26th Canadian Congress of Applied Mechanics (CANCAM 2017)

Progress in Canadian Mechanics Volume 2

Victoria, Canada
29 May – 1 June 2017

Editors:

Y. Shi
B. Nadler

ISBN: 978-1-5108-5678-3

Printed from e-media with permission by:

Curran Associates, Inc.
57 Morehouse Lane
Red Hook, NY 12571

proceedings.com

Some format issues inherent in the e-media version may also appear in this print version.

Printed by Curran Associates, Inc. (2018)

Western University – Department of Mechanical and Materials Engineering
Spencer Engineering Building
London ON N6A 5B9
Canada

Phone: 519-661-2136
Fax: 519-661-3020

mme@uwo.ca

Additional copies of this publication are available from:

Curran Associates, Inc.
57 Morehouse Lane
Red Hook, NY 12571 USA
Phone: 845-758-0400
Fax: 845-758-2633
Email: curran@proceedings.com
Web: www.proceedings.com

CANADIAN CONGRESS OF APPLIED MECHANICS

CANCAM

CONGRÈS CANADIEN DE MÉCANIQUE APPLIQUÉE

Progress in Canadian Mechanics

Volume 2

Proceedings of the 26th Canadian Congress of Applied Mechanics (CANCAM 2017)
Victoria, British Columbia, May 29 – June 1, 2017

Editors: **Shi, Y.**

Nadler, B.

PREFACE

It is with great pleasure that we welcome you to our celebration of the 26th Canadian Congress of Applied Mechanics (CANCAM) at the University of Victoria, Victoria, BC, Canada.

CANCAM is a non-profit organization established in 1975 with the sole purpose of promoting mechanics in Canada through the organization of a bi-annual event called "Canadian Congress of Applied Mechanics". The first CANCAM was held at Université Laval in 1967. Throughout the past fifty years, CANCAM has excelled in showcasing creativity, in providing an ideal environment for technical exchange and networking, and most importantly in highlighting the latest advances in the field of mechanics, with applications to mechanical, civil, materials, aerospace engineering and related areas. To address the latest achievements, CANCAM 2017 offers a rich and comprehensive menu for professionals.

Victoria, an island destination, is one of the most beautiful cities in Canada. There are countless opportunities to explore heritage architecture, colorful gardens, traditions like afternoon tea, and outdoor adventure. We'd love to share our beautiful city and all it offers with you and your family. Welcome to join us at CANCAM 2017 and help make it yet another wonderful event filled with research discussions, social and networking opportunities and fostering new connections.

Taking this opportunity, we sincerely thank all the individuals who have contributed to the organization and support of CANCAM 2017. Special thanks are extended to our colleagues in the organizing committee, session chairs, and the student volunteers for their warm, thoughtful service to all participants. We also would like to express our high appreciation and gratitude to distinguished plenary speakers, and all of the authors and participants. All of your contribution and support will make CANCAM 2017 a fruitful and memorable event!

We gratefully acknowledge the generous sponsorships from High Speed Imaging, UVic, and CANCAM. We hope you enjoy your time at the congress in Victoria.

Yang Shi, P.Eng., FASME, FCSME, FIEEE
Professor, General Chair of CANCAM 2017

Organizing Committee

Honorary Chair: Clarence W. De Silva (University of British Columbia)
General Chair: Yang Shi (University of Victoria)
Program Chair: Ben Nadler (University of Victoria)
Publications Chair: Brad Buckham (University of Victoria)
Student Activities Chair: Mohsen Akbari (University of Victoria)
Publicity Chair: Keivan Ahmadi (University of Victoria)

CANCAM Central Committee

President: Jim Bugg (University of Saskatchewan)
Past President: Clarence W. De Silva (University of British Columbia)
Vice-President: J. Maciej Floryan (Western University)
Secretary: Arun Misra (McGill University)
Treasurer: Leszek Sudak (University of Calgary)
Member: Donald Bergstrom (University of Saskatchewan)
Member: Remus Tutunea-Fatan (Western University)
Member: Yang Shi (University of Victoria)
Member: Ben Nadler (University of Victoria)

Plenary Speakers

Dr. Janet A. W. Elliott, FCIC
Professor and Tier I Canada Research Chair in Thermodynamics
Department of Chemical and Materials Engineering
Adjunct Professor Laboratory Medicine and Pathology
University of Alberta

Dr. Elliott obtained her B.A.Sc. in Engineering Science (Engineering Physics Option) and her M.A.Sc. and Ph.D. in Mechanical Engineering at the University of Toronto. She has been a Visiting Professor at the Massachusetts Institute of Technology and at the Oxford Centre for Collaborative Applied Mathematics.

Dr. Elliott currently serves as Associate Editor of the journal Cryobiology, on the Editorial Advisory Board of The Journal of Physical Chemistry, and on the American Chemical Society Division of Colloid & Surface Chemistry Executive Committee. She has previously served as a member of the Physical Sciences Advisory Committee for the Canadian Space Agency and on the Board of Directors of the Canadian Society for Chemical Engineering.

Dr. Elliott's research has been recognized nationally in science and engineering by the Canadian Society for Chemical Engineering Syncrude Canada Innovation Award (2008), the Natural Sciences and Engineering Research Council Doctoral Prize (1998), the Canadian Council of Professional Engineers Young Engineer Achievement Award (2001), the Canadian Institute for Advanced Research Young Explorer's Prize (2002) and Time Magazine's Canadians Who Define the New Frontiers of Science (2002). Dr. Elliott has also received provincial and University awards including the University of Alberta Teaching Unit Award (2004, 2016). As one student put it, "She could convince rocks to study thermodynamics."

Dr. Afzal Suleman, FCAE, AFAIAA, FRAeS
Professor and Canada Research Chair in Computational and Experimental Mechanics
Department of Mechanical Engineering
University of Victoria

Afzal Suleman is a Professor, Canada Research Chair in Computational and Experimental Mechanics, and Director of the Center for Aerospace Research at the University of Victoria, BC, Canada. Afzal Suleman obtained his BSc (Honours) and MSc in Aeronautical Engineering from Imperial College, University of London (UK) followed by a PhD in Aerospace Engineering from the University of British Columbia in Canada (1992). He was awarded the National Academy of Sciences Fellowship at Wright-Patterson Air Force Research Labs, U.S. Air Force (1992-1994). He is Fellow of the Canadian Academy of Engineers, Fellow of the Royal Aeronautical Society, Associate Fellow of AIAA, and Fellow of the Lisbon Royal Academy of Sciences. He has served as a National Delegate at UN-COPUOS (1999-2005) and is currently a Panel Member and National Delegate at the NATO AVT Panel (1998-Present). Also, he is currently a member of the Government of Canada Space Advisory Board and the Canadian Armed Forces Defence Advisory Board.

Dr. Saeid Habibi, FASME, FCSME
Professor
Department of Mechanical Engineering
McMaster University

Saeid Habibi is a Senior NSERC Industrial Research Chair and a Professor in the Department of Mechanical Engineering at McMaster University. He obtained his Ph.D. in Control Engineering from the University of Cambridge, U.K. His academic background includes research into intelligent control, state and parameter estimation, fault diagnosis and prediction, Variable Structure Systems, actuation systems, mechatronics and fluid power. The application areas for his research have included automotive, aerospace, and robotics. He has developed his own theory related to estimation and target tracking referred to as the Smooth Variable Structure Filter (SVSF).

Dr. Habibi has held a number of managerial and industrial positions. He is presently the Director of the Centre for Mechatronic and Hybrid Technology (CMHT) at McMaster University. He was the Chair of the Department of Mechanical Engineering from 2008 to 2013. He also spent a number of years in industry as a Project Manager and Senior Consultant for Cambridge Control Ltd, U.K., where he was involved in automotive and aerospace related projects. He then moved to AlliedSignal Aerospace Canada (presently part of Honeywell) where managed their Systems Engineering Department. Dr. Habibi has been the recipient of a number of awards including the 2012 Best Paper Prize from IEEE Transportation Electrification Conference, 2 corporate awards for his contributions to the AlliedSignal Systems Engineering, and the Institution of Electrical Engineers (IEE) F.C. Williams best paper award for his contribution to the Variable Structure Systems theory. He is a Professional Engineer and an ASME Fellow.

Dr. Rodney Herring
Associate Professor
Department of Mechanical Engineering
University of Victoria

Dr, Herring obtained his BASc in 1980 and MASc in 1982 from the University of Windsor and his PhD in 1985 from the University of Birmingham, UK. His personal research working experience involves 17 years performing industrial research including 7 years in the Canadian automotive industry (Tecumseh Racing & Chryslers, Windsor), 5 years in the nuclear industry (Chalk River Nuclear Laboratories & Ontario Hydro) and 5 years in the US aerospace industry (Lockheed-Martin, Baltimore, MD), as well as, 5 years performing basic research involving condensed matter physics (Exploratory Research for Advanced Technology (ERATO), Saitama, Japan), 7 years in space research (Canadian Space Agency, Montreal) and 15 years in academia (UVic, Victoria).

Dr. Sadik Dost, FCAE
Professor and Canada Research Chair in Semiconductor Crystal Growth
Department of Mechanical Engineering
University of Victoria

Dr. Dost is Professor and Director of the Crystal Growth Laboratory at the University of Victoria. He joined UVic as Professor of Mechanical Engineering in 1989 after serving as a faculty member at the University of Calgary for nine years. He has obtained his Ph.D. in Istanbul Technical University in the area of Continuum Mechanics in 1974. He graduated from Karadeniz Technical University with a degree of Dip. Ing. in Civil Engineer in 1969.

During his tenure at UVic, Dr. Dost has served as Founding Director of a university research center---centre for Advanced Materials and Related Technology (CAMTEC) ---from 1992 to 1997, and also as Chair of the Department of Mechanical Engineering for one year in 1993 and for six years from 1997 to 2003 until his appointment as Canada Research Chair (CRC) in 2003 who served until the end of 2016. He has also served in a large number of University Committees, and has held various potions in learned and professional bodies. He is a fellow of the Canadian Society for Mechanical Engineering, the Engineering Institute of Canada, and Canadian Academy of Engineering. He is a registered professional engineer in BC, and has served in professional and learned bodies at various capacities and also on advisory boards and committees in federal and provincial government bodies including the Canadian Space Agency.

His current research combines experimental and theoretical study of crystal growth of semiconducting bulk single crystals from the liquid phase with particular focus on materials like Cadmium Zinc Telluride (CZT) and Silicon Germanium (SiGe) that are needed for a variety of applications mainly for solar energy conversion, medical imaging, and infra-red and security devices. His research programs involve extensive industrial collaborations. He has collaborated and carried out joint research projects with various research groups from Canada, Japan, Turkey, and Europe.

Dr. Dost is well known internationally for his research, and has published extensively; with well over 200 journal articles, 3 books, 10 book chapters, and a large number of conference publications. With the support of government and industry, he has established a leading edge crystal growth research program in Canada, and a unique state-of-the-art crystal growth laboratory (CGL). In 2009, Dr. Dost has been named as the recipient of the Craigdarroch Gold Medal for Career Achievement in Research by the University of Victoria, and has also received the 2009 University of Victoria Legacy Awards for Research. The award honours recipients "whose energy and dedication have contributed significantly to the quality and reputation of the University of Victoria."

Table of Contents

Plenary Lectures

Topology Optimization

Vibration of Structures

Multicomponent Materials

Robotics and Control Systems

Fluid Mechanics

Computational Solid Mechanics

Mechatronics and Control Systems

Vibration and Dynamics

Thermodynamics and Heat Transfer

Biomechanics and Biomaterials

Nanostructures and MEMS

Renewable Energy Systems 2

Fluid Mechanics 2

Fluid Mechanics and Heat Transfer

Plenary Lectures

Proceedings of the 26th CANCAM
Victoria, British Columbia, Canada, May 29 – June 1, 2017

THERMODYNAMICS AND BIOMECHANICS IN CRYOPRESERVATION: HOW MATH CAN CHANGE LIVES

Janet A. W. Elliott
Department of Chemical and Materials Engineering
University of Alberta
Edmonton, Alberta, Canada
janet.elliott@ualberta.ca

ABSTRACT

Cryobiology is the study of life at low temperatures. Cryopreservation (using temperatures down to that of liquid nitrogen at –196 °C) is the only way to preserve long-term viability and function of mammalian cells for research and medical use. While many cells in suspension are routinely cryopreserved and used clinically and in research, the vast majority of tissues cannot be adequately cryopreserved today. Articular cartilage is the smooth white tissue covering the ends of bones allowing them to glide over one another in articulating joints such as the knee. The lack of a cryopreservation protocol was limiting the joint-saving transplants of this tissue. We have introduced an osmotic virial equation approach for multisolute solution thermodynamics in biology, improving thermodynamics and transport modeling in cryopreservation. In collaborative research spanning 15 years, we combined mathematical modelling with biological experiments to develop the first successful protocol to cryopreserve intact human articular cartilage on bone with high cell viability and function. A key component of this work was a biomechanical model of the transport of thermodynamically non-dilute amounts of cryoprotectant in the linear elastic porous medium of cartilage including water movement, the role of ions, the spatially varying tissue shrinking and swelling, and the spatially inhomogeneous initial state of the tissue. Our collaborative, interdisciplinary research group allows students to do graduate work in engineering and medical sciences. This work was funded by NSERC, CIHR, Alberta Innovates, the Edmonton Orthopedic Research Committee and University of Alberta.

Proceedings of the 26th CANCAM
Victoria, British Columbia, Canada, May 29 – June 1, 2017

UNMANNED AIR SYSTEMS: A TOOL IN THE FLIGHTWORTHINESS EVALUATION OF NOVEL AIRCRAFT CONFIGURATIONS

Afzal Suleman
Department of Mechanical Engineering
University of Victoria
Victoria, British Columbia, Canada
suleman@uvic.ca

ABSTRACT

The research program aims to improve the performance of complex aerospace engineering systems through advances in mathematical and computational models, and experimental methods that incorporate multidisciplinary analysis, design optimization and subscale UAV model flight testing for the synthesis of optimal and novel aircraft designs. The design and development of physical flight test platforms provide a low-cost opportunity to evaluate flight worthiness of new and unconventional aircraft configurations. The presentation will outline some of the experimental UAV flight test programs for aeroelastic evaluation of joined-wing and high-aspect ratio aircraft configurations in collaboration with OEMs. The UAV based flight test programs enable designers to retrieve quantifiable data and to provide a qualitative assessment of the aircraft handling qualities. It provides new perspectives that may lead to identification of design issues early in the development process thus avoiding expensive re-designs at the detailed design phase of the full scale transport aircraft.

THERMODYNAMICS AND BIOMECHANICS IN CRYOPRESERVATION: HOW MATH CAN CHANGE LIVES

Saeid Habibi
Department of Mechanical Engineering
McMaster University
Hamilton, Ontario, Canada
habibi@mcmaster.ca

ABSTRACT

Concerns related to the impact of fossil fuels on climate change, population health and the environment have resulted in a momentous shift in the global automotive strategy. Electrification of automobiles is a key strategy in reducing transportation-related greenhouse gas emissions as well as fossil fuel consumption. An equally important feature for the car of the future is being autonomous capable. The latter has the potential to change our mode of urban transportation and can be an enabler for fully electric vehicles. Saeid Habibi, Director of the Centre for Mechatronics and Hybrid Technologies at McMaster University. Dr. Habibi will present emerging trends that will impact the car of the future and will provide an overview of some of the related research currently being conducted at the McMaster University.

Proceedings of the 26th CANCAM
Victoria, British Columbia, Canada, May 29 – June 1, 2017

CLEANING AND MONITORING EARTH

Rodney Herring
Department of Mechanical Engineering
University of Victoria
Victoria, British Columbia, Canada
rherring@uvic.ca

ABSTRACT

As is now well-known, increased CO2 in the atmosphere and the oceans are contributing to global warming. Additionally, acidification of the ocean will kill oxygen producing phytoplankton such as diatoms that live in glass houses and responsible for producing ~25% of the Earth's oxygen. With this in mind, research projects were initiated to help reduce CO2 emissions and monitor the Earth's atmosphere, oceans and landmasses. Nanotechnology using visible light activated iron (Fe) doped titania (TiO2) photocatalyst crystals of ~5 nm dia. have been invented to cheaply, effectively and efficiently clean 100 % of the organics from waste water converting them to CO2 for easy sequestration or conversion to other useful forms, as well as, converting the waste water's ammonia to nitrites and nitrates (fertilizer) useful for agriculture benefiting Earth by helping to clean its oceans and atmosphere. The Fe doped TiO2 nanocrystals, which are not perfect, are cheaply produced using the sol gel process, effectively applied with respect to UV activated photocatalysts and bioreactors by using visible light that passes through water and glass strongly adhering to its surface for no loss of the nanocrystals to the environment. Because the Fe doped TiO2 nanocrystals are not poisoned by the wastewater, are self-cleaned and not consumed in the cleaning process, wastewater facilities based on this technology will have a long life with low maintenance. Many other significant applications of this nanotechnology, which just requires water and visible light, are being investigated including keeping surfaces clean of biological materials useful for marine and hospital environments.

Additionally, methods are actively being developed at UVic to monitor the Earth's atmosphere, oceans and landmass using a new type of camera based on software defined radio/sonar (SDR/S) to measure the power, energy and direction of movement of the Earth's atmosphere as well as its three-dimensional temperature, pressure and composition. This presentation will focus on SDR's monitoring of Earth with some references to its SDS applications.

THE ROLE OF CONTINUUM MECHANICS IN CRYSTAL GROWTH AND SOLIDIFICATION

Sadik Dost
Department of Mechanical Engineering
University of Victoria
Victoria, British Columbia, Canada
sdost@uvic.ca

ABSTRACT

Growth of semiconductor single crystals under electric and magnetic fields is of interest to increase and better control of crystal growth rate, to suppress and control the adverse effect of natural convection and to obtain better mixing in the growth melt (liquid solution) for better crystal uniformity, which all are favorable conditions for a prolonged growth of high quality crystals. To this end, in parallel to well-designed experiments, modeling is essential to shed light on various aspects of these growth processes and also to better understand the transport phenomena involved.

In this talk the models developed over the years, based on "electromagnetic interactions", will be presented for "solution crystal growth" conducted under electric and magnetic fields. Basic and constitutive equations of a binary electromagnetic continuum mixture are specialized for two important solution growth techniques—Liquid Phase Electroepitaxy (LPEE) and Travelling Heater Method (THM). As an application, an LPEE growth of GaAs bulk crystals under a strong static magnetic field is considered. Experimental results that have shown that the growth rate under an applied static magnetic field is also proportional to the applied magnetic field and increases with the field intensity level, are predicted from these models. The contribution of a third-order material constant in LPEE is also predicted from these models. The prediction of increasing growth rate in THM growth under rotating magnetic fields from modeling was verified by experiments.

Industrial applications of these growth processes require seed crystals to initiate growth. To this end the growth techniques of Liquid Phase Diffusion (LPD) and Vertical Gradient Freezing (VGF) are utilized to obtain the needed seed crystals. In this direction at UVic we have designed and built 25mm LPD and 75mm VGF systems to grow Silicon Germanium (SiGe) and Cadmium Zinc Telluride (CZT) crystals that are essential materials for efficient solar energy conversion and medical imaging and infra-red device applications. Samples of grown SiGe and CZT crystals and their characteristics will also be presented.

Topology Optimization

Proceedings of the 26th CANCAM
Victoria, British Columbia, Canada, May 29 – June 1, 2017

LARGE SCALE TOPOLOGY OPTIMIZATION UTILIZING ELEMENT DELETION & REFINEMENT

Christopher Woischwill, Stephen Roper, Daozhong Li, Chris Carrick, Il Yong Kim[1]

Department of Mechanical and Materials Engineering
Queen's University, Kingston, ON, Canada
[1]kimiy@queensu.ca

ABSTRACT

This paper presents a methodology that reduces the computational cost of density-based topology optimization studies and permits the solution of problems that are too large to be solved with existing methods. The proposed methodology uses a "coarse-to-fine" outer loop approach where topology results from a coarse finite element mesh are used to generate finer meshes of a smaller size in subsequent outer loops. After each topology optimization study, only high-density elements and low-density elements that are required to resolve numerical issues are refined and brought forward to the next outer loop; all other unused elements are deleted. This results in far fewer elements being used when compared to existing brute force approaches (i.e. refining all elements without any element deletion) while maintaining the same mesh refinement. The advantages of this methodology over existing brute force approaches are demonstrated with linear static and modal analysis examples.

KEYWORDS: Topology optimization, large scale topology optimization, model reduction, computational efficiency

INTRODUCTION

Density-based topology optimization is well-established in theory and has been used in a wide range of industry applications [1]–[3]. In this approach, a design space is discretized into finite elements and each element is given a prescribed initial density. Optimization is then utilized to determine the optimal density of each element (normalized between 0 and 1) which in effect finds the optimal distribution of material within a designable region subject to loads, boundary conditions, and structural performance constraints. Commercial topology optimization solvers use the Solid Isotropic Material with Penalization (SIMP) method, which penalizes the stiffness of intermediate and low-density elements to discourage unphysical results [4].

However, the aforementioned discretization process can lead to a large number of design variables in larger problems, which exponentially increases both the solve time as well as computational memory requirements. In most topology optimization problems, simply decreasing the resolution of the mesh (and thus decreasing the number of finite elements) is undesirable since the accuracy of the finite element analysis decreases [5]. Furthermore, larger finite elements might not be able to accurately represent small structural members (especially sheet metal structures). It follows that there is a need to develop a methodology and computational tool that both reduces the computational cost of topology optimization studies and enables the solution of large and comprehensive problems demanded by industry.

This paper proposes a methodology titled Large Scale Topology Optimization (LSTO) which reduces the computational cost of topology optimization problems by automatically deleting and refining elements in a coarse-to-fine approach. In the following sections, the methodology used by the tool to strategically create new topology optimization problems with finer meshes is described. Then, three numerical examples are used to compare the LSTO methodology to existing brute force approaches.

PROCESS

The proposed LSTO methodology and computational tool uses an outer loop approach as shown below in Fig. 1.

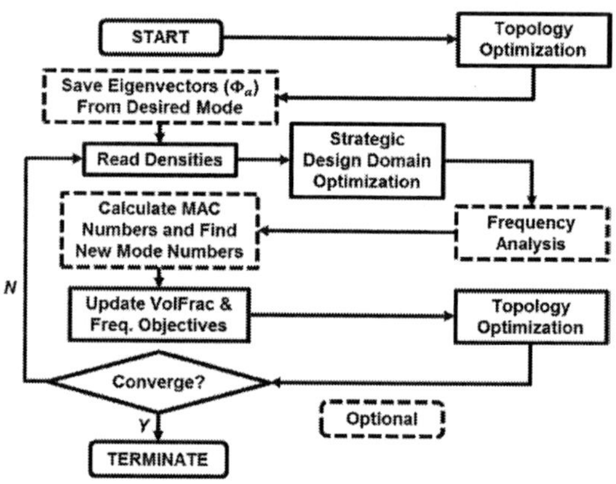

Figure 1: Large Scale Topology Optimization.

Each LSTO study begins with topology optimization using a coarse mesh. The results from this initial topology optimization study are then used to generate the design space of the following outer loop: elements with a density greater than a specified threshold are "kept" and are used in the following outer loop, while all other elements are only brought forward to the next outer loop if they are needed to resolve numerical issues. Processes shown in dashed borders are optional and are only utilized if specific mode shapes need to be tracked throughout multiple outer loops.

ELEMENT DELETION & REFINEMENT

After each run of topology optimization, elements which had a final element density less than a user-specified threshold are marked as "void" and are only brought forward to the design space in the following outer loop if they resolve at least one of three numerical issues: (1) pinched loadpaths, (2) hinged elements, and (3) isolated elements.

Pinched loadpaths are strands of finite elements which produce small and jagged loadpaths which are unsuitable for topology optimization. Pinched loadpaths are corrected by keeping all "void" elements surrounding these loadpaths and using them as designable space in the following outer loop.

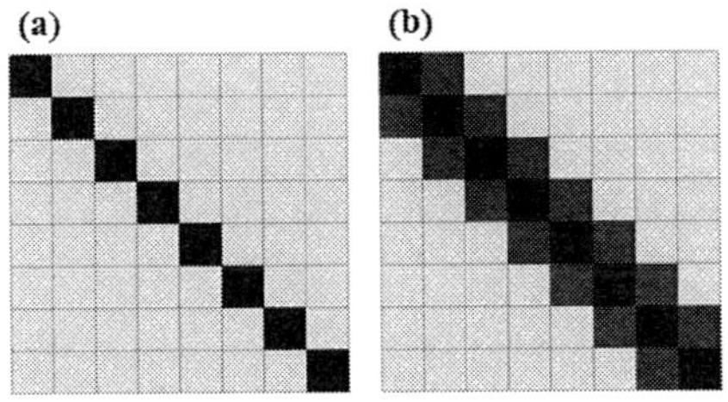

Figure 2: (a) Pinched loadpath; (b) corrected structure.

Hinged elements are defined as elements which can have undesirably large rotations relative to adjacent elements [6]. Examples include two-dimensional elements which share only a single node with other elements and three-dimensional elements which share only a single node or edge with other elements. Since this results in the stiffness matrix becoming nearly singular, these elements produce many numerical issues in finite element analysis including very large displacements. This issue is resolved by keeping all "void" elements adjacent to hinged elements and using them in the following outer loop.

Isolated elements are elements, or groups of elements, which are not contiguously connected to any other elements or boundary conditions. These elements produce many of the same numerical issues as the hinged elements including very large displacements. All isolated elements are deleted and do not return in the design space used in the following outer loop regardless if they are "kept" or "void".

All "void" elements which were not used to resolve numerical issues are deleted. The remaining elements are then refined to the desired mesh size and saved to a finite element model to be used in the following outer loop.

MODE TRACKING

A modal analysis of a structure using the finite element method returns several eigenvalues and eigenvectors which are physically interpreted as the natural frequencies and corresponding mode shapes of the system. Since the magnitudes of these eigenvalues and eigenvectors are sensitive to the refinement of the mesh, it is possible that these mode shapes are reordered as the mesh refinement is changed. For example, it is possible that the first mode shape (the mode shape corresponding to the smallest frequency) found at a coarse mesh refinement is a pumping motion, while the first mode shape found at a finer mesh refinement is a completely different twisting motion. This can become problematic since commercial topology optimization solvers can only accept mode numbers as optimization parameters (e.g. "maximize the frequency of Mode 2") [7].

Oftentimes in industry it is important to maximize the frequency of a particular mode shape. Hence, LSTO must be able to track any number of mode shapes across its multiple outer loops. This is accomplished by calculating the MAC numbers between target mode shapes, as given by Eq. (1)

$$\mathrm{MAC}(\Phi_a, \Phi_b) = \frac{|\Phi_a^T \Phi_b|^2}{(\Phi_a^T \Phi_a)(\Phi_b^T \Phi_b)} \quad (1)$$

where Φ_a, Φ_b represent two unique eigenvectors [8]. The MAC number is a real number between 0 and 1; when $\mathrm{MAC}(\Phi_a, \Phi_b) = 1$, Φ_a and Φ_b are the same mode shape.

Before the first outer loop, the user can specify which mode numbers (e.g. "the 4th and 6th modes") are to be tracked across multiple outer loops. After the remeshing process in each outer loop, a normal modal analysis is conducted and all calculated eigenvectors are compared to the targeted eigenvectors from the first outer loop through MAC numbers. The new targeted modes are then identified and the optimization parameters for the next topology optimization run are updated to the newly-identified mode numbers.

2D EXAMPLE: LOADED KNEE STRUCTURE

To demonstrate the capability of the proposed methodology, a classical topology optimization problem – the loaded knee structure shown in Fig. 3 – is considered. The structure is subjected to a single point load of magnitude P and is fixed in all degrees of freedom at the top of the structure. The objective of this study is to determine the optimal structure within the designable region that minimizes compliance (sum of element strain energy) while only using 30% of the available volume in the structure. In this model, the Young's Modulus (E) is 210 GPa, the density (ρ) is 7850 kg/m^3, the Poisson's ratio (ν) is 0.3, L is 500 mm, and P is 100 N.

The designable space is discretized into square first-order quadrilateral finite elements where the average mesh size (defined as the length of one element) is L/400 or 1.25 mm. Each quadrilateral element has a thickness of 1 mm into the page. The proposed method begins with a coarser mesh size (10 mm) in its first outer loop, and sequentially deletes and refines elements until it arrives at an average mesh size of 1.25 mm. To ensure that loadpaths are not prematurely deleted, LSTO begins with a large volume fraction constraint (45%), and reduces the volume fraction constraint each outer loop until it arrives at a volume fraction of 30% at the final level of mesh refinement.

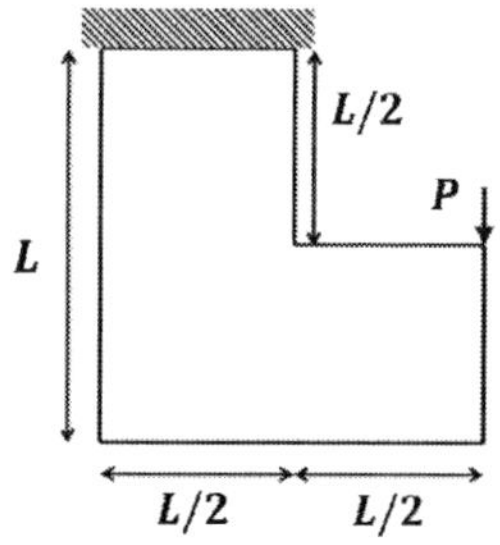

Figure 3: Loaded knee structure.

Table 1 below compares the number of designable elements, the mass of the optimal structure, the final objective value of optimization (total strain energy), and CPU time at each level of mesh refinement.

Table 1: Optimized results comparison (2D).

Mesh Size [mm]	No. of Elements	Mass [kg]	Obj. [J]	CPU Time [s]
Existing Approach (Brute Force)				
1.25	120,000	0.442	3.083	2,240
Proposed Method (LSTO)				
10.00	1,875	0.662	2.111	45
5.00	4,200	0.588	2.290	47
2.50	14,191	0.515	2.555	100
1.25	49,352	0.422	2.917	495

The optimized structures obtained from both methods at the final level of mesh refinement are shown below in Fig. 4.

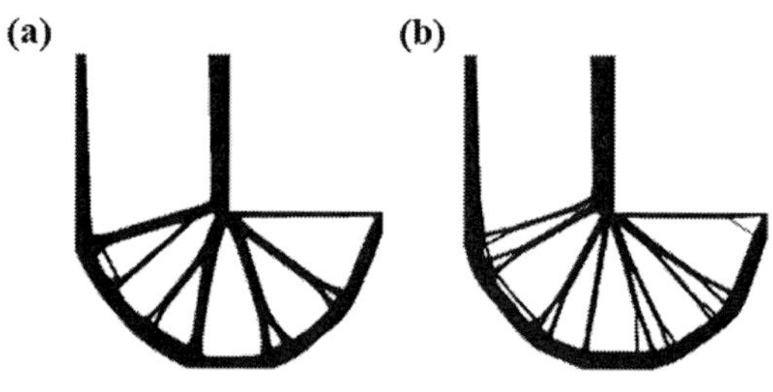

Figure 4: Topology results: (a) brute force and (b) LSTO.

While both methods produced similar optimized topology results, the proposed method used only 687 seconds of CPU time to complete the study (192 seconds to complete the first three levels of mesh refinement and 495 seconds to complete the final level of mesh refinement) while the existing approach used 2,240 seconds of CPU time to complete the study at the same mesh size. By deleting elements between each outer loop, the proposed LSTO method reduced computational time by over 69% while also reducing the objective by over 5% compared to the brute force approach.

3D EXAMPLE: CONTROL ARM (LINEAR STATIC)

The second numerical example considers a three-dimensional model of an automotive control arm as shown below in Fig. 5, which is adapted from an exemplary model packaged with Altair HyperWorks. The design space is discretized into first-order hexahedral finite elements and all regions except for the bushings and their attachments (shown in black) are designable. Three linear static loadcases, each of which apply a 1000 N force to the front bushing and use the displacement constraints applied at the locations shown, are considered. The objective of this study is to determine the optimal structure using a mesh size of 1.5 mm while using only 30% of the available design space. Symmetry and draw direction manufacturing constraints are also considered. The optimization objective is a simple weighted sum of the compliance for each of the three loadcases. In this model: E = 200 GPa, ρ = 7850 kg/m^3, and ν = 0.3.

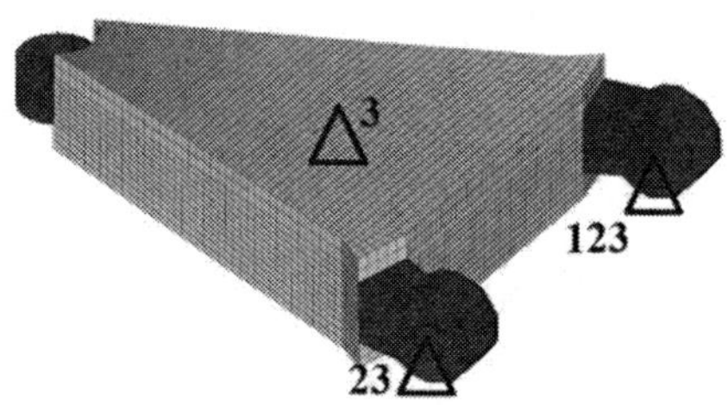

Figure 5: Automotive control arm.

LSTO begins with a mesh size of 6.0 mm and uses three outer loops to reach a final mesh size of 1.5 mm. As done with the previous numerical example, LSTO uses larger volume fraction constraints in earlier outer loops to ensure loadpaths are not prematurely deleted. Table 2 below compares the number of hexahedral elements, the mass of the optimal structure, the optimal objective value, and CPU time at each level of mesh refinement.

Table 2: Optimized results comparison (3D – linear static).

Mesh Size [mm]	No. of Elements	Mass [kg]	Obj. [J]	CPU Time [s]
Existing Approach (Brute Force)				
1.5	2,145,024	N/A	N/A	N/A
Proposed Method (LSTO)				
6.0	33,516	21.647	0.012	1,184
3.0	167,102	19.415	0.014	5,415
1.5	1,230,492	17.183	0.020	193,173

No solution at the final mesh size could be obtained with the brute force approach; the solver became hung during optimization and had to be manually terminated before the job reached convergence. The proposed LSTO approach, however, successfully obtained results at the final mesh refinement level using 193,173 s of CPU time. The optimized structure obtained using LSTO is shown below in Fig. 6.

Figure 6: Optimized topology from LSTO.

3D EXAMPLE: CONTROL ARM (MODAL)

The third numerical example considers the same control arm introduced in the last section but with a fourth modal analysis loadcase and a new optimization objective formulation. The objective of this study is to minimize the weighted sum of compliance for the three static loadcases in addition to the inverse of a pumping mode eigenvalue using a fixed modal analysis. The "pumping mode" is a mode shape where the control arm pivots about the rear bushings and the front bushing pumps up and down vertically. Each term is weighted equally and normalized in each iteration to ensure that no loadcase dominates optimization. Mode tracking is enabled in this example, and hence the pumping mode shape will be tracked in all outer loops even if the modes are reordered when the designable volume changes.

Again, LSTO begins with a mesh size of 6.0 mm and uses three outer loops to reach the final mesh size of 1.5 mm. Table 3 below compares the number of hexahedral elements, the mass of the optimal structure, the optimal objective value, and CPU time at each mesh refinement level.

Table 3: Optimized results comparison (3D – modal).

Mesh Size [mm]	No. of Elements	Mass [kg]	Obj.	CPU Time [s]
Existing Approach (Brute Force)				
1.5	2,145,024	N/A	N/A	N/A
Proposed Method (LSTO)				
6.0	33,516	21.647	18.66	4,603
3.0	177,965	19.415	26.94	28,609
1.5	1,278,808	17.183	28.11	815,493

Similarly, no solution at the final mesh size could be obtained with the brute force approach since the solver became hung during optimization and had to be manually terminated before the job reached convergence. However, the final level of mesh refinement could be considered using the proposed LSTO method. The optimized structure obtained at the final level of mesh refinement using LSTO is shown below in Fig. 7. The pumping mode was successfully tracked from Mode 3 in the first outer loop, to Mode 5 in the second outer loop, and then to Mode 2 in the final outer loop.

Figure 7: Optimized topology from LSTO.

CONCLUSIONS

A methodology and computational tool which performs large scale topology optimization using an automated coarse-to-fine approach has been developed. The tool includes modules that improve mesh quality (including the deletion of hinged and isolated elements) and track mode shapes across multiple design spaces. It has been demonstrated that the LSTO approach can both solve problems that cannot be solved using existing approaches as well as solve problems significantly faster than existing methods without compromising solution quality.

Future improvements to be made to the LSTO approach include support for different element types (e.g. triangular and tetrahedral elements), support for different types of optimization (e.g. size, shape, and composites), and support for new types of problems (e.g. thermal and acoustics).

ACKNOWLEDGMENTS

This research was funded by Automotive Partnership Canada and General Motors of Canada. Technical advice and direction were gratefully received from Joe Moore, Balbir Sangha, Manish Pamwar, Derrick Chow, and Chandan Mozumder, at General Motors.

REFERENCES

[1] C. Li and I. Y. Kim, "Topology, size and shape optimization of an automotive cross car beam," *Proc. Inst. Mech. Eng. Part D J. Automob. Eng.*, vol. 229, no. 10, p. 954407014561279, Sep. 2014.

[2] C. Li, I. Y. Kim, and J. Jeswiet, "Conceptual and detailed design of an automotive engine cradle by using topology, shape, and size optimization," *Struct. Multidiscip. Optim.*, vol. 51, no. 2, pp. 547–564, Feb. 2015.

[3] N. Cristello and I. Y. Kim, "Multidisciplinary design optimization of a zero-emission vehicle chassis considering crashworthiness and hydroformability," *Proc. Inst. Mech. Eng. Part D J. Automob. Eng.*, vol. 221, no. 5, pp. 511–526, 2007.

[4] M. P. Bendsøe and N. Kikuchi, "Generating optimal topologies in structural design using a homogenisation method," *Comput. Methods Appl. Mech. Eng.*, vol. 71, no. 2, pp. 197–224, 1988.

[5] I. Y. Kim and B. M. Kwak, "Design space optimization using a numerical design continuation method," *Int. J. Numer. Methods Eng.*, vol. 53, no. 8, pp. 1979–2002, 2002.

[6] G. H. Yoon, Y. Y. Kim, M. P. Bendsoe, and O. Sigmund, "Hinge-free topology optimization with embedded translation-invariant differentiable wavelet shrinkage," *Struct. Multidiscip. Optim.*, vol. 27, no. 3, pp. 139–150, 2004.

[7] M. Eldred, V. Venkayya, and W. Anderson, "Mode Tracking Issues in Structural Optimization," *Aiaa J.*, vol. 33, no. 10, pp. 1926–1933, 1995.

[8] T. S. Kim and Y. Y. Kim, "MAC-based mode-tracking in structural topology optimization," *Comput. Struct.*, vol. 74, no. 3, pp. 375–383, 2000.

Proceedings of the 26th CANCAM
Victoria, British Columbia, Canada, May 29 – June 1, 2017

PRACTICAL APPLICATIONS FOR MULTI-MATERIAL DESIGN UTILIZING SIZE AND TOPOLOGY OPTIMIZATION

Stephen Roper, Daozhong Li, Chris Carrick, Christopher Woischwill, Il Yong Kim[1]

Department of Mechanical and Materials Engineering
Queen's University, Kingston, ON, Canada
[1]kimiy@queensu.ca

ABSTRACT

Multi-material design optimization is a rapidly advancing field, with current research focusing on theory and tool development. This paper examines the combination of conventional optimization strategies with modern techniques to develop advanced system and component-level design concepts. Here, multi-material size and topology optimization are considered in two numerical studies of a high-performance Formula SAE vehicle, with results indicating significant benefits and future potential for multi-material applications in global design frameworks.

KEYWORDS: size optimization, topology optimization, finite element analysis, multi-material, single-material, design

INTRODUCTION

In practical structural applications, optimization is typically used to minimize some objective function by adjusting the physical parameters of a given system within a design space [1]. In many mechanical engineering problems, this can be achieved by parametrizing and manipulating a series of geometric design variables such as gauge thickness or cross sectional properties of beams or bars. This method of size optimization is widely used in structural mechanics, and leaves the global geometry of the system unchanged. Coupled with the finite element method, there are multiple examples of size optimization that usually consider the design of truss structures under displacement or stress constraints [2] [3]. While size optimization is a well-known field in numerical analysis, it is typically restricted to a single-material design space. This even applies to commercial computer-aided engineering (CAE) software, which generally only support single material optimization and neglects material selection in the design loop.

To address these limitations multi-material design has emerged as a prominent field, with current research focusing on theory and tool development such as density-based topology optimization and heuristic approaches. Here, simple academic case studies have been examined with some extension into practical problems, such as automotive component design for lightweighting and improved mechanical performance [4] [5]. Current state-of-the-art applications implement a minimum compliance problem, subject to volume or mass constraints to obtain the optimum material selection and existence within a design space.

While the previous research conducted in this field successfully applies multi-material optimization techniques within isolated studies, its combination with conventional strategies and implementation within global design frameworks are often neglected. That is, by adopting techniques from their respective fields, the combination of single and multi-material optimization strategies can be used in parallel to address both component and system-level designs. This paper focuses on the practical application of multi-material design, coupling conventional and modern optimization methods within large system-based, and small component-based structures. An example of each is provided considering the development of a high-performance Formula SAE competition vehicle. Considering three-dimensional geometries, complex loading scenarios and various optimization problem statements, the three objectives are:

(a) System-level multi-material size optimization, to minimize mass subject to displacement constraints of a Formula SAE space frame chassis.
(b) Component-level multi-material topology optimization, to minimize compliance subject to volume fraction constraints of a Formula SAE suspension upright.
(c) Optimization performance comparisons, to demonstrate the significance of coupled multi-material design approaches.

The culmination of this work are multi-material design concepts at the system and component level in practical automotive applications. Analysis was conducted using a custom MATLAB program (multi-material size optimization) and in-house solvers (multi-material topology optimization) interfacing with commercial FE solvers.

MULTI-MATERIAL SIZE OPTIMIZATION

The Queen's University Formula SAE Team designs and builds a new high-perform racing vehicle each year to compete in national and international competitions. In order to gain competitive advantage on the track, mass reductions are desired to improve vehicle dynamics, acceleration, and braking. To improve these overall performance metrics, this investigation applies nonlinear constrained programming techniques and the finite element method (FEM) to solve a structural optimization problem, minimizing mass subject to displacement constraints. Here, material and geometric design variables are combined to consider: (a) beam cross sectional design, and (b) material type. The initial baseline design is a rigid spaceframe structure composed of hollow circular tubing of varying inner diameters and thicknesses.

Here, single-material size optimization (SMSO) is first presented to consider only geometric design variables. Next, multi-material size optimization (MMSO) is implemented to determine both the optimum tube geometry and material selection throughout the structure.

OPTIMIZATION PROBLEM STATEMENT

From the MMSO problem definition, the optimal solution must reveal the tube geometry and material type for each frame element in the structure to produce the most lightweight design. In this study, displacement constraints at the load nodes act as the performance criteria, and reflect the current 2017 all-steel baseline design. To determine the optimum structure three primary design variables are considered: inner tube radius (r_i) tube thickness (t_i), and material type, represented here by density (ρ_i).

The tube design variables were assigned reasonable bounds to reflect manufacturing and supplier availability, as well as competition guidelines. In this case, the upper bounds for both r_i and t_i were assigned 130% of the initial baseline design; the lower bounds were set to the minimum requirement for structural members in the SAE competition. For practical design considerations, the material selection variable should be an integer representing either steel or aluminum; however, to simplify the optimization problem statement a relaxed version of this design variable was established, making it continuous between the two material alternatives. Additionally, this design variable is the 0-1 normalization of physical density, signifying an aluminum (0) or steel (1) element.

This forms the multi-variable non-linear programming problem in Eq. (1), where i is the number of elements in the structure, and j is the number of nodes with explicitly assigned displacement constraints:

$$\begin{aligned} \text{minimize:} \quad & \sum(\rho_i(\rho_{ST}-\rho_{AL})+\rho_{AL})A_iL_i \qquad (1) \\ \text{such that:} \quad & \boldsymbol{K(\rho,r,t)u=F} \\ & u_j \le \bar{u} \\ & \underline{r} \le r_i \le \bar{r} \\ & \underline{t} \le t_i \le \bar{t} \\ & \boldsymbol{0 \le \rho \le 1} \\ & r_i, t_i \in \mathbb{R}^2 \end{aligned}$$

To relate the material density design variable to mechanical properties (Young's Modulus) required for calculating optimization responses a transformation function is implemented; in this case, the Solid Isotropic Material with Penalization approach proposed by Bendsøe and Sigmund is used [6]. Developed initially for topology optimization, this approach is widely utilized for its conceptual simplicity and practicality for actively eliminating intermediate density elements and forcing convergence of a continuous variable to their upper or lower bounds. In a two-phase design space, the Young's Modulus is related to material density by:

$$E(\rho)=(\rho)^p E_1+[1-(\rho)^p]E_2 \qquad (2)$$

where E_1 and E_2 represent steel and aluminum respectively. A penalty factor (p) is set to actively penalize the mechanical properties of intermediate density elements, and help produce a more physically meaningful discretized result.

FINITE ELEMENT METHOD

To solve this structural optimization problem, the finite element method (FEM) was applied to discretize the chassis into individual elements and develop the mathematical representation as shown in Figure 1. This was achieved through the application of a two-node, 3D finite frame element, which combines simple FE bar and beam elements to account for six degrees of freedom: three displacements (u_x, u_y, u_z) and three rotations (θ_x, θ_y, θ_z) in a global Cartesian coordinate system [7]. This element type assumes isotropic material properties with constant cross sectional properties and is suitable for representing the welded spaceframe structure for the Formula SAE chassis, capable of resisting in-plane bending and transvers shear forces. The load case considered in this study is global bending of the frame.

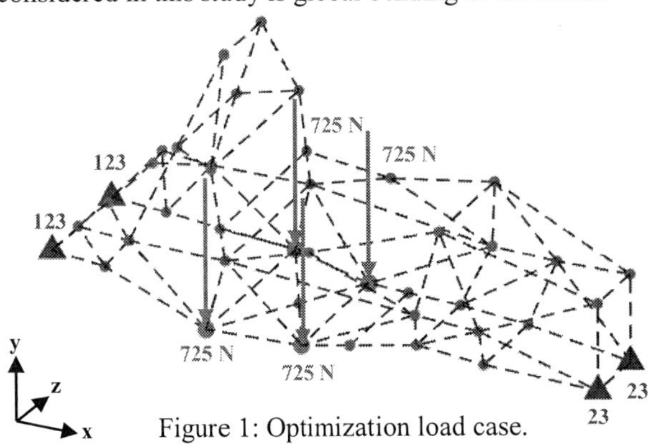

Figure 1: Optimization load case.

The finite element method is applied to the simplified 3D space frame in Figure 1, resulting in 43 distinct nodes and 97 elements. Here, curved structural elements are simplified into straight connections and non-structural components are removed entirely. Each element is subject to both changes in geometric and mechanical properties and during each iteration the local and global stiffness matrices are recalculated to determine the system response (nodal displacement) in the constraint function.

IMPLEMENTATION AND RESULTS

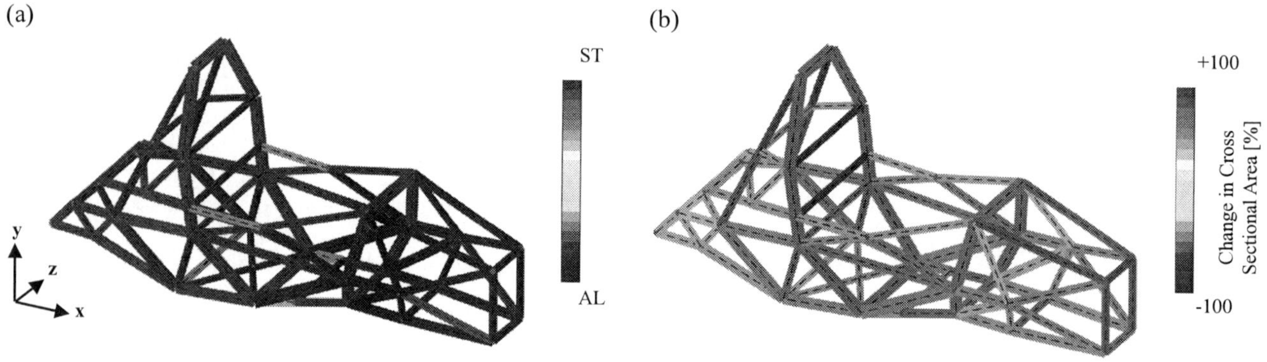

Figure 2: Results, multi-material size optimization: (a) material distribution and (b) change in tube cross sectional area.

In order to solve the nonlinear constrained optimization problem the fmincon local solver in MATLAB was utilized. This platform readily allowed a method for programming the finite element method (with nonlinear constraints) and objective function while providing flexibility for the optimization algorithm; in this analysis, the Sequential Quadratic Programming (SQP) solver was utilized. The final MMSO results from the study are shown in Figure 2, with comparisons to SMSO in Table 1.

By examining Figure 2(a), the optimum chassis design is composed of multiple steel and aluminum elements, with a limited number of intermediate densities. The steel material placement is appropriate considering the bending load case applied to the chassis, with most of the weaker aluminum elements in the front bulk head. Figure 2b) also indicates the change in tube properties for the final optimized result, which shows similar trends to the material distribution. Here, the rear tubes tended to increase in size, while elements near the front bulkhead decreased and roll hoops remained non-designable; as expected, the aluminum tubes near the rear of the chassis show a large increase in cross sectional area.

Table 1: Summary of results.

	Baseline	$SMSO^{ST}$	$SMSO^{AL}$	MMSO
$Mass_{OPT\,[kg]}$	26.48 n/a	21.78 (-17.7%)	16.39 (-38.1%)	17.98 (-32.1%)
$Status_{OPT\,[-]}$	n/a	Feasible	Infeasible	Feasible

The results in Table 1 demonstrates the advantage of a multi-material chassis design, which achieved significant mass reductions through strategic selection of tube material and geometric properties. This shows a 32.1% improvement in mass savings compared to the baseline design and 14.4% compared to conventional all-steel optimization approach. While the $SMSO^{AL}$ result would suggest a significant opportunity for mass savings, this design is infeasible under the optimization constraints and convergence criteria. In fact, given the geometric bounds, an all-aluminum design is incapable of matching the baseline displacement constraints.

MULTI-MATERIAL TOPOLOGY OPTIMIZATION

Multi-material topology optimization offers another method for implementing mixed-material design. This tool is typically applied at the component level, and can be used to determine the optimum material distribution and type within a predefined design space. In this case, MMTO is applied to an upright for the front suspension system on the Queen's Formula SAE vehicle, with four load cases applying bending and torsional stresses. Unlike the system-level design where MMSO was utilized to minimize mass, here, MMTO is implemented to improve component stiffness.

OPTIMIZATION PROBLEM STATEMENT

As an extension from conventional topology optimization, MMTO is typically implemented using a density approach and also uses the SIMP interpolation function described in Equation 2. Here, the objective is to minimize compliance subject to a mass fraction constraint as shown in Eq. (3)

$$\begin{aligned} \text{minimize:} \quad & C(\boldsymbol{u}) = \boldsymbol{F}^T \boldsymbol{u} \\ \text{such that:} \quad & \boldsymbol{Ku} = \boldsymbol{F} \\ & \sum W_i(\rho_i^1, \rho_i^2) \le \overline{W} \\ & \mathbf{0} < \boldsymbol{\rho} \le \boldsymbol{\rho}^1 \le \mathbf{1} \\ & \mathbf{0} \le \boldsymbol{\rho}^2 \le \mathbf{1} \end{aligned} \tag{3}$$

where ρ_i^1 is the material existence design variable, either void (0) or solid (1); ρ_i^2 is the material selection variable with aluminum (0) and steel (1); and W_i is element mass calculated using Eq. (4) and element volume V.

$$W(\rho^1, \rho^2) = [\rho^1(\rho_{AL} + \rho^2(\rho_{ST} - \rho_{AL}))]V \tag{4}$$

Here, MMTO is solved using a gradient based optimization algorithm (Method of Moving Asymptotes) coupled with commercial FE solvers to handle complex geometries and load cases for practical design applications.

IMPLEMENTATION AND RESULTS

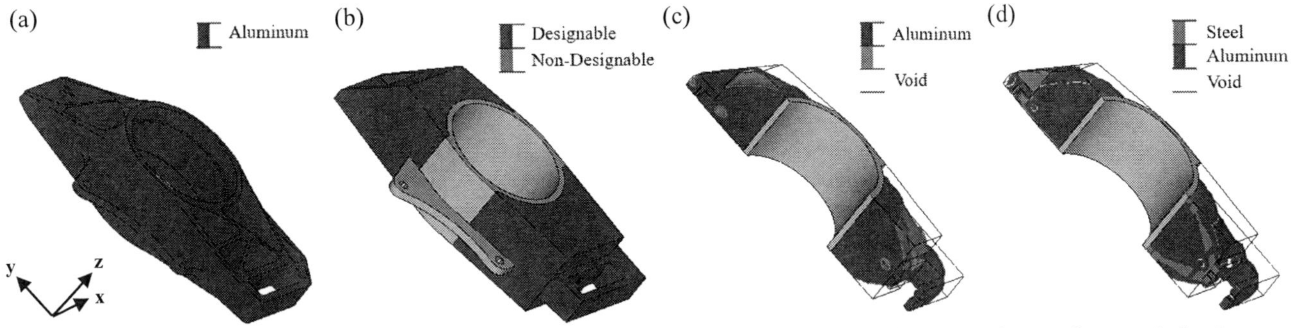

Figure 3: Multi-material topology optimization: (a) original upright design, (b) design space for topology optimization, (c) SMTO material distribution (section view), (d) MMTO material distribution and material selection (section view).

This study was completed using the geometry shown in Figure 3(a), which shows the baseline design for the front upright on the current vehicle. In order to obtain a suitable optimized design, a design domain was created in Figure 3(b), with designable and non-designable regions. Single-material topology optimization was then performed to represent a conventional design approach for improving component performance. With a mass fraction constraint of 15%, the optimum design achieved a compliance of 576.6 mJ and mass of 1.181 kg. Next MMTO was performed using an identical problem statement, achieving a static compliance of 446.5 mJ, indicating a 22.6% improvement from the all-aluminum design in Figure 3(c). Here, Figure 3(d) shows how steel is utilized to stiffen critical connection points around the non-designable regions, and areas of the applied load to resist the bending and torsional load cases. These results are summarized in Table 2.

Table 2: Summary of SMTO and MMTO results.

	$SMTO^{AL}$	MMTO
$Compliance_{OPT\ [mJ]}$	576.6 (n/a)	446.5 (-22.6%)
$Mass_{OPT\ [kg]}$	1.181 (n/a)	1.181 (n/a)

CONCLUSIONS

The practical applications of multi-material optimization were studied in both system and component-level design for a high-performance Formula SAE vehicle. The analysis from multi-material size optimization revealed significant advantages for incorporating aluminum and steel members in a space frame chassis design with mass savings of 32.1% compared to the baseline design. At the component-level, multi-material topology optimization was applied to improve stiffness while satisfying a user defined mass target. In this case, MMTO utilized a minimum compliance problem statement and outperformed the conventional SMTO approach by 22.6%.

These findings indicate the vast potential for multi-material design in practical applications when coupled with conventional and modern tools. From these case studies it is clear that selecting the optimum material type, distribution and geometry simultaneously provides significant benefit to global optimization objectives, and offers a compelling method for creating non-intuitive and competitive designs.

ACKNOWLEDGMENTS

This research was funded by Automotive Partnership Canada and General Motors of Canada. Technical advice and direction were gratefully received from Joe Moore, Balbir Sangha, Manish Pamwar, Derrick Chow, and Chandan Mozumder, at General Motors.

REFERENCES

[1] I. Y. Kim and B. M. Kwak, "Design space optimization using a numerical design continuation method," *International Journal of Numerical Methods,* vol. 53, pp. 1979 - 2002, 2002.

[2] J. A. Bennett, "Automated Design of Truss and Frame Geometry," *Computers and Structures,* vol. 8, pp. 717-721, 1978.

[3] K. Svanberg, "Optimization of Geometry in Truss Design," *Computer Methods in Applied Mechanics and Engineering,* vol. 28, pp. 63-80, 1981.

[4] C. Li, I. Y. Kim and J. Jeswiet, "Conceptual and detailed design of an automotive engine cradle by using topology, shape, and size optimization," *Structural and Multidisciplinary Optimization,* vol. 51, no. 2, pp. 547 - 564, 2015.

[5] C. Li and I. Y. Kim, "Topology, size and shape optimization of an automtoive cross car beam," *Journal of Automtoive Engineering,* vol. 229, no. 10, pp. 1361 - 1378, 2015.

[6] M. Bendsøe and O. Sigmund, Topology Optimization: Theory, Methods, and Applications, New York: Springer, 2003.

[7] A. Ferreira, MATLAB Codes for Finite Element Analysis, Porto: Springer, 2009.

MULTI-MATERIAL LIGHTWEIGHT DESIGN BY TOPOLOGY OPTIMIZATION

Daozhong Li, Stephen Roper, Christopher Woischwill, Christopher Carrick, Il Yong Kim[1]

Department of Mechanical and Materials Engineering
Queen's University, Kingston, ON, Canada
[1]kimiy@queensu.ca

ABSTRACT

This paper describes the mathematical foundation for multi-material topology optimization (MMTO), which extends the classical material interpolation models and updates the sensitivity expressions. This provides a method for readily performing multi-material topology optimization to simultaneously determine both material selection and distribution within a given design space. Also presented are two numerical examples demonstrating current capability to solve a minimum compliance problem.

KEYWORDS: topology optimization, multi-material, SIMP

NOMENCLATURE

In this paper, the top-right subscript of a quantity that is relevant to materials indicates the material choice. Important symbols are listed below, while other quantities not in this list follow standard mathematical conventions.

ρ^j	one design variable relevant to the *j*th material, generally in any design cell
ρ_e^j	one design variable relevant to the *j*th material in the *e*th design cell
$\boldsymbol{\rho}^j$	all design variables relevant to the *j*th material
$(\rho^j)^p$, $(\rho_e^j)^p$	powers of ρ^j and ρ_e^j with the component p
E^j	original elastic modulus of the *j*th material
W^j	original element weight of the *j*th material, generally in any element

INTRODUCTION

With the homogenization method, the classical topology optimization (TO) problem was converted to the problem of analyzing a "porous" structure [1]. The homogenization concept has been further developed in many different approaches, including density and level set [2], among others [3]. In 2001, Eschenauer and Olhoff [4] provided a comprehensive review of this field, which was also summarized by Bendsøe and Sigmund [5] covering research before 2003.

In the density approach [6], the design variables (which are termed densities) are used to determine material properties through an interpolation function. For example, the well-known Solid Isotropic Material with Penalization (SIMP) method adopts a power function as the relationship between density and material properties. Another material interpolation model, Rational Approximation of Material Properties" (RAMP) [3], [7], was developed to alleviate some issues associated with low-density elements in the SIMP approach.

Single material topology optimization (SMTO) with isotropic materials is a mature field and has been widely applied in many industry applications. For example, Kim and his colleagues published several papers on SMTO design optimization and topology optimization in automotive structures [8]–[13], where they also investigated damping and large-scale problems [13], [14]. Multi-material topology optimization (MMTO) however remains under preliminary development. Here, only limited case studies have been performed, such as MMTO in the design of multiphysics actuators [15] by solving a minimum compliance problem with weight constraints. Other examples apply the level-set approach [16]–[18] to MMTO, which generates clear and crisp definitions of the optimized geometry by iso-contours of the level-set functions, however this approach suffers from slow convergence, objective oscillations, local minima and strong dependency on the initial guess [2].

This paper introduces the MMTO methodology based on the SIMP interpolation model and examines its integration with the general mathematical expression of the standard topology optimization problem. This includes practicle discussions of numerical technqiues such as checkerboard filtering, as well as sample case studies to demonstrate the advantages of MMTO.

MULTI-MATERIAL INTERPOLATION

The SIMP method interpolates Young's Modulus using Equation 1:

$$E^{(1,2)}(\rho) = (\rho)^p E^1 + [1-(\rho)^p]E^2 \qquad (1)$$

where $\rho \in [0,1]$ is the artificial density design variable; $E^{(1,2)}$ are the interpolated elastic moduli of two materials; E^1 and E^2 are the original elastic moduli of two materials; and p is the penalty factor to penalize intermediate density elements (typically $p=3$). For single material topology optimization (SMTO), $E^2=0$ represents void material (i.e. no material existence).

The SIMP function shown in Equation 1 can be extended for M materials as shown in the recursive equation below:

$$\begin{aligned} &E^{(1,\cdots,M)}(\rho^1,\rho^2,\cdots,\rho^{M-1}) \\ &=(\rho^1)^p E^{(1,\cdots,M-1)}+[1-(\rho^1)^p]E^M \\ &=(\rho^1)^p\{(\rho^2)^p E^{(1,\cdots,M-2)}+[1-(\rho^2)^p]E^{M-1}\} \\ &\quad +[1-(\rho^1)^p]E^M \\ &=\cdots \end{aligned} \tag{2}$$

where p is the penalty factor and $E^1>E^2>\cdots>E^M\geq 0$. Correspondingly, the relationship of the original weights (or physical densities) of the M materials is $W^1>W^2>\cdots>W^M\geq 0$. In both research and practical applications, materials are typically sorted in this way such that the heavier materials are completely replaced by a lighter alternative if the lighter material is stronger (without considering any non-mechanical constraints such as cost). That is, weaker and heavier materials should not exist as a viable material option.

TOPOLOGY OPTIMIZATION PROBLEMS

In multi-material topology optimization, for M materials, the design variable vector $\boldsymbol{\rho}$ is explicitly written as $\boldsymbol{\rho}^1,\boldsymbol{\rho}^2,\cdots,\boldsymbol{\rho}^M$. Thus, the two-material minimum compliance problem under volume fraction constraints will be written as Equation 3:

$$\begin{aligned} \text{min.}\quad & C(\boldsymbol{u})=\boldsymbol{F}^T\boldsymbol{u} \\ \text{s.t.}\quad & g_1(\boldsymbol{\rho}^1)=\Sigma_e \rho_e^1 V_e/\Sigma_e V_e\leq\overline{\Delta_1} \\ & g_2(\boldsymbol{\rho}^1,\boldsymbol{\rho}^2)=\Sigma_e \rho_e^1\rho_e^2 V_e/\Sigma_e V_e\leq\overline{\Delta_2} \\ & \boldsymbol{Ku}=\boldsymbol{F}\ \text{(for linear static problems)} \\ & \forall e,\rho_e^j\in[0,1], j=1,2 \end{aligned} \tag{3}$$

where ρ_e^j represents any design variable in the eth design cell, $\overline{\Delta_1}$ and $\overline{\Delta_2}$ are the volume fraction limits which must satisfy $\overline{\Delta_1}\geq\overline{\Delta_2}$, V_e is the volume of the eth design cell; $\boldsymbol{u}$ is the displacement vector, $\boldsymbol{F}$ is the external force vector, and $\boldsymbol{K}$ is the global stiffness. The sensitivities are calculated per Equations 4.1 – 4.3:

$$\frac{\partial C}{\partial\rho_e^j}=-\boldsymbol{u}^T\frac{\partial C}{\partial\rho_e^j}\boldsymbol{u}=-\frac{C_e}{E_e}\frac{\partial E_e}{\partial\rho_e^j} \tag{4.1}$$

$$\frac{\partial g_1}{\partial\rho_e^1}=\frac{V_e}{\Sigma_i V_i}\qquad \frac{\partial g_1}{\partial\rho_e^2}=0 \tag{4.2}$$

$$\frac{\partial g_2}{\partial\rho_e^1}=\frac{\rho_e^2 V_e}{\Sigma_i V_i}\qquad \frac{\partial g_2}{\partial\rho_e^2}=\frac{\rho_e^1 V_e}{\Sigma_i V_i} \tag{4.3}$$

where C_e, E_e are the compliance and current Young's modulus of the design cell, respectively, and $\boldsymbol{K}_e$ is the element stiffness in the global level.

CHECKERBOARD FILTERS

In practical applications one common numerical instability is the checkerboard phenomenon, as shown in Figure 1. Checkerboard structures are not only difficult to physically interpret, but also overpredict stiffness. To obtain a checkerboard-free structure, a mesh-independent filter can be applied, as described in Equation 5.1 and 5.2.

Figure 1: Checkerboard phenomenon.

$$\left(\frac{\partial\Phi}{\partial\rho_i}\right)^*=\frac{\sum_{k=1}^{N}\left(H_k\rho_k\frac{\partial\Phi}{\partial\rho_k}\right)}{\rho_i\sum_{k=1}^{N}H_k} \tag{5.1}$$

$$H_k=\max_k\{r-dist(k,i),0\} \tag{5.2}$$

The sensitivity of the ith design cell, $\frac{\partial\Phi}{\partial\rho_i}$, is modified to $\left(\frac{\partial\Phi}{\partial\rho_i}\right)^*$, which is a weighted sum of the sensitivities of neighboring elements (the ith design cell is excluded) within a circle of a given radius, r, where $dist(k,i)$ is the center-to-center distance between the ith design cell and its kth neighbour. N is the number of neighbours. For the multi-material problem, Eq. (5) should be extended to Eq. (6) so that the filter will be applied to $\boldsymbol{\rho}^1$ and $\boldsymbol{\rho}^2$, respectively.

$$\left(\frac{\partial\Phi}{\partial\rho_i^j}\right)^*=\frac{\sum_{k=1}^{N}\left(H_k\rho_k^j\frac{\partial\Phi}{\partial\rho_k^j}\right)}{\rho_i^j\sum_{k=1}^{N}H_k} \tag{6}$$

NUMERICAL STUDIES

Two finite element models are generated to show the advantage of the MMTO method. For simplicity, we assign two isotropic materials: steel and aluminum with E^1=210GPa, E^2=70GPa, ρ_{phy}^1=7.8g/cm^3, ρ_{phy}^2=2.7g/cm^3, and Poisson's ratio v=0.3. All problems are solved using the Method of Moving Asymptotes (MMA) with the SIMP interpolation scheme ($p=3$). The MMTO results are discretized by the elastic modulus rule; for example, when the elastic modulus of an element is less than $E_{crit}=5\%\cdot E^2$, the element is set as void. Similarly, if the elastic modulus of the element is larger than E^2, the element is set as steel, and when the elastic modulus is between E_{crit} and E^2, the element is set as aluminum.

For the 2D model, the SMTO-steel minimum compliance problem uses one volume fraction constraint ($\overline{\Delta_1}\le 30\%$), while MMTO requires an extra volume fraction constraint ($\overline{\Delta_2}\le 15\%$). For the 3D model, we use $\overline{\Delta_1}\le 20\%$ and $\overline{\Delta_2}\le 10\%$.

CASE 1: MBB BEAM LIGHTWEIGHT DESIGN

The first model, as shown in Figure 2, is a Messerschmitt-Bölkow-Blohm (MBB) beam with 1mm thickness. The model is composed of 8,000 design cells. A 1,000N upward nodal force is located at the midpoint of the top edge. The objectives and iteration histories of the two problems are shown in Table 1.

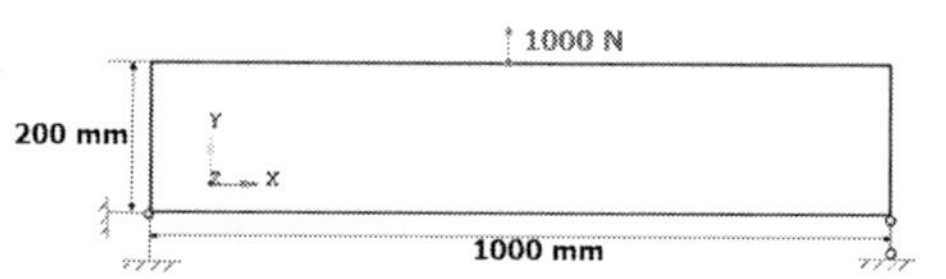

Figure 2: MBB beam design domain

Table 1 shows that the MMTO solution is lighter than the SMTO result, but has a larger compliance compared to the all-steel design. The material distribution of both SMTO and MMTO is shown in Fig. 3. The multi-material layout under the bending load case confirms expectations: the strong material (steel) is located on the top and bottom edges.

Table 1: Topology Optimization History of the MBB Beam

	Compliance$_{OPT}$ [mJ]	Mass$_{OPT}$ [kg]	No. of Iterations
SMTO^{STEEL}	653.113	0.468 (-70.00%)	27
MMTO	1031.884	0.304 (-80.51%)	32

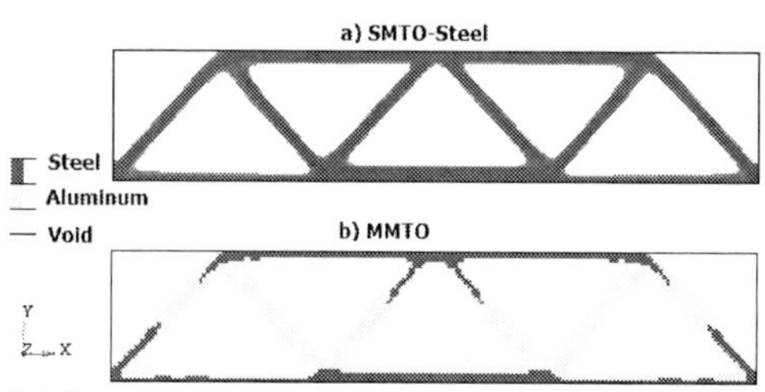

Figure 3: Material distributions of the MBB beam: a) SMTO-steel solution; b) MMTO solution.

CASE 2: 3D BOX LIGHTWEIGHT DESIGN

The second numerical example is the 3D box as shown in Figure 4, with a domain consisting of 38,400 design cells. The right side is fixed and a 1,000 N force along the +Z axis is located at the center of the left side. The objectives and iteration histories of the two problems are shown in Table 2.

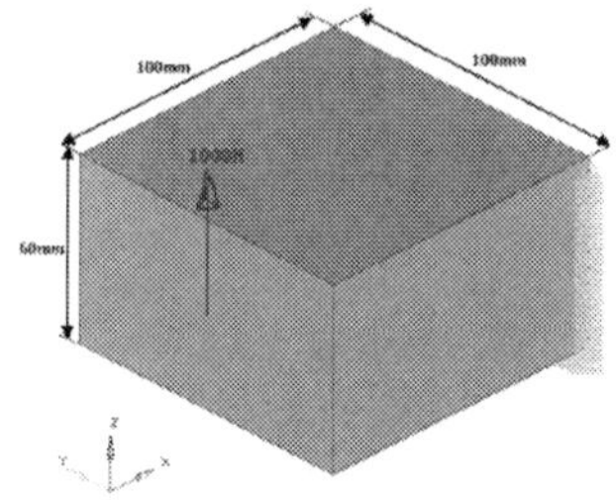
Figure 4: Design domain of the box.

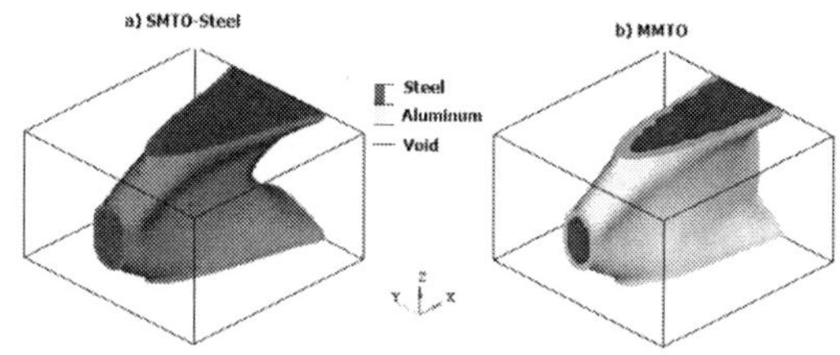

Figure 5: Material distributions of the 3D box: a) SMTO-steel solution; b) MMTO solution.

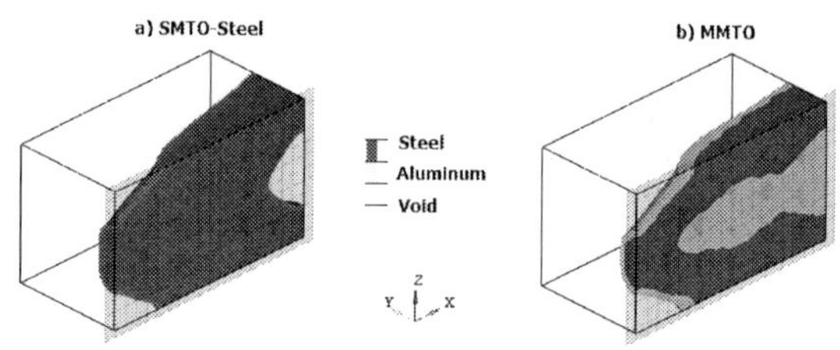

Figure 6: Sectional views of material distributions of the 3D box: a) SMTO-steel solution; b) MMTO solution.

Similar to the MBB case, Table 2 shows that the MMTO result is lighter than the SMTO result, but is also more compliant than the SMTO design. Figures 5 and 6 demonstrate the material distribution, where the high-stress regions form two bars connecting from the fixed end of the box to the location of the external force.

Table 2: Topology Optimization History of 3D Box

	Compliance$_{OPT}$ [mJ]	Mass$_{OPT}$ [kg]	No. of Iterations
SMTO^{STEEL}	8.306	1.394 (-80.14%)	20
MMTO	11.376	0.612 (-91.28%)	18

From both cases, the volume fraction limits are critical to the final solutions. However, those values are selected arbitrarily by users. It is difficult to give proper limits because neither compliance nor fractions directly relate to the performance indices. Thus, the minimum compliance problem under volume fractions is typically used to find load paths in practical applications.

CONCLUSIONS

This paper presents the multi-material lightweight design method of solving minimum compliance problems with multi-material interpolation models. The numerical examples prove that MMTO can determine the best material choices and distribution. Nevertheless, the benefit of MMTO over SMTO seems unclear: weights decrease, but compliances also increase. The reason is that neither the compliance nor the arbitrary volume fractions directly relate to the performance indices in practical applications. Thus, the MMTO minimum compliance problem subject to volume fractions can be used to find load paths and determine the material selection and distribution in practical applications. To obtain lighter designs that satisfy performance requirements, the authors will continue to research lightweight multi-material design by solving minimum weight problems under performance constraints instead.

ACKNOWLEDGMENTS

This research was funded by Automotive Partnership Canada and General Motors of Canada. Technical advice and direction were gratefully received from Joe Moore, Balbir Sangha, Manish Pamwar, Derrick Chow, and Chandan Mozumder, at General Motors.

REFERENCES

[1] M. P. Bendsøe and N. Kikuchi, "Generating optimal topologies in structural design using a homogenization method," *Comput. Methods Appl. Mech. Eng.*, vol. 71, no. 2, pp. 197–224, Nov. 1988.

[2] N. P. van Dijk, K. Maute, M. Langelaar, and F. van Keulen, "Level-set methods for structural topology optimization: a review," *Struct. Multidiscip. Optim.*, vol. 48, no. 3, pp. 437–472, Sep. 2013.

[3] O. Sigmund and K. Maute, "Topology optimization approaches," *Struct. Multidiscip. Optim.*, vol. 48, no. 6, pp. 1031–1055, Dec. 2013.

[4] H. A. Eschenauer and N. Olhoff, "Topology Optimization of Continuum Structures: A review," *Appl. Mech. Rev.*, vol. 54, no. 4, pp. 331–390, 2001.

[5] M. P. Bendsøe and O. Sigmund, *Topology Optimization: Theory, Methods and Applications*, 2nd ed. Springer, 2003.

[6] M. P. Bendsøe, "Optimal shape design as a material distribution problem," *Struct. Optim.*, vol. 1, no. 4, pp. 193–202, Dec. 1989.

[7] M. Stolpe and K. Svanberg, "An alternative interpolation scheme for minimum compliance topology optimization," *Struct. Multidiscip. Optim.*, vol. 22, no. 2, pp. 116–124, Sep. 2001.

[8] I. Y. Kim and B. M. Kwak, "Design space optimization using a numerical design continuation method," *Int. J. Numer. Methods Eng.*, vol. 53, no. 8, pp. 1979–2002, Mar. 2002.

[9] C. Li, I. Y. Kim, and J. Jeswiet, "Conceptual and detailed design of an automotive engine cradle by using topology, shape, and size optimization," *Struct. Multidiscip. Optim.*, vol. 51, no. 2, pp. 547–564, Feb. 2015.

[10] N. Cristello and I. Y. Kim, "Multidisciplinary design optimization of a zero-emission vehicle chassis considering crashworthiness and hydroformability," *Proc. Inst. Mech. Eng. Part D J. Automob. Eng.*, vol. 221, no. 5, pp. 511–526, Jan. 2007.

[11] C. Li and I. Y. Kim, "Topology, Size and Shape Optimization of an Automotive Cross Car Beam," *Proc. Inst. Mech. Eng. Part D, J. Automob. Eng.*, vol. 229, pp. 1361–1378, 2015.

[12] N. Cristello and I. Y. Kim, "Multi-objective shape optimisation of an automotive universal joint assembly," *Int. J. Heavy Veh. Syst.*, vol. 16, no. 3, pp. 271–293, 2009.

[13] S. Y. Kim, I. Y. Kim, and C. K. Mechefske, "A new efficient convergence criterion for reducing computational expense in topology optimization: reducible design variable method," *Int. J. Numer. Methods Eng.*, vol. 90, no. 6, pp. 752–783, May 2012.

[14] S. Y. Kim, C. K. Mechefske, and I. Y. Kim, "Optimal damping layout in a shell structure using topology optimization," *J. Sound Vib.*, vol. 332, no. 12, pp. 2873–2883, Jun. 2013.

[15] O. Sigmund, "Design of multiphysics actuators using topology optimization - Part II: Two-material structures," *Comput. Methods Appl. Mech. Eng.*, vol. 190, no. 49–50, pp. 6605–6627, 2001.

[16] G. Allaire, F. Jouve, and A.-M. Toader, "Structural optimization using sensitivity analysis and a level-set method," *J. Comput. Phys.*, vol. 194, no. 1, pp. 363–393, Feb. 2004.

[17] J. A. Sethian and A. Wiegmann, "Structural Boundary Design via Level Set and Immersed Interface Methods," *J. Comput. Phys.*, vol. 163, no. 2, pp. 489–528, Sep. 2000.

[18] M. Y. Wang, X. Wang, and D. Guo, "A level set method for structural topology optimization," *Comput. Methods Appl. Mech. Eng.*, vol. 192, no. 1–2, pp. 227–246, Jan. 2003.

TOPOLOGY OPTIMIZATION USING PERIDYNAMICS METHOD

A. Abdelhamid
Department of Mechanical Engineering, University of Victoria, British Columbia, Canada,
aahmed@uvic.ca

A. Sohouli
MIT-Portugal Program, IDMEC-CCTAE, Instituto Superior Técnico, University of Lisbon, Portugal,
abdolrasoul.sohouli@ist.utl.pt

N. Tofighi
Department of Mechanical Engineering, University of Victoria, British Columbia, Canada,
ntofighi@uvic.ca

M. Yildiz
Faculty of Engineering and Natural Sciences, Sabanci University, Istanbul, Turkey
meyildiz@sabanciuniv.edu

A. Suleman
Department of Mechanical Engineering, University of Victoria, British Columbia, Canada,
suleman@uvic.ca

ABSTRACT

This paper presents a topology optimization method based on the peridynamics method. An efficient heuristic topology optimization method is used to avoid calculation of the gradients of the objective function and constraints. The strain energy is selected as the objective function and it is optimized based on a modified solid isotropic material with penalization approach. The results show the robustness and suitability of the proposed peridynamics method in the topology optimization problem.

KEYWORDS: Topology optimization, Peridynamics, Meshless method, Heuristic optimization, Sensitivity analysis

INTRODUCTION

Peridynamics (PD) is a numerical method in the computational mechanics field that has been proposed to predict deformations with discontinuities. The original PD, later named as bond-based PD theory, is a meshless method that was introduced by Silling [1], and subsequently generalized as a state-based PD theory by Silling et al. in [2] and represented as a nonlocal theory. The core concept behind PD is its reformulation of the Classical Continuum Mechanics (CCM) equations to better-suited forms. PD does not consider stress or strain, and the non-locality existing in the PD integro-differential equation of motion establishes the relationship between CCM and molecular dynamics for the objective of enabling long-range forces and fracture problem solutions. The nonlinear integro-differential equation of motion in the PD theory contains spatial integrals and time differentiation in order to permit discontinuities and it is usually unfavorable to be solved analytically.

The presentation of PD models for functionally graded materials (FGMs) in the literature is infrequent. But in [3], a bond-based PD model for FGMs was. PD formulations for the composite type materials were modified to consider the material property difference in the design domain and a conical bond function was also introduced.

The classical topology optimization may be classified as gradient and non-gradient methods. As the name suggests, calculation of the gradients of the objective function and constraints are required in gradient based methods. Optimality Criteria (OC), SLP, SQP, CONLIN and MMA are a few examples of gradient methods. More commonly used methods in non-gradient methods are Evolutionary Structural Optimization (ESO) method, Bidirectional ESO (BESO), Genetic Algorithm (GA), Simulated Annealing Optimization (SAO). Svanberg and Werme [4] reported that continuous design variables are more successful for minimization of the objective function in the topology optimization methods. Thus, a non-gradient method with continuous design variables is the most suitable option for topology optimization using PD due to easy implementation and efficiency. The Solid Isotropic Material with Penalization (SIMP) is a widely used approach in topology optimization presented by Rozvany and Zhou [5].

In this study, a modified SIMP approach is used with PD. The strain energy is minimized with the volume fraction constraint where the materials are defined for the particles. A heuristic and non-gradient optimization method introduced by [6] is implemented in the PD formulation.

PERIDYNAMICS THEORY

In PD theory, a body or a material is discretized to a number of volumetric material points. Each material point interacts with its neighboring points within a finite distance denoted as the "horizon" (δ). The deformation of point (x_k) is dependent on the total deformation of the neighboring points or "family members" H_{x_k}. Each interaction is donated as a "bond", and represented as the position vector between point x_k and a family member x_j in the initial condition and between y_k and y_j after deformation. Furthermore, due to the influence of H_{x_k} on point x_k, force density vectors appears to be the force exerted by point x_j on point x_k as t_{kj}, vice versa.

Each interaction of a material point with its neighboring points within a finite distance develops a scalar-valued micropotential, w_{kj}, and w_{jk} vice versa, that depends on both the stretch between interacting points and the property of the material. It is defined as follows:

$$w_{kj} = \frac{\delta}{|x_j - x_k|} \quad (1)$$

The summation of micropotentials developed from the interactions of material points with each of the family members within the horizon is denoted as the strain energy density, W_k, as defined by:

$$W_k = b\sum_{j=1}^{N} w_{kj}\left(|y_j - y_k| - |x_j - x_k|\right)V_j \quad (2)$$

where V_j is the volume of material point j. Considering the conical effects of the family members [3], a PD parameter b is defined for two-dimensional plate as:

$$b = \frac{6E}{\pi\delta^4(1-v)}\left(1 - \frac{x_j - x_k}{\delta}\right) \quad (5)$$

where E is the average modulus of elasticity of any two interacting points according to [3]:

$$E = \frac{E_k + E_j}{2} \quad (6)$$

By using the principle of virtual work and solving the Lagrange's equation L, the PD equations of motion at point x_k can be derived as:

$$\frac{d}{dt}\left(\frac{\partial L}{\partial \dot{u}_k}\right) - \left(\frac{\partial L}{\partial u_k}\right) = 0 \quad (7)$$

$$L = T - U \quad (8)$$

where T represents the kinetic energy, and U represents the potential energy of the body [7]. The integral equation of motion may be defined considering a horizon H for a material point k interacting with neighbors $j = 1,..,N$ as:

$$\rho(k)\ddot{u}(k,t) = \int_H \left(t_{kj}(u_j - u_k, x_j - x_k, t) - t_{jk}(u_k - u_j, x_k - x_j, t)\right) dH + b(k,t) \quad (9)$$

where b is the body force, ρ is the material density, and ü is the acceleration. t_{kj} is one of pairwise force density vectors and it is defined as:

$$t_{kj}(u_j - u_k, x_j - x_k, t) = \frac{1}{V_j}\frac{\partial W_k}{\partial(|y_j - y_k|)}\frac{y_j - y_k}{|y_j - y_k|} \quad (10)$$

There are no spatial derivatives on PD equation of motion. In this case, constraint conditions are not needed on the integro-differential equation. But, such conditions can only be introduced by proposing constraints on displacement and velocity fields of an imaginary material layer along the boundary of the real material. Based on numerical experiments from [2], it was suggested that to ensure accuracy, the extent of the imaginary boundary layer is equal to a horizon size. In the PD equations of motion, the application of external loads is different from the classical continuum theory as the boundary traction does not appear. In the PD theory, external loads should be applied as body forces in the real material. Hence, any extent of an imaginary boundary layer should be very small and close to the boundary. The introduced extend of the boundary layer for external load will have a depth of a "grid size" (Δ).

TOPOLOGY OPTIMIZATION

The statement of topology optimization problem is modified for the PD method. Thus, the minimum strain energy problem with a material volume constraint may be stated as:

$$\begin{cases} \min W \\ \text{such that} \begin{cases} \sum_{k=1}^{N_t} \rho_k V_k = \bar{V} \\ 0 \le \rho_i \le 1 \end{cases} \end{cases} \quad (11)$$

where ρ_k is the density or the design variable, $\bar{V}$ is the volume fraction constraint and N_t is the total number of PD particles. The compliance of the structure is represented by C. The material property of each particle is defined based on the modified SIMP approach:

$$E(\rho_k) = E_{min} + \rho_k^P(E_0 - E_{min}) \quad (12)$$

Here, E_0 and E_{min} are the elastic modulus of the solid material and elastic modulus of the void material, respectively. P is the penalization coefficient. The optimal layout is found using a non-gradient introduced in [6], which it is modified for the PD approach. In the first step, the particle density is distributed according to the compliance of the particle by:

$$\rho_k^{opt} = \frac{RM}{\sum_{k=1}^{N_t} C_k^q} C_k^q \quad (13)$$

Where ρ_k^{opt} is the optimal particle density in the current iteration. RM is the remaining material amount, q is the proportion exponent and C_i is the compliance of the particle. Then, the particle density is updated for the next step as:

$$\rho_k^{new} = \alpha\rho_k^{prev} + (1-\alpha)\rho_k^{opt} \tag{14}$$

Here ρ_k^{new} is the new particle density in the next iteration, ρ_k^{prev} is the particle density in the last iteration. The history coefficient α is used to control the effect of the last iteration on the current iteration. A density filtering for the particle i is defined as:

$$\hat{\rho}_k = \frac{1}{\sum_{j=1}^{N}\xi_{kj}}\left(\sum_{j=1}^{N}\xi_{kj}\rho_k^{new}\right) \tag{15}$$

where assuming filter radius R, the filter weights ξ_{kj} are calculated as follows:

$$\xi_{kj} = 1 - \frac{|x_k - x_j|}{R} \tag{16}$$

Then, Heaviside projection operator is employed to obtain a distinct layout as [8]:

$$A_i = 1 - \frac{\hat{\rho}_i}{\eta} \tag{17}$$

$$B_i = \frac{\hat{\rho}_i - \eta}{1-\eta} \tag{18}$$

$$\rho_i = \begin{cases} \eta[exp(-\beta A_i) - A_i exp(-\beta)] & if\ 0 < \hat{\rho}_i \le \eta \\ (1-\eta)[1 - exp(-\beta B_i) + B_i exp(-\beta)] + \eta & if\ \eta < \hat{\rho}_i \le 1 \end{cases} \tag{19}$$

In which β and η are the projection parameters. In this study, the penalization coefficient is 3 and η is 0.5. A continuation projection scheme is applied to achieve a distinct optimal topology. Thus, β starts at one and increases in increments of 0.15 with 20 iterations until the design variables attain a constant value sin the optimization.

NUMERICAL EXAMPLE

A well-known case study is selected to test the proposed topology optimization approach. A cantilevered beam with a width to length ratio of two under a concentrated load is shown in Figure 1. The design domain is discretized into 60×30 uniformly distributed particles. The elastic modulus of the solid material and elastic modulus of the void material are selected as 10^7 GPa and 10^4 GPa, respectively. A PD horizon size of 4 particles is used for this case.

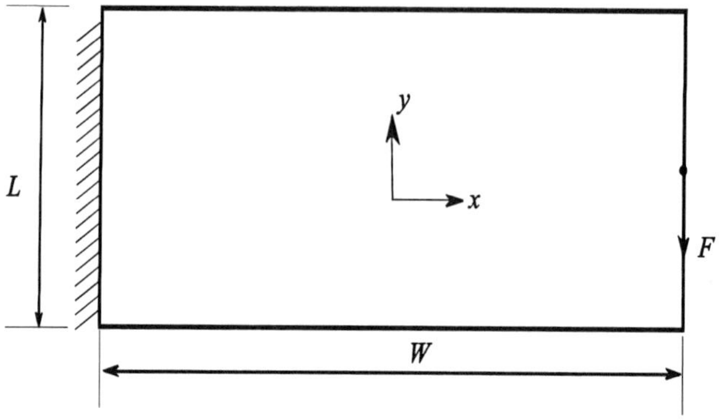

Figure 1: Geometry and loading of cantilever beam.

The optimal result for filter radius of 1.5 patches is shown in Figure 2. The material volume constraint is 50% of the design domain. The topology result of the proposed method shows very good agreement with the FE method

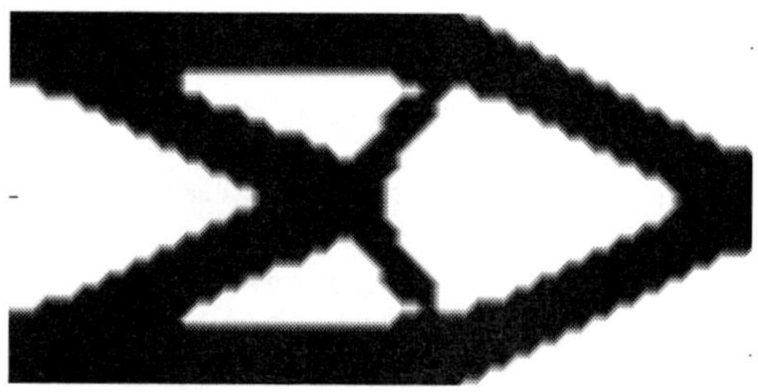

Figure 2: Optimal topology by the PD method.

CONCLUSIONS

In this study, an alternative topology optimization approach in PD was developed. A heuristic optimization method is applied to avoid depravities calculations. A continuation projection scheme has been used to map non-distinct valued design variables into discrete values. It has been shown that the proposed approach is a very effective and robust approach to find optimal topologies.

ACKNOWLEDGMENTS

The authors would like to acknowledge the NSERC Canada Research Chair and Discovery Grant Programs.

REFERENCES

1. Silling, S.A., 2000. Reformulation of elasticity theory for discontinuities and long-range forces. *Journal of the Mechanics and Physics of Solids*, *48*(1), pp.175-209.

2. Silling, S.A., Epton, M., Weckner, O., Xu, J. and Askari, E., 2007. Peridynamic states and constitutive modeling. *Journal of Elasticity*, *88*(2), pp.151-184.

3. Cheng, Z., Zhang, G., Wang, Y. and Bobaru, F., 2015. A peridynamic model for dynamic fracture in functionally graded materials. *Composite Structures*, *133*, pp.529-546.

4. Svanberg, K. and Werme, M., 2007. Sequential integer programming methods for stress constrained topology optimization. Structural and Multidisciplinary Optimization, 34(4), pp.277-299.

5. Rozvany, G.I.N. and Zhou, M., 1991. Applications of the COC algorithm in layout optimization. In Engineering optimization in design processes, pp. 59-70.

6. Biyikli, E. and To, A.C., 2015. Proportional Topology Optimization: A New Non-Sensitivity Method for Solving Stress Constrained and Minimum Compliance Problems and Its Implementation in MATLAB. PloS one, 10(12), p.e0145041.

7. Madenci, E. and Oterkus, E., 2014. *Peridynamic theory and its applications* (Vol. 17). New York: Springer.

8. Li, L. and Khandelwal, K., 2015. Volume preserving projection filters and continuation methods in topology optimization. Engineering Structures, 85, pp.144-161.

Proceedings of the 26th CANCAM
Victoria, British Columbia, Canada, May 29 - June 1, 2017

ANT COLONY OPTIMIZATION ALGORITHM APPLIED TO TOPOLOGY OPTIMIZATION OF AIRCRAFT STRUCTURES

M. Bras* and A. Suleman
University of Victoria, Victoria B.C., Canada

R. Diogo and A. C. Marta
Instituto Superior Técnico, Universidade de Lisboa, Lisbon, Portugal

ABSTRACT

A structural Topology Optimization (TO) of the cross-section of a wing-box is carried out using an Ant Colony Optimization (ACO) method. ACO is a meta-heuristic biologically influenced algorithm that has been proven to be useful to solve NP-hard combinatorial optimization problems in an expedite way. Its application to solving topology optimization has been introduced recently.

The algorithm is first used with a cantilever plate example and the results compared with the literature. The algorithm is then coupled with an external finite element (FEM) solver to perform a topology optimization of the cross-section of a wing box. The external aerodynamic loads are computed from a computational fluid dynamics (CFD) analysis for a specific flight condition and applied to the structural model.

The advantages of using ACO to discrete TO problems are demonstrated and the results of the optimization are discussed.

Keywords: heuristic methods, ant colony optimization, topology optimization, aircraft structures

INTRODUCTION

Ant Colony Optimization (ACO) is a combinatorial optimization method inspired by the foraging behavior of ants. When exploring the surrounding environment and upon finding food, ants leave behind a trail of pheromones to mark the path. If other ants find such path, they are likely to follow it and stop travelling at random. Colorni, Dorigo, and Maniezzo [2] developed the Ant System (AS) as a novel nature-inspired metaheuristic algorithm to solve combinatorial optimization problems. Different variations of the AS algorithm were developed, such as Elitist Ant System (EAS) [3], $\mathcal{MAX}-\mathcal{MIN}$ Ant System ($\mathcal{MM}$AS) [9, 10], and Rank-based Ant System (RAS) [1]. The implementation of the AS is listed in Algorithm 1.

A common combinatorial optimization problem is the Travelling Salesman Problem (TSP) that requires the Hamiltonian cycle with the least total distance a salesman can take through each city. To apply ACO to the TSP problem, ants are simulated moving around a graph of connected nodes such that

- the ant can choose one city to travel to at a time. Each city is chosen according to a probability function that considers the town distance and the amount of pheromones on the path connecting the towns;
- previously visited towns cannot be visited again by the agent, to guarantee a Hamiltonian cycle;
- upon completing a tour, the agent lays down a pheromone trail on the path travelled according to the *pheromone update rule.* [4]

When applied to Topology Optimization (TO) problems, metaheuristic methods, such as the ACO method, present advantages when compared to gradient-based algorithms. One of these is the ability to use a wider variety of objective functions. A metaheuristic algorithm does not require the objective function to be differentiable, as the search process is stochastic and not based on the gradient-based. Additionally, the discrete nature of a TO problem makes the ACO method straightforward to implement and no modifications need to be made to the problem. In contrast, a gradient-based algorithm requires the optimization variables to be continuous and modification to the TO problem.

*Corresponding author. E-mail: mbras@uvic.ca

Algorithm 1 Ant System algorithm

initialize Ant System:
- create n_a ants (colony dimension);
- distribute ants randomly across the towns;
- deposit an initial amount of pheromones, τ_{ij}, on the graph edges;
- define the maximum number of iterations n_t;

```
for n = 1, ..., n_t do
    for k = 1, ..., n_a do
        while not Hamiltonian tour do
            choose next town, using the transition probability rule;
        end while
    end for
    distribute pheromones τ_ij on the edges;
end for
```

Gradient-based algorithms, however, are faster and require less computational work. In ACO, multiple finite element analysis need to be performed at the end of each iteration, which can be problematic if the domain under analysis is large and complex. Furthermore, with the increase in the number of optimization variables, finding a solution using a metaheuristic algorithm can be challenging, as a result of the stochastic search process.

Successful sizing optimization problems have been solved using ACO methods, including AS and its variant EAS [5, 6]. ACO has also been successfully used in TO problems [7], and has been found useful in the search for innovative solutions to structural problems [8].

TOPOLOGY OPTIMIZATION USING ACO ALGORITHM

In order to use the ACO method to solve a TO problem, some changes need to be made to the original algorithm. The optimization problem can be stated as

$$\begin{aligned} &\text{minimize: } U(\mathbf{u}) \\ &\text{subjected to: } \frac{V_s(\mathbf{u})}{\overline{V}_s} \leq 1, \text{ and physical constraints,} \end{aligned} \tag{1}$$

where U is the strain energy, $\mathbf{u}$ is the displacement field, $V_s(\mathbf{u})$ is the available material volume and $\overline{V}_s$ is the total material volume.

The physical constraints require the element where the load is applied to and the support elements to be present in the final design, and that the structure must be a connected structure, i.e a continuous path that connects all the elements with the physical constraints must be present in the final design.

The objective function to be minimized is the strain energy $U(\mathbf{u})$. It needs to be discretized so it can have a representation across the discrete domain. After discretization of the domain, the strain energy function becomes

$$U(\mathbf{u}) = \frac{1}{2}\sum_{e=1}^{N}\int_{V_e} \varepsilon^T(\mathbf{u}) D_e \varepsilon(\mathbf{u}) dV, \tag{2}$$

where V_e is the element volume, ε is the element strain, D_e is the element constitutive matrix and N is the number of finite elements.

The optimization process starts with the initialization of the AS, which requires a finite element analysis of the structure. Afterwards, the solution search process starts. A finite element analysis is performed at the end of each solution and the pheromone matrix is updated. The solution search process ends when the maximum volume fraction defined has been achieved and the boundary conditions have been met.

Unlike the TSP, where the ACO method used a transition probability, there is now an *element transition rule* instead. This rule corresponds to the probability of an element being selected as the next move for an ant [7]

$$P_i = \frac{[\tau_i(t)]^\alpha}{\sum\limits_{j=1}^{N}[\tau_j(t)]^\alpha}, \tag{3}$$

where $\tau_i(t)$ is the pheromone trail in element i at iteration t and $\tau_j(t)$ is the pheromone trail of a neighbor element j at the same iteration t. Element j must be a neighbor element in order to be used in the next iteration of the optimization process. Parameter α is used to control the relative weight of the pheromone trail. Care should be taken when choosing its value as it can lead to premature convergence to non-optimal solutions.

In 2D, a neighbor is defined as any element that shares one edge with another element and can have a maximum of 4 neighbors. In 3D, a neighbor is any element that shares one face with another element and can have a maximum of 6 neighbors. Irrespective of the problem dimension, an ant can only contain neighbor elements in its path.

The pheromone intensity $\Delta\tau_i^k$ laid by ant k on element i is given by

$$\Delta\tau_i^k = \frac{(U_i^k)^\lambda}{\sum\limits_{j=1}^{N}(U_j^k)^\lambda}, \tag{4}$$

where U_i^k is the strain energy of element i of the solution obtained by ant k, λ is a parameter used to tune the influence of the strain energy in the algorithm, helping with its convergence.

After determining the increment in pheromones for every element from all ants' solutions, the pheromone matrix can be updated. The pheromone matrix is updated according to the *pheromone update rule* [8]

$$\tau_{ij}(t+n) = \rho\,\tau_{ij}(t) + \sum_{k=1}^{n_a}\Delta\tau_{ij}, \tag{5}$$

where ρ is the evaporation rate and n_a is the dimension of the colony.

In order to improve the results generated by the ACO method in the TO problem, a noise cleaning filter is introduced during the pheromone update process [7] to prevent the formation of small members in the structure.

The noise cleaning filter modifies the strain energy of the element with the strain energy of the neighbor elements. The modified strain energy is

$$\hat{U}_i = \frac{\sum_{e=1}^{n_i} H_e U_e}{\sum_{e=1}^{n_i} H_e}, \tag{6}$$

where H_e is given by

$$H_e = V_e[r_{\text{min}} - r(i,e)],\ e \in \{1, 2, \ldots, n^i\}, \tag{7}$$

with V_e being the volume of element e, r_{min} the minimum allowed structure member size, $r(i,e)$ the distance between element i, whose strain energy is being modified, and element e, a neighbor element that satisfies $r(i,e) \leq r_{\text{min}}$, and n_i the number of elements that satisfy the last condition.

OPTIMIZATION OF CANTILEVER PLATE

An example TO problem was used to benchmark the ACO algorithm. The example is a 2D cantiliver plate subjected to a point load in its extremity and constrained at two points. The results were compared with the results obtained by Kaveh et al. [7]. The values of Young's modulus and Poisson's ratio for each element were chosen to be the same as used by the authors, 7.9×10^9 Pa and 0.30, respectively. Table 1 summarizes the parameters of the algorithm, which were chosen to minimize the strain energy of the final structure.

Table 1: Parameters used in the ACO algorithm.

Parameter	n_a	n_t	ρ	α	λ	r_{min}
Value	15	30	0.3	1	2.2	1.2

The best solution of the ones depicted in Figure 1 resulted in a strain energy of 4.875 J, compared to 4.89 J for the the best solution found by Kaveh et al. [7], with a volume fraction of 0.45. The solution displays some resemblances, such as the overall shape and the central structural member, which increases the stiffness of the structure. However, the solution obtained does not display symmetry.

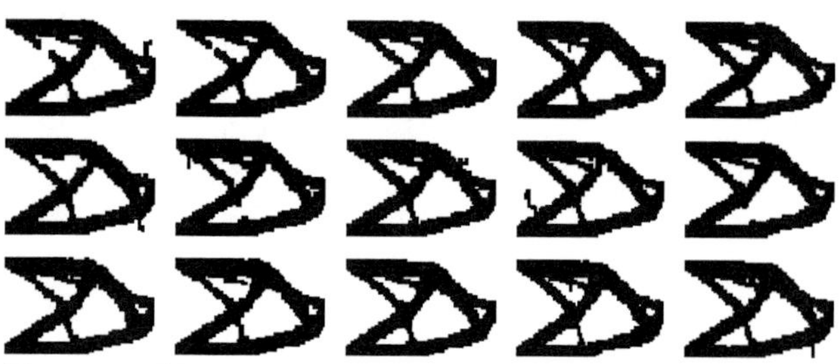

Figure 1: Plate topologies found by a colony of 15 ants.

OPTIMIZATION OF WING CROSS-SECTION

An optimization of a wing box cross-section was performed in this section. The finite element analysis was conducted for the entire wing box domain, while the optimization was carried out for the cross-section only, whose topology is repeated along the spanwise dimension.

A pressure distribution calculated from a CFD analysis was applied to the external surfaces of the wing. The top and bottommost layer of elements were constrained to be always contained in the optimization solutions. Also, at least one path connecting the upper and lower surfaces was requied to be present in the final topology so that a contiguous body was present at all times.

Aluminium was selected for the wing box material with a density, Young's modulus and Poisson's ratio of $2.81\,\text{g}\,\text{cm}^{-3}$, 71.7 GPa and 0.33, respectively.

The finite element mesh of the wing box contained 1152 elements at the cross-section, repeated in 26 spanwise layers for a total of 29 952 elements. These values were determined after a mesh convergence analysis.

During the solution process, the ant's path defines the shape of the cross-section at the wing root. Each cross-section element chosen by the ant aggregates a set of structural elements with it. This set correspondes to the spanwise elements behind the selected element. If an element does not belong to the ant's path, the corresponding spanwise row of elements disappears as well.

The pheromone update rule is given by Equation 5. A relation between the strain energy of a cross-section element i, $U_i^{2D}(\mathbf{u})$, and each element j of the aggregate, $U_{ij}^{3D}(\mathbf{u})$, is given by

$$U_i^{2D}(\mathbf{u}) = \sum_{j=1}^{N} U_{ij}^{3D}(\mathbf{u}), \tag{8}$$

where N is the number of spanwise elements.

The optimization of the wing box cross-section was performed using the parameters listed in Table 1. Two volume fractions of 0.35 and 0.45 were compared. The evolution of strain energy with the optimization process is presented in Figure 2 for the two volume fractions.

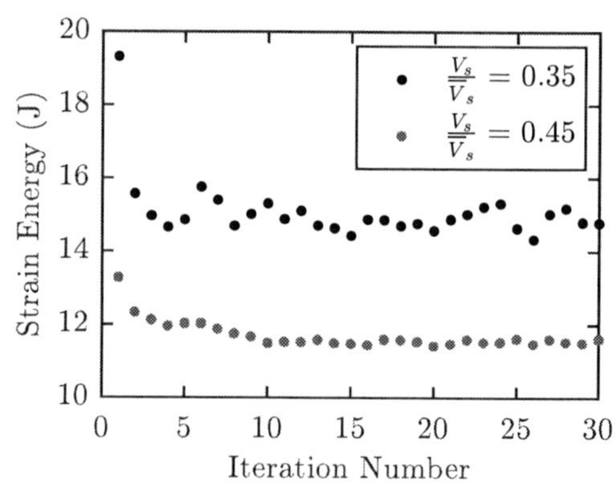

Figure 2: Minimum strain energy evolution throughout the wing cross-section optimization process.

Figure 3a depicts the topology of the wing box cross-section created by the ants' solution for a volume fraction of 0.35. It can be seen that the ant's path contains mainly the elements closer to the bottom right and upper left corners. In the middle area of the wing box cross-section, few elements are active. On both sides of the wing box the majority of solutions displays a connection between the upper and lower skin, resulting in a lower strain energy for these solutions.

The solutions obtained using a volume fraction of 0.45 present one or two members that connect the bottom and upper skin approximately at the mid-chord. The accumulation of elements in the upper left and lower right corners is more evident with the increased volume fraction. This is depicted in Figure 3b. The presence of elements in these regions stiffens the structure in torsion and is a consequence of the aerodynamic torsional moment. The lower and upper skins have also been reinforced in many of the solutions found, displaying usually two layers of elements close to the skin.

(a) $\frac{V_s}{V_s} = 0.35$

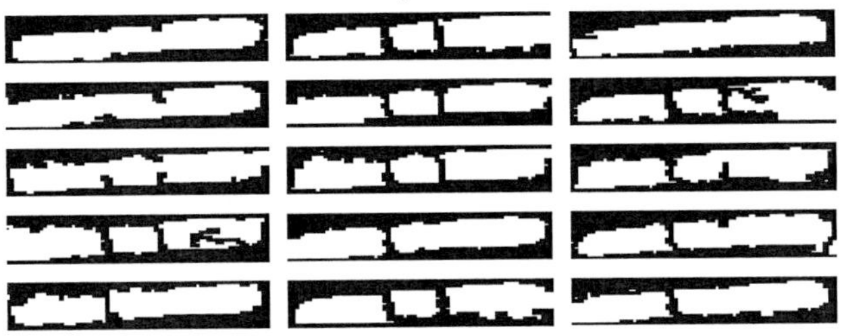

(b) $\frac{V_s}{V_s} = 0.45$

Figure 3: Wing cross-section topologies found by a colony of 15 ants for two volume fraction values.

Figure 4a depicts the solution with the lowest strain energy, 14.783 J, for a volume fraction of 0.35. With the increase of the volume fraction to 0.45, the strain energy decreased to 11.617 J, 21.42 % lower. This is depicted in Figure 4b.

(a) $\frac{V_s}{V_s} = 0.35$

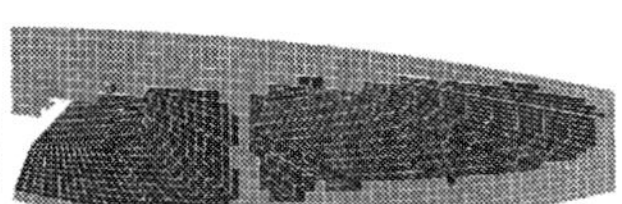

(b) $\frac{V_s}{V_s} = 0.45$

Figure 4: Best wing cross-section topologies found for two volume fraction values.

CONCLUSIONS

A topology optimization using an ant colony optimization method was first conducted for a cantilever plate example and later for the cross-section of a wing box.

The cantilever plate example served as a benchmark of the algorithm and the strain energy values agreed well with the ones presented by Kaveh et al. [7].

For the wing box cross-section optimization, a finite element model of the wing was created and an aerodynamic load case was computed using CFD. A constant cross-section topology along the spanwise direction of the wing was considered for the optimization problem. Two volume fractions of 0.35 and 0.45 were studied which resulted in minimum strain energy values of 14.783 J and 11.617 J, respectively.

It was possible to observe an accumulation of elements in the upper left and lower right corners of the cross-section that stiffened the structure as a consequence of the aerodynamic torsional moment.

This work demonstrated the advantages of the ACO method, including its flexibility to deal with discrete problems as well as its ability to work with non-differentiable objective functions, making it a good candidate for TO problems.

REFERENCES

[1] Bernd Bullnheimer, Richard F. Hartl, and Christine Strauss. "An Improved Ant System Algorithm for the Vehicle Routing Problem". In: *Annals of Operations Research* 89 (1999), pp. 319–328.

[2] Alberto Colorni, Marco Dorigo, and Vittorio Maniezzo. "Distributed Optimization by Ant Colonies". In: *Proceedings of the First European Conference on Artificial Life* (1992), pp. 134–142.

[3] Marco Dorigo, Vittorio Maniezzo, and Alberto Colorni. "The Ant System: Optimization by a Colony of Cooperating Agents". In: *IEEE Transactions on Systems, Man, and Cybernetics-Part B* 26.1 (1996), pp. 1–13.

[4] Marco Dorigo and Thomas Stützle. *Ant Colony Optimization.* The MIT Press, 2004.

[5] O. Hasançebi and S. Çarbaş. "Ant Colony Search Method in Pratical Structural Optimization". In: *International Journal of Optimization in Civil Engineering* 1 (2011), pp. 91–105.

[6] A. Kaveh and S. Shojaee. "Optimal Design of Skeletal Structures Using Ant Colony Optimization". In: *International Journal for Numerical Methods in Engineering* 70.5 (2007), pp. 563–581.

[7] A. Kaveh et al. "Structural Topology Optimization Using Ant Colony Methodology". In: *Engineering Structures* 30.9 (2008), pp. 2559–2565.

[8] Guan-Chun Luh and Chun-Yi Lin. "Structural Topology Optimization Using Ant Colony Optimization Algorithm". In: *Applied Soft Computing* 9.4 (2009), pp. 1343–1353.

[9] Thomas Stützle and Holger H. Hoos. "MAX-MIN Ant System". In: *Future Generation Computer Systems* 16.8 (2000), pp. 889–914.

[10] Thomas Stützle and Holger H. Hoos. "MAX-MIN Ant System and local search for the traveling salesman problem". In: *Proceedings of 1997 IEEE International Conference on Evolutionary Computation (ICEC '97).* IEEE, 1997, pp. 309–314.

Vibration of Structures

Proceedings of the 26th CANCAM
Victoria, BC, Canada, May 29 - June 1, 2017

ELECTROMECHANICAL RESPONSE OF CANTILEVERED CARBON NANOTUBE RESONATORS BASED ON A NEW NONLINEAR MODEL

Hamed Farokhi
Department of Mechanical Engineering
McGill University
Montreal, QC, Canada H3A 0C3
hamed.farokhi@mail.mcgill.ca

Michael P. Paidoussis
Department of Mechanical Engineering
McGill University
Montreal, QC, Canada H3A 0C3
michael.paidoussis@mcgill.ca

Arun K. Misra
Department of Mechanical Engineering
McGill University
Montreal, QC, Canada H3A 0C3
arun.misra@mcgill.ca

ABSTRACT

This paper investigates the nonlinear dynamical behaviour of cantilevered carbon nanotube (CNT) resonators, employing a new nonlinear model for the electrostatic load as well as a new nonlinear size-dependent continuum model for the cantilevered CNT. All sources of nonlinearity, i.e. geometrical and inertial nonlinearities as well as nonlinearities associated with damping, small-scale, and electrostatic load are taken into account. The continuous model of the system is discretized employing the Galerkin method and solved numerically utilizing a continuation technique.

Keywords: carbon nanotube resonator, cantilevered nanoresonator, electrostatic load distribution, nonlinear dynamics, finite element analysis

INTRODUCTION

Practical applications of nanoelectromechanical systems (NEMS) are becoming more feasible as new fabrication technologies emerge. These nanoscale devices have significantly high operating frequency and low power consumption, and they can be implemented as nanoresonators and nanosensors. Carbon nanotubes (CNTs) are among suitable materials to be used as nanoscale resonators, since they have the desired electrical and mechanical properties.

A large amount of research has been carried out on the dynamical behaviour of CNT resonators employing continuum models. For instance, Wu et al. [2] examined the resonant frequency and mode shapes of a cantilevereded single-walled carbon nanotube analytically and numerically. Poot et al. [1] contributed to the field by investigating the transverse oscillation in suspended carbon nanotubes and analyzed the electrostatic force and the gate tuning. Ouakad and Younis [3] investigated the nonlinear dynamic response of doubly clamped and cantilevereded CNT resonators employing the method of multiple scales. Further investigations on this topic were conducted by Hajnayeb and Khadem [4], who developed a model for nonlinear vibrations of double-walled clamped-clamped CNT resonators; they used a Taylor series approximation for the electrostatic load and employed the method of multiple scales, along with a single-mode approximation, to examine the system behaviour.

Eichler et al. [5] performed an experimental investigation on the damping mechanisms in carbon nanotube and graphene resonators; they showed that the damping in such systems depends strongly on the amplitude of the motion and concluded that a *nonlinear amplitude-dependent* modelling of the damping is required.

All of the valuable studies in the literature, which examined the behaviour of an electrostatically actuated CNT resonator based on a continuum model, consider the CNT to be *infinitely long* and hence do not take into account the end effects associated with the finite length of the CNT resonator. Moreover, they considered a linear viscous damping mechanism to model the damping in such systems. In the present study, these limitations are addressed by developing a new nonlinear electrostatic load model which accounts for end effects, and also developing a new nonlinear size-dependent con-

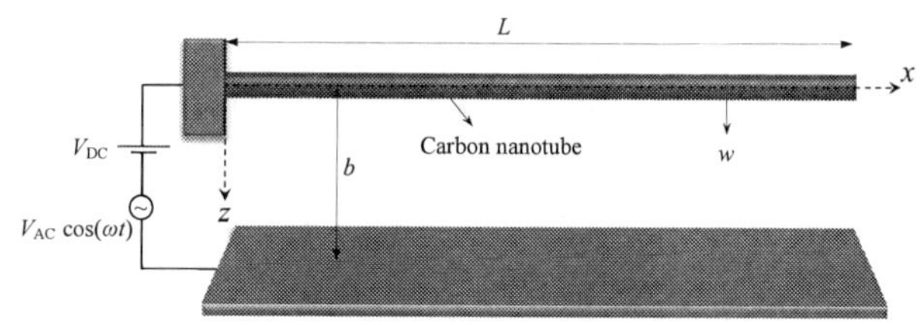

Figure 1: Schematic of a cantilevered CNT resonator.

tinuum model which accounts for all sources of nonlinearity, i.e. geometric, inertial, damping, and size-dependent ones.

NEW ELECTROSTATIC LOAD MODEL

Figure 1 shows a schematic of a cantilevered CNT resonator positioned at a distance b from the gate underneath. The CNT is of length L, outer radius R, cross-sectional area A, second moment of area I, and Young's modulus E. x and z represent the axial and transverse directions, respectively; $u(x,t)$ is the longitudinal displacement, in the x direction, and $w(x,t)$ is the transverse one, in the z direction.

In this section, a new electrostatic load model (which is in fact a modification of the model in the literature) is developed for cantilevered CNT resonators which accounts for end effects due to finite length of the CNT. 3D finite element simulations are conducted employing COMSOL Multiphysics software to obtain the electrostatic load distributions on various CNT resonators in both straight and deflected configurations. A block-shape electrode is considered at the clamped end to model the clamped-end effect.

Based on the results obtained for various cases via 3D finite element simulations, the following model is proposed for obtaining the electrostatic load distribution (per unit length) on cantilevered CNT resonators:

$$F_e^{(\text{cant.})} = \frac{\pi \epsilon V^2}{b} \times \left[\frac{1}{\sqrt{\left(1-\frac{w}{b}\right)\left(2S+1-\frac{w}{b}\right)}\left[\cosh^{-1}\left(1+\frac{1-\frac{w}{b}}{S}\right)\right]^2} + \hat{F}_0 (C_1 F_{\text{fit}} + C_2) \right], \quad (1)$$

in which $S = R/b$,

$$\hat{F}_0 = \frac{1}{\sqrt{(2S+1)}\left[\cosh^{-1}\left(1+\frac{1}{S}\right)\right]^2},$$

$$F_{\text{fit}} = \frac{-0.562\left(\frac{x}{L}\right)^4 + 1.8555\left(\frac{x}{L}\right)^3 - 1.23\left(\frac{x}{L}\right)^2 - 0.066\left(\frac{x}{L}\right)}{\left(\frac{x}{L}\right)^3 - 1.067\left(\frac{x}{L}\right)^2 + 0.0728\left(\frac{x}{L}\right) - 0.00743},$$

$$C_1 = 1 - 0.075 R_c, \quad (2)$$

$$C_2 = (1.6L_c - 1.8b_c)\left(\frac{x}{L}\right)^2 - 1.6(L_c - b_c)\left(\frac{x}{L}\right) - 1,$$

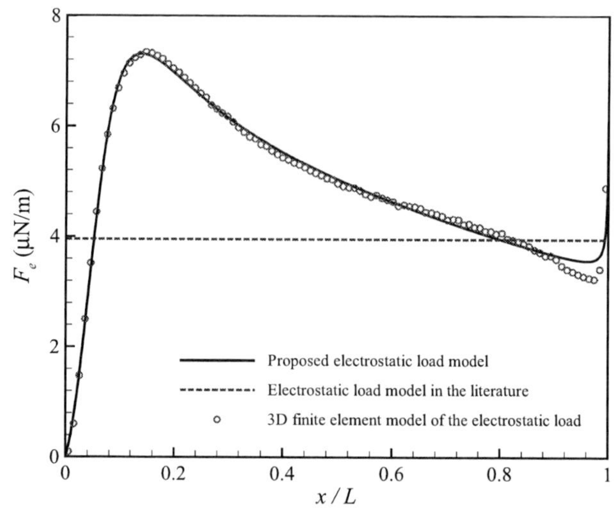

Figure 2: Electrostatic force distribution on a cantilevered CNT resonator predicted by different models; $L = 3000$ nm, $b = 300$ nm, and $R = 5$ nm.

and

$$R_c = \frac{R - R_0}{R_0}, \quad b_c = \frac{b - b_0}{b_0}, \quad L_c = \frac{L - L_0}{L_0},$$
$$L_0 = 3000 \times 10^{-9} \text{ m}, \quad b_0 = 300 \times 10^{-9} \text{ m},$$
$$R_0 = 5 \times 10^{-9} \text{ m}. \quad (3)$$

The proposed model works best for cantilevered CNT resonators in the range of

$$2000 \text{ nm} < L < 4000 \text{ nm}, \quad 200 \text{ nm} < b < 500 \text{ nm},$$
$$1 \text{ nm} < R < 10 \text{ nm}. \quad (4)$$

A comparison of different models (i.e. the model in the literature, the proposed model, and the 3D finite element model) is shown in Fig. 2, for a cantilevered CNT resonator in the trivial equilibrium configuration. The purpose of this figure is to show that even when $w(x,t) = 0$, the electrostatic force varies along the length due to the end effects. As seen in Fig. 2, the model in the literature predicts a constant electrostatic force along the length of the cantilevered CNT resonator; however, the 3D finite element simulation shows that the clamped end has a substantial effect on the electrostatic force distribution. Furthermore, it is also seen that the proposed analytical model predicts almost the same electrostatic force distribution as the finite element model. Quantitatively, the model in the literature predicts the total force with 19.4 % error, while the proposed model does so with less than 1 % error.

Figure 3 shows the electrostatic force distribution on deflected cantilevered CNT resonators for various deflection amplitudes based on three models: the finite element model, the proposed model, and the model in the literature. In particular, the deflected configuration of the cantilevered CNT resonator is modelled as $w = C_1\phi_1$; ϕ_1 is the first eigenfunction for transverse motion of a cantilevered (clamped-free) beam.

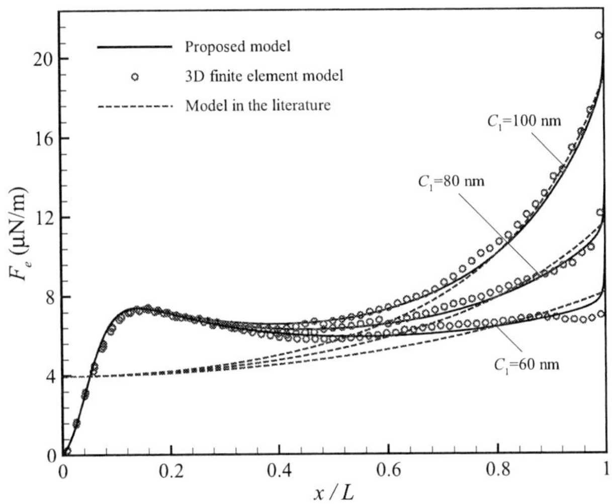

Figure 3: Electrostatic force distribution on the cantilevered CNT resonator for different amplitudes of deflection; L = 3000 nm, b = 300 nm, and R = 5 nm.

When a CNT resonator is in a deflected configuration, *two factors* affect the electrostatic load distribution along the length. The first factor is the transverse displacement, i.e. $w(x,t)$. The electrostatic load increases with increasing deflection amplitude; it also increases more at the tip of the resonator since that corresponds to the largest transverse displacement. As seen in Fig. 3, *all three models* are capable of detecting the load variations due to deflection. The second factor is the effect of clamped and free ends; it is seen that only the *proposed model* and the *3D finite element model* are capable of capturing the end effects. Quantitatively, the proposed model predicts the total force distribution with less that 1% error for the three deflection amplitudes shown here; the model in the literature, on the other hand, predicts the total force with 15% error when C_1 = 60 nm, 14% error when C_1 = 80 nm, and 13% error when C_1 = 100 nm.

NONLINEAR CONTINUOUS MODEL

Employing Hamilton's principle together with the modified couple stress theory and the Kelvin-Voigt dissipative model, the dimensionless equation of motion of the cantilevered CNT resonator can be obtained as

$$
\begin{aligned}
&\frac{\partial^2 w}{\partial t^2}+\alpha^2\frac{\partial}{\partial x}\left\{\frac{\partial w}{\partial x}\int_1^x\int_0^x\left[\frac{\partial w}{\partial x}\frac{\partial^3 w}{\partial x\partial t^2}+\left(\frac{\partial^2 w}{\partial x\partial t}\right)^2\right]\mathrm{d}x\,\mathrm{d}x\right\}\\
&+(1+l_s)\frac{\partial^4 w}{\partial x^4}+\eta_s(1+l_s)\frac{\partial^5 w}{\partial x^4\partial t}+\alpha^2\frac{\partial}{\partial x}\left[\left(\frac{\partial w}{\partial x}\right)^2\frac{\partial^3 w}{\partial x^3}\right.\\
&\left.+\frac{\partial w}{\partial x}\left(\frac{\partial^2 w}{\partial x^2}\right)^2\right]+\eta_s\alpha^2\frac{\partial}{\partial x}\left[\left(\frac{\partial w}{\partial x}\right)^2\frac{\partial^4 w}{\partial x^3\partial t}\right.\\
&\left.+\frac{\partial w}{\partial x}\frac{\partial^2 w}{\partial x\partial t}\frac{\partial^3 w}{\partial x^3}+\frac{\partial^2 w}{\partial x\partial t}\left(\frac{\partial^2 w}{\partial x^2}\right)^2+2\frac{\partial w}{\partial x}\frac{\partial^2 w}{\partial x^2}\frac{\partial^3 w}{\partial x^2\partial t}\right]\\
&+\frac{1}{4}l_s\alpha^2\frac{\partial}{\partial x}\left[\left(\frac{\partial w}{\partial x}\right)^2\frac{\partial^3 w}{\partial x^3}+\frac{\partial w}{\partial x}\left(\frac{\partial^2 w}{\partial x^2}\right)^2\right]\\
&+\frac{1}{4}\eta_s l_s\alpha^2\frac{\partial}{\partial x}\left[\left(\frac{\partial w}{\partial x}\right)^2\frac{\partial^4 w}{\partial x^3\partial t}+\frac{\partial w}{\partial x}\frac{\partial^2 w}{\partial x\partial t}\frac{\partial^3 w}{\partial x^3}\right.\\
&\left.+2\frac{\partial w}{\partial x}\frac{\partial^2 w}{\partial x^2}\frac{\partial^3 w}{\partial x^2\partial t}\right]-\frac{1}{4}l_s\frac{\alpha^2}{\beta}\frac{\partial}{\partial x}\left[\left(\frac{\partial w}{\partial x}\right)^2\frac{\partial^5 w}{\partial x^5}\right.\\
&\left.+3\frac{\partial w}{\partial x}\left(\frac{\partial^3 w}{\partial x^3}\right)^2+4\frac{\partial w}{\partial x}\frac{\partial^2 w}{\partial x^2}\frac{\partial^4 w}{\partial x^4}\right]\\
&-\frac{1}{4}\eta_s l_s\frac{\alpha^2}{\beta}\frac{\partial}{\partial x}\left[\left(\frac{\partial w}{\partial x}\right)^2\frac{\partial^6 w}{\partial x^5\partial t}+\frac{\partial w}{\partial x}\frac{\partial^2 w}{\partial x\partial t}\frac{\partial^5 w}{\partial x^5}\right.\\
&\left.+6\frac{\partial w}{\partial x}\frac{\partial^3 w}{\partial x^3}\frac{\partial^4 w}{\partial x^3\partial t}+4\frac{\partial w}{\partial x}\frac{\partial^2 w}{\partial x^2}\frac{\partial^5 w}{\partial x^4\partial t}+4\frac{\partial w}{\partial x}\frac{\partial^3 w}{\partial x^2\partial t}\frac{\partial^4 w}{\partial x^4}\right]\\
&-\frac{\xi(V_{\mathrm{DC}}+V_{\mathrm{AC}}\cos(\Omega t))^2}{\sqrt{(1-w)(2S+1-w)}\left[\cosh^{-1}\left(1+\frac{1-w}{S}\right)\right]^2}\\
&-\frac{\xi(V_{\mathrm{DC}}+V_{\mathrm{AC}}\cos(\Omega t))^2}{\sqrt{(2S+1)}\left[\cosh^{-1}\left(1+\frac{1}{S}\right)\right]^2}(C_1F_{\mathrm{fit}}+C_2)=0, \qquad (5)
\end{aligned}
$$

in which the following dimensionless quantities are used

$$
\begin{aligned}
&w^*=\frac{w}{b},\ x^*=\frac{x}{L},\ \gamma=\frac{m_p}{\rho AL},\ \alpha=\frac{b}{L},\ l_s=\frac{\mu A l^2}{EI},\\
&\beta=\frac{AL^2}{I},\ \xi=\frac{\pi\epsilon L^4V_0^2}{EIb^2},\ t^*=\frac{t}{\tau},\ \Omega=\omega\tau,\ \eta_s=\frac{\eta}{\tau},\\
&S=\frac{R}{b},\ V_{\mathrm{DC}}^*=\frac{V_{\mathrm{DC}}}{V_0},\ V_{\mathrm{AC}}^*=\frac{V_{\mathrm{AC}}}{V_0}, \qquad (6)
\end{aligned}
$$

and the asterisk notation is dropped for convenience. $\tau=\sqrt{\rho AL^4/EI}$ and V_0 is a reference voltage (set to 1 throughout this study).

The continuous model obtained in Eq. (5) is reduced into a high-dimensional discretized model employing the Galerkin technique and solved by means of a continuation technique.

NONLINEAR DYNAMICAL RESULTS

In this section, the nonlinear dynamical response of the CNT resonator is studied. The numerical calculations are performed for a multi-walled CNT resonator of L = 2500 nm, b = 390 nm, outer radius R_o = 5 nm and inner radius R_i = 0.5 nm.

Figure 4 shows the AC frequency-amplitude curves for two models of the cantilevered CNT resonator, one based on the proposed electrostatic load model and the other based on the model in the literature which is for infinitely long CNT resonators; l_s = 0.0 for both models. As seen in sub-figure (a), when V_{DC} = 1.5 and V_{AC} = 0.024, the proposed model predicts a much larger peak amplitude of vibrations and also a stronger nonlinear softening behaviour, compared to the model in the literature. In particular, there is 27% difference between the peak amplitude predicted by the two models.

By changing the DC and AC voltages to 1.85 and 0.0053, respectively, a new AC frequency-response diagrams is obtained and plotted in sub-figure (b). As seen, for this case, the differences between the two models become much more significant compared to the previous case. The proposed model predicts a much larger vibration amplitude overall as well as a much stronger nonlinear softening behaviour. Quantitatively, the proposed model predicts the peak resonance amplitude with 29% difference from the model in the literature; furthermore, there is 18% difference between the nonlinear resonance frequency predicted by the two models.

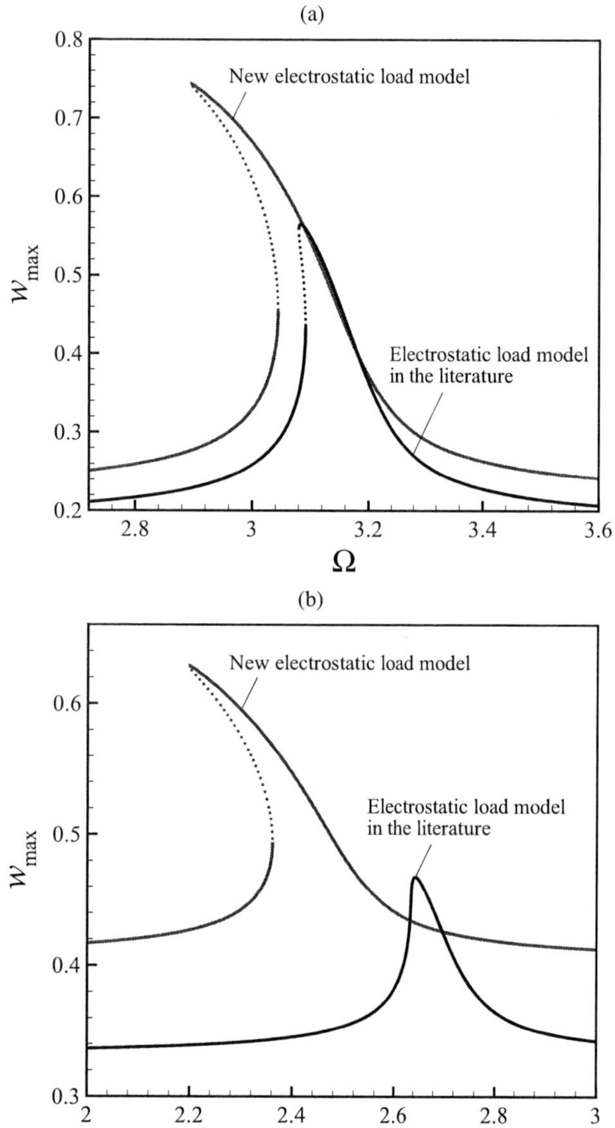

Figure 4: Comparison between AC frequency-amplitude curves of the cantilevered CNT resonator obtained based on the new electrostatic load model and the model in the literature: the maximum transverse displacement at $x = 1.0$ when (a) $V_{DC} = 1.5$ and $V_{AC} = 0.024$, and (b) $V_{DC} = 1.85$ and $V_{AC} = 0.0053$.

CONCLUSIONS

A new electrostatic load model and a new nonlinear continuous model have been developed for cantilevered CNT resonators. Examining the electrostatic load distribution on a cantilevered CNT resonator employing different electrostatic load models shows that the proposed model predicts a load distribution very similar to that obtained via 3D finite element analysis. In particular, for both straight and deflected CNT resonators, the proposed model predicts the total electrostatic load with less than 1% error; however, the model in the literature predicts the total electrostatic load with almost 20% error for the straight CNT resonator and with more than 13% error for the deflected CNT resonator.

A comparison between the AC frequency-amplitude diagrams of the CNT resonator obtained employing the new electrostatic load model and the model in the literature shows that the proposed more accurate model predicts a much larger overall amplitude of vibration as well as a stronger nonlinear behaviour. The difference becomes even more significant at larger values of the DC voltage.

Hence, the conclusions can be reached that the end-effects on the electrostatic load distribution of cantilevered CNT resonators are significant and must be taken into account.

ACKNOWLEDGMENTS

The authors gratefully acknowledge the financial support from the Natural Sciences and Engineering Research Council of Canada (NSERC) and the Fonds de recherche du Québec – Nature et technologies (FRQNT).

REFERENCES

[1] M. Poot, B. Witkamp, M. A. Otte, and H. S. J. van der Zant, "Modelling suspended carbon nanotube resonators," *Physica Status Solidi (B)*, vol. 244, no. 11, pp. 4252–4256, 2007.

[2] D. H. Wu, W. T. Chien, C. S. Chen, and H. H. Chen, "Resonant frequency analysis of fixed-free single-walled carbon nanotube-based mass sensor," *Sensors and Actuators A: Physical*, vol. 126, no. 1, pp. 117–121, 2006.

[3] H. M. Ouakad and M. I. Younis, "Nonlinear dynamics of electrically actuated carbon nanotube resonators," *Journal of Computational and Nonlinear Dynamics*, vol. 5, no. 1, pp. 011009–011009, 2009.

[4] A. Hajnayeb and S. E. Khadem, "Nonlinear vibration and stability analysis of a double-walled carbon nanotube under electrostatic actuation," *Journal of Sound and Vibration*, vol. 331, no. 10, pp. 2443–2456, 2012.

[5] A. Eichler, J. Moser, J. Chaste, M. Zdrojek, I. Wilson-Rae, and A. Bachtold, "Nonlinear damping in mechanical resonators made from carbon nanotubes and graphene," *Nature nanotechnology*, vol. 6, no. 6, pp. 339–342, 2011.

Proceedings of the 26th CANCAM
Victoria, British Columbia, Canada, May 29 – June 1, 2017

NONLINEAR VIBRATION ANALYSIS OF A VISCOELASTIC CRACKED PLATE

Shamim Mashrouteh
Faculty of Engineering and Applied Science
University of Ontario Institute of Technology
Oshawa, Ontario, Canada
Shamim.Mashrouteh@uoit.ca

Davood Younesian
Center of Excellence in Railway Transportation
Iran University of Science and Technology
Tehran, 16846-13114, Iran
Younesian@iust.ac.ir

Ebrahim Esmailzadeh
Faculty of Engineering and Applied Science
University of Ontario Institute of Technology
Oshawa, Ontario, Canada
Ezadeh@uoit.ca

ABSTRACT

Vibration analysis of a rectangular orthotropic plate with an inclined part-through crack surrounded by a nonlinear Winkler-Pasternak elastic medium is performed in this study. The crack is located at the center of the plate with an arbitrary angle of orientation and is modeled using a simplified line-spring model. The Galerkin procedure is applied to derive the time-dependent modal coordinate of the system and the method of multiple time-scales is utilized to obtain the frequency response of the forced vibration system. The external resonance of the first four natural vibration modes are investigated, and a parametric sensitivity study is carried out to examine the effect of different parameters on the vibration behavior of the system.

Keywords: Orthotropic plate; Nonlinear vibration; Cracked plate; Winkler-Pasternak foundation.

INTRODUCTION

Vibration analysis of flexible structures such as beams, shafts, and plates is of high importance since they are widely being used in building, aviation, automation and ship building industries [1]. In real life applications, imperfections in flexible structures are highly probable and two of the most common of these imperfections are in surface and in-depth cracks. Therefore, it is essential to be able to predict and analyze the vibration behavior of such flexible plates which are mostly affected by crack imperfection [2].

In order to formulate the crack terms in the governing equations of motion of plates, the line-spring model has been mostly considered in the literature [3-5]. Recently, a new function named "Modified Comparison Functions" for analyzing the vibration characteristic of a simply supported rectangular cracked plate is introduced [6]. The elastic behavior of the plate at the crack location is modeled as a line-spring with a varying stiffness along the crack. In order to increase the accuracy of the introduced function, the authors considered the following assumptions: (a) the crack has an influence on the mode shapes of its neighborhood, (b) its maximum effect occurs at the crack location, and (c) this effect will be vanished at an adequate distance from the crack.

An important point missing in their studies is that the crack may not be always parallel to the side edges and could be oriented by an angle with respect to the coordinates of the plate [7, 8]. Ismail and Cartmell [7] presented an analytical approach for the forced vibration analysis of a plate containing a surface crack of variable angular orientation. They modeled the plate when subjected to a harmonic excitation based on the classical plate theory and considered the effect of different boundary conditions. They applied the Berger formulation to derive the governing equation of motion and utilized the multiple time-scales method to investigate the amplitude-frequency equation. They found that the vibration behavior and the nonlinear natural frequency of the cracked plate can be highly affected by the orientation of the crack.

The orthotropic behavior of a plate, which is the result of either using materials with orthotropic stress-strain relations in nature or by a simplification applied to composite relations, has been ignored in the previously cited references. The importance of orthotropic plates and their various applications leads to a wealthy literature. Though researchers have investigated the effect of the crack on the vibration behavior of orthotropic plates [9, 10], in all referenced articles the cracks are assumed to be perpendicular to the edges of the plate and the effect of the inclined crack on an orthotropic plate is ignored.

The lack of adequate knowledge on the effect of oriented crack on the nonlinear vibration behavior of orthotropic plate led the authors to conduct an investigation in which forced vibration analysis of a rectangular orthotropic plate with an inclined crack were studied. The inclined crack is assumed to be located at the center of the plate and has a thickness through the plate. It is assumed that the plate has simply supported boundary conditions on all sides and is subjected to a harmonic excitation with a concentrated force.

The Kirchhoff plate theory has been utilized to derive the governing equation of motion and the Galerkin method has been applied to obtain the ordinary differential equation of motion for each vibration modes. The nonlinear frequency response of the system for each mode has been obtained by applying the method of multiple time-scales [11-13]. Effect of different crack angles on the frequency response of the first four modes of the system has been investigated numerically and discussed in details.

MATHEMATICAL MODELING

The governing differential equation of motion for forced vibration of a rectangular orthotropic plate with an inclined crack embedded in nonlinear Winkler-Pasternak foundation is derived by utilizing an equilibrium principle on the plate elements based on the Kirchhoff plate theory as [14]:

$$D_x \frac{\partial^4 w}{\partial x^4} + 2D_{xy} \frac{\partial^4 w}{\partial x^2 \partial y^2} + D_y \frac{\partial^4 w}{\partial y^4} + \rho h \frac{\partial^2 w}{\partial t^2} = -\mu \frac{\partial w}{\partial t} - k_1 w - k_3 w^3 + k_p \left(\frac{\partial^2 w}{\partial x^2} + \frac{\partial^2 w}{\partial y^2} \right) + F_x \frac{\partial^2 w}{\partial x^2} + \bar{F}_y \frac{\partial^2 w}{\partial y^2} + \frac{\partial^2 \bar{M}_y}{\partial y^2} + 2\frac{\partial^2 \bar{M}_{xy}}{\partial x \partial y} + 2\bar{F}_{xy} \frac{\partial^2 w}{\partial x \partial y} + Q_z \tag{1}$$

Parameters $w(x,y,t)$, ρ, μ, k_1, k_3, and k_p are the respective displacement, mass density, damping coefficient, linear and nonlinear Winkler elastic foundation coefficients, and Pasternak elastic foundation coefficient [15, 16]. Symbols D_x and D_y are the flexural rigidities in x and y directions, and D_{xy} is the effective torsional rigidity for a uniform thickness of the orthotropic plate.

Moreover, F_x, $\bar{F}_y$ and $\bar{F}_{xy}$ are the in-plane forces per unit for an orthotropic plate with inclined crack and have been defined based on the line spring model [17].

The Galerkin procedure [18] is employed to obtain the approximate solution of Eq. (1). According to this technique, for a plate with a simply-supported boundary condition at all edges, the displacement of plate written as:

$$w(x,y,t) = \sum_{i=1}^{m} \sum_{j=1}^{n} \left(A_{ij} \sin\left(\frac{i\pi x}{lx}\right) \sin\left(\frac{j\pi y}{ly}\right) \psi_{ij}(t) \right) \tag{2}$$

Furthermore, the external harmonic force is assumed to be applied to the plate at point (x_0, y_0) is

$$Q_z = q_0 \cos(\Omega t) \delta(x - x_0) \delta(y - y_0) \tag{3}$$

Substituting Eqs. (2) and (3) into Eq. (1) and applying the orthogonality principle and performing few mathematical simplifications, the ordinary differential equation of motion for the first four vibration modes can be obtained as:

$$\ddot{\psi}_{11} + \omega_{11}^2 \psi_{11} = -2\mu\dot{\psi}_{11} - \alpha_{11}\psi_{11}^3 - \alpha_{12}\psi_{11}\psi_{12}^2 - \alpha_{13}\psi_{11}\psi_{21}^2 - \alpha_{14}\psi_{11}\psi_{22}^2 - \alpha_{15}\psi_{22}\psi_{11}^2 - \alpha_{16}\psi_{22}\psi_{12}^2 - \alpha_{17}\psi_{22}\psi_{21}^2 - \alpha_{18}\psi_{22}^3 + \alpha_{f1} q(t) \tag{4-a}$$

and

$$\ddot{\psi}_{12} + \omega_{12}^2 \psi_{12} = -2\mu\dot{\psi}_{12} - \alpha_{21}\psi_{12}\psi_{11}^2 - \alpha_{22}\psi_{12}^3 - \alpha_{23}\psi_{12}\psi_{21}^2 - \alpha_{24}\psi_{12}\psi_{22}^2 - \alpha_{25}\psi_{21}\psi_{11}^2 - \alpha_{26}\psi_{21}\psi_{12}^2 - \alpha_{27}\psi_{21}^3 - \alpha_{28}\psi_{21}\psi_{22}^2 + \alpha_{f2} q(t) \tag{4-b}$$

$$\ddot{\psi}_{21} + \omega_{21}^2 \psi_{21} = -2\mu\dot{\psi}_{21} - \alpha_{31}\psi_{21}\psi_{11}^2 - \alpha_{32}\psi_{21}\psi_{12}^2 - \alpha_{33}\psi_{21}^3 - \alpha_{34}\psi_{21}\psi_{22}^2 - \alpha_{35}\psi_{12}\psi_{11}^2 - \alpha_{36}\psi_{12}^3 - \alpha_{37}\psi_{12}\psi_{21}^2 - \alpha_{38}\psi_{12}\psi_{22}^2 + \alpha_{f3} q(t) \tag{4-c}$$

and

$$\ddot{\psi}_{22} + \omega_{22}^2 \psi_{22} = -2\mu\dot{\psi}_{22} - \alpha_{41}\psi_{22}\psi_{11}^2 - \alpha_{42}\psi_{22}\psi_{12}^2 - \alpha_{43}\psi_{22}\psi_{21}^2 - \alpha_{44}\psi_{22}^3 - \alpha_{45}\psi_{11}^3 - \alpha_{46}\psi_{11}\psi_{12}^2 - \alpha_{47}\psi_{11}\psi_{21}^2 - \alpha_{48}\psi_{11}\psi_{22}^2 + \alpha_{f4} q(t) \tag{4-d}$$

where ω_{mn} is the natural frequency of the system and corresponds to each vibration mode.

SOLUTION PROCEDURE

The multiple time-scales method is utilized to investigate the amplitude-frequency response of the thin orthotropic plate with an inclined crack embedded on an elastic foundation. Based on the published articles on the use of this method to obtain the frequency response of the nonlinear systems, an approximate solution of the problem in terms of different time scales can be obtained as [19]:

$$\psi_{mn} = u_i(t,\varepsilon) = \varepsilon u_{i1}(T_0, T_2) + \varepsilon^3 u_{i3}(T_0, T_2) + O(\varepsilon^3) \qquad i = 1, 2, \ldots, m \times n \tag{5}$$

By expanding the derivatives with respect to time t in terms of the partial derivatives with respect to T_n, when the frequency of the external force gets closer to the natural frequency of the ith mode of the system, $\Omega \approx \omega_{mn}$, by assuming $\Omega = \omega_{mn} + \varepsilon^2 \sigma_i$, and separating the same orders of ε, the following set of equations can be obtained:

$$D_0^2 u_{i1} + \omega_{mn}^2 u_{i1} = 0 \tag{6-a}$$

$$D_0^2 u_{i3} + \omega_{mn}^2 u_{i3} = -2D_0 \left(D_2 u_{i1} + \mu u_{i1} \right) - \left(\alpha_{ii} u_{i3}^3 + \delta_i \right) + \alpha_{fi} \cos\left(\left(\omega_{mn} + \varepsilon^2 \sigma_i \right) t \right) \tag{6-b}$$

The approximate solution of Eq. (6-a) can be expressed as:

$$u_{i1} = A_i(T_2) \exp(i\omega_{nm} T_0) + \bar{A}_i(T_2) \exp(-i\omega_{nm} T_0) \tag{7-a}$$

$$A_i = \frac{1}{2} a_i \exp(j\theta_i) \tag{7-b}$$

By substituting Eqs. (7-a) and (7-b) into Eq. (6-b) and subtracting the secular terms, the frequency response relationship of the system for the external resonance of the ith mode of the system can be obtained as:

$$\sigma_i = \frac{3\alpha_{ii} a_i^2}{8\omega_{mn}} \pm \sqrt{\left(\frac{\alpha_{fi} q_0}{2\omega_{mn} a_i} \right)^2 - \mu^2} \tag{7-b}$$

NUMERICAL RESULTS AND DISCUSSIONS

The frequency response of a thin orthotropic plate with an inclined crack embedded in the nonlinear Winkler-Pasternak foundation has been illustrated in this section. The angle of the crack with respect to the x-axis is one the most important parameters in this study and its effect on the frequency response is investigated in this section. The numerical values of the parameters used in this study are presented in Table 1.

Table 1. Numerical values of physical parameters of the system

Parameter	*Numerical value*	*Parameter*	*Numerica l value*
E_x	$208GPa$	ν_x	0.23
E_y	$18.9GPa$	ν_y	0.0208
G	$5.7GPa$	μ	0.08

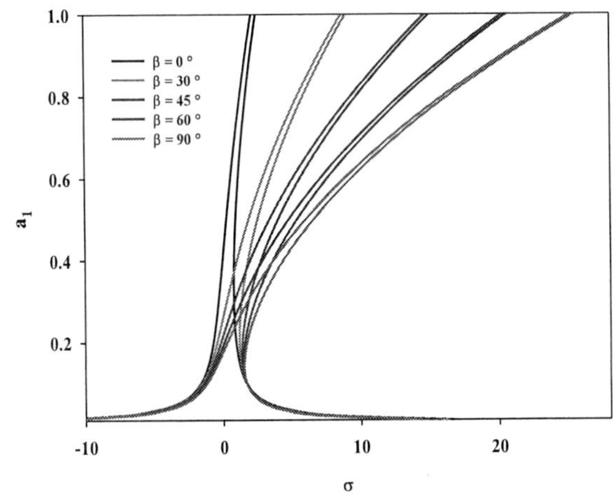

Figure 1: Effect of the crack angle on the frequency response curve of the first mode

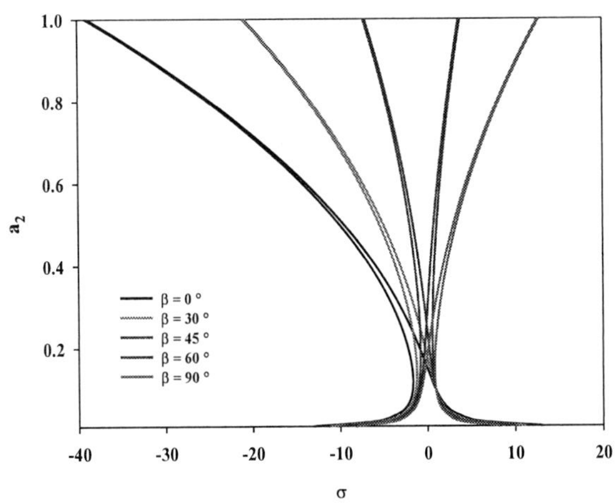

Figure 2: Effect of the crack angle on the frequency response curve of the second mode

The effect of different crack angles on the frequency response of the orthotropic plate for the first four natural modes of the system are illustrated in Figures 1 to 4. Considering the first mode of the system in Figure 1, while the crack is parallel to the x-axis, the vibration behavior shows linear behavior. As the crack angle increases with respect to the x-axis the nonlinearity of the vibration behavior increases and tends toward the nonlinear hardening zone.

The frequency response for the second mode has been illustrated in Figure 2. It can be seen that as the crack angle increases, the vibration behavior of the second mode changes from the linear status to nonlinear softening.

Figure 3 shows that the crack angle has less influence on the frequency response of the third mode in comparison with the other vibration modes of the system, and as the crack angle increases, the nonlinearity of the frequency response would tend to increase. The frequency response of the fourth vibration mode has been illustrated in Figure 4. It can be seen that the system behavior tend to change toward a hardening nonlinear zone as the crack angle increases.

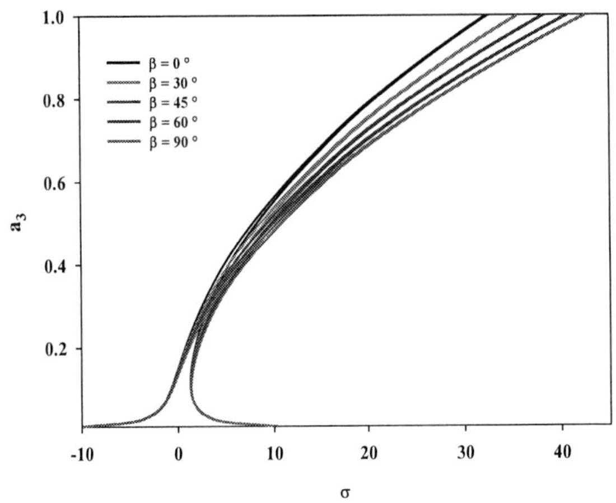

Figure 3: Effect of the crack angle on the frequency response curve of the third mode

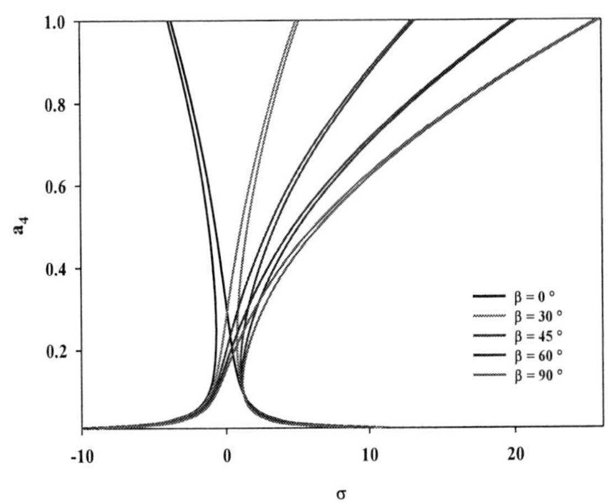

Figure 4: Effect of the crack angle on the frequency response curve of the fourth mode

Comparing these four figures, one could easily conclude that the second and the fourth modes of the system are more sensitive to the variation of the crack angle while the third mode shows less sensitivity to any crack angle changes.

CONCLUSIONS

Nonlinear forced vibration analysis of an orthotropic rectangular plate with an inclined crack embedded in a nonlinear Winkler-Pasternak elastic foundation has performed in this study. The crack was assumed to be located at the center of the plate at an angle with respect to the side edge of the plate. The Kirchhoff plate theory and the simplified line-spring model were employed to develop the governing partial differential equation of motion of the system. The Galerkin procedure was used to rewrite the equation of motion of the system as the time-dependent modal coordinates and the method of multiple time-scales was utilized to perform vibration analysis and to find the frequency response curves. The external resonance investigation indicates that the crack angle has a significant effect on the vibration behavior of the system. Although any increase in the crack angle makes the vibration behavior of the second mode change from the linear to nonlinear softening, the same changes in the crack angle cause the fourth mode behavior to change from a linear to nonlinear hardening. Finally, it is found that the third mode of the system shows less sensitivity to any crack angle variation.

ACKNOWLEDGMENTS

The financial support provided by the Natural Science and Engineering Research Council (NSERC) of Canada to complete this research is gratefully acknowledged

REFERENCES

1. Dimarogonas, A.D. 1996, "Vibration of cracked structures: a state-of-the-art review". *Engineering Fracture Mechanics*, **55**(5): 831-857.
2. Yu, S. 2002, "Free flexural vibration of rectangular plates having single cracks". *The 23rd International Congress of Aeronautical Sciences (ICAS)*, Toronto, Canada, pp. 421.1-421.8.
3. Rice, J. and Levy, N. 1972, "The part-through surface crack in an elastic plate", *Journal of Applied Mechanics*, **39**(1): 185-194.
4. Rice, J.R. 1972, "Some remarks on elastic crack-tip stress fields", *International Journal of Solids and Structures*, **8**(6): 751-758.
5. Joshi, P., Jain, N. and Ramtekkar, G. 2014, "Analytical modeling and vibration analysis of internally cracked rectangular plates", *Journal of Sound and Vibration*, **333**(22): 5851-5864.
6. Khadem, S. and Rezaee, M. 2000, "Introduction of modified comparison functions for vibration analysis of a rectangular cracked plate", *Journal of Sound and Vibration*, **236**(2): 245-258.
7. Ismail, R. and Cartmell, M. 2012, "An analysis of the effects of the orientation angle of a surface crack on the vibration of an isotropic plate", *Journal of Physics: Conference Series*, **382**(1): 1-6.
8. Huang, C., Leissa, A. and Li, R. 2011, "Accurate vibration analysis of thick, cracked rectangular plates", *Journal of Sound and Vibration*, **330**(9): 2079-2093.
9. Gupta, A., Jain, N.K., Salhotra, R., and Joshi, P.V. 2015, "Effect of microstructure on vibration characteristics of partially cracked rectangular plates based on a modified couple stress theory", *International Journal of Mechanical Sciences*, **100**: 269-282.
10. De Jong, F.B.P. 2007, "Renovation techniques for fatigue cracked orthotropic steel bridge decks" TU Delft, Delft University of Technology, The Netherland.
11. Nayfeh, A.H., Mook, D.T. and Lobitz, D.W. 1974, "Numerical-perturbation method for the nonlinear analysis of structural vibrations", *AIAA Journal*, **12**(9): 1222-1228.
12. Maccari, A. 1999, "The asymptotic perturbation method for nonlinear continuous systems", *Nonlinear Dynamics*, **19**(1): 1-18.
13. Zhang, W. 2001, "Global and chaotic dynamics for a parametrically excited thin plate", *Journal of Sound and vibration*, **239**(5): 1013-1036.
14. Ventsel, E. and Krauthammer, T. 2001, *Thin Plates and Shells Theory: Analysis, and Applications*, CRC press.
15. Civalek, Ö. 2006, "Harmonic differential quadrature-finite differences coupled approaches for geometrically nonlinear static and dynamic analysis of rectangular plates on elastic foundation", *Journal of Sound and Vibration*, **294**(4): 966-980.
16. Shih, Y.-S. and Blotter, P. 1993, "Non-linear vibration analysis of arbitrarily laminated thin rectangular plates on elastic foundations", *Journal of Sound and Vibration*, **167**(3): 433-459.
17. Ismail, R. and Cartmell, M. 2012, "An investigation into the vibration analysis of a plate with a surface crack of variable angular orientation", *Journal of Sound and Vibration*, **331**(12): 2929-2948.
18. Sadri, M. and Younesian, D. 2014, "Nonlinear free vibration analysis of a plate-cavity system", *Thin-Walled Structures*, **74**: 191-200.
19. Nayfeh, A.H. and Mook, D.T. 2008, *Nonlinear Oscillations*. John Wiley and Sons.

Proceedings of the 25th CANCAM
London, Ontario, Canada, May 31 – June 4, 2015

FEA AND EXPERIMENT RESEARCH OF LOWER ARM B

Zhihong Liu[1], Huigang Wang[1]

[1]**Northwestern Polytechnical University**
Xi'An, Shann'Xi, China
Liuzh9706@gmail.com
Wanghg74@nwpu.edu.cn

TianYang Chen[2], Xin Zhang[2]

[2]**F.Tech R&D North America**
Troy, Ohio, USA
Email: tchen@ftech-rd.com
xzhang@ftech-rd.com

ABSTRACT

Lower arm B is an important part in the vehicle, which bears spring load and damper load or as stable bar in some cases. Study LAB's dynamic performance will give sight to control noise and vibration in cabin for the vehicle designer, because LAB make a lot of contributions on the suspension assembly which will result in noise and vibration in cabin. In this paper, first the FEA model of LAB is updated through modal test results, and second the modal test is improved further by using the pretest analysis results. At last, the natural frequency, damping of LAB is gained for further study in the suspension assembly.

KEYWORDS: Lower arm B, Suspension system, Modal experiment, MAC, FEM analysis

1. INTRODUCTION

Car's suspension system is responsible for smoothing out the ride and keeping the car in control. The suspension system also provides comfort for passengers to limiting the impact of particular road conditions to not only the car, but the passengers riding inside.

In the vehicle suspension system, lower control arm (LAB, referred to Fig.1) is a critical part in the rear suspension system, which usually designed to bear spring load and damper load or stabi-bar load in some cases. Thus, the LAB dynamic characteristic will make a lot of contributions to the suspension assembly. In order to get better understand the dynamic performance and NV performance of the rear suspension system, study the lower arm B's dynamic features can give us an whole insight about the suspension's NV performance.

Figure 1: A Typical multi-link Rear Suspension with LAB

2. FEA SIMULATION AND MODAL TEST

LAB is made of steel, the weight is 2.333kg; its geometry property is given in Fig.2-3, dimension unit is mm.

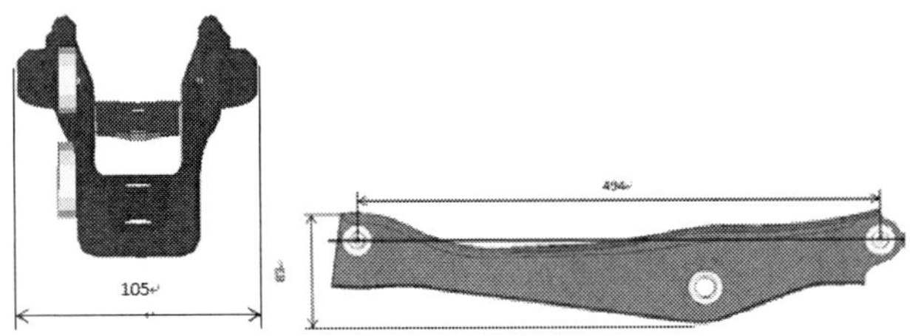

2-1 width 2-2 length and height

Figure 2: Geometry size of Lower arm B

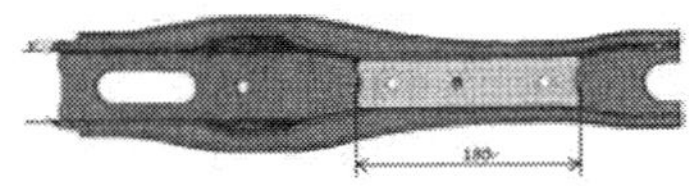

Figure 3: Welded Patch in Lower arm B

2.1 FE Analysis of Lower arm B

LAB is created in Nastran, as shown in Figure 4. The arm body and patch are shell mesh, with mesh size of 7mm and thickness of 2.6mm, 2.0mm respectively. Patch welding is modeled with RBE3 elements. Nuts are modeled as solid elements. All material property is steel. Mode shapes of LAB under free-free boundary condition (BC) is given in Fig.5.

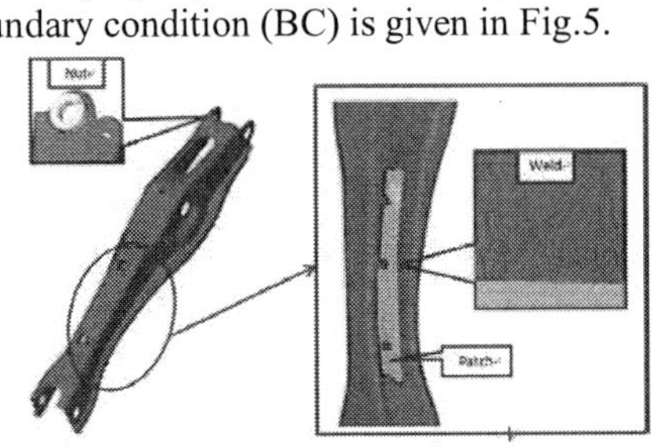

Figure 4: FE mesh of Lower arm B

2.2 Experiment Analysis of lower arm B

In this section, intuitive test setup and pretest analysis are used for getting the optimal test geometry, measurement points and driving points from an existing FE model. Thereafter, modal assurance criterion (MAC) is used to evaluate the new quality levels.

The lower arm B is supported by bungee cords vertically to mimic the free-free boundary condition applied in FEA. And the correlation results based on this setup is given in Fig.6.

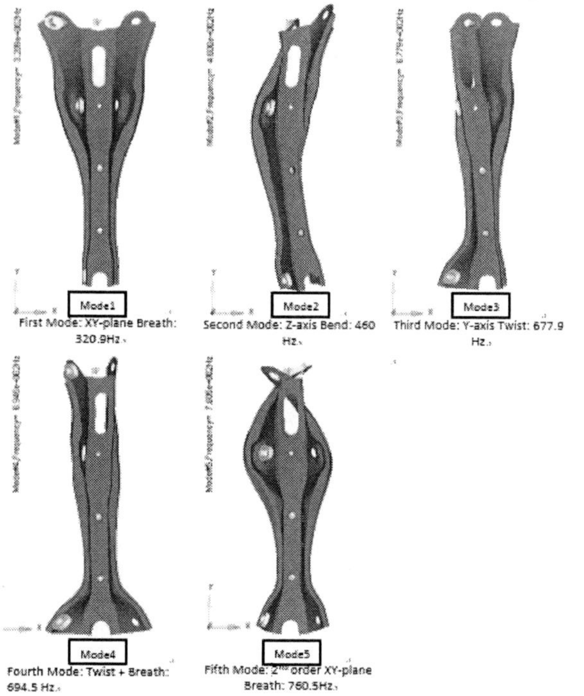

Fig.5 Mode shapes of LAB

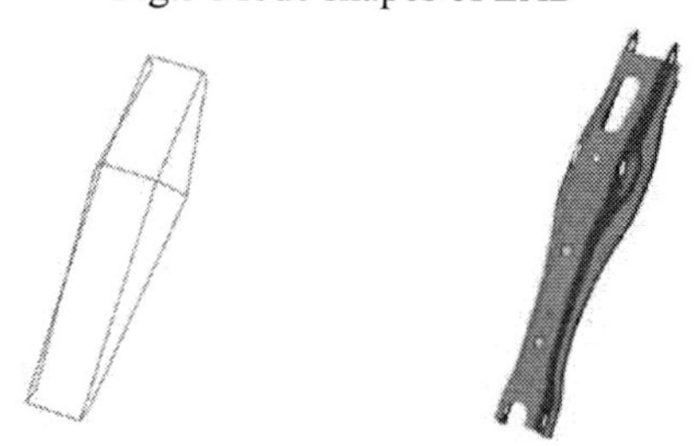

Fig.6 geometrical comparison the between testing wire frame model (right) and CAD model (left)

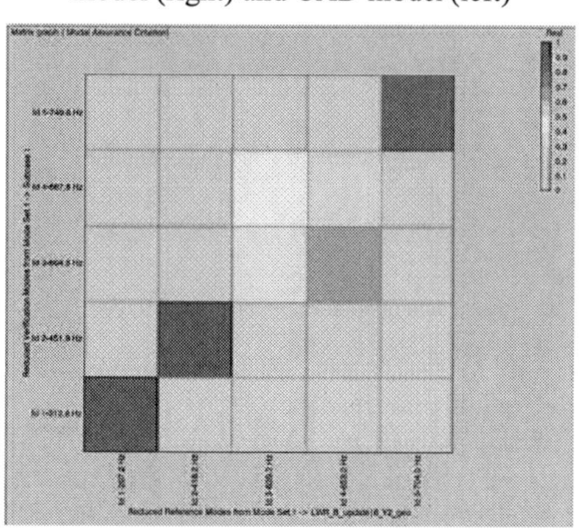

Figure 7 MAC Value Results

From Fig.6, we can see the arm geometry can be represented relatively well. The correlation result is shown in Figure **7**. The mode 1, 2 and 5 are correlated relatively accurately but we still missing mode 3 and 4.

Further optimization of the testing setup is necessary. Note that there's a big patch welded on the back of the arm, as shown in Figure 3. However, in the test Layout, we didn't capture this geometry (meaning no accelerometers installed around that area). Actually, according to the correlation results, this welded patch feature has much contribution on the modal results. Then we made our final accelerometer setup as shown in Figure 8, we added 4 measurement points to the patch welding area, which are attached to both sides of the arm, as shown in Figure 9.

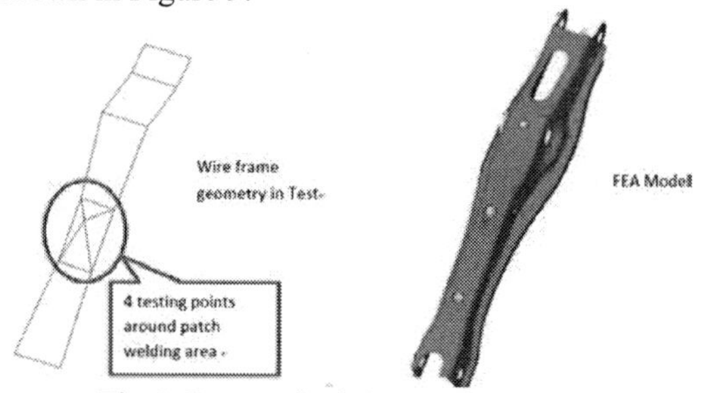

Fig.8 Geometrical Comparison for Final Setup

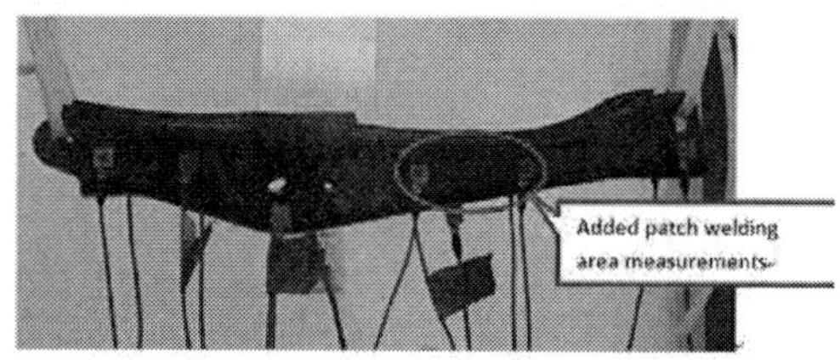

Fig.9 On-site testing setup

Figure 10 gives out the correlation results based on this setup. We can see the mode 3 and 4 are well correlated this time. MAC value for all modes is at least 0.85, meaning the off-diagonal terms are below 0.2, which is considered an acceptable correlation in the engineering application.

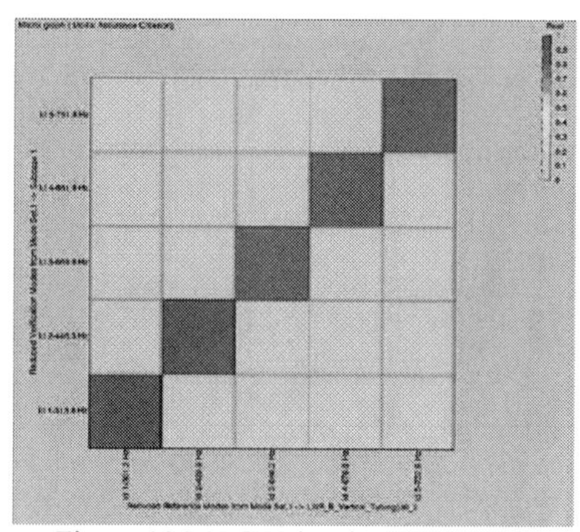

Figure 10 MAC Value for Final Setup

2.3 Influence factors in the Modal test for Lower arm B

There are 3 factors which will influence the test results are discussed in this section. First is the hanging mode: vertical VS horizontal; Second is the driving points in the test; Third is the additional mass needed considered in the test.

(1) Hanging pattern's influence on the mode test for the LAB. Hanging pattern is given from fig.11-12; and the resulting test data are given from table 1- table 2.

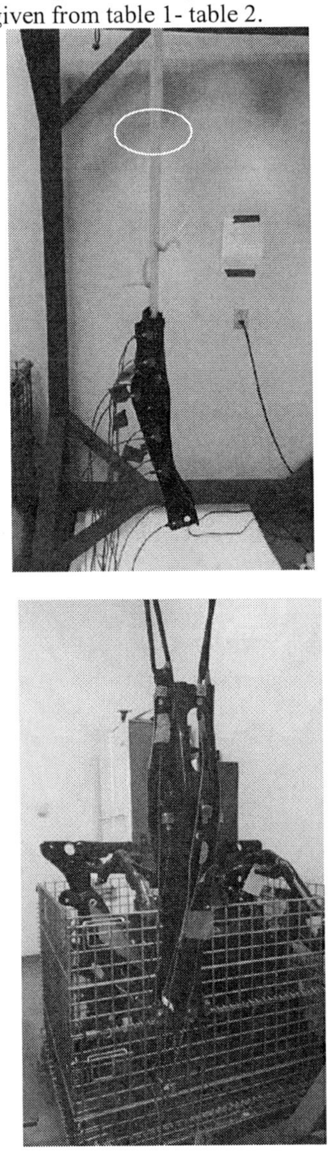

Fig.11Vertical hanging: Elastic tubing(upper) VS Bugee cord (lower)

Seen the test data from table 1-2, we can see that horizontal hanging pattern is better than the vertical's for the test of LAB, and using elastic tubing is more approximating the free-free boundary.

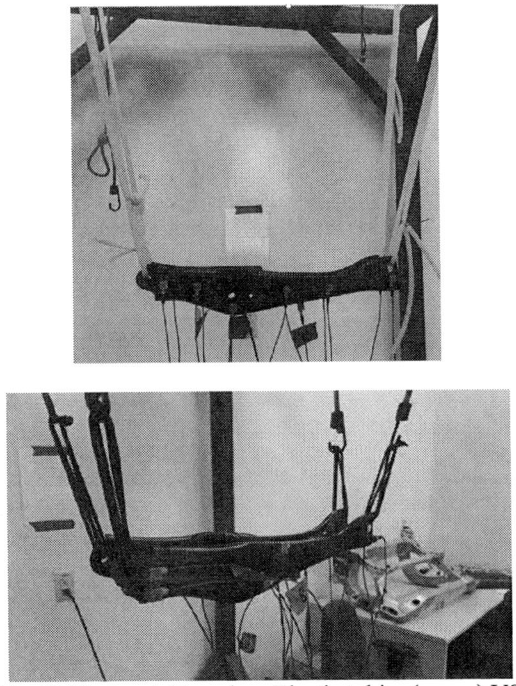

Fig.12 Horizontal hanging: Elastic tubing(upper) VS Bugee cord(lower)

Table 1: Modes from Vertical (upper) and horizontal (lower) Suspension with Bungee Cord Testing

Mode	un-damped Frequency (Hz)	Modal Damping (%)
1	297.2356	0.3693
2	418.1576	0.5533
3	639.228	0.4369
4	652.9684	0.7001
5	704.0253	0.8006

Mode	un-damped Frequency (Hz)	Modal Damping (%)
1	298.9999	0.2266
2	426.8763	0.4864
3	650.882	0.1587
4	671.9143	0.4327
5	722.296	0.4556

Table 2: Modes from Vertical (Upper) and horizontal (Lower) Suspension with Elastic Tubing Testing

Mode	un-damped Frequency (Hz)	Modal Damping (%)
1	301.2	0.3226
2	429.9	0.5489
3	646.2	0.533
4	676.8	0.2395
5	732.9	0.364

Mode	un-damped Frequency (Hz)	Modal Damping (%)
1	301.398	0.4385
2	428.5777	0.4811
3	643.4306	0.5914
4	672.9246	1.0576
5	732.3264	0.7225

(2) Derivation of Driving Point's Location

In order to excite all the interested modes, 3 driving points are needed in the test. The 3 points are near the location of accelerometers for the assurance of good coherence at these locations and all the driving points are impacted in Global X direction.

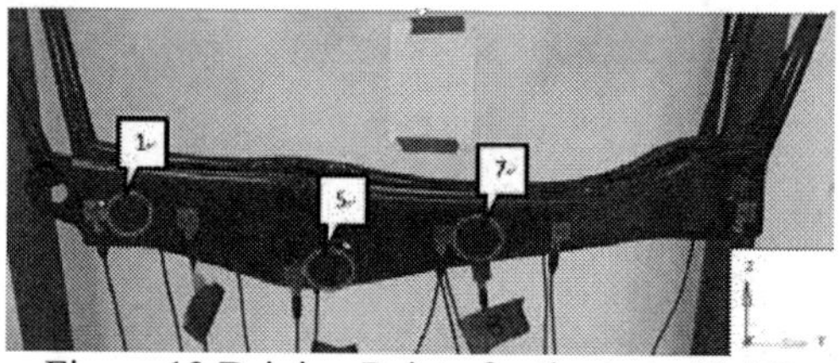

Figure 13 Driving Points for the test of LAB

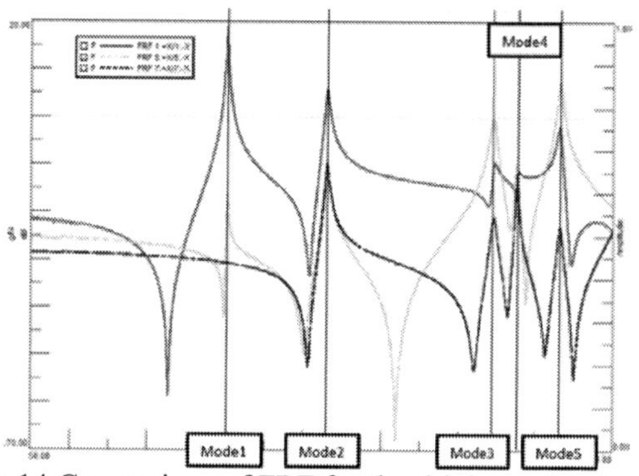

Figure 14 Comparison of FRF for the driving points location

(3) The accelerometer mass

Table 3 shows the correlation summary between FEA and modal test. We can see the MAC value is reasonably accepted but frequency difference is obvious.

In the test, the accelerometer is ICP tri-axial accelerometer model 356A15. According to the manufacture catalog, the weight of each accelerometer is 10.5 gram, the totally weight of used accelerometers is 0.1365 Kg which occupied about 6% of the gross weight of the arm. Thus, this weight is better to be considered in the FEA simulation for correlation purpose. After applying the accelerometer mass, we compare the part weight between FEA and Test

Comparing Table 3 and Table 4, we can find out with added accelerometer mass, the MAC remains almost the same, however the frequency difference is reduced remarkably. Frequency discrepancy among the first 5 modes is below 5%, which is accurate enough for engineering purpose.

Table 3: Correlation Results without including accelerometer mass in FEA model

Mode	f/Hz FEA	f/Hz Modal Testing	Frequency Difference (%)	MAC Value
1	320.9	301.2	6%	0.98
2	460	429.9	7%	0.96
3	677.9	646.2	5%	0.89
4	694.5	676.8	3%	0.89
5	760.5	732.9	4%	0.85

Table 4: Correlation Results including accelerometer mass in FEA model

Mode	f/Hz FEA	f/Hz Modal Testing	Frequency Difference (%)	MAC Value
1	313.8	301.2	4%	0.98
2	446.3	429.9	4%	0.96
3	669.8	646.2	4%	0.83
4	681.9	676.8	1%	0.86
5	751.8	732.9	3%	0.85

3 CONCLUSIONS

In this paper, different hanging style, transducer and wire's mass are taken into account in the test which help gain more accurate test results in actual application of engineering;

FEA model of lower arm B is updated according to the modal test data which means we will reduce the in even for a small part we need to consider the model explicit.

Because of the success gained with FEA and Modal test in the project, the technique will be used for predicting noise and vibration problems related to rear suspension assembly.

ACKNOWLEDGMENTS

I would like to express my sincere gratitude to Ohio State Universtity and F.tech R&D North America, Inc. who provide me the opportunity to work in this project.

REFERENCES

1. J.E.Mottershead,M.I. Friswell, A.1993 "Model updating in structural dynamics: a survey", JSV, 167, pp347-375.
2. Ewins, D. J., A. 2007, "Modal Testing: Theory and Practice", Research Studies Press Ltd., London.
3. Crocker, M.J., A. 2007, "Handbook of Noise and Vibration Control", John Wiley & Sons, Inc.
4. LMS Application Note, Substructuring Overview and Theory, LMS International, www.lmsna.com

RESEARCH ON DYNAMIC BEHAVIORS OF ELECTROMECHANICAL COUPLING MAIN DRIVE SYSTEM UNDER FRICTION

Liu Shuang
department of electrical engineering
Yanshan University
Qinhuangdao, Hebei province, China,
shliu@.ysu.edu.cn

Li Shuo
department of electrical engineering
Yanshan University
Qinhuangdao, Hebei province, China,
shuoli@stumail.ysu.edu.cn

Wu Zheng
department of electrical engineering
Yanshan University
Qinhuangdao, Hebei province, China,
994897898@qq.com

ABSTRACT

In this paper, electromechanical coupling system is regarded as research object. According to the theory of electromagnetic field, the method of energy, lumped mass method and Lagrange-Maxwell equations, the nonlinear dynamic model of the electromechanical main transmission system driven by AC synchronous motor has been established. Secondly, the lubrication friction's influence of the electromechanical coupling nonlinear dynamic model of main drive system has been considered. Taking lubricated friction coefficient as changing parameter, the Hopf bifurcation of the system is studied from the view of smooth dynamics. At last, the correctness of the results is verified by numerical simulation. The simulation shows that, with the change of the coefficient of friction, lubrication system near the bifurcation point may appear the phenomenon such as volatility and instability, and something can be learned that the stability of the system dynamic behavior is influenced by both the lubricated friction coefficient and the system initial conditions.

KEYWORDS: Electromechanical coupling drive system; Lubrication friction; Hopf bifurcation; Non-linear;

INTRODUCTION

Because of Mechatronics, electromechanical coupling system was born, and now it has been the foundation of industrial technology development. However, the stability of the system will be affected by many factors, and these have affected the safety of system operation seriously. Thus study of nonlinear dynamic characteristics of electromechanical coupling system has been paid more and more attention. Zhang[1]considering the structure of elastic coupling and sets up a kind of electromechanical coupling model of rolling mill systems. What's more, he puts forward a method to control the electromechanical coupling vibration. Through the establishment of sheet mill and the oil film thickness between the model in the process of rolling, Wang[2] provides a theoretical basis for determining the lubrication friction torque. What's more, through the analysis of the system fault, Shi Yang[3-4] puts forward the control method to improve the system stability. However, the analysis above mainly reveals the transmission mechanism of the complex dynamic behavior from the perspective of machinery. Therefore, from the perspective of electromechanical coupling this article studied dynamics behavior of friction in mechanical and electrical drive system.

1. THE NONLINEAR DYNAMIC MODEL OF THE ELECTROMECHANICAL MAIN TRANSMISSION SYSTEM

In order to simplify the model and be accordant with the practical meaning, according to the lumped-mass method, the transmission system of AC synchronous motor is abstracted as a two-mass relative rotation system including motor side and load side. The model is shown in Fig. 1.

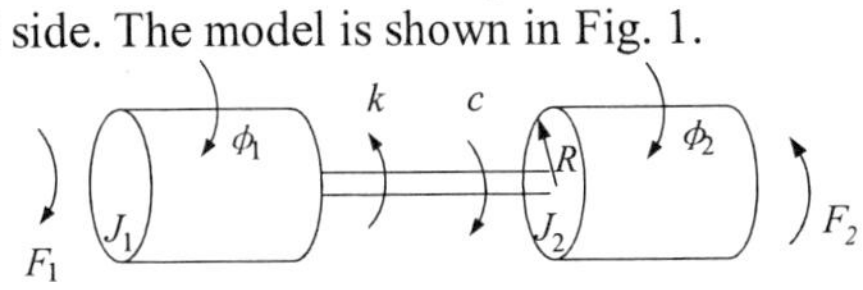

Figure 1: The two-mass rotation system dynamic model

The energy of the air-gap field between AC synchronous motor stator magnetic can be expressed by the following formula:

$$W_e = \frac{RL}{2}\int_0^{2\pi} \Lambda(\alpha,t)F^2(\alpha,t)d\alpha \qquad (1\text{-}1)$$

Where J_1、J_2 respectively expresses the motor side of rotary inertia and the load side of rotary inertia; $\varphi_i(i=1,2)$ and $\dot{\varphi}_i(i=1,2)$ are the angle and speed of the motor side and load side; K is the torsional rigidity of the connecting shaft in transmission system ; C is the damping force of connecting shaft; L is the equivalent length for the rotor; R is the average radius of the rotor.

Therefore the unit air-gap permeance between the rotor and the stator can be deduced as:

$$\Lambda(\alpha,t) = \Lambda_0[1+\frac{1}{2}(x^2+y^2)+x\cos\alpha+y\sin\alpha+ \qquad (1\text{-}2)$$
$$\frac{1}{2}(x^2-y^2)\cos 2\alpha + xy\sin 2\alpha]$$

Where $\Lambda_0 = \mu_0/\sigma$, $\sigma = k_\mu\delta_0$. μ_0 is the air permeability coefficient, δ_0 is the air-gap size, k_μ is the magnetic saturation.

According to the theory of motor, the air-gap of base wave magnetic motive force in the rotor and the stator is:

$$F(\alpha,t) = F_{sm}\cos(\omega t-\alpha) + F_{jm}\cos(\omega t + p\varphi_1 - \alpha - \psi_1 - \psi_2) \quad (1\text{-}3)$$

Where F_{sm} is the amplitude of magnetic motive force in the stator and F_{jm} is the rotor amplitude of magnetic motive force. $\psi_1 + \psi_2$ is the lagging phase angle of the rotor magnetic field potential compared to the stator magnetic field potential.

By substituting $L = T + W_e - V$ into the Lagrange-Maxwell function, then the function are:

$$\frac{d}{dt}\frac{\partial L}{\partial \dot{\varphi}_i} - \frac{\partial L}{\partial \varphi_i} + \frac{\partial F}{\partial \dot{\varphi}_i} = Q_i \quad (i=1,2) \quad (1\text{-}4)$$

Then

$$\begin{cases} J_1\ddot{\varphi}_1 + k(\varphi_1-\varphi_2) + c(\dot{\varphi}_1-\dot{\varphi}_2) = T_1 \\ J_2\ddot{\varphi}_2 - k(\varphi_1-\varphi_2) - c(\dot{\varphi}_1-\dot{\varphi}_2) = T_2 \end{cases} \quad (1\text{-}5)$$

Where T_1 is the electromagnetic torque, T_2 is the load torque. From (1-1) to (1-3) the electromagnetic expression can be obtained as:

$$T_1 = \frac{\partial W_e}{\partial \varphi_1} = b_0 + b_1\varphi_1 + b_2\varphi_1^2 + b_3\varphi_1^3 \quad (1\text{-}6)$$

Where

$$b_0 = -\pi RL\Lambda_0 F_{jm}F_{sm}\sin(\psi_1+\psi_2)$$

$$b_1 = -\pi RL\Lambda_0 F_{jm}F_{sm}\cos(\psi_1+\psi_2)$$

$$b_2 = -\frac{1}{2}\pi RL\Lambda_0 F_{jm}F_{sm}\sin(\psi_1+\psi_2)$$

$$b_3 = -\frac{1}{6}\pi RL\Lambda_0 F_{jm}F_{sm}\sin(\psi_1+\psi_2)$$

By substituting (1-6) into (1-5) the dynamic model of electromechanical transmission system can be obtained as follows:

$$\begin{cases} J_1\ddot{\varphi}_1 + k(\varphi_1-\varphi_2) + c(\dot{\varphi}_1-\dot{\varphi}_2) = b_0 + b_1\varphi_1 + b_2\varphi_1^2 + b_3\varphi_1^3 \\ J_2\ddot{\varphi}_2 - k(\varphi_1-\varphi_2) - c(\dot{\varphi}_1-\dot{\varphi}_2) = T_2 \end{cases} \quad (1\text{-}7)$$

From the equation (1-7), the equations show that mechanical and electrical drive system model is a complex high-dimensional nonlinear model. The characteristics of the load torque T_2 will determine the dynamics characteristics of the entire transmission system.

2. THE LUBRICATION FRICTION'S INFLUENCE OF THE ELECTROMECHANICAL COUPLING NONLINEAR DYNAMIC MODEL

The load torque T_2, contains two parts. One is the external disturbance excitation, the other is friction between the load and load torque.

$$T_2 = T_0 + T_r \quad (2\text{-}1)$$

Where

$$T_0 = fR\cos(\omega t) \quad (2\text{-}2)$$

In the actual working process of transmission system, due to the spraying of lubricants between the load side and the load, T_r is a kind of Lubrication friction.

In order to simplify the process of dynamics research, the effect of viscosity coefficient η and roll speed v_r on oil film thickness are emphasized. While modeling, the oil film thickness model can be simplified as follows:

$$\varepsilon = b_0\eta v_r + c_0 \quad (2\text{-}3)$$

Where b_0 and c_0 are the corresponding constant coefficient.

So the friction coefficient can be expressed as:

$$\mu = a_0 + ae^{-b\eta v_r + c} \quad (2\text{-}4)$$

Where b and c are the corresponding friction coefficient.

According to (2-4), the friction torque of the load side can be represented as:

$$T_f = P\mu R = PRa_0 + PRae^{-b\eta v_r + c} \quad (2\text{-}5)$$

The equation(1-7) is the transmission system dynamics model. Assume the state variables:

$$T_1 = T_{10} + \Delta T_1,\ T_2 = T_{20} + \Delta T_2,\ \varphi_1 = \varphi_{10} + \theta_1$$

$$\varphi_2 = \varphi_{20} + \theta_2,\ \dot{\varphi}_{10} = \dot{\varphi}_{20},\ \ddot{\varphi}_{10} = \ddot{\varphi}_{20} = 0 \quad (2\text{-}6)$$

Where T_{10}, T_{20} respectively expresses the constant composition of the electromagnetic torque in motor side and the load side torque; $\Delta T_1, \Delta T_2$ respectively represents the nonlinear section of electromagnetic torque and the disturbance in load torque; $\varphi_{10}, \varphi_{20}$ respectively expresses the torsion angle between the motor side and load caused by electromagnetic torque constant T_{10} and constant load torque T_{20}; θ_1, θ_2 respectively expresses the torsion angle caused by the nonlinear electromagnetic torque and torque disturbance .

The friction torque (2-5) can be derived using Taylor's expansion at the steady speed of $\dot{\varphi}_{20}$ ($v_r = R\dot{\varphi}_2$), reserve the first two parts and substitute (2-6) into (1-7), then the equation is as follows:

$$\begin{cases} J_1\ddot{\theta}_1 + c(\dot{\theta}_1-\dot{\theta}_2) + k(\Delta\varphi+\theta_1-\theta_2) = T_{10} + \Delta T_1 \\ J_2\ddot{\theta}_2 - c(\dot{\theta}_1-\dot{\theta}_2) - k(\Delta\varphi+\theta_1-\theta_2) = -T_{20} - \Delta T_2 \end{cases} \quad (2\text{-}7)$$

Where $T_{10} = b_0$, $\Delta T_1 = b_1\theta_1 + b_2\theta_1^2 + b_3\theta_1^3$

$$\Delta T_2 = -\mu_0 Pb\eta R^2\dot{\theta}_2 + \frac{1}{2}\mu_0 Pb^2\eta^2R^3\dot{\theta}_2^2 + fR\cos\omega t$$

$T_{20} = PR(a_0 + \mu_0)$ $\mu_0 = e^c \cdot \mu_0 = e^c$

When the system is running under constant torque, there are two equations $\Delta T_1 = \Delta T_2 = 0$, $\theta_1 = \theta_2 = 0$. By substituting them into (2-7), the equation is written as follows:

$$\begin{cases} K\Delta\varphi = T_{10} \\ K\Delta\varphi = T_{20} \end{cases} \quad (2\text{-}8)$$

By substituting (2-8) into (2-7):

$$\begin{cases} J_1\ddot{\theta}_1 + c(\dot{\theta}_1 - \dot{\theta}_2) + k(\theta_1 - \theta_2) = b_1\theta_1 + b_2\theta_1^2 + b_3\theta_1^3 \\ J_2\ddot{\theta}_2 - c(\dot{\theta}_1 - \dot{\theta}_2) - k(\theta_1 - \theta_2) = -fR\cos\omega t \\ +\mu_0 Pb\eta R^2\dot{\theta}_2 - \frac{1}{2}\mu_0 Pb^2\eta^2 R^3\dot{\theta}_2^2 \end{cases} \quad (2\text{-}9)$$

Where the equations (2-9) are the influence caused by friction lubrication to electromechanical coupling system.

3. THE HOPF BIFURCATION OF THE SYSTEM

In order to simplify the calculation, assume the dimensionless variables:

$$t^* = \frac{\omega}{\Omega}t, \quad y_1 = \frac{\theta_1}{\theta_c}, \quad y_2 = \frac{\theta_2}{\theta_c}, \quad \omega_1^2 = \frac{K\Omega^2}{J_1\omega^2},$$

$$\omega_2^2 = \frac{K\Omega^2}{J_2\omega^2}, \quad \xi_1 = \frac{c\Omega}{J_1\omega}, \quad \xi_2 = \frac{c\Omega}{J_2\omega}, \quad k_1 = \frac{b_1\Omega^2}{J_1\omega^2},$$

$$k_2 = \frac{b_2\theta_c\Omega^2}{J_1\omega^2}, \quad k_3 = \frac{b_3\theta_c^2\Omega^2}{J_1\omega^2}, \quad F = -\frac{f\Omega^2}{J_2\varphi_c\omega^2},$$

$$\mu_1 = \frac{\mu_0 Pb\eta R^2\Omega}{J_2\omega}, \quad \mu_2 = \frac{\mu_0 Pb^2\eta^2 R^3\theta_c}{2J_2}.$$

Then the dynamic model can be simplified as follows:

$$\begin{cases} \ddot{y}_1 + \xi_1(\dot{y}_1 - \dot{y}_2) + \omega_1^2(\ddot{y}_1 - \ddot{y}_2) = k_1 y_1 + k_2 y_1^2 + k_3 y_1^3 \\ \ddot{y}_2 - \xi_2(\dot{y}_1 - \dot{y}_2) - \omega_2^2(y_1 - y_2) = F\cos\omega t + \mu_1\dot{y}_2 - \mu_2\dot{y}_2^2 \end{cases} \quad (3\text{-}1)$$

When $F = 0$, the system is a kind of autonomous system. Assume the state variables: $x_1 = y_1$, $x_2 = \dot{y}_1$, $x_3 = y_2$, $x_4 = \dot{y}_2$, $a_1 = k_1 - \omega_1^2$, $b_1 = -\xi_1$, $c_1 = \omega_1^2$, $d_1 = \xi_1$, $m_2 = k_2$, $m_3 = k_3$, $a_2 = \omega_2^2$, $b_2 = \xi_2$, $c_2 = -\omega_2^2$, $d_2 = \mu_1 - \xi_2$, $n_2 = -\mu_2$, the equation (3-1) can be written as follows, and the equilibrium point is at the origin.

$$\begin{cases} \dot{x}_1 = x_2 \\ \dot{x}_2 = a_1x_1 + b_1x_2 + c_1x_3 + d_1x_4 + m_2x_1^2 + m_3x_1^3 \\ \dot{x}_3 = x_4 \\ \dot{x}_4 = a_2x_1 + b_2x_2 + c_2x_3 + d_2x_4 + n_2x_4^2 \end{cases} \quad (3\text{-}2)$$

For equation (3-2), its jacobian matrix at the origin is:

$$A(x_0, b) = \begin{Bmatrix} 0 & 1 & 0 & 0 \\ a_1 & b_1 & c_1 & d_1 \\ 0 & 0 & 0 & 1 \\ a_2 & b_2 & c_2 & d_2 \end{Bmatrix} \quad (3\text{-}3)$$

The characteristic equation of A is written as follows:

$$\lambda^4 + p_1\lambda^3 + p_2\lambda^2 + p_3\lambda + p_4 = 0 \quad (3\text{-}4)$$

Where $p_1 = -b_1 - d_2$, $p_2 = b_1d_2 - b_2d_1 - a_1 - c_2$, $p_3 = a_1d_2 - a_2d_1 - b_1c_2 - b_1c_1$, $p_4 = a_1c_2 - a_2c_1$。

Then Hurwitz criterion is used to solve the Hopf bifurcation point of the system.

Assume that the friction coefficient $b = b^*$ is the Hopf bifurcation points in the system. For the equation (3-4), the condition of Hopf bifurcation can be expressed as:

$$\begin{cases} p_i(b^*) > 0 \quad (i = 1,2,3,4) \\ \Delta_i'(b^*) > 0 \quad (i = 1,2) \\ \Delta_3'(b^*) = 0 \\ \frac{d\Delta_3'(b)}{db}\Big|_{b=b^*} = 0 \end{cases} \quad (3\text{-}5)$$

Where Δ_i' is the Hurwitz determinant of the system $\Delta_1' = p_1$, $\Delta_2' = \begin{vmatrix} p_1 & 1 \\ p_3 & p_2 \end{vmatrix}$,

$$\Delta_3' = \begin{vmatrix} p_1 & 1 & 0 \\ p_3 & p_2 & p_1 \\ 0 & p_4 & p_3 \end{vmatrix}.$$

Select the variables as follows: $a_1 = -25$, $b_1 = -2$, $c_1 = 13$, $d_1 = 2$, $m_2 = 4.6$, $m_3 = 15.8$, $a_2 = 19.5$, $b_2 = 3$, $d_2 = 17.42b - 3$, $c_2 = -19.5$, $n_2 = -61.33b^2$.

From the bifurcation condition (3-5), the Hopf bifurcation point of the system is $b^* = 0.01533$, and the system contains a pair of imaginary roots.

Assume that the imaginary roots of the system are $\pm\omega_0 i$, U and W are the corresponding eigenvectors of the matrix, then the equations are written as follows:

$$UA = \omega_0 iU, \quad AW = \omega_0 iW, \quad UW = 1 \quad (3\text{-}6)$$

Assume the variable:

$$\rho = \mathrm{Re}(-Uf_{xxx}WWW^* + 2Uf_{xx}WA^{-1}(0)f_{xx}WW^* + Uf_{xx}W^*[A(0) - 2i\omega_0 I]^{-1}f_{xx}WW)$$

Where W^* is the conjugate complex numbers of W.

$$A(0) = A(x_0, b)\big|_{b=b^*}$$

$$f_{xxx}WWW^* = \left(\frac{\partial}{\partial x}\left(\frac{\partial}{\partial x}\left(\frac{\partial f(x,b)}{\partial x}\times W\right)\times W\right)\times W^*\right)\Bigg|_{b=b^*, x=x_0}$$

From the sign of ρ, the stability of the periodic solution of Hopf bifurcation can be judged .

By substituting $b^* = 0.01533$ into calculation, $\rho = -0.078 < 0$, the periodic solution of the system at the bifurcation point is not asymptotically stable, the Hopf bifurcation belongs to a kind of subcritical bifurcation.

4. THE NUMERICAL SIMULATION

The numerical simulation of the dynamic behavior is carried out near the bifurcation point $b^* = 0.01533$ in the system.

When $b = 0.01148$, it is less than the bifurcation point. As shown in Fig.4-1, the speed of load side gradually decreases with time, the system is convergent and eventually tends to be stable.

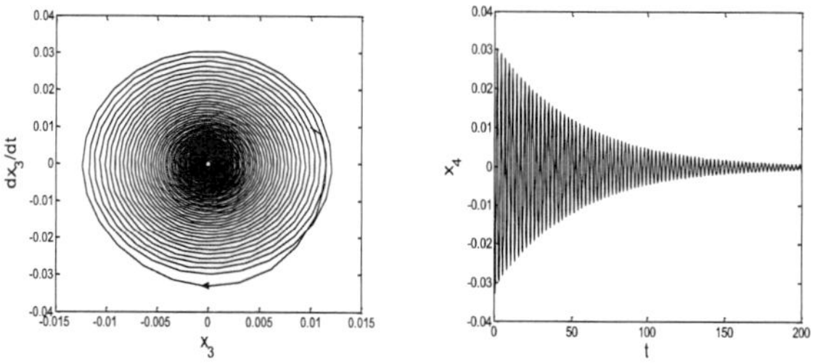

(a) The phase orbits (b) The Time history diagram

Figure 4-1, when $b = 0.01148$, The figures are as shown.

When $b = 0.01492$, it approaches the bifurcation point, at this time subcritical bifurcation occurs, the movement of the system trajectory will be affected by the initial conditions. As shown in Fig.4-2, when the initial point is far from the equilibrium point, the trajectory is gradually divergent from a limit cycle, and the system is unstable. As shown in Fig.4-3, when the initial point is close to the equilibrium point, the system trajectory is convergent and the system tends to be stable.

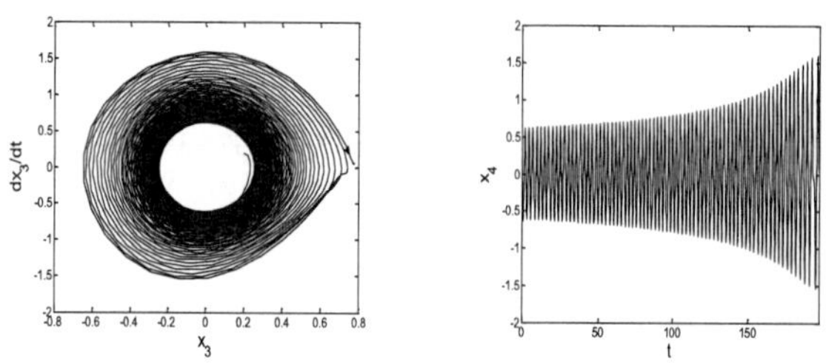

(a) The phase orbits (b) The Time history diagram

Figure 4-2, when $b = 0.01492$, the initial condition is $[0.2\ 0.2\ 0.2\ 0.2]$. The figures are as shown.

When $b = 0.02296$, it is less than the bifurcation point. As shown in Fig.4-4, the load side speed gradually diverges with time, the system emerges self-oscillation phenomenon.

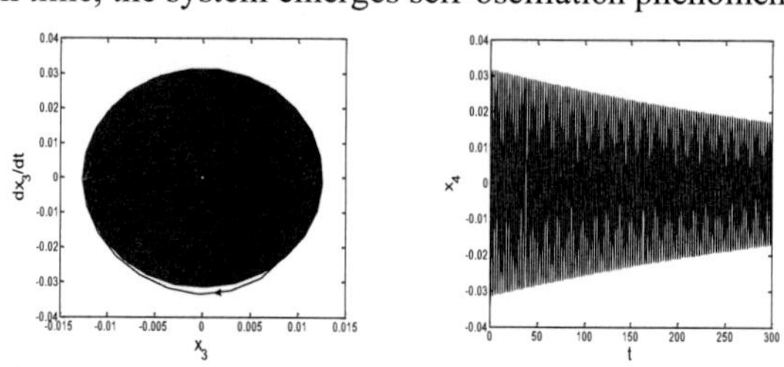

(a) The phase orbits (b) The Time history diagram

Figure 4-3, when $b = 0.01492$, the initial condition is $[0.01\ 0.01\ 0.01\ 0.01]$. The figures are as shown.

In the analysis above, the system has complicated dynamic behavior. The friction coefficient and the initial condition of the system are directly related to the stability of the system, and the system may have violent concussion and even instability near the bifurcation point. At this time, the system has more complex dynamic phenomena.

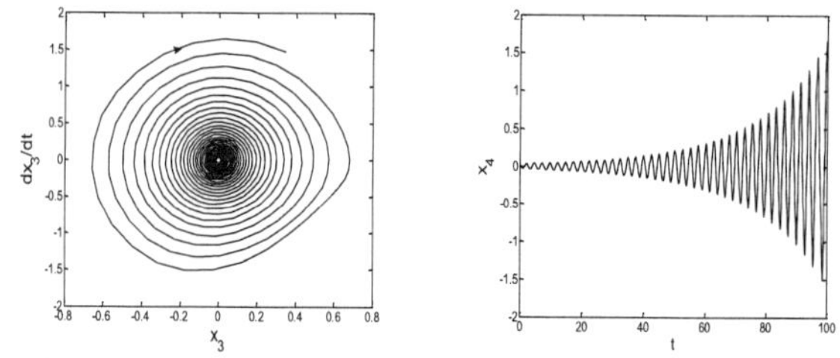

(a) The phase orbits (b) The Time history diagram

Figure 4-4, when $b = 0.02296$, The figures are as shown.

5. CONCLUSIONS

Firstly the dynamic model of electromechanical main transmission system is established, and on the basis of it, the influence of lubrication friction on electromechanical main transmission system is considered. What's more, taking lubricated friction coefficient as changing parameter, the Hopf bifurcation of the system is studied from the view of smooth dynamics. In the end, the correctness of the results is verified by numerical simulation. The simulation shows that, with the change of the coefficient of friction, lubrication system near the bifurcation point may appear the phenomenon such as volatility and instability. In order to ensure the safe operation of the whole system, these phenomena should be avoided in the process of actual production.

ACKNOWLEDGMENTS

This paper is supported by National Natural Science Foundation of China (No. 61673334).

REFERENCES

1. Zhang, R. C., & Zhuo, C. L., 2015. Study on the Vibration Characteristics of the Electromechanical Coupling in the Main Drive System of Rolling Mill with Impacts of the Magnetic Parameters. Machinery Design & Manufacture(8), 128-132.
2. Wang, y. l., & Shen J. F., 1980. Lubrication theory foundation. Lubrication theory foundation. Machinery Design & Manufacture.
3. Huang, J., Shi, Y., & Zhang, X. 2015. Active fault tolerant control systems by the semi - markov model approach. International Journal of Adaptive Control & Signal Processing, 28(9), 833-847.
4. Zhang, H., Shi, Y., & Mehr, A. S. 2012. Stability and stabilization in switched discrete-time systems. International Journal of Adaptive Control & Signal Processing, 26(26), 991–1012.

INTERACTION OF PITCH AND HEAVE DOF`S IN STALL FLUTTER

Dominique Poirel (poirel-d@rmc.ca), Luba Goyaniuk and Azémi Benaissa
Department of Mechanical and Aerospace Engineering
Royal Military College of Canada
Kingston, Ontario, Canada

ABSTRACT

The stall flutter behavior of a pitch-heave aeroelastic system is experimentally investigated at Reynolds numbers ranging from 60,000 to 140,000. Self-sustained large amplitude oscillations, in the order of 30° to 50° in pitch and up to 0.6 in heave (non-dimensionalized by the chord length), are observed. They occur for various combinations of pitch and heave spring stiffness coefficients. It is concluded that the LCOs are pitch dominated, and that the heave response, for the most part, is driven by the pitch acting as a forcing mechanism. For frequency ratio values close to one, however, there appears to be a feedback relationship from the heave to the pitch. In this regime the system response is dictated by the coupling between pitch and heave rather than by a forcing in pitch as seen over frequency ratios well below or above 1.

KEYWORDS: stall flutter, negative aerodynamic damping, LCO, pitch-heave coupling, lock-in

INTRODUCTION

Stall flutter is an aeroelastic phenomenon normally associated with dynamic stall which describes the lag between the aerodynamic loads and the oscillatory motion of an airfoil in and out of stall (McCroskey, 1981). For a given set of conditions, this lag can cause negative aerodynamic damping which then enables stall flutter to occur (Fung, 1955).

Preliminary results of one-degree-of-freedom (1DOF) stall flutter observed in the Aeroelasticity Lab at RMC were reported for various locations of the pitch rotation axis, for instance for an axis located aft of the quarter chord point (Peristy *et al.*, 2014). In this work, we report additional results on stall flutter and we focus on the interaction between the pitch and heave degrees of freedom.

EXPERIMENTAL SET-UP

The basic setup is comprised of the archetypical NACA0012 wing mounted elastically via pulleys and springs on a free rotating axis in pitch, which can also translate in heave so as to permit 2DOF motion. The wing is rigid with uniform chord length and its surface is smooth. Its aspect ratio is 3.9, but the flow can be considered as quasi-2D because of endplates located along the wind tunnel walls. See Fig. 1.

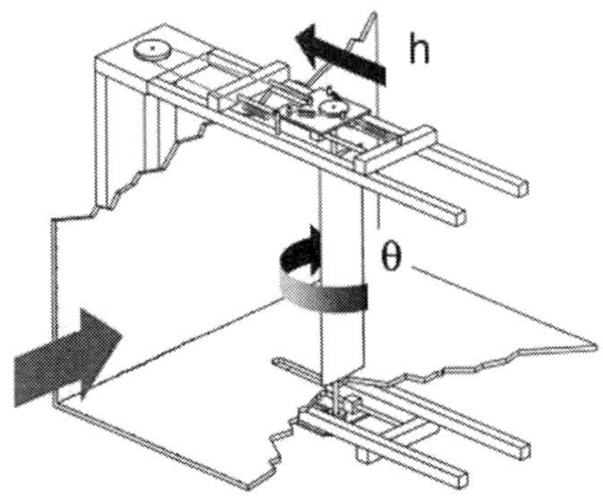

Figure 1: Flutter rig

For this study, the basic parameters which are varied are the frequency ratio, $\overline{\omega} = \omega_h / \omega_\theta$, and the Reynolds number, Re_c, where c is the chord. The flow is considered incompressible as the Mach number is lower than 0.05. The turbulence intensity of the freestream is also small at 0.16%. The wing is aligned vertically to isolate the response from gravity. The rotation axis is located at 0.35c aft of the leading edge. The chord is 0.156 m long.

RESULTS AND DISCUSSION

Results are presented for three Reynolds numbers and various values of frequency ratio, in addition to the pitch only case. A typical response in time of the pitch and heave is presented in Fig. 2, followed by the histogram in Fig. 3 and the power spectrum in Fig. 4.

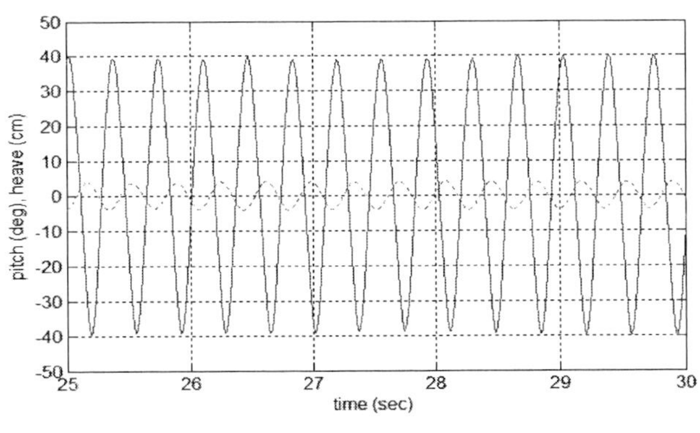

Figure 2: Pitch (——) and heave (- - - -) time responses; $Re_c = 87{,}000$, $\overline{\omega} = 0.82$

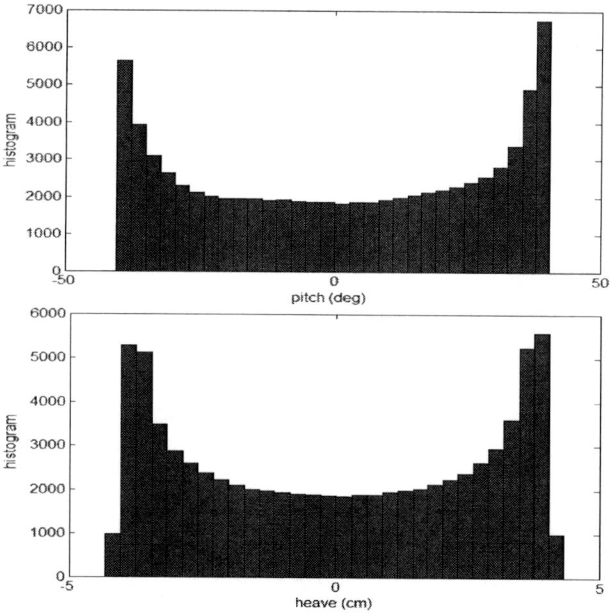

Figure 3: Pitch (top) and heave (bottom) histograms; $Re_c = 87{,}000$, $\overline{\omega} = 0.82$

Both DOFs are well behaved exhibiting simple harmonic motion. This dynamics is representative of all frequency ratios away from one. However, as the frequency ratio gets closer to one, a second frequency appears and a beating phenomenon is observed. This is exemplified in Figs 5 and 6 for a frequency ratio of 0.92, for the time response and the frequency content, respectively.

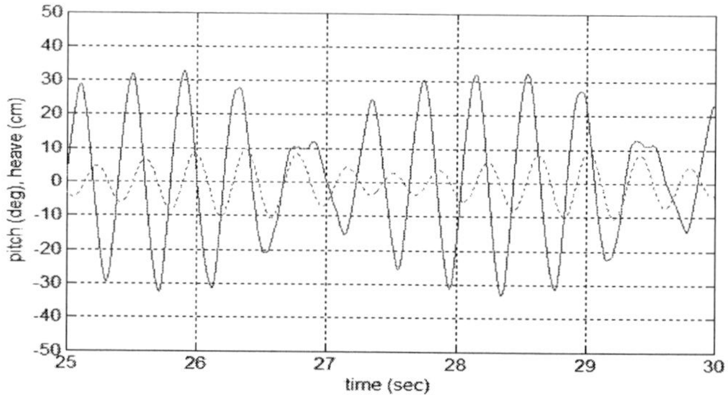

Figure 5: Pitch (——) and heave (- - - -) time responses; $Re_c = 87{,}000$, $\overline{\omega} = 0.92$

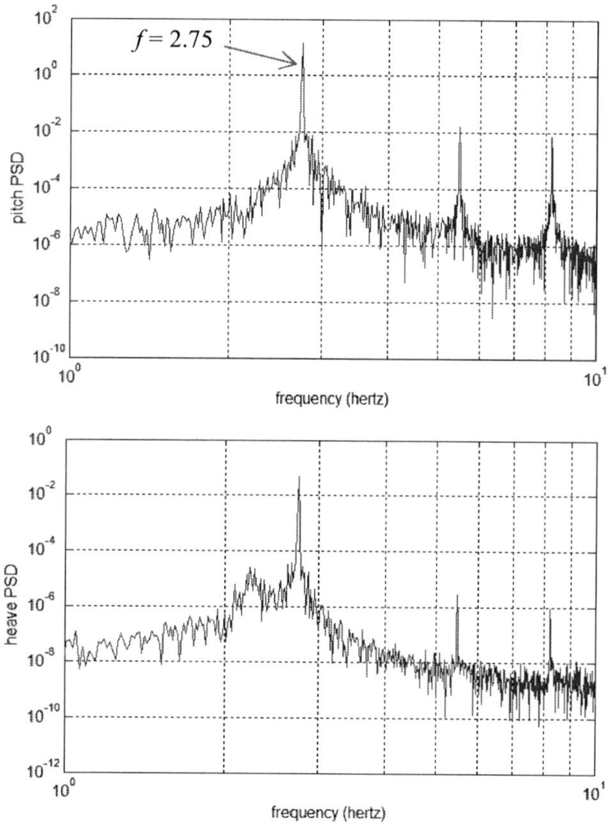

Figure 4: Pitch (top) and heave (bottom) PSDs; $Re_c = 87{,}000$, $\overline{\omega} = 0.82$

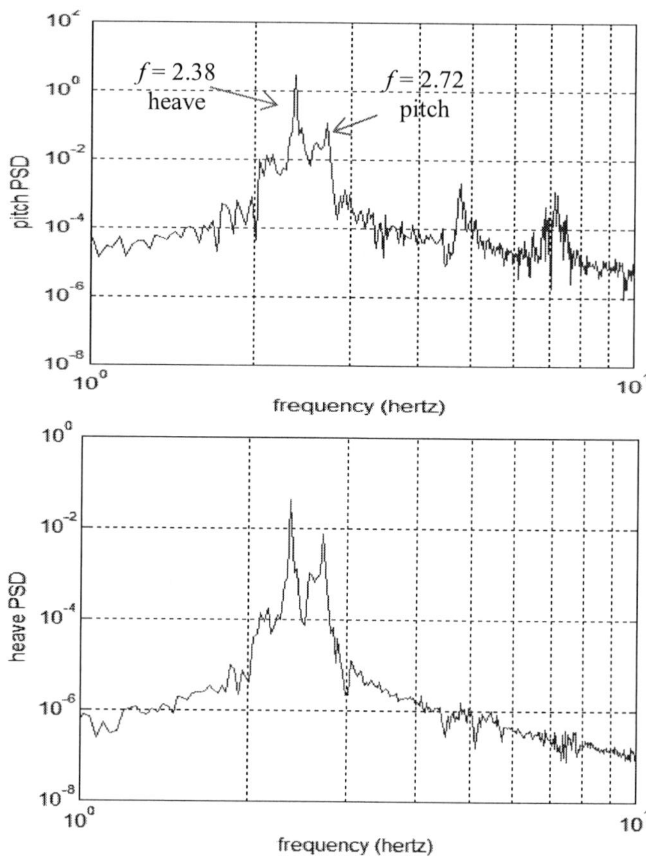

Figure 6: Pitch (top) and heave (bottom) PSDs; $Re_c = 87{,}000$, $\overline{\omega} = 0.92$

The compiled pitch and heave LCO amplitudes, as a function of frequency ratio, are shown in Figs. 7 and 8, respectively. The LCO frequency is shown in Fig. 9.

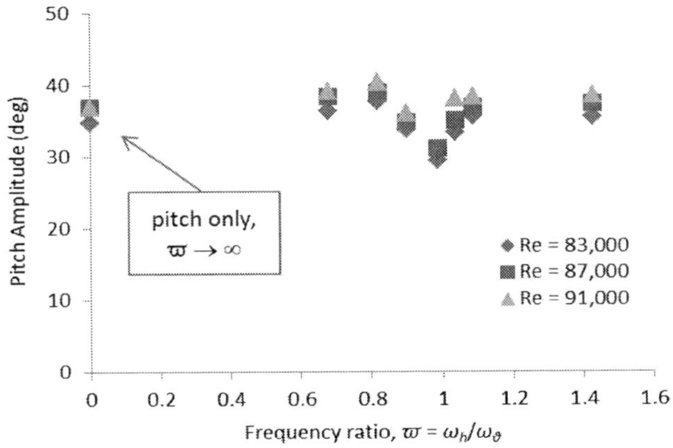

Figure 7: Pitch amplitude vs frequency ratio

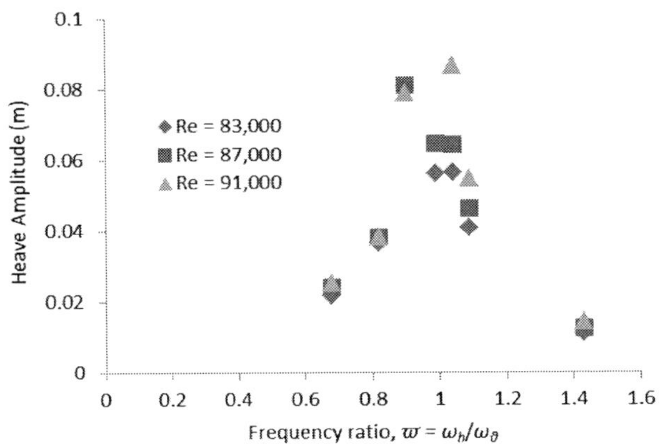

Figure 8: Heave amplitude vs frequency ratio

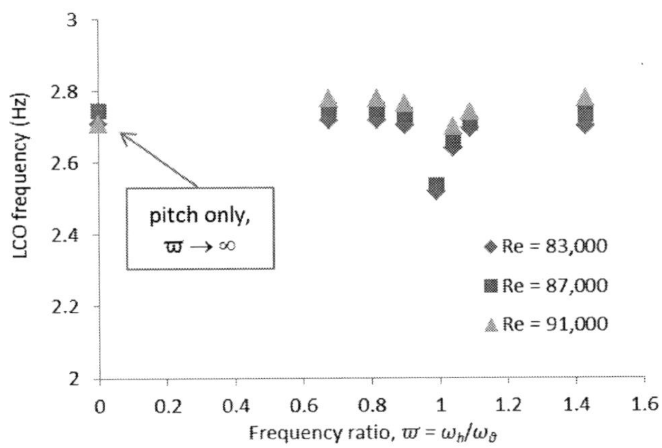

Figure 9: LCO frequency vs frequency ratio

Note that the 1DOF pitch only amplitude and frequency values are shown on the *y*-axis aligned with $\overline{\varpi} = 0$; this is solely to make the figures more readable as in reality the 1DOF pitch is represented by $\overline{\varpi} \to \infty$ since the heave stiffness is effectively infinite.

In comparison with the 1DOF pitch only case, the introduction of the heave DOF does not alter meaningfully the LCO dynamics for frequency ratios away from unity. Both the pitch amplitude (Fig. 5) and frequency (Fig. 7) remain close to the 1DOF problem in these regimes. It is clear that the heave motion does not interact with the pitch mechanism.

On the other hand, when the uncoupled structural natural frequency of the heave approaches the uncoupled structural natural frequency in pitch, a significant change in dynamics is observed. This event is characterized by a drop in pitch amplitude, a drop in in LCO frequency and an increase in heave amplitude. The increase in heave amplitude is akin to resonance since the frequency of the "forcing" (pitch) is close to the natural frequency of the system being forced (heave). However, the change in pitch dynamics suggests that there is a strong (two-way) coupling between both DOF's for $\overline{\varpi} \approx 1$ which resembles lock-in of the pitch to the heave.

CONCLUSIONS

Experimental results of stall flutter in 2DOF (pitch-heave) have shown that for a pitch dictated flutter phenomenon, the heave behaves generally as a forced response. However, for frequency ratio values close to one the feedback relationship from the heave to the pitch becomes important. In this regime the system response is dictated by the coupling between pitch-heave, rather through a forcing mechanism in pitch only.

ACKNOWLEDGEMENTS

The first and third authors acknowledge financial support from NSERC and DND. The second author acknowledges financial support from NSERC:Create:SERA.

REFERENCES

1. McCroskey, W. 1981, "The phenomenon of dynamic stall", NASA TM-81264.
2. Fung, Y.C. 1955, *An Introduction to the Theory of Aeroelasticity*, John Wiley & Sons, Inc., New York
3. Peristy, L., Poirel, D. and Yuan, W. 2014, "Characterization of the aeroelastic behaviour of a NACA0012 airfoil at transitional Reynolds numbers", *Proceedings of the Canadian Society of Mechanical Engineering (CSME) International Congress*, Toronto, Canada.

Multicomponent Materials

MONITORING DAMAGE ACCUMULATION EFFECTS ON THE ELASTIC PROPERTIES OF COMPOSITES

Cagatay Yilmaz
Faculty of Engineering and Natural Sciences, Integrated Manufacturing Technologies Research and Application Center, Sabanci University, 34956, Istanbul, Turkey
cagatayyilmaz@sabanciuniv.edu

Afzal Suleman
Department of Mechanical Engineering, Center for Aerospace Research
University of Victoria, Victoria, BC, V8W 3P6 CANADA
suleman@uvic.ca

Mehmet Yildiz
Faculty of Engineering and Natural Sciences, Integrated Manufacturing Technologies Research and Application Center, Sabanci University, 34956, Istanbul, Turkey
meyildiz@sabanciuniv.edu

ABSTRACT

Damage accumulation in Glass Fiber Reinforced Polymer (GFRP) composites is monitored based on Poisson's ratio and elastic modulus measurement for two different fiber stacking sequences subjected to both quasi-static and quasi-static cyclic tensile loadings. The sensor systems utilized to measure strain include a dual-extensometer, a biaxial strain gage and a novel embedded-biaxial Fiber Bragg Grating (FBG) sensor. These sensors are used concurrently to measure biaxial strain whereby the evolution of Poisson's ratio and elastic modulus as a function of the applied axial strain is evaluated. It is observed that each sensor system measures a non-constant Poisson's ratio and elastic modulus, which is a sign of damage accumulation under the applied tensile loading. An Acoustic Emission (AE) system is also utilized to capture elastic stress waves emitted by damage. The acquired AE signals are distinguished by analyzing frequency contents to the different damage types. The accumulation of different damage types is presented along with damaged elastic properties.
KEYWORDS: Poisson's ratio, FBG, glass fiber, elastic modulus

INTRODUCTION

In literature, there are several approaches developed and investigated to understand the accumulation of damage and damage state of composite structures under static and dynamic loading conditions. For example, Highsmith and Reifsnider studied stiffness reduction as a damage indicator in composite materials [1]. Several researchers examined the reduction of Poisson's ratio as a damage indicator [2,3] since Poisson's ratio $\upsilon_{xy} = -\varepsilon_y / \varepsilon_x$ embodies both axial and transversal strain (ε_x and ε_y, respectively) information and is affected by the transverse cracks formed as a result of applied longitudinal strain [4]. Smith et al.[5] modelled the evolution of Poisson's ratio using a shear-lag theory under static loading conditions, correlated their analytic model with experimental results, and indicated that Poisson's ratio decreases as the transverse crack density increases.

Novel embedded biaxial FBG sensor is also used to make a comparison between the surface mounted electrical strain sensor systems and embedded optical strain sensor. FBG is a section of a single mode optical fiber which contains a periodic variation of refractive index formed holographically on the core of the optical fiber along the fiber direction, and acts like a stop-band filter [6]. An FBG sensor reflects a small portion of the incoming electromagnetic spectrum while enabling the passage of the others. The center wavelength of the reflected portion of the incident electromagnetic spectrum is referred to as the Bragg wavelength, $\lambda_B = 2n_{eff}\Lambda$ where Λ is grating period and n_{eff} is the effective refractive index of the FBG sensor. When subjected to strain or temperature variations, the grating period and the effective refractive index of the FBG sensor change, thereby causing a shift in the Bragg wavelength λ_B. The change in the Bragg wavelength can be coupled to external effects, namely, temperature and strain through the following equation $\Delta\lambda_B / \lambda_B = (\alpha + \xi)\Delta T + (1 - \rho_e)\varepsilon$ where α and ξ are the thermal expansion and thermo-optic coefficients of the fiber core, respectively. Here, $\Delta\lambda_B$ is the shift in the Bragg wavelength, ΔT is the change in the temperature of the grating region and ρ_e denotes effective photo-elastic constant of the fiber core, which is taken as 0.22 in this work, and ε is the axial strain of the grating region. For a constant temperature, one can write the previous relation as $\Delta\lambda_B / \lambda_B = (1 - \rho_e)\varepsilon$.

In this study, we have investigated the effect of transverse cracks on the reduction of Poisson's ratio and as a function of imposed axial strain or stress. To this end, composite specimens with two different ply stacking sequences (i.e., uniaxial laminates, and laminates with two 90^0 off-axis plies) are instrumented with a biaxial extensometer and strain gage and then subjected to a quasi-static loading, thereby shedding light on the intimate relation between the reduction in Poisson's ratio and elastic modulus, and the transverse cracking. Given that, in angle ply composites, the deformation in the lateral direction is enhanced, which in turn can easily obscure the effect of transverse cracks on the lateral deformation, we limit our study to laminates with 0^0 and 90° stacking orientations. A novel approach for embedding biaxial FBG sensor into a composite laminate is proposed and utilized for multi-axis strain acquisition as well as for structural health monitoring of composites through referring to the reduction in Poisson's ratio. Specimens with extensometer, strain gage and embedded biaxial FBG sensor are subjected to quasi-static cyclic loading to study the reduction in Poisson's ratio under cyclic loading. It is shown that there can be notable variation in the values of lateral strain recorded by extensometer, strain gage and FBG sensor due to the difference in the gage length of these strain sensors. Referring to the reduction in Poisson's ratio, it is shown that composite specimens with a higher number of off-axis plies are more prone to the formation of transverse cracking. This study contributes to the state of the art in the field of composites materials testing and it is the first reported study on the measurement of the decline in Poisson's ratio for composites with different fiber stacking sequences using embedded biaxial FBG sensors.

MATERIALS AND METHODOLOGY

Composite Materials

As a reinforcement, 330 gsm unidirectional (0o) E-glass stitched fabric (Metyx, Turkey) was used. As a matrix material, Araldite LY 564 epoxy and XB3403 hardener system (Hunstman) was used. Composite laminates with two different stacking sequences, namely, $(0)_6$ and $[90/90/0]_s$, were manufactured by vacuum infusion method as shown in Figure 1 (a)-(b). Figures 1 (a) and (b) respectively show the water-heated metal curing table and the vacuum bagging setup. Composite plates were cured at 75 °C for 15 hours. Having completed the curing process, laminates were cut into the test specimen according to ASTM D3039 standard using water-cooled diamond circular saw.

Equipment and Test

Specimens were subjected to the quasi-static monotonic and qquasi-static cyclic tensile tests using Zwick Z100 universal testing machine with a load cell of $\pm 100\ kN$ under the constant rate of crosshead displacement (2 mm/min). The quasi-static cyclic tensile test is performed through loading-unloading cycles with 50 MPa increment in stress with respect to previous maximum stress. Epsilon 3542 with fixed gage length of 25 mm and Epsilon 3575 with a controllable gauge length were respectively used as axial and transverse extensometers. The signals of extensometers were collected by National Instruments NI SCXI-1000 main chassis with NI SCXI-1520 card at a sampling frequency of 1000 Hz. Mistras PCI- 2 acoustic emission (AE) hardware and an AEwinPCI2-4 software with a sampling rate of 2 MHz.were used to the AE data.

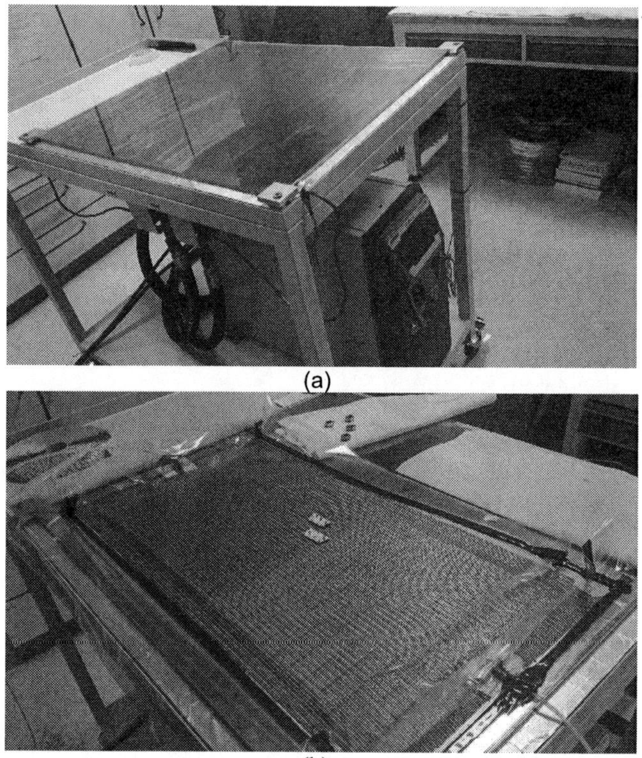

(a) (b)

Figure 1: a) water-heated curing table, and b) vacuum bagging process

RESULT AND DISCUSSION

In order to assess the mechanical properties and evolution of Poisson's ratio of composite plates with different stacking sequences, test specimens were subjected to quasi-static tensile loading until failure and all the data collected during the tests were processed and tabulated in Table 1 along with the fiber volume fractions. During the quasi-static tensile tests of all composite coupons, both axial and transversal strains were recorded by extensometers. Biaxial strain gages were attached only to a single coupon for each plate to investigate the effect of sensor type (i.e., gage length, measurement location) on the strain measurement. Figure 2(a) yields stress-strain curves for composite coupons having different stacking sequences. One can see from Table 1 that, as the number of 0° layer increases, so do the strength and elastic modulus of composites, as expected. It is observed that the axial strain at failure does not differ significantly for composites with three different stacking sequences. Considering the iso-strain condition, strain of fibers, composite and matrix must be same at break, namely, $\varepsilon_c = \varepsilon_f = \varepsilon_m$. The fact that the failure of the composite takes place at the ultimate strain of the fibers parallel to the loading direction explains why composites with different stacking orientations have nearly the same axial strain. On the other hand, as seen in both Figure 2(b) and Table 1, on increasing the amount of 90° fibers (perpendicular to loading), the ultimate transverse strain of the composite decreases due to the fact that 90° fibers increase the elastic modulus of the composite in the transversal direction.

Table 1: Mechanical properties of composites with three different stacking configurations (strain measurements are based on extensometer).

	G6U1 $[0]_6$	G6B1 $[90/90/0]_s$
Strength (σ_u) MPa	716	310
Elastic Modulus (GPa)	28	15
Axial Strain (ε_x) ($\mu\varepsilon$)	26175	25739
Transverse Strain (ε_y) ($\mu\varepsilon$)	5363	1334
Fiber Volume Fraction (V_f) (%)	41.3	42
Maximum Poisson's ratio	0.253	0.113
Minimum Poisson's ratio	0.203	0.051
Change (%) in Poisson's ratio	19	54.3

Poisson's ratios for three different stacking sequences are calculated using experimental data collected by dual extensometers during quasi-static tensile tests and are also given in Table 1.

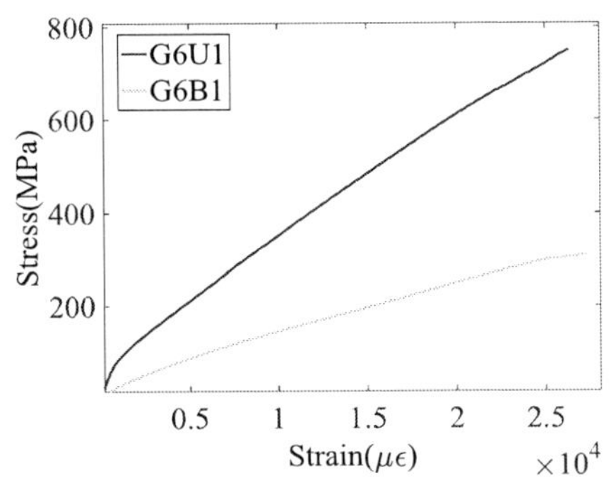

(a)

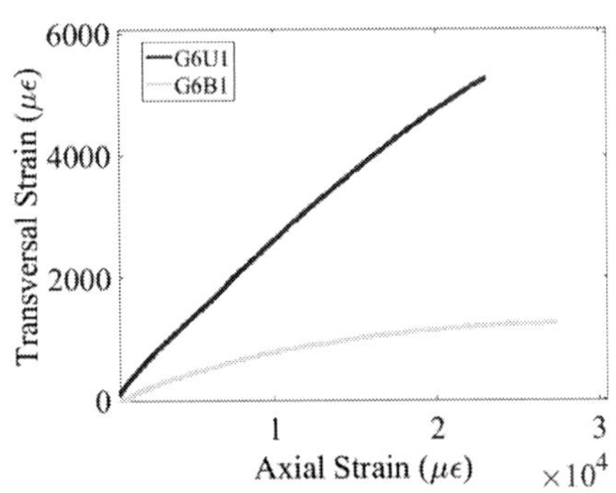

(b)

Figure 2: (a) A representative stress-strain (extensometer) curve, and (b) the variation of extensometer based transverse strain as a function of axial strain for composite with three different stacking sequences

Collected strain data were processed to be able to calculate major Poisson's ratio (υ_{xy}) and longitudinal elastic modulus (E_{xx}). Figure 3 shows the normalized values of Poisson's ratio and elastic modulus for the stacking sequence of $(0)_6$ and $[90/90/0]_s$. It is clear that change in the Poisson's ratio is much higher than that in elastic modulus with respect to applied strain for two different stacking sequences. This result also clearly indicates that stacking configuration notably affects the evolution of Poisson's ratio and the elastic modulus trend. When the number of off-axis plies increase, the level of the reduction in Poisson's ratio and elastic modulus raises.

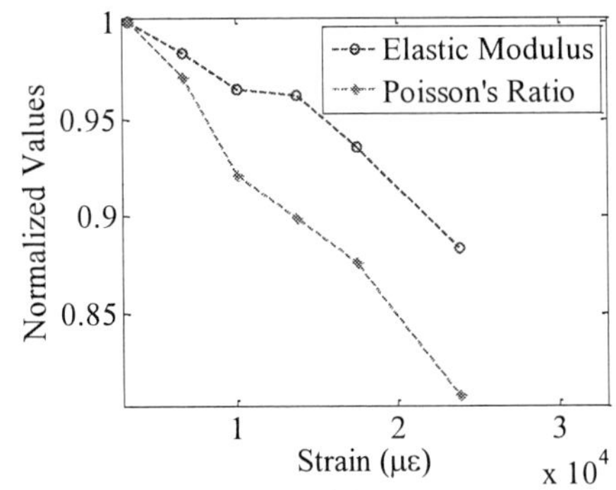

(a)

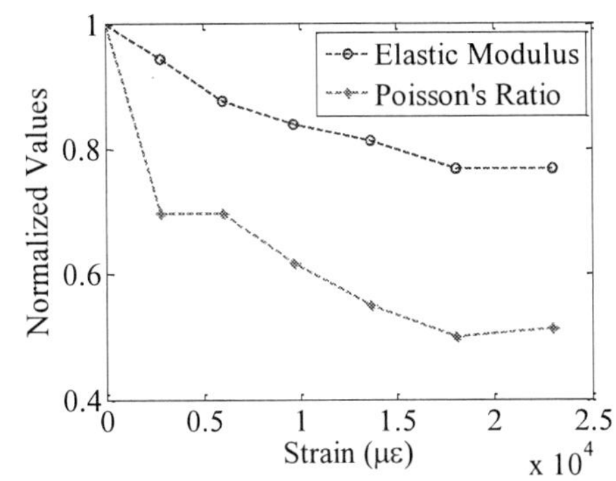

(b)

Figure 3: Normalized elastic modulus and Poisson's ratio with respect to applied strain for (a) unidirectional (G6U1), and (b) biaxial (G6B1) composites

Figure 4 shows the progressive change of Poisson's ratio for G6B1 stacking sequence with respect applied axial strain. The change in Poisson's ratio in Figure 4 is analyzed in embedded biaxial FBG system which also indicates a decreasing trend as well.

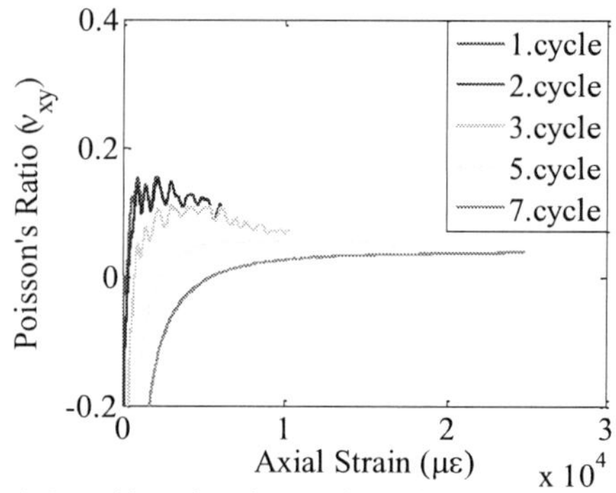

Figure 4: Tracking the change in Poisson's ratio with biaxial FBG sensor for G6B1 laminate

Figure 5 shows cumulative transverse crack AE signals which are converted by AE hardware by capturing the elastic stress waves released by transverse cracks. As seen in Figure 3, G6U1 and G6B1 plate show different evolution trend and percent change in their measured elastic properties so does their transverse crack signal. The G6B1 plate has the higher decrease in Poisson's ratio and elastic modulus when compared to G6U1 plate. The cumulative transverse crack signal of G6B1 plate's is higher than the G6U1 plate.

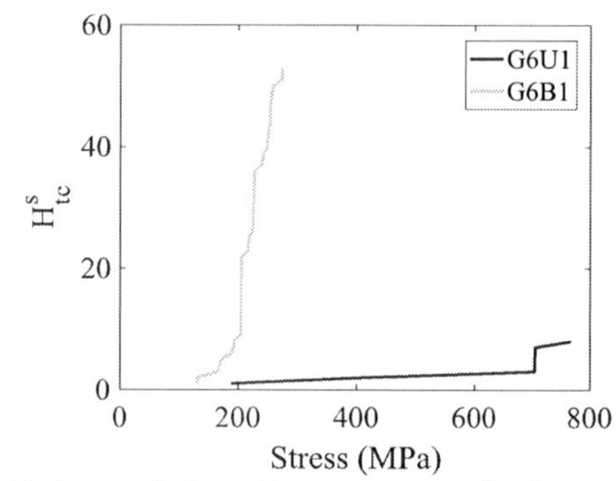

Figure 5: Accumulation of transverse cracks for two different stacking sequence with applied stress.

CONCLUSIONS

The paper presents the effect of stacking sequence on the evolution of elastic modulus and Poisson's ratio under damage. When the number of off-axis plies increase, a higher reduction in both elastic modulus and Poisson's ratio is observed. Furthermore, referring to the results of the quasi-static cyclic test on two different stacking sequences, it is noted that the reduction in Poisson's ratio is higher than that in elastic modulus~~. It is also observed that the change in the Poisson's ratio is greater than that of elastic modulus for the stacking sequence studied~~. The AE transverse crack signal also overlaps with percent change in Poisson's ratio and it indicates that as the Poisson's ratio decreases, the AE exhibit more transverse crack signals .

ACKNOWLEDGMENTS

The authors gratefully acknowledge the funding provided by The Scientific and Technological Research Council of Turkey (TUBITAK), and Ministry of Science, Industry and Technology of Turkey for the project, 112M357, and 01307.STZ.2012-1, respectively. The authors also gratefully acknowledge the funding provided to first author by the Scientific and Technological Research Council of Turkey (TUBITAK) under the 2214-A International Research Fellowship Program for PhD Students for tenure at the Center of Aerospace Research at the University of Victoria.

REFERENCES

1. A.L. Highsmith, K.L. Reifsnider, Stiffness-reduction mechanisms in composite laminates, Damage in composite materials, 775 (1982) 103-117.
2. F. Gao, L. Boniface, S. Ogin, P. Smith, R. Greaves, Damage accumulation in woven-fabric CFRP laminates under tensile loading: Part 1. Observations of damage accumulation, Composites science and technology, 59 (1999) 123-136.
3. M. Surgeon, E. Vanswijgenhoven, M. Wevers, O. Van Der Biest, Transverse cracking and Poisson's ratio reduction in cross-ply carbon fibre-reinforced polymers, Journal of Materials Science, 34 (1999) 5513-5517.
4. 4.W. Van Paepegem, I. De Baere, E. Lamkanfi, J. Degrieck, Monitoring quasi-static and cyclic fatigue damage in fibre-reinforced plastics by Poisson's ratio evolution, International Journal of Fatigue, 32 (2010) 184-196.
5. P.A. Smith, J.R. Wood, Poisson's ratio as a damage parameter in the static tensile loading of simple crossply laminates, Composites Science and Technology, 38 (1990) 85-93.
6. C.J. Keulen, E. Akay, F.F. Melemez, E.S. Kocaman, A. Deniz, C. Yilmaz, T. Boz, M. Yildiz, H.S. Turkmen, A. Suleman, Prediction of fatigue response of composite structures by monitoring the strain energy release rate with embedded fiber Bragg gratings, Journal of Intelligent Material Systems and Structures, (2014) 1045389X14560358.

EIGENFREQUENCY OPTIMIZATION OF VARIABLE STIFFNESS LAMINATED COMPOSITES

A. Sohouli
MIT-Portugal Program, IDMEC-CCTAE, Instituto Superior Técnico, University of Lisbon, Portugal,
abdolrasoul.sohouli@ist.utl.pt

M. Yildiz
Faculty of Engineering and Natural Sciences, Advanced Composites & Polymer Processing Laboratory, Sabanci University, Istanbul, Turkey,
meyildiz@sabanciuniv.edu

A. Suleman
Department of Mechanical Engineering, University of Victoria, British Colombia, Canada,
suleman@uvic.ca

ABSTRACT

This paper presents an optimization framework for the design of vibrating composite structures with material constraints subject to external excitation frequencies to avoid resonance in a given interval. A new decoupled discrete material optimization technique is employed where the design variables are the piecewise patch orientations and materials of the fiber reinforced composite. A sequential approximation optimization algorithm is used for the maximization of fundamental and higher order eigenfrequencies. To overcome the non-differentiability issues of the eigenfrequency constraints, the Kreisselmeier–Steinhauser function is used in the optimization framework. The numerical results show that the proposed framework is suitable for material and fiber orientation optimization of composite laminates structures with eigenfrequency constraints.

KEYWORDS: Decoupled discrete material optimization, Composite structures, Eigenfrequency constraints, Kreisselmeier–Steinhauser function

INTRODUCTION

Composite materials are desirable in engineering structures due to directional stiffness and high strength to weight ratio properties. The composite laminated structures may be composed of different materials such as Fiber Reinforced Polymers (GFRP/CFRP) bonded together by resin. Thus, the design of composite structures must be optimized for material and fiber orientation to achieve a desired structural performance. Different optimization methods have been developed to find the best desirable performance of composite structures in engineering applications. Ghiasi [1, 2] provided a review of the optimization of laminated composite. Discrete Material Optimization (DMO) is an efficient method to design composite structures and was introduced by Stegman and Lund [3], as a generalization of multi-phase topology optimization problems [4]. DMO distributes a set of discrete candidate materials over a structural domain by means of material interpolation techniques. The candidate materials could be fiber reinforced composite materials with different orientations or different materials in the optimization process. When the number of design variables is high, DMO is computationally expensive and may not converge to an optimal design. This problem is very critical when the design variables in the optimization problem are material, orientation and topology optimization, simultaneously. A scheme is proposed herein to reduce the number of design variables by using a multi-level or decoupled analysis of material and orientation optimization. The objective of this study is the development of an optimization framework of the composite structures for the maximization of the minimum and higher order eigenfrequencies with a multi-material and multi-fiber orientation variables. The maximization of the minimum eigenfrequency design of laminated composite plates was investigated by Lund and Stegmann [5], Pedersen [6] and Blasques [7]. Multiple eigenfrequencies further complicates the problem. Therefore, the Kreisselmeier–Steinhauser function [8] is used to overcome this problem during the optimization process. The sequential approximate optimization (SAO) is usually used in this kind of problem in order to overcome high computational cost of optimization process. In this study, a continuation exponential approximation method is used which the convexity of the approximation is increased as penalization coefficient increases. The optimization framework is developed based on a commercially available finite element software for the decoupled DMO scheme.

DECOUPLED DISCRETE MATERIAL OPTIMIZATION

Discrete Material Optimization is inspired from structural topology optimization research where instead of one material, a set of discrete candidate materials over the structural domain is optimized according to an objective function. This set of discrete materials can consist of different materials, one material at different orientations or a combination of different materials and orientations. If the set of the materials is defined per each element of the design domain, the total number of candidate materials is the number of elements multiplied by the number of candidate materials. The number of design variables is high, and this is a critical disadvantage of DMO. These design variables can be reduced by merging several design variables from elements into a single variable for a patch. This strategy is called patch design that covers several elements of the structure. In the classical DMO schemes [3], the constitutive matrix E_p for each patch is defined as:

$$E_p = E_0 + \sum_{c=1}^{nc} w_{c,p}\left(E_{c,p} - E_0\right) \qquad \forall p \qquad (1)$$

Here E_0 is a low stiffness material to ensure positive definiteness for E_p. $E_{c,p}$ is the constitutive tensor for the cth candidate material and the pth patch, nc is the number of candidate materials. $w_{c,p}$ is the weights of the cth candidate material and the pth patch. The objective of DMO technique is set to one for a single weight and to zero for all other weights. The number of design variables is an issue in Discrete Material Optimization even when employing the patch design. A two-stage method named Decoupled Discrete Material Optimization (DDMO) is herein proposed to address this drawback. The optimization step is implemented at two levels, namely the material optimization level, and the fiber orientation optimization level. The number of candidate materials nc is divided into the number of candidate fiber orientations no and the number of candidate materials nm. Two types of weight functions are defined to achieve multi-level optimization. In the material optimization step, the formulation of the constitutive matrix E_p for each patch is given as:

$$E_p = E_0 + \sum_{m=1}^{nm} \varrho_{m,p}\left(\Delta E_{m,p}\right) \qquad \forall p \qquad (2)$$

$$\Delta E_{m,p} = \begin{cases} \sum_{o=1}^{nm} \widetilde{w}_{o,p}\left(E_{m,o,p} - E_0\right) & for \quad m \in M_A \\ E_{m,p} - E_0 & Otherwi \end{cases} \qquad (3)$$

Here $\varrho_{m,p}$ is the material weight of the mth candidate material and the pth patch. $E_{m,o,p}$ is the constitutive tensor for the mth candidate material, the oth candidate fiber orientation and the pth patch. M_A is set of materials that its fiber orientation should be also optimized in the orientation optimization step, the rest of materials are in the set of M_I. The constitutive tensor in the set of M_I for the mth candidate material and the pth patch is represented by $E_{m,p}$. $\widetilde{w}_{o,p}$ is the rescaled fiber orientation weight of the oth candidate orientation and the pth patch and given by:

$$\widetilde{w}_{o,p} = \frac{w_{o,p}}{\sum_{o=1}^{no} w_{o,p}} \qquad (4)$$

Where $w_{o,p}$ is the fiber orientation weight of the oth candidate orientation and the p^{th}patch. The rescaled fiber orientation weights are considered constant in the material optimization phase, and the material optimization weights are optimized in this step. In the orientation optimization of DDMO, the constitutive matrix E_p for each patch is represented as:

$$E_p = E_0 + \zeta_p \sum_{o=1}^{no} w_{o,p}\left(E_{o,p} - E_0\right) \qquad \forall p \qquad (5)$$

$$\zeta_p = \frac{\sum_{m \epsilon M_A} \varrho_{m,p}}{\sum_{m=1}^{nm} \varrho_{m,p}} \qquad (6)$$

Here ζ_p is the rescaled material weight of the pth patch. $E_{o,p}$is the constitutive tensor of one of the candidate materials in the set of M_A, the oth candidate fiber orientation, and the pth patch. In the orientation optimization step, the fiber orientation weights are optimized by considering rescaled material weight as constant values.

INTERPOLATION FUNCTIONS

A local density filtering operator $\phi(X)$ is used to map from the design variable space to the density variable space for each and same candidate material or fiber orientation in the DDMO, similar to [5]. The filtering is defined as:

$$\hat{x}_{c,p} = \phi_{c,p}(x) = \frac{1}{\sum_{k \epsilon N_{c,p}} \delta_k}\left(\sum_{k \epsilon N_{c,p}} \delta_k x_{c,p}\right) \qquad (5)$$

Where $N_{c,p}$ is an index set of local neighborhood the pth patch and the cth candidate material or fiber orientation, given by:

$$N_{c,p} = \{k \,|\, \|X_k - X_{c,p}\| \leq R\} \qquad (6)$$

Here, X_k and $X_{c,p}$ are the positions vector of the centroid of the patches, and R is the prescribed filter radius. The filter weights, δ_k, are calculated as follows:

$$\delta_k = 1 - \frac{\|X_k - X_{c,p}\|}{R} \qquad (7)$$

A projection operator of the Heaviside approximation is applied in order to map non-discrete valued design variables into discrete values by the following equation:

$$\bar{x}_{c,p} = \frac{\tanh(\beta\eta) + \tanh(\beta(\hat{x}_{c,p} - \eta))}{\tanh(\beta\eta) + \tanh(\beta(1 - \eta))} \qquad (8)$$

Where β and η are the filter parameters to control the threshold projection limit and the steepness of the Heaviside approximation. The sum of projected $\bar{x}_{c,p}$ may not be equal to one because of the nonlinear projection operator. Hence, a rescale function is used as follows:

$$\tilde{x}_{c,p}\left(\bar{x}_{c,p}\right) = \frac{\bar{x}_{c,p}}{\sum_{c=1}^{nc} \bar{x}_{c,p}} \qquad (9)$$

Subscript c is used as the general variable that could be m as the candidate material or o as the candidate fiber orientation. In this study, the same weight function is used for the material and orientation optimization step. Thus, subscript c is changed to m and o for the material optimization and the orientation optimization step, respectively. Using the filtering scheme and projection operator as described above, the constitutive properties of a given patch as a design domain is defined using the RAMP scheme [4], and presented as follows:

$$\varrho_{m,p}(\tilde{x}_{m,p}) = \frac{\tilde{x}_{m,p}}{1+p_e(1-\tilde{x}_{m,p})} \quad (10)$$

$$w_{o,p}(\tilde{x}_{o,p}) = \frac{\tilde{x}_{o,p}}{1+p_e(1-\tilde{x}_{o,p})} \quad (11)$$

Here nc is changed to nm and no for the material and the orientation optimization. The penalization coefficient is represented by p_e to obtain a distinct result.

PROBLEM FORMULATION

The objective function consists in minimizing the compliance to solve for the optimal material, orientation and topology of composite plates. The optimization problem can be stated as:

$$\text{Objective:} \quad Max\ \underline{KS}(x)$$

$$\text{Subject to: } f_1(x) = \sum_{p=1}^{np}\sum_{c=1}^{nc}\hat{x}_{c,p} \le \overline{m}$$

$$\sum_{j=2}^{m} f_j(x) = \sum_{c=1}^{nc} x_{c,p} = 1; \qquad \forall\ (p) \quad (12)$$

$$x_{c,p} \in [0;1]; \qquad \forall\ (c,p)$$

where $f_1(X)$ and $\sum_{j=2}^{m} f_j(x)$ are the mass and equality weight constraints, respectively. For simplicity, the weights are defined in $w_{c,p}(x_{c,p})$ form where c and p are the candidate material and patch number, respectively. $\underline{KS}(x)$ is the Kreisselmeier–Steinhauser (KS) function to optimize the eigenfrequencies of the composite structures. The KS function is used as a differentiable function. The set of the eigenfrequencies may be expressed by $\omega = \{\omega_1, \dots, \omega_n\}$, and $\underline{\omega}$ is a subset of ω as $\underline{\omega} = \{\omega_1, \dots, \omega_{ng}\}$. $\underline{\omega}$ is ordered such that $\omega_1 \le \cdots \le \omega_{ng}$. The subset of $\underline{\omega}$ is a group of eigenfrequencies to maximize. Thus, the KS function may be defined by:

$$\underline{KS}(x) = \omega_{ng}(x) - \frac{1}{\beta_s} ln\left(\sum_{g=1}^{ng} e^{-\beta_s(\omega_g(x)-\omega_{ng}(x))}\right) \quad (13)$$

where the parameter $\beta_s \ge 1$ is such that the function $\underline{KS}(x)$ will tend to the minimum of $\underline{\omega}$ as β_s increases. In this study, the Sequential Approximate Optimization (SAO) techniques are used to assure a reasonably fast optimization. The basic concept in SAO techniques is to construct approximate analytical sub-problems successively at the design point for the objective function and constraints. The primal approximate sub-problem at X^k is expressed as:

$$min -\hat{f}_0^k(X)$$

$$subject\ to\ \hat{f}_j^k(X) \le 0, \quad j = 1,2,\cdots,m. \quad (14)$$

$$\alpha_i^k \le x_i \le \beta_i^k, \qquad i = 1,2,\cdots,n$$

where $\hat{f}_j^k(X), j = 0,1,\cdots,m$ are the approximated functions of the original objective and constraint functions, respectively. n is the number of design variables, X is the vector of the design variables and α_i^k and β_i^k are the move limits. In this study, a convex exponential approximation is used to approximate the primal sub-problem [9]. The algorithm searches for the new feasible designs vector X^{k+1} such that the new objective function follows $-\hat{f}_0^{k+1} < -\hat{f}_0^k$ while satisfying the constraint function. In the exponential approximation method, the objective and constraint functions in the form of primal sub-problem are given by [9]:

$$\hat{f}_j^k(X) = f_j(X^k) + \sum_{i=1}^{n}\left(\left[\left(\frac{x_i}{x_i^k}\right)^{a_p} - 1\right]\left(\frac{x_i^k}{a_p}\right)\left(\frac{\partial f_j}{\partial x_i}\right)_{x_i=x_i^k}\right) \quad (15)$$

$$\alpha_i^k \le x_i \le \beta_i^k$$

where j is used for the sake of notational brevity; which $j = 0$ is the approximated objective function and $1 \le j \le m$ is for the approximated function, and k in the number of iteration. The parameter a_p is an exponential parameter to ensure that the approximation is convex and it is linked to the penalty parameter with $a_p = -2p$, and it is updated by increasing the penalization coefficient. Thus, the constrained optimization problem Eq. (14) is solved using the Augmented Lagrangian Method (ALM) [10]. The optimization solver is L-BFGS-B [11], which is a limited-memory quasi-Newton code for bound-constrained optimization. The computational framework for DDMO is established by programing an interface between a commercial finite element code and the MATLAB. In this study, all the cases are analyzed using ABAQUS with S4 shell elements.

NUMERICAL EXAMPLE

A fully clamped square plate is selected to demonstrate the application of the framework to optimize the lowest eigenfrequency of the group of eigenfrequencies $\{\omega_1, \dots, \omega_5\}$ shown in Figure 1. β_s is chosen to be one in order to optimize the lowest eigenfrequencies. The one-layer plate is divided into 400 elements and the candidate materials are the unidirectional carbon fiber reinforced polymer (CFRP) and the unidirectional glass fiber reinforced polymer (GFRP). The material properties of the material candidates used in this study are given in Table 1. The candidate orientations are four fiber orientations, $\theta = \{-45^\circ, 0^\circ, 45^\circ, 90^\circ\}$, where the fiber orientations are shown in Figure 1.

Table 1: Material Properties of the candidates.

Property	Unit	CFRP	GFRP
Longitudinal modulus E_1	GPa	112.30	38.55
Transverse modulus E_2	GPa	8.03	12.06
In-plane shear modulus G_{12}	GPa	3.62	4.52
Transverse shear modulus G_{23}	GPa	2.90	4.52

Major Poisson's ratio υ_{12}	-	0.27	0.28
Minor Poisson's ratio υ_{23}	-	0.39	0.33
Mass density ρ	$^{Kg}/_{m^3}$	1455	1815

The filter radius is 1.5 patches in the filtering scheme, and the allowable CFRP material is 50 % of the beam material in the material constraint.

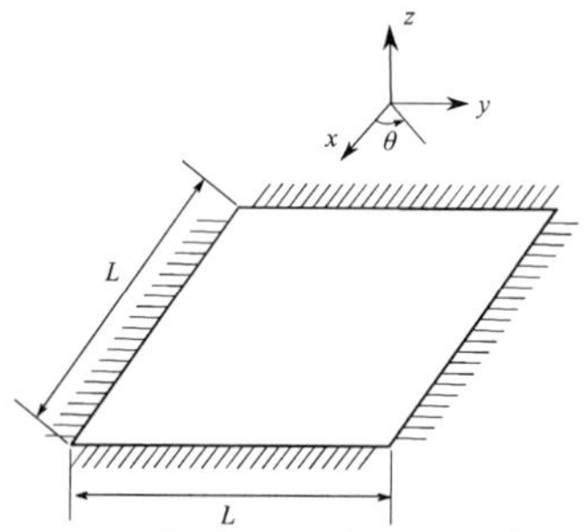

Figure 1: Fully clamped plate under central concentrated load

The optimal materials and fiber orientations of the plate is shown in Figure 2. In the figure illustrated, blue and black colors represent the GFRP and CFRP, respectively.

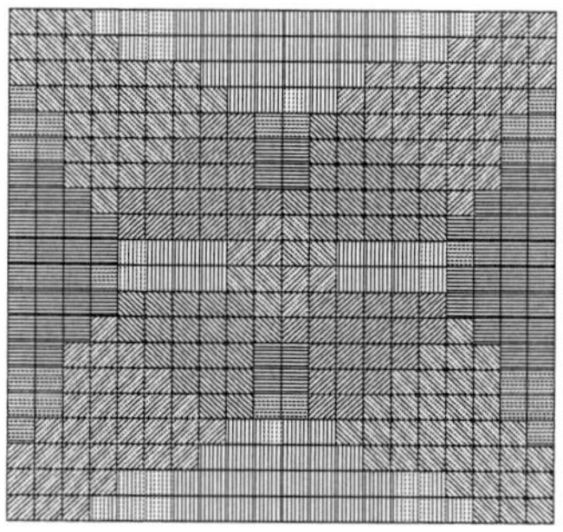

Figure 2: Optimal material and fiber orientation of the plate.

CONCLUSIONS

In this paper, a novel computational framework employing a decoupled Discrete Material Optimization approach has been applied to maximize the fundamental and higher order eigenfrequencies of laminated hybrid composite structures. To overcome multiple eigenfrequencies, the Kreisselmeier–Steinhauser function was used. It has been shown that the optimization framework is an efficient design tool in the design of laminated composites.. The framework can be also used to maximize the gap between consecutive eigenfrequencies.

ACKNOWLEDGMENTS

The authors would like to acknowledge the Graduate Fellowship from the Fundação para a Ciencia e Tecnologia through the MIT-Portugal program and the NSERC Canada Research Chair and Discovery Grant Programs.

REFERENCES

1. Ghiasi H., Pasini D., and Lessard L., 2009, "Optimum stacking sequence design of composite materials Part I: Constant stiffness design", Composite Structures, Review vol. 90, no. 1, pp. 1-11.
2. Ghiasi H., Fayazbakhsh K., Pasini D., and Lessard L., 2010, "Optimum stacking sequence design of composite materials Part II: Variable stiffness design", Composite Structures, Review vol. 93, no. 1, pp. 1-13.
3. J. Stegmann, E. Lund, 2005 , "Discrete material optimization of general composite shell structures", International Journal for Numerical Methods in Engineering, Article vol. 62, no. 14, pp. 2009-2027.
4. S. N. Sorensen, E. Lund, 2013 , "Topology and thickness optimization of laminated composites including manufacturing constraints", Structural and Multidisciplinary Optimization, Article vol. 48, no. 2, pp. 249-265.
5. Lund E. and Stegmann, J., 2006, "Eigenfrequency and buckling optimization of laminated composite shell structures using discrete material optimization", IUTAM Symposium on Topological Design Optimization of Structures, Machines and Materials, pp. 147-156.
6. José Pedro Blasques, "Multi-material topology optimization of laminated composite beams with eigenfrequency constraints"
7. Blasques, J.P., 2014. "Multi-material topology optimization of laminated composite beams with eigenfrequency constraints", Composite Structures, 111, pp.45-55.
8. Kreisselmeier G, Steinhauser R., 1979, "Systematic control design by optimizing a vector performance index", International federation of active controls symposium on computer-aided design of control systems, Zurich, Switzerland.
9. Li L. and Khandelwal K., 2015, "Volume preserving projection filters and continuation methods in topology optimization", Engineering Structures, Article vol. 85, pp. 144-161.
10. Vanderplaats G. N. A, 1984, Numerical optimization techniques for engineering design: with applications, McGraw-Hill, New York.
11. Byrd R. H., Lu P. H., Nocedal J., and Zhu C. Y., 1995, "A Limited Memory Algorithm for Bound Constrained Optimization", Siam Journal on Scientific Computing, Article vol. 16, no. 5, pp. 1190-1208.

THE EFFECT OF ALKALIZATION ON PROPERTIES OF RAFFIA PALM FIBER

Opeoluwa Fadele
Mechanical Engineering
University of Saskatchewan
Saskatoon, Saskatchewan, Canada
oef265@mail.usask.ca

Ikechukwuka N. A. Oguocha
Mechanical Engineering
University of Saskatchewan
Saskatoon, Saskatchewan, Canada
iko340@mail.usask.ca

Akindele Odeshi
Mechanical Engineering
University of Saskatchewan
Saskatoon, Saskatchewan, Canada
ago145@mail.usask.ca

Majid Soleimani
Biological Engineering
University of Saskatchewan
Saskatoon, Saskatchewan, Canada
mas233@mail.usask.ca

ABSTRACT

Natural fibers used in fabricating polymer matrix composites are often given some form of chemical treatment to improve interfacial bonding to the matrix. This paper presents the results of a study carried out to determine the effect of alkali (NaOH) treatment on properties of natural fibers derived from the raffia palm tree (*Raphia farinifera*). Morphological, thermal and chemical structural changes in the fibers were investigated using scanning electron microscopy (SEM), Fourier transform infrared spectroscopy (FTIRS) and differential scanning calorimetry (DSC). It was found that treating the fibers with 10wt.% NaOH for 5 h at 60°C increased the degradation temperature of both hemicellulose and cellulose by 7% and 3%, respectively, and the cellulose content by 22%. It was also found that although hemicellulose and lignin were removed after alkali treatment, the chemical functionalities of cellulose remained unaffected.

KEYWORDS: Raffia palm fibers, scanning electron microscopy, infrared spectroscopy, alkali treatment, thermal analysis.

INTRODUCTION

Increasing environmental awareness throughout the world has greatly impacted materials engineering and design. Renewed interest in the utilization of natural materials addresses ecological issues such as recyclability and environmental safety. Currently, synthetic fibers such as glass, carbon and aramid are used widely in polymer matrix composites (PMCs) because of their high specific stiffness and specific strength [1]. Although the synthetic fiber reinforced polymer composites (FRPCs) possess superior mechanical strength properties, they have got some drawbacks such as high cost, poor recyclability and non-biodegradability. Consequently, research efforts over the last few years have focused on development of viable and cost-effective natural fibers (NFs) reinforced polymer composites (NFRPCs) [2]. The use of natural fibers in polymer composites is beneficial because they are cheap, sustainable, biodegradable, and non-toxic compared to the synthetic fibers. However, NFs are hydrophilic in nature, which reduces their compatibility with hydrophobic polymer matrices. The presence of a natural waxy substance on the fiber surface and hydroxyl groups lead to ineffective fiber-polymer matrix bonding. High moisture absorption of the fiber causes swelling within the polymer matrix which leads to dimensional instability [3]. Thus, it is very challenging to obtain a strong bond between NFs and polymer matrices due to the difference in chemical structure between the two constituents. The problem, in most cases, is ineffective stress transfer through the fiber-matrix interface [1]. To enhance the fiber-matrix adhesion and, consequently, the properties of NFs, various physical and chemical treatments are applied. Among them is the treatment with sodium hydroxide (NaOH). Previous studies of alkali treatment have revealed that its effectiveness depends mainly on the fiber type, concentration of the alkali used, the soaking temperature and duration [4]. The present study focuses on optimizing the surface chemistry of raffia palm fibers for enhanced interfacial adhesion with polymers. The initial results obtained are presented in this report.

MATERIALS AND METHODS

The raw raffia palm fibers (RPFs) used in this work were obtained from southern Nigeria. For ease of treatment, processing and characterization, the remnant binders were manually removed and the resulting raw fibers were subsequently cleaned in 2% detergent solution to remove oily substances and other impurities on the fiber surface. After cleaning, the fibers were dried in an oven maintained at 70°C for 24 h and cooled to room temperature before being ground and sieved using a 1-mm screen. Fig. 1 shows photographs of the as-cleaned and ground fibers. The diameter of the cleaned but unground fibers was determined using a micrometer screw gauge. Twenty-eight fibers were chosen at random and diameter measurements were taken at ten locations along each fiber due to non-uniformity in diameter along the fiber length. The average fiber diameter obtained is 1.53 ± 0.29 mm. The density of the raffia palm fibers was measured using a gas (nitrogen) pycnometer. The average density obtained is 1.50 ± 0.01 g/cm^3.

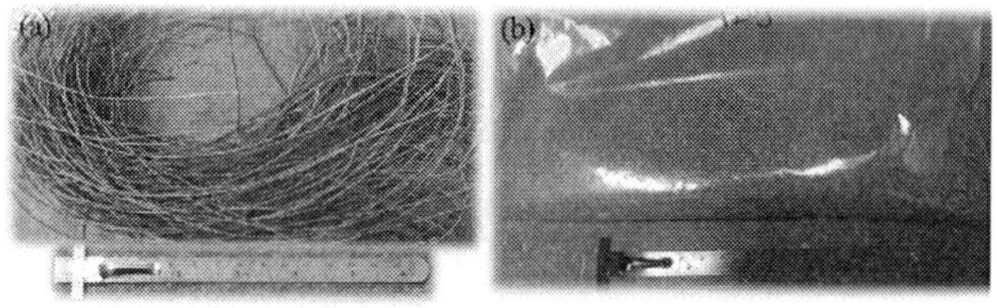

Figure 1: Images of raffia palm fiber (a) cleaned and washed with 2% detergent (b) ground fiber.

For chemical treatment, ground RPFs were immersed in 10% w/v aqueous NaOH solution at 60°C for 5 h, with a liquid-to-material ratio of 10 to 1. After the immersion, the fibers were initially washed in laboratory water and finally in distilled water to ensure that no NaOH was left. Subsequently, the fibers were dried at 60°C for 24 h. The chemical composition (cellulose, hemicellulose and lignin content) of RPFs before and after alkali treatment was determined by acid digestion using an Ankom 200 Fiber Analyzer™. The standard acid detergent fiber (ADF), neutral detergent fiber (NDF) and acid detergent lignin (ADL) procedures were followed in this analysis. 0.5 g of ground fibers was used and each measurement was run in triplicates. The cellulose content was calculated from the difference between the ADF and ADL readings; the hemicellulose content was based on the difference between the NDF and ADF reading and the lignin content was the ADL reading. The results shown in Table 1 for each constituent represent the average (dry matter basis) of the three runs.

Table 1. Chemical composition of raffia fiber.

Fiber	Cellulose (wt.%)	Hemicellulose (wt.%)	Lignin (wt.%)
Untreated fiber	53.36	12.32	23.78
Alkaline treated fiber	65.29	12.66	21.89

Microstructural examination of untreated and alkaline treated RPFs was conducted using a JEOL JSM-6010LV scanning electron microscope at an accelerating voltage of 5 kV. The micrograph of the untreated RPF in Fig. 2(a) shows the presence of wax, oil and surface impurities. Although waxes and oils provide a protective layer to the surface of the fibers, they lead to weak bonding with polymer if not removed [5]. On the other hand, the micrograph of a typical alkali treated fiber shown in Fig. 2(b) shows a much cleaner surface although it felt rougher when touched.

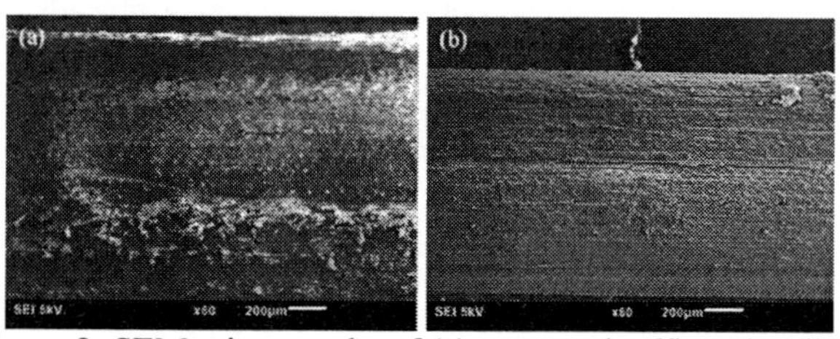

Figure 2: SEM micrographs of (a) untreated raffia palm fiber, and (b) raffia palm fiber treated with hot 10wt% NaOH.

Fourier transform-infrared spectrometry (FTIRS) was carried out in an inVia Raman microscope equipped with a Smiths IlluminantIR™ accessory that allows Fourier transform infrared (FTIR) spectra to be acquired using an all reflective objective (ARO) and a diamond attenuated total reflection (ATR) objective. The spectra were obtained with an accumulation of 512 scans and a resolution of 4 cm^{-1}.

Thermal analysis of the fibers was carried using a model 2910 V4.4E (TA Instruments, USA) modulated differential scanning calorimeter (MDSC). Each scan was performed in an open aluminum pan under argon gas from room temperature to 390°C at a heating rate of 5°C/min. Approximately 10 mg of ground RPFs was used and the measurement was repeated twice to ensure reproducibility.

RESULTS AND DISCUSSION

The extent of chemical changes associated with alkali treatment of RPFs was analyzed using Fourier transform-infrared spectroscopy (FTIRS). Typical spectra obtained from this test are shown in Fig.3 for untreated and treated fibers, with the absorbance peaks of interest clearly identified.

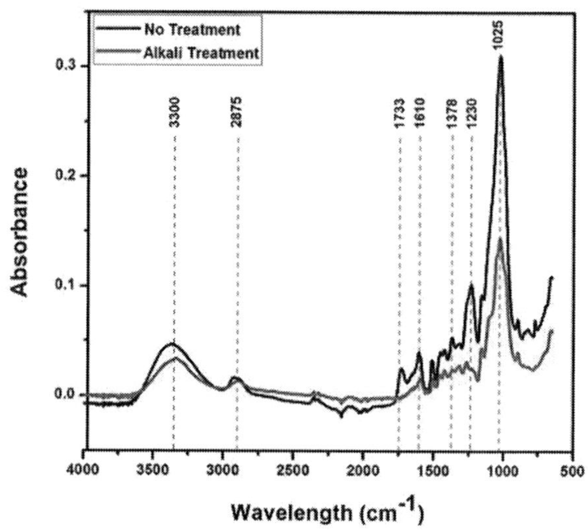

Figure 3: FTIRS spectra of untreated and alkali treated RPFs.

The spectrum of alkaline treated RPFs is similar to that of the untreated fiber. However, after NaOH treatment, there were reductions in certain bands in comparison with the untreated fiber. The large absorption band in the region of 3650–3000 cm^{-1} refers to ν–OH groups and represents the hydrophilicity of the NFs [6]. The peak at 3300 cm^{-1} corresponds to ν–OH stretching, which was found to decrease after alkali treatment. The absorption peak at 1733 cm^{-1} is attributed to the ν(C=O) stretching of methyl ester and carboxylic acid in pectin [7]. The presence of this band at 1733 cm^{-1} before NaOH treatment indicates the presence of pectin which disappears after alkali treatment. The absorbance peak of NFs in the region of 1610 cm^{-1} attributed to ν(C=C) stretching shows the presence of fatty acids [8,9]. This peak disappears after alkali treatment, thus confirming the removal of any traces of oils. The spectra also exhibited weak absorption peak at 1378 cm^{-1} attributed to ν(C=C) stretching indicates the presence of lignin [8]. Nevertheless, Tibolla et al. [6] reported that vibrations around 2875 cm^{-1} attributed to C−H vibration of lignin did remain but reduced, as seen in the spectra of alkali treatment of RPFs. The disappearance of the peak around 1230 cm^{-1} is attributed to the ν(C=O) stretching after alkalization and indicates a significant reduction of hemicellulose materials [8]. A significant reduction in absorption peak at 1025 cm^{-1} is attributed to ν(C−O−C) stretching band, and it is believed to be caused by the degradation of hemicellulose in alkali treated RPFs [10].

RPFs exhibit a thermal behaviour comparable to other natural fibers commonly used in polymeric composites as shown in Fig. 4. The DSC curve for RPFs shows an endothermic peak at around 125°C−140°C, which is due to evaporation of water, and exothermic peaks at higher temperatures, which can be attributed to decomposition hemicellulose, cellulose and lignin [11].

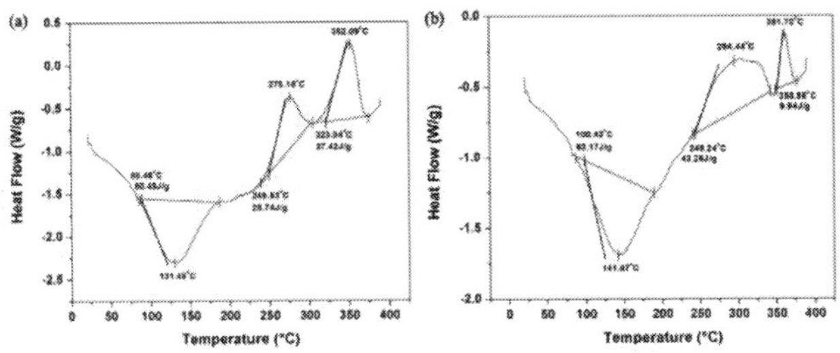

Figure 4: DSC thermograms obtained for (a) untreated and dried raffia fiber (b) alkaline treated and dried raffia fiber.

The structural changes in the fiber due to the alkaline treatment, as revealed by FTIR spectroscopy are likely to have direct effect on the thermal degradation characteristics of NFs [7]. The DSC curves as shown in Fig. 4(a-b) reveals that RPFs became less hydrophilic after alkali treatments, as there was an overall increase in the amount of heat generated, T_g values, degradation temperature of hemicellulose, cellulose and lignin. Similar results of an increase in the decomposition temperature of cellulose from 357°C−367°C for hemp fiber was reported by Hao [12]. The increased cellulose content after alkali treatment led to increase in the endothermic peak, enthalpy values and thermal properties of raffia fibers as presented in Table 2. The increase in thermal properties of alkaline treated RPFs corresponds with the disappearance of the C=O band in the FTIR spectra of alkali treated fibers. From FTIR results, it was apparent that this band disappearance, significantly affected the thermal degradation of RPFs. This was probably because more non-cellulosic material was removed and high degree of crystalline structural order was obtained, which required a higher degradation temperature [7]. Initial degradation of RPFs started at 130°C with a broad endotherm, which indicates the vaporization of water molecules in the fiber [13], and suggest that the fibers experienced glass transition stage at this peak temperatures [12]. The second decomposition stage (i.e. the two exothermic peaks) is due to the degradation of cellulosic substances such as hemi-cellulose and cellulose [14]. The degradation process involves two distinct stages. The low temperature decomposition (first exothermic peak) observed in the temperature range of 250°C−300°C, can be attributed to the thermal degradation of hemicelluloses - the least thermally stable lignocellulosic component.

Table 2. Reaction enthalpies obtained for untreated and treated raffia fibers obtained from DSC analysis.

Sample	Enthalpy (J/g)	Endothermic (°C)	Hemi-cellulose DT (°C)	Cellulose DT (°C)
Untreated fiber	60.45	131.48	275.18	352.09
Alkaline treated fiber	63.17	141.87	294.44	361.70

The second decomposition process (second exothermic peak) is observed in the temperature range of 300°C−400°C and can be attributed to the decomposition of cellulose [15]. Deepa et al. [16] obtained similar results for the thermal properties of banana fiber with the degradation of the hemi-cellulose around 280°C−290°C followed by decomposition of cellulose around 340°C−360°C. Structurally lignin has a complex composition of aromatic rings with several branches and hence its decomposition generally occurs at an high rate within the whole temperature range from ambient to temperatures higher than 700°C [13]. No exothermic or endothermic reactions are observed between 180 and 250°C, suggesting that RPFs are stable within these temperatures. Aziz and Ansell [5] reported that chemical treatment increases the thermal stability of both kenaf and hemp fibers. Thus, it can then be concluded from these results that the thermal stability of RPFs increases after chemical treatment. The DSC results proved that RPFs have enhanced thermal properties, making them less prone to degradation and therefore can be processed at higher temperatures when compared with their non-treated counterparts.

CONCLUSIONS

1. Alkali treatment changed both the morphological and chemical properties of raffia palm fiber. It removed surface impurities on the fiber, which is expected to provide good interfacial adhesion with polymer.
2. There were observable structural changes in non-cellulosic components of RPFs after alkali treatment as revealed by FTIRS. Certain chemical groups were removed upon alkalization, thus rendering fewer hydroxyl groups in the fiber available for reactions.
3. DSC results showed that thermal degradation of RPFs depended mainly on the cellulose and non-cellulosic components present of the fibers.

REFERENCES

[1] M. M. Kabir, H. Wang, K. T. Lau, and F. Cardona, "Chemical treatments on plant-based natural fibre reinforced polymer composites: An overview," *Compos. Part B Eng.*, vol. 43, no. 7, pp. 2883–2892, 2012.

[2] G. R. Arpitha, M. R. Sanjay, and B. Yogesha, "Review on Comparative Evaluation of Fiber Reinforced Polymer Matrix Composites," *Adv. Eng. Appl. Sci. An Int. J.*, vol. 4, no. 4, pp. 44–47, 2014.

[3] 2014 Jayamani et al., "The Effect of Natural Fibres Mercerization on Natural Fibres / Polypropylene Composites : A Study of Thermal Stability , Morphology and Infrared Spectrum," *Aust. J. Basic Appl. Sci.*, vol. 8, no. 15, pp. 332–340, 2014.

[4] R. G. Elenga, P. Djemia, D. Tingaud, T. Chauveau, J. G. Maniongui, and G. Dirras, "Effects of Alkali Treatment on the Microstructure, Composition, and Properties of the Raffia textilis Fiber," *Bioresour. Peer-reviewed online J.*, vol. 8, no. 2, pp. 2934–2949, 2013.

[5] S. H. Aziz and M. P. Ansell, "The effect of alkalization and fibre alignment on the mechanical and thermal properties of kenaf and hemp bast fibre composites: Part 1 - polyester resin matrix," *Compos. Sci. Technol.*, vol. 64, no. 9, pp. 1219–1230, 2004.

[6] H. Tibolla, F. M. Pelissari, and F. C. Menegalli, "Cellulose nanofibers produced from banana peel by chemical and enzymatic treatment," *LWT - Food Sci. Technol.*, vol. 59, no. 2P2, pp. 1311–1318, 2014.

[7] S. Ouajai and R. A. Shanks, "Composition, structure and thermal degradation of hemp cellulose after chemical treatments," *Polym. Degrad. Stab.*, vol. 89, no. 2, pp. 327–335, 2005.

[8] L. Y. Mwaikambo and M. P. Ansell, "Chemical modification of hemp, sisal, jute, and kapok fibers by alkalization," *J. Appl. Polym. Sci.*, vol. 84, no. 12, pp. 2222–2234, 2002.

[9] C. H. Chan, C. H. Chia, S. Zakaria, I. Ahmad, and A. Dufresne, "Production and characterisation of cellulose and nano- crystalline cellulose from kenaf core wood," *BioResources*, vol. 8, no. 1, pp. 785–794, 2013.

[10] I. Udoetok, L. Wilson, and J. Headley, "Quaternized Cellulose Hydrogels as Sorbent Materials and Pickering Emulsion Stabilizing Agents," *Materials (Basel).*, vol. 9, no. 8, p. 645, 2016.

[11] M. A. Martins, J. D. C. Pessoa, P. S. Gonçalves, F. I. Souza, and L. H. C. Mattoso, "Thermal and mechanical properties of the acai fiber/natural rubber composites," *J. Mater. Sci.*, vol. 43, no. 19, pp. 6531–6538, 2008.

[12] A. Hao, "Mechanical and Thermal Properties of Kenaf Polypropylene Nonwoven Composites," *Phd Thesis; Univ. Texas*, p. 177, 2013.

[13] S. S. Saravanakumar, A. Kumaravel, T. Nagarajan, P. Sudhakar, and R. Baskaran, "Characterization of a novel natural cellulosic fiber from Prosopis juliflora bark," *Carbohydr. Polym.*, vol. 92, no. 2, pp. 1928–1933, 2013.

[14] S. Wong, R. Shanks, and A. Hodzic, "Interfacial improvements in poly (3-hydroxybutyrate)-flax fibre composites with hydrogen bonding additives," *Compos. Sci. Technol.*, vol. 64, no. 9, pp. 1321–1330, 2004.

[15] M. Tajvidi and a. Takemura, "Thermal Degradation of Natural Fiber-reinforced Polypropylene Composites," *J. Thermoplast. Compos. Mater.*, vol. 23, no. 3, pp. 281–298, 2010.

[16] B. Deepa *et al.*, "Structure, morphology and thermal characteristics of banana nano fibers obtained by steam explosion," *Bioresour. Technol.*, vol. 102, no. 2, pp. 1988–1997, 2011.

3D PRINTED SELF-FILLING MICROWELL ARRAYS FOR CELL SPHEROID FORMATION

Amir Seyfoori
Breast cancer research center, ACECR
Tehran, Iran, Department of Mechanical Engineering, Laboratory for Innovation in MicroEngineering (LiME), University of Victoria, 3800 Finnerty Rd., Victoria, BC, Canada, V8P 2C5
Klm.1985@yahoo.com

Mahdi Rezayati Charan
Breast cancer research center, ACECR,
Tehran, Iran
m.rezayati.ch@gmail.com

Neda Jalili
Breast cancer research center, ACECR,
Tehran, Iran
nedajalili67@yahoo.com

Mahdi Rahmanian
Breast cancer research center, ACECR,
Tehran, Iran
mrahmanian1@gmail.com

Leyla Farahmand
Breast cancer research center, ACECR,
Tehran, Iran,
laylafarahmand@gmail.com

Mohsen Akbari*
Department of Mechanical Engineering, Laboratory for Innovation in MicroEngineering (LiME), University of Victoria, 3800 Finnerty Rd., Victoria, BC, Canada, V8P 2C5
makbari@uvic.ca

ABSTRACT

Formation of cell spheroids in microwell arrays is a promising approach for high-throughput screening of chemotherapeutic agents. This method offers the advantage of better mimicking the complexities of tumors as compared to conventional two-dimensional (2D) monolayers. Integrating microwell arrays within microfluidic systems enables performing multiple assays in short periods with small cell and reagent consumption. However, using these technologies to their full potential is hindered by the inability to seed the cells within the wells with high yield and reproducibility. This issue is mainly caused by the current limitations in the microfabrication strategies that are restricted to planar designs; an inherent limitation of current soft lithography techniques. Herein, we use three-dimensional printing to overcome this challenge and develop microwell arrays with non-planar channels for guiding the cells to microwells. We characterized the fabrication process and demonstrated the ability to culture liver carcinoma cells (HepG2) in microwells and form cell spheroids. We envision that the proposed innovative approach opens avenues of opportunities for high-throughput three-dimensional cell culture for drug screening.

KEYWORDS: 3D printing, Microwell array, Microfluidic, 3D culturing, Microfabrication

INTRODUCTION

Traditional 2D cell culture methods show different kinds of limitation in providing a proper microenvironment to simulate the cell function and signaling pathways as it occurs in a specific tissue. To solve these problems, applying 3D culturing seem to be a good alternative for common 2D cell culture methods [1]. Among the methods of 3D cell culturing systems like natural extracellular matrix (ECM) matrigel or scaffold based synthetic gels, microwell have some advantages such as potential of scaling up in high-throughput cancer studies. Another key point of using microwell systems instead of 2D culture plates is the possibility of forming cell spheroids with specific size and shape [2]. Recently, variety of methods based on soft lithography have been developed to construct mirowells for 3D cell culturing with diverse shape and sizes. Polydimethylsiloxane (PDMS), poly (methyl methacrylate) (PMMA) and polyethylene glycol (PEG) are the most well-known materials which are used to fabricate mold based microwells with controlled geometry [3]. Since mold fabricating by photolithographic methods with defined geometries like round and conical bottom for suitable cell spheroid formations in microwells is so complicated, more feasible methods are also used to form microwells made out of polycarbonate including laser sintering and laser micro-ablation [3]. Another feasible method which has been recently used to fabricate microdevices is additive manufacturing using 3D printing. Steriolithography-based 3D printing is a powerful technique through which mold-less or mold-based microdevices with complicated geometries can be generated. Accordingly, by Dynamic Light Projection (DLP) 3D printer and with the use of photo-crosslinkable materials complex mold designs for the replication process is achievable. In the present study, for the first time DLP based 3D printing mold containing an array of self-filling microwells connecting to cell carrying channels are replicated by agarose to generate the hepatocellular carcinoma (HepG2) cell spheroids with different sizes.

2. EXPERIMENTAL

2.1. Mold 3D printing and microwell fabrication

In this work, new design of microwell arrays was fabricated using home-made DLP 3D printing set-up with the X-Y resolution of 30 µm. As illustrated in Fig. 1, the mold was designed by Solid Work software and 3D printed using 3DM castable high resolution resin. Followed by 3D printing, replica molding was conducted using 2% agarose in PBS at high temperature to create the versatile microwell geometries of spherical and rectangular shape. Bovine serum albumin (BSA) was added and solved to agarose solution when its temperature was decreased to 40°C to alter the miceowell surface non-adherent and ready for cell spheroid formation. After physical cross-linking of the agarose hydrogel by the time, mold template was removed and microwell array was developed. In this study, for apprising the effect of microwell dimension on the final size of formed spheroids, 3 different sizes of 250, 500 and 750 µm were fabricated.

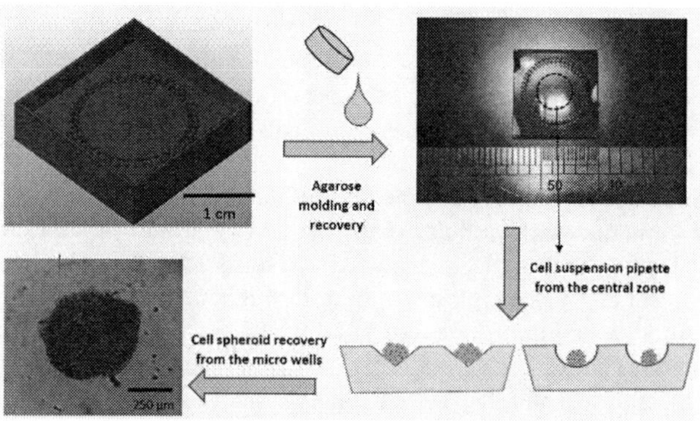

Figure1: Schematic view of the total process from mold printing to cell spheroid recovery.

2.2. Scanning electron microscopy (SEM)

Imaging of the 3D printed molds was conducted by FE-SEM (Hitachi S-4800) to measure the final sizes of the fabricated molds and their difference with the designed one. Moreover, fabricated microwells were imaged by optical microscope (Leica) and the total difference between primary design and final device was extracted.

2.3. HepG2 cell culture and spheroid formation

Human hepatoblastoma cells (HepG2 cells) were cultured at 37 °C in a 5% CO_2 humidified incubator in 89% DMEM culture medium supplemented with 10% FBS, and 1% penicillin–streptomycin. All microwell devices were placed in 6 well culture plates after fabrication, washed with ethanol for 2 hours, and kept under UV light to get sterilized. HepG2 cell suspension with the concentration of 10^6 cells /ml were pipetted at the central inlet of the microdevice and then they were allowed to fill the microwells through the fabricated channels with the aid of their gravity force as well as primary velocity. Microwells were kept in incubator for 10 min to allow for complete well filing without any cell residue in the channel and then they were kept in the fresh culture medium for 7 days to get the spheroids with different sizes. Spheroid sizes was analyzed by optical microscopy and subsequently ImageJ software. Moreover, different channel slopes effect on the self-filling ability of the microwell design was investigated by cell track measurement in different time intervals.

3. Results and discussion

DLP 3D printer was used to fabricate the modified mold design containing microchannels to have a conductive role on the cells to access the microwells. As shown in Fig. 2, channels with the width of 300µm connected the central inlet of cell pipetting zone to the different size and shapes of microwells. In this way, cells will reach the microwells with same amount of the primary concentration at the entrance zone. Both of microwell shapes with spherical and rectangular cross-section were filled with HepG2 cell lines completely. Tapered wells with round bottom shape have been reported as a suitable design for providing the proper shear stress and flow velocity applied on cells in comparison to flat cylindrical one. In this way cell seeding will be conducted uniformly and the viability of the cells will be improved [4]. These designs are not so feasible to be fabricated by the conventional lithographic methods while using a 3D printer, fabrication of such a complicated design can be achieved. Due to natural shrinkage of the resin followed by curing under UV exposure, there is a definite size difference between the designed and printed specific parameters. As shown in Fig. 2B and 2C, the size reduction due to resin curing is approximately circumvented by size increment which is occurred by agarose molding. This increment after molding may be attributed to the hydrogel physical characters of the agarose material like low viscosity and the tapered design of the well which can pull the hydrogel to the side and make the final feature larger.

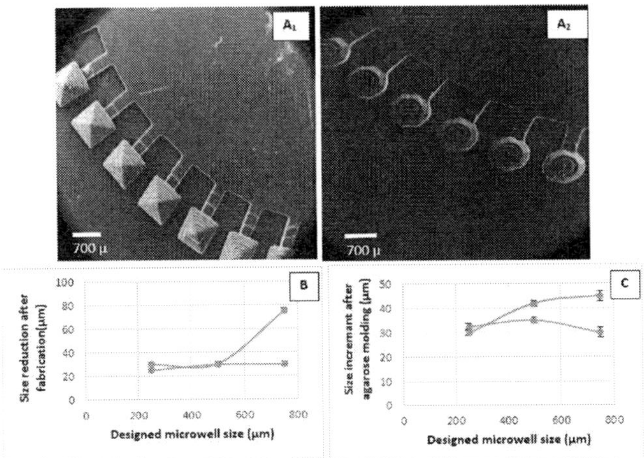

Figure 2: SEM images of 3D printed molds A_1) rectangular and A_2) spherical. B) Size reduction due to resin shrinkage after printing. C) Size increment after agarose molding.

Three-dimensional cell culture is a promising approach that enables mimicking the complexities of tumor microenvironment by recapitulating the cell-cell and cell-drug interactions as well as nutrient and drug mass transport *in vitro*. Moreover, 3D cultured cells tune the cell-cell interactions so that biological events such as cell differentiation, growth, viability and protein expression is altered in a way which is so different with 2D culture condition. Cell spheroids are amongst the most promising 3D culturing methods through which drug screening analysis are performed [5]. In our study, due to the specific shape and geometry of microwells, trapped cells in each of them formed spheroids after about 7 days. Regarding to Fig. 3, by apprising the cell spheroid sizes during their growth, we can see that both of the growth and cell infusion process is occurred simultaneously and as a consequence a final spheroid size is obtained. As seen in Fig. 3, roundness of the cells is more apparent after 7 days in spheroidal microwells in comparison to rectangular one. This may be related to the higher surface energy of the wall effect for rectangular shape as well as less end cross-section area which is efficient for cell settle down compared to spherical microwell shape.

In order to investigate the effect of channel slope on the self-filling ability of the cells, 2 different slops containing 0 °and 20° were designed and it was showed that the velocity of the cell track inside the micro channels strongly depended on the gravity, which was introduced by applying slope to the channel so that cell trajectory velocity for the flat and 20° channels was obtained about 0.01 μm/s and 0.3μm/s, respectively.

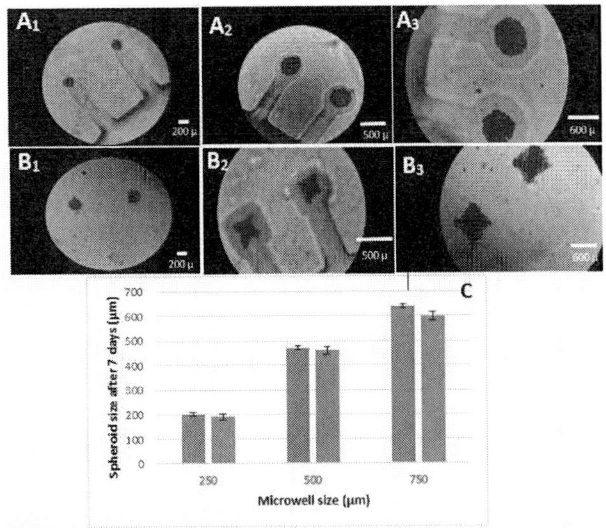

Figure 3: Optical microscope of cell spheroids after 7 days seeding. A_1&B_1)250, A_2&B_2) 500, A_3&B_3) 750 μm for spherical and rectangular wells respectively. C) Final sizes of spheroids after 7 days.

CONCLUSIONS

HepG2 cell spheroids were generated using different sizes and shapes of agarose micowells which were fabricated by DLP 3D printer method followed by replica molding. Self-filling property of this design was obtained by the microchannels which connected the pipetting zone to the end wells. Spherical shape microwell formed cell spheroids with better roundness which can be used as a simple tumor model for further drug screening analysis.

ACKNOWLEDGMENTS

M.A. would like to thank the Natural Sciences and Engineering Research Council of Canada (NSERC) and Canadian Foundation for Innovation for supporting this work

REFERENCES

1. Gong, X., o Lin, C., Cheng, J., Su, J., Zhao, H., Liu, T., Wen, X., Zhao, P. 2015, "Generation of Multicellular Tumor Spheroids with Microwell-Based Agarose Scaffolds for Drug Testing". *PLOS ONE*, 10(6). pp. e0130348.

2. Kim, S., Lee, G., Park,J. 2013, "Microwell Fabrication Methods and Applications for Cellular Studies", *Biomed Eng Lett, 3,pp.* 131-137.

3. Kang, D., Kim, H., Kim, P., Suh, K. 2014, "Poly (ethylene glycol) (PEG) Microwells in Microfluidics: Fabrication Methods and Applications", *BioChip J.*, 8(4). pp. 241-253

4. Selimovi, S. Piraino, F., Bae, H., Khademhosseini, A. 2011, "Microfabricated polyester conical microwells for cell culture applications", *Lab on a chip*, 11(14), pp. 2325-32.

5. Mikhail, A.;, Eetezadi, S., Allen, C. 2013, "Multicellular Tumor Spheroids for Evaluation of Cytotoxicity and Tumor Growth Inhibitory Effects of Nanomedicines in Vitro: A Comparison of Docetaxel-Loaded Block Copolymer Micelles and Taxotere", 8(4), pp. e62630.

INVESTIGATING ENERGY TRANSFER MECHANISM IN THE HIGH-ENERGY MECHANICAL ALLOYING PROCESS TO SYNTHESIZE NANOSTRUCTURED AL_2O_3-NI(CR) PARTICLES

Lipika Nath
Mechanical Engineering
University of New Brunswick
Fredericton, NB, Canada
lnath@unb.ca

Gobinda Saha
Mechanical Engineering
University of New Brunswick
Fredericton, NB, Canada
gsaha@unb.ca

ABSTRACT

High-energy mechanical alloying (HE-MA) is an efficient method for producing nanocomposite powder particles suitable for application in high velocity oxy-fuel (HVOF) coatings development process. In the first phase of this project, a detailed design of experiment (DoE) was carried out to study the optimal parameters required for processing microcrystalline Al_2O_3-Ni(Cr) blended mix for (1) the reduction of the alumina crystal grain size to a nanometric level, and (2) reinforcing the Al_2O_3 nano-grains into the micron-sized Ni(Cr) alloy binder [1]. Now in the second phase, the aim is to study the ball-to-powder energy transfer phenomena during the HE-MA process with a new design of experiment (DoE). High-energy mechanical alloying is accomplished in a planetary mill. This is a significant work, as energy evolved during alloying has a tangible effect on progressive grain size refinement. Additionally, energy works as a catalyst for binding grains in composite particles. The anticipated relationship when identified will play a key role in optimizing the design composition of novel composites which are then used as feedstock for coatings, applicable to many industries, including energy and environment, industrial manufacturing, and aerospace.

Throughout this project, various batch specimens of powders are collected and characterized to study the particle morphologies, cross-sectional microstructures, crystal structures, and phase formation using scanning electron microscopy (SEM), high-resolution transmission electron microscopy (HR-TEM), and X-ray diffraction (XRD) techniques, respectively. Energy released due to perfect elastic collision and perfect inelastic collision are calculated using the 'kinetic energy model' first proposed by Burgio *et al.* [2]. Moreover, the aim of this present work is to find out the correlation between input energy and end product and having a good understanding of the HE-MA process in terms of repeated fracturing and cold welding by varying milling parameters e.g., milling time, milling aid, ball to powder ratio (BPR) and weight composition of precursor powder particles.

KEYWORDS: Nanocomposite, High-energy mechanical alloying (HE-MA), Wear, HVOF, Cermet.

1.0 INTRODUCTION

One of the most significant and detrimental problems of mechanical components of automobiles, power generation units construction equipment, aircraft engine, chemical processing equipment is high temperature wear and corrosion. As both wear and corrosion are surface phenomena, application of surface coatings with materials consist of high temperature corrosion resistance properties is an effective way to avoid this problem. The thermal spraying is one of the most successful of all the advanced coating methodologies because of the wide range of coating materials and substrates to which it can be applied. The HVOF technique is one of the most industrially-used thermal spray techniques as it produces an extremely dense coating of higher bond strength and higher hardness together with lower porosity [3-5]. However, the properties of coatings are highly influenced by the properties and morphology of the underlying feedstock powder. High-energy mechanical alloying (HE-MA) is an effective method to synthesize nanostructured feedstock powder particles as it features a strong mechanical energy transfer into the processed powder particles [6-7]. During HE-MA process, the powder particles are subjected to high energy impact. In a vibratory or planetary mill, energy transfer occurs due to collision of balls, charged with powder, and against the vial walls. In this process, the dissipated energy from an impact gets transformed mainly into heat, resulting in a temperature rise of both the ball and the plate. A small fraction of it is stored in the material as structural disorder. The quantitative measures of energetics in high-energy mechanical alloying (HE-MA) process attracted attention for a long time. In the first phase of this study, nanostructured Al_2O_3-Ni(Cr) was manufactured using HE-MA process [1]. It became clear that (1) energy evolution plays a key role in HE-MA process from elastic deformation to perfectly plastic deformation of the powder particles, (2) only α-Al_2O_3 can be selected for precursor material instead of α-Al_2O_3 and γ-Al_2O_3 mixture as at high temperature (>350°C) γ-Al_2O_3 is converted into α-Al_2O_3, and (3) to obtain nanostructured Al_2O_3-Ni(Cr) powder particles, more than 24 hours milling time is not required.

After getting these significant findings, it was decided to conduct further analysis in this work with a new design of experiment (DoE). The primary goal of this current research work is to quantify the amount of energy consumption to obtain a desired powder particle microstructure, followed by examining its impact on the formation of nanocrystalline grain reinforced ceramic metallic nanocomposites (CMNC) during different milling stages. Milling parameters e.g milling time (1h, 2h, 3h, 6h, 9h, 12h), ball to powder ratio (BPR) (5:1 and 10:1) and weight composition of α-Al_2O_3 and Ni(Cr) alloy (75 wt% of α-Al2O3 blended with 25 wt% of Ni(Cr) and 50 wt% of α-Al2O3 blended with 50 wt% of Ni(Cr)) are varied in the DoE.

2.0 EXPERIMENTAL PROCEDURES

2.1 Design of Material

Nanostructured α-Al_2O_3 grains were blended with micron-sized Ni(Cr) alloy in a planetary ball mill. After the HE-MA process, α-Al_2O_3 nano-grains were embedded into micron-sized Ni(Cr) alloy binder as shown in Fig. 1.

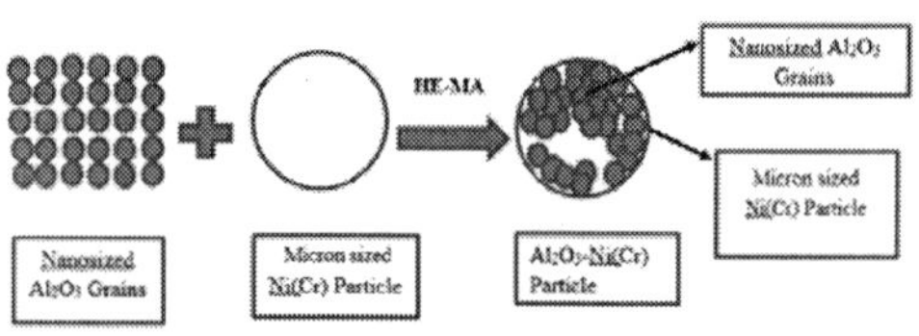

Fig.1: Schematic showing α-Al_2O_3 grains reinforced in micron-sized Ni(Cr) alloy binder.

2.2 Materials and Method

Two commercial grade precursor materials were used in this study: metal oxide ceramic grains (α-Al_2O_3 from Atlantic Equipment Engineers, NJ) and metal alloy particle (Ni(Cr) from Atlantic Equipment Engineers, NJ). Their physical characteristics are shown in Table 1.

Precursor Powder	Mean Particle Size
α-Al_2O_3	50 nm, >99.5% purity, wet chemical synthesis
Ni80/ Cr20 Alloy	44 µm, gas atomized

Table 1: Precursor particle characteristics.

2.3 Design of Experiment (DoE)

Pre-mix (blended) powders were mechanically-alloyed and sampled using a DoE, specified in Table 2 below.

Sample Designation	Mixture (wt. %)		Ball to Charge Ratio		Milling aid	
	α-Al_2O_3	Ni(Cr)	5:1	10:1	Stearic acid	Cyclohexane
S1	50	50	Yes		Yes	
S2	50	50		Yes	Yes	
S3	50	50	Yes			Yes
S4	50	50		Yes		Yes
S5	75	25	Yes		Yes	
S6	75	25		Yes	Yes	
S7	75	25	Yes			Yes
S8	75	25		Yes		Yes

Table 2: Design of experiment (DoE) of HE-MA process.

The high-energy mechanical alloying was done in a planetary ball mill and the samples were taken out for characterization and testing at intervals 1h, 2h, 3h, 6h, 9h and 12h. The rotational speed of the mill was 1200 rpm.

3.0 References

1. Nath, L., Saha, G. C., 2017, "Synthesis and characterization of nanocrystalline Al_2O_3-Ni(Cr) particles using high-energy mechanical alloying process", *Surface & Coatings Technology*, In Press.
2. N. Burgio, A. Iasonna, M. Magini, S. Martelli and F. Padella, 1991, "Mechanical alloying of the Fe–Zr system. Correlation between input energy and end products", *Nuovo Cimento,* vol 59, pp. 459–476.
3. Zhao, L., Zwick, J., & Lugscheider, E., 2003, "The influence of milling parameters on the properties of the milled powders and the resultant coatings," *Surface and Coatings Technology*, vol. 168, no. 2-3, pp. 179-185.
4. Parker, D. W., & Kutner, G., L., 1991, "HVOF-Spray Technology-Poised for Growth," *Journal of Advanced Materials and Processes*, vol. 4, pp. 68-74.
5. Saha, G. C., Khan, T. I., & Glenesk, L. B., 2008, "Development of Wear Resistant Nanostructured Duplex Coatings by High Velocity Oxy -Fuel Process for Use in Oil Sands Industry," *Journal of Nanoscience and Nanotechnology*, vol.8, pp. 1–8.
6. C. Suryanarayana, 2001, "Mechanical alloying and milling," *Progress in Materials Science,* vol. 46, no. 1-2, pp. 1-184.

7. Soni, P., R., 2001 " Mechanical alloying: fundamentals and applications," *Cambridge International Publishing*.

Turbulence

NUMERICAL SIMULATION OF TURBULENT FLOW OVER A FORWARD FACING STEP

Ali Nematollahi
Department of Mechanical Engineering
University of Manitoba
Winnipeg, Manitoba, Canada
nematola@myumanitoba.ca

Mark F. Tachie
Department of Mechanical Engineering
University of Manitoba
Winnipeg, Manitoba, Canada
mark.tachie@umanitoba.ca

ABSTRACT

In this study, a numerical simulation was performed to investigate the effect of Reynolds number on the flow characteristics of a two-dimensional forward-facing step using commercial computational fluid dynamics (CFD) code (ANSYS CFX, release 16.2) and different turbulence models. The test cases were investigated for Reynolds numbers ranging from 2040 to 13000. The turbulence models employed are the standard k-ε, baseline (BSL) and an omega-based second moment closure (BSL Reynolds stress model). The numerical results were compared with recent measured data obtained using a particle image velocimetry (PIV) system. Unlike the standard k-ε and BSL Reynolds stress models, the BSL model predicted the length of the recirculation region well. In addition, the Reynolds stress model correctly reproduced negative values of Reynolds shear stress in the vicinity of the leading edge of the step and positive values at the center of recirculation region, in conformity with the experimental results.

KEYWORDS: forward-facing step (FFS); Reynolds number; turbulence models; computational fluid dynamics (CFD)

INTRODUCTION

Flow separation and reattachment caused by a sudden change in geometry are encountered in many engineering and industrial applications such as combustors, gas and wind turbines and pipe systems. Such flow phenomena usually generate noise and structural vibrations and may also reduce the efficiency of fluid-thermal engineering systems. Over the past decades, significant research efforts have been made to understand separated and reattached turbulent flows produced from obstacles such as forward and backward-facing steps with the ultimate goal of designing high performance fluid engineering systems. Because of their relative complexity (caused by flow reversal, adverse and favorable pressure gradient and streamline curvature in these flows), separated and reattached turbulent flows are also used as acute test cases to assess the predictive performance of engineering turbulence models [1]. While numerical simulation of a backward-facing step (BFS) has been performed extensively in the past [2], the capability of different turbulence models to reproduce the flow characteristics over a forward-facing step (FFS) has not been systematically assessed. The pertinent features of flow over a FFS and the Cartesian coordinate system adopted in the present study are shown in Figure 1. The x-coordinate is aligned with the streamwise direction, and the y-coordinate is aligned with the wall-normal direction; $x = 0$ corresponds to the leading edge of the step and $y = 0$ is on top of the upstream wall. Generally, the flow field over a FFS can be classified into the following three distinct regions (as shown in Fig. 1): (*i*) the upstream region, (*ii*) the recirculation region, which spans from the first separation point in front of the step to the reattachment point on top of the step, (*iii*) the recovery region, which is the region beyond the reattachment point where a new boundary layer redevelops. Parameters such as Reynolds number (Re_h), blockage ratio (h/H), aspect ratio (W/h where W is the channel width), freestream turbulence intensity (Tu), and the ratio of the approach boundary layer thickness to step height (δ/h) may significantly affect the reattachment length (L_r).

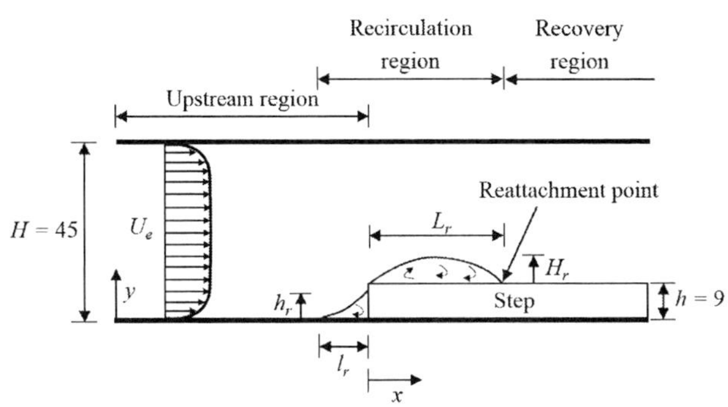

Figure 1: Salient flow features of FFS (all dimensions in millimeters)

Turbulent flow over FFS has been studied quite extensively using both experimental [3] and numerical techniques [4-7]. The effect of Reynolds number on the reattachment length, mean and turbulent statistics in the recirculation region on top of a FFS was investigated by Essel et al. [3] experimentally. A wide range of Reynolds numbers (2040 to 9130) was considered. It was observed that the reattachment length increased gradually by increasing Reynolds number up to $Re_h = 6380$ and after that it became independent of Reynolds number. Zhang et al. [4] performed a numerical simulation at $Re_h = 50000$ to examine the capability of standard k-ε model to reproduce the flow properties over a FFS. Their results showed that the k-ε model can predict the mean velocities and static pressure well but the turbulent kinetic energy (*tke*) is under-predicted. Also, the reattachment length was under-estimated by 15%. Gasset et al. [5] used RNG k-ε model to simulate the flow over a FFS with the Reynolds number of 50000. It was found that the RNG k-ε model under-estimated the reattachment length. Addad et al. [6] carried out a numerical simulation for the flow over FFS using large eddy simulation (LES) at Re_h = 170000. They compared the reattachment length, mean and turbulent quantities and a good agreement between their results and LDA data was obtained. Hattori and Nagano [7] used direct numerical simulation (DNS) to investigate the effect of Reynolds number (900, 1800 and 3000) on the mean and turbulent statistics. The counter-gradient diffusion phenomena (CDP), defined as the opposite sign between the Reynolds shear stress and the mean velocity gradient, between the streamwise mean velocity and Reynolds shear stress was observed near the corner of the step.

The objective of the present study is to assess the capability of two two-equation models (standard k-ε and baseline (BSL)) and one Reynolds stress model (BSL Reynolds stress model) available in the commercial CFD code, ANSYS CFX to predict the turbulent characteristics of FFS. This objective is achieved by comparing the numerical results with the experimental data of Essel et al. [3]. The paper structure is as follows; Section 2 describes the governing equations, boundary conditions, numerical procedure and computational domain and grid generation while Section 3 summarizes the main findings.

GOVERNING EQUATIONS

The flow is considered to be two-dimensional, steady state and turbulent and the fluid is incompressible. The Reynolds-Averaged Navier-Stokes (RANS) equations for continuity and momentum conservation are given by

Continuity:

$$\frac{\partial U_j}{\partial x_j} = 0 \tag{1}$$

Momentum:

$$\frac{\partial}{\partial x_j}(\rho U_i U_j) = -\frac{\partial P}{\partial x_i} + \frac{\partial}{\partial x_j}\left(\mu \frac{\partial U_i}{\partial x_j} - \rho\overline{u_i u_j}\right) \tag{2}$$

where U_i and u_i denotes the mean and fluctuating velocity components in the x_i direction, ρ is the density, μ is the dynamic viscosity, P is the pressure and $\overline{u_i u_j}$ is the Reynolds stress tensor. The two-equation models (standard k-ε and BSL models) assumed eddy-viscosity relationship to relate the Reynolds stresses in Equation (2) to the mean velocity gradients and the turbulent viscosity, given by

$$-\rho\overline{u_i u_j} = \mu_t \left(\frac{\partial U_i}{\partial x_j} + \frac{\partial U_j}{\partial x_i}\right) - \frac{2}{3}\rho\delta_{ij}k \tag{3}$$

where μ_t is the eddy viscosity, δ_{ij} is the Kronecker delta and k is the turbulent kinetic energy (*tke*).

On the other hand, Reynolds stress models solve transport equations for all the components of Reynolds stress tensor. These models do not use the eddy viscosity hypothesis, instead a separate transport equation is solved for each of the six Reynolds stress components of $\rho\overline{u_i u_j}$. In principle, the Reynolds stress models are more suited to complex turbulent flows because the individual Reynolds stresses are directly computed. It has been suggested, however, that the predictive performance of Reynolds stress models is not always superior to results from two-equation models [8]. The BSL Reynolds stress model adopted in the present study is an ω-based model. The advantage of the ω-equation in the ANSYS CFX is that it allows for a more accurate near wall treatment with an automatic switch from a wall function to a low Reynolds-number formulation based on the grid spacing.

BOUNDARY CONDITIONS

The boundary conditions considered at various boundaries are as follows. At the inlet, water with an approach velocity, U_e, ranging from 0.227 *m/s* (Re_h = 2040) to 1.444 *m/s* (Re_h = 13000) and upstream turbulence intensity (*Tu*) of 5% enters the solution domain uniformly. At the outlet, zero static pressure was applied. The symmetry condition was used for the symmetry planes and no-slip ($U = 0$) and no penetration ($V = 0$) boundary conditions were imposed on the solid walls.

NUMERICAL METHOD

The problem under consideration is two-dimensional flow domain and a turbulent flow regime. The simulations were performed by the computational fluid dynamics software ANSYS CFX. Only a brief overview of pertinent information is provided here. Further details of the implementation can be found in the CFX manuals [8]. In order to solve the governing equations in CFX, the fluid domain is subdivided into a finite number of cells and the equations are changed into algebraic form via a discretization process in which the Finite Volume Method (FVM) is used. The root-mean-square (RMS) criterion with a residual target value of 1.0×10^{-6} was used for checking the convergence of the solutions and the equations are solved using double precision.

FLOW DOMAIN AND GRID GENERATION

The configuration and dimensions of the two-dimensional domain, which are chosen based on the experimental work by Essel et al. [3], are shown in Figure 1. A structured grid was used for meshing the computational domain. In order to achieve grid-independent solutions, grid convergence tests were conducted by comparing the reattachment length, (L_r/h), streamwise variation of skin friction coefficient (C_f), and velocity profiles in the recirculation region (not shown) for the Re_h = 9,130. Five different grids (57200, 115600, 207400, 307400 and 420400 elements) were considered and the results indicated that the optimal grid size is 307,400 elements. This grid size was employed in all the simulations. It may be noted that for the two ω-based models (BSL and BSL Reynolds stress models), finer grids were used in close proximity to the step and walls in order to accurately resolve the flow properties within the viscous sub-layer.

RESULTS AND DISCUSSION

REATTACHMENT LENGTH

The variation of the reattachment length, L_r, over the step with Re_h is shown in Figure 2. The experimental results from Essel et al. [3] are included for comparison. The reattachment point was determined as the position where the streamwise skin friction coefficient (C_f) on the step becomes zero. The standard k-ε model under-predicted the reattachment length while the BSL Reynolds stress model over-predicted the values of L_r. However, there is a good agreement between the results obtained from the BSL model and the experimental data for $4030 \leq Re_h \leq 9130$. The experimental results indicated that there are two distinct regimes. In the first regime, the reattachment length increases with Re_h (for $Re_h < 6380$) and in the second region L_r is independent of Re_h (for $Re_h \geq 6380$). In the first region, all the turbulence models failed to accurately predict the rapid increase in the reattachment length of the experimental results. However, in the second region, the models showed that the L_r is independent of Re_h which is in agreement with the experiment.

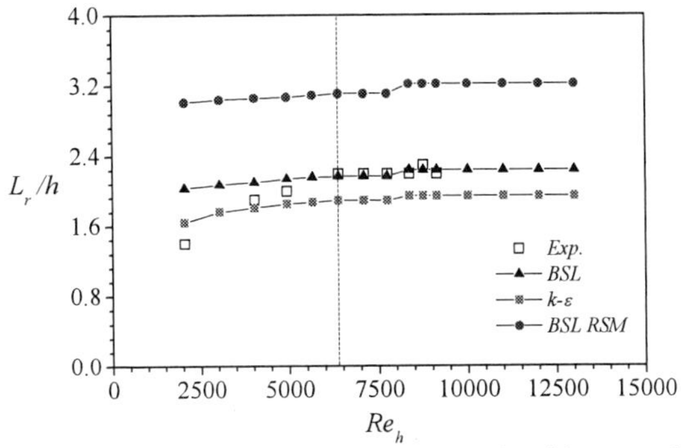

Figure 2: Variation of reattachment length with Reynolds number

MEAN VELOCITIES

Figure 3 shows the contour plots of the streamwise mean velocity obtained from the turbulence models for $Re_h = 2040$ and 8360. The experimental results are included for comparison. Since the contour plots of BSL and BSL Reynolds stress models are qualitatively similar, only the results for the BSL model are presented. The approach boundary layers deflect upward as it approaches the step and accelerate from the leading edge of the step due to mass conservation. In each case, distinct recirculation region is observed on top of the step. The size of the recirculation bubble on the top of the step increases as the Reynolds number increases. On top of the step, U_{max} increased to a local peak value that tends to increase with Re_h. Similar to the experimental results the position of the maximum velocity is located above the recirculation region, however, the magnitude of the maximum velocity predicted by the turbulence models is higher than in the experiment with a maximum deviation of 17%.

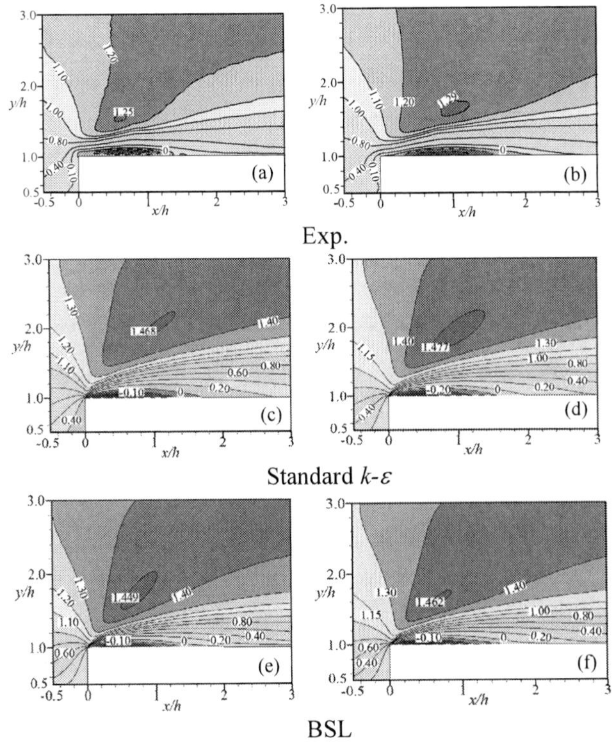

Figure 3: Contour plots of streamwise mean velocity for Re_h = 2040 (a,c,e) and Re_h = 8360 (b,d,f)

Profiles of streamwise and wall-normal mean velocities obtained at selected x-locations in the recirculation region ($x' = x/L_r = 0,\ 0.5, 1.0$) and one step height downstream of reattachment point (x^*) for $Re_h = 8360$ for the three turbulence models are shown in Figure 4. The local maximum velocity, U_e, is used as the velocity scale. As it is shown, the shape of the velocity profiles changes as streamwise distance increases. This is because of the severe adverse pressure gradient and momentum deficit caused by separation. All the models are able to predict the streamwise mean velocity well but BSL model performed better than other models.

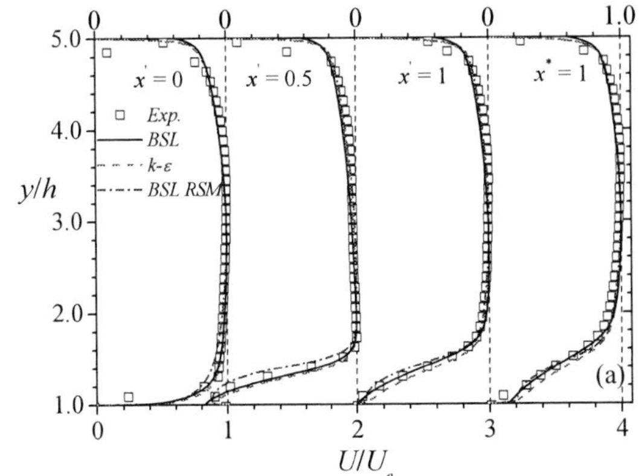

Figure 4: Profiles of streamwise mean velocity

REYNOLDS STRESSES

The profiles of the Reynolds stresses at selected streamwise locations for $Re_h = 8360$ are shown in Figure 5. At the leading edge of the step and close to the wall, the peaks in the normal stresses are larger than the profiles further downstream (Figure 5a and 5b). As the streamwise distance increases, the Reynolds stresses reduces. This can be associated with the breakdown of large scale structures generated from separation. The turbulence model is able to predict the Reynolds stresses fairly well. At the leading edge of the step, high negative peaks of Reynolds shear stress are noticeable in the experimental results while a very small negative value is predicted by the turbulence model (Figure 5c). Also, this negative value becomes positive at the center of recirculation region which is consistent with the experimental results.

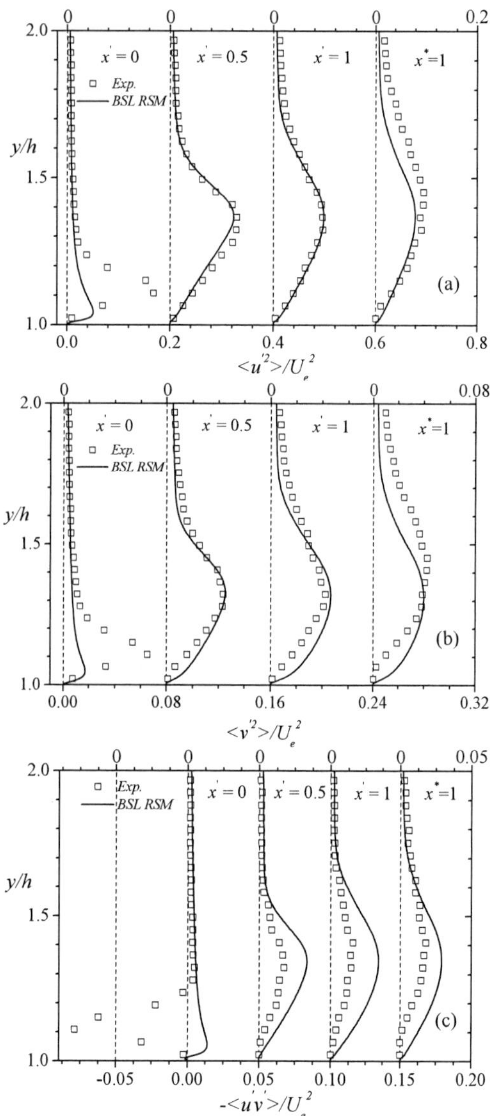

Figure 5: Profiles of Reynolds stresses

CONCLUSIONS

The flow over a forward facing step is studied numerically using CFD software ANSYS CFX. The standard k-ε, BSL and BSL Reynolds stress turbulence models were employed and the numerical results were compared with available measured data. The effects of Reynolds number (ranging from 2040 to 13000) on the separation-reattachment lengths and flow structure associated with the forward facing step flow are considered. The standard k-ε and BSL Reynolds stress models under-predicted and over-estimated the reattachment length, respectively while the BSL model performed better for a wide range of Reynolds numbers. Furthermore, among the models considered, the BSL model showed a better agreement with the measured mean velocities. The negative values of Reynolds shear stress at the leading edge of the step was predicted by the BSL Reynolds stress model and a good agreement with the experimental results was obtained.

ACKNOWLEDGMENT

The financial support of this work by the Natural Sciences and Engineering Research Council of Canada (NSERC-DG) to Mark F. Tachie is gratefully acknowledged.

REFERENCES

1. Craft, T., 1998, “Developments in a low-Reynolds-number second-moment closure and its application to separating and reattaching flows”. *International Journal of Heat and Fluid Flow*, 19, pp. 541–548.
2. Le, H., Moin, P., Kim, J., 1997, “Direct numerical simulation of turbulent flow over a backward-facing step”. *Journal of Fluid Mechanics*, 330, pp. 349–374.
3. Essel, E.E., Nematollahi, A., Thacher, E.W., Tachie, M.F., 2015, “Effects of upstream roughness and Reynolds number on separated and reattached turbulent flow. *Journal of Turbulence*, 16, pp. 872–899.
4. Zhang, C., 1994, “Numerical predictions of turbulent recirculating flows with a k-ε model”. *Journal of Wind Engineering and Industrial Aerodynamics*, 51, pp. 177–201.
5. Gasset, N., Poitras, G.J., Gagnon, Y., Brothers, C., 2005, “Study of atmospheric boundary layer flows over a coastal cliff”. *Wind Engineering*, 29, pp. 3–24.
6. Addad, Y., Laurence, D., Talotte, C., Jacob, M.C., 2002, “Large eddy simulation of a forward-backward facing step for acoustic source identification”, 5th *International Symposium on Engineering Turbulence Modelling and Experiments*, Elsevier.
7. Hattori, H., Nagano, Y., 2010, “Investigation of turbulent boundary layer over forward-facing step via direct numerical simulation”. *International Journal of Heat and Fluid Flow*, 31, pp. 284–294.
8. CFX, Help Files, ANSYS Inc., 2015.

ASSESSMENT OF CFD PREDICTIONS OF A PLANAR TURBULENT OFFSET JET

Ali Nematollahi
Department of Mechanical Engineering
University of Manitoba
Winnipeg, Manitoba, Canada
nematola@myumanitoba.ca

Mark F. Tachie
Department of Mechanical Engineering
University of Manitoba
Winnipeg, Manitoba, Canada
Mark.Tachie@umanitoba.ca

Scott J. Ormiston
Department of Mechanical Engineering
University of Manitoba
Winnipeg, Manitoba, Canada
Scott.Ormiston@umanitoba.ca

ABSTRACT

A comparative study of mean flow and turbulence characteristics of a two-dimensional turbulent offset jet issuing from a finite width nozzle into stagnant surroundings has been performed using different turbulence models in the computational fluid dynamics (CFD) software ANSYS CFX. The Reynolds number considered for the present study is 15000 based on the jet inlet mean velocity and the jet width, and the offset height ratios are varied between 3 and 15. The standard k-ε, baseline (BSL) and QI Reynolds stress models are used for the numerical simulations. The computational results are compared with available experimental data in the literature in order to test the predictive capability of the different turbulence models. The results showed that the reattachment length increased almost linearly with increasing offset height ratio. More importantly, all three models predicted the reattachment length reasonably well. It is also observed that the streamwise variation of the normalized skin friction coefficient obtained from the standard k–ε and QI Reynolds stress models deviated considerably from the experimental results because of the application of wall functions, while a very good agreement has been obtained for the BSL model.

KEYWORDS: offset jet; turbulence models; computational fluid dynamics (CFD)

INTRODUCTION

Offset jets are used in many engineering devices for cooling and mixing applications. Some of these applications include mixing processes in boilers, cooling of gas turbine blades, injection systems, heating and air conditioning applications and heat exchangers. Offset jets are formed when a jet is discharged at a certain distance from a parallel wall boundary. The offset jet possesses characteristic features of both a free jet (close to the nozzle exit) and a wall jet (after the jet impinges on the wall). This makes the flow field of an offset jet relatively more complicated than for free and wall jets.

The salient features of an offset jet and the nomenclature used in the present study are shown in Figure 1, where H is the jet center line distance from the wall and b is the jet width. The offset height ratio (OR) is defined as the ratio of the distance from the jet center line to the wall to the jet width, i.e. OR = H/b. The x-coordinate is aligned with the streamwise direction and the y-coordinate is aligned with the wall-normal direction.

Generally, the flow field of an offset jet can be classified into the following three distinct regions (as shown in Fig. 1): the recirculation region, the impingement region and the wall jet region.

When the jet discharges from the nozzle, it deflects towards the wall because of the pressure difference across the jet caused by unbalanced entrainment which is called the Coanda effect [1]. Beyond the impingement region, the flow behaves like a wall jet flow regardless of the offset height ratio [2]. Accurate prediction of the flow characteristics, particularly in the impingement region, is quite difficult. This can be partly attributed to the complex nature of the flow field. Within the impingement region, for example, part of the fluid is deflected backward (forming a recirculation region) due to an adverse pressure gradient dominating in the recirculation region and the rest of the fluid moves in the forward streamwise direction (forming the wall jet region). Because the flow is also subjected to a large strain rate and streamline curvature in the recirculation and impingement regions [3], an accurate prediction using standard engineering turbulence models is challenging.

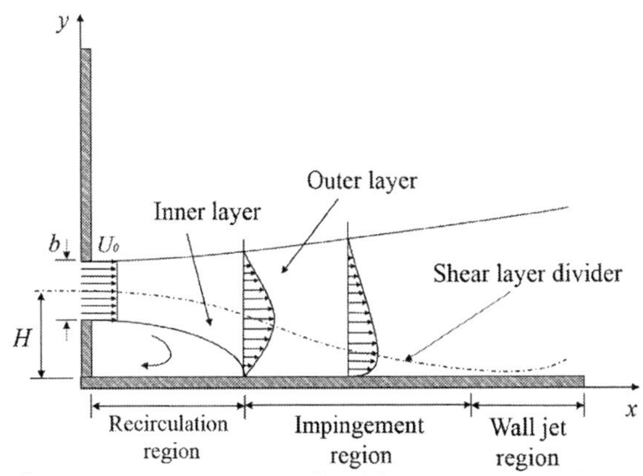

Figure 1: Salient features of offset jets

The turbulent offset jet has been studied in the past experimentally and numerically. Pelfrey and Liburdy [3] investigated the mean flow characteristics of a turbulent offset jet with an offset height ratio of 7.0 and Reynolds number of 15000 using single component laser Doppler anemometry (LDA). It was concluded that this flow cannot be modeled as a thin shear layer flow because the ratio of curvature to shear strain rate is significantly high in the recirculation region. They also remarked that for large Reynolds number (Re > 10000), the impingement distance becomes independent of Re.

Nasr and Lai [4] studied the mean flow and turbulent statistics for a small offset height ratio of 2.125 and Reynolds number of 11000 experimentally and numerically. Three different turbulence models (standard *k-ε*, RNG *k-ε* and Reynolds stress models) were used. It was found that the standard *k-ε* model performed better than other models for a small offset height ratio. Rathore and Das [5] performed a comparative study to examine the capability of three turbulence models (the standard *k-ε* and two low-Reynolds number *k-ε* models (Launder and Sharma (LS) and Yang and Shin (YS) models) in predicting the flow field and turbulence characteristics of an offset jet. A wide range of offset ratios (3, 5, 7, 9, and 11) was considered at *Re* = 15000. The results were compared with available experimental data in the literature. It was found that the low-Reynolds number models performed better than the standard *k-ε* model in predicting the skin friction coefficient due to their ability in resolving the entire boundary layer including the viscous sub-layer.

The first objective of the present study is to assess the capability of two two-equation models (standard *k-ε* and baseline (BSL) models) and an epsilon-based Reynolds stress (RS) model (QI Reynolds stress model) available in the commercial CFD code, ANSYS CFX to predict the flow characteristics of a turbulent offset jet. This objective is achieved by comparing the numerical results with available experimental data of Pelfrey and Liburdy [3]. The second objective is to investigate the effect of offset height ratio on reattachment length and mean velocities of a two-dimensional offset jet.

GOVERNING EQUATIONS

The flow is considered to be two-dimensional, steady-state, and turbulent and the fluid is incompressible. The Reynolds-Averaged Navier-Stokes (RANS) equations for continuity and momentum conservation are given by

Continuity:

$$\frac{\partial U_j}{\partial x_j} = 0 \tag{1}$$

Momentum:

$$\frac{\partial}{\partial x_j}(\rho U_i U_j) = -\frac{\partial P}{\partial x_i} + \frac{\partial}{\partial x_j}\left(\mu \frac{\partial U_i}{\partial x_j} - \rho\overline{u_i u_j}\right) \tag{2}$$

where U_i and u_i denotes the mean and fluctuating velocity components in the x_i direction, ρ is the density, μ is the dynamic viscosity, P is the pressure, and $\overline{u_i u_j}$ is the Reynolds stress tensor. The two-equation models (standard *k-ε* and BSL models) assumed eddy-viscosity relationship to relate the Reynolds stresses in Equation (2) to the mean velocity gradients and the turbulent viscosity, given by

$$-\rho\overline{u_i u_j} = \mu_t\left(\frac{\partial U_i}{\partial x_j} + \frac{\partial U_j}{\partial x_i}\right) - \frac{2}{3}\rho\delta_{ij}k \tag{3}$$

where μ_t is the eddy viscosity, δ_{ij} is the Kronecker delta, and k is the turbulent kinetic energy (*tke*). On the other hand, Reynolds stress models solve transport equations for all the components of Reynolds stress tensor. These models do not use the eddy viscosity hypothesis, instead a separate transport equation is solved for each of the six Reynolds stress components of $\rho\overline{u_i u_j}$. In principle, the Reynolds stress models are more suited to complex turbulent flows because the individual Reynolds stresses are directly computed. It has been suggested, however, that the predictive performance of Reynolds stress models is not always superior to results from two-equation models [6]. The QI (Quasi-Isotropic) Reynolds stress model adopted in the present study is based on the *ε*-equation and was developed by Launder and Reece [7]. In this model, the pressure-strain correlation is linear.

BOUNDARY CONDITIONS

The boundary conditions considered at various boundaries are as follows. At the inlet, air with an approach velocity, U_0, of 18.54 *m/s* (*Re* = 15000) and an upstream turbulence intensity (*Tu*) of 5% enters the solution domain uniformly. At the outlet, zero static pressure was applied. At the top, an opening boundary condition with zero relative pressure was prescribed. The symmetry condition was used for the two *x-y* planes in the computational domain and no-slip ($U = 0$) and no penetration ($V = 0$) boundary conditions were imposed on the solid walls.

NUMERICAL METHOD

The simulations were performed by the CFD software ANSYS CFX. In order to solve the governing equations in CFX, the fluid domain is subdivided into a finite number of cells and the equations are changed into algebraic form via a discretization process in which the Finite Volume Method (FVM) is used. The root-mean-square (RMS) criterion with a residual target value of 1.0×10^{-6} was used for checking the convergence of the solutions and the equations are solved using double precision.

FLOW DOMAIN AND GRID GENERATION

A structured, non-uniform Cartesian grid was used for meshing the computational domain. The domain size and grid convergence tests were conducted by comparing the reattachment length, streamwise skin friction coefficient and velocity profiles at different axial positions. Because the maximum difference between the reattachment lengths for different domain sizes was less than 1%, the domain size of 75×60 was adopted for this study. The grid convergence tests were conducted for each turbulence models for OR = 7.0. Four different grids were considered for the standard *k-ε* and QI RS models and the results indicated that the optimal grid sizes are 238,000 and 312,000 elements, respectively. For the BSL model, 770,400 elements were considered to be enough. It may be noted that for the omega-based model (BSL), finer grids were used in the jet entrance region and at the walls. This was done in order to accurately resolve the flow properties within the viscous sub-layer while a coarser grid was used near the wall for the epsilon-based models (the standard k-ε and QI Reynolds stress models) due to the application of a wall function.

VALIDATION OF NUMERICAL RESULTS

The numerical results are validated with the experimental results of Pelfrey and Liburdy [3] for OR = 7.0 and Reynolds number of 15000. The streamwise mean velocity profiles at different downstream locations (x/b = 3, 6, 9, 12, 15 and 17) for the three turbulence models are shown in Figure 2. These locations are within the recirculation region (x/b = 3, 6, 9), impingement region (x/b = 12) and wall jet region (x/b = 15, 17). The results show that the standard k-ε and BSL models are able to predict the streamwise mean velocity reasonably well while the QI RSM model failed to predict the velocity profiles, especially in the wall jet region. Also, the ε-based models (the standard k-ε and QI Reynolds stress models) are not able to accurately capture the velocity profiles near the wall due to the application of the wall function. In general, the BSL model performed better than the other models, especially near the wall.

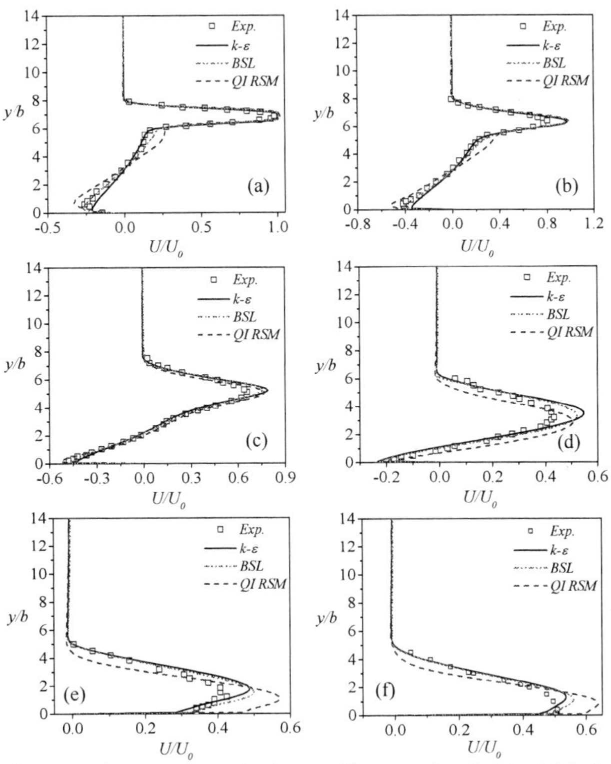

Figure 2: Streamwise velocity profiles at x/b = 3 (a), 6 (b), 9 (c), 12 (d), 15 (e), and 17 (f) for OR = 7.0

RESULTS AND DISCUSSION

MEAN VELOCITY CONTOURS

Figure 3 shows the contour plots of the streamwise mean velocity (U) obtained from the turbulence models for OR = 3 and 15. The contour plots are similar for all the models. Due to asymmetric in entrainment, the jet deflects towards the wall. A high density of contours is observed at the exit of the jet where the jet interact with the quiescent fluid immediately. The existence of a large recirculation bubble with characteristic negative U-velocity can be noticed. Pelfrey and Liburdy [3] reported a maximum backward velocity of $0.41U_0$ for OR = 7.0. The corresponding value was $0.42U_0$, $0.47U_0$ and $0.53U_0$ for the standard k-ε, BSL and QI RS models, respectively.

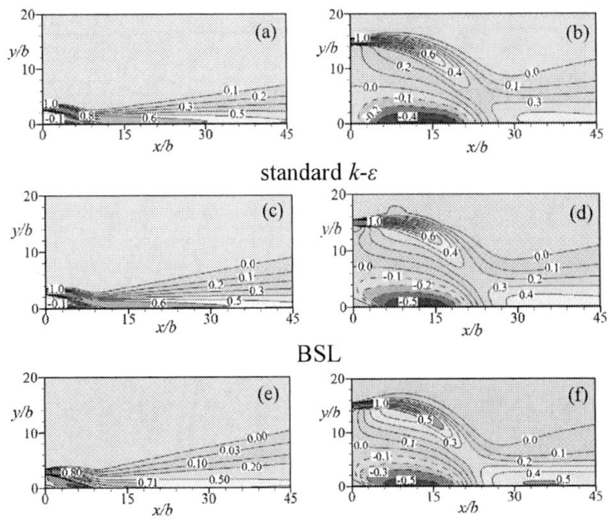

Figure 3: Contour plots of streamwise mean velocity for OR = 3 (a,c,e) and OR = 15 (b,d,f)

REATTACHMENT LENGTH

The reattachment length is one of the most important parameters of separated and reattached flows. The reattachment point was determined as the streamwise position where the streamwise skin friction coefficient (C_f) on the wall becomes zero. The comparison of the reattachment length with the experimental results of Holland and Liburdy [8], Perry [9] and the correlation of Nasr and Lai [10] ($L_r/b = 2.632(H/b)^{0.851}$) is shown in Figure 4. All the models are able to predict the reattachment length reasonably well. The present results predicted a reattachment length of 13.48, 13.23 and 12.15 for OR = 7.0 for the standard k-ε, BSL and QI RS models, respectively. The experimental results of Pelfrey and Liburdy [3] showed the normalized reattachment length (L_r/b) of 13.0. The percentage differences between the numerical and experimental results are 3.6%, 1.7% and 6.5% for the standard k-ε, BSL and QI RS models, respectively. Also, the present results are compared with the experimental data of Holland and Liburdy [8] who used the same conditions as Pelfrey and Liburdy [3] with offset ratios of 3, 7 and 11. The percentage difference between the numerical results and the experimental data of Holland and Liburdy [8] are 5.1%, 4.5% and 10.5% for OR = 3.0 and the corresponding errors for OR = 11 are 7.2%, 3.0% and 5.4% for the standard k-ε, BSL and QI RS models, respectively. All the models show a deviation with the correlation proposed by Nasr and Lai [10] for OR > 8.0. In addition, the numerical results obtained from BSL model show a good agreement with the experimental results of Perry [9] who investigated the effect of offset ratio (ranging from 1 to 10) on the reattachment length for $Re = 13000$.

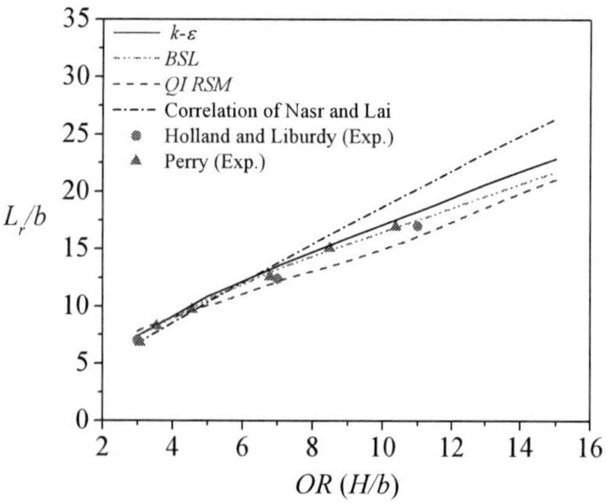

Figure 4: Variation of the reattachment length with OR

STREAMWISE SKIN FRICTION COEFFICIENT

The distribution of the skin friction coefficient (C_f) normalized by the maximum skin friction coefficient ($C_{f,max}$) for OR = 7.0 is shown in Figure 5. The skin friction coefficient is negative in the recirculation region (not shown) and becomes zero at the reattachment point. After the reattachment point, the skin friction coefficient increases rapidly due to the conversion of static pressure into velocity. In the wall jet region, the skin friction coefficient decreases gradually. The standard k-ε and QI RS models failed to accurately predict the variation of skin friction coefficient due to the application of wall function. The BSL model predicted the skin friction coefficient relatively accurately compared with the experimental data. Although there are some deviations after the reattachment point, (i.e., $(x-L_r)/b < 3$), a very good agreement was obtained in the wall jet region for the BSL model.

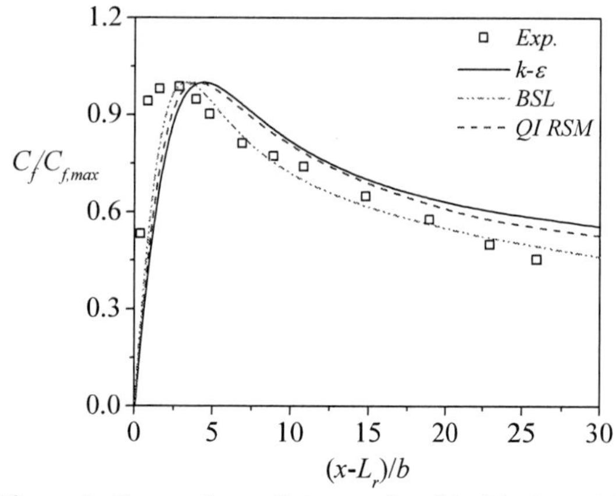

Figure 5: Comparison of streamwise skin friction with experimental results of Pelfrey and Liburdy [3] for OR = 7.0

CONCLUSIONS

A numerical study of a turbulent offset jet has been carried out using CFD software ANSYS CFX. The standard k-ε, BSL, and QI Reynolds stress turbulence models were employed and the numerical results were compared with available measured data to assess the capability of the models. The effects of offset ratios (ranging from 3 to 15) on the flow characteristics were investigated for $Re = 15000$. All the models are able to predict the reattachment length well. The BSL model, however, performed better than other models. Unlike the BSL model, the ε-based models (the standard k–ε and QI Reynolds stress models) could not predict the streamwise variation of the normalized skin friction coefficient accurately due to the application of wall function. To sum up, for the range of offset ratios and Reynolds number considered in this study, among the models considered, the BSL model performed better than standard k-ε and QI Reynolds stress models. The performance of other turbulence models should be considered in future studies.

ACKNOWLEDGMENT

The financial support of this work by the Natural Sciences and Engineering Research Council of Canada (NSERC-DG) to Mark F. Tachie is gratefully acknowledged.

REFERENCES

1. Tritton, D.J., 1977, "Physical Fluid Dynamics", Von Norstrand Reinhold, UK, pp. 284–286.
2. Rajaratnam, N., Subramanya, N., 1968, "Plane Turbulent Reattached Wall Jets". *ASCE Journal of Hydraulics Div.*, 94, pp. 95–112.
3. Pelfrey, J.R.R., Liburdy, J.A., 1986, "Mean Flow Characteristics of a Turbulent Offset Jet". *Transaction ASME Journal Fluids Engineering*, 108, pp. 82–88.
4. Nasr, A., Lai, J.C.R., 1998, "A turbulent Plane Offset Jet with Small Offset Ratio". *Experiments in Fluids*, 24, pp. 47–57.
5. Rathore, S.K., Das, M.K., 2013, "Comparison of two low-Reynolds number turbulence models for fluid flow study of wall bounded jets". *International Journal of Heat and Mass Transfer*, 61, pp. 365–80.
6. CFX, Help Files, ANSYS Inc., 2015.
7. Launder, B.E., Reece, G.J., Rodi, W., 1975, "Progress in the developments of a Reynolds-stress turbulence closure". *Journal of Fluid Mechanics*, 68, pp.537-566.
8. Holland, J.T., Liburdy, J.A., 1990, "Measurements of the thermal characteristics of a turbulent offset jets". *International Journal of Heat and Mass Transfer*, 33, pp. 69–78.
9. Perry, C.C., 1967, "Two-dimensional jet reattachment", PhD dissertation, University of Michigan.
10. Nasr, A., Lai, J.C.S., 1997, "Comparison of flow characteristics in the near field of two parallel plane jets and an offset jet". *Physics of Fluids*, 9, pp. 2919–2931.

TURBULENCE MODELING OF A SINGLE-PHASE EJECTOR WORKING WITH HFO REFRIGERANTS

S. Croquer, Y. Fang, S. Poncet*
Département de génie mécanique
Université de Sherbrooke
Sherbrooke, Québec, Canada
*Sebastien.Poncet@USherbrooke.ca

Z. Aidoun
CanmetEnergy
Natural Resources Canada
Varennes, Québec, Canada

Y. Bartosiewicz
Inst. Mech., Materials, Civil Eng.
Université Catholique de Louvain
Louvain-la-Neuve, Belgique

ABSTRACT

The present work reports a numerical analysis of a supersonic ejector in single-phase condition using R134a as well as novel HydroFluoroOlefins (HFO) refrigerants R1234yf and R1234ze(E) as working fluids. Comparisons were made regarding the ejector performances under varying operating conditions and refrigerant mixture proportions. A comparison of the ejector exergetic performance and flow structure was carried out. R1234yf appears to be a good candidate for drop-in replacement of R134a in a real ejector, while using R1234ze(E) would induce modifications of some components of the refrigeration cycle due to its thermodynamic properties.

Keywords: HFO refrigerants, R134a, single-phase ejector, numerical modeling

INTRODUCTION

The constant increase in energy consumption and decrease in natural resources necessitate a more efficient use of energy. Coupled with the political will to reduce the greenhouse gas emissions, national initiatives develop to increase the system efficiency and replace refrigerants with a high environmental impact. In that context, a renewed interest recently appeared to incorporate ejectors in refrigeration systems to replace the expansion valve or reduce the compressor load [1]. Most of the refrigeration or air-conditioning systems work with HFC refrigerants, mainly R134a and R245fa. Their high ozone depletion potential (ODP) and Global Warming Potential (GWP) demonstrate their direct impact on the environment in case of releasing to the surroundings. In 2006, the European Union initiated the F-gas regulation to completely phase out gases with fluorine starting from 2017 [2].
A new class of synthetic fluids, named HydroFluoroOlefins (HFO), has been recently developed and appeared as promising alternatives. Among them, R1234yf and R1234ze(E) are good candidates for drop-in replacement of R134a in ejector refrigeration systems. These are two environmentally friendly refrigerants because of their zero ODP and very low GWP. The researches on HFO are recent and mostly focus on their thermodynamic properties. Wang [3] and Mota-Babiloni et *al.* [4] reviewed heat pump refrigeration and air-conditioning systems working with R1234yf and R1234ze(E), respectively. For ejector refrigeration systems, the drop-in replacement of R134a by HFOs received few attention. Lawrence and Elbel [5] performed a coupled experimental and thermodynamical analysis of a real air-conditioning system for a small vehicle. R1234yf was found to be more affected by the separator inefficiency compared to R134a, while it gained more benefit from the two-phase ejector. Very recently, Smierciew et *al.* [6] investigated experimentally an ejector refrigeration system working with R1234ze(E). It was shown that this refrigerant may be competitive for ejector cycle driven by low temperature heat source. The present work is the first to evaluate the potentiality of drop-in replacement of R134a by two HFOs in an existing single-phase supersonic ejector.

OPERATING CONDITIONS

The ejector geometry corresponds to the ejector developed by Garcia del Valle et *al.* [7]. The pressure and temperature at the primary and secondary inlets and outlet of the ejector used in the experiments are summarized in Table 1. These values were obtained for R134a and include a $10K$ overheat at both inlets. In order to substitute R134a by R1234yf and R1234ze(E), two options are considered: (i) keeping the operating conditions unchanged or (ii) maintaining the same primary inlet temperature $T_{primary}$ and adjusting the pressures along the ejector to achieve the $10K$ overheat and same pressure ratios $P_{out}/P_{secondary}$. The operating conditions for the latter approach are summarized in Table 2.

Table 1: Operating conditions for the ejector corresponding to the experiments of Garcia del Valle et $al.$ [7].

Operation Point	$T_{primary}$ $[^oC]$	$T_{secondary}$ $[^oC]$	$P_{primary}$ $[kPa]$	$P_{secondary}$ $[kPa]$	P_{out} $[kPa]$
1	89.35	20.05	2598.0	414.6	757.2
2	94.35	20.05	2888.8	414.6	826.6
3	99.15	20.05	3188.1	414.6	897.1

Table 2: Operating conditions for the ejector adjusted for the HFO refrigerants R1234yf and R1234ze(E), keeping the same primary inlet temperature, pressure ratios and overheat as in the experiments of Garcia del Valle et $al.$ [7].

Operation Point	$T_{primary}$ $[^oC]$	**R1234yf** $P_{primary}[kPa]$	$P_{secondary}[kPa]$	$P_{out}[kPa]$
1'	89.35	2487.2	396.9	724.9
2'	94.35	2753.9	395.3	788.0
3'	99.15	3028.8	393.9	852.3
Operation Point	$T_{primary}$ $[^oC]$	**R1234ze(E)** $P_{primary}[kPa]$	$P_{secondary}[kPa]$	$P_{out}[kPa]$
1'	89.35	1980.5	316.1	577.2
2'	94.35	2203.5	316.2	630.5
3'	99.15	2432.5	316.3	684.5

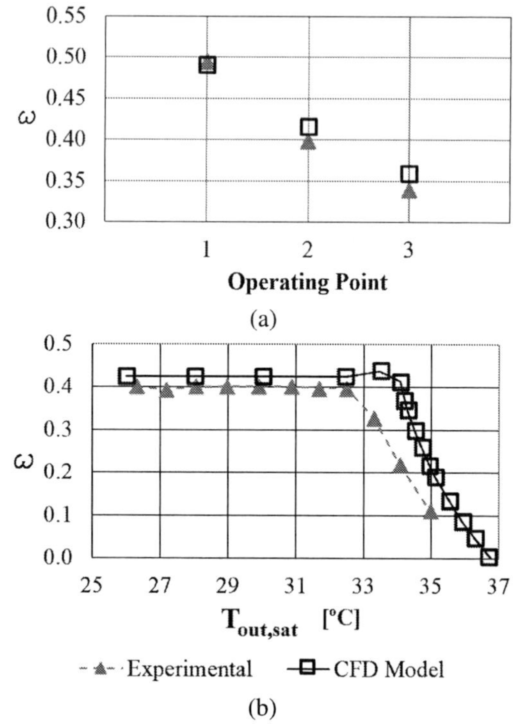

Figure 1: Validation of the ejector CFD model versus the experimental data of Garcia del Valle et $al.$ [7]: (a) Entrainment ratio for three inlet conditions; (b) Entrainment ratio versus outlet saturation temperature for fixed inlet conditions corresponding to OP2. Results obtained for R134a.

NUMERICAL MODELING

The governing equations for a steady state, compressible, 2D axisymmetric flow are solved using a finite volume approach. The advective and diffusive terms are discretized using second-order upwind and second-order central difference schemes, respectively. Gradients are evaluated using a least-square procedure. The resulting system of algebraic equations is solved using a full coupled pressure-based solver. Turbulence effects are modeled using the $k-\omega$ SST model. Fluid properties are evaluated using the NIST Database. The present model is fully described in [8] and showed better performances compared to other combinations of turbulence and thermodynamic modelings.

A structured grid composed of 650000 elements is sufficient here to get grid independent solutions, while guaranteeing a wall coordinate y^+ lower than 1. Total pressure and temperature are imposed at the inlets and static pressure is fixed at the outlet. Walls are considered as adiabatic with no-slip.

The CFD model is validated by comparing the predicted entrainment ratio to the experimental results of Garcia del Valle et $al.$ [7] for pure R134a. The entrainment ratio ω relates the secondary to primary mass flows. Figure 1a shows the change in ω for conditions given in Table 1. At OP1, the deviation is within 2% of the experimental value, increasing to about 5% for OP3. Figure 1b compares the experimental [7] and calculated ejector operating curves for OP2. The difference between the predicted and experimental critical point temperature is less than 0.5%, showing that the model is thus capable of correctly estimating the entrainment ratio and the critical operating point of the ejector.

EJECTOR PERFORMANCE ANALYSIS

Table 3 shows the variation in entrainment ratio when using R134a or R1234yf at fixed operating conditions (OP1 to OP3). The change from R134a to R1234yf implies a slight decrease of 4% to 6.9%. For R1234ze(E), single-phase operation is maintained only at OP1, and hence is not considered here. The effect of changing the working fluid while keeping $T_{primary}$ and the same pressure ratios is also shown for the three pure refrigerants (OP1' to OP3'). A slight decrease of ω is also observed when changing from R134a to R1234yf, although the difference is weaker compared to the previous cases. On the other hand, using R1234ze(E) leads to an increase in ω of about 3.5% in average.

Figure 2a presents the entrainment ratio as a function of the mass fraction of R134a in both HFO-R134a mixtures. For a mixture with 20% of R134a, the R1234yf mixture provides a higher entrainment ratio than the R1234ze(E) mixture at OP1. By increasing the percentage of R134a in both mixtures, ω increases. When the proportion of R134a reaches 60%, the

Table 3: Comparison in terms of entrainment ratio for R134a, R1234yf and R1234ze(E) and six operating conditions.

	R134a	R1234yf	R1234ze(E)
OP1	0.5126	0.4919 (-4.0%)	-
OP2	0.4364	0.4165 (-4.6%)	-
OP3	0.3804	0.3543 (-6.9%)	-
OP1'	0.5126	0.5044 (-1.6%)	0.5201 (+3.1%)
OP2'	0.4364	0.4297 (-1.5%)	0.4442 (+3.4%)
OP3'	0.3804	0.3752 (-1.4%)	0.3907 (+4.1%)

two mixtures have the same value of ω. Increasing the R134a in the mixture can improve ω for R1234ze(E) from 0.45 to 0.5 and for R1234yf from 0.48 to 0.5. The differences being not quite large, R1234yf appears as a better potential substitute of R134a in future refrigeration systems than R1234ze(E).

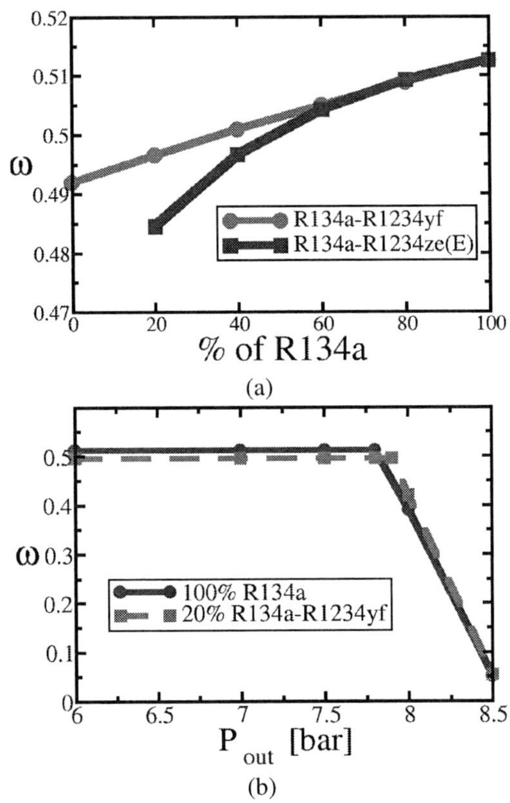

Figure 2: (a) Variation in ejector entrainment ratio with the percentage of R134a for mixtures of R134a-R1234yf and R134a-R1234ze(E); (b) Variation of entrainment ratio versus outlet pressure between R134a and mixture R134a-R1234yf. Results obtained at OP1.

Figure 2b compares the entrainment ratio ω as a function of the outlet pressure for pure R134a and a mixture composed of 20% R134a - 80% R1234yf. The R134a provides a slightly higher entrainment ratio (+4.0%) at on-design conditions compared to the R1234yf-R134a mixture. However, the critical outlet pressure is slightly higher for the R134a-R1234yf mixture ($P_{out,c} = 7.9$ bars, +1.3% compared to the case with pure R134a) expanding the range of on-design conditions. The limit critical outlet pressure remains the same ($P_{out,lc} = 8.55$ bars) for both fluids.

Figures 3a and b show iso-contours of the density gradient and Mach number along the ejector for 40% of R134a in a R134a-R1234ze(E) mixture at OP1. The iso-contours of the density gradient enable to visualize more easily the intensity and the structure of the shock train as well as relaxation waves, which appear in the divergent part of the primary nozzle. A train of oblique relaxation waves appears at the throat of the primary nozzle and gets progressively weaker when moving to the outlet. A second train of oblique relaxation waves appear also at the outlet section of the primary nozzle. Reflexions with the wall of the constant section area may be also observed. From the contours of the density gradient, the train of oblique shock-waves is also clearly visible in the second half of the constant section together with Mach disks. Figure 3b presents the Mach number distribution for the same case and highlights in particular the sonic line (white line) for which $Ma = 1$. It confirms two former results: for the primary fluid, the sonic condition is obtained just after the throat and the mixing between the primary and secondary fluids is mainly due to the shear stress between the two flows as demonstrated by Croquer et $al.$ [9]. Note that similar behaviors have been obtained for the other mixtures. Increasing the percentage of R134a in the mixture just shifts the shock train upstream in the constant section area for R1234yf but it does not change its position for R1234ze(E). The intensity of the shock does not vary monotonously with the proportion of HFO (not shown here).

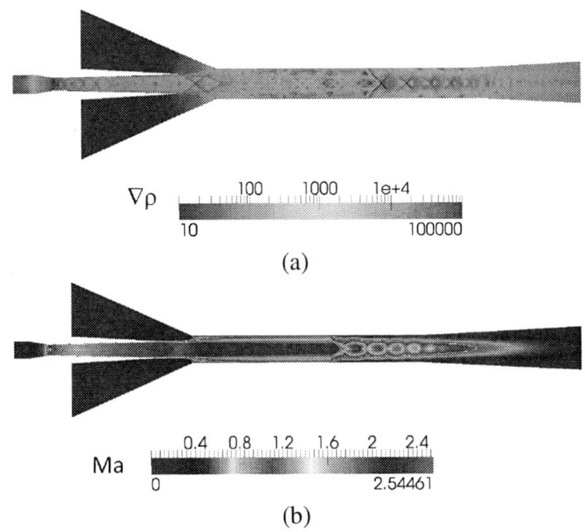

Figure 3: Iso-contours along the ejector of the (a) density gradient $\nabla\rho$ and (b) Mach number Ma for 40% of R134a in a R134a-R1234ze(E) mixture. Results obtained at OP1.

An interesting thermodynamic quantity to help improving the design of future ejectors is the ejector exergetic efficiency η_{χ}. CFD can easily provide more accurate values of η for each part of the ejector, which are necessary for the development of more accurate 1D models. The ejector exergetic efficiency is defined as the exergy ratio between the outlet and inlets as:

$$\eta_\chi = \frac{\dot{\chi}_{out}}{\dot{\chi}_{primary,in} + \dot{\chi}_{secondary,in}} \quad (1)$$

where $\dot{\chi}$ is the total stream exergy defined as:

$$\dot{\chi} = \dot{m}[(h - h_o) - T_o\,(s - s_o)] \quad (2)$$

where the index 0 refers to the dead state, defined here as a saturated vapor at the evaporator temperature and h refers to the total enthalpy (including the kinetic energy).
The exergy efficiency of the ejector is shown in Table 4 for all the studied operating conditions. In general, the ejector shows better performance with R134a whatever the conditions, with a minimal value at OP1 of 56% (resp. 55%) for R134a (resp. R1234yf). The exergy efficiency for both fluids increases by increasing the primary inlet temperature. At OP3, the maximal exergy efficiency reaches 59.5% (resp. 58.7%) for R134a (resp. R1234yf). When the same pressure ratios are kept constant, the ejector shows the worst performance with R134a than with any of the HFOs. For R1234yf and R1234ze(E), the exergy efficiency increases respectively by 2% and 1% in average for OP1' to OP3'. The maximum value (60.8%) is obtained with R1234yf at OP3' (60.1% for R1234ze(E)), whereas the minimum value (56.1%) is observed for R134a at OP1'. Overall, the exergetic efficiency of the ejector lies in the range $54.7\% - 60.8\%$, and no clear tendency was observed among all the considered operating conditions and fluids.

Table 4: Ejector exergy efficiency.

	Exergy Efficiency η_χ [−]		
	R134a	R1234yf	R1324ze(E)
OP1	0.5606	0.5475 (-2.4%)	-
OP2	0.5785	0.5719 (-1.1%)	-
OP3	0.5952	0.5874 (-1.3%)	-
OP1'	0.5606	0.5700 (+1.6%)	0.5648 (+0.7%)
OP2'	0.5785	0.5900 (+2.0%)	0.5842 (+1.0%)
OP3'	0.5952	0.6084 (+2.2%)	0.6010 (+1.0%)

CONCLUSIONS

A numerical analysis was carried out on the effect of substituting the working fluid of an ejector (R134a) with new low-GWP HFO refrigerants (R1234yf and R1234ze(E)) for different operating conditions. Results were obtained via a validated CFD model of the supersonic ejector.

Given the resemblance in properties among the three fluids, no noticeable differences in terms of the ejector performances were observed. Ejector entrainment ratio varied within a 5% range for all the considered operating conditions and refrigerants without significant changes in the flow structure. Moreover, the on-design operating limiting pressure for R1234yf is slightly larger than for R134a. As a result, the R1234yf can expand the range of on-design conditions and appears then as a good substitute of HFC. When the primary inlet temperature and pressure ratios across the ejector are kept, an increase of about 2 points in entrainment ratio is observed with R1234ze(E) relative to R134a.

ACKNOWLEDGMENTS

This work is part of the NSERC chair on industrial energy efficiency established at Université de Sherbrooke in 2014 with the support of Hydro-Québec, Ressources Naturelles Canada and Rio Tinto Alcan. All calculations have been done using the computational resources of Compute Canada.

REFERENCES

[1] Besagni, G., Mereu, R. and Inzoli, F. 2016, "Ejector refrigeration: a comprehensive review," *Renewable and Sustainable Energy Reviews*, Vol(53), pp. 373-407.

[2] The National Archives 2006, "Regulation (EC) No 842/2006 of the European Parliament and the Council of 17 May 2006 on certain fluorinated greenhouse gases."

[3] Wang, C.C. 2014, "System performance of R-1234yf refrigerant in air-conditioning and heat pump system - an overview of current status," *Applied Thermal Engineering*, Vol(73), pp. 1412-1420.

[4] Mota-Babiloni, A., Navarro-Esbri, J., Molès, F., Barragán Cervera, A., Peris, B. and Verdú, G. 2016, "A review of refrigerant R1234ze(E) recent investigations," *Applied Thermal Engineering*, Vol(95), pp. 211-222.

[5] Lawrence, N. and Elbel, S. 2012, "Experiment and analytical investigation of automative ejector air-conditioning cycles using low-pressure refrigerants," *International Refrigeration and Air Conditioning Conference*, Purdue, USA, pp. 1-10.

[6] Smierciew, K., Gagan, J., Butrymowicz, D., Lukaszuk, M. and Kubiczek, H. 2017, "Experimental investigation of the first prototype ejector refrigeration system with HFO-1234ze(E)," *Applied Thermal Engineering*, Vol(110), pp. 115-125.

[7] García del Valle, J., Saíz Jabardo, J.M., Castro Ruiz, F. and San José Alonso, J.F. 2014, "An experimental investigation of a R134a ejector refrigeration system," *International Journal of Refrigeration*, Vol(46), pp. 105-113.

[8] Croquer, S., Poncet, S. and Aidoun, Z. 2016, "Turbulence modeling of a single-phase R134a supersonic ejector. Part 1: Numerical benchmark," *International Journal of Refrigeration*, Vol(61), pp. 140-152.

[9] Croquer, S., Poncet, S. and Aidoun, Z. 2016, "Turbulence modeling of a single-phase R134a supersonic ejector. Part 2: Local flow structure and exergy analysis," *International Journal of Refrigeration*, Vol(61), pp. 153-165.

ANALYTICAL SOLUTION OF DRAG REDUCTION BY CHEMICAL ADDICTIVES IN TURBULENT PIPE FLOW

Xin Zhang, Xili Duan, Yuri Muzychka
Faculty of Engineering and Applied Science
Memorial University of Newfoundland
St. John's, NL, Canada
xz4476@mun.ca

ABSTRACT

Drag reduction induced by drag-reducing agents (DRAs) has been studied for many years, but few studies can explain the mechanism of this phenomenon. Based on the FENE-P (Finitely Extensible Nonlinear Elastic-Peterlin) model, this study develops a new mathematical model to predict the upper limit of the drag reduction with polymers in a turbulent pipe flow. This model assumes that all the vortex structures disappear with addition of DRAs in turbulent flow. The maximum drag reduction by a given DRA at a given concentration can be predicted directly via several easy to measure parameters, i.e., bulk velocity of the fluid, geometry of the pipe and relaxation time of the DRA.

KEYWORDS: Drag reduction, analytical solution, turbulent flow, polymer, pipeline

INTRODUCTION

Adding a small amount of polymers can significantly reduce flow friction in pipelines, an important phenomenon first found by Toms in 1948 [1]. This critical finding has been applied in many fields, including the transportation of oil and gas, particularly in oil and gas pipelines in Alaska [2]. By adding several ppm (weight based) of drag-reducing agents (DRAs), an extra 1300m^3/h of oil was transported compared to pipelines without drag-reducing polymers. This increase accounted for more than one percent of U.S. oil consumption at that time.

Given the benefits of DRAs in the oil and gas industry, a large amount of research has been completed in this area. Lioumbas et al. [3] have studied how the non-ionic surfactant affects the flow pattern in slightly inclined pipelines. Karami et al. [4] examined the drag reduction caused by many parameters in a crude oil pipeline, i.e. diameter of pipelines, relative roughness, flow rate of oil, concentration of DRAs (this complex DRA contained several types of DRA) and temperature. In this study, the drag reduction increases with decreasing diameter and increasing temperature. Different from the previous drag reduction study, in this research, researchers use frequency of the interfacial waves instead of pressure drop to indicate the drag reduction, where a few additives can influence the flow pattern.

In addition to experimental studies, numerical simulation is another widely-used method to study drag reduction. Many researchers attempt to use simulation to validate the experimental data and explain the mechanism. Resende et al. [5] have proposed a new Low-Reynolds-number model to predict drag reduction. The main improvement of this model is a modified eddy viscosity compared to previous work by Pinho et al. [6]. Iaccarino et al. [7] introduce a Reynolds-averaged model of drag reduction based on an elliptic relaxation model.

Many numerical simulations have attempted to validate the mechanism of drag reduction, but the predicted results have not matched experimental data. This is a result of limitations of the mathematical models in commercial software, which cannot consider the complex interaction between polymers and fluids. Some researchers have tried to develop mathematical models to clarify the mechanism. Barenblatt [8] has developed a model to predict the velocity profile in drag reduction by a high-molecular-weight polymer. This model is obtained using the assumption that only velocity in the streamwise direction is taken into consideration and velocity in the other two directions are zero. In addition to this assumption, the concentration of drag-reducing polymer remains constant in turbulent flow. This model is reasonable, but the important Navier-Stokes equation (NS equation) is not included, which undermines the application in prediction of drag reduction in both experiment and real industry.

In summary, even though many works have attempted to explain the mechanism of drag reduction and establish a model to validate the mechanism, few can relate the experimental data and theory, especially when the DRAs' viscosity and elasticity are incorporated. In this work, a new solution of drag reduction based on the Navier-Stokes equation and some basic assumptions is presented to provide an analytical solution to predict the maximum drag reduction when all the turbulent structures disappear. Moreover, this model is also validated by previous experiment data, which makes the model more reliable than the former ones.

MODEL FORMULATION

In a 2-D turbulent flow, the governing equation (with Reynolds transformation) for drag reduction by DRAs (including polymers and surfactants) is as follows [9]:

$$\frac{\partial \vec{u}}{\partial t}+(\vec{u}\cdot\nabla)\vec{u}=-\nabla p+\frac{\beta}{Re_b}\nabla^2\vec{u}+\frac{1-\beta}{Re_b}\nabla\vec{T} \quad (1)$$

where $\vec{u}$ is the velocity vector in turbulent flow; ∇p is the pressure drop along a certain direction; $\vec{T}$ is the stress tensor brought by DRAs; Re_b is the Reynolds number based on bulk velocity, defined as $\frac{dU_b\rho}{\mu}$; β is the ratio of the viscosity of solvent to the total viscosity after polymer is added at zero shear rate. β is defined as follows:

$$\beta=\frac{\mu_S}{\mu_S+\mu_{P,0}} \quad (2)$$

where μ_S and $\mu_{P,0}$ represent the viscosity of solvent and polymers added in at zero shear rate. Tensor $\vec{T}$ is defined by Bird et al. [10] in the FENE-P model as:

$$\vec{T}=\frac{1}{W_i}\left(\frac{\vec{c}}{1-\frac{\vec{c}^2}{l^2}}-\vec{I}\right) \quad (3)$$

where Wi is Weissenberg number, defined as:

$$W_i=\lambda\dot{r} \quad (4)$$

where λ is the relaxation time, from the disorder state to the equilibrium state of the polymer; l is the maximum length of the polymer in the solution; $\dot{r}$ is the shear rate in the turbulent flow. I is a unit tensor; In equation (3), $\vec{c}$ is conformation tensor of the polymer, defined as:

$$\vec{c}=QQ \quad (5)$$

For tensor $\vec{c}$, the governing equation is as following:

$$\frac{\partial \vec{c}}{\partial t}+(\vec{u}\cdot\nabla)\vec{c}=\vec{c}\cdot(\nabla\vec{u})+(\nabla\vec{u})^T\cdot\vec{c}-\vec{T} \quad (6)$$

As many unknown terms are contained in the governing equations, some reasonable assumptions are made to simplify the equations above. Previous publications have proved that DRAs can reduce the fluctuation velocity or Reynolds stress [11]. If all the fluctuation velocities in two directions, x and y, disappear, the fluctuation velocity and the Reynolds stress are 0. Meanwhile, other terms, i.e. $\frac{\partial \vec{c}}{\partial t}$ and $\frac{\partial \vec{u}}{\partial t}$ are also 0. Kostic [12] showed that there was a laminarization trend of turbulent flows after DRA was added. If all turbulent structures disappear, then the turbulent flow becomes laminar. This assumption has also been partially proven by experiments. Shao et al. [13] used PIV to indicate that fluctuating velocity (Reynolds stress) decreases when DRAs are added. In the extreme situation, it can be seen that all turbulent structures disappear, and the drag reduction is larger than the measured drag reduction.

As shown by Barenblatt [8], u_y can be considered negligible as u_x is much more important in pipelines. Based on this assumption, u_y is 0. Based on the former two assumptions, equations (1), (3) and (6) can be simplified as follows:

$$\frac{dp}{dx}=\frac{\beta}{Re}\frac{d^2\left(\frac{u_x}{U_b}\right)}{dy^2}+\frac{1-\beta}{R_e}\frac{dT}{dx} \quad (7)$$

$$\frac{u_x}{U_b}\frac{dc}{dx}=-T=-\frac{1}{W_i}\left(\frac{c}{1-\frac{c^2}{l^2}}-1\right) \quad (8)$$

Other assumption involves approximation of the velocity profile to a second order polynomial distribution in the laminarized turbulent flow with DRAs by Kostic [12].

$$\frac{u_x}{U_b}=ay^2+by+c \quad (9)$$

In a pipeline, three boundary conditions can be used to obtain the exact equation.

$$u_x(R)=u_x(-R)=0 \quad (10)$$

$$U_b=\frac{\int_{-R}^{R}u_x}{2R} \quad (11)$$

Therefore, the velocity profile is as following:

$$\frac{u_x}{U_b}=-\frac{3}{2R^2}y^2+\frac{3}{2} \quad (12)$$

According to equation (7), the drop pressure without polymers can be represented as

$$\frac{dp}{dx}=\frac{1}{Re}\frac{d^2(\frac{u_x}{U_b})}{dy^2} \quad (13)$$

Based on the definition of drag reduction ($DR\%$)

$$DR\%=\frac{\left(\frac{dp}{dx}\right)_{without}-\left(\frac{dp}{dx}\right)_{with}}{\left(\frac{dp}{dx}\right)_{without}} \quad (14)$$

$\left(\frac{dp}{dx}\right)_{with}$ and $\left(\frac{dp}{dx}\right)_{without}$ represent the drop pressure with and without polymers, respectively. Using equations (7) and (11), equation (12) can be written as:

$$DR\% = \frac{\left(\frac{dp}{dx}\right)_{without} - \left(\frac{dp}{dx}\right)_{with}}{\left(\frac{dp}{dx}\right)_{without}} = (1-\beta)\left(1 - \frac{\frac{dT}{dx}}{\frac{d^2\left(\frac{u_x}{U_b}\right)}{dy^2}}\right) \quad (15)$$

Using the chain rule, $\frac{dT}{dx}$ can be written as:

$$\frac{dT}{dx} = \frac{dT}{dc}\frac{dc}{dx} = \frac{1}{\frac{u_x}{U_b}W_i^2} \frac{\left(1+\frac{c^2}{l^2}\right)\left(1-c-\frac{c^2}{l^2}\right)}{\left(1-\frac{c^2}{l^2}\right)^3} \quad (16)$$

Once u_x is replaced by bulk velocity, the equation for drag reduction (*DR*) can be shown as:

$$DR\% = \frac{\left(\frac{dp}{dx}\right)_{without} - \left(\frac{dp}{dx}\right)_{with}}{\left(\frac{dp}{dx}\right)_{without}} = (1-\beta)\left(1-\frac{2R^2}{3W_i^2}\right)\frac{\left(1+\frac{c^2}{l^2}\right)^2\left(1-c-\frac{c^2}{l^2}\right)}{\left(1-\frac{c^2}{l^2}\right)^3}$$

(17)

To further simplify Eq. (17), the confirmation tensor and the maximum length of the polymer need to be investigated. *C* and *l* are less than 1 according to Eq. (18):

$$l = na = 10^5 \times 154 \times 10^{-12} \text{ pm} = 1.54 \times 10^{-5} \text{ m} \quad (18)$$

Thus, the final equation to predict the maximum *DR* is:

$$DR = (1-\beta)(1-\frac{2R^2}{3W_i^2}) \quad (19)$$

VALIDATION AND DISCUSSION

Before validating this model, several actions should be taken. The Weissenberg number can be replaced by polymer's relaxation time. According to the definition of the Weissenberg number and shear rate, the Weissenberg number can be expressed as:

$$W_i = \lambda\dot{r} = \lambda\frac{8U_b}{d} \quad (20)$$

The definition of relaxation time is [14]:

$$\lambda = \frac{\mu}{G} \quad (21)$$

where *G* is the polymer modulus measured by rheometer, which represents the polymer elasticity with a unit, Pa.

A previous publication by Japper-Jaafar et al. [15] is used to validate the model. Details can be seen in the original paper. The major result given by this model is that the DR by Eq. (19) should be larger than the DR given experimentally, since this model provides the upper limit of drag reduction. The validation results are shown in Fig. 1. In this figure, DR_{exp} and DR_{cal} represent the DR by Eq. (19) and the experimental DR.

$$DR_{exp} < DR_{cal} \quad (22)$$

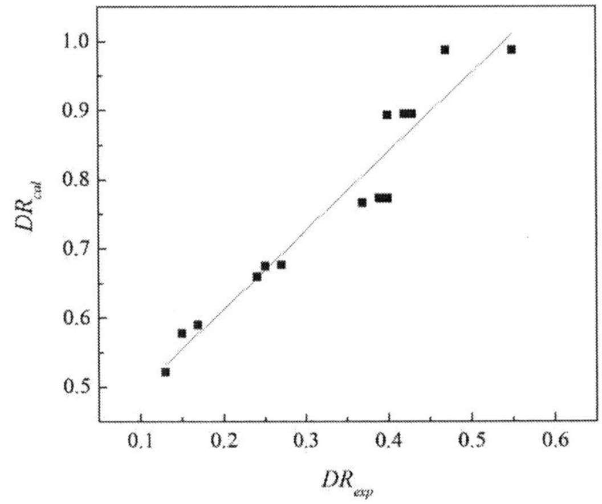

Figure 1: *DR* predicted by Eq. (19) and *DR* in the experiment

In Fig. 1, DR_{cal} is greater than DR_{exp} over the whole range. So, it is reasonable to conclude that the initial aim of this model is achieved. In addition to the qualitative analysis above, Fig. 1 shows a linear relationship between DR_{cal} and DR_{exp}. This straight line is called the "calibration curve", and can be used to predict the drag reduction.

Another method to predict drag reduction using the current model is shown in Fig. 2. For this purpose, a coefficient, α, can be introduced and defined as,

$$\alpha = \frac{DR_{exp}}{DR_{cal}} \quad (22)$$

In Fig. 2, the approximate range of α is from 0.2 to 0.6. This parameter can be used to predict the drag reduction before design or chemical addition in pipelines.

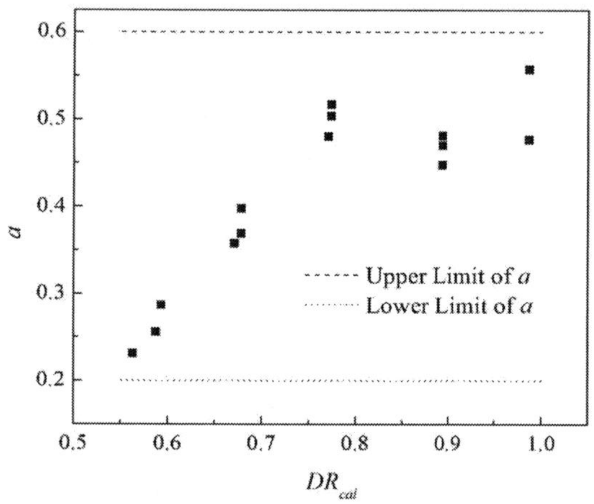

Figure 2: An empirical method to predict *DR*

CONCLUSIONS

In this work, a new model for drag reduction by DRAs is proposed, which is supported by previous experiment data. This model has the potential to be applied to predict the upper limit of drag reduction with chemical additives in turbulent pipe flows. This model could also be used to explain the mechanism of drag reduction by DRAs.

ACKNOWLEDGMENTS

Financial support for this research by Research and Development Corporation (RDC) and the Natural Sciences and Engineering Research Council of Canada (NSERC) is gratefully appreciated.

REFERENCES

1. Toms, B. A. 1948, "Some observations on the flow of linear polymer solutions through straight tubes at large Reynolds numbers", *1st International Congress on Rheology*, Amsterdam, Holland, pp. 135-141.
2. Burger, E. D., Munk, W. R., and Wahl, H. A. 1982, "Flow increase in the Trans Alaska Pipeline through use of a polymeric drag-reducing additive", *Journal of Petroleum Technology*, Vol(34), pp. 377-386.
3. Lioumbas, J. S., Kolimenos, C., and Paras, S. V. 2009, "Liquid layer characteristics in gas–liquid flow in slightly inclined pipes: Effect of non-ionic surfactant additives", *Chemical Engineering Science*, Vol(64), pp. 5162-5172.
4. Karami, H. R., and Mowla, D. 2012, "Investigation of the effects of various parameters on pressure drop reduction in crude oil pipelines by drag reducing agents", *Journal of Non-Newtonian Fluid Mechanics*, Vol(177), pp. 37-45.
5. Resende, P. R., Kim, K., Younis, B. A., Sureshkumar, R., and Pinho, F. T. 2011, "A FENE-P k–ε turbulence model for low and intermediate regimes of polymer-induced drag reduction", *Journal of Non-Newtonian Fluid Mechanics*, Vol(166), pp. 639-660.
6. Pinho, F. T., Li, C. F., Younis, B. A., and Sureshkumar, R. 2008, "A low Reynolds number turbulence closure for viscoelastic fluids", *Journal of Non-Newtonian Fluid Mechanics*, Vol(154), pp. 89-108.
7. Iaccarino, G., Shaqfeh, E. S., and Dubief, Y. 2010, "Reynolds-averaged modeling of polymer drag reduction in turbulent flows", *Journal of Non-Newtonian Fluid Mechanics*, Vol(165), pp. 376-384.
8. Barenblatt, G.I., 2008. "A mathematical model of turbulent drag reduction by high-molecular-weight polymeric additives in a shear flow", *Physics of Fluids*, Vol(20), p. 091702.
9. Housiadas, K.D. and Beris, A.N., 2004, "An efficient fully implicit spectral scheme for DNS of turbulent viscoelastic channel flow", *Journal of Non-Newtonian Fluid Mechanics*, Vol(122), pp. 243-262.
10. Bird, R., Curtiss, C., Armstrong, R. and Hassager, O. 1987, *Dynamics of Polymer Liquids Vol. 2 Kinetic Theory*. Wiley, New York.
11. Warholic, M.D., Heist, D.K., Katcher, M. and Hanratty, T.J., 2001, "A study with particle-image velocimetry of the influence of drag-reducing polymers on the structure of turbulence", *Experiments in fluids*, Vol(31), pp. 474-483.
12. Kostic, M., 1994. "The ultimate asymptotes and possible causes of friction drag and heat transfer reduction phenomena", *Journal of Energy*, Vol(16), pp. 01-14.
13. Shao, X.M., Lin, J.Z., Wu, T. and Li, Y.L., 2002. Experimental research on drag reduction by polymer additives in a turbulent pipe flow, *Canadian Journal of Chemical Engineering*, Vol(80), pp. 293-298.
14. Malkin, A.I., Malkin, A.Y. and Isayev, A.I. 2006, *Rheology: concepts, methods, and applications*. ChemTec Publishing, Toronto.
15. Japper-Jaafar, A., Escudier, M.P. and Poole, R.J., 2009. Turbulent pipe flow of a drag-reducing rigid "rod-like" polymer solution, *Journal of Non-Newtonian Fluid Mechanics*, Vol(161), pp. 86-93.

A MATHEMATICAL MODEL FOR LARGE EDDY SIMULATION OF TWO-PHASE TWO-WAY COUPLED TURBULENT DILUTED BUBBLY FLOWS

Mohammad Haji Mohammadi
School of Engineering
University of British Columbia, Okanagan
Kelowna, BC, Canada
mhajit@alumni.ubc.ca

Joshua Brinkerhoff
School of Engineering,
University of British Columbia, Okanagan
Kelowna, BC, Canada
Joshua.Brinkerhoff@ubc.ca

ABSTRACT

An efficient mathematical formulation for turbulent, two-phase, two-way coupled diluted bubbly flows is presented based on large eddy simulation and a purely Eulerian (two-fluid approach) framework. Through a filtering process, the microscopic governing equations are converted to macroscopic equations via spatial averaging of solution variables, leading to a loosely coupled form suitable for dilute turbulent bubbly flow mixtures. The resulting mathematical model is based on four distinct sets of equations, namely mixture momentum balance, pressure Poisson equation, and momentum and mass balances of gas phase, accompanied by an equation of state. This mathematical formulation provides an efficient numerical procedure for two-way coupling of bubbly flows at low gas holdups. The formulation is validated against a fully turbulent bubble column test case via comparison to available experimental results.

KEYWORDS: Large eddy simulation, Eulerian-Eulerian, bubbly flows, turbulence, Smagorinsky model, bubble column

INTRODUCTION

Bubbly flows are encountered in a broad range of environmental, natural, and industrial problems. For industrial applications with millions of bubbles, employing the direct numerical simulation (DNS) [10] and Eulerian-Lagrangian (E-L) tracking [6] method is not feasible due to excessive computational demands. Eulerian-Eulerian (E-E) methods have shown promising results in terms of the prediction accuracy of flow fields and computational cost for industrial applications. E-E methods are based on converting the microscopic equations for both of the phases to macroscopic set of equation by averaging the equations (integrating on computational cells) in space. Separate momentum and mass balance equations have to be solved for each phase. The averaging process introduces terms known as force terms and stress terms in the momentum equation, hence modeling of these terms is inevitable. Modeling of local turbulence has been a major challenge in bubbly flows. Reynolds-averaged Navier-Stokes (RANS)-based models are often based on the assumption of isotropic turbulence, despite the fact that bubbly flows show high anisotropy in velocity fluctuations especially in the gravitational direction [3]. Large-eddy simulation (LES) resolves the larger scales while eddies of smaller scales and the interaction of bubbles with the surrounding liquid turbulence in bubbly flows are modeled. In LES, this ad-hoc assumption is not made for resolved scales, and isotropic turbulence is only assumed for the unresolved scales.

Fewer works have been reported on the large eddy simulation of bubbly flows than RANS calculations, since LES of bubbly flows are more recent. The earliest work on the large eddy simulation of bubble plumes were performed by Smith and Mileli [8]. They employed LES to simulate the liquid phase and Random Dispersion Model (RDM) for the gas phase to describe the turbulent dispersion of bubbles in a two-phase buoyant plume. A relatively coarse cylindrical grid was used to simulate the axisymmetric cylindrical bubble column. Similarly, Deen at. al. [1] carried out LES of a bubbly flow with the commercial code ANSYS CFX. They employed the constant Smagorinsky model for eddy viscosity evaluation and compared their results with a k-ϵ turbulence model. Even though wall models were not used for the LES, their results proved improvement for LES compared to k- ϵ model. In a work by Dhotre et. al. [4] E-E LES of a bubble column was carried out with a Smagorinsky model. In their work, they varied the model constant Smagorinsky coefficient value and found that model constant of C_s=0.12 yields solutions similar to the LES solution with the Germano wall model procedure.

Most of the works with the aim of numerical simulation of bubbly flows are carried out with commercial CFD packages that may not be based on efficient numerical algorithms. Moreover, the continuous advances in the computational power may allow for evolution of higher accurate numerical algorithms/schemes. In this work, we have developed an efficient mathematical model for the large eddy simulation of turbulent bubbly flows at low gas holdups. The filtered form of equations are employed and some simplifying assumptions, based on the work of Sokolichin et al. [9], are made that enables efficient LES of bubbly flows.

GOVERNING EQUATIONS

Defining the filtering operation

$$f(x) = \int_{\Omega} G(r, x) f(x - r, t) d\Omega \quad [1]$$

and applying it to the component weighted microscopic Navier-Stokes equations yields

$$\frac{\partial}{\partial t}(\alpha^k \rho^k) + \nabla \cdot (\alpha^k \rho^k \tilde{u}^k) = \Gamma^k \quad [2]$$

$$\frac{\partial}{\partial t}(\alpha^k \rho^k \tilde{u}^k) + \nabla \cdot (\alpha^k \rho^k \tilde{u}^k \tilde{u}^k) = \nabla \cdot [\alpha^k(\tilde{\Pi}^k - \tau^k)] + \alpha^k \rho^k g + M^k + \sigma^k \quad [3]$$

where G is the filter function, α^k is the filtered void fraction of phase k, ρ^k is the filtered density, $\tilde{u}^k$ is the filtered velocity field of phase k, Γ^k is the mass transfer term, σ^k is the interfacial momentum source term, τ^k is the subgrid stress tensor for unresolved scales that needs to be modeled for both phases and Π is the total molecular stress. M^k is the interfacial momentum source term due to interfacial forces, which represents the rate of drain or gain of momentum through the interface by the movement of two phases sharing an interface. The interfacial momentum source term is the most important term that needs to be modeled to quantify the bubble plume behavior. In this work, we have considered the four main components of the interfacial momentum term as the drag force, lift force, virtual mass force, and turbulent dispersion force. The mathematical expressions of these forces are provided and explained in detail by Drew and Passman [5]. Compared to other terms, some terms may be small in two phase bubbly flows and are neglected, i.e. $\Gamma^k = 0$ and $\sigma^k = 0$.

Because equations (2) and (3) are typically very stiff, we employ the mixture equations to solve for the liquid phase. The mixture equations are derived by summing up the mass and momentum balance equations, as below

$$\frac{\partial}{\partial t}(\rho_m) + \nabla \cdot (\rho_m u_m) = 0 \quad [4]$$

$$\frac{\partial}{\partial t}(\rho_m u_m) + \nabla \cdot (\rho_m u_m u_m) = -\nabla \cdot \tau_m - \nabla p + \rho_m g - \nabla \cdot \left[\alpha_G \rho_G \left(1 - \alpha_G \frac{\rho_g}{\rho_m}\right) u_r u_r\right] \quad [5]$$

where u_r is the relative or slip velocity defined as $u_r = u_G - u_L$. The mixture variables are mixture density, $\rho_m = \alpha_G \rho_G + \alpha_L \rho_L$, mixture momentum, $\rho_m u_m = \alpha_G \rho_G u_G + \alpha_L \rho_L u_L$, and mixture shear stress $\tau_m = \alpha_G \tau_G + \alpha_L \tau_L$.

TURBULENCE

The turbulence in the two phase bubbly flows is modeled as

$$\tilde{\tau}^k = -\mu_{eff}\left[\nabla \tilde{u}^k + (\nabla \tilde{u}^k)^T - \frac{2}{3} I (\nabla \cdot \tilde{u}^k)\right] \quad [6]$$

where μ_{eff} is the effective turbulent viscosity. Unlike turbulence in single phase flows, the turbulence in two phase bubbly flows is the combination of shear-induced turbulence and bubble-induced turbulence. The shear-induced turbulence is due to the local action of eddies on the surrounding flow field, typically defined as below

$$\tau_{ij}^k = -2\mu_t^k \tilde{S}_{ij}^k + \frac{1}{3}\delta_{ij}\tau_{rr}^k \quad [7]$$

where μ_t^k is called the turbulent viscosity evaluated by the Smagorinsky model,

$$\mu_t^k = \rho^k (C_s \tilde{\Delta})^2 \sqrt{2\tilde{S}_{ij}\tilde{S}_{ij}} \quad [8]$$

and C_s is the Smagorinsky coefficient, calculated by the dynamic procedure and $\tilde{\Delta}$ is the grid filter size. The bubble-induced turbulence is a result of interaction of bubbles with turbulent eddies. In this work we employ the model of Milleli et. al. [11],

$$\mu_{BI} = C_s \tilde{\Delta} \rho_L \alpha_G |u_G - u_L| \quad [9]$$

where u_G and u_L are the gas and liquid velocity fields, respectively.

MODEL SIMPLIFICATION

As mentioned earlier, equations (2) and (3) are harder to solve numerically than single phase flows since they have to be solved in a coupled manner. As a result, the pressure equation is non-linear and requires iterations to converge. There are certain valid assumptions for dilute bubbly flows that can make couple equations in a looser manner. Owing to the fact that $\rho_L \gg \rho_G$, the terms containing the ratio ρ_G / ρ_L can be ignored. Also for bubbly flows where the gas holdup is small, the mixture quantities can be approximated by the liquid quantities, i.e. $\rho_m \approx \rho_L$, $u_m \approx u_L$. This approximation is similar to the Boussinesq approximation in single phase flows. Hence, the mixture equations can be simplified to

$$\nabla . (\tilde{u}\_L) = 0 \quad [10]$$

$$\frac{\partial}{\partial t}(\rho_L \tilde{u}_L) + \nabla . (\rho_L \tilde{u}_L \tilde{u}_L) = -\nabla p - \nabla . \tau_L + (1 - \alpha_G)\rho_L g. \quad [11]$$

The simplified mixture momentum has an extra term, compared to single-phase problems. The last term on the right hand side represents the gravitational force, which is proportional to the local gas holdup. This term can be assimilated by the buoyancy effect in free convection problems. The divergence free condition set for the liquid phase greatly reduces the computational cost of the solution, since single phase flow codes can be readily employed, with some minor change for the buoyancy force, to compute the liquid phase flow field.

To account for the turbulent mixing effect in the mass balance equation of the gas phase, Sokolichin [4] introduced the turbulence dispersion effect with the eddy diffusivity $D_{turb} = \nu_L^t / S_c$, where S_c is the Schmidt number taken as unity for bubbly flows. Hence the gas mass balance equation incorporating the turbulent dispersion effect reads as

$$\frac{\partial}{\partial t}(\alpha_G \rho_G) + \nabla . (\alpha_G \rho_G u_G) = \frac{1}{Sc} \nabla . (\nu_L^t \rho_G \nabla \alpha_G) \quad [12]$$

To calculate the gas density, an equation of state is required. The ideal gas law is used to calculate the gas density,

$$\rho_G = \frac{p}{RT_0} \quad [13]$$

where R is the specific gas constant and T_0 is the absolute temperature in Kelvin.

SOLUTION PROCEDURE

Marching in time yields the time-accurate solution of the governing equations. The total procedure for each time step to converge is divided into 6 distinct steps:

1. Calculate the shear- and bubble-induced turbulence employing equations 8 and 9.

2. The liquid phase equations (10, 11) are coupled using the fractional step method to solve for the pressure and liquid velocity field.
3. The gas density field is calculated from equation 13.
4. Equation 3 is solved to obtain the gas velocity field.
5. Equation 12 is solved to calculate the gas void fraction.
6. Convergence of void fraction is assessed. The convergence criterion is $\|(\alpha_G)_{fresh}-(\alpha_G)_{initial}\|_\infty<10^{-6}$. If converged, proceed to next time step, otherwise return to step 1 and repeat with the updated void fraction field.

The above solution procedure is done iteratively, i.e. the iteration is required to satisfy the above mentioned criterion. In our work, converged solutions were obtained within 2 iterations per time step. All of the above equations are discretized via second-order schemes in space and time, except the pressure which is discretized based on second-order in space and first-order accurate in time. Fractional step method is employed to couple the pressure and liquid velocity fields where the generalized minimal residual method (GMRES) is used to solve the pressure correction equation. Central differencing scheme is employed for the advective term in the liquid momentum equation. Due to the inherent discontinuous nature of void fraction field, a proper scheme should be employed for the gas advective scheme. In this work, we have used the second-order total variation diminishing (TVD) scheme with the superbee flux limiter. The same flux limiter is also employed for the discretization of the advective term in the gas mass-balance equation. The Crank-Nicolson scheme is used for both gas mass and momentum equations to discretize the equations in time.

TEST CASE AND RESULTS

The problem under consideration is the bubble column test case, illustrated schematically in Figure 1. The test case is a benchmark problem in the two phase bubbly flow field which has been studied extensively by others [3] to validate different numerical methods. The bubble column has a height of H = 0.45 m with square cross section of W×D = 0.15×0.15 m^2. Air is introduced into the bubble column through a W×D = 3×3 cm^2 sparger at the bottom. The sparger has a 7×7 holes of 1 mm diameter which is positioned at the center of the cross section at the bottom of the bubble column. A 31x31x101 spatial grid is employed for the simulation. This grid satisfies the grid resolution set by Mileli et al. [7]. The superficial gas velocity of u_s=4.9x10^{-3} m/s is imposed at the bottom plane which corresponds to the gas jet velocity of u_g=2.806 m/s from the sparger with the void fraction of α_G = 0.0427. A constant time step size of Δt=0.001 which corresponds to CFL number of 0.5 is used for all time steps. The simulation was conducted using two computational nodes with 12 processors per node.

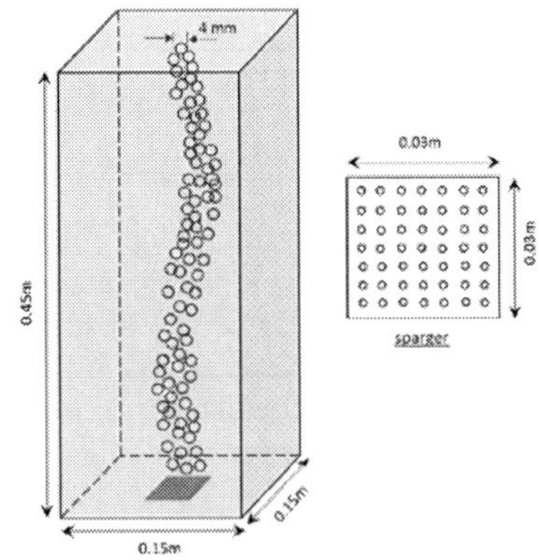

Figure 1: Schematic of the bubble column test case

Figure 2 shows the instantaneous gas axial velocity contours superimposed on the liquid velocity vectors. Large and small vortices in the liquid phase are detectable moving through the bubble column while interacting with each other. To validate the results of our mathematical formulation, the time-averaged results are compared with experimental data reported by Deen et al. [2]. Time-averaging procedure should be initiated after the flow in the bubble column reaches statistically stationary turbulence which is measured by the surrogate of the mean kinetic energy of flow field, i.e. total kinetic energy of the resolved flow. Our simulation showed that stationary turbulence is achieved after 20 seconds of the physical time. The flow fields are then time-averaged for another 250 seconds.

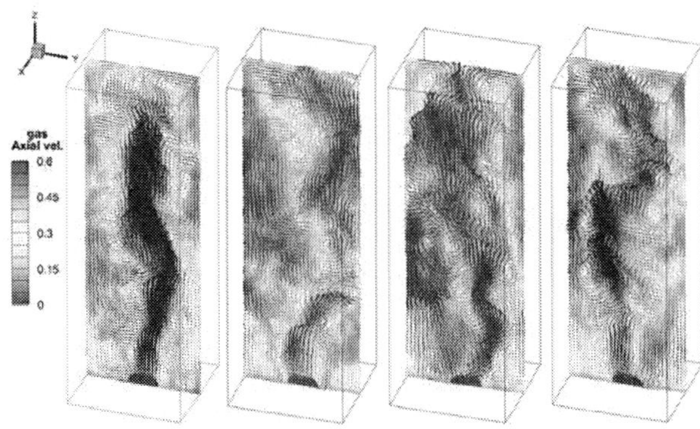

Figure 2: Instantaneous gas axial velocity contours superimposed on the liquid velocity vectors after t=55, t=70, t=85 and t=100 seconds of simulation

The time-averaged gas and liquid axial velocities are shown in Figure 3 for a cross-sectional plane through the bubble column at x=0.075 m and z=0.25 m. To quantitatively evaluate the performance and accuracy of the proposed method, a comparison is made with the experimental data reported by Deen et al. [2] and other results in the literature for the axial velocity component of both gas and liquid phases. Figure 3 shows an excellent agreement between our present results and the experimental data. Our results demonstrate improvement in predicting the flow field in bubbly flows.

To study the behavior of the second-order turbulence statistics, the turbulence kinetic energy (TKE) is plotted in Figure 4 at the middle plane at x=0.075m and at different heights of the column. The maximum TKE level is achieved at about half the column height and decays gradually as the plume travels upward toward the top free surface. This phenomenon occurs mainly due to the slip boundary condition imposed at the top face of the bubble column. Also the contours at different heights clearly shows that the spread of the turbulence occurs in circular shape, except very close to the sparger, in spite of the release of bubble from a square-shaped sparger.

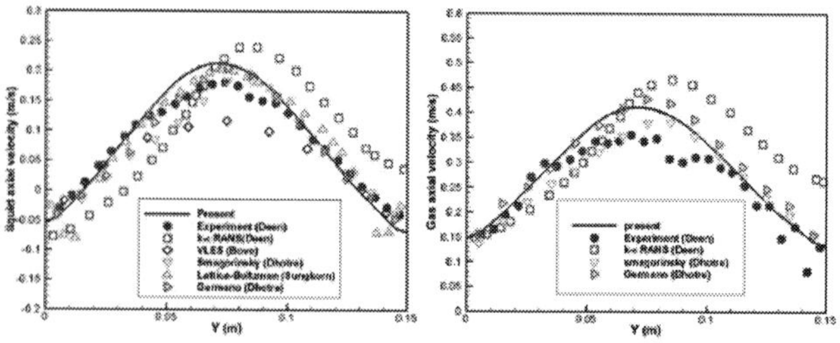

Figure 3: comparison of the gas and liquid axial velocity component on the middle plane x=0.075m at the height z=0.25m with experimental data of Deen et al. [2] and other reported results in the literature.

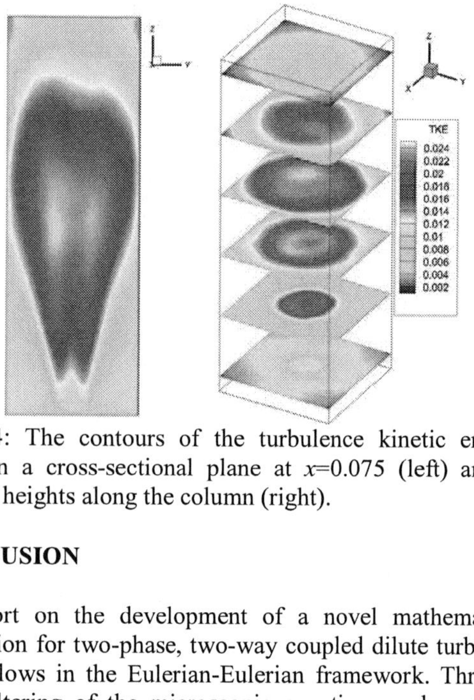

Figure 4: The contours of the turbulence kinetic energy (TKE) in a cross-sectional plane at x=0.075 (left) and at different heights along the column (right).

CONCLUSION

We report on the development of a novel mathematical formulation for two-phase, two-way coupled dilute turbulent bubbly flows in the Eulerian-Eulerian framework. Through single filtering of the microscopic equations and applying assumptions valid for dilute bubbly flows, the stiff set of governing equations are converted to a system of equations that are much easier to solve. The performance of the method is shown by simulating a turbulent problem for a standard benchmark test case of a bubble column. The time-averaged liquid and gas phase velocities achieve very good agreement with experimental measurements, suggesting that the mathematical formulation accurately captures the physical behavior of the liquid and gas flow field.

REFERENCES

1. Deen, N. G., Solberg, T. and Hjertager, B. H. 2001. Large eddy simulation of the gas–liquid flow in a square cross-sectioned bubble column, Chemical Engineering Science 56, 21: 6341–6349.
2. Deen, N. G., Hjertager, B. H. and Solberg, T. 2000. Comparison of PIV and LDA measurement methods applied to the gas-liquid flow in a bubble column, 10th international symposium on applied of laser techniques to fluid mechanics, Lisbon (Portugal), Vol. 38.
3. Dhotre, M., Deen, N., Niceno, B., Khan, Z. and Joshi, J. 2013. Large eddy simulation for dispersed bubbly flows: a review, International Journal of Chemical Engineering, vol. 2013, Article ID 343276, 22 pages.
4. Dhotre, M., Niceno, B. and Smith, B. 2008. Large eddy simulation of a bubble column using dynamic sub-grid scale model, Chemical Engineering Journal, 136: 337–348.
5. Drew, D. A. and Passman, S. L. 1999. Theory of multicomponent fluids, Springer.
6. Lau, Y., Bai, W., Deen, N. and Kuipers, J. 2014. Numerical study of bubble breakup in bubbly flows using a deterministic Euler–Lagrange framework. Chemical Engineering Science, 108: 9–22.
7. Milelli, M., Smith, B. and Lakehal, D. 2001. Large-eddy simulation of turbulent shear flows laden with bubbles, Direct and Large-Eddy Simulation IV, pp. 461–470.
8. Smith, B. and Milelli, M. 1998. An investigation of confined bubble plumes, Third International Conference on Multiphase Flow, 8–12.
9. Sokolichin, A., Eigenberger, G. and Lapin, A. 2004. Simulation of buoyancy driven bubbly flow: established simplifications and open questions, AIChE Journal, 50: 24–45.
10. Unverdi, S. O. and Tryggvason, G. 1992. A front-tracking method for viscous, incompressible, multi-fluid flows, Journal of Computational Physics, 100: 25.

Elasticity/Plasticity

ESHELBY STRESS AND ENERGY FLOW VECTOR FOR UNIFORM BODIES

Mawafag F. Alhasadi
Graduate Programme in
Mechanical Engineering
The University of Calgary
Calgary, Alberta, Canada
mfalhasa@ucalgary.ca

Marcelo Epstein
Department of Mechanical and
Manufacturing Engineering
The University of Calgary
Calgary, Alberta, Canada
mepstein@ucalgary.ca

Salvatore Federico
Department of Mechanical and
Manufacturing Engineering
The University of Calgary
Calgary, Alberta, Canada
salvatore.federico@ucalgary.ca

ABSTRACT

The theory of configurational forces is presented here within a general framework including thermo-elasticity, volumetric growth inertial effects. Inspired by the seminal works by Eshelby [1, 3] on configurational forces and by Noll [9] on material uniformity, we obtain a differential identity for the *modified Eshelby stress*, involving the *torsion* of the connection induced by the *material isomorphism*. This results in a generalisation of the work by Epstein and Maugin [4].

Moreover, we show that Eshelby's variational approach can be used to formulate not only the balance of material momentum, but also the balance of energy, as the temporal counterpart of the theory of configurational forces. The analogue of differential identity for the modified Eshelby stress is made an identity for a *modified energy flow vector*. Eventually, this leads to concluding that the driving force for material evolution of inhomogeneities is the *dynamical Eshelby stress*, while the driving force for volumetric growth is a novel *dynamical Mandel stress*.

Keywords: material uniformity, Eshelby stress, energy flow vector, volumetric growth, canonical balance of momentum, canonical balance of energy.

INTRODUCTION

In his pioneering study on the mechanics of materials with inhomogeneities, Eshelby [1] introduced the mechanical counterpart of Maxwell's energy-momentum tensor, which is now commonly referred to as the Eshelby stress. The integral of the divergence of the Eshelby stress over a region containing a defect represents the *material* or *configurational* force acting on the defect necessary to cause the *change in configuration* needed to displace the defect. In a subsequent work, Eshelby [3] showed that, for an elastic body, the Eshelby stress arises naturally from a variational approach, as the proper measure of stress in the *material* balance of linear momentum.

In another fundamental work, Noll [9] defined a uniform body as a body in which every point is made of the same material: in this definition, the energy at each pair of points can be related by a linear isomorphism (i.e., an invertible linear map) $\boldsymbol{P}$, called *material isomorphism*. Noll [9] also showed that the torsion of the connection induced by the material isomorphism captures the presence of inhomogeneities in the body. We recall that, given a connection (also referred to as covariant derivative) with Christoffel symbols Γ^A_{BC}, the torsion is the third-order tensor $\mathbf{Tor}$ with components $\mathrm{Tor}^A{}_{BC} = \Gamma^A_{BC} - \Gamma^A_{CB}$. In the case of Noll's material isomorphism $\boldsymbol{P}$, the Christoffel symbols are given by $\Gamma^A_{BC} = P^A{}_\alpha\,(\boldsymbol{P}^{-1})^\alpha{}_{B,C}$ (in the notation of [4]).

Later, Epstein and Maugin [4] succeeded in finding a natural relation between Noll's and Eshelby's theories by deriving an identity involving a "modified" Eshelby stress and the torsion of the connection of the material isomorphism, for a uniform body with inhomogeneities, in the static case, in the absence of body forces and without growth.

The objective of this research is to obtain a generalised differential identity for the modified Eshelby stress in a uniform body in a general thermoelastic framework, involving inertial

IN MEMORY OF PROF. GAETANO GIAQUINTA (1945-2016)

effects, body forces and volumetric growth. This identity involves, aside from the modified Eshelby stress and the torsion of the connection arising from the material isomorphism, thermal, growth and linear momentum terms. Then, an equivalent construction is employed to derive the differential identity of the modified energy flow vector for a uniform thermoelastic body whose Lagrangian explicitly depends on time.

THEORY OF CONFIGURATIONAL FORCES

Here, we derive the inhomogeneity (configurational) force following the basic idea presented in Eshelby's work [3]. We assume to have a thermoelastic body with Lagrangian density function

$$\begin{aligned}\mathcal{L}(X,t) &= \hat{\mathcal{L}}(\phi(X,t),\dot{\phi}(X,t),\boldsymbol{F}(X,t),\Theta(X,t),X,t),\\ \mathcal{L} &= \hat{\mathcal{L}}\circ(\phi,\dot{\phi},\boldsymbol{F},\Theta,\mathcal{X},\tau),\end{aligned} \tag{1}$$

where $\mathcal{L}$ is the Lagrangian density *field* (per unit reference volume), and $\hat{\mathcal{L}}$ is the corresponding *constitutive function*, whose arguments are the values of the configuration map ϕ (with components ϕ^a), of the velocity $\dot{\phi}$, of the deformation gradient $\boldsymbol{F}$ (with components $F^a{}_A = \phi^a{}_{,A}$) and of the absolute temperature $\Theta > 0$ is the absolute temperature, and the point X and time t. If we use the material identity map $\mathcal{X}$ (such that $\mathcal{X}(X,t) = X$, in components, $\mathcal{X}^A(X,t) = X^A$) and the time map τ (such that $\tau(X,t) = t$), we can omit the argument (X,t) everywhere, use the function composition symbol "$\circ$", and still express the explicit dependence of the Lagrangian density on the point X and time t. We shall use this notation throughout this work for all constitutive functions.

The Lagrangian density is defined as $\mathcal{L} = K - W$, i.e., the difference of the kinetic energy density and the potential energy density. The kinetic energy density (per unit reference volume) is

$$K = \hat{K}\circ(\dot{\phi},\mathcal{X},\tau) = \tfrac{1}{2}\,\hat{\rho}_R\circ(\mathcal{X},\tau)\,\dot{\phi}.\dot{\phi}, \tag{2}$$

where $\rho_R = \hat{\rho}_R\circ(\mathcal{X},\tau)$ denotes the referential mass density, which could depend explicitly not only on the point X, but also on time t, in the case of volumetric growth. The potential energy density (per unit reference volume) is assumed to be given by the sum of an elastic energy and a potential of the external body forces, i.e.,

$$\begin{aligned}W &= \hat{W}\circ(\phi,\boldsymbol{F},\Theta,\mathcal{X},\tau)\\ &= \hat{W}_{\text{el}}\circ(\boldsymbol{F},\Theta,\mathcal{X},\tau) + \hat{W}_{\text{ext}}\circ\phi.\end{aligned} \tag{3}$$

The first Piola-Kirchhoff stress $\boldsymbol{T}$ (with components $T_a{}^A$), the entropy density (per unit reference volume) S, and the external body force density (per unit reference volume) $\boldsymbol{f}$ (with components f_a) are given by

$$T_a{}^A = -\frac{\partial\hat{\mathcal{L}}}{\partial F^a{}_A}\circ(\phi,\dot{\phi},\boldsymbol{F},\Theta,\mathcal{X},\tau), \tag{4a}$$

$$S = \frac{\partial\hat{\mathcal{L}}}{\partial\Theta}\circ(\phi,\dot{\phi},\boldsymbol{F},\Theta,\mathcal{X},\tau), \tag{4b}$$

$$f_a = \frac{\partial\hat{\mathcal{L}}}{\partial\phi^a}\circ(\phi,\dot{\phi},\boldsymbol{F},\Theta,\mathcal{X},\tau). \tag{4c}$$

Assuming no mass flux but a non-zero mass source density Π, and zero "irreversible" momentum density, the balance of mass and balance of linear momentum read [5]

$$\partial_t\rho_R = \Pi, \tag{5a}$$

$$\partial_t(\rho_R\,g_{ab}\dot{\phi}^b) - \Pi\,g_{ab}\,\dot{\phi}^b = T_a{}^A{}_{|A} + f_a, \tag{5b}$$

where the vertical bar refers to the covariant differentiation with respect to the Levi-Civita connection of the material metric $\boldsymbol{G}$, and $\boldsymbol{g}$ is the spatial metric tensor. Note that, since there we assume to have no "irreversible" momentum density, the balance of linear momentum reduces to the classical one by use of the balance of mass.

Let us now take the material gradient of the Lagrangian density, i.e., the partial derivatives with respect to X^A,

$$\mathcal{L}_{,A} = \left[\hat{\mathcal{L}}\circ(\phi,\dot{\phi},\boldsymbol{F},\Theta,\mathcal{X},\tau)\right]_{,A}. \tag{6}$$

The "explicit" gradient (usually denoted $(\partial\mathcal{L}/\partial X^A)|_{\text{expl}}$, in the literature) gives the *material inhomogeneity force* or *configurational force*

$$\mathfrak{F}^{\text{inh}}_A = \frac{\partial\hat{\mathcal{L}}}{\partial\mathcal{X}^A}\circ(\phi,\dot{\phi},\boldsymbol{F},\Theta,\mathcal{X},\tau) \tag{7}$$

The final expression of the material gradient (6) is

$$\mathfrak{F}^{\text{inh}}_A = \dot{\mathcal{P}}_A + \mathcal{Z}_A - S\,\Theta_{,A} - \mathfrak{E}_A{}^B{}_{|B}, \tag{8}$$

where $\mathcal{Z}_A = \Pi\,g_{ab}\,\dot{\phi}^b\,F^a{}_A$, $\mathcal{P}_A = \rho_R\,\dot{\phi}^b\,g_{ab}\,F^a{}_A$ is the *material momentum*, and $\mathfrak{E}_A{}^B$ is the *dynamic Eshelby stress*

$$\mathfrak{E}_A{}^B = F^a{}_A\frac{\partial\hat{\mathcal{L}}}{\partial F^a{}_B}\circ(\phi,\dot{\phi},\boldsymbol{F},\Theta,\mathcal{X},\tau) - \mathcal{L}\,\delta_A{}^B, \tag{9}$$

Equation (8) is called the *balance of material momentum* within the general thermoelastic framework.

MATERIAL UNIFORMITY

A body is uniform if all points are made of the same material, i.e., if all point are pairwise materially isomorphic [9, 4]. In other words, the body is uniform if there exits a material isomorphism $\boldsymbol{P}(X,t) : T_{X_0}\mathcal{B}\to T_X\mathcal{B}$ from a fixed point X_0, called the *archetypal material point* to each point X of the body. For a uniform body, the dependence of the Lagrangian density $\hat{\mathcal{L}}$ on the point X is only through $\boldsymbol{P}$, i.e.,

$$\hat{\mathcal{L}}\circ(\phi,\dot{\phi},\boldsymbol{F},\Theta,\mathcal{X},\tau) = J_{\boldsymbol{P}}^{-1}\,\hat{\Lambda}\circ(\phi,\dot{\phi},\boldsymbol{F}\boldsymbol{P},\Theta,\tau) \tag{10}$$

where $\hat{\Lambda}$ is the Lagrangian density per unit volume of archetype, and $J_{\boldsymbol{P}} = \det(\boldsymbol{P})$.

We consider now $\hat{\mathcal{L}}$ as a function of the variables ϕ, $\dot{\phi}$, $\boldsymbol{F}$, $\boldsymbol{P}$, Θ, $\mathcal{X}$, and τ, and define the constitutive function $\widetilde{\mathcal{L}}$ such that

$$\begin{aligned}\hat{\mathcal{L}}\circ(\phi,\dot{\phi},\boldsymbol{F},\Theta,\mathcal{X},\tau) &= J_{\boldsymbol{P}}^{-1}\,\hat{\Lambda}\circ(\phi,\dot{\phi},\boldsymbol{F}\boldsymbol{P},\Theta,\mathcal{X},\tau)\\ &= \widetilde{\mathcal{L}}\circ(\phi,\dot{\phi},\boldsymbol{F},\boldsymbol{P},\Theta,\mathcal{X},\tau).\end{aligned} \tag{11}$$

Now, we differentiate $\hat{\mathcal{L}}$ with respect to $\boldsymbol{F}$ and, after some simple manipulation we obtain the identity

$$\frac{\partial \hat{\Lambda}}{\partial(\boldsymbol{F}\boldsymbol{P})} \circ (\#) = J_{\boldsymbol{P}} \frac{\partial \hat{\mathcal{L}}}{\partial \boldsymbol{F}} \circ (\#)\, \boldsymbol{P}^{-T}. \tag{12}$$

where from now on we use (#) in place of the appropriate arguments, for the sake of brevity. Now, we differentiate $\widetilde{\mathcal{L}}$ with respect to $\boldsymbol{P}$ to estimate the energy expenditure caused by a change in the material isomorphism $\boldsymbol{P}$:

$$\frac{\partial \widetilde{\mathcal{L}}}{\partial \boldsymbol{P}} \circ (\#) = \frac{\partial J_{\boldsymbol{P}}^{-1}}{\partial \boldsymbol{P}} \boldsymbol{P}^{-T} \hat{\Lambda} \circ (\#) + J_{\boldsymbol{P}}^{-1} \frac{\partial \hat{\Lambda}}{\partial(\boldsymbol{F}\boldsymbol{P})} \circ (\#) \boldsymbol{F}^{T}. \tag{13}$$

Substituting the identity (12) into Eq. (13) we have

$$\frac{\partial \widetilde{\mathcal{L}}}{\partial \boldsymbol{P}} \circ (\#) = \left(\boldsymbol{F}^{T} \frac{\partial \hat{\mathcal{L}}}{\partial \boldsymbol{F}} \circ (\#) - \mathcal{L}\, \boldsymbol{I}^{T} \right) \boldsymbol{P}^{-T} = \mathfrak{E}\, \boldsymbol{P}^{-T}. \tag{14}$$

Since $\hat{\mathcal{L}} \circ (\#) = \widetilde{\mathcal{L}} \circ (\#)$, also the material gradients are equal, i.e., $\left[\hat{\mathcal{L}} \circ (\#)\right]_{,A} = \left[\widetilde{\mathcal{L}} \circ (\#)\right]_{,A}$, and we obtain

$$\frac{\partial \hat{\mathcal{L}}}{\partial \mathcal{X}^{A}} \circ (\#) = \frac{\partial \widetilde{\mathcal{L}}}{\partial P^{B}{}_{\alpha}} \circ (\#)\, P^{B}{}_{\alpha,A} + \frac{\partial \widetilde{\mathcal{L}}}{\partial \mathcal{X}^{A}} \circ (\#). \tag{15}$$

Now, use of Eq. (7) and Eq. (14) yields

$$\mathfrak{F}^{\text{inh}}_{A} = \mathfrak{E}_{B}{}^{C} (\boldsymbol{P}^{-T})_{C}{}^{\alpha}\, P^{B}{}_{\alpha,A} + \frac{\partial \widetilde{\mathcal{L}}}{\partial \mathcal{X}^{A}} \circ (\#), \tag{16}$$

where $-(\boldsymbol{P}^{-T})_{C}{}^{\alpha}\, P^{B}{}_{\alpha,A}$ is the Christoffel symbol Γ^{B}_{CA} of the connection induced by $\boldsymbol{P}$ [2]:

$$\Gamma^{B}_{CA} = P^{B}{}_{\alpha} (\boldsymbol{P}^{-1})^{\alpha}{}_{C,A} = -(\boldsymbol{P}^{-T})_{C}{}^{\alpha}\, P^{B}{}_{\alpha,A}. \tag{17}$$

By applying the definition of configurational force (8) and the uniformity condition (10), Eq. (16) becomes

$$\mathfrak{E}_{A}{}^{B}{}_{|B} = \mathfrak{E}_{B}{}^{C}\, \Gamma^{B}_{CA} + \dot{\mathcal{P}}_{A} + \mathcal{Z}_{A} - S\, \Theta_{,A}. \tag{18}$$

The mass source term Π in $\boldsymbol{\mathcal{Z}} = \Pi\, \dot{\phi}\, \boldsymbol{g}\, \boldsymbol{F}$ can be expressed in terms of $\boldsymbol{P}$ via [5]

$$\Pi = -\rho_R\, \text{tr}(\dot{\boldsymbol{P}}\boldsymbol{P}^{-1}) = -\rho_R\, \text{tr}(\boldsymbol{L_P}), \tag{19}$$

where $\boldsymbol{L_P} = \dot{\boldsymbol{P}}\boldsymbol{P}^{-1}$ is the "inhomogeneous velocity gradient". Therefore, Eq. (18) becomes

$$\mathfrak{E}_{A}{}^{B}{}_{|B} = \mathfrak{E}_{B}{}^{C}\, \Gamma^{B}_{CA} + \dot{\mathcal{P}}_{A} + (\boldsymbol{L_P})^{A}{}_{A}\, \mathcal{P}_{A} - S\, \Theta_{,A}. \tag{20}$$

As in the static case presented by Epstein and Maugin [4], the sought differential identity applies to the *modified* dynamical Eshelby stress

$$\widetilde{\mathfrak{E}}_{A}{}^{B} = J_{\boldsymbol{P}}\, \mathfrak{E}_{A}{}^{B}, \tag{21}$$

and reads

$$\begin{aligned} \widetilde{\mathfrak{E}}_{A}{}^{B}{}_{;B} = {} & \widetilde{\mathfrak{E}}_{A}{}^{C}\, (\mathbf{Tor})^{B}{}_{CB} + \widetilde{\mathfrak{E}}_{B}{}^{C}\, (\mathbf{Tor})^{B}{}_{CA} + \dot{\widetilde{\mathcal{P}}}_{A} \\ & + (\boldsymbol{L_P})^{A}{}_{A}\, \widetilde{\mathcal{P}}_{A} - \widetilde{S}\, \Theta_{,A}. \end{aligned} \tag{22}$$

where the semi-colon ";" denotes the $\boldsymbol{P}$-covariant derivative, i.e., the covariant derivative with respect to the Christoffel symbols (17), $\widetilde{\mathcal{P}}_{A} = J_{\boldsymbol{P}}\, \mathcal{P}_{A}$ is modified linear momentum tensor, $\widetilde{S} = J_{\boldsymbol{P}}\, S$ is the modified referential entropy density, and $\mathbf{Tor}$ is the torsion of the $\boldsymbol{P}$-connection.

We remark that, in the homogeneous case, when $P^{\alpha}{}_{A} = \delta^{\alpha}{}_{A}$, Eq. (22) and Eq. (20) take the *same* form, coinciding with that found by Maugin [8]. Note also that, in the homogeneous case, there cannot be any mass source, since $\Pi = -\rho_R\, \text{tr}(\boldsymbol{L_P}) = 0$. Eq. (22) and Eq. (20) are the two possible forms of the balance of material momentum when uniformity is taken into consideration.

CANONICAL BALANCE OF ENERGY

We show that Eshelby's approach [3] for deriving the canonical balance of momentum Eq. (8) can be used here to formulate the *balance of energy*. In this case, the derivative of the Lagrangian has to be taken with respect to *time*. The canonical equation of balance of energy is derived here within the same general framework of the previous sections.

In addition to the balance of mass (5a) and the balance of linear momentum (5b), we add the balance of energy [6] in the presence of the volumetric growth source Π, i.e.,

$$\rho_R\, \partial_t(e \circ (\phi, \tau)) = \partial_t E - \Pi \frac{E}{\rho_R} = T_{a}{}^{A}\, \dot{F}^{a}{}_{A} - Q^{A}{}_{|A}, \tag{23}$$

where $\boldsymbol{Q}$ denotes the material heat flux, $e \circ (\phi, \tau)$ is the internal energy per unit mass, seen as a function of the material point X, and $E = \rho_R[e \circ (\phi, \tau)]$ is the internal energy per unit reference volume.

In the time derivative of Lagrangian density,

$$\partial_t \mathcal{L} = \partial_t \left[\hat{\mathcal{L}} \circ (\phi, \dot{\phi}, \boldsymbol{F}, \Theta, \mathcal{X}, \tau)\right]. \tag{24}$$

let us substitute Eqs. (5a), (5b) and (4a) into Eq. (24), so that we obtain the *explicit time derivative* (which would be denoted $(\partial \mathcal{L}/\partial t)|_{\text{expl}}$, in the classical notation)

$$\frac{\partial \hat{\mathcal{L}}}{\partial \tau} \circ (\#) = \Pi\, \dot{\phi}.\dot{\phi} - S\, \dot{\Theta} - \text{Div}(\boldsymbol{\mathcal{R}}) - \partial_t(K + W), \tag{25}$$

where $\boldsymbol{\mathcal{R}}$ is called the *energy flow vector* [2] and is given by

$$\boldsymbol{\mathcal{R}} = -\dot{\phi}\, \boldsymbol{T} = -\boldsymbol{T}^{T} \dot{\phi}. \tag{26}$$

By substituting (23) and (25) into Eq. (24), we have

$$\Theta\, \partial_t S + \text{Div}\,(\boldsymbol{Q}) - \Pi \frac{E}{\rho_R} = \hat{h}^{\text{ext}}, \tag{27}$$

where we defined the *external* source of energy

$$\hat{h}^{\text{ext}} = \boldsymbol{f}\, \dot{\phi} - \frac{\partial \hat{W}}{\partial \tau} \circ (\phi, \boldsymbol{F}, \Theta, \mathcal{X}, \tau). \tag{28}$$

Eq. (27) is called the *canonical balance of energy* in the general framework.

"UNIFORMITY" WITH RESPECT TO TIME

Let us consider the usual Lagrangian density (1) and let us this time impose the same uniformity condition (10), but *on time* rather than on the point. This amounts to defining, at every point X, the isomorphism $\boldsymbol{P}(X,t) : T_X\mathcal{B} \to T_X\mathcal{B}$, which maps the tangent space of X into *itself* and *evolves* with time. In other words, the dependence of $\hat{\mathcal{L}}$ on X and t is again only through $\boldsymbol{P}$ but, this time, *each point is its own archetype* [7].

Now, we consider once again $\hat{\mathcal{L}}$ along with its "version" $\widetilde{\mathcal{L}}$, in which the dependence on X and t has been replaced by a dependence on $\boldsymbol{P}$, see Eq. (11). Since $\hat{\mathcal{L}}\circ(\#) = \widetilde{\mathcal{L}}\circ(\#)$, also the time derivatives are equal, i.e., $\partial_t\big[\hat{\mathcal{L}}\circ(\#)\big] = \partial_t\big[\widetilde{\mathcal{L}}\circ(\#)\big]$. Thus, we obtain

$$\frac{\partial \hat{\mathcal{L}}}{\partial \tau} \circ (\#) = \mathfrak{E}_A{}^C\, \dot{P}^A{}_\alpha (P^{-1})^\alpha{}_C = \mathfrak{E}_A{}^C\, (\boldsymbol{L_P})^A{}_C. \quad (29)$$

Now, we note that (19) and (2) imply

$$\Pi\; \dot{\phi}.\dot{\phi} = -\mathrm{tr}(\boldsymbol{L_P})\; (2\,K), \quad (30)$$

substitute Eq. (25) into Eq. (29), and obtain the divergence of the energy flow vector $\boldsymbol{\mathcal{R}}$ as

$$\begin{aligned} \mathcal{R}^A{}_{|A} = &- 2\,K(\boldsymbol{L_P})^A{}_A - S\,\dot{\Theta} - \partial_t(K+W) \\ &- \mathfrak{E}_A{}^C\,(\boldsymbol{L_P})^A{}_C. \end{aligned} \quad (31)$$

We now rescale the energy flow vector $\mathcal{R}^A$ as

$$\widetilde{\mathcal{R}}^A = J_{\boldsymbol{P}}\,\mathcal{R}^A, \quad (32)$$

which satisfies the differential identity

$$\begin{aligned} \widetilde{\mathcal{R}}^A{}_{;A} = &\,\widetilde{\mathcal{R}}^A\,(\mathbf{Tor})^B{}_{AB} - 2\,J_{\boldsymbol{P}}\;\mathrm{tr}(\boldsymbol{L_P})\;(K) - \widetilde{S}\,\dot{\Theta} \\ &- J_{\boldsymbol{P}}\;\partial_t(K+W) - \widetilde{\mathfrak{E}}_A{}^C\,(\boldsymbol{L_P})^A{}_C. \end{aligned} \quad (33)$$

Thus, the *canonical balance of energy* in the general framework of a uniform body reads

$$\Theta\;\partial_t S + Q^A{}_{|A} - \Pi\,\frac{E}{\rho_R} = \mathfrak{E}_A{}^C\,(\boldsymbol{L_P})^A{}_C. \quad (34)$$

With Eq. (34), the Clausius-Duhem inequality [6] reads

$$\mathfrak{E}_A{}^C\,(\boldsymbol{L_P})^A{}_C + \Pi\,\frac{W}{\rho_R} - \frac{1}{\Theta}\,Q^A\,\Theta_{,A} \geq 0. \quad (35)$$

Note that, if Π vanishes, $\boldsymbol{L_P}$ is *traceless* and describes *pure remodelling*:

$$\mathfrak{E}_A{}^C\,(\boldsymbol{L_P})^A{}_C - \frac{1}{\Theta}\,Q^A\,\Theta_{,A} \geq 0. \quad (36)$$

Equation (35) can also take the form

$$\mathfrak{M}_A{}^C\,(\boldsymbol{L_P})^A{}_C + K\,(\boldsymbol{L_P})^A{}_A + \frac{1}{\Theta}\,Q^A\,\Theta_{,A} \leq 0, \quad (37)$$

where tensor $\boldsymbol{\mathfrak{M}}$, with components

$$\mathfrak{M}_A{}^B = -F^a{}_A\,\frac{\partial \hat{\mathcal{L}}}{\partial F^a{}_B} \circ (\phi, \dot{\phi}, \boldsymbol{F}, \Theta, \mathcal{X}, \tau), \quad (38)$$

is the *dynamical Mandel stress*.

SUMMARY

The theory of configurational forces can be presented within the general framework of material uniformity in *both* point and time. Not only the modified Eshelby stress, but also the modified energy flux obeys a particular differential identity involving the torsion arising from the uniformity isomorphism. Also, the canonical balance of energy derived via Eshelby's variational approach can be seen as the temporal counterpart of the theory of inhomogeneity. Finally, it has been deduced that the *dynamical Mandel stress* is the dual of the "inhomogeneity velocity gradient" in the presence of volumetric growth, and this duality describes the dissipation due to the motion of the inhomogeneities.

ACKNOWLEDGMENTS

Libyan - North American Scholarship Programme [MFA], Natural Sciences and Engineering Research Council of Canada through the NSERC Discovery Programme [ME, SF].

REFERENCES

[1] Eshelby, J. D., 1951, "The force on an elastic singularity", *Philosophical Transactions of the Royal Society A*, Vol. 244A, pp. 87-112.

[2] Eshelby, J. D., 1970, "Energy relations and the energy-momentum tensor in continuum mechanics", in *Inelastic Behaviour of Solids*, Kanninen, M. F., et al., Eds., pp. 77-115, McGraw-Hill, New York.

[3] Eshelby, J. D., 1975, "The elastic energy-momentum tensor", *Journal of Elasticity*, Vol. 5, pp. 321-335.

[4] Epstein, M., Maugin, G. A., 1990, "The Energy Momentum Tensor and Material Uniformity in Finite Elasticity", *Acta Mechanica*, Vol. 83, pp. 127-133.

[5] Epstein, M., Maugin, G. A., 2000, "Thermomechanics of volumetric growth in uniform bodies", *International Journal of Plasticity*, Vol. 16, pp. 951-978.

[6] Epstein, M., 2007, *Material Inhomogeneities and Their Evolution*, Springer, Berlin.

[7] Epstein, M., 2015, "Mathematical characterization and identification of remodeling, growth, aging and morphogenesis", *Journal of the Mechanics and Physics of Solids*, Vol. 84, pp. 72-84.

[8] Maugin, G. A., 2006, "On canonical equations of continuum thermomechanics", *Research Mechanics Communications*, Vol. 33, pp. 705-710.

[9] Noll, W., 1967, "Materially uniform bodies with inhomogeneities", *Archive for Rational Mechanics and Analysis*, Vol. 27, pp. 1-32.

Proceedings of the 25th CANCAM
London, Ontario, Canada, May 31 - June 4, 2015

AN ATOMIC-INFORMED MODEL FOR SPALL FAILURE IN MAGNESIUM

Mauricio Ponga
Department of Mechanical Engineering
University of British Columbia
Vancouver, BC, Canada
mponga@mech.ubc.ca

David Funes Rojas
Department of Mechanical Engineering
University of British Columbia
Vancouver, BC, Canada
dfunesrojas@alumni.ubc.ca

ABSTRACT

We develop an atomic-informed model for dynamic failure induced by spallation in Hexagonal Close-Packed (HCP) Magnesium (Mg) single crystals. The model is based on the maximum strain energy density that the material containing nanovoids can store before emitting cracks due to cavitation. The critical value for strain energy density is obtained directly from multiscale simulations of nanovoid growth under hydrostatic tensile loads. The model is computationally efficient, predicts the spall behavior of Mg single crystals and hardening coefficient in close agreement with experiments.

Keywords: Spallation, Magnesium

INTRODUCTION

Predictive modelling and simulation of materials under extreme and dynamic loads remain a challenging endeavor despite the significant research efforts made in the fields of atomic simulations, multiscale modeling of materials, continuum mechanics including finite element methods, etc. In the context of dynamic loads, the failure often happens due to spallation. Spall failure is characterized by the rapid thermo-mechanical nucleation, growth, and coalescence of nanovoids in a narrow plane - the so-called *spall plane* - where tensile shock waves collide together and produce a hydrostatic tensile stress state. The hydrostatic tensile stress state is characterized by exceedingly large stresses that produces the catastrophic failure of the material due to the rapid propagation of cracks [1–3].

Many efforts have been carried out in the past to understand and predict spall behavior in metals. However, due to the complexity of this phenomenon, the development of predictive micromechanics models without prescription of *ad-hoc* coefficients is challenging. Here, we develop a model for dynamic failure induced by spallation in HCP Mg single crystals using solely atomic-scale information of nanovoid growth. The model is based on the maximum strain energy density that the material containing a small porosity can store before failing. The critical value of strain energy density is obtained directly from molecular dynamic simulations carried out by [4,5] for Mg single crystals at the same strain rates and porosities. As such, the model needs only two parameters that are directly obtained from atomic calculations bypassing the need to prescribe *ad-hoc* coefficients and/or parameters. The model is computationally efficient, predicts the spall behavior of Mg single crystals and hardening coefficient in close agreement with experiments [6].

DESCRIPTION OF THE MODEL

Ponga *et al.* [4] and Grégoire and Ponga [5] have studied nanovoid growth by dislocation emission in Mg single crystals using atomic-scale simulations. In their analysis, they have found that the failure of nanovoids happens in a brittle fashion, where nanovoids emit a limited amount of dislocations before the catastrophic failure of the sample is reached. The peak stress values obtained by [4,5] are summarized in Figure 1. The results obtained by Ponga *et al.* [4] and Grégoire and Ponga [5] suggested the possibility of modeling the dynamic behavior of Mg by considering the brittle nature of the failure observed in these simulations.

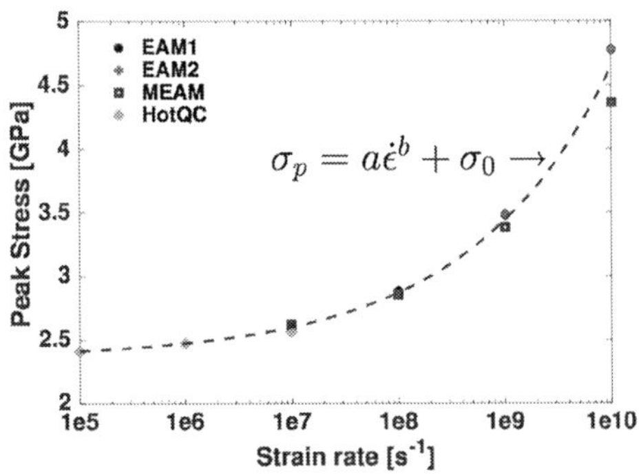

Figure 1: Relevant data collected from Refs. [4] and [5]. The evolution of the peak stress with strain rate is asymptotic for slow to moderate strain rates.

An important result obtained in Refs. [4,5] is that for moderate strain rates $\dot{\epsilon}_0 < 10^7$ s^{-1}, the peak stress observed in nanovoid simulations tends asymptotically to a certain value. This indicates that the hardening experienced by the increment of the strain rate is small for moderate strain rates. Based on these two main observations, we develop a simplified model for dynamic behavior of HCP-Mg.

In our model, we assume that the macroscopic Mg-single crystal contains a void porosity sufficiently small ($p < 0.01$ %) like the one simulated by Ponga *et al.* [4] and Grégoire and Ponga [5]. This is consistent with experimental measurements of void porosity in metals. Under these conditions, the characteristic length of nanovoids is much smaller than the length scale of the crystal. Thus, the real material with a complex spatial distribution of nanovoids is replace by a defect-free material with equivalent properties. In particular, the *equivalent material* is required to have the same elastic properties than Mg whereas the fracture properties are obtained from the maximum strain energy density that the material containing nanovoids can store before emitting cracks catastrophically. This parameter, usually defined as the critical strain energy density is directly computed from the atomic scale simulations of Ponga *et al.* [4] and Grégoire and Ponga [5] as,

$$W_{\text{critical}} = \int_0^{\epsilon_f} \sigma d\epsilon \tag{1}$$

where W_{critical} is the critical strain energy density used in our model, ϵ_f is the maximum strain before the material emits cracks and σ is the mean stress. This critical toughness is used in the macroscopic model within the context of peridynamics. In the next section, we briefly summarize the peridynamic theory and we show how the critical strain energy density are used in the peridynamics context. The motivation for using the proposed model within the peridynamics framework is that is more suitable for modelling cracks nucleation and propagation than classical continuum mechanics models like Finite Elements Methods (FEM).

PERIDYNAMICS THEORY

Classic continuum models assume that the deformation is twice continuously differentiable, which makes unable to compute spontaneous nucleation and propagation of cracks. The peridynamics theory proposed by Silling [7] eschews this issue by using a nonlocal integral formulation of the classical continuum mechanics problem. This approach upgrades some of the issues that arise when formulating fracture problems with continuum-based methods that rely on continuous fields to evaluate gradients, i.e., FEM. We also note that the choice of peridynamics is completely arbitrary, and the model is perfectly valid in other frameworks.

Let $\mathcal{R}$ be the three-dimensional domain where the body is defined and $\mathbf{x}$ be a position vector of a material particle in the reference configuration, such that $\mathbf{x} \in \mathcal{R}$. The particle in the position $\mathbf{x}$ interacts with the particles inside an open ball, a subdomain denoted as $\mathcal{R}_x$, with $\mathcal{R}_x \in \mathcal{R}$, centered in $\mathbf{x}$ with radius δ. Beyond δ the particle $\mathbf{x}$ does not interact with others. The displacements of the particles are denoted as $\mathbf{u}$, and their relatives displacements and positions with $\boldsymbol{\mu} = \mathbf{u}' - \mathbf{u}$ and $\boldsymbol{\xi} = \mathbf{x}' - \mathbf{x}$, respectively. The peridynamics equation of motion of any particle, $\mathbf{x}$, is given by

$$\rho\ddot{\mathbf{u}} = \int_{\mathcal{R}_x} \mathbf{f}(\mathbf{u}(\mathbf{x}'(t)), \mathbf{u}(\mathbf{x}(t)), \mathbf{x}'(t) - \mathbf{x}(t)) dV_{\mathbf{x}'} + \mathbf{b}(\mathbf{x}, t)$$

$$\mathcal{R}_x \in \mathcal{R} \quad t \geq 0 \tag{2}$$

Here, $\mathbf{b}$ is the body force; $\ddot{\mathbf{u}}$ is the acceleration, and ρ is the density in the reference configuration. The vector field $\mathbf{f}$ is known as *pairwise force function*, which defines the force interactions between pairs of particles and quantifies the amount of force exerted by particle $\mathbf{x}'$ on the particle $\mathbf{x}$. Its units are force per unit of volume squared and it must satisfy the next conditions: *linear admissibility condition* (Newton's third law), *angular admissibility condition* (conservation of angular momentum) and it must be zero beyond the region $\mathcal{R}_x$, see [7–9] for more details. The choice of the pairwise function is not unique and it can be change accordingly to the type of material that one is interested in simulating.

In the peridynamics theory, a material is called *microelastic* if the pairwise function satisfies the following condition

$$\int_\Gamma \mathbf{f}(\boldsymbol{\eta}, \boldsymbol{\xi}) d\boldsymbol{\eta} = 0; \quad \forall \boldsymbol{\xi} \neq 0; \quad \forall \text{ closed curve } \Gamma \in \mathcal{R} \tag{3}$$

The pairwise function, $\mathbf{f}$, is defined by a *micropotential* which depends on the relative displacement vector $\boldsymbol{\eta}$ only through the scalar distance between the deformed points, y.

$$w(\boldsymbol{\eta}, \boldsymbol{\xi}) = \hat{w}(y, \boldsymbol{\xi}) \quad \forall \boldsymbol{\eta}, \boldsymbol{\xi} \quad y = |\boldsymbol{\eta} + \boldsymbol{\xi}| \tag{4}$$

Consequently, the interaction between two particles, called *bond*, can be interpreted as a nonlinear spring. Now, differentiating with respect to y, we obtain

$$\mathbf{f}(\boldsymbol{\eta}, \boldsymbol{\xi}) = \frac{\boldsymbol{\eta} + \boldsymbol{\xi}}{|\boldsymbol{\eta} + \boldsymbol{\xi}|} \hat{f}(y, \boldsymbol{\xi}); \qquad \hat{f}(y, \boldsymbol{\xi}) = \frac{\partial \hat{w}}{\partial y} \tag{5}$$

f is the *scalar bond force*.

The value of the expression above is determined by the mode of failure. Ponga *et al.* [4] and Grégoire and Ponga [5] suggested that Mg fails in a brittle mode; hence, we adopt the *Prototype Microelastic Brittle* (PMB) model in peridynamics. The PMB states that the bonds will fail if a critical value, namely the *critical stretch* s_0, is reached. In the linearized version of the model the *scalar bond force*, $\hat{f}$, splits into two contributions: a linear scalar function, g, proportional to the bond stretch s through the constant c; and a boolean function μ, which adds to the model history dependence by becoming zero when the critical stretch is achieved. The boolean function μ is the key of peridynamics that allows for crack nucleation and propagation.

$$\hat{f}(y(t), \boldsymbol{\xi}) = g(s(t, \boldsymbol{\xi}))\mu(t, \boldsymbol{\xi}) \tag{6}$$

$$g(s) = sc \quad \forall s \quad \text{with} \quad c = \text{constant} \tag{7}$$

$$\mu(t, \boldsymbol{\xi}) = \begin{cases} 1 & s(t', \boldsymbol{\xi}) < s_0 \quad 0 \leq t' \leq t \\ 0 & \text{otherwise} \end{cases} \tag{8}$$

where s_0 depends on the minimum stretch among all bonds connected to a given point on the material. The function μ let us define the *local damage*, ϕ, at the point $\mathbf{x}$ and at the time t,

$$\phi(\mathbf{x}, t) = 1 - \frac{\int_{\mathcal{R}_x} \mu(\mathbf{x}, t, \boldsymbol{\xi}) dV_\xi}{V_\xi} \tag{9}$$

It goes from zero to one, as the damage in the subdomain $\mathcal{R}_x$ increases. The constant c can be related with the bulk modulus from classical theory - the strain energy density is $W_c = 9ks^2/2$ - by evaluating the energy and the comparison with the classical continuum theory. In peridynamics, the energy per unit volume in the body at a given point (i.e., the local strain energy density) is given by:

$$W = \frac{1}{2} \int_{\mathcal{R}_\mathbf{x}} w(\boldsymbol{\eta}, \boldsymbol{\xi}) dV_\xi = \frac{\pi c s^2 \delta^4}{4} \tag{10}$$

we obtain a relationship between the constant c and the bulk modulus, k:

$$c = \frac{18k}{\pi \delta^4} \tag{11}$$

Finally, we proceed to compute the critical bond stretch value s_0 in the PMB model. In the classical peridynamic theory, s_0 is obtained by forcing the *equivalent material* to have the same energy release rate. While this is useful for cracks propagation, it is not useful when the cracks are propagating due to nanovoid growth and coalescence. Instead, we force the *equivalent material* to have the same critical strain energy density to those obtained by Ponga *et al.* and Gregorie and Ponga using MD simulations. This is performed by requiring that the strain energy density in the peridynamic model Eqn. 10 equals its MD counterpart given by Eqn. 1.

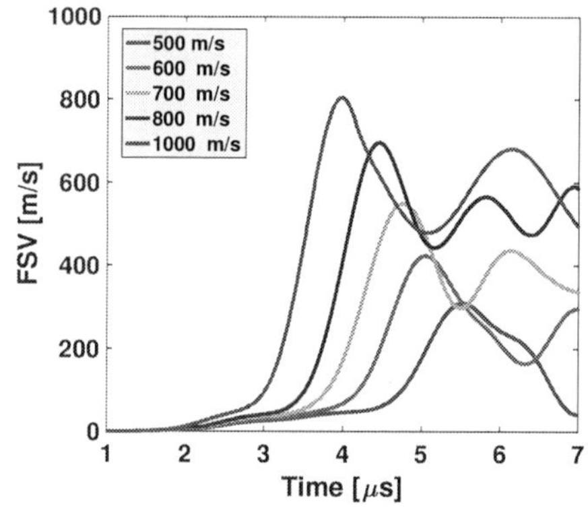

Figure 2: Experimental values of spall strength of Mg single crystals and the values computed with the model

$$W^{\text{PD}}_{\text{critical}} = \frac{\pi c s_0^2 \delta^4}{4} = W^{\text{MD}}_{\text{critical}} \rightarrow s_0 = \frac{2}{\delta^2}\sqrt{\frac{W^{\text{MD}}_{\text{critical}}}{\pi c}} \tag{12}$$

Finally, the simulations are carried out with the peridynamics implementation in the LAMMPS code [10].

RESULTS AND DISCUSSION

A plate impact experiment is simulated next. We consider a cylindrical projectile of diameter $\phi = 40$ mm and thickness $t_p = 2$ mm traveling at a velocity v_p towards a target. The target is a cylinder of diameter $\phi_t = 40$ mm and thickness $t_t = 10$ mm. Five different impact velocities are considered, i.e., $v_{p1} = 500$ m/s, $v_{p2} = 600$ m/s, $v_{p3} = 700$ m/s, $v_{p4} = 800$ m/s and $v_{p5} = 1000$ m/s . Both projectile and target are discretized using a cubic mesh with size $l_0 = 0.5$ mm given a total of 200,000 nodes with volume $V_0 = 8 \times 10^{-10}$ m^3. The density of the material is set to $\rho_0 = 1738$ Kg/m^3. The free surface velocity is measured in the central node on the free surface of the target in the opposite side of the impact.

Figure 2 shows the free surface velocity for the five cases considered. Let us now describe qualitatively the evolution of the free surface velocity. First, the velocity is zero until $t_1 = 2\mu$s, where the projectile is traveling towards the target. Then, an elastic wave is propagated through the target increasing the velocity and reaching the Hugoniot Elastic Limit (HEL) around $t_2 = 3.6\mu$s which is consistent with the longitudinal speed of sound in the material. The HEL is highlighted in the inset of figure 2. After this point, there is a second stage of plastic deformation, where the free surface velocity is increased at a much larger rate than in the elastic regime. Eventually, the free surface velocity increases almost at the same value than the impact velocity and a pull back is observed. The pullback is characterized by a sudden drop in the free surface velocity. As time goes by, the impact velocity is reached again and a second pull back is observed. The qualitatively behavior

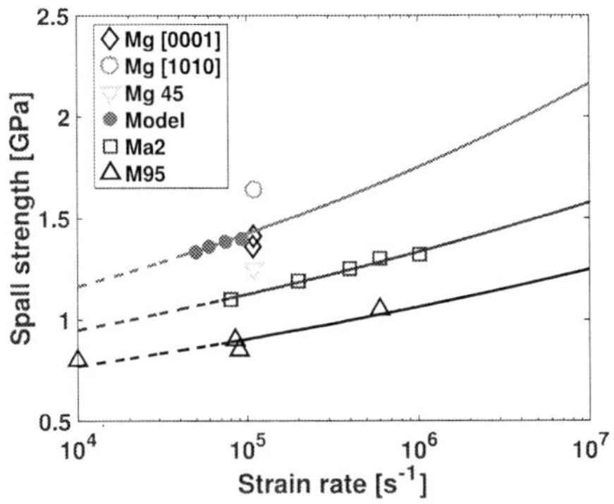

Figure 3: Comparison different experiments obtained from Ref. [6] and the developed model. Ma2 and M95 are polycrystalline Mg samples.

of the free surface velocity is in agreement with experimental measurements of spall strength in metals [6, 11–13].

Once the free surface velocity has been obtained, the spall strength is calculated as

$$\sigma_{\text{spall}} = \frac{1}{2}\rho_0 c_l \Delta u_{pb} \quad (13)$$

here c_l and ρ_0 are the longitudinal speed of sound in Magnesium and its reference density, respectively and Δu_{pb} is the difference between the peak and the first pull back of the free surface velocity.

Now, let us compare the model with experimental results [6] obtained for Mg single crystals. The results obtained with our model (red dots figure 3) predict the spall stress in well agreement with experimental results for the range of strain rate between $\dot{\epsilon} = 10^4 - 10^5$ s^{-1}. The tendency of our results predicted with the fitted red curve in dashed line shows very similar hardening coefficients that the one obtained for Mg polycrystalline samples (Ma2 and M95) observed in experiments. This fact lends credence to the model and opens the possibility to use it for poly-crystalline samples of Mg-alloys.

The current micromechanics model for dynamic behavior of Mg single crystal does not directly consider twinning which is most likely to occur in Mg. Even though twinning is an important deformation mechanism, we argue that it is activated at very small stresses much before the nanovoids start to emit dislocation and growth. As such, we posse that twinning should happen at very low stresses and do not affect the failure by nanovoid growth and coalescence. The obtained results that ignore twinning seem to support this simplification.

In conclusion, we developed a simple and reliable tool for simulating dynamic behavior of Mg single crystals where nanovoid growth and coalescence are the main mechanisms of failure. Currently, the authors are extending the work to high temperature impact tests and Mg polycrystalline alloys.

ACKNOWLEDGMENTS

We gratefully acknowledge the support from the Natural Sciences and Engineering Research Council of Canada (NSERC) through the Discovery Grant under Award Application Number RGPIN-2016-06114 and the support of Compute Canada through the Westgrid consortium for giving access to the supercomputer grid.

REFERENCES

[1] Stevens *et al.*, 1972, "Spall fracture in aluminum monocrystals: a dislocation-dynamics approach", *Journal of Applied Physics* Vol(43) pp 4922-7.

[2] Meyers and Aimone, 1983, "Dynamic fracture (spalling) of metals" *Prog. Mater. Sci.* Vol(28) pp 10-96.

[3] Becker *et al.*, 2002, "Understanding metal-failure dynamics" *Ind. Heat.*

[4] Ponga *et al.*, 2016, "Dynamic behavior of nano-voids in magnesium under hydrostatic tensile stress", *Modelling and Simulation in Materials Science and Engineering*, Vol(24), 065003.

[5] Grégoire and Ponga, (2017), "Nanovoid failure in Magnesium under dynamic loads", *Preprint submitted to Acta Materialia.*

[6] Kanel *et al.*, 2014, "Shock response of magnesium single crystals at normal and elevated temperatures", *Journal of Applied Physics*, Vol(116), 143504.

[7] Silling, 2000, "Reformulation of elasticity theory for discontinuities and long-range forces", *Journal of the Mechanics and Physics of Solids*, Vol(48) pp. 175-209.

[8] Silling and Bobaru, 2004, "Peridynamic modeling of membranes and fibers", *International Journal of Non-Linear Mechanics*, Vol(40) pp. 395-409.

[9] Silling and Askari, 2005, "A meshfree method based on the peridynamic model of solid mechanics", *Computer and Structures*, Vol(83) pp. 1526-1535.

[10] Rahman *et al.*, 2014, "PDLAMMPS - made easy".

[11] Garkushina *et al.*, 2014, "Effect of Structural Factors on Mechanical Properties of the Magnesium Alloy Ma2-1 under Quasi Static and High Strain Rate Deformation Conditions", *Physics of the Solid State*, Vol(57), No. 2, pp. 337-343.

[12] Garkushin *et al.*, 2012, "The resistance to deformation and fracture of magnesium ma2-1 under shock-wave loading at 293 k and 823 k of the temperature", *AIP Conference Proceedings.*

[13] Kanel *et al.*, 1996, "Spall fracture properties of aluminum and magnesium at high temperatures", *Journal of Applied Physics*, Vol(79), 8310.

Buckling Paradox and Anisotropic Plastic Plate Bifurcation

Suresh Shrivastava
Department of Civil Engineering and Applied Mechanics
McGill University
Montreal, Quebec, Canada H3A 2K6
Email: suresh.shrivastava@mcgill.ca

ABSTRACT

Anisotropic elastic/plastic plate bifurcation is investigated. The pre-bifurcation loading is biaxial $\sigma_{vv} = \alpha\sigma_{\xi\xi}$, $-1 \le \alpha \le 1$, with ξv axes along the sides of a simply supported rectangular plate. The ξv axes form an angle β from the xy principal axes of anisotropy. Analytical variational method is used together with Hill's incremental theory of anisotropic strain-hardening plasticity. Bifurcation stresses are determined for the anisotropic, and the classical isotropic incremental and deformation theories of plasticity, for equibiaxial compression $(\alpha = 1)$, equal compression and tension $(\alpha = -1)$, and uniaxial $(\alpha = 0)$ cases. The plastic plate buckling paradox is examined for each of the cases.

KEYWORDS: anisotropic plastic bifurcation, isotropic plastic bifurcation (incremental, deformation), plate buckling paradox

1. INTRODUCTION

The plate plastic buckling paradox originated from the work of Handelman and Prager [1]. They found that the bifurcation stresses predicted by the isotropic Mises incremental theory, the "correct" theory of strain hardening plasticity, were absurdly higher than those from the "incorrect" deformation theory of plasticity. Experiments generally favour the predictions of the deformation theory. Hence, the plastic plate buckling paradox: a correct theory yields the wrong results, while an incorrect one gives the right results. Onat and Drucker [2] explained the paradox by showing that by taking "unavoidable" geometric imperfection into account in a nonlinear growth analysis of a plate, the incremental theory gives maximum loads matching the deformation theory bifurcation loads. As the growth analysis is quite complicated, and can only be done numerically, efforts have been made to lower the bifurcation loads of the incremental theory by various legitimate means, but without success.

Here, a bifurcation analysis for plate buckling is performed by considering the anisotropic behaviour of the plate material. For this purpose Hill's incremental theory [3] for anisotropic strain hardening of sheet metals is used. The x principal axis of anisotropy is taken as the rolling direction; y is the transverse principal axis. These anisotropy axes are assumed to remain fixed, uninfluenced by the loading of the plate for bifurcation. Plane stress conditions are assumed to prevail. The loading is biaxial $\sigma_{vv} = \alpha\sigma_{\xi\xi}, -1 \le \alpha \le 1$, with ξv axes parallel to the sides of the rectangular plate, simply supported at $\xi = 0, a$ and at $v = 0, b$. The ξ axis is at an angle $+\beta$ (which may or may not be zero) measured counter clock wise from the x principal axis of anisotropy. Three cases are treated: (1) equibiaxial, (2) equal compression tension, and (3) uniaxial compression. The anisotropic bifurcation stresses are compared with those from the isotropic incremental and deformation theories.

2. ANISOTROPIC CONSTITUTIVE RELATIONS

The stresses along the anisotropy axes are

$$\begin{aligned} \sigma_{xx} &= \sigma_{\xi\xi}\cos^2\beta + \sigma_{vv}\sin^2\beta \\ \sigma_{yy} &= \sigma_{\xi\xi}\sin^2\beta + \sigma_{vv}\cos^2\beta \\ \sigma_{xy} &= (\sigma_{\xi\xi 1} - \sigma_{vv})\cos\beta\sin\beta \\ \sigma_{yx} &= (\sigma_{\xi\xi} - \sigma_{vv})\cos\beta\sin\beta \end{aligned} \quad (1)$$

The anisotropic yield criterion of Hill [3] is

$$2f(\sigma_{ij}) = (G+H)\sigma_{xx}^2 + (F+H)\sigma_{yy}^2 - 2H\sigma_{xx}\sigma_{yy} + 2N\sigma_{xy}^2 = 1 \quad (2)$$

where

$G+H=\frac{1}{X^2}, F+H=\frac{1}{Y^2}, 2H=\frac{1}{X^2}+\frac{1}{Y^2}-\frac{1}{Z^2}, 2N=\frac{1}{T^2}$.

X, Y, Z and T are the current direct and shear yield stresses. For isotropy $Z=Y=\sqrt{3}T=X$.

Normality rule gives the plastic strain increments: $d\varepsilon_{ij}^P = d\lambda \frac{\partial f}{\partial \sigma_{ij}}$ where $d\lambda = dW^P = \sigma_{ij} d\varepsilon_{ij}^P$. Explicitly:

$$\begin{aligned} d\varepsilon_{xx}^P &= dW^P\{(G\sigma_{xx}+H(\sigma_{xx}-\sigma_{yy})\}, \\ d\varepsilon_{yy}^P &= dW^P\{(F\sigma_{yy}-H(\sigma_{xx}-\sigma_{yy})\}, \\ d\varepsilon_{zz}^P &= -dW^P(G\sigma_{xx}+F\sigma_{yy}) \\ d\varepsilon_{xy}^P &= 2dW^P N\sigma_{xy}, d\varepsilon_{yz}^P=0, d\varepsilon_{zx}^P=0 \end{aligned} \quad (3)$$

The elastic increments, assumed isotropic, are

$$d\varepsilon_{ij}^E = \frac{1+\nu}{E} d\sigma_{ij} - \frac{\nu}{E} d\sigma_{kk}\delta_{ij},\ d\varepsilon_{ii}^E = \frac{1-2\nu}{E} d\sigma_{kk} \quad (4a)$$

$$d\sigma_{ij} = \frac{E}{1+\nu}\{d\varepsilon_{ij}^E + \frac{\nu}{(1-2\nu)} d\varepsilon_{kk}^E \delta_{ij}\} \quad (4b)$$

The total strain increments, elastic plus plastic, are

$$d\varepsilon_{ij} = \frac{1+\nu}{E} d\sigma_{ij} - \frac{\nu}{E} d\sigma_{kk}\delta_{ij} + dW^P \frac{\partial f}{\partial \sigma_{ij}} \quad (5)$$

Now, as defined by Hill, the equivalent stress and the equivalent plastic strain increments are:

$$\sigma_{eq} = \sqrt{\frac{3}{2}\frac{h^2}{(F_0+G_0+H_0)}}, d\varepsilon_{eq}^P = \sqrt{\frac{2}{3}\frac{(F_0+G_0+H_0)}{h^2}} dW^P \quad (6)$$

where F_0, G_0, H_0 and N_0 are related to the initial yield stresses in direct and shear stresses, and

$$h = \sqrt{(F_0+H_0)\sigma_{yy}^2 + (G_0+H_0)\sigma_{xx}^2 - 2H_0\sigma_{xx}\sigma_{yy} + 2N_0\sigma_{xy}^2} \geq 1 \quad (7)$$

is the non-dimensional hardening function, assumed common for all yield stresses:

$$X = hX_0, Y = hY_0, Z = hZ_0, T = hT_0 \text{ or,}$$

$$F_0 = h^2F, G_0 = h^2G, H_0 = h^2H, N_0 = h^2N. \quad (8)$$

Also

$$dW^P = \sigma_{mn} d\varepsilon_{mn}^P = \sigma_{eq} d\sigma_{eq} \frac{d\varepsilon_{eq}^P}{d\sigma_{eq}} = \frac{d(\sigma_{eq}^2)}{2}\frac{(\lambda-1)}{E} \quad (9)$$

$\lambda = E/E_t$, E and E_t are the elastic and tangent moduli from a uniaxial stress-strain curve of the material. Hence

$$d\varepsilon_{ij} = \frac{(1+\nu)d\sigma_{ij} - \nu d\sigma_{kk}\delta_{ij}}{E} + \frac{3(\lambda-1)dh^2}{8Eh^2(F_0+G_0+H_0)}\frac{\partial h^2}{\partial \sigma_{ij}} \quad (10)$$

where $\frac{\partial f}{\partial \sigma_{ij}} = \frac{1}{2h^2}\frac{\partial h^2}{\partial \sigma_{ij}}$ has been substituted.

For plane stress behaviour $d\sigma_{zz}=0$, one has

$$d\varepsilon_{zz} = -\frac{\nu}{E}(d\sigma_{xx}+d\sigma_{yy}) - (d\varepsilon_{xx}^P + d\varepsilon_{yy}^P). \quad (11)$$

The above relations can be inverted to express $d\sigma_{ij}$ in terms of $d\varepsilon_{ij}$. The results can be expressed as:

$$\begin{Bmatrix} d\sigma_{xx} \\ d\sigma_{yy} \\ d\sigma_{xy} \\ d\sigma_{yx} \end{Bmatrix} = \begin{bmatrix} C_{11} & C_{12} & C_{13} & C_{14} \\ C_{21} & C_{22} & C_{23} & C_{24} \\ C_{31} & C_{32} & C_{33} & C_{34} \\ C_{41} & C_{42} & C_{43} & C_{44} \end{bmatrix} \begin{Bmatrix} d\varepsilon_{xx} \\ d\varepsilon_{yy} \\ d\varepsilon_{xy} \\ d\varepsilon_{yx} \end{Bmatrix}, C_{ji} = C_{ij} \quad (12)$$

The current incremental moduli C_{ij} depend on the stresses at bifurcation, $\lambda = E/E_t$, and the anisotropic parameters F_0, G_0, H_0, N_0. Referred to the $\xi\nu$ axes:

$$\begin{Bmatrix} d\sigma_{\xi\xi} \\ d\sigma_{\nu\nu} \\ d\sigma_{\xi\nu} \\ d\sigma_{\nu\xi} \end{Bmatrix} = [R] \begin{Bmatrix} d\sigma_{xx} \\ d\sigma_{yy} \\ d\sigma_{xy} \\ d\sigma_{yx} \end{Bmatrix}, [R]=[R(\beta)], [R]^{-1}=[R]^T=[R(-\beta)], \quad (13)$$

$$[R] = \begin{bmatrix} \cos^2\beta & \sin^2\beta & \cos\beta\sin\beta & \cos\beta\sin\beta \\ \sin^2\beta & \cos^2\beta & -\cos\beta\sin\beta & -\cos\beta\sin\beta \\ -\cos\beta\sin\beta & \cos\beta\sin\beta & \cos^2\beta & -\sin^2\beta \\ -\cos\beta\sin\beta & \cos\beta\sin\beta & -\sin^2\beta & \cos^2\beta \end{bmatrix} \quad (14)$$

where, as defined before, β is the angle from the x principal axis of anisotropy to the ξ axis of loading. Thus

$$\begin{Bmatrix} d\sigma_{\xi\xi} \\ d\sigma_{\nu\nu} \\ d\sigma_{\xi\nu} \\ d\sigma_{\nu\xi} \end{Bmatrix} = \begin{bmatrix} B_{11} & B_{12} & B_{13} & B_{14} \\ B_{12} & B_{12} & B_{12} & B_{12} \\ B_{12} & B_{12} & B_{12} & B_{12} \\ B_{12} & B_{12} & B_{12} & B_{12} \end{bmatrix} \begin{Bmatrix} d\varepsilon_{\xi\xi} \\ d\varepsilon_{\nu\nu} \\ d\varepsilon_{\xi\nu} \\ d\varepsilon_{\nu\xi} \end{Bmatrix}, [B]=[R][C][R]^T \quad (15)$$

$B_{ji} = B_{ij}$ are too complicated to express explicitly.

3. CLASSICAL ISOTROPIC THEORIES

The incremental stress strain relations based upon the J_2 (Mises) yield criterion are given by [4]:

$$\sigma_{\xi\xi} = B'\varepsilon_{\xi\xi} + C'\varepsilon_{vv} \quad \sigma_{vv} = C'\varepsilon_{\xi\xi} + D'\varepsilon_{vv} \qquad (16a)$$

$$\sigma_{\xi v} = 2F'\varepsilon_{\xi v}, \sigma_{v\xi} = 2F'\varepsilon_{v\xi} \qquad (16b)$$

where

$$B' = \frac{E\{(\lambda + 3 + 3e + 4\lambda\alpha(\alpha - 1)\}}{\Delta},$$

$$D' = \frac{E\{(\alpha^2(\lambda + 3 + 3e) - 4\lambda(\alpha - 1)\}}{\Delta},$$

$$C' = \frac{E\{2(1+\alpha^2)(\lambda - 1 + 2v) + \alpha(5 + 3e - 5\lambda - 4v)\}}{\Delta},$$

$$F' = \frac{E}{(2 + 2v + 3e)},$$

$$\Delta = (1 + \alpha^2 - \alpha)\{\lambda(5 + 3e - 4v) - (1 - 2v)^2\} + 3\alpha(1 - \lambda + e)(1 - 2v). \qquad (17)$$

The parameters $\lambda = E/E_t, e = E/E_s - 1$ where $E_t = d\sigma/d\varepsilon$ and $E_s = \sigma/\varepsilon$ are the tangent and secant moduli at the equivalent stress level σ'_{eq} which for the biaxial state of stress $\sigma'_{vv} = \alpha\sigma'_{\xi\xi}$, is

$$\sigma'_{eq} = \sqrt{3J_2} = \sqrt{(\sigma'_{\xi\xi})^2 - \sigma'_{\xi\xi}\sigma'_{vv} + (\sigma'_{vv})^2} = \sqrt{1 - \alpha + \alpha^2} \times \sigma'_{\xi\xi} \qquad (18)$$

Above are the current moduli of the J_2 deformation theory. The current moduli of the J_2 incremental theory follow by substituting $e=0$ in the above equations. Moduli for the elastic theory are obtained by substituting $e=0$ and $\lambda = 1$. Thus, the constitutive relations incorporate the two isotropic strain-hardening plasticity theories, and also the purely elastic behaviour. We write:

$$\begin{Bmatrix} d\sigma_{\xi\xi} \\ d\sigma_{vv} \\ d\sigma_{\xi v} \\ d\sigma_{v\xi} \end{Bmatrix} = \begin{bmatrix} B'_{11} & B'_{12} & B'_{13} & B'_{14} \\ B'_{21} & B'_{22} & B'_{23} & B'_{24} \\ B'_{32} & B'_{32} & B'_{33} & B'_{34} \\ B'_{41} & B'_{42} & B'_{43} & B'_{44} \end{bmatrix} \begin{Bmatrix} d\varepsilon_{\xi\xi} \\ d\varepsilon_{vv} \\ d\varepsilon_{\xi v} \\ d\varepsilon_{v\xi} \end{Bmatrix} \qquad (19)$$

where $B'_{ji} = B'_{ij}$ are the isotropic moduli in (19) in lieu of the anisotropic moduli in (15).

4. STRESS STRAIN CURVE

The material is taken to be 2024S-T3 Aluminum alloy, used in aerospace industry, with a stress strain curve:

$$\varepsilon = \frac{\sigma}{76{,}500} + 0.002(\frac{\sigma}{300})^7 \qquad (20)$$

where σ is in MPa units [5] and $E = 76{,}500$ MPa. This is taken as the required relation between $\sigma_{eq} = \sigma$ and $\varepsilon^P_{eq} = \int d\varepsilon^p_{eq}$. Yielding begins around 170 MPa. The representation is experimentally valid up to $\sigma = 310$ MPa. Poisson's ratio is taken as 0.33.

For accounting anisotropic plasticity $X_0 = 170$ MPa is taken. Other yield stresses Y_0, Z_0, T_0 are given appropriate values in terms of X_0. The angle β between the $x-\xi$ axes is taken here to be just 0^0 and 22.5^0.

4. FORMULATION FOR BIFURCAION

Let $w(\xi, v)$ be the bifurcation transverse z displacement. Invoking the Kirchhoff kinematic hypothesis for thin plates, the incremental strains (symbol d omitted hereon) due to bifurcation are

$$\varepsilon_{\xi\xi} = -zw_{\xi\xi}, \varepsilon_{vv} = -zw_{vv}, \varepsilon_{\xi v} = -zw_{\xi v}, \varepsilon_{v\xi} = -zw_{v\xi} \qquad (21)$$

The stress increments are therefore

$$\begin{aligned} \sigma_{\xi\xi} &= -z(B_{11}\varepsilon_{\xi\xi} + B_{12}\varepsilon_{vv} + B_{13}\varepsilon_{\xi v} + B_{14}\varepsilon_{v\xi}) \\ \sigma_{vv} &= -z(B_{21}\varepsilon_{\xi\xi} + B_{22}\varepsilon_{vv} + B_{23}\varepsilon_{\xi v} + B_{24}\varepsilon_{v\xi}) \\ \sigma_{\xi v} &= -z(B_{31}\varepsilon_{\xi\xi} + B_{32}\varepsilon_{vv} + B_{33}\varepsilon_{\xi v} + B_{34}\varepsilon_{v\xi}) \\ \sigma_{v\xi} &= -z(B_{41}\varepsilon_{\xi\xi} + B_{42}\varepsilon_{vv} + B_{43}\varepsilon_{\xi v} + B_{44}\varepsilon_{v\xi}) \end{aligned} \qquad (22)$$

As per Shanley, plastic bifurcation under increasing load requires B_{ij} to be constants over the thickness. Hence

$$\begin{aligned} M_{\xi\xi} &= \frac{t^3}{12}(B_{11}w_{\xi\xi} + B_{12}w_{vv} + B_{13}w_{\xi v} + B_{14}w_{v\xi}), \\ M_{vv} &= \frac{t^3}{12}(B_{21}w_{\xi\xi} + B_{22}w_{vv} + B_{23}w_{\xi v} + B_{24}w_{v\xi}), \\ M_{\xi v} &= \frac{t^3}{12}(B_{31}w_{\xi\xi} + B_{32}w_{vv} + B_{33}w_{\xi v} + B_{34}w_{v\xi}). \\ M_{v\xi} &= \frac{t^3}{12}(B_{41}w_{\xi\xi} + B_{42}w_{vv} + B_{43}w_{\xi v} + B_{44}w_{v\xi}). \end{aligned} \qquad (23)$$

Now, the principle of virtual work requires

$$\int_0^a \int_0^b (M_{\xi\xi}\delta w_{\xi\xi} + M_{\nu\nu}\delta w_{\nu\nu} + M_{\xi\nu}\delta w_{\xi\nu} + M_{\nu\xi}\delta w_{\nu\xi} - P_1 w_\xi \delta w_\xi - \alpha P_1 w_\nu \delta w_\nu) d\xi d\nu = 0 \quad (24)$$

where virtual work of axial loads P_1 and $P_2 = \alpha P_1$ per unit length has been accounted. Substitution of $M_{\xi\xi}$ etc. renders the virtual work equation as

$$\frac{t^3}{12}\int_0^a \int_0^b \{(B_{11}w_{\xi\xi} + B_{12}w_{\nu\nu} + B_{13}w_{\xi\nu})\delta w_{\xi\xi} + (B_{21}w_{\xi\xi} + B_{22}w_{\nu\nu} + B_{23}w_{\xi\nu} + B_{24}w_{\nu\xi})\delta w_{\nu\nu} + (B_{31}w_{\xi\xi} + B_{32}w_{\nu\nu} + B_{33}w_{\xi\nu} + B_{34}w_{\nu\xi})\delta w_{\xi\nu} + (B_{41}w_{\xi\xi} + B_{42}w_{\nu\nu} + B_{43}w_{\xi\nu} + B_{44}w_{\nu\xi})\delta w_{\nu\xi} - P_1(w_\xi \delta w_\xi + \alpha w_\nu \delta w_\nu)\} d\xi d\nu = 0 \quad (25)$$

4. DEERMINING THE BIFURCATION STRESS

Galerkin procedure is used. The displacement w and its variation δw are taken of the same form, satisfying the simple support conditions around the edges. The functions

$$w = \Sigma A_i \sin\frac{m_i \pi \xi}{a} \sin\frac{n_i \pi \nu}{b},$$
$$\delta w = \Sigma \delta A_i \sin\frac{m_i \pi \xi}{a} \sin\frac{n_i \pi \nu}{b} \quad (26)$$

are such functions, where m_i, n_i are the number of half buckles in ξ, ν directions. A_i are the amplitudes and δA_i are their variations. The terms are such that $m_i + n_i$ is equal to $2,3,4,5,6,7,8$, giving 28 terms. When substituted in the above virtual work equation and integrated, the arbitrary character of δA_i yields a system of linear homogeneous equations of a generalized eigenvalue problem:

$$\frac{t^3}{12}[K_1]\{A\} - P_1[K_2]\{A\} = \{0\} \quad (27)$$

For a chosen $\sigma_{\xi\xi}$, the problem can be cast as

$$[[K_1'] - \frac{b^2}{t^2}[K_2]]\{A\} = \{0\} \quad (28)$$

where $[K_1'] = \frac{b^2}{12\sigma_{\xi\xi}}[K_1]$ depends on $\sigma_{\xi\xi}$, moduli B_{ij} or B'_{ij} and $r = \frac{a}{b}$. The eigenvalue is $\frac{b^2}{t^2}$ and {A} is the corresponding eigenvector (relative amplitude vector).

Mathematica programs were constructed to obtain the critical values of $\frac{t}{b}$ versus $\sigma_{\xi\xi}$ for the present incremental anisotropic theory of plasticity of Hill, and for the two classical isotropic theories.

5. EQUIBIAXIAL COMPRESSION LOADING

The results are given for rectangular plates with $a/b = 5$. The equibiaxial compressive stresses result in zero shear stress everywhere. The graphs below show gradual divergence between the incremental and deformation theory. At $t/b \approx 0.0927$ the critical stress is 310 MPa for the incremental theory, whereas for the deformation theory it is around 286 MPa which is lower by 7.7%. This is not large difference and can be made still smaller by taking the transverse shear effects in the incremental theory. For circular SS plate [6] this difference is about 1%. This means that for this equibiaxial loading the paradox is practically non-existent. For a slight anisotropy $X_0 = Y_0 = 170$ MPa $Z_0 = 0.9X_0$ the critical stress is around 290 MPa, somewhat higher than that from the deformation theory and lower than the incremental theory. No experimental data are available for this loading.

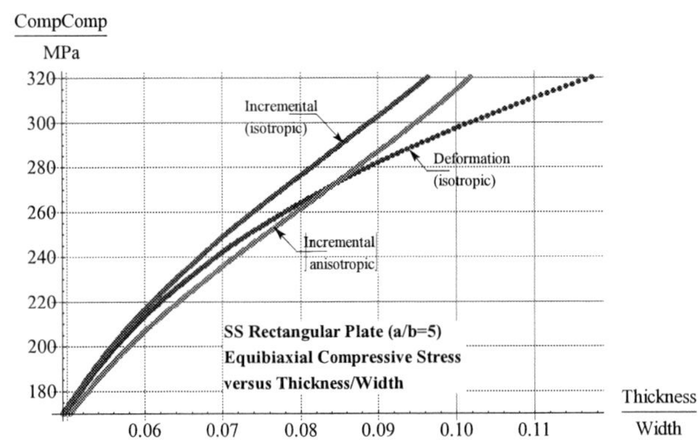

Fig 1: Critical Stress $\sigma_{\xi\xi}$ in Equibiaxial Loading

6. EQUAL COMPRESSION TENSION LOADING

Figure 2 below shows, for $a/b = 1$ square plates, the extreme divergence between the bifurcation stresses predicted by the isotropic incremental and deformation theories, confirming the plastic plate buckling paradox for this case of equal compression and tension loading. The results of the incremental theory exhibit increasing stiffness with increasing t/b ratio, while those of the deformation theory show markedly softening behaviour. It seems that the result of the incremental theory are too high and those of the deformation theory too low. In the absence of any experimental results for such loading, it is conjectured that the true values are midway between the results of the two theories.

Graphs obtained by choosing

$$X_0 = 170, Y_0 = 1.3 \times X_0, Z_0 = 1.35 \times X_0, T_0 = X_0 / \sqrt{3}\ \text{MPa}$$

are shown for $\beta = 0$ and $\beta = 22.5^\circ$. These graphs lie between the results of the two other theories.

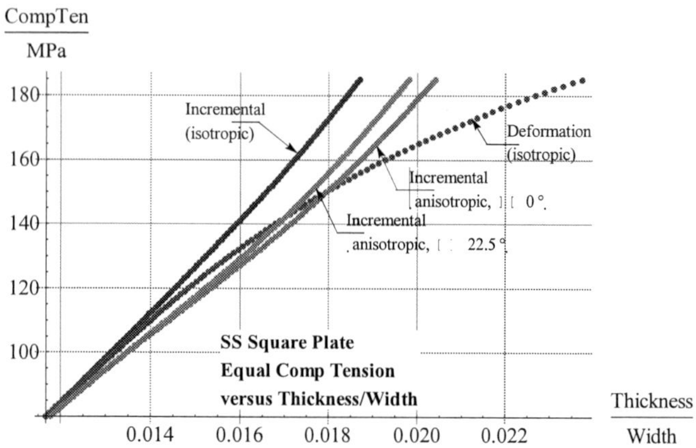

Fig. 2: Critical Stress $\sigma_{\xi\xi}$ in Equal Comp-Tension

The above anisotropic values are not unique. They were chosen by trial and error to obtain graphs midway between the graphs of the two isotropic theories, keeping in mind that the yield stresses should not be very dissimilar.

7. UNIAXIAL COMPRESSION LOADING

This case, for long plates, is the one most discussed, and has many experimental studies generally favoring the results of the deformation theory. It was also the case dealt by Handleman and Prager [1], and the paradox was discovered.

Figure 3, drawn for $a/b = 5$, shows the wide divergence between the results of the incremental and the deformation theories for the uniaxial loading. The anisotropic graphs are for the following values:

$$X_0 = 170, Y_0 = 1.6 \times 170, Z_0 = 1 \times 170, T_0 = 1.35 \times 170\ \text{MPa},$$
and $\beta = 0^\circ, \beta = 22.5^\circ$.

The graphs for these values are quite close to the critical stresses predicted by the deformation theory, especially for $\beta = 0^\circ$.

This holds true for the entire range of stress in plastic strain-hardening behaviour. For $\beta = 22.5^\circ$ a slightly stiffer behaviour is predicted than for $\beta = 0$.

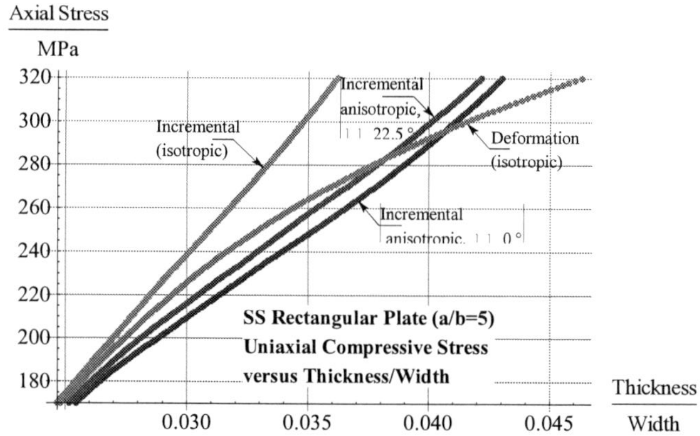

Fig. 3: Critical Stress $\sigma_{\xi\xi}$ in Uniaxial Compression

8. CONCLUSION

It is clear that the incremental theory, by considering the anisotropic plastic behaviour of the plate material, predicts bifurcation stresses which are close to the isotropic deformation theory results, which in turn are close to the available experimental values.

REFERENCES

[1] Handleman G. H. and Prager. W. 1949. Plastic buckling of a rectangular plate under edge thrusts: *NACA Report* 946.

[2] Onat E. T. and Drucker. D. C. 1953. Inelastic instabilities and incremental theories of plasticity: *Journal of the Aeronautical Sciences*, vol. 20, pp. 181-186.

[3] Hill R. 1950. The Mathematical Theory of Plasticity, *Oxford University Press*, London.

[4] Shrivastava S. and Mayemba B. M. 2013. Plastic Buckling Paradox in Biaxial loading. *Proceedings of the 25th Cancam, Western University*, London, Ontario.

[5] Seide P. and Stowell E. Z. 1948. Elastic and plastic buckling of simply supported solid-core sandwich plates in compression: *NACA Technical Report*, NACA 967.

[6] Bahous A. 2014. Plastic buckling of circular plates on a partial elastic foundation: *M. Eng. Thesis*, McGill University.

Proceedings of the 26th CANCAM
Victoria, BC, Canada, May 29 – June 1, 2017

TEMPERATURE DEPENDENT HYPERELASTIC CONSTITUTIVE MODELING USING MULTIPLICATIVE DECOMPOSITION OF DEFORMATION GRADIENT

Rohan Thakkar
Department of Mechanical Engineering
York University
Toronto, Ontario, Canada
rohan33@yorku.ca

Aleksander Czekanski
Associate Professor
Department of Mechanical Engineering
York University
Toronto, Ontario, Canada
alex.czekanski@lassonde.yorku.ca

ABSTRACT

In this paper, temperature dependent hyper-elastic constitutive law has been formulated by decomposing the total deformation into its thermal and mechanical parts. An analytical stress-strain relation has been derived for incompressible Mooney-Rivlin solid under uniaxial tensile load. Obtained results have been compared with experimental test data. Section-2 is dedicated to the formulation of multiplicative decomposition of deformation gradient (MDD) for linear isotropic thermos-elasticity. Temperature dependent material law for uniaxial tension is derived in section 3. Results and Conclusions are stated in sections 4 and 5 respectively.

Keywords: Multiplicative decomposition, finite strain thermoelasticity, temperature dependent hyperelasticity

1 INTRODUCTION

Temperature dependent constitutive laws are well discussed problems in the framework of finite strain thermoelasticity. In this classical approach, Helmholz's free energy is expressed as a function of mechanical strains, temperature and entropy. In the next step, stress-strain relations are correlated with thermodynamic entropy by enforcing energy conservation law in constitutive formulation of deformation. Finally, the coupled heat equation is determined by applying Fourier's law of heat conduction. Only two configurations of a continua are considered for developing temperature dependent stress-strain relations: initial undeformed state at reference temperature, and final stressed configuration at current temperature [1-2].

An alternative and simplified approach to formulate coupled constitutive relation is to decompose the deformation into its thermal and mechanical components [2-3]. Contrary to the classical thermodynamic theory, the total deformation is divided into three compositions. **I)** the initial undeformed zero stress state at reference temperature **II)** an inter-mediate stress-free formation at current temperature and the **III)** the deformed shape at current temperature. The intermediate con-ceptual configuration at current temperature is derived by isothermal destressing the final deformed configuration. In other words, the overall deformation is decomposed in a two steps: in the first step the body undergoes only thermal expansion and then during the second step deforms mechanically.

Despite gaining attention in elastoplastic deformation, MDD approach is less explored for thermoelastic problems. Vujosevic and Lubarda [2] compared the results obtained by classical approach with MDD method. They demonstrated thermodynamic validity of the approach by showing similarity between the entropy expressions for MDD and classical method. Hartman [4] demonstrated that for isotropic and pure volumetric temperature evolution $F=F_\theta F_M$ and $F=F_M F_\theta$ decomposition formulations give similar final results with different intermediate configurations. Darijani [5] suggested an explicit analytical solution for the induced thermoelastic stress distribution in thick walled thermoelastic cylinder vessel under thermal and mechanical loading using MDD approach.

2 FOUNDATION OF MDD METHOD

Let us consider an isotropic body $\mathcal{B}_0$ at reference temperature θ_0 in the stress free state. We are interested in the deformed configuration $\mathcal{B}$ at the temperature field θ. It is convenient to analyze the motion of the body into two consecutive stages: First the body $\mathcal{B}_0$ goes through thermal expansion when temperature changes from θ_0 to θ and transforms into an intermediate configuration $\mathcal{B}_\theta$. Next, it deforms under mechanical load and transfigures into final $\mathcal{B}$ state. This two-step deformation is shown in Figure 1 (right). The whole motion can be expressed by decomposition of deformation gradient and Jacobi into its mechanical and thermal parts as:

$$F = \frac{\partial x}{\partial X} = \frac{\partial x}{\partial x_\theta}\frac{\partial x_\theta}{\partial X} = F_M F_\theta \qquad (1)$$

$$J = \det(F) = \det(F_M F_\theta) = J_M J_\theta \qquad (2)$$

As the present analysis is restricted to thermally isotropic materials, F_θ is a diagonal tensor given by:

$$F_\theta = \lambda_\theta I \qquad (3)$$

Here λ_θ is thermally induced stretch,

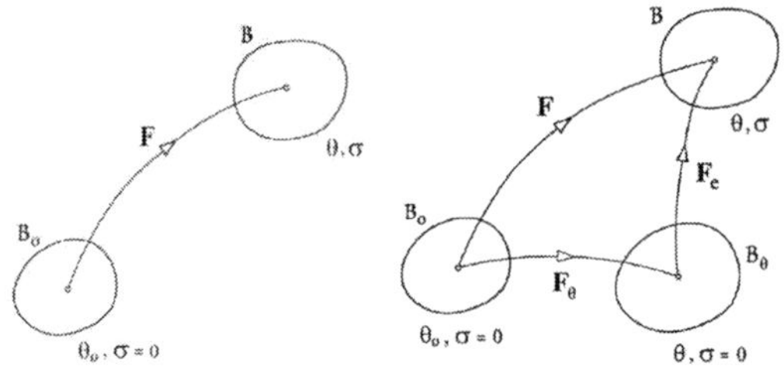

Figure 1 Classical and Multiplicative decomposition of deformation gradient based thermoelasticity

approximated using following the ansatz [6]:

$$\lambda_\theta = exp\left(\int_{\theta_0}^{\theta} \propto d\theta\right) = exp(\propto(\theta - \theta_0)) \qquad (4)$$

Where, $\propto$ is the coefficient of linear thermal expansion.

3 MOONEY-RIVLIN HYPERELASTICITY

Stress-strain relationship for hyperelastic material is derived from strain energy density function.

$\sigma = \frac{dW}{dI}$, where $W= W(I_1, I_2, I_3)$ is strain energy density function. I_1, I_2 and I_3 are invariants of B tensor and expressed by:

$$I_1 = \mathrm{tr}(B), I_2 = (\mathrm{tr}(B))^2 - \mathrm{tr}(B^2), I_3 = \det(B) \quad (5)$$

Moony-Rivlin (MR) strain energy density W is given by:

$$W = C_1(I_1 - 1) + C_2(I_2 - 1) + D(j - 1) \qquad (6)$$

In eq. (6), C_1, C_2 and D are the material constants. An incompressibility condition can be imposed by enforcing $j = 1$. As a result, an incompressible Moony-Rivlin energy density is expressed by:

$$W = C_1(I_1 - 1) + C_2(I_2 - 1) \qquad (7)$$

Following hyperelastic constitutive frame-work, the Cauchy stress is given by:

$$\sigma = -pI + 2(C_1 + C_2 I_1)B - 2C_2 B \cdot B \qquad (8)$$

Here $B = FF^T$- is the Left Cauchy-Green tensor and p is a Lagrange multiplier which is derived by imposing a boundary condition.

4 MDD FOR MR HYPERELASTICITY

In this section, temperature dependent stress-strain relation for incompressible MR solid under simple tension has been derived by applying MDD method. For incompressible material J_M=1. As result, for this particular problem from eq. (1) and (2), we can state;

$$J = J_\theta = \det F_\theta = \exp\left(3\propto(\theta - \theta_0)\right) \tag{9}$$

Because the value of $\propto$ is generally of the order of 10^{-3}, we can further simplify eq.(9) as follows:

$$\propto \ll 0, \ \rightarrow J \approx 1 + 3\propto(\theta - \theta_0) \approx 1 \tag{10}$$

For uniaxial extension, mechanical part of the deformation gradient can be written as

$$F_M = \lambda_M \, e_1 \otimes e_1 + \lambda_M{}^{-0.5} \, e_2 \otimes e_2 + \lambda_M{}^{-0.5} e_3 \otimes e_3 \tag{11}$$

Following eq. (1) the overall deformation gradient is calculated as:

$$F = F_M F_\theta = \lambda_\theta \begin{pmatrix} \lambda_M \, e_1 \otimes e_1 + \lambda_M^{-0.5} \, e_2 \otimes e_2 \\ +\lambda_M{}^{-0.5} e_3 \otimes e_3 \end{pmatrix} \tag{12}$$

Consequently, the temperature dependent left Cauchy-Green deformation tensor becomes

$$B = FF^T = \lambda_\theta^2 \begin{pmatrix} \lambda_M^2 \, e_1 \otimes e_1 + \lambda_M^{-1} \, e_2 \otimes e_2 \\ +\lambda_M^{-1} e_3 \otimes e_3 \end{pmatrix} \tag{13}$$

By substituting B in (8), we can derive the expression for Cauchy stress in terms of thermal and mechanical stretch given by:

$$\sigma = -pI + \{2\lambda_\theta^2\lambda_M^2(C_1 + C_2 I_1 - C_2\lambda_\theta^2\lambda_M^2)\} \, e_1 \otimes e_1 + \{2\lambda_\theta^2\lambda_M^{-1}(C_1 + C_2 I_1) - 2C_2\lambda_\theta^4\lambda_M^{-2}\} \, e_2 \otimes e_2 + \{2\lambda_\theta^2\lambda_M^{-1}(C_1 + C_2 I_1) - 2C_2\lambda_\theta^4\lambda_M^{-2}\} \, e_3 \otimes e_3 \tag{14}$$

In fact, for the uniaxial deformation, the thickness of the specimen in the e_3 direction is much smaller than the length and width. Hence, the sample is in a plane stress and therefore, $\sigma_{33} \approx 0$. By substituting this condition in above equation, the expression for the Lagrange multiplier p becomes:

$$p = 2\,\lambda_\theta^2\lambda_M^{-2}\{(C_1 + C_2 I_1)\lambda_M - C_2\lambda_\theta^2\} \tag{15}$$

Finally, by substituting p in eq. (14), we can reach to temperature dependent incompressible Mooney-Rivlin material law as:

$$\sigma = 2\lambda_\theta^2[\{\lambda_M^2(C_1 + C_2 I_1 - C_2\lambda_\theta^2\lambda_M^2)\} - \{\lambda_M^{-2}(C_1 + C_2 I_1 - C_2\lambda_\theta^2\lambda_M^2)\}] \, e_1 \otimes e_1 \tag{16}$$

5 RESULTS

We have considered experimental data reported by Xu et. al [7] for comparing derived analytical results. In their research work, uniaxial tensile tests on natural rubber have been carried out at temperatures ranging from 297K to 388K in an environment chamber. The deformation in the specimens has been recorded using a non-contact optical measurement system. Prior to the test, the specimens were subjected to 10-12 cycles of loading up to 0.6 strain for minimizing Mullin's effect.

Figure 2 compares the experimental data with the analytical results. The required material constants C_1 and C_2 are identified using the room temperature test data at 297K. Whereas $\propto$ is determined from higher temperature test data. From the plots it is observed that analytical results are in excellent agreement with experimental data for the constant values of: C_1=0.175 Mpa, C_2=0.081 MPa and $\propto$=1.85×10^{-3}/K.

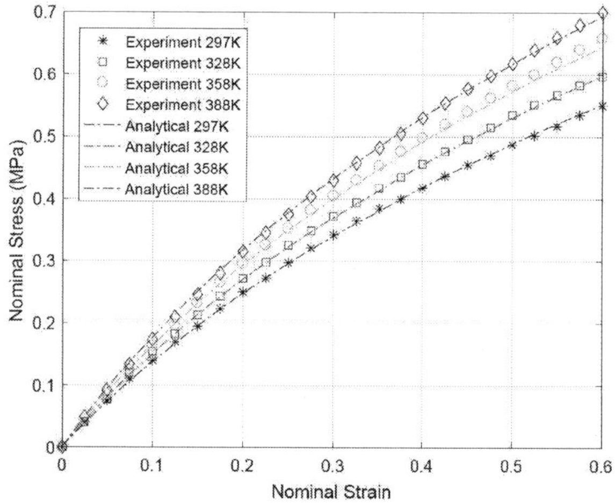

Figure 2 Comparison of experimental data with analytical results: C_1=0.175 Mpa, C_2=0.081 MPa and $\propto=1.85\times10^{-3}$ /K

6 CONCLUSIONS

To summarize, an incompressible Mooney-Rivlin material law has been extended to incorporate temperature dependency by exploiting multiplicative decomposition of deformation gradient approach. Analytical stress-strain relations are derived for uniaxial tension mode. An excellent agreement between numerical results and experimental data reported by Xu et. al has been obtained.

Following these encouraging initial results, the conducted work will be extended for higher uniaxial strain as well as for different modes of deformations. For that purpose, tests will be carried out on natural rubber for a range of temperatures. A better form of hyper-elastic strain energy function will be utilized for characterizing higher strains efficiently. The developed model will be implemented into commercial finite element code via user defined subroutine to analyze complex geometries and loadings.

ACKNOWLEDGMENTS

The authors would like to acknowledge Dr. Taisiya Sigaeva for her assistance during the research. The grants provided by Lassonde School of Engineering and NSERC/Quanser Chair in Design Engineering are thankfully acknowledged.

REFERANCES

[1] C. Truesdell and W. Noll, *The Non-Linear Field Theories of Mechanics*. Berlin, Heidelberg: Springer Berlin Heidelberg, 2004.

[2] L. Vujosevic and V. a. Lubarda, "Finite-strain thermoelasticity based on multiplicative decomposition of deformation gradient," *Theor. Appl. Mech.*, vol. 28, no. 28–29, pp. 379–399, 2002.

[3] R. Stojanovic, S. Djuric, and L. Vujosevic, "On fnite thermal deformations," *Arch. Mech. Stosow.*, vol. 16, pp. 103–108, 1964.

[4] S. Hartmann, "Comparison of the multiplicative decompositions $F = F_\theta F_M$ and $F = F_M F_\theta$ in finite strain thermo-elasticity," *Tech. Rep. Ser.*, 2012.

[5] H. Darijani, "Finite Deformation Thermoelasticity of Thick-Walled Cylindrical Vessels Based on the Multiplicative Decomposition of Deformation Gradient," *J. Therm. Stress.*, vol. 35, no. 12, pp. 1143–1157, Jan. 2012.

[6] G. A. Holzapfel, "Nonlinear Solid Mechanics: A Continuum Approach for Engineering," *John Wiley Sons*, p. 470, 2000.

[7] X. Li, T. Bai, Z. Li, and L. Liu, "Influence of the temperature on the hyper-elastic mechanical behavior of carbon black filled natural rubbers," *Mech. Mater.*, vol. 95, pp. 136–145, 2016.

A higher-order elastic beam model for indentation problems

Min Li
[a] **Department of Mechanical Engineering**
University of Alberta
Edmonton, AB T6G 2G8, Canada
[b] **Nanjing University of Aeronautics and Astronautics, Nanjing 210016, China**
Contact person : mli5@ualberta.ca

C. Q. Ru
Department of Mechanical Engineering
University of Alberta
Edmonton, AB T6G 2G8, Canada

ABSTRACT

A higher-order elastic beam model is applied to investigate indentation of a thin beam by a rigid indenter. Different from all existing methods, the present model is a Kerr-type simple differential relation between the applied pressure and the normal deflection of the pressured surface of elastic beam. The contact-zone width-displacement relation and the load-displacement relation are obtained in explicit form and illustrated with numerical examples.
KEYWORDS: Indentation, Beam, Kerr model, Contact

INTRODUCTION

Indentation of an elastic beam by a rigid indenter is of both scholarly and practical interest [1-3]. Since classical beam model fails for the indentation problem due to highly non-uniform pressure in the contact zone, Keer & Miller [1] approached this problem by a superposition of an infinite layer plane-strain solution with a pure bending solution of beam. On the other hand, Kim et al. [2] investigated the contact problem by adopting an asymptotic method combined with finite element method. Unlike all existing methods, the present method is based on a single differential relation between the contact pressure and the normal deflection of the pressured surface of elastic beam, derived by Ru [4] using the Kerr model [5], which holds both inside and outside the contact zone. Efficiency of the present method will be demonstrated by comparing its predictions with some known basic results available in literature (such as Keer & Miller [1]).

PROBLEM FORMULATION

As shown in Fig. 1, for a suspended elastic thin beam supported at its two ends, the Kerr-type relation [5] between the (downward) vertical deflection $w(x)$ of its upper surface and the pressure $p(x)$ exerted on the upper surface is given by Ru [4]

$$p-\frac{1}{3}h^2\frac{d^2}{dx^2}p+\frac{1}{15}h^4\frac{d^4}{dx^4}p=D\frac{d^4}{dx^4}w, \tag{1}$$

where $p(x)$ is uniform in the direction perpendicular to the x-w plane, h is the thickness of the beam, D is the bending rigidity given by $D=\frac{Eh^3}{12(1-\nu^2)}$, and E and ν are its Young's modulus and Poisson' ratio. The differential equation applies both inside and outside the contact zone provided that the beam is sufficiently thin compared to the size of contact zone. In particular, when $p(x)$ is uniform (e.g. outside the contact zone, where $p(x)=0$), the relation reduces to the classical elastic beam model. Therefore, the present model can be considered as a simple higher-order beam model which includes two additional higher-order terms of $p(x)$ relevant for highly non-uniform pressure $p(x)$ inside the contact zone.

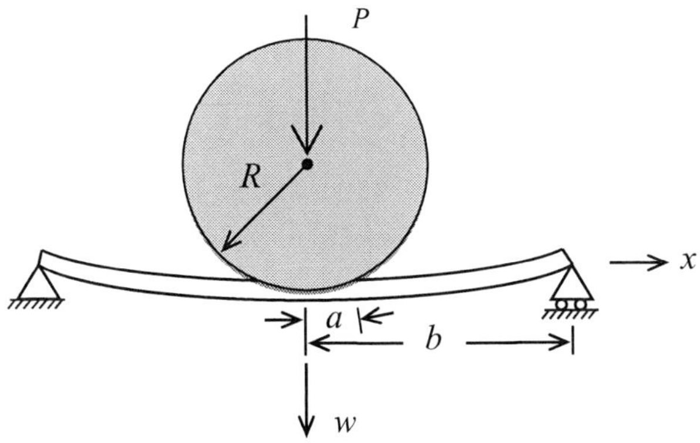

Figure 1: Beam indented by a rigid cylinder indenter

Let us first assume that the indenter remains contact with the beam at the center and the half width of contact-zone be a. Thus, in the contact zone ($-a\le x\le a$), the symmetric surface deflection $w(x)$ of the upper surface is identical to that of the surface of the cylinder indenter and has the following form

$$w(x)=\sqrt{R^2-x^2}-R+\delta,\ -a\le x\le a, \tag{2}$$

where δ is the (downward) displacement of the indenter. In general, the radius of the indenter can be assumed to be much larger than the half width of contact-zone (namely $R>>a$).

Then, due to the symmetry of the problem, we consider only the deflection $w(x)$ outside the contact zone in right zone $a\le x\le b$. The substitution of $p(x)=0$ into Eq. (1) gives the governing equation outside the contact zone

$$\frac{d^4}{dx^4}w = 0 \quad (a \le x \le b), \tag{3}$$

with the boundary conditions:

$$\text{clamped case: } w\big|_{x=b} = 0, \; \frac{dw}{dx}\bigg|_{x=b} = 0; \tag{3}$$

$$\text{simply supported case: } w\big|_{x=b} = 0, \; \frac{d^2w}{dx^2}\bigg|_{x=b} = 0. \tag{4}$$

On the other hand, the continuity of surface deflection and its slope for smooth indentation at $x=a$ give

$$w\big|_{x=a} = \delta - \frac{a^2}{2R}, \; \frac{dw}{dx}\bigg|_{x=a} = \frac{-a}{R}. \tag{5}$$

a). For the clamped case, by employing the boundary condition and the continuity condition, we obtain

$$w(x) = \frac{(x-b)^2}{(a-b)^3}\left[\left(\frac{ab}{R} - 2\delta\right)(x-b) + 3\delta(a-b) - \frac{a\left(a^2+ab-2b^2\right)}{2R}\right]. \tag{6}$$

With the continuity of the second derivatives of the deflection at $x=a$, we have

$$w''(a) = \frac{-1}{R}, \tag{7}$$

from which a simple relation between the half contact-zone width a and the displacement δ is obtained

$$\frac{\delta R}{h^2} = \frac{b(b+2a)}{6h^2}. \tag{8}$$

On the other hand, the total applied load can be written as

$$P = \int_{-a}^{a} p(x)dx = 2V\big|_{x=a} = 2Dw'''\big|_{x=a}, \tag{9}$$

where $w'''\big|_{x=a}$ is calculated by the deflection outside the contact zone at $x=a$. Then, Eqs. (7), (9) and (10) give the relation between the indentation load P and the displacement δ

$$P = \frac{4Db}{R(a-b)^2} = \frac{16Db^3}{R(6\delta R - 3b^2)^2}. \tag{10}$$

b). For the simply supported case, in a similar way, we have

$$\frac{\delta R}{h^2} = \frac{-a^2+2ab+2b^2}{6h^2}, \tag{11}$$

$$P = \frac{2D}{R(b-a)} = \frac{2D}{R(\sqrt{3b^2-6\delta R})}. \tag{12}$$

In Fig. 2-3, the efficiency of the proposed method is validated by comparing its predictions with some known results of Keer & Miller [1].

CONCLUSIONS

A simple new method is developed to solve the indentation problem of a beam, and the merits of the proposed method are demonstrated by some easy-to-use explicit formulas and numerical results (more details are omitted here).

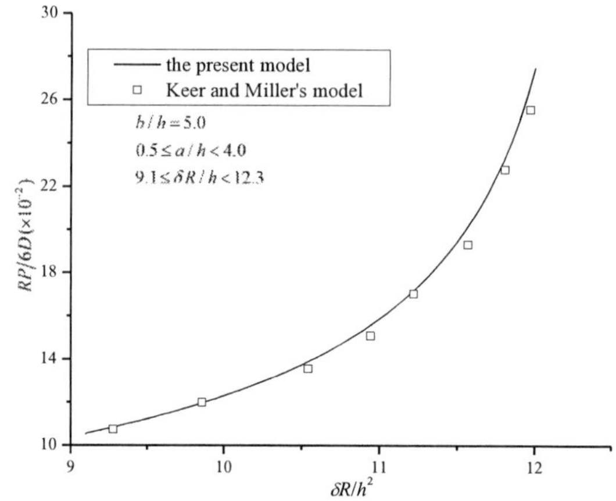

Figure 2: Load-displacement: simply supported ends

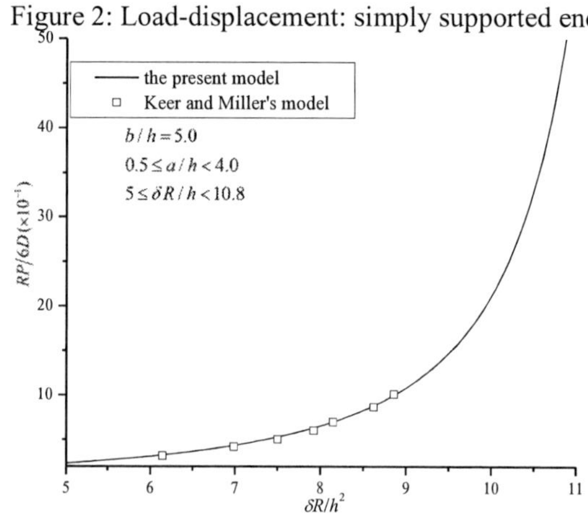

Figure 3: Load-displacement: clamped ends

ACKNOWLEDGMENTS

The support of Natural Science and Engineering Research Council of Canada (NSERC-RGPIN204992) and the China Scholarship Council are appreciated.

REFERENCES

1. Keer, L.M. and Miller, G.R. 1983, “Smooth indentation of finite layer”, J Eng Mech, 109, 706-717.
2. Kim, J.H., Ahn, Y.J., Jang, Y.H. and Barber, J.R. 2014, “Contact problems involving beams”, Int J Solids Struct, 51, 4435-4439.
3. Thredgold, J.M., Lucas, S.K. and Howlett, P.G. 2006, “On the contact of a rigid sphere and a thin plate”, Math Comput Modell, 43, 119-131.
4. Ru, C.Q. 2002, “Surface instability of an elastic thin film interacting with a suspended elastic plate”, J Appl Mech, 69, 97-103.
5. Kerr, A.D. 1984, “On the formal development of elastic foundation models”, Ing Arch, 54, 455-464.

Advanced Manufacturing 1

DEVELOPMENT OF WC-Co TOOL BY ADDITIVE MANUFACTURING

Sanjay Kumar
Department of Mechanical Engineering,
York University, Toronto, Canada
Email: sanjaykr@yorku.ca,

Aleksander Czekanski
Department of Mechanical Engineering,
York University, Toronto, Canada
Email: alex.czekanski@lassonde.yorku.ca

ABSTRACT

WC-Co is a well-known material for conventional tooling but is not yet suitable for additive manufacturing (AM). Fabricating it successfully by selective laser sintering (SLS) will pave the way for its acceptance and commercialization. The present work is an attempt to make complex 3D parts from WC-Co without using any post-processing (heat treatment, infiltration) and any additional material (binder). It is aimed to do comprehensive study of the process parameters so that the advantage of AM could be realized for hard metals. In order to find the best parameters for making parts encompassed with microfeatures, various process parameters are varied, and their effects are studied by observing microstructures and by performing hardness tests. The finding that microfeatures could be fabricated using WC-Co will herald renewed interest in investigating hardmetals using SLS for manufacturing complex hard tools, molds and wear-resistance parts.

KEYWORDS: selective laser sintering, WC-Co, tool, additive manufacturing

1.0 INTRODUCTION

WC-Co is used extensively for fabricating wear-resistant parts, cutting tools, molds, mining parts for excavation, electronic packaging and nozzles by traditional manufacturing processes. These processes have limitations such as restricted design, difficult post processing (machining) and high cost for tools comprising micro features [1]. Additive manufacturing (AM) is natural solution to various ills, but due to lack of process improvement and unavailability of commercial machines for hardmetals, shaping WC-Co by AM is not as straightforward as is for metals, polymers and low melting point composites.

WC-Co has been tried to be processed using ink jet printing [2] or binder jetting based 3D printing process [3]. But these processes require extensive post-processing (oven treatment) to remove binder and subsequently to infiltrate porous parts for increasing density. This compromises the strength, size and design of final parts manufactured.

Laser-based additive manufacturing such as selective laser sintering (SLS), selective laser melting (SLM) and Laser engineered net shaping (LENS) promises to make stronger parts and is preferable over ink jet printing and 3D printing. Xiong et al used LENS to make simple 3D blocks from WC-Co, but there is no further attempt to use LENS for making complex parts [4]. Majority of works on this material has been done by SLS and SLM as these processes have an edge over other AM process for making complex parts from high melting point material. Wang et al were able to make a simple 3D part using SLS of WC-Co (comprising 91%WC) but the part was porous [5]. Kumar attempted to solve various manufacturing problems for making 3D parts and did wear analysis on fabricated molds. However, parts needed to be infiltrated with bronze for improving densification and strength [6]. Gu et al fabricated WC-Co parts comprising WC from 18 wt. % [7] to 45 wt. % [8-10]. They made simple blocks out of WC-Co with the help of Cu as another binder. Gu and Meiners fabricated in-situ WC using a composition of W-Ni-graphite. It furnished a composite comprising of WC/Ni_2W_4C, W and Ni. However, the amount of WC in the composite was not determined [11]. Khymrov et al got crack-free WC-Co parts but the percentage of WC was low (25%) and fabricated parts were smaller. By increasing WC to 50%, cracks could not be avoided [12]. Uhlmann et al made 83WC-17Co parts comprising of cooling channels but fine micro-cracks were observed in all samples [13]. Past works show that making a WC-Co part by AM is easy if the amount of WC is low, shape of the part is simple and size is small. Purity of WC-Co is compromised if it is either infiltrated or processed with the help of another materials (binder, Cu, Ni).
The present research is an attempt to make complex 3D parts from WC-Co without taking any help of any post-process and without using any additional material in the WC-Co mixture. It is intended to study the limitation of the process so that the advantage of AM could be realized for hard metals. In order to find best parameters for making part comprising of microfeatures, various process parameters are varied and their effects are studied.

2.0 EXPERIMENTAL

2.1 Material & Machine

A pre-alloyed powder (size from 15 to 45 μm) consisted of 83 wt. % WC and 17 wt. % Co is selected for experimentation. The powder is made by sintering and agglomeration and is of near-spherical shape as shown in Fig.1. Each powder comprises of several grains of WC of approximate size of 5 μm bound together by Co. Agglomerated powder is preferred over mixture of constituents as it allows uniform distribution of binder (Co) in an SLS process during powder deposition and processing, which subsequently helps achieve maximum densification with minimum amount of Co.

Fig. 1. Image of WC/Co powder

An AM system, EOS M 280, made for processing customized metals is used for processing powders. Before processing, chamber is vacuumed and filled up with nitrogen gas. In order to use a small area of powder bed platform for making parts, adaptive plates are used. The plate is made up of 1045 cold rolled steel of thickness 6 mm and size 10 mm X 10 mm and acts as a substrate on which parts are fabricated. Polishing of fabricated parts is accomplished using following steps in sequence: 20 min plane grinding using water, 20 min fine grinding using liquid containing 9 μm diamond particle, 10 min polishing using liquid containing 3 μm diamond particle and 5 min cloth polishing using liquid containing 0.05 μm aluminum oxide particle. Polished parts are observed using optical microscope (Zeiss Imager.M2), and hardness test is done using Rockwell hardness tester (using a load of 150 kg). Some selected samples are observed using scanning electron microscope (Hitachi SU3500). Wear test is done by Bruker NanoSurfaces UMT Tribolab while wear volume is measured by Bruker 3D optical microscope. Parameters for the wear test are: stroke length 5 mm, frequency 20 Hz, time 2 hours, applied load 20 N, alumina ball of size 9.525 mm as counterbody.

2.2 Experiments

In order to process powders, it is intended to supply enough laser energy to melt only cobalt and leave WC intact. Because higher laser energy causes melting of WC and subsequent formation of M_6C compound by chemical reaction with Co. M_6C has lower hardness and wear-resistance than WC and decreases the mechanical property of WC/Co.

Scan speed, hatch spacing and laser power are three parameters varied for doing experiments with an aim to make parts of minimum size of 10 mm dia. X 5 mm height. Scan speed is the speed at which laser beam scans the powder bed while hatch spacing is the distance between two adjacent scans. For preliminary experiments, scan speed is varied from 800 to 1400 mm/s, laser power is varied from 225 to 370 W while hatch spacing and layer thickness are kept constant as 0.1mm and 0.04 mm respectively. Samples fabricated corresponding to the highest laser power of 370 W and lower values of scan speed were the best (shown in Fig. 2). Inferring from these results, further experiments were carried out varying laser power from 270 to 370 W, scan speed from 900 to 1100 mm/s while keeping hatch spacing (Hs) constant at 0.1, 0.07 and 0.04 mm. Results from these experiments showed that samples fabricated corresponding to hatch spacing of 0.04 mm were remained intact whiles others were displaced due to weak attachment with the base plate (shown in Fig. 3). In order to increase the adhesion, further experiments were done: by increasing the base plate temperature to 200°C, varying laser power from 270 to 370 W, scan speed from 600 to 800 mm/s, using hatch spacing of 0.04 mm. This helped make taller samples upto the height of 12 mm.

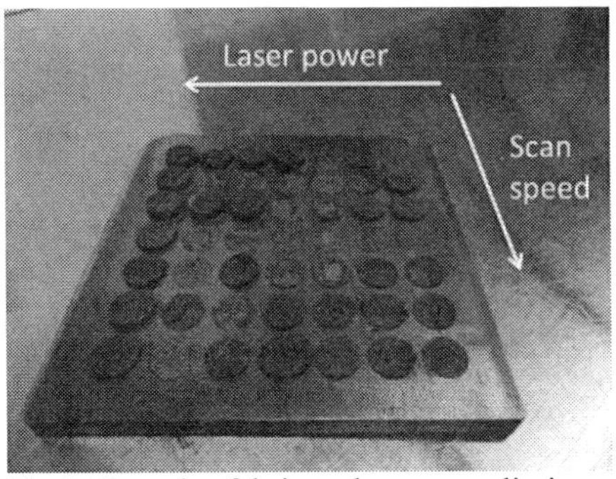

Fig 2- Samples fabricated as per preliminary experiments

Fig 3- Samples of 5 mm height fabricated at various hatch spacing

3.0 RESULTS & DISCUSSION

Microstructures of all samples at all parameters showed cracks (shown in Fig. 4). The size and width of cracks decreased for samples fabricated at higher base plate temperature.

Hardness test showed results from 53 to 60 HRC for the samples fabricated without increasing the base plate temperature while the hardness for samples fabricated at increased base plate temperature varied from 54 to 58 HRC with less variation both 1) among different samples and 2) in each sample. This is because the samples are built higher and are kept at higher temperatures for longer duration. This anneals the sample and increases uniformity in hardness values.

Fig 4: SEM image of a sample

Microstructural analysis showed that the lowest scan speed (600 mm/s) and the highest laser power (370 W) gave fewer pores and considered the best parameters. But, higher laser power caused Co to deplete from samples, therefore, lower laser power was favored for samples showing same amounts of pores and cracks. It gave the optimized parameter as shown in Table 1.

Process parameter	Value
Laser power (W)	270
Scan speed (mm/s)	500
Layer thickness (mm)	0.04
Hatch spacing (mm)	0.04
Preheating (°C)	200

Table 1: Values for optimized parameters

Wear Test

A sample found to be best (made using scan speed 600 mm/s, laser power 370 W, hatch spacing 0.04 mm, preheating 200°C) was selected for wear test. For comparison, a sample made with the best tooling material (EOS maraging steel MS1) was taken. It was found that the wear volume for WC-Co and maraging steel was 0.003 mm^3 and 0.697 mm^3 respectively showing far better wear resistance of WC-Co than the best commercial tooling material available

Model

In order to test optimization of parameters for making complex parts, a model (design shown in Fig. 5) comprising micro features is fabricated using optimized parameters (as per Table 1). The model consists of 1) holes of diameter 0.5, 1, 2, 4 mm and each of depth 2 mm, 2) cylindrical pins of diameter 0.5, 1, 2, 4 mm and each of height 5 mm and 3) thin walls of height 10 mm and thickness 2 mm making angle 90°, 75°, 60°, 45°, 30°, 15° with the substrate plane. The model fabricated is shown in Fig. 6 while measured dimensions of model features are given in Table 2.

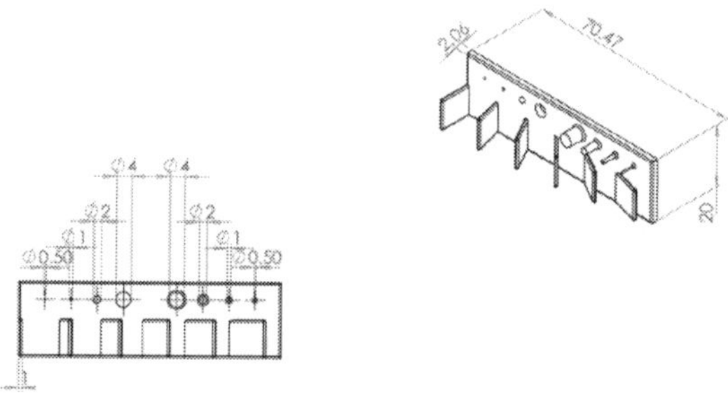
Fig 5- Design of a model comprising micro features.

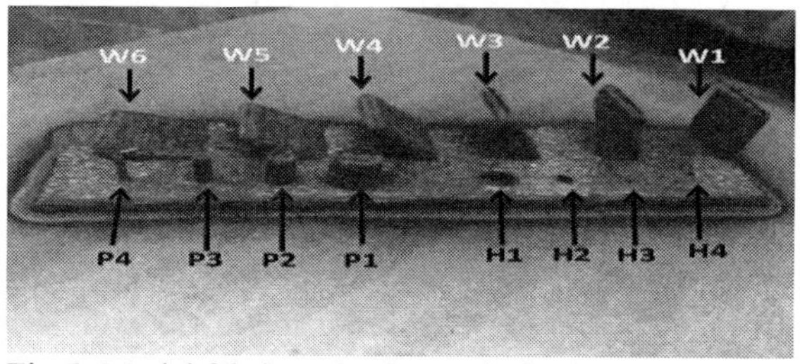

Fig 6- Model fabricated using optimized process parameters.

Features	**Dimension of CAD file**		**Dimension achieved**	
Cylindrical pin	**Diameter (mm)**		**Diameter (mm)**	
P1	4		4.81	
P2	2		2.72	
P3	1		1.58	
P4	0.5		1.10	
Hole	**Diameter (mm)**		**Diameter (mm)**	
H1	4		3.46	
H2	2		1.63	
H3	1		Could not be fabricated	
H4	0.5		Could not be fabricated	
Wall	**Thickness (mm)**	**Angle (°)**	**Thickness (mm)**	**Angle (°)**
W1	2	90°	1.95	90°
W2	2	75°	1.98	76°
W3	2	60°	2.03	61°
W4	2	45°	2.15	46°
W5	2	30°	2.45	30°
W6	2	15°	-	-

Table 2- Dimension of micro features of model and fabricated part

Table 2 shows cylindrical pins from 0.5 to 4 mm diameter that could be fabricated with a tolerance of 0.6 to 0.8 mm. Holes from 2 to 4 mm diameter with tolerance 0.4 to 0.6 mm could be fabricated with a tolerance of 0.5 mm. These tolerances are quite high in comparison to the previous works done on metals [14-15]. However, previous works are done for materials customized for the machine and it was expected that machine would furnish better tolerance for its customized material.

In addition to it, in the present work, parameters are optimized for bulk and not for boundary. At the boundary area, powders get adhered to the built edge leading to an increase in dimension (as shown in Fig. 4). This resulted in enhancement of pin diameter and constriction of hole. This is the reason smaller holes of diameter 0.5 and 1 mm could not be produced.

Thin walls bent at angles from 30 to 90° could be fabricated with a tolerance of 1° while tolerance for its thickness was about 0.1 mm. This is comparable to past works done for metals [14-15]. With an increase in bent of wall, thickness of walls increased and at high bent (15°), angle could not be measured. The reason could be attributed to the same powder-adherence phenomenon responsible for increase in pin diameter and decrease in hole diameter.

CONCLUSIONS

A part comprising of various micro features is fabricated showing that complex tools and molds could be fabricated with AM process. Hardness obtained for final samples are better than the existing tool steels. Microstructures have shown that cracks are unavoidable. It shows the limitation of present powder bed systems as the temperature of the base plate (or processing chamber) of the systems could not be increased more than that used in the present research. Further research is required to find whether a complex tool having cracks will fare better than the crack-free tools made from tool steel.

ACKNOWLEDGMENTS

We would like to thank Mohawk College's CFI and NSERC funded Additive Manufacturing Resource Centre for supporting this project. The grant provided by Lassonde and NSERC Chair in Design Engineering is thankfully acknowledged. Technical assistance provided by Mr. David Marcinkiewicz during wear test is acknowledged.

REFERENCES

1. Brookes, K. J. A. 2015, "3D-printing style additive manufacturing for commercial hardmetals", Metal Powder Report, Vol. 70, Issue 3, pp. 137-140.
2. Benichou, A., Laufer, L. 2016, "Tungsten carbide-cobalt ink composition for 3D inkjet printing", US Patent, publication number 20160236372.
3. Kernan, B. D., Sachs, E. M., Oliveira, M. A., Cima, M. J. 2007, "Three dimensional printing of tungsten carbide-10 wt % cobalt using a cobalt oxide precursor", International Journal of Refractory Metals & Hard Materials, Vol. 25, pp. 82-94.
4. Xiong, Y., Smugeresky, J. E., Ajdelsztajn, L., Schoenung, J. M. 2008, "Fabrication of WC-Co cermets by laser engineered net shaping", Materials Science and Engineering A, Vol. 493, pp. 261-266.
5. Wang, X. C., Laoui, T., Bonse, J., Kruth, J. P., Lauwers, B., Froyen, L. 2002, "Direct selective laser sintering of hard metal powders: experimental study and simulation", International Journal of Advanced Manufacturing Technology, Vol. 19, pp. 351-357.
6. Kumar, S. 2009, "Manufacturing of WC-Co moulds using SLS machine", Journal of Materials Processing Technology, Vol. 209, pp. 3840-3848.
7. Gu, D., Shen, Y., Dai, P., Yang, M. C. 2006, "Microstructure and property of sub-micro WC-10%Co particulate reinforced Cu matrix composites prepared by selective laser sintering", Trans. Nonferrous Met. Soc. China, Vol. 16, pp. 357-362.
8. Gu, D., Shen, Y. 2006, "WC-Co particulate reinforcing Cu matrix composites prepared by direct laser sintering", Materials Letters, Vol. 60, pp. 3664-3668.
9. Gu, D., Shen, Y., Zhao, L., Xiao, J., Wu, P., Zhu, Y. 2007, "Effect of rare earth oxide addition on microstructures of ultra-fine WC-Co particulate reinforced Cu matrix composites prepared by direct laser sintering", Materials Science and Engineering A, Vol. 445-446, pp. 316-322.
10. Gu, D., Shen, Y. 2008, "Direct laser sintered WC-10Cu/Co nanocomposites", Applied Surface Science, Vol. 254, pp. 3971-3978.
11. Gu, D., Meiners, W. 2010, "Microstructure characteristics and formation mechanisms of in situ WC cemented carbide based hardmetals prepared by selective laser melting", Materials Science and Engineering A, Vol. 527, pp. 7585-7592.
12. Khmyrov, R. S., Safronov, V. A., Gusarov, A. V. 2016, "Obtaining crack-free WC-Co alloys by selective laser melting", Physics Procedia, Vol. 83, pp. 874-881.
13. Uhlmann, E., Bergmann, A., Gridin, W. 2015, "Investigation on additive manufacturing of tugsten carbide-cobalt by selective laser melting", Procedia CIRP, Vol. 35, pp. 8-15.
14. Kruth, J.-P., Vandenbroucke, B., Van Vaerenbergh. J., Mercelis, P. 2005, "Benchmarking of different SLS/SLM processes as rapid manufacturing techniques", International Conference on Polymers & Molds Innovation (PMI), Ghent, Belgium April 20-23.
15. Li, Y., Anam, M. A. 2014, "An investigation of standard test part design for additive manufacturing", SFF Symposium proceedings, pp. 901-922.

Proceedings of the 26th CANCAM
Victoria, British Columbia, Canada, May 29 – June 1, 2017

FABRICATION OF INEXPENSIVE FILAMENTS FOR FUSED DEPOSITION MODELING

Aleksander Czekanski
Department of Mechanical Engineering,
York University, Toronto, Canada
alex.czekanski@lassonde.yorku.ca

Sanjay Kumar
Department of Mechanical Engineering,
York University, Toronto, Canada
sanjaykr@yorku.ca

ABSTRACT

Fused Deposition Modeling (FDM), has a largest number of Additive Manufacturing (AM) systems sold worldwide because of its ability to make inexpensive 3D products. Nonetheless, its application is limited due to the limited range of the strength of the product achievable. The present work is an effort to expand the range by developing filaments using recycled polymer. The recycled material is a discarded powder (polyamide) collected from Selective Laser Sintering (SLS) systems. Polyamide (PA12) is mixed with various percentages of tungsten carbide, and the resulting powder mixture is fed into an extruder to fabricate filaments. Cost analysis besides various characterizations such as DSC, MFI, SEM and tensile tests are performed. It is found that the fabricated filaments have better and comparable tensile strength to the existing filaments, and could be economically manufactured. Moreover, the strategy presented here will give new impetus to recycling and help make AM self-sustainable.

KEYWORDS: fused deposition modeling, additive manufacturing, selective laser sintering, recycling.

1.0 INTRODUCTION

Fused Deposition Modeling (FDM) is a material extrusion based additive manufacturing (AM) process in which a product is fabricated by melting polymer based filaments and depositing molten materials onto a platform. FDM is famous among all AM techniques because it is office-friendly, and offers inexpensive machines, open source software and feedstock materials. However, it is not acceptable for critical engineering applications because of the lack of robust feedstock materials furnishing stronger mechanical properties. ABS (Acrylonitrile Butadiene Styrene) and PLA (Poly Lactic Acid) are mostly used in FDM while other commercially available materials are polyimide, polycarbonate, nylon, polyvinyl alcohol, high density polyethylene, polyetheretherketone, high impact polystyrene, etc. [2]. In past, efforts were made to develop composite filaments by mixing various types of polymers such as ABS, Polypropylene, Nylon with Fe, Cu, Al, Al_2O_3, carbon fibre etc. [2-8]. It was found that carbon fibre and Al_2O_3 improve the mechanical property of composites while Fe, Cu and Al improve both mechanical and thermal properties. Filaments with higher mechanical and thermal properties are useful for tooling applications [9]. Developing nanocomposite by adding 4 to 5% nano particles (graphene, clay) to ABS has greatly increased its tensile strength and thermal stability [10-11]. An investigation is also undertaken to utilize nylon waste to make low-cost filaments, which promises to enable wider adoption of FDM systems [12-15].

Recycling waste materials generated by another AM process for making FDM filaments will help not only to reduce the cost of filaments but also to make AM more viable. Selective laser sintering (SLS) is a powder bed based AM process in which a product is fabricated by depositing powder layers on a platform, and scanning them by laser beam as per CAD file. The layer deposited covers the whole area of the platform irrespective of the size of the desired product. Consequently, even if a tiny product needs to be made, a large amount of powders is required. The fraction of heap of powders, which remain on the platform and do not contribute to the mass of the product still go through thermal cycles because of the heat generated by scanning action, and attached heaters. This leads to the deterioration of the properties of the powders, and the property of the product made from such degraded powders is no longer guaranteed [16-17]. There are numerous studies on the degradation of polymer powders and its effect on final properties. In order to maintain required property, various strategies to mix fresh and used powders have been suggested [18-19]. As per recommendation of EOS, used powders are mixed with 30-50% of unused powders, and can be used few more times before powders turn into waste. These waste powders are high-value wastes, and can not be equated with low value plastic wastes (bottle, bag, packaging etc.) generated in everyday life because producing powders is a time-taking and energy-intensive process. Consequently, if these waste powders are not again used as powders then the energy-efficiency of SLS is seriously compromised. An efficient plan needs to be contemplated for this purpose. In the past, there are many works on used SLS powders but there is no work presented on the processing of waste SLS powders. This work is an attempt to utilize the waste SLS powders by developing filaments for FDM. This will link two processes (SLS & FDM) and help make AM more sustainable.

In this study, discarded SLS powders are mixed with various amounts of tungsten carbide. The resulting powder mixture is processed in a single screw extruder to make filaments, which are subsequently characterized to find their properties.

2.0 EXPERIMENT

2.1 Material Used

PA2200 Balance is a polyamide 12 SLS material offered by EOS GmbH [20]. Its average grain size is 56 µm and bulk density is 0.45 g/cm^3. The part fabricated by SLS will provide tensile modulus (ASTM D638), tensile strength and Elongation at Break of 1700 MPa, 48 MPa and 24% respectively. Waste powders are discarded PA 2200 powders which are no longer suitable to be used in an SLS system. These powders during processing have gone through a number of thermal cycles and mechanical handling, which has increased their molecular weights and changed their properties. Hereon, these waste powders are referred to as PA. WC (tungsten carbide) powder of average size 5 µm and density 15.63 g/cm^3 is purchased from GTP PA, USA. PA is mixed with WC by taking its 12.5, 25, 37.5 and 50 weight percentages. WC is selected primarily because of its hardness and strength, which will arrest the flow of polymer chain and increase strength.

The high density of WC will drive it to settle at the bottom of the container both in powder and molten states during mixing and processing. In order to circumvent this fact, small size of WC is selected. Moreover, the powders are mixed in a hot plate stirrer. For mixing, powders are put in a beaker and a stir bar is used to mix it. By changing stirring speed, powders are thoroughly mixed.

2.2 Filament Fabrication

A single screw extruder (nozzle diameter 1.75 mm) named Filastruder (as shown in Fig.1) is used to make filaments from powder mixtures [21]. A temperature of 180°C is set at Filastruder and the powder mixture is continuously poured in a hopper for making filaments.

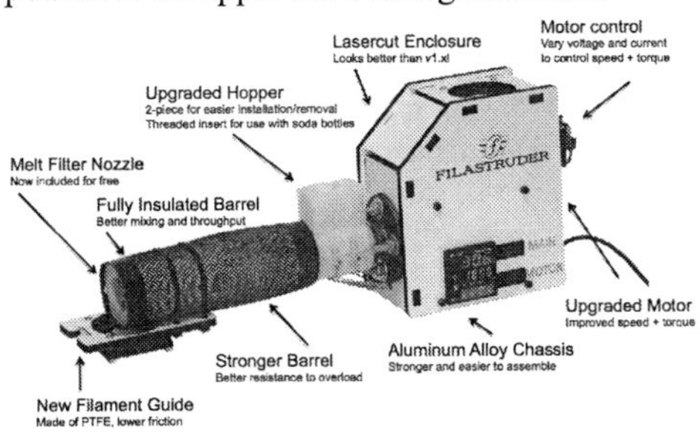

Fig. 1- Filastruder

2.3 Characterization

Differential scanning calorimetry (DSC) and Melt flow index (MFI) are evaluated for some of powder samples. DSC is done to know the glass transition temperature, melting point and heat of reaction for both powder mixtures and processed materials. MFI is performed to assess whether molten filament will flow well through nozzle during part fabrication in FDM systems.

To determine the strength of the fabricated filaments, tensile tests are conducted. Scanning electron microscopy (SEM) is carried out to observe WC powder distribution in a powder mixture and filaments. Microstructures of filaments are observed to know the type of fracture.

Tensile Test

The tensile test of the filament is performed as per ASTM D3822 using an MTS machine of capacity 10 kN, model LPS.104 procured from MTS Systems Corp., MN, USA. For this test, a filament of uniform diameter of length 17.6 cm is cut from the bunch of filaments. 76 mm is fixed as the gage length (as shown in Fig 2a) while 50 mm on both sides of the filament is tightened under a grip (as shown in Fig 2b). There is no much work carried out in this type of testing but there is one instance where the same gage length is taken [19]. Authors select this gage length so that comparisons could be made, if required. Selection of this gage length also conforms to ASTM D3822 standard. Both sides of the filament are padded with cotton before being fitted in the fixture of the machine [Fig 2a]. This ensures that the filament shall not be snapped at the end. Various extension rates such as 10 mm/min, 100 mm/min and 181.4 mm/min are tried during preliminary tests and it is found that 10 mm/min gives consistent result. Consequently, for final experiments, the extension rate of 10 mm/min is adopted.

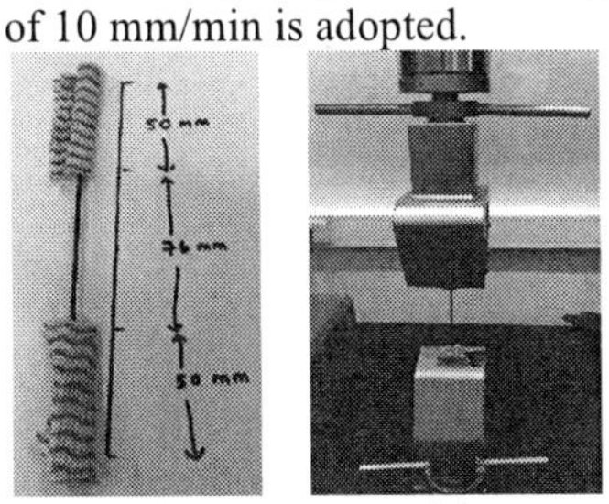

(a) (b)

Fig 2(a)- Filament padded with cloth, (b) Tensile test setup

3.0 RESULTS & DISCUSSION

DSC results show that the thermal cycle during SLS has increased the molecular weight of PA and changed its crystallization and melting point from 138°C [EOS] to 144°C and 186°C [EOS] to 189°C respectively. Moreover, with an addition of 12.5%WC, glass transition temperature increases from 47.74 to 49.04°C. It shows a decrease in chain flexibility with an increase in WC. This is advantageous from the point of view of FDM as it leads to a decrease in the shrinkage of FDM parts [11].

All filaments show high level of ductility (>200%). Because of their ductile nature, it was relatively easy to perform tensile test of filaments. In case of low ductile material (commercial PLA filament), snapping at the end was common. The mechanism of failure observed during test was breaking of a skin of the filament and its subsequent thinning.

During production of filaments, because of the low thermal conductivity of PA (~0.1 W/mK), the filaments were solidified faster along periphery rather than uniformly across the diameter of the filament resulting in creation of skin and bulk with different properties. After breaking of the skin, the bulk started to stretch leading to a break and separation.

Table 1 provides mechanical properties obtained for various filaments. It shows that the tensile strength and modulus increase with an increase in WC percentages. Increasing WC leads to an increase in agglomeration which may act as a stress concentrator and decrease the strength [7]. Despite this fact, properties are higher showing some agglomerated particles acted as a consolidated WC particle responsible for an increase in the strength. Properties of composite filaments are higher than that obtained for ABS or nylon based composite showing that tested composite filament has potential to fulfill requirements for fabricating high-strength filaments for FDM [15].

Filament Composition	Young's Modulus (MPa)	Tensile strength (MPa)	Elongation at break (%)
100%PA	700	25	Did not break
87.5%PA + 12.5%WC	800	30	700
75%PA + 25%WC	920	40	400
62.5%PA + 37.5%WC	1170	45	300
50%PA + 50%WC	1300	60	260

Table 1- Mechanical properties of various filaments tested as per ASTM D3822

Melt flow index (MFI) results shows an increase in MFI with an increase in amount of WC. Increase in MFI is a relatively fair situation for FDM as molten materials will not be clogged in nozzle because of an increase in WC percentages. Consequently, selection of maximum amount of WC to be added to PA will not be decided from flow-ability point of view but from its effect on the brittleness and the ductility of fabricated parts. However, melt-flow volume rate (MVR) calculated from MFI has shown acute changes (from 0 to 50% type material) indicating that translating a high-strength filament into a high-strength 3D product may not be possible by a given range of process parameters available in a given FDM system.

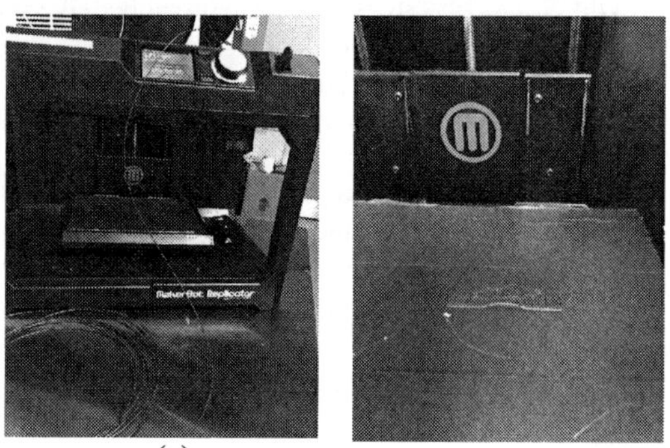

(a) (b)

Fig 3- (a) fabrication of part by filament, (b) tensile test part

In order to check the suitability of the working of filaments in an FDM machine, filaments were manually introduced into the nozzle of an FDM machine (MakerBot Replicator 2.0 as shown in Fig. 3a) to make parts [22]. There is no clogging of the filament noticed. Deposition with higher WC loaded filament was found relatively easy because of its higher MFI. A tensile test part fabricated using PA + 37.5%WC composite filament is shown in Fig 3b.

Cost analysis

Nylon based filaments are expensive (upto $100 per kg) [23] because products obtained from them furnish high toughness and less variation in strengths (with build orientation) in comparison to amorphous material based filaments (as shown in Table 2). In Table 2, ULTEM-1010 is ABS based material while Nylon 12 is nylon-based material supplied by Stratasys [23]. Table shows properties furnished by samples fabricated in both build orientations: vertical (V) and horizontal (H).

Material	Tensile strength (MPa) (ASTM D638)	Flexural strength (MPa) (ASTM D790)	Tensile Elongation at break (%) (ASTM D638)
ULTEM-1010	81 (H) 29 (V)	144 (H) 77 (V)	3.3 (H) 1.3 (V)
Nylon 12	48 (H) 44 (V)	69 (H) 59 (V)	30 (H) 5 (V)

Table 2- Mechanical properties offered by FDM materials offered by Stratasys [23]

Filastruder (for producing filaments) costs $300. Waste PA was procured at no charge. WC costs $40 per kg. For fabricating 1 kg filament containing 25 weight % WC, materials cost would be $10 for fabricating. Assuming other (transportation, storing, inventory) costs as $5, a robust filament will cost less than $20 per kg. If one needs to pay for waste PA, the cost will increase upto $30 per kg. The cost is still far lower than the present market rate of $100 per kg. Consequently, the present research substantiates a methodology to create high-value filaments cost-effectively.

There are a number of polymers for SLS but 95% of polymers consumed are polyamide [17]. It is an expensive SLS powder and costs about $100 per kg. A SLS system contains about 20 kg of powder (more or less depending upon the size and the model). Approximately, 25% of powders gets wasted by degradation during fabrication. A busy SLS system can generate about 50 kg waste powder per month. If a company is using a number of SLS systems for mass customization, the waste powder generated will be 100s of kg. Presently, there is no mechanism to utilize these high-value wastes, which add up to the burden of waste management. If these wastes will be turned into filaments for FDM as demonstrated in the present research, this will not only generate revenue but also contribute to a sustainable manufacturing practice.

CONCLUSIONS

It is demonstrated that a feedstock material (filament) for FDM can be fabricated from a mixture of waste SLS powder and tungsten carbide. Filament provides adequate mechanical property and is compatible with FDM machine. Though, materials analysis showed that for fabricating a high-strength part, further testings from various different FDM systems are required.

Fabrication cost of filament is exceedingly low while mass customization has propelled generation of SLS waste powders yet to be properly recycled. Converting these powders into robust inexpensive filaments would not only make AM self-sustainable but also contribute the efficient utilization of high-value resources.

ACKNOWLEDGMENTS

Anubis Manufacturing Consultants, Mississauga, Canada is thankfully acknowledged for providing PA 2200 powder. The grant provided by Lassonde School of Engineering and NSERC Chair in Design Engineering is thankfully acknowledged.

REFERENCES

1. Hunt, E., Zhang, C., Anzalone, N. et al, 2015, "Polymer recycling codes for distributed manufacturing with 3D printers", Resources, Conservation and Recycling, 97, 24-30.
2. Hwang, S., Reyes, E. I., Moon, K. S., Rumpf, R. C., Kim N. S., 2015, "Thermo-mechanical characterization of metal/polymer composite filaments and printing parameter study for fused deposition modeling in the 3D printing process", Journal of Electronic Materials, 44 (3), 771-778.
3. Kalita, S. J., Bose, S., Hosick, H. L. et al, 2003, "Development of controlled porosity polymer-ceramic composite scaffolds via fused deposition modeling", Materials Science and Engineering C 23, 611-620.
4. Nikzad, M., Masood, S. H., Sbarski, I. et al, 2009, "A study of melt flow analysis of an ABS-iron composite in fused deposition modeling process", Tsinghua Science and Technology, 14(1), 29-37.
5. Nikzad, M., Masood, S. H., Sbarski, I., 2011, "Thermo-mechanical properties of a highly filled polymeric composites for fused deposition modeling", Materials and Design, 32, 3448-3456.
6. Roberson, D., Shemelya, C. M., MacDonald, E., Wicker, R., 2015, "Expanding the applicability of FDM-type technologies through materials development", Rapid Prototyping Journal, 21(2), 137-143.
7. Torrado, A. R., Shemelya, C. M., English, J. D. et al, 2015, "Characterizing the effect of additives to ABS on the mechanical property anisotropy of specimens fabricated by material extrusion 3D printing", Additive Manufacturing, 6, 16-29.
8. Zhong, W., Li, F., Zhang, Z. et al, 2001, "Short fibre reinforced composites for fused deposition modeling", Materials Science and Engineering A 301, 25-130.
9. Masood, S. H., Song, W. Q., 2004, "Development of new metal/polymer materials for rapid tooling using fused deposition modeling", Materials and Design, 25, 587-594.
10. Dul, S., Fambri, L., Pegoretti, A., 2016, "Fused deposition modeling with ABS-graphene nanocomposites", Composites: Part A, 85, 181-191.
11. Weng, Z., Wang, J., Senthil, T. et al, 2016, "Mechanical and thermal properties of ABS/montmorillonite nanocomposites for fused deposition modeling 3D printing", Materials and Design, 102, 276-283.
12. Baechler, C., De Vuono, M., Pearce, J. M., 2013, "Distributed recycling of waste polymer into RepRap feedstock", Rapid Prototyping Journal, 19 (2), 118-125.
13. Boparai, K. S., Singh, R., Singh, H., 2016, "Experimental investigations for development of Nylon 6-Al-Al2O3 alternative FDM filament", Rapid Prototyping Journal, 22 (2), 217-224.
14. Boparai, K. S., Singh, R., Singh, H., 2016, "Process optimization of single screw extruder for development of Nylon 6-Al-Al2O3 alternative FDM filament", Rapid Prototyping Journal, 22(4), 766-776.
15. Boparai, K. S., Singh, R., Singh, H., 2016, "Development of rapid tooling using fused deposition modeling: a review", Rapid Prototyping Journal, 22(2).
16. Goodridge, R. D., Hague, R. J. M., Tuck, C. J., 2010, "Effect of long-term aging on the tensile properties of a polyamide 12 laser sintering material", Polymer Testing, 29(4), 483-493.
17. Goodridge, R. D., Tuck, C. J., Hague, R. J. M., 2012, "Laser sintering of polyamide and other polymers", Progress in Materials Science, 57(2), 229-267.
18. Dotchev, K., Yussof, W., 2009, "Recycling of polyamide 12 based powders in the laser sintering process", Rapid Prototyping Journal, 15(3), 192-203.
19. Wudy, K., Drummer, D., Kuhnlein, F., Drexler, M., 2014, "Influence of degradation behavior of polyamide 12 powders in laser sintering process on produced parts", AIP Conference Proceedings 1593, 691, 691-695.
20. EOS (www.eos.info).
21. Filastruder (www.filastruder.com).
22. MakerBot (www.makerbot.com).
23. Stratasys (www.stratasys.com).

Predicting the Mechanical Properties in Fused Deposition Modelling of Elastomers using Artificial Neural Network

A. Czekanski
Department of Mechanical Engineering
York University
Toronto, Ontario, Canada
alex.czekanski@lassonde.yorku.ca

M.S. Chaudhry
Department of Mechanical Engineering
York University
Toronto, Ontario, Canada
salman01@yorku.ca

ABSTRACT

Fused Deposition Modelling has proved to be the most promising fabricating technique in additive manufacturing technology. Mechanical properties of parts fabricated from fused deposition modelling technique are greatly influenced by the selection of process parameters. In order to enhance the mechanical properties of a FDM fabricated part, it is necessary to study the influence of process parameters on the resulting mechanical performance. The main focus of the study is to characterize the influence of initial process parameter on the mechanical performance of the thermoplastic polyurethane under quasi-static strain rate. In this study the effect of part percentage infill, layer height and raster orientation on the mechanical properties of a FDM part is evaluated. Furthermore, statistical analysis revealed the relative importance and presence of significant interactions between the input parameters. Moreover, artificial neural networking (ANN) approach was implemented to evaluate the regression model between the input factors and resulting strength in the samples. Predicted results from the trained neural network model were found to be in good agreement with experimental values.

KEYWORDS: Mechanical Characterization, FDM, ANN, Elastomers

INTRODUCTION

Fused deposition modeling is a process in 3-D printing in which a thermoplastic material is extruded in layers to create a three dimensional object (Figure 1). In this process, the design of the product is first realized within computer aided design (CAD) software. From a solid model it is exported to STL tessellated format. This faceted model is then sliced into parallel horizontal cross-sections. In this process, the build material used, such as acrylonitrile butadiene styrene (ABS) is in the form of flexible filament [1]. Through a temperature controlled extrusion nozzle, each layer is melted and deposited on to the build plate in the form of fibers (Figure 1). The finished 3-D part takes the form of a vertically stacked laminated composite with a network of fibers and voids. Bonding between adjacent fibers is achieved by thermally driven diffusion welding [2]. The overall bond strength is strongly affected by the envelope temperature and variations in convective conditions within the building chamber [3]. During fabrication, bottom layers are susceptible to deformation due to residual stresses caused by uneven rapid heating and cooling cycles. This localized contraction and expansion affects not only the mechanical performance but also causes distortions during processing [4]. Residual stress induced deformation has shown dependence on the stacking section length [5].

FDM process produces parts with unique characteristics. The material is deposited in a way that results in directional anisotropic behavior. Moreover, overall quality and performance of a 3-D printed part is also affected by the build parameters such as printing direction (raster angle), percentage amount of infill and layer thickness as illustrated in Figure 2 and Figure 3. Recently, a huge amount of attention was given to analyzing and characterizing their effects [6] [7]. Rayegani and Onwubolu [8] modelled a functional relationship between the build parameters and tensile strength using group method of data handling (GMDH). The predicted valued were a very close match for samples with different orientation, and build parameters. A similar study done by Sood et al. [2] investigated the influence of layer thickness, raster angles, infill width and air voids on the bonding and distortion within the part and its mesostructured configuration. Furthermore, they investigated the effect of these parameters on the compressive strength of the specimens and drew conclusions about the importance of fiber-to-fiber bonding [9]. Compressive strength is also affected by the anisotropic behavior due to build direction, showing a reduced strength in transverse direction as compared to axial. [10].

The most common thermoplastic polymers used in FDM based 3-D printers are Polylactic acid (PLA) and Acrylonitrile butadiene styrene (ABS). With the rapid growth in consumer-based 3-D printing industry, new advanced materials for this process are being developed. One of the most interesting additions to the family of FDM based printing material is thermoplastic polyurethane (TPU), which demonstrates high flexibility and rubber like characteristics. While a lot of attention has been devoted to understanding the performance of process parameters on popular polymers like ABS, in the case of new FDM materials however, very little work has been done in terms of characterization and optimization of printing parameters. Moreover, absence of such guidelines and standards for new materials is a matter of concern.

The aim of this work is to study the effect of printing parameters such as percentage infill, layer height and raster angle on the mechanical properties of 3D printed TPU samples. This study aims to cover this knowledge barrier by incorporating 3-D printed TPU samples at quasi-static and high strain rate

EXPERIMENTAL PROCEDURE

In this study, build parameters including sample orientation and layer thickness are considered at different infill percentages. TPU material used in this study was received in the form of filament rolls from Ninja Flex. This material is desirable due to its useful mechanical properties, such as high flexibility and high elongation at break as compared to traditional FDM materials. A commercial entry level FDM based 3-D printer was used for fabricating the samples (Makerbot 2). Some modifications were made to the extruder to work with flexible filament using the custom profiles settings. The extrusion temperature was kept constant at 230°C for all the prints and samples.

Table 1 The printing parameters of thermoplastic polyurethane 3D printed samples

Control Factors		Fixed Factors	
Percentage Infill (%)	100		
	90		
	80		
Layer Height (mm)	0.1	Part Orientation	Flat
	0.2	No. of Shells	2
	0.3	Nozzle Diameter	0.4
	0.4	Coarseness	0.0001 mm
		Part Orientation	Horizontal
Raster Angle (0)	0/90		
	45/45		

Quasi-static tension tests were conducted using a commercial load frame (MTS Criterion 43) in accordance to ASTM D638 Standard Test Methods for Tensile Properties of Plastic [11]. The tests were carried out at a crosshead speed of 0.05 mm/min and data was collected at 50 Hz. Extension values were recorded using a MTS Laser Extensometer having accuracy up to 0.001mm. From the extension and load values recorded from tension test stress/strain curves were generated for further analysis.

RESULTS AND DISCUSSION

Quasi-Static Tensile

To carry out quasi-static tensile tests, experiment was designed by applying full factorial approach in design of experiment (DOE). This methodology allows to account for every possible combination of parameters. Table 2, provides the outcomes of three test runs for each possible combination of process parameter. Figure 4(a), illustrates the categorical mean tensile strength for each build parameters at different levels. It can be noted from the main effect plot that the most significant factor effecting the tensile strength of 3D printed TPU samples is the layer height. As the layer height of the sample increases from 0.1 mm – 0.4 mm the resulting TPU sample is strengthened by 73.5%. For every 0.1 mm increase in the layer height the tensile strength of the specimen is increased by 25%. At 0.1 mm of layer height due to more number of layers present in the sample, there is an increased distortion effect due to high temperature gradient in the bottom layers. Fewer number of layers reduces the number of heating and cooling cycles minimizing the distortion effect. Layer height can be seen as a trade-off between the mechanical strength and the overall surface smoothness and finish of the 3d printed parts.
It can also be noted from the figure that the raster angle or the arrangement of fibers in the sample also affects the tensile strength of the specimen with a reduction of almost 12% in tensile strength when the fibers are rotated from 0°/90° to +45°/-45° orientation. At 0°/90° raster orientation some of the fibers are aligned in the loading direction. The sample fails when the fibers carrying the load breaks. At a raster orientation of +45°/-45°, the sample breaks when the thermally fused bonds between the fibers fails. Furthermore, it is interesting to

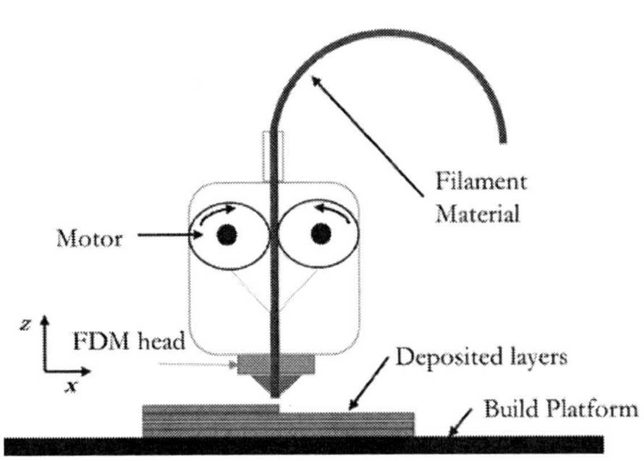

Figure 2 Fused Deposition Modelling Process

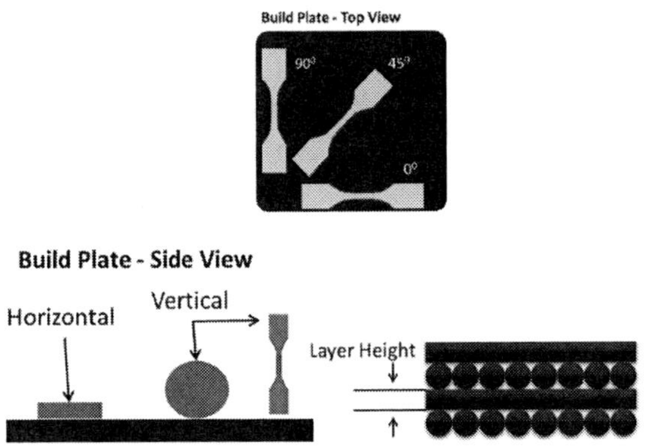

Figure 1 Visualization of sample orientation and layer height.

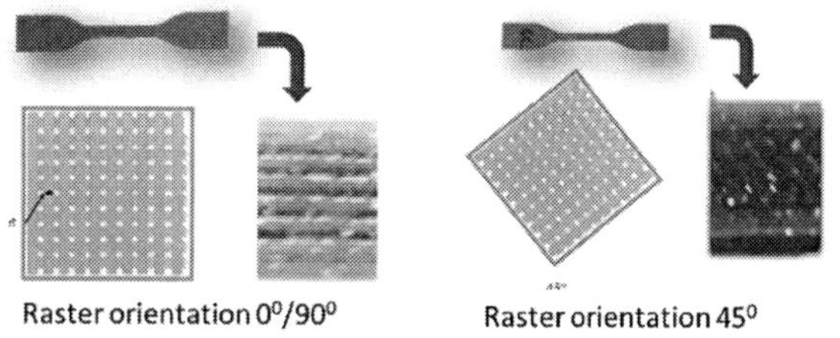

Figure 3 Different raster angle as a result of thermoplastic polyurethane part printed in different orientations.

see a decrease in the tensile strength as the percentage infill in the samples are increased. The probable reason for this behavior can be justified by taking into account the thermal degradation of material due to the presence of heated extruder head for extended period of times on a region at high infill.

Infill percentage at 100% results in no or negative airgaps between the adjacent layers resulting in restricted heat dissipation and accumulation of extra stress. This resulted in approx. 23.5% reduction in tensile strength, from 22.19 MPa at 80% infill to 16.99 MPa at 100% infill. Furthermore, possible interaction between the printing parameter can be deduced from interaction plot. Examining Figure 5 reveals a visible interaction between the height and percentage infill in the sample. It is also evident from the figure that raster angle has no combined effect with either the layer height or the percentage infill in the sample.

Next in this study the levels of printing parameters were used as input and the experimental tensile strength as output to develop an artificial neural network (ANN) model. The trained and validated ANN model with a total of 10 neurons in the hidden and one neuron in the output layer was validated against the experimental means for each combination of factors present in Table 2. Figure 4(b) illustrates the comparison between the experimental and the model results. The ANN model provides a very close agreement, and can be used as a valuable tool as a predictive model when dealing with large number of printing parameters with multiple levels. However, the ANN modelling process does not provide any information about the effect of each factor on the predicted output tensile strength.

CONCLUSIONS

In this study, the effect of 3d printing parameters on the mechanical tensile strength at quasi-static is analyzed. The effect of infill percentage, layer height and raster angle on the mechanical properties and the interactions between these factors have been studied in detail. The experiment was designed using the full factorial approach in DOE, in which every possible combination of printing factors was taken into account. The resulting tensile strength at quasi-static rates were evaluated for three samples for each combination of factor, to conduct further statistical analysis. Results revealed that at quasi-static rates, the most significant factor effecting the ultimate tensile strength of the 3d printed sample is the layer height. A change in the layer height form 0.1 mm to

Table 2 Input process parameter and corresponding experimental and predicted tensile strength

Printing Parameter			Tensile Strength (MPa)				
Percentage Infill (%)	Layer Height (μm)	Raster Angle (⁰)	Test 1	Test 2	Test 3	Mean	Predicted
100	100	0/90	16.03	16.84	16.85	16.57	16.03
100	100	45/45	11.98	13.07	10.80	11.95	11.19
100	200	0/90	19.36	17.61	18.52	18.50	18.31
100	200	45/45	14.09	14.22	14.50	14.27	14.66
100	300	0/90	17.96	16.79	18.78	17.84	17.89
100	300	45/45	20.16	20.25	20.15	20.19	21.32
100	400	0/90	19.04	18.55	18.85	18.81	18.88
100	400	45/45	18.85	15.39	19.31	17.85	18.23
90	100	0/90	14.75	13.80	12.73	13.76	11.51
90	100	45/45	10.73	11.24	7.97	9.98	11.51
90	200	0/90	13.33	16.01	14.32	14.55	14.97
90	200	45/45	12.57	15.17	13.97	13.91	14.95
90	300	0/90	30.61	26.19	31.42	29.41	28.28
90	300	45/45	28.60	26.41	25.16	26.72	28.15
90	400	0/90	30.02	29.02	33.02	30.68	30.58
90	400	45/45	28.81	28.69	28.00	28.50	30.58
80	100	0/90	21.16	16.01	19.62	18.93	18.19
80	100	45/45	17.32	17.33	14.24	16.30	18.19
80	200	0/90	21.20	28.57	27.46	25.74	27.93
80	200	45/45	18.69	17.72	20.20	18.87	19.48
80	300	0/90	17.28	21.31	19.11	19.23	19.60
80	300	45/45	19.15	19.36	18.18	18.90	19.61
80	400	0/90	30.46	30.63	28.59	29.89	29.65
80	400	45/45	27.48	32.35	29.21	29.68	29.65

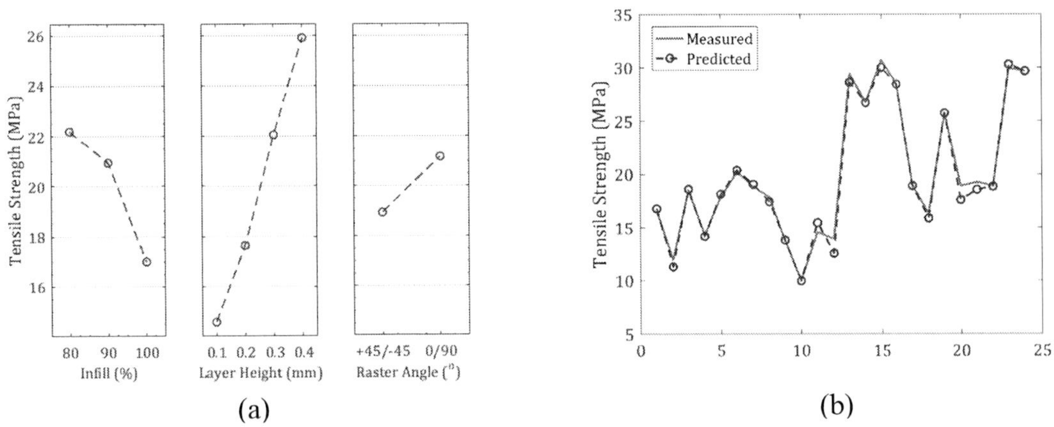

Figure 4 Main effect plot of each factor for quasi-static tensile test, (b) Experimental vs predicted tensile strength

0.4 mm, results in less amount of layer and reduced distortion due to thermal contraction increasing the tensile strength by 73.5%. Layer height is found to be the most significant factor followed by percentage infill with raster orientation found to be the least contributing factor effecting the tensile strength at quasi-static strain rates. For percentage infill, maximum tensile strength resulted in the samples with 80 percent infill which is reduced by 23.5% at 100% infill level. However, a strong interaction between the infill percentage and layer height was also found with a contribution of ~30% towards the tensile strength in specimen. The error contribution in the results is evaluated to be 1.98%. The large amount of error can be attributed to the distortion samples resulting in in different degree of thermal gradient. This results in a weak interlayer bonding which is the primary mode of failure under loading. Next in the study a nonlinear regression model was developed using artificial neural network (ANN). This model proved to be a very useful tool to deal with the nonlinear relation of the printing parameter with the mechanical properties of FDM fabricated parts. The experimental results with the model predicted results are presented in the study for quasi-static tensile strength.

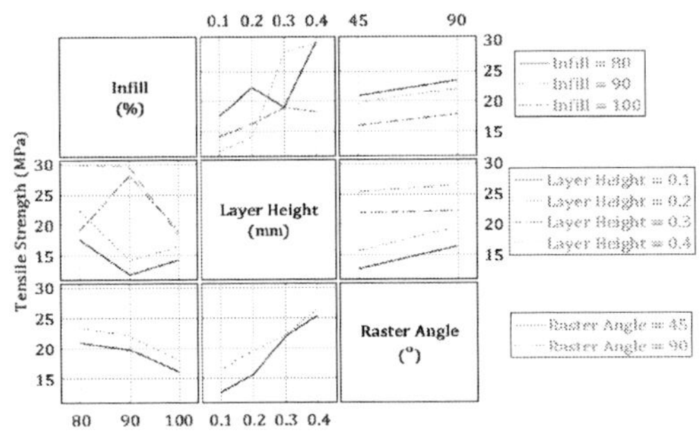

Figure 5 Interaction plot for quasi-static tensile test results.

REFE RENCES

[1] Upcraft S and Fletcher R., "The Rapid Prototyping Technologies," *Assembly Automat,* vol. 23, no. 4, pp. 318-330, 2003.

[2] Sood AK, Ohdar RK and Mahapatra SS, "Parametric Appraisal of Fused Deposition Modelling Process Using the Grey Taguchi," *Proc Inst Eng B J Eng Manuf,* vol. 224, no. 1, pp. 135-145, 2010.

[3] Q. Sun, "Effect of Processing Conditions on The Bonding Quality of FDM Polymer Filaments," *Rapid Prototyping Journal,* vol. 14, no. 2, pp. 72-80, 2007.

[4] Zhang Y and Chou K, "A Paramteric Study of Part Distortion in Fused Deposition Modelling Using Three-Dimensional Finite Element Analysis," *J Eng Manuf,* vol. 228, no. 8, pp. 959-967, 2008.

[5] Wang TM, Xi JT and Jin Y, "A Model Research for Prototype Warp Deofrmation in the FDM Process," *Int J Adv Manuf Technol,* vol. 33, no. 11-12, pp. 1087-1096, 2007.

[6] Constance Ziemian, Mala Sharma and Sophia Ziemian, "Anisotropic Mechanical Properties of ABS Parts Fabricated by Fused Deposition Modelling," in *Mechanical Engineering*, InTech.

[7] Lee B.H., Abdullah J. and Khan Z., "Optimization of Rapid Prototyping Parameters for Production of Flexible ABS Object," *Journal of Materials Processing Technology,* vol. 30, no. 10, pp. 54-61, 2005.

[8] Rayegani F. and Onwubolu G.C., "Fused Deposition Modelling (FDM) Process Parameter Prediction and optimization Using Group Method for Data Handling (GMDH) and Differential Evolution (DE)," *The International Journal Of Advanced Manufacturing Technology,* 2014.

[9] Anoop K. Sood, Raj K. Ohdar and Siba S. Mahapatra, "Experimental Investigation and Empirical Modeling of FDM Process for Compressive Strength Improvement," *Jouranl of Advanced Research,* vol. 3, pp. 81-90, 2012.

[10] Lee C, Kim S, Kim H and Ahn S, "Measurement of Anisotropic Compressive Strength of Rapid Prototyping Parts," *Journal of Materilas Processing Technology,* vol. 187, pp. 627-630, 2007.

[11] ASTM Standard D638, "Standard test methods for tensile properties of plastics," *ASTM International, West Conshohocken, PA,* 2010.

FROM TUNNELING TO HOPPING: A COMPREHENSIVE INVESTIGATION OF TEMPERATURE COEFFICIENT OF RESISTANCE IN CARBON NANOTUBE/POLYMER NANOCOMPOSITES

Z. H. Zhu
Department of Mechanical Engineering
York University
Toronto, Ontario, Canada
gzhu@yorku.ca

S. Gong
School of Metallurgy and Environment
Central South University
Changsha, Hunan, China
gongshen011@163.com

ABSTRACT

Temperature coefficient of resistance (TCR) of carbon nanotubes (CNT)/polymer nanocomposites is an essential property in sensor applications. In this study, we developed a CNT/polymer composite with zero-TCR, which showed nearly constant resistance with less than 3% deviation. Meanwhile, a high TCR (98% resistance change ratio) was observed at glass transition of polymer matrix. The underlying transport mechanisms were investigated using a multiscale CNT percolation network model together with Landauer-Büttiker formula and Arrhenius law. It is revealed that thermally assisted tunneling and thermally activated hopping are core mechanisms that affect TCR in different temperature ranges. The findings will benefit the design of future sensors made from CNT/polymer nanocomposites.

KEYWORDS: Carbon Nanotubes, Polymer Nanocomposite, Temperature Coefficient of Resistance, Tunneling, Hopping.

INTRODUCTION

The induced electrical conductivity and piezoresistivity of Carbon nanotubes polymer nanocomposites can be potentially used in next generation nanoelectromechanical sensors with high accuracy and sensitivity [1-3]. However, these electrical properties are sensitive to the inevitable environmental temperature and the mechanisms underlying the temperature coefficient of resistance (TCR) of the CNT/polymer nanocomposites should be well understood to enable the practical use of these nanocomposites.

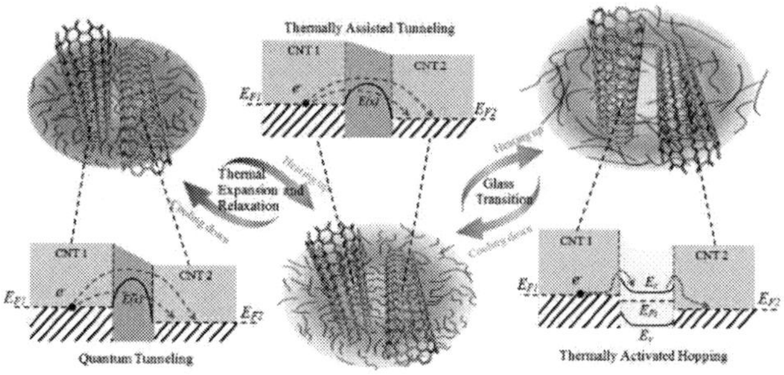

Fig. 1: Effects of thermal expansion and glass transition. [4]

RESULTS AND DISCUSSION

A. Zero- or enhanced TCR of CNT/polymer nanocomposites

The purpose of current work is to control or tailor the TCR of CNT/polymer nanocomposites for better design of advanced materials based on the following two ways. The first is to minimize the temperature effect to obtain zero-TCR in certain temperature range, and the second is to achieve high NTC of resistance with a wide temperature range by introducing glass transition. Fig. 1 shows effects of thermal expansion and glass transition on the intertube space and potential barrier height. Three electron transport mechanisms dominated in different temperature regimes. In the limit of zero temperature only quantum tunneling is important. At high temperatures above glass temperature, the electron escape over the barrier is due to thermally activated hopping. In the intermediate regime, the main mechanism is the thermally assisted tunneling. Because the thermal expansion and glass transition are reversible, the intertube space and potential barrier are restored when the temperature is decreased.

As shown in Figure 2a, conductive behaviors of CNT/polymer nanocomposites for temperatures ranging from 293 to 473 K with various CNT loadings have been investigated. It shows the glass tansition temperature T_g is about 383 K and the percolation threshold is about 0.12 wt.%. At low temperature range (from room temperature to T_g), the nanocomposites exhibited NTC properties at low CNT loadings and PTC properties at high CNT loadings, respectively. In the mid range, the nanocomposites exhibited zero-TCR when the CNT loadings were about 3 wt.%. The experimental results indicate that the CNT loading plays a significant role in determining the characteristics of TCR at low temperatures. As the temperature exceeds T_g, the increasement of conductivity in nanocomposites of all CNT loadings were observed. Next, the temperature effect on the resistance change ratio with respect to the room temperature was investigated and the results are shown in Figure 2b. The nanoomposites with CNT loadings close to the percolation threshold exhibit the highest resistance change ratio. The resistance decreased by 78% as the temperature increased from 293 K to 383 K, and then further decreased by 98% when the temperature reached 473 K. It should be noted that the trends of resistance change ratio in the nanocomposites with CNT loadings of 4 and 5 wt.% are different. The resistances increased first and then reversed to decrease as the temperature arised.

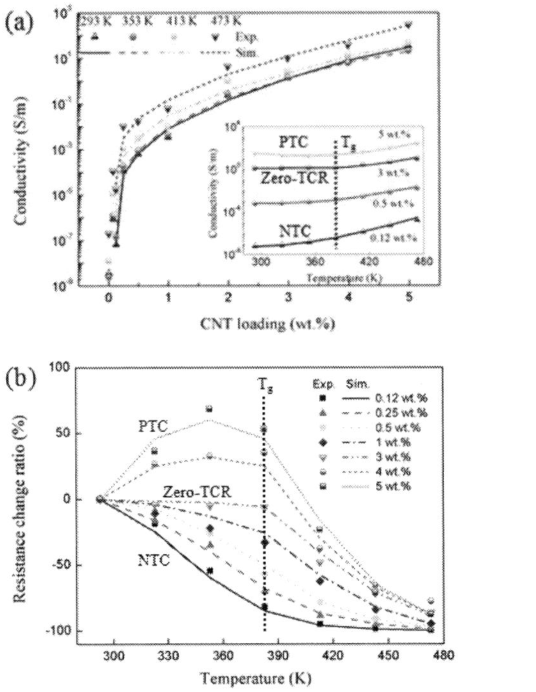

Fig. 2. Experimental and modeling results of zero- and enhanced TCR of CNT/polymer nanocomposites. (a) Conductivity behavior of the CNT/polymer nanocomposite at various CNT loadings and temperatures. (b) Temperature behavior of the resistance change ratio (normalized to the values of room temperature).

B. Mechanisms of TCR

As shown in Figure 3a, the contributions of thermally assisted tunneling (mechanism A), thermal expansion of matrix (mechanism B) and thermally activated hopping (mechanism C) are calculated by Monte Carlo simulations based on the multi-scale percolation network model. Results show that the hopping is the dominant mechanism when temperature is above T_g, while the tunneling and thermal expansion are of equal importance when temperature is below T_g. Mechanism A and B have the opposite effect on the TCR properties, thus a zero-TCR is possible to achieve in CNT/polymer nanocomposites. The contribution of thermal expansion is quite sensitive to the sparseness of CNT network, since increase of intertube distances causes determination of tunneling trasnports at crossed CNT junctions, leading to the decrease in total conductivity. When the CNT loading is about 3 wt. %, these two mechanisms (A and B) offset each other perfectly and the resistances of nanocomposites are temperature-independent for temperatures ranging from 293 to 383 K.

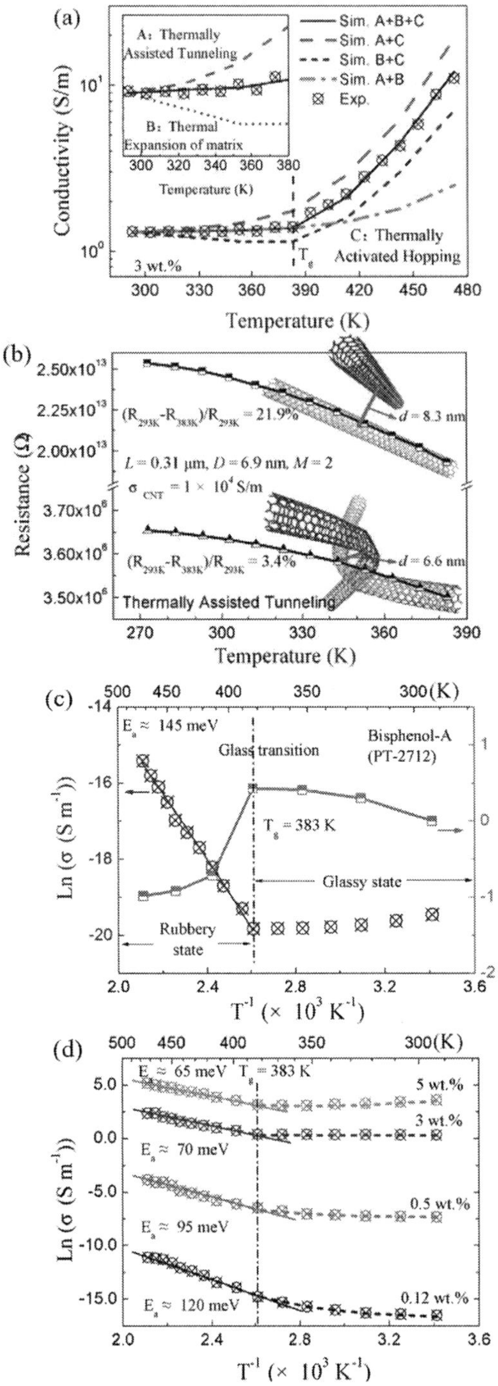

Fig.3 Mechanisms of zero- and enhanced TCR of CNT/polymer nanocomposites. Thermally assisted tunneling (A), thermal expansion of matrix (B) and thermally activated hopping (C) are the three core mechanisms. a) Simulation results show the contributions of these three mechanisms. Zero-TCR below T_g is determined by mechanism A and B, while TCR above T_g is mainly affected by mechanism C. b) Resistance change ratio on two types of junctions contributed by thermally assisted tunneling. The effect of thermally assisted tunneling on the junctions formed by undistorted

CNTs is much larger (more than 6 times) than that on the junctions formed by distorted CNTs. c) and d) Ln(σ) versus 1/T for the polymer matrix and the composites with different CNT loadings. The thermal activation energy E_a are calculated from linear fitting of experiment data.

EXPERIMENT SECTION

Raw Materials: Thin multi-wall carbon nanotubes NC7000, purchased from NANOCYL were 9.5 nm in average diameter and 1.5 μm in length. The epoxy resin and curing agent used were Bisphenol-A PT-2712 and PT-2712 Part B3, purchased from Aeropoxy. These two parts were mixed at room temperature with a resin/hardener mixing weight ratio of 100/22.

CNT/polymer Composite Synthesis: the CNTs were mixed into the part-A of epoxy resin by mechanical stirring for 30 minutes and bath sonication (Sonicor SC-40) for 6 hours at the room temperature. Then CNT/resin mixture was mixed with the amine hardener by stirring for 30 minutes and bath sonication for 1 hours with iced water to prevent the premature solidification of epoxy. After the mixing process, the CNT/polymer mixture was degassed in a vacuum oven (Fisher Scientific Model 281A) at room temperature for 30 minutes before it was injected into moulds. Four electrical probes were inserted into samples with equal spacing of 1.5 mm between probles in order to measure the volumetric electrical conductivity by four-point four probe method. The CNT/polymer composites were cured at room temperature for 24 hours and followed by several heat-cool-heat cycles ranging from room temperature to 200°C to remove possible thermal history.

Conductivity measurements: The electrical conductivity of the CNT/polymer composite samples was measured by the four-point probe method as per ASTM F1529-97. The constant DC current was supplied by a Keithley 6221 power supply and the voltage drop over two inner probes was measured by a Keithley 6517B high resistance meter.

All samples were also characterized in a two-point contact geometry. The electrical resistivity was measured with a digital multimeter when it was lower than 10^7 Ω. A high-resistance meter was used when the resistivity exceeded. Different applied voltages were used on different samples, depending on the level of resistivity of the sample. The two series of experimental results do not have significant differences.

CONCLUSIONS

An effective approach is developed for CNT/polymer nanocomposites with zero-TCR and enhanced TCR. It is found that the zero-TCR below T_g could be achieved by matching the opposite contributions of thermally assisted tunneling and thermal expansion of matrix. On the other hand, the NTC of composites above T_g were enhanced because the transport mechanism at CNT junctions experienced a transition from tunneling to hopping.

ACKNOWLEDGMENTS

This work is supported by the Canadian Space Agency and the Natural Sciences and Engineering Research Council of Canada (NSERC).

REFERENCES

1. Amjadi, M., Kyung, K.-U., Park, I. and Sitti, M. 2016, "Stretchable, Skin-Mountable, and Wearable Strain Sensors and Their Potential Applications: A Review", *Adv. Funct. Mater.*, Vol. 26, pp. 1678–1698.
2. Zhang, J., Liu, X., Neri, G. and Pinna, N. 2016, "Nanostructured Materials for Room-Temperature Gas Sensors", *Adv. Mater.*, Vol. 28, pp. 795–831.
3. He, C., Zhao, N., Shi, C., Liu, E. and Li, J. 2015, "Fabrication of Nanocarbon Composites Using In Situ Chemical Vapor Deposition and Their Applications" *Adv. Mater.*, Vol. 27, pp. 5422–5431.
4. Gong S., Zhu Z.H., Li, Z. 2017, "Electron tunnelling and hopping effects on temperature coefficient of Resistance of Carbon Nanotube/Polymer Nanocomposites", *Physical Chemistry Chemical Physics*, Vol. 19, pp. 5113-5120.

THE EFFECT OF QUENCHING MEDIUM ON THE MICROSTRUCTURE AND MECHANICAL BEHAVIOUR OF INCONEL 625 WELDED ALLOY

Anas M. Atieh
Industrial Engineering Department
German Jordanian University
Amman, Jordan
anas.atieh@gju.edu.jo

Abdulaziz AlHazaa
Physics & Astronomy Department
College of Science
King Saud University
Riyadh, Saudi Arabia
aalhazaa@ksu.edu.sa

Atta G. Attaelmanan
College of Science
University of Sharjah
Sharjah, UAE
aattaelmanan@sharjah.ac.ae

ABSTRACT

Nickel-chromium based super alloys Inconel 625 have diversified applications in petrochemical and marine industries. Inconel 625 has high corrosion resistance and high temperature strengths. In general, this alloy is a difficult material to machine. In this study, the Inconel 625 has been TIG welded and heat treated for one hour at 900°C and then quenched using a medium of oil, water, and air to enhance the surface mechanical characteristics of the alloy. The results showed that no significant change in microstructure observed for the post-weld heat treated alloy. However, when the Inconel 625 was heat treated to 900°C and then quenched in water, the hardness of the alloy increased from around 300 HV to 450 HV. On the other hand, the tensile strength was decreased for all heat treated samples quenched in oil, water and air compared to non-heat treated welded samples.

KEYWORDS: Inconel 625, heat treatments, quenching, tensile, hardness

INTRODUCTION

For most ferrous alloys, it is well established that a quenching medium (e.g. water or oil) can be effectively used to enhance the surface mechanical properties, hardness and wear resistance. In the past decade, the application of heat treatment showed significant improvement of alloys mechanical characteristics due to surface properties enhancements [1–3].

Inconel 625 is a nickel chromium molybdenum alloy that features an outstanding corrosion resistance in a wide range of corrosive media. It is commonly used for its high strength and excellent weldability by different joining processes such as tungsten inert gas (TIG) welding [4]. Inconel 625 is heavily used in potash extraction industry in Jordan to store chemicals and in the pipelines, where TIG welding is used to join chemicals pipelines. However, very little research work has concentrated on the effect of post-weld heat treatment characteristics of this alloy. Cortial et al. studied the effect of heat treatment between 600 and 1000 °C and holding time of 8 hours on the mechanical and corrosion characteristics of welded Inconel 625 alloy. The work suggested that beyond 850°C the ratio between alloying elements Nb, Mo, Si, Cr, Ni and Fe approximate to 1 between the inter-dendritic spaces and dendritic cores, and the composition heterogeneity disappeared at 1000°C [5]. Furthermore, it was reported that between 750 and 950 °C and a holding time of 8 hours, a catastrophic reduction of mechanical properties of the Inconel 625 alloy was achieved. This was attributed to the precipitation of intermetallic compound δ Ni_3(Nb, Mo, Cr, Fe, Ti) phase. This research work concluded that the higher the heat treatment temperature the weaker the mechanical strengths [5].

In this study, the quenching medium of oil, water, and air has been used as a method of altering the mechanical characteristics of welded Inconel 625 alloy. The effect on microstructure, hardness and strength have been investigated with unifying all testing parameters and altering quenching regimes only.

EXPIREMENTAL PROCEDURE

Inconel 625 plates were brought from Arab potash company (Dead see, Jordan) in the dimension of 20x10x1 cm. The elemental compositional analysis is shown in table1.

Table 1:Inconel 625 composition in weight %

Ni	Cr	Mo	Nb	Fe	C	Si	Al	Ti	Mn
61	21.5	9.0	3.6	2.0	0.05	.20	.20	.20	.20

The edges of the Inconel 625 plates were cut into V-shape, then brought into contact for TIG welding using Inconel N82/125/10ts filler wire. TIG welding was carried out by 5 upside passes and 3 downside passes to overcome the plate bending and crack formation. After TIG welding, Inconel 625 plates were cut and machined into 4 sets of testing specimens to ensemble each test. The set included two dog-bone shaped specimen for tensile testing, 2 microstructure specimens and 2 hardness specimens. Testing specimens was prepared according to ASTM standards ASTM E3-11, ASTM E8/E8M-15a, and E384-11. Same number of specimens were prepared for each of the 3 quenching medium; air, oil and water and one control set with no quenching treatment.

In order to compare the changes in surface hardness before and after quenching media treatment, hardness measurements were obtained using Tru-Blu universal hardness tester (Tru-Blu U/10 version F139, CA, USA). 10 Kgs load was applied with 15 second indentation time. The hardness measures were taken along the thickness starting from the base metal through the weld zone then ending at the other end.

Tensile testing was performed on 25 KN Testometric (FS300 CT, England) universal tensile testing machine at room temperature and constant cross-head speed of 0.5 mm/min.

The microstructure of the welded sample was observed using electron microprobe analyzer SEM (VEGA 3, TESCAN, Czech Republic) equipped with energy dispersive X-ray spectroscopy (EDS) for compositional analysis.

RESULTS AND DISCUSION

Figure 1 shows optical micrograph of the alloy (a) quenched in air and (b) quenched in oil. There was no significant difference in microstructure of the Inconel 625 alloy after the quenching in air or oil

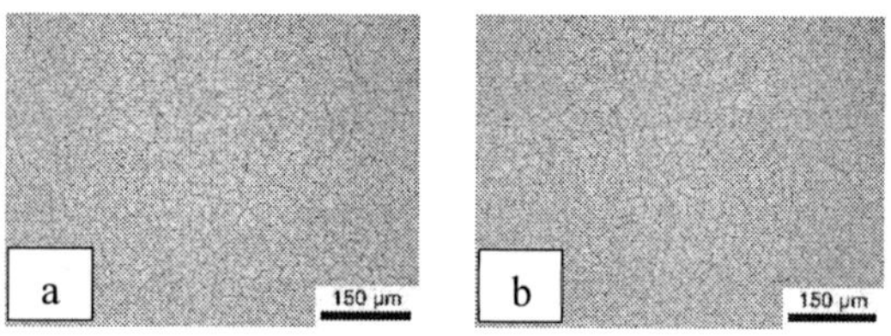

Figure 1:optical micrograph for (a) air and (b) oil quenched

Similar observation was seen for SEM micrograph and EDS analysis obtained for all quenching medias as seen in Figure 2.

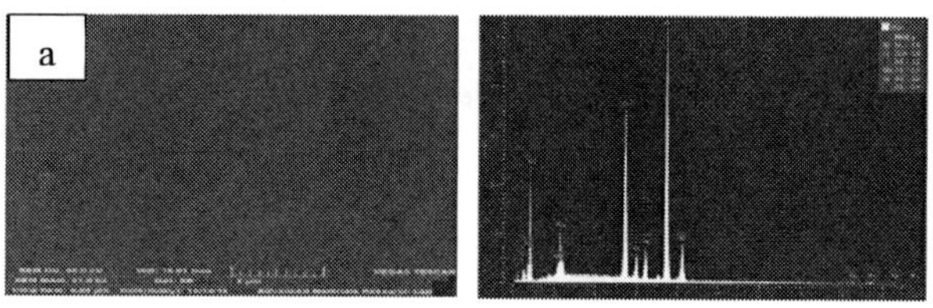

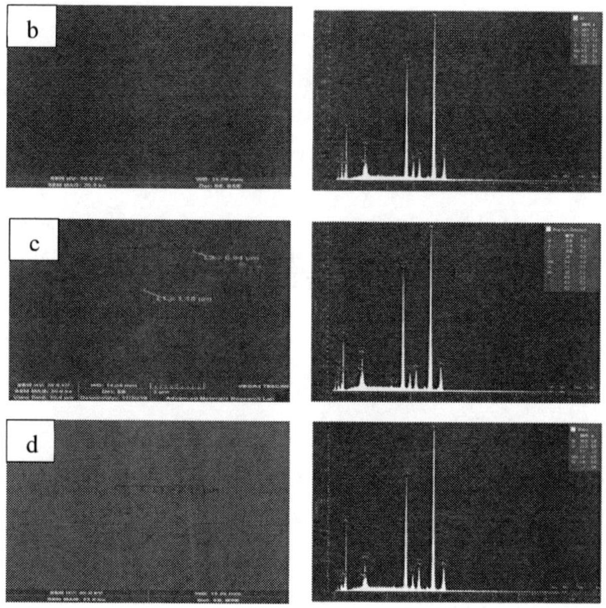

Figure 2: SEM micrographs and EDs analysis for (a) Non-quenched (b) air quenched (c) oil quenched (d) water quenched

However, the hardness measurements taken across the joint interface revealed that Inconel 625 samples quenched in water showed higher hardness of 450 VHN compared to all other samples quenched in oil, air or non-quenched with an average of 280 VHN.

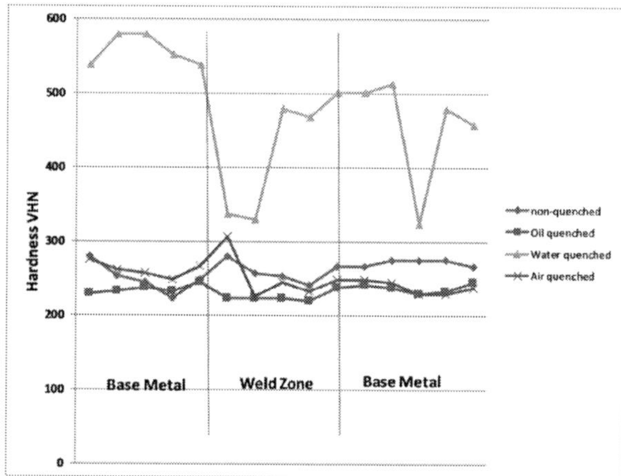

Figure 3: Hardness profile

on the other hand, ultimate tensile strength UTS results showed that the non-quenched and water quenched may results in higher tensile strength (Avg. 700 MPa) while oil and air quenching deteriorate the strength to 400 and 600 MPa. Similar mechanical behavior was reported for other alloys such as Al 6061 and NST-37-2 steel where water quenched specimen has higher UTS and hardness value compared to oil and air quenched ones [6,7].

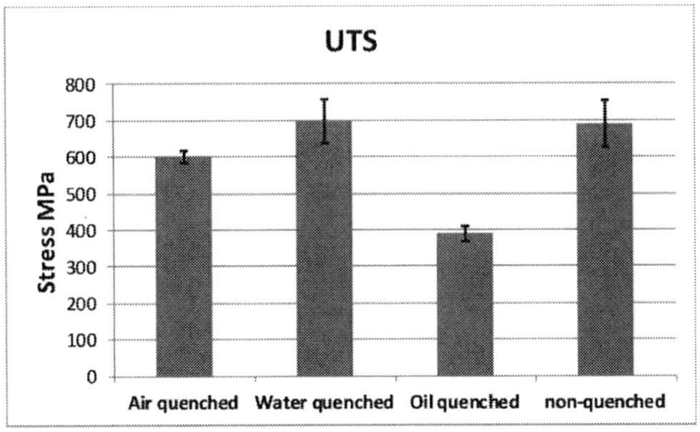

Figure 4: Ultimate tensile strength

Inconel 625 alloy derives most of its strength from solid solution strengthening [8]. It consists of a homogeneous, single, solid-solution strengthened γ phase. When subjected to constant elevated temperatures from 700 to 900°C for at least one hour, an δ phase which is an incoherent equilibrium phase will precipitate out of γ [9].

The microstructure of Inconel 625 consists of a face-centered cubic matrix phase (γ) strengthened with substitutional elements like tungsten, rhenium, and tantalum that cause strain in the lattice due to their larger atomic radii relative to the dominant element of this alloy i.e. nickel. Other phases such as body-centered tetragonal precipitate (γ′′) with a composition of Ni_3Nb can be brought out of the γ matrix [10]. Another possible phase is orthorhombic soft δ phase.

It is expected that the differences in mechanical strength between the water quenched and oil/air quenched could be attributed due to the transformation of the strong γ phase to the softer δ phase which form at the high heat treatment temperature of 900°C for one hour and inter-looking the new δ phase when water quenching medium was used.

CONCLUSIONS

Microstructure analysis and mechanical tests were performed on INCONEL 625 specimens to interpret the effect of heat treatment followed by three different quenching mediums: air, oil and water on weldment properties of Inconel 625 alloys. Tensile testing and hardness results has shown that water quenched have the highest UTS value of 700 MPA while oil and air quenching deteriorate the mechanical properties. No significant difference was observed on the microstructure of the alloy specimens treated by the three different quenching mediums.

ACKNOWLEDGMENTS

The authors gratefully acknowledge the SEM/EDS analysis performed by Mr Mohamed Shameer at the Center for Advanced Material Research, University of Sharjah. Contribution in the experimental testing conducted at the German Jordanian University by Aya Abo Daqa, Karmel Enaim, and Faisal Falah are also appreciated.

REFERENCES

[1] Molinari, A., Pellizzari, M., Gialanella, S., Straffelini, G., and Stiasny, K. H., 2001, "Effect of Deep Cryogenic Treatment on the Mechanical Properties of Tool Steels," J. Mater. Process. Technol., **118**(1–3), pp. 350–355.

[2] Huang, J. ., Zhu, Y. ., Liao, X. ., Beyerlein, I. ., Bourke, M. ., and Mitchell, T. ., 2003, "Microstructure of Cryogenic Treated M2 Tool Steel," Mater. Sci. Eng. A, **339**(1–2), pp. 241–244.

[3] Feng, C., and Khan, T. I., 2008, "The Effect of Quenching Medium on the Wear Behavious of a Ti-6Al-4V Alloy," J. Mater. Sci., **43**(2), pp. 788–792.

[4] "www.Specialmetals.com."

[5] Cortial, F., Corrieu, J.-M., and Chantal Vernot-Loier, 1994, "Heat Treatments of Weld Alloy 625: Influence on the Microstructure, Mechanical Properties and Corrosion Resistance," Miner. Met. Mater. Soc., pp. 859–870.

[6] Abubakre, O. K., Mamaki, U. P., and Muriana, R. A., 2009, "Investigation of the Quenching Properties of Selected Media on 6061 Aluminum Alloy," J. Miner. Mater. Charact. Eng., **8**(4), pp. 303–315.

[7] Fadare, D. A., Fadara, T. G., and Akanbi, O. Y., 2011, "Effect of Heat Treatment on Mechanical Properties and Microstructure of NST 37-2 Steel," J. Miner. Mater. Charact. Eng., **10**(3), pp. 299–308.

[8] Ryan Lewis, Megan McCabe, Jennifer Rowe, Regan Rumph, Bryce Simmons, J. W., 2017, *Delta Phase Transformation in Inconel 625.*

[9] Eiselstein, H. L., and Tillack, D. J., 1991, "The Invention and Definition of Alloy 625," Superalloy 718,625 and Various Derivatives," Miner. Met. Mater. Soc.

[10] Donachie, M. J., and Donachie, S. J., 2002, "Understanding Superalloy Metallurgy," *Adapted from Superalloys: A Technical Guide, 2nd Edition, Approved by the ASM Handbook Committee for Addition to the ASM Handbook Supplements, 2011.*

Robotics and Control Systems

Proceedings of the 26th CANCAM
Victoria, BC, Canada, May 29 - June 1, 2017

WALL-FOLLOWING MOTION PRIMITIVES FOR A THREE-WHEELED CLEANING ROBOT

David J. Yoon, Steven L. Waslander
Mechanical and Mechatronics Engineering
University of Waterloo
Waterloo, Ontario, Canada
j32yoon@uwaterloo.ca, stevenw@uwaterloo.ca

Kenneth Lee, Pablo Molina
Avidbots Corporation
Waterloo, Ontario, Canada
kenneth@avidbots.ca, pablo@avidbots.ca

ABSTRACT

Commercial floor scrubbing persists as a manual task, despite the repetitive nature and controlled environment. Specifically for cleaning the boundaries of a floor, detailed mopping by hand is a commonly used method due to the difficulties of driving large floor scrubbing machines near obstacles and corners. This work proposes a method that allows autonomous floor scrubbing machines with holonomic drive platforms to accomplish wall cleaning into and around wall corners in a complete fashion, which otherwise is very difficult to achieve. Three motion primitives are defined to cover a large proportion of straight wall and corner combinations. The same strategy is usable to clean around obstacles of arbitrary shape. Experimental testing on a custom omni-directional floor scrubbing machine is presented, validating the method.

Keywords: Mobile Robotics, Industrial Cleaning, Floor Scrubber, Motion Primitives, Omni-directional

INTRODUCTION

Floor cleaning is a large industry and is ripe for automation. Substantial effort has been put into floor care for both commercial application [1, 2] and research. Specifically in the industry of floor scrubbing, autonomy can replace costly operators and provide detailed cleaning reports from accumulated data.

Much work has been done on automated cleaning. An early attempt by Hofner and Schmidtt [3] generates parallel cleaning tracks by referencing a known planar map. This works well for rectilinear rooms but more was required for cleaning close to wall boundaries and corners. As an improved approach to covering rooms with more intricate designs, Choset [4] decomposes the map into multiple convex cells, each easier to complete coverage upon. The simplest approach to achieving coverage of a cell is a back-and-forth motion. Other motions for coverage have been investigated in literature, such as variations of spiral motions [5, 6]. There are several more works completed on coverage, these examples being only a few. However, key features that they do not address are the complications that arise from large floor scrubbing machines. More specifically, we refer to considering the direction of travel and the shape of the robot's body, which critically affect cleaning around intricate wall contours and obstacles - the primary focus of this work.

Coverage with the purpose of cleaning is not complete by simply the passing of the robot's planar footprint. Rather, what matters is the coverage area and direction of travel of the scrubbing head, the cleaning mechanism that rotates a brush onto the floor. Most designs only include one vacuum at the back of the scrubbing head. Thus, forward movement is the most effective for scrubbing and picking up excess cleaning fluid. Lateral motions may be permissible, but backward movement will leave a large puddle of cleaning fluid.

The robot's body is commonly taken to be of circular shape, which is typical of a domestic vacuum robot, but not of an industrial floor scrubbing machine. Having a circular body has the advantage of spot rotations that allow changes in orientation without collision. An elongated body will have difficulty in achieving scrubbing head coverage around obstacles and corners using a conventional steering drive mechanism or a two-wheel differential drive. Though a path can be planned in open space, coverage completeness degrades around intricate contours and obstacles.

Previous literature suggested a different approach, where wall coverage is handled separately from general floor coverage [5]. A dedicated wall following behaviour adds padding around walls and obstacles, allowing regular coverage to be completed without the additional complexities. Several works have been completed on autonomous wall following. The simplest approach would be a variation of feedback control on a proximity sensor reading from the wall. Others have applied more sophisticated control strategies, some examples being fuzzy incremental control [7] on sonar data, as well as nonlinear control laws based on sonar and odometry [8]. Toibero et al. [9] introduced a control switching strategy that selects among three different controllers depending on the type of wall contour. However, these approaches were developed for domestic robots with a circular footprint. This disregard leads to collisions as the desired distance away from the wall is tuned to be lower for elongated machine bodies.

The predominant method to date of handling the detailed contours of walls, corners, and obstacles is not through the floor scrubbing machine, but mopping by hand. In order to transfer this task to the machine, we propose a hardware alteration of the drive system to solve the complexities of wall coverage. Previous literature has touched upon omni-directional cleaning vehicles [10, 11], but have not gone into detail about wall coverage. Our contribution is a novel approach to wall coverage with an omni-directional drive platform. It involves trajectories that exactly cover wall corners of various angles that can only be accomplished with a holonomic drive system.

Using a known map, high level path planning is completed offline by the placement of waypoints. Trajectories are developed based on a collection of motion primitives to connect the waypoints and a Linear Quadratic Regulator (LQR) [12] control strategy is used for tracking. These motion primitives consist of trajectories to handle straight walls and corners, similar to the categorization of Toibero et al. [9]. In this work, we exclude the possibility of curved surfaces or non-static obstacles, as the former can be tracked by matching the curvature, and the latter can be avoided by waiting for them to pass. Our contributions include an omni-directional robot design for floor cleaning, the definition of motion primitives that plan exact coverage of wall corners with the robot footprint, a system that is capable of wall following in most rooms, and demonstrations of it working in practice. A custom made omni-directional floor scrubber was used for experimental testing. Experimental results from a mock room are presented, based on mapping data of a physical hallway, demonstrating the ability to cover any arrangement of straight walls that the robot can physically fit into.

PROBLEM FORMULATION

A. Cleaning Environment

The cleaning environment defined in this work is comprised of a set of sequential vertices and edges. Each vertex, v $\in \mathbb{R}^2$, is a wall corner in the environment. Each edge, e $\in \mathbb{R}^{2\times 2}$, is a collision-free line segment that connects adjacent vertices, such that e_i is defined by v_i and v_{i+1}. For the case of a protruding obstacle or wall segment of non-rectilinear geometry, a discretized arrangement of vertices and edges can be defined to cover the surface.

The wall is maintained to always be on the right side of the robot with respect to its heading, thus defining the direction of wall coverage. Setting the robot's planar footprint to be a fixed rectangle, the front and back corners that are adjacent to the wall are what dictate the wall following motion. The objective of complete coverage is achieved by these two corners maintaining contact with the edges e_{i-1} and e_i at all times during the cleaning of vertex v_i. Each vertex is a resting point for the robot as the existence of a vertex is coupled with the requirement to stop and turn to the subsequent edge. The intersection of the rectangular footprint and an edge must be avoided as it may mean collision, depending on the how close the vertices were placed to the real wall or obstacle.

For the wall following task, each vertex is further defined as a waypoint. The ith waypoint is defined as a position and orientation on a planar surface in the global frame and using a right handed coordinate system, $W_i = [x_i^w \ y_i^w \ \theta_i^w]^T \in SE(2)$. The heading of each waypoint is defined by the angle of the subsequent edge, with respect to the previous edge. In other words, W_i has a heading θ_i^w which points along e_i, defined relative to the edge e_{i-1}. Moving to a waypoint is complete once the front adjacent corner of the footprint lies on top of the waypoint.

A wall corner's type is defined by its waypoint heading θ_i^w. Outer corners have a heading less than 0 degrees, while inner corners have a heading greater than 0 degrees (assuming the vehicle cleans to the right of the wall). If the heading θ_i^w is equal to 0 degrees, it can be ignored. It is assumed that the angle of an inner corner will never exceed 90 degrees because it is not possible for a rectangular footprint to fit into the tight space available. If such a case were to appear during waypoint selection, a processing step is completed by sectioning out the acute angle, resulting in a right angle corner and an obtuse angle corner. Another possibility is to create two obtuse angle corners. An additional assumption is the length of the edge e_i is greater than the width of the robot's footprint, to ensure a turn can be completed before the next corner is reached. Special care is taken to stop an inner corner turn early for when the length of the edge e_i is shorter than the length of the robot, allowing for such corners in implementation.

B. Motion Kinematics

Figure 1a illustrates the robot's wheel arrangement in its local body frame. To achieve holonomic motion, each of the three wheels is an omni-wheel that produces a linear velocity in the direction perpendicular to its axis. The linear wheel velocities are represented as v_1, v_2, and v_3. As shown in Figure 1a, values θ_2, θ_3, L_1, L_2, and L_3 are measurements required to define the wheel positions with respect to the robot's frame of reference.

The input of the robot at time t is $u_t^b = [v_x^b \ v_y^b \ w^b]^T \in \mathbb{R}^3$, where v_x^b is the linear velocity along the x-axis of the robot frame, v_y^b is the linear velocity along the y-axis of the robot frame, and w^b is the angular velocity about the origin. The relationship between the wheel velocities and the command velocities [13] is defined as

$$\begin{bmatrix} v_1 \\ v_2 \\ v_3 \end{bmatrix} = \begin{bmatrix} 0 & 1 & L1 \\ -\sin\theta_2 & \cos\theta_2 & L2 \\ -\sin\theta_3 & \cos\theta_3 & L3 \end{bmatrix} \begin{bmatrix} v_x^b \\ v_y^b \\ w^b \end{bmatrix}. \quad (1)$$

The state of the robot at time t is $x_t = [x_1 \ x_2 \ x_3]^T \in \mathbb{R}^2$, where x_1 and x_2 define the robot's position and x_3 defines the heading in the global frame. Let $R_{x_3} \in \mathbb{SO}(2)$ be a rotation matrix defined by the robot heading x_3. Then the relationship between u_t^b and the command velocity vector with respect to the global frame u_t^g is

$$u_t^g = R_{x_3} u_t^b. \quad (2)$$

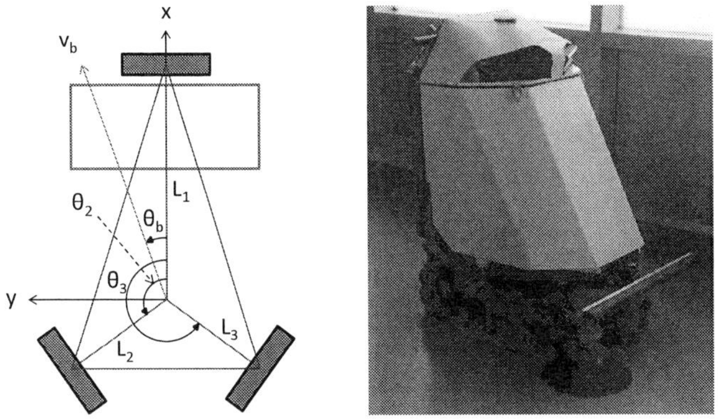

(a) Omni-drive footprint. (b) Image of experimental platform.

Figure 1: a) The three grey rectangles are the wheels and the blue rectangle is the footprint of the scrubbing head. Each wheel is perpendicular to its axle. b) Experimental omni-directional sweeper-scrubber platform built by Avidbots Corporation. The rotary brush at the front corner overcomes the physical limitations of the scrubbing head footprint.

Therefore the linear motion model of the robot without disturbances and a time step of dt is defined as

$$x_t = x_{t-1} + u_t^g \, dt. \tag{3}$$

A trajectory is defined as a time sequential list of desired states and command inputs, $\pi = [x_{t_0}, u_{t_0}^g]...[x_{t_f}, u_{t_f}^g]$ for the time interval $t_0...t_f$.

TRAJECTORY GENERATION AND CONTROL

Three types of trajectories were developed: a straight trajectory, an outer corner trajectory, and an inner corner trajectory. The three trajectories are motion primitives that connect the waypoints W_i together to achieve wall cleaning for any combination of connected straight edges that satisfy the stated assumptions on the corner angles and edge lengths.

A. Straight Trajectory

The straight trajectory involves a translation without angular rotation. The robot must start moving from waypoint W_i at zero initial velocity and end at W_{i+1} at zero final velocity. The robot is to move along its heading, therefore the trajectory is a sequence of collinear poses spaced according to a jerk constraint.

Piecewise position curves were created for linear displacement along the path [14]. The curves are constant in velocity, acceleration, or jerk depending on the section of the piecewise curve. The trajectory $\pi_{straight}$ is then approximated from the linear displacement profile. We apply maximum jerk until maximum acceleration is reached, holding the acceleration constant until maximum velocity is reached, and then holding the velocity constant until the position is halfway through the path. The sequence is then repeated in reverse to symmetrically reach the end of the path. The formulation differs when maximum values of the kinematic elements are not met, which depends on the length of the path and the maximum values set. These cases were identified and handled appropriately.

B. Outer Corner Trajectory

Complete coverage around an outer corner is achieved by having the scrubbing mechanism turn about its front corner that is adjacent to the wall. The robot is assumed to start with its front corner already on top of the waypoint that defines the corner, from its previous motion. The turn is complete once the heading of the robot is aligned with the succeeding wall.

The trajectory π_{open} is a rotation about one of the footprint's corners, but with command velocities and states with respect to the command centre of the robot. The command centre is defined as the intersection of the wheel axles in Figure 1a. A piecewise position curve, as used in the straight trajectory, was applied to form the angular displacement w of the robot. The required sequence of control inputs $u_{t0}^g...u_{tf}^g$ of π_{open} is then generated according to the kinematic relationship of velocities

$$v_B = v_A + r_{BA} \times w, \tag{4}$$

where v_B is the velocity vector of a point B on a rigid body, v_A is the velocity vector of a point A on the same rigid body, and r_{BA} is a displacement vector from point B to A. Thus the resulting computation of v_B at each time step defines $u_{t0}^g...u_{tf}^g$ of π_{open}.

The sequence of required states $x_{t0}...x_{tf}$ of π_{open} is computed by observing that the command centre is displaced as a circular locus of points about the point of spot rotation with radius r_{BA}.

C. Inner Corner Trajectory

Inner corner coverage is completed by having the robot's rectangular footprint fit into the corner and transition out without collision. In what follows, we describe coverage of a perpendicular inner corner, but the same concepts are applicable for any inner corner between 0 and 90 degrees.

Assume the robot starting pose places the top right corner directly over the corner waypoint, so that the robot is fit exactly into the available space. The points of interest are the front and back corners of the footprint that lie on the wall edges. These two points combined with the command centre of the robot create a rigid triangle. Let points A, B and C represent the vertices of the rigid triangle as shown in Figure 2. Additionally, let the two sides of the wall that define the corner be two separate line segments that intersect each other at the corner vertex O. Set the path of vertex A to be constrained to the line segment that it already resides on and set the path of vertex B to be constrained to the remaining line segment. By translation of triangle ABC along the line segments, the path of vertex C is a locus of points defined by an ellipse with its centre on vertex O [15].

The ellipse defines the spatial path of the trajectory π_{inner}. A challenge was faced in determining the motion profile as the

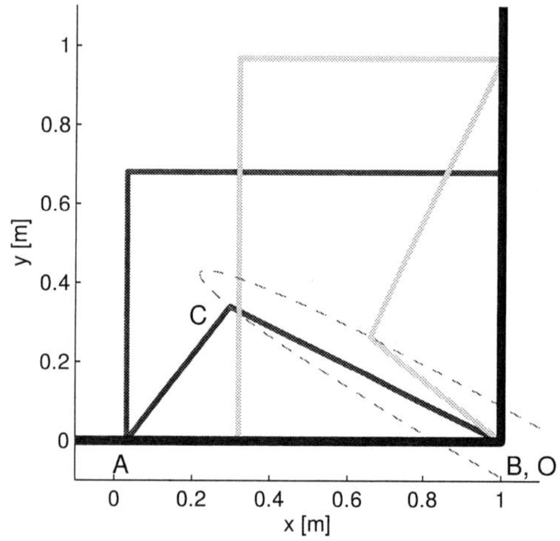

Figure 2: Locus of points that define the elliptical trajectory. The blue and green rectangles are the start and end positions of the robot footprint. Their respective triangles define the start and end of the elliptical path.

translational and rotational velocities are dependent on each other. The motion profile is created by first defining the sequence of translational velocities v_x^g, v_y^g. The magnitude of the translational velocities were set proportional to the curvature of the ellipse, setting the magnitude smaller for smaller radii of curvature. The sequence of rotational velocities are then computed by computing the orientation of the robot throughout the trajectory and numerically approximating the angular velocity with the orientation at the previous time step.

D. Controller

The LQR control strategy was used to track the generated trajectories with the motion primitives incorporated as feedforward inputs for tracking. The motion model given in Equation (3) is nonlinear. Since the only nonlinear component in the equation is the rotational term, a simple feedback linearization was accomplished by formulating the control inputs in the global frame. The formulated commands are then transformed to the body frame for use. Although this transformation can be problematic, as it relies heavily on accurate knowledge of the vehicle orientation, in practice, accurate pose estimation is a critical component of any autonomous cleaning system.

IMPLEMENTATION AND RESULTS

A. Simulation

Figure 3 shows a progression of inner and outer corner trajectories at 30 degree intervals. The trajectories were generated with a time step of 0.1 seconds and every fifth trajectory state is shown. The direction of the arrows define the heading of the robot. These trajectories were generated without considering a tolerance distance away from the wall. As such, these trajectories provide complete coverage with zero distance between the robot and wall.

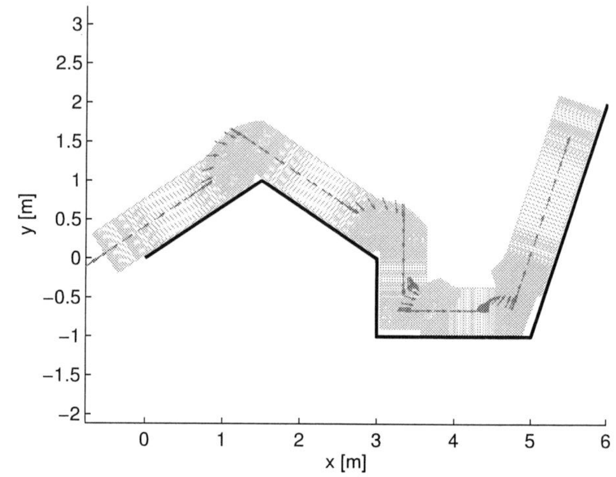

Figure 4: Example scrubbing head coverage. Minor white spaces are due to the scrubbing head not encompassing the entire front portion of the robot footprint.

An example of the scrubbing head coverage is shown in Figure 4. This example demonstrates that the coverage path nearly covers the entire path of straight walls, inner corners, and outer corners. The scrubbing head footprint realistically does not cover the entire front of the robot footprint, thus leaving a few areas of white space.

B. Hardware Validation

The complete implementation was tested on hardware using a custom built autonomous floor scrubbing machine. The entire software was implemented using the Robot Operating System (ROS) framework [16]. Localization of the robot was accomplished using a planar laser scanner with the adaptive Monte Carlo localization algorithm from the ROS library. The machine, made by Avidbots Corporation, is a sweeper scrubber hybrid machine. It is approximately 1 meter in length and 1.2 meters in height. Onboard computation is performed on an Intel 4th generation i5 NUC. A laser scanner and camera sensors are located at the bottom and top of the machine, respectively. Figure 1b shows an image of the experimental platform. The scrubbing head is located behind the front wheel, underneath the body. A rotary brush is attached to the side that will be adjacent to walls during wall following.

The testing environment was built based on a portion of a hallway that was mapped with laser scanner data. Figure 5 shows a screen capture from Rviz, the visualization tool provided by ROS, of the testing environment with the computed wall-following trajectory. The light green rectangle in the top-left corner is the footprint of the robot, demonstrating both the scale and how relatively close the robot travels to the walls. The desired trajectory shown was generated based on a 15 centimeter tolerance away from any obstruction. This tolerance was necessary to accommodate the inaccuracy of the trajectory tracking, preventing collision. It should be noted that this means the experimental implementation is even more conservative than the simulated coverage shown in Figure 4.

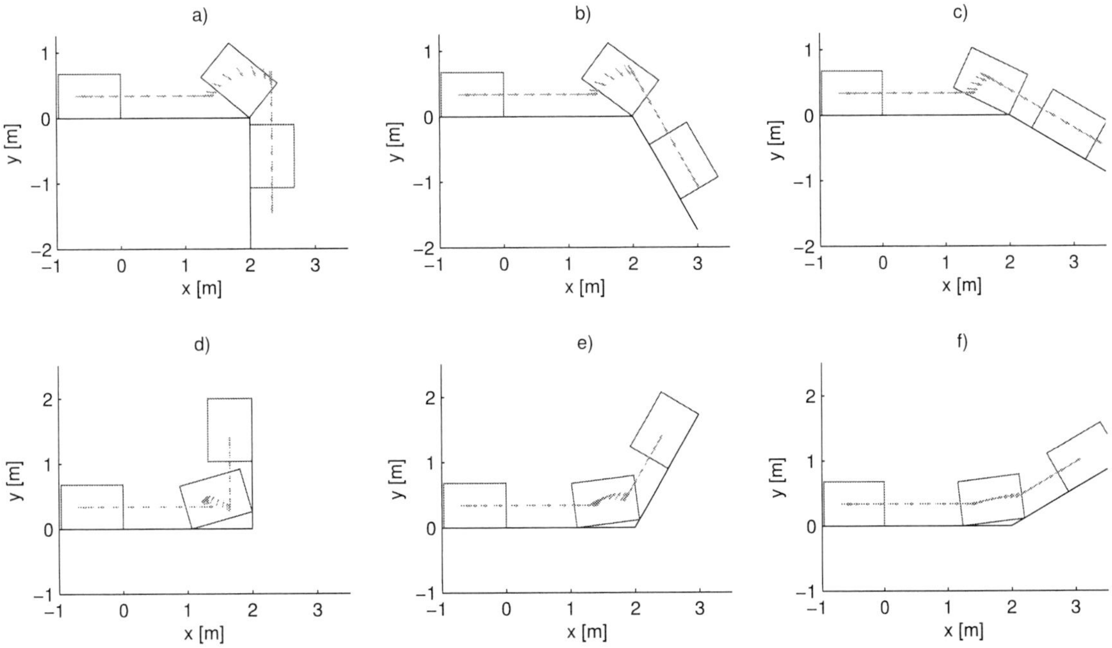

Figure 3: a) -90 degree outer corner, b) -60 degree outer corner, c) -30 degree outer corner, d) 90 degree inner corner, e) 60 degree inner corner, f) 30 degree inner corner. Three instances of the robot footprint are shown in progression for each trajectory.

The issue of conservative spacing is resolved with the use of the side rotary brush.

It should be noted that a precise external reference of the robot's path against the desired trajectory is not available, as no such indoor positioning system was available during testing. As a substitute, Figure 6 and 7 shows close-ups of the robot's path relative to the desired trajectory, as reported by the localization algorithm. The desired trajectory is shown without the orientation for a cleaner visualization. Inspection of Figure 6 shows overshoot on the scale of 10 centimeters for both the start and end of each turn. Likewise, Figure 7 shows the same amount of overshoot for an inner corner turn. Despite the inaccuracies in trajectory tracking, the wall following algorithm proves to be successful when provided a 15 centimeter safety offset relative to the wall.

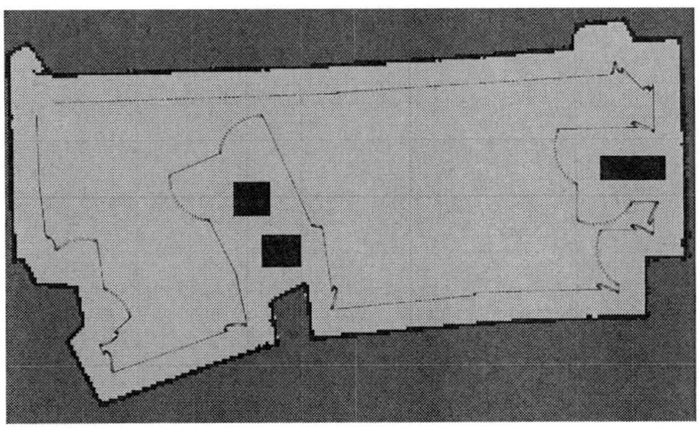

Figure 5: Map of the testing environment and desired trajectory, taken from the ROS visualization tool, Rviz. The green rectangle is the robot footprint.

CONCLUSIONS

Performing wall following as a separate procedure from regular floor coverage enables close cleaning to walls and a simplification of the coverage cleaning problem. This work presents a method of wall cleaning for straight wall contours that handles the difficulties of maneuvering around obstacles with an elongated-body machine by using a holonomic drive system. The wall following algorithm uses a combination of three motion primitives to cover straight trajectories and various angles of wall corners. With the current algorithm, walls of arbitrary curvature or obstacles of arbitrary shape can be handled by a discretized selection of straight edges.

Future work includes automating the high level planner for waypoint selection, dynamic re-planning for obstacles not known a priori, additional motion primitives to handle curved contours, and improvements to the control strategy for trajectory tracking. The current waypoint selection process is completed manually because the environment is assumed to be static. Automating the process will be more user-friendly to the machine operators. Dynamic re-planning for moveable obstacles, however, is a necessary improvement that will contribute towards full autonomy. The current method of handling curved contours, a discretization of straight edges, is inefficient. The addition of new motion primitives will improve speed and efficiency. Lastly, the experimental results show that there is room for improvement in trajectory tracking. Improved tracking performance will allow the robot to clean faster and closer to walls.

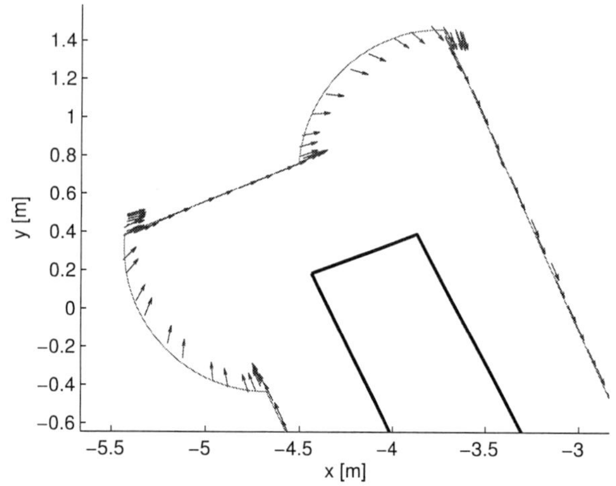

Figure 6: Robot path as reported by localization plotted against desired path (solid line) of two outer corners in succession. The arrow directions represent the orientation of the robot.

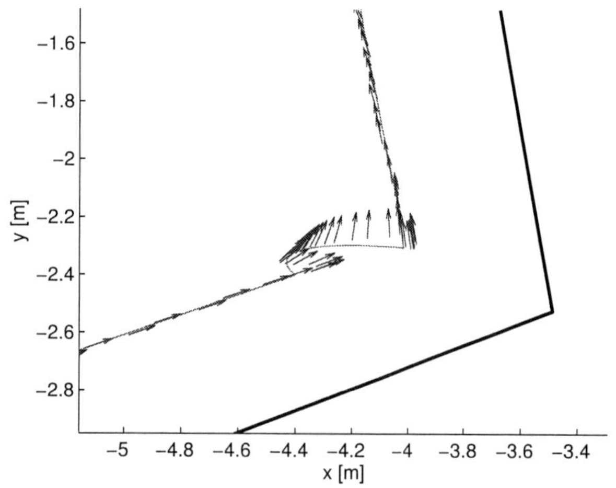

Figure 7: Robot path as reported by localization plotted against desired path (solid line) of one inner corner. The arrow directions represent the orientation of the robot.

ACKNOWLEDGMENTS

The authors would like to thank Keshav Iyengar for developing sections of the outside turn generator, and helping validate the system in simulation and hardware. Also, we'd like to thank Faizan Sheikh and members of Avidbots Corporation for supporting this project.

REFERENCES

[1] "iRobot Corporation," http://www.irobot.com/, accessed: 2015-09-06.

[2] "Intellibot Robotics," http://www.intellibotrobotics.com/, accessed: 2015-09-06.

[3] C. Hofner and G. Schmidt, "Path planning and guidance techniques for an autonomous mobile cleaning robot," in *IEEE/RSJ/GI International Conference on Intelligent Robots and Systems (IROS)*, vol. 1, 1994, pp. 610–617.

[4] H. Choset, "Coverage of known spaces: The boustrophedon cellular decomposition," *Autonomous Robots*, vol. 9, no. 3, pp. 247–253, 2000.

[5] T.-k. Lee, S.-H. Baek, S.-Y. Oh, and Y.-H. Choi, "Complete coverage algorithm based on linked smooth spiral paths for mobile robots," in *11th International Conference on Control Automation Robotics & Vision (ICARCV)*, 2010, pp. 609–614.

[6] E. Gonzalez, O. Alvarez, Y. Diaz, C. Parra, and C. Bustacara, "Bsa: a complete coverage algorithm," in *IEEE International Conference on Robotics and Automation (ICRA)*, 2005, pp. 2040–2044.

[7] D. Hanafi, Y. M. Abueejela, and M. F. Zakaria, "Wall follower autonomous robot development applying fuzzy incremental controller," *Intelligent Control and Automation*, vol. 4, pp. 18–25, 2013.

[8] R. Carellia and E. O. Freire, "Corridor navigation and wall-following stable control for sonar-based mobile robots," *Robotics and Autonomous Systems*, vol. 45, pp. 235–247, 2003.

[9] J. M. Toibero, F. Roberti, and R. Carelli, "Stable contour-following control of wheeled mobile robots," *Robotica*, vol. 27, no. 1, pp. 1–12, 2009.

[10] X. Gao, K. Li, Y. Wang, G. Men, D. Zhou, and K. Kikuchi, "A floor cleaning robot using swedish wheels," in *IEEE International Conference on Robotics and Biomimetics (ROBIO)*, 2007, pp. 2069–2073.

[11] W.-P. Chen, S.-S. Chen, L. K. Wang, S.-P. Cheng, S.-C. Chang, and M.-F. Lu, "A design of large space automatic cleaning robot with omni-directional wheels," in *International Conference on Fuzzy Theory and Its Applications (iFUZZY)*, 2013, pp. 140–145.

[12] K. Ogata and Y. Yang, "Modern control engineering," 1970.

[13] R. Siegwart, I. R. Nourbakhsh, and D. Scaramuzza, *Introduction to autonomous mobile robots*. MIT press, 2011.

[14] L. Biagotti and C. Melchiorri, *Trajectory planning for automatic machines and robots*. Springer, 2008.

[15] H. Dorrie, *100 great problems of elementary mathematics*. Dover Publications, 1965.

[16] "Open Source Robotics Foundation," http://http://www.ros.org/, accessed: 2015-09-06.

Proceedings of the 26th CANCAM
Victoria, BC, Canada, May 29 - June 1, 2017

FAULT-TOLERANT CONTROL FOR QUADROTOR HELICOPTER WITH PARAMETRIC UNCERTAINTY BASED ON INTEGRAL SLIDING MODE CONTROL

Ban Wang
Department of Mechanical and Industrial Engineering
Concordia University
Montral, Quebec, Canada
npuwangban@gmail.com

Youmin Zhang
Department of Mechanical and Industrial Engineering
Concordia University
Montral, Quebec, Canada
ymzhang@encs.concordia.ca

ABSTRACT

This paper proposes an adaptive fault-tolerant control strategy for a quadrotor helicopter with parametric uncertainties and disturbances based on an integral sliding mode control strategy. Sliding mode control is known as a robust method to maintain system performance and keep it insensitive to disturbances. With the help of the proposed on-line adaptive scheme, both uncertain model parameters and actuator faults can be compensated without the knowledge of the uncertainty bound and fault information. In this case, there is no need to increase the discontinuous control gain for accommodating parametric uncertainties and actuator faults, which may stimulate the chattering effect, compared to the conventional sliding mode control. The effectiveness of the proposed control strategy is validated through simulation of a payload transportation application using a quadrotor helicopter subject to actuator faults.
Keywords: Actuator fault, adaptive fault-tolerant control, integral sliding mode control, parametric uncertainty.

INTRODUCTION

Over the last decades, unmanned aerial vehicles (UAVs) have been widely used by commercial industries, research institutes and military sectors, such as payload transportation, forest fire detection and fighting, surveillance, border patrol, remote sensing, and aerial mapping, etc. [1, 2, 3, 4]. Recently, with the development of micro-electromechanical systems and onboard computer, more and more small scale UAVs are available in the market which further broadens the applications of UAVs. For many applications, in order to accomplish the specific task, different sensors and measurement systems need to be incorporated with the assigned UAV to make it fully functional to the specified mission. In this sense, a UAV can be regarded as an aerial sensor or payload carrier. The weights of the onboard sensors or payloads may vary depending on different applications which brings a big challenge for the onboard controller. Moreover, UAVs are usually employed to work in complex and hazardous environment, such as urban surveillance, fire fighting, and special military actions [5]. These situations may seriously threaten the reliability and safety of UAVs and the expensive onboard instruments/payloads. Especially, for those applications carried out in urban area, the failure of UAV may put people's life in danger. Therefore, the reliability and survivability of UAVs are becoming another big concern which is normally beyond what a conventional controller can provide [6]. It turns out that it will be beneficial to have a UAV system with the capability of tolerating certain parametric uncertainties and actuator faults without imperiling itself and its surroundings. On the other hand, among those different types of UAVs, multirotor helicopters have drawn more and more attention in both industrial and academic communities in view of their special ability of vertical take-off and landing and affordable price [7]. As an example of multirotor helicopters, the quadrotor helicopter is a relatively simple and easy-to-fly system. Many control strategies have been proposed to maintain a good control performance of the quadrotor helicopter, including linear quadratic regulator (LQR) control [8], dynamical programming [9], differential flatness approach [10], and model predictive control [11], to name a few. Some of the approaches focus on path planning for payload transportation application without being affected by payload swing, and some are interested in fault-tolerant control to ensure the safety of the quadrotor helicopter in the presence of actuator fault. However, among these studies, the total take-off weight are assumed constant throughout the entire flight phase. Moreover, in order to make the control strategy more useful for real applications, the capability of tolerating certain parametric uncertainties, some level of loss of control effectiveness faults and unknown disturbances should be achieved within one control scheme.

Sliding mode control (SMC) is an approach to design robust control systems dealing with large uncertainties with discontinuous control strategy which demonstrates invariance to so-called matched uncertainties while on a reduced-order sliding surface [12, 13]. The property of insensitivity to parametric uncertainty makes sliding mode as one of the most promising control approaches. However, in order to use this conventional SMC, the bound of uncertainty is always needed at the design stage. For some applications, the payload can be added at any time with unknown mass and actuator fault might also occur with unknown magnitude. Therefore, it is hard to obtain the exact uncertainty bound in advance. This stimulates a new control strategy which incorporates adaptive scheme into SMC to compensate uncertainties and disturbances without knowing the exact uncertainty bounds.

In this paper, an adaptive control strategy is proposed to simultaneously accommodate parametric uncertainties and actuator faults for a quadrotor helicopter based on an integral sliding mode control approach. With the synthesized adaptive scheme, the proposed control strategy can adaptively generate appropriate control signals to compensate parametric uncertainties and actuator faults simultaneously without merely relying on the robust discontinuous control strategy. Since the proposed adaptive scheme is synthesized in both continuous and discontinuous control parts, the control effort is significantly reduced for accommodating uncertainties and faults, which means larger uncertainties and faults could be tolerated within this control scheme, and the chattering effect can also be suppressed.

PROBLEM FORMULATION

Problem Statement

Consider a nonlinear affine system with parametric uncertainty and disturbance subject to actuator fault:

$$\begin{aligned} \dot{x}(t) &= f(x(t),t) + g(x(t),t)v(t) + d(t) \\ v(t) &= B_u H(t) u(t) \end{aligned} \quad (1)$$

where $x(t) \in \mathbb{R}^n$ is the state vector, $u(t)$ is the control input and $v(t)$ is the intermediate virtual control. The vectors $f(x(t),t) \in \mathbb{R}^n$ and $g(x(t),t) \in \mathbb{R}^n$ are the nonlinear functions containing parametric uncertainties. The bounds of these uncertainties are unknown *a priori*. $d(t)$ represents disturbance which is unknown but bounded, i.e., $|d(t)| \leq D_d$. $B_u \in \mathbb{R}^{n\times m}$ is the control effectiveness matrix. $H(t) = diag(h_1(t), \cdots, h_m(t))$ represents the control effectiveness level of the actuators, where $h_j(t)_{(j=1,\ldots,m)}$ is a scalar satisfying $0 \leq h_j(t) \leq 1$. If $h_j(t) = 1$, the jth actuator works perfectly. Otherwise, the jth actuator suffers some level of fault with a special case $h_j(t) = 0$ denoting the complete failure of the jth actuator.

For expressing parametric uncertainties of the system, let $\hat{f}(x(t),t)$ and $\hat{g}(x(t),t)$ represent the estimated parts of $f(x(t),t)$ and $g(x(t),t)$, respectively. Therefore, the estimated errors $\tilde{f}(x(t),t)$ and $\tilde{g}(t)$ can be obtained as follows:

$$\begin{aligned} \tilde{f}(x(t),t) &= f(x(t),t) - \hat{f}(x(t),t) \\ \tilde{g}(x(t),t) &= g(x(t),t) - \hat{g}(x(t),t) \end{aligned} \quad (2)$$

For simplicity of expression, the notation (t) is omitted in the following sections, e.g., $x(t)$ is expressed as x.

System Formulation

Since the quadrotor helicopter is an underactuated system, only the attitude and height controllers are directly related to the actuators. Thus, the state vector is defined as:

$$\begin{aligned} x &= [z \ \dot{z} \ \phi \ \dot{\phi} \ \theta \ \dot{\theta} \ \psi \ \dot{\psi}]^T \\ &= [x_1 \ x_2 \ x_3 \ x_4 \ x_5 \ x_6 \ x_7 \ x_8]^T \end{aligned} \quad (3)$$

With this state vector, the dynamical equations of the quadrotor helicopter can be resolved into the following subsystems. Height subsystem: $\dot{x}_1 = x_2, \dot{x}_2 = f_1(x) + g_1 v_1 + d_1$ with $f_1(x) = -g$, $g_1 = \cos\phi\cos\theta/m$ and $d_1 = -K_3/m\dot{z}$; Roll subsystem: $\dot{x}_3 = x_4, \dot{x}_4 = f_2(x) + g_2 v_2 + d_2$ with $f_2(x) = x_6 x_8 (I_{yy} - I_{zz})/I_{xx}$, $g_2 = 1/I_{xx}$ and $d_2 = -I_r\dot{\theta}\Omega/I_{xx} - K_4 L\dot{\phi}/I_{xx}$; Pitch subsystem: $\dot{x}_5 = x_6, \dot{x}_6 = f_3(x) + g_3 v_3 + d_3$ with $f_3(x) = x_4 x_8 (I_{zz} - I_{xx})/I_{yy}$, $g_3 = 1/I_{yy}$ and $d_3 = I_r\dot{\phi}\Omega/I_{yy} - K_5 L\dot{\theta}/I_{yy}$; Yaw subsystem: $\dot{x}_7 = x_8, \dot{x}_8 = f_4(x) + g_4 v_4 + d_4$ with $f_4(x) = x_4 x_6 (I_{xx} - I_{yy})/I_{zz}$, $g_4 = 1/I_{zz}$ and $d_4 = -K_6\dot{\psi}/I_{zz}$. Then, each subsystem can be written as a single-input nonlinear system with the help of the intermediate virtual control given by:

$$\begin{aligned} \dot{x}_{2i-1} &= x_{2i} \\ \dot{x}_{2i} &= f_i(x) + g_i v_i + d_i \end{aligned} \quad (4)$$

where $i = 1,2,3,4$ represents each subsystem.

CONTROL STRATEGY

In this section, an integral sliding mode control based adaptive fault-tolerant control strategy is designed to accommodate parametric uncertainties and actuator faults for quadrotor helicopter. The schematic of the proposed control strategy is described in Fig. 1.

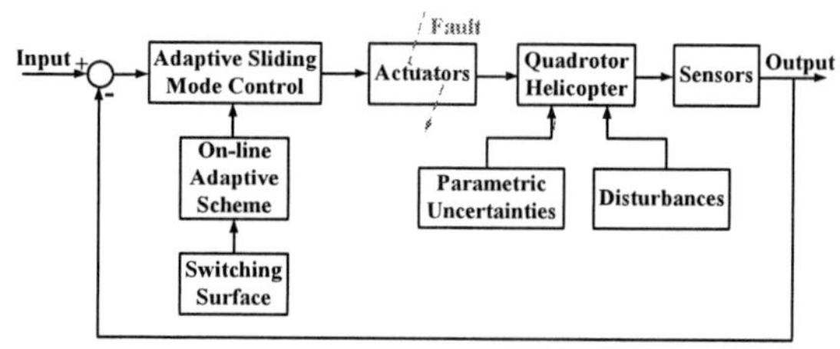

Figure 1: Schematic of the proposed control strategy

Integral Sliding Mode Control

The integral sliding surface for the system is defined by the following set:

$$S_i = \{x \in \mathbb{R}^n : \ \sigma_i(x) = 0\} \quad (5)$$

For $\forall i = 1, 2, 3, 4$, denoting x^d_{2i-1} and x^d_{2i} as the desired trajectories, the tracking errors can be defined as $\tilde{x}_{i1} = x_{2i-1} - x^d_{2i-1}$ and $\tilde{x}_{i2} = x_{2i} - x^d_{2i}$. The switching function σ_i is then defined as:

$$\sigma_i = \tilde{x}_{i2} + k_{i2}\tilde{x}_{i1} + k_{i1}\int_{t_0}^{t} \tilde{x}_{i1}(\tau)d\tau - k_{i2}\tilde{x}_{i1}(t_0) - \tilde{x}_{i2}(t_0) \quad (6)$$

Then, the next problem is to design an appropriate control law to make the sliding surface attractive. The control law is designed in the following form:

$$v_i = v_{i0} + v_{i1} \tag{7}$$

where v_{i0} is the continuous nominal control part to stabilize the ideal system, and v_{i1} is the discontinuous control part for compensating the perturbations and disturbances in order to ensure the sliding motion.

The equivalent control v_{i0} is designed by equalizing $\dot{\sigma}_i = 0$, and the disturbance d_i is omitted in this case as shown below:

$$v_{i0} = g_i^{-1}(\dot{x}_{2i}^d - k_{i2}\tilde{x}_{i2} - k_{i1}\tilde{x}_{i1} - f_i(x)) \tag{8}$$

The discontinuous control part, which is synthesized to reject the disturbance d_i, is designed as:

$$v_{i1} = -g_i^{-1}k_{ci}sign(\sigma_i) \tag{9}$$

where k_{ci} is a positive high gain which makes the sliding surface attractive.

However, in order to account for the disturbance, the control discontinuity is increased which may lead to control chattering. One can remove this condition by smoothing the control discontinuity in a thin boundary layer neighboring the switching surface [14]. The boundary layer can be formulated as follows:

$$\bar{B} = \{\tilde{x}_{i1}, \tilde{x}_{i2}, |\sigma_i| \leq \Phi_i\} \tag{10}$$

where Φ_i is the boundary layer thickness with positive value.

Accordingly, the developed control law becomes:

$$v_i = g_i^{-1}(\dot{x}_{2i}^d - k_{i2}\tilde{x}_{i2} - k_{i1}\tilde{x}_{i1} - f_i(x)) - g_i^{-1}k_{ci}sat(\sigma_i/\Phi_i) \tag{11}$$

where the *sat* function is defined as:

$$sat(\sigma_i/\Phi_i) = \begin{cases} sign(\sigma_i) & if \quad |\sigma_i| > \Phi_i \\ \sigma_i/\Phi_i & if \quad |\sigma_i| < \Phi_i \end{cases} \tag{12}$$

Synthesized Adaptive Control Strategy

With the consideration of the parametric uncertainties and actuator faults, the above developed control law in (11) could be rewritten as:

$$\begin{aligned} v_i &= \hat{g}_i^{-1}(\dot{x}_{2i}^d - k_{i2}\tilde{x}_{i2} - k_{i1}\tilde{x}_{i1} - \hat{f}_i(x)) - \hat{g}_i^{-1}k_{ci}sat(\sigma_i/\Phi_i) \\ u &= B_{ui}^+ v_i = B_{ui}^T(B_{ui}B_{ui}^T)^{-1}v_i \end{aligned} \tag{13}$$

Denoting $\hat{\chi}_i = g_i^{-1}\hat{f}_i(x)$ and $\hat{\Upsilon}_i = \hat{g}_i^{-1}$, the above control law can be rearranged as:

$$v_i = -\hat{\chi}_i + \hat{\Upsilon}_i(\dot{x}_{2i}^d - k_{i2}\tilde{x}_{i2} - k_{i1}\tilde{x}_{i1} - k_{ci}sat(\sigma_i/\Phi_i)) \tag{14}$$

In order to compensate parametric uncertainties and actuator faults, the estimated parameters $\hat{\chi}_i$ and $\hat{\Upsilon}_i$ should be updated on-line. Such that, the on-line adaptive schemes for estimating the uncertain parameters are designed as:

$$\dot{\hat{\chi}}_i = \sigma_{\Delta i} \tag{15}$$

$$\dot{\hat{\Upsilon}}_i = (-\dot{x}_{2i}^d + k_{i2}\tilde{x}_{i2} + k_{i1}\tilde{x}_{i1} + k_{ci}sat(\sigma_i/\Phi_i))\sigma_{\Delta i} \tag{16}$$

where $\sigma_{\Delta i}$ is the measurement of the algebraic distance between the current state and the boundary layer defined in (17). It features $\dot{\sigma}_{\Delta i} = \dot{\sigma}_i$ outside the boundary layer and $\sigma_{\Delta i} = 0$ inside the boundary layer.

$$\sigma_{\Delta i} = \sigma_i - \Phi_i sat(\sigma_i/\Phi_i) \tag{17}$$

Theorem 1: Consider a nonlinear affine system (1) with parametric uncertainties, actuator faults and bounded disturbances. Given the designed integral sliding surface (6), by employing the feedback control law (14) and updated by (15) and (16), the sliding motion will be achieved and maintained inside the boundary layer (10) regardless of the parametric uncertainties, actuator faults and disturbances with the discontinuous gain chosen as $k_i \geq \eta_i + D_i$.

Proof of Theorem 1: Consider the following Lyapunov candidate function:

$$V_i = \frac{1}{2}\left[\sigma_{\Delta i}^2 + \Upsilon_i^{-1}(\hat{\chi}_i - \chi_i)^2 + \Upsilon_i^{-1}(\hat{\Upsilon}_i - \Upsilon_i)^2\right] \tag{18}$$

Then, the derivative of the selected Lyapunov candidate function would be:

$$\begin{aligned} \dot{V}_i =& \sigma_{\Delta i}\dot{\sigma}_{\Delta i} + \Upsilon_i^{-1}(\hat{\chi}_i - \chi_i)\dot{\hat{\chi}}_i + \Upsilon_i^{-1}(\hat{\Upsilon}_i - \Upsilon_i)\dot{\hat{\Upsilon}}_i \\ =& \sigma_{\Delta i}(\Upsilon_i^{-1}\chi_i - \Upsilon_i^{-1}\hat{\chi}_i + \Upsilon_i^{-1}\hat{\Upsilon}_i(\dot{x}_{2i}^d - k_{i2}\tilde{x}_{i2} - k_{i1}\tilde{x}_{i1} \\ & - k_{ci}sat(\sigma_{\Delta i}/\Phi_i)) + d_i - \dot{x}_{2i}^d + k_{i2}\tilde{x}_{i2} + k_{i1}\tilde{x}_{i1}) \\ & + \Upsilon_i^{-1}(\hat{\chi}_i - \chi_i)\dot{\hat{\chi}}_i + \Upsilon_i^{-1}(\hat{\Upsilon}_i - \Upsilon_i)\dot{\hat{\Upsilon}}_i \\ =& \Upsilon_i^{-1}(\hat{\chi}_i - \chi_i)(\dot{\hat{\chi}}_i - \sigma_{\Delta i}) + (\Upsilon_i^{-1}\hat{\Upsilon}_i - 1)[\dot{\hat{\Upsilon}}_i + (\dot{x}_{2i}^d \\ & - k_{i2}\tilde{x}_{i2} - k_{i1}\tilde{x}_{i1} - k_i sat(\sigma_{\Delta i}/\Phi_i))\sigma_{\Delta i}] + d_i\sigma_{\Delta i} \\ & - k_{ci}sat(\sigma_{\Delta i}/\Phi_i)\sigma_{\Delta i} \end{aligned} \tag{19}$$

Substituting Eqs. (15) and (16) into Eq. (19) leads to:

$$\begin{aligned} \dot{V}_i =& - k_{ci}sat(\sigma_{\Delta i}/\Phi_i)\sigma_{\Delta i} + d_i\sigma_{\Delta i} \\ \leq& - (\eta_i + D_i)sat(\sigma_{\Delta i}/\Phi_i)\sigma_{\Delta i} + d_i\sigma_{\Delta i} \\ \leq& - \eta_i|\sigma_{\Delta i}| \end{aligned} \tag{20}$$

Thus, the system satisfies the standard $\eta-$reachability condition, and the performance of the system is maintained with the proposed control law in the presence of parametric uncertainties, actuator faults and disturbances.

SIMULATION RESULTS

In order to validate the performance of the proposed adaptive control strategy for accommodating parametric uncertainties and actuator faults, a simulation based on a quadrotor helicopter subject to mass variation and actuator fault is demonstrated. The quadrotor helicopter is firstly lifted to 0.8 m to pick up a payload with weight of 0.6 kg. With this payload, it is lifted to 2 m and commanded to follow a set of predesigned trajectory. During this phase, an actuator fault is injected at 50 sec. At the end, the payload is dropped from the air to the targeted area. During the whole flight phase, the mass of the payload and the level of actuator fault are both unknown to the onboard controller. With the synthesized adaptive scheme, the expected ability to estimate the uncertain parameters adaptively and maintain the tracking performance of the quadrotor helicopter throughout the entire mission are shown in Figs. 2(a) and 2(b) for demonstrating the achieved tracking performance of pitch and height, respectively. Compared to the commonly used linear quadratic regulator (LQR) controller [8], the proposed adaptive control strategy shows a better tracking performance with the ability to estimate the total mass of the quadrotor helicopter and to compensate the actuator fault. The estimation of the total mass of the quadrotor helicopter is shown in Fig. 2(c).

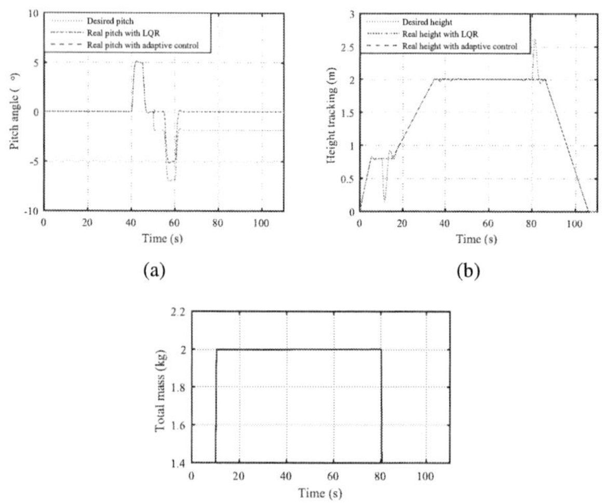

(a) (b)

(c)

Figure 2: Simulation results

CONCLUSIONS

In this paper, an adaptive strategy based on integral sliding mode control is proposed to accommodate parametric uncertainties, actuator faults and disturbances for a quadrotor helicopter. With the help of the developed online adaptive scheme, the uncertain parameters can be adaptively estimated without the knowledge of the uncertainty bound and actuator fault information to maintain tracking performance of the quadrotor helicopter. The effectiveness of the proposed control strategy are validated through a simulation scenario where the quadrotor helicopter is subject to mass variation and actuator fault during the entire flight phase.

References

[1] C. Yuan, Y. M. Zhang, and Z. X. Liu, "A survey on computer vision based technologies for automatic forest fire detection using UAVs and remote sensing techniques," *Canadian Journal of Forest Research*, vol. 45, no. 7, pp. 783–792, 2015.

[2] F. A. Goodarzi, D. Lee, and T. Lee, "Geometric control of a quadrotor UAV transporting a payload connected via flexible cable," *Int. Journal of Control, Automation and Systems*, vol. 13, no. 6, pp. 1486–1498, 2015.

[3] Z. Li, Y. Liu, R. Walker, R. Hayward, and J. Zhang, "Towards automatic power line detection for a UAV surveillance system using pulse coupled neural filter and an improved hough transform," *Machine Vision and Applications*, vol. 21, no. 5, pp. 677–686, 2010.

[4] H. Xiang and L. Tian, "Development of a low-cost agricultural remote sensing system based on an autonomous unmanned aerial vehicle (UAV)," *Biosystems Engineering*, vol. 108, no. 2, pp. 174–190, 2011.

[5] E. Semsch, M. Jakob, D. Pavlicek, and M. Pechoucek, "Autonomous UAV surveillance in complex urban environments," in *Proc. of the IEEE/WIC/ACM International Joint Conferences on Web Intelligence and Intelligent Agent Technology*, vol. 2, September 2009, pp. 82–85.

[6] Y. M. Zhang and J. Jiang, "Bibliographical review on reconfigurable fault-tolerant control systems," *Annual Reviews in Control*, vol. 32, no. 2, pp. 229–252, 2008.

[7] N. V. Hoffer, C. Coopmans, A. M. Jensen, and Y. Chen, "A survey and categorization of small low-cost unmanned aerial vehicle system identification," *Journal of Intelligent & Robotic Systems*, vol. 74, no. 1-2, p. 129, 2014.

[8] B. Wang, K. A. Ghamry, and Y. M. Zhang, "Trajectory tracking and attitude control of an unmanned quadrotor helicopter considering actuator dynamics," in *Proc. of the 35th Chinese Control Conference (CCC)*, July 2016, pp. 10 795–10 800.

[9] I. Palunko, R. Fierro, and P. Cruz, "Trajectory generation for swing-free maneuvers of a quadrotor with suspended payload: A dynamic programming approach," in *Proc. of the IEEE International Conference on Robotics and Automation (ICRA)*, 2012, pp. 2691–2697.

[10] K. Sreenath, T. Lee, and V. Kumar, "Geometric control and differential flatness of a quadrotor UAV with a cable-suspended load," in *Proc. of the 52nd IEEE Conference on Decision and Control (CDC)*, December 2013, pp. 2269–2274.

[11] B. Yu, Y. M. Zhang, I. Minchala, and Y. H. Qu, "Fault-tolerant control with linear quadratic and model predictive control techniques against actuator faults in a quadrotor UAV," in *Proc. of the Conference on Control and Fault-Tolerant Systems (SysTol)*, October 2013, pp. 661–666.

[12] R. Hess and S. Wells, "Sliding mode control applied to reconfigurable flight control design," *Journal of Guidance, Control, and Dynamics*, vol. 26, no. 3, pp. 452–462, 2003.

[13] C. Edwards, H. Alwi, and M. T. Hamayun, "Fault tolerant control using integral sliding modes," in *Proc. of the European Control Conference (ECC)*, 2014, pp. 2291–2296.

[14] J.-J. Slotine and S. S. Sastry, "Tracking control of nonlinear systems using sliding surfaces, with application to robot manipulators," *International Journal of Control*, vol. 38, no. 2, pp. 465–492, 1983.

HOVERING QUADROTOR CONTROL IN A WINDY ENVIRONMENT USING NEURAL NETWORK

Yintao Zhang
Department of Mechanical and Industrial Engineering
Concordia University
Montreal, Quebec, Canada
zhangyintao23@gmail.com

Youmin Zhang
Department of Mechanical and Industrial Engineering
Concordia University
Montreal, Quebec, Canada
ymzhang@encs.concordia.ca

ABSTRACT

In recent years, quadrotor UAVs have obtained an increasing range of applications. This paper proposes a neural network controller for quadrotors hovering in a windy environment. Firstly, the quadrotor model in a windy environment is established. Secondly, a PID controller is set up for quadrotor hovering. After that, a controller based on back propagation neural network is designed. Simulation results show that neural network controller has a better performance and efficiency than a conventional PID controller.

KEYWORDS: Quadrotor hovering, PID control, back propagation neural network

INTRODUCTION

Unmanned Aerial Vehicles (UAVs) have gained enormous commercial potential during the last years. In fact, the essential needs of surveillance, remote sensing, monitoring, military and security applications put UAVs in the centerpiece of researchers and engineers.

From a control engineer's point of view, the most significant ability of neural networks is its universal approximation to any nonlinear function, which can be used to deal with nonlinear systems [1]. In order to control these nonlinear systems, one may need to construct the system model. Although some control methods are not dependent on system model, such as PID controller, some other controllers can only be designed when a precise system model is built. However, in real systems, it is a nontrivial task to know the exact system model and its parameters.

To solve this problem, neural networks have been extensively studied in the past decades, since they have many excellent properties such as parallel distributed processing, learning and adaption, data fusion, multi-input multi-output processing.

The research results demonstrate that neural network is effective in identification and control of nonlinear systems with unknown dynamics due to its powerful nonlinear approximation property. For instance, in [2-3], feedforward neural networks (FNNs) were utilized to identify the system models. In [4], an adaptive feedforward neural network tracking controller was proposed for a robotic manipulator with input dead zone and output constraint. In [5], an adaptive predictor incorporated with a high-order neural network (HONN) observer was proposed to obtain the future predictions of system states, which was used in the control design to circumvent the input delay and nonlinearities. An adaptive neural network control method was investigated in [6] to stabilize a class of uncertain nonlinear strict-feedback systems with full-state constraints.

In this study, back propagation neural network is used in the controller design of a hovering quadrotor. Firstly, a quadrotor hovering model in windy environment is built. After that, a PID controller is set up. Then a controller based on neural network is designed. Finally, simulation results of two controllers are given.

MODELING

The state equations of a typical quadrotor can be given below [7].

$$\begin{bmatrix} \ddot{x} \\ \ddot{y} \\ \ddot{z} \\ \dot{p} \\ \dot{q} \\ \dot{r} \\ \dot{\phi} \\ \dot{\theta} \\ \dot{\psi} \end{bmatrix} = \begin{bmatrix} \frac{1}{m}\sum_{i=1}^{4} F_{T_i} \cdot \begin{bmatrix} S_\theta C_\psi C_\phi + S_\psi S_\phi \\ S_\theta S_\psi C_\phi - C_\psi S_\phi \\ C_\theta C_\phi \end{bmatrix} + \begin{bmatrix} 0 \\ 0 \\ g \end{bmatrix} \\ qr\frac{I_y - I_z}{I_x} + q\omega_r\frac{J_r}{I_x} + \frac{l\left(-F_{T_2} + F_{T_4}\right)}{I_x} \\ pr\frac{I_z - I_x}{I_y} - p\omega_r\frac{J_r}{I_y} + \frac{l\left(F_{T_1} - F_{T_3}\right)}{I_y} \\ pq\frac{I_x - I_y}{I_z} + \frac{-M_{Q_1} + M_{Q_2} - M_{Q_3} + M_{Q_4}}{I_z} \\ \begin{bmatrix} 1 & \tan\theta\sin\phi & \tan\theta\cos\phi \\ 0 & \cos\phi & -\sin\phi \\ 0 & \sin\phi/\cos\theta & \cos\phi/\cos\theta \end{bmatrix}\begin{bmatrix} p \\ q \\ r \end{bmatrix} \end{bmatrix} \tag{1}$$

where S stands for *sin* and C stands for *cos*, other parameters and variables have the normal meaning.

The rotor aerodynamics analysis is shown in Figure 1. v_1 is the induced velocity caused by rotor turning, v_w is wind speed and v_t is the total induced velocity.

$$\vec{v}_t = \vec{v}_1 + \vec{v}_w \tag{2}$$

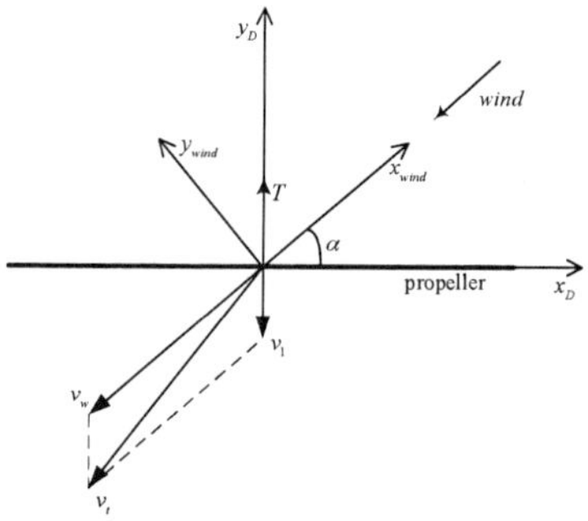

Figure 1: Rotor aerodynamics

According to the rotor slipstream theory, the rotor lift force in no-wind environment [8-9] is

$$T_{no} = 2\rho A v_1^2 \tag{3}$$

where ρ is air density, A is the effective area of a rotor disk. Also, the force is proportional to the squared rotor rotating speed.

The rotor lift force with wind effect can be expressed as [7],

$$T = 2\rho\pi R^2 v_1 \sqrt{v_1^2 + v_w^{\,2} + 2v_1 v_w \sin\alpha}, |\alpha| \le 90^\circ \tag{4}$$

where α is the angle between wind and propeller.

PID CONTROLLER DESIGN

In this study, it is assumed that the quadrotor is hovering in a constant windy environment. Compared to the quadrotor, the weight, rotation and heeling moment of the each rotor are too small so that they could be ignored.

The structure of PID controller is shown in Figure 2.

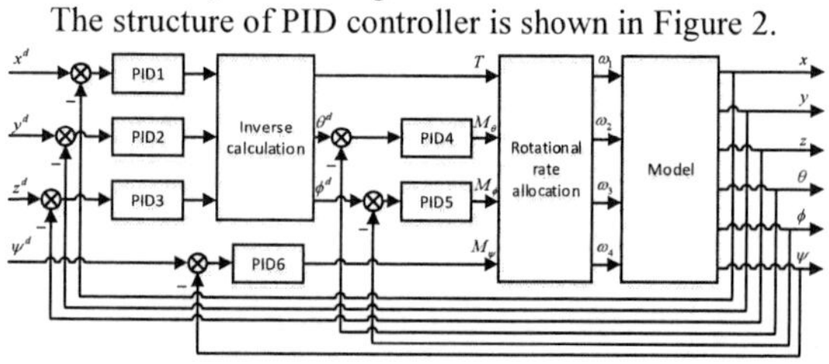

Figure 2: PID control system

A. Position PID Controller Design

The equations of the first three PID controllers are shown as follows.

$$\begin{cases} \ddot{x}^d = k_{px}\left(x^d - x\right) - k_{dx}\dot{x} + k_{ix}\int\left(x^d - x\right) \\ \ddot{y}^d = k_{py}\left(y^d - y\right) - k_{dy}\dot{y} + k_{iy}\int\left(y^d - y\right) \\ \ddot{z}^d = k_{pz}\left(z^d - z\right) - k_{dz}\dot{z} + k_{iz}\int\left(z^d - z\right) \end{cases} \tag{5}$$

After that, the inverse calculation should be done to obtain the total thrust, desired pitch angle and desired roll angle. The specific equations are shown below.

$$T = \frac{m\left(\ddot{z}^d - g\right)}{C_\theta C_\phi} \tag{6}$$

$$\begin{cases} \theta^d = \arctan\left(\dfrac{\ddot{x}^d \cdot C_\psi + \ddot{y}^d \cdot S_\psi}{\ddot{z}^d - g}\right) \\ \phi^d = \arcsin\left(\dfrac{\ddot{x}^d \cdot S_\psi - \ddot{y}^d \cdot C_\psi}{\sqrt{\left(\ddot{x}^d\right)^2 + \left(\ddot{y}^d\right)^2 + \left(\ddot{z}^d - g\right)^2}}\right) \end{cases} \tag{7}$$

B. Pose Controller Design

The rest three controllers are designed to calculate three angular accelerations.

$$\begin{cases} \ddot{\theta}^d = k_{p\theta}\left(\theta^d - \theta\right) - k_{d\theta}\dot{\theta} + k_{i\theta}\int\left(\theta^d - \theta\right) \\ \ddot{\phi}^d = k_{p\phi}\left(\phi^d - \phi\right) - k_{d\phi}\dot{\phi} + k_{i\phi}\int\left(\phi^d - \phi\right) \\ \ddot{\psi}^d = k_{p\psi}\left(\psi^d - \psi\right) - k_{d\psi}\dot{\psi} + k_{i\psi}\int\left(\psi^d - \psi\right) \end{cases} \tag{8}$$

The integral terms on the posture loop will result in the instability of the system [10], so the integral parameters need to be zero and the three controllers are degenerated to PID control channels.

The equations of the moment around *x*, *y*, and *z* axes are

$$\begin{cases} M_\phi = I_x\ddot{\phi}^d - \dot{\theta}\dot{\psi}\left(I_y - I_z\right) \\ M_\theta = I_y\ddot{\theta}^d - \dot{\phi}\dot{\psi}\left(I_z - I_x\right) \\ M_\psi = I_z\ddot{\psi}^d - \dot{\theta}\dot{\phi}\left(I_x - I_y\right) \end{cases} \tag{9}$$

After that, we can obtain the rotation rate of each propeller. Finally, the six variables of current state could be generated by setting the four rotation rates as the input of quadrotor model designed previously.

NEURAL NETWORK CONTROLLER DESIGN

The transmission of biological neuron signals is a complicated electrochemistry process through synapses. In artificial neural network, this process is simplified to a set of numbers which could be updated by a particular training rule; and this set of numbers is stored in the neuron thresholds and link values. The neuron network in this study consists of three parts (as shown in Figure 3): the input layer, the hidden layers and the output layer. The input signal enters the network through input layer. After a complex calculation in hidden layers, the output signal is generated in the output layer. Comparing this output signal to the desired output, the output error could be obtained. Then the error signal is transferred backwards. In this back propagation process, the error is apportioned to every layer and every neuron by a particular algorithm. Then the values of each unit could be calibrated and a new neural network is generated. Repeat this process until the total error meets the demand, then the final neural network is obtained and the neural network controller of this study is accomplished.

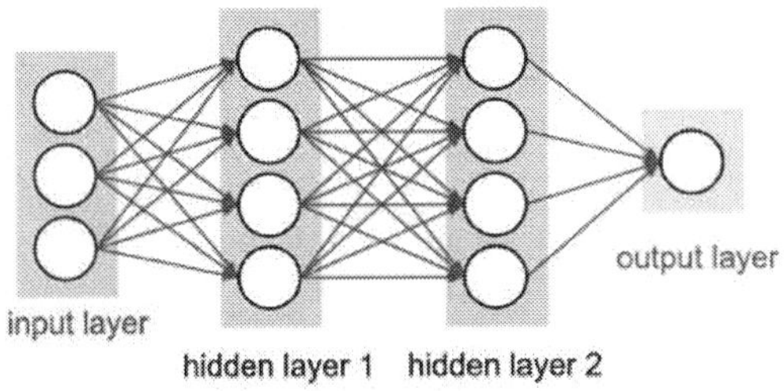

Figure 3: Neural network

In this study, the artificial neural network is designed as a controller. So it is assumed that the neural network has i input neurons, j output neurons and h neurons in hidden layers. The specific steps of the neural network controller design are as follows.

Step 1. Initialization

Normalize all the inputs and outputs of the training sample to a range from 0 to 1. It is convenient to set, calculate and calibrate the values of the neural network if all the numbers are between 0 and 1.

Then set the value of all parameters and variables, such as max training number of times, the desired error, the learning rate, all the link values and neuron thresholds and so on.

Step 2. Forward calculation

Input a new training sample into the neural network, and this input signal would be transferred to the output layer within a complex calculation.

$X = [x_1, x_2, \cdots, x_i]$ is the input. $Y = [y_1, y_2, \cdots, y_j]$ is the system output and $D = [d_1, d_2, \cdots, d_j]$ is the desired output.

The input signal has i variables, so the input layer has i neurons. The output value of ith neuron in input layer is

$$S_i = f(x_i) \tag{10}$$

where $f(x)$ stands for the activation function. Eq.(11) is the activation function in this study.

$$f(x) = \frac{1}{1+e^{-ax}} \tag{11}$$

where $a = 1.7159$.

The output value of hth neuron in hidden layers is expressed in Eq.(12). θ_i is the threshold of ith neuron. w_{hi} is the link value between two neurons, h and i.

$$S_h = f\left(\sum w_{hi} S_i + \theta_i\right) \tag{12}$$

In this way, the system output could be obtained at the output layer. So the output error could be obtained by Eq.(13). The total error of all samples could be obtained by Eq.(14).

$$E = \sum \left(D_j - S_j\right)^2 \tag{13}$$

$$E_{total} = \sum_{k=1}^{K} E_k = \sum_{k=1}^{K} \sum \left(D_j - Y_j\right)^2 \tag{14}$$

Compare the total error to the desired error e. If the total error is bigger, then go to the next step. If the total error is smaller, then the final neural network is generated.

Step 3. Back propagation

The main purpose of back propagation is to calibrate the link values and neuron thresholds by transferring the output error backward.

Firstly, calibrate the link value and neuron threshold of output layer.

$$\begin{cases} w_{jh}' = w_{jh} + \Delta w_{jh} = w_{jh} + \eta \delta_j^Y S_h \\ \theta_j' = \theta_j + \Delta\theta_j = \theta_j + \eta \delta_j^Y \\ \delta_j^Y = S_j(1-S_j)(D_j - S_j) \end{cases} \tag{15}$$

where $\eta = 0.1$ is the learning efficiency.

Then, calibrate the link value and neuron threshold of hidden layers.

$$\begin{cases} w_{hi}' = w_{hi} + \Delta w_{hi} = w_{hi} + \eta \delta_h^H S_i \\ \theta_h' = \theta_h + \Delta\theta_h = \theta_h + \eta \delta_h^H \\ \delta_h^H = f'\left(\sum w_{hi} S_i + \theta_i\right) \sum \delta_j^Y w_{jh} \end{cases} \tag{16}$$

In this way, all the link values and neuron thresholds would have been modified.

Step 4. Algorithm circulation

Test the number of training samples. If it reaches the maximum number, then the training section ends. If not, test the number of training times. If the training time reaches the maximum number, then the training section ends. Or go back to Step 2.

SIMULATION AND ANALYSIS

The simulation environment and PID controller parameters are shown in TABLE I and Table II.

TABLE I. SIMULATION ENVIRONMENT PARAMETERS

Air density (kg/m^3)	1.29	Blade radius (m)	0.1
Weight (kg)	1.6	Arm length (m)	0.3
Rotation inertia around x_b, y_b (kg • m^2)	0.03	Rotation inertia around z_b (kg • m^2)	0.04
Tension ratio	1.55e-5	Reaction moment ratio	2.82e-7

TABLE II. SIMULATION ENVIRONMENT PARAMETERS

Position control channel		
k_{px}, k_{py}, k_{pz}	k_{ix}, k_{iy}, k_{iz}	k_{dx}, k_{dy}, k_{dz}
0.3	0.1	0.9
Pose control channel		
$k_{p\phi}, k_{p\theta}, k_{p\psi}$	$k_{d\phi}, k_{d\theta}, k_{d\psi}$	
15	5	

As is known, quadrotor has a symmetrical structure, which means the motions along the axis x_n and y_n have the same characteristics, and the longitudinal motion is separated from the lateral motion, a constant lateral wind with the azimuth of 53.1° in the navigation coordinate system, ten meters per second, is assumed to simulate for a convenient description of the problem. Furthermore, the Euler angles are limited between -10° and 10° so that the wind influence on the rotor lift forces can be regarded to be constant.

Figures 4 and 5 illustrate that the PID controller and neural network controller both achieve satisfactory control effects and when the aircraft reaches the hovering point, the displacements along the two axes are all equal to zero. Obviously, the neural network controller has a better performance with faster and smoother responses than that of PID controller in this study.

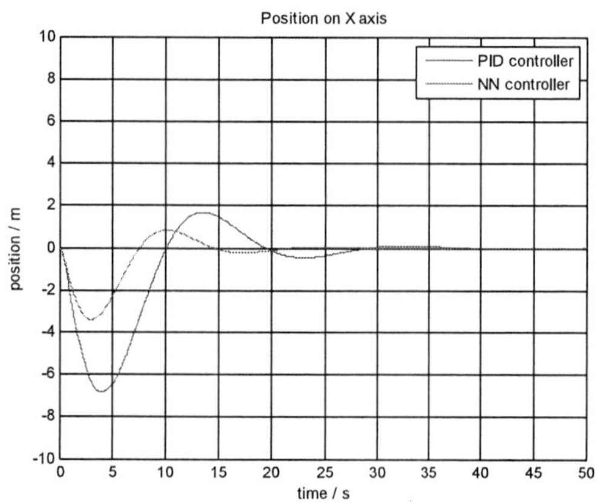

Figure 4: Position displacement along X axis

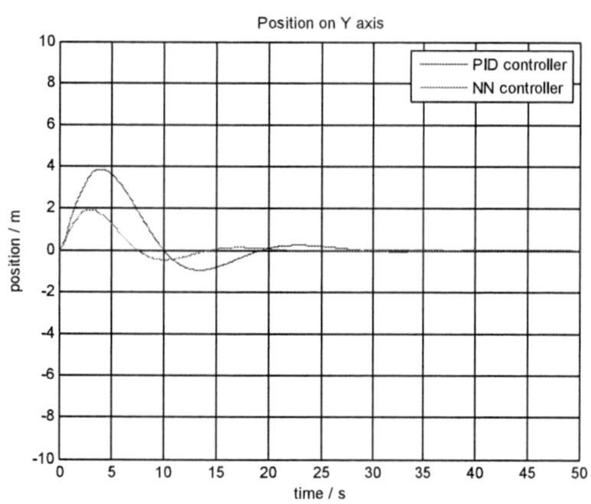

Figure 5: Position displacement along Y axis

CONCLUSIONS

In this paper, a neural network based solution for hovering quadrotor in windy environment is investigated and proposed. The simulation results show that the proposed method has faster and smoother responses than that of PID controller in a constant windy environment.

REFERENCES

1. K. Hunt, D. Sbarbaro, R. bikowski, and P. Gawthrop, "Neural networks for control systemsa survey," *Automatica*, vol. 28, no. 6, pp. 1083-1112, 1992.
2. S. Lu and T. Basar, "Robust nonlinear system identification using neural-network models," *IEEE Transactions on Neural Networks*, vol. 9, no. 3, pp. 407-429, May 1998.
3. A. Alessandri, R. Bolla, M. Gaggero, and M. Repetto, "Modeling and identification of nonlinear dynamics for freeway traffic by using information from a mobile cellular network," *IEEE Transactions on Control Systems Technology*, vol. 17, no. 4, pp. 952-959, Jul. 2009.
4. W. He, A. O. David, Z. Yin, and C. Sun, "Neural network control of a robotic manipulator with input dead zone and output constraint," *IEEE Transactions on Systems, Man, and Cybernetics: Systems*, vol. 46, no. 6, pp. 759-770, Jun. 2016.
5. "Adaptive neural network predictive control for nonlinear pure feedback systems with input delay," *Journal of Process Control*, vol. 22, no. 1, pp. 194-206, 2012.
6. Y.-J. Liu, J. Li, S. Tong, and C. L. P. Chen, "Neural network control-based adaptive learning design for nonlinear systems with full-state constraints," *IEEE Transactions on Neural Networks and Learning Systems*, vol. 27, no. 7, pp. 1562-1571, Jul. 2016.
7. Y. L. He, Y. M. Chen, M. F. Zhou, "Modeling and control of a quadrotor helicopter under impact of wind disturbance," *Journal of Chinese Inertial Technology*, vol. 21, no.5, pp. 624-630, 2013.
8. B. W. McCormick, *Aerodynamics of V/STOL flight,* Academic Press, 1967.
9. A. Nagaty, S. Saeedi, C. Thibault, M. Seto and H. Li, "Control and navigation framework for quadrotor helicopters," *Journal of Intelligent & Robotic Systems*, vol. 70, no.1-4, pp. 1-12, 2013.
10. B. Erginer, E. Altuğ, "Modeling and PD control of a quadrotor VTOL vehicle," *IEEE Intelligent Vehicles Symposium*, pp. 894-899, 2007.

Proceedings of the 26th CANCAM
Victoria, British Columbia, Canada, May 29 - June 1, 2017

NONLINEAR MODEL PREDICTIVE TRACKING CONTROL OF WHEELED MOBIL ROBOTS USING MODIFIED C/GMRES ALGORITHM

Yuanyang Pei, Jian Pan
School of Electrical and Electronic Engineering
Hubei University of Technology
Wuhan, China
p_p_pyy@163.com, hgd_teacher@163.com

Kunwu Zhang, Yang Shi
Department of Mechanical Engineering
University of Victoria
Victoria, BC, Canada
kunwu@uvic.ca, yshi@uvic.ca

ABSTRACT

In this brief, a nonlinear model predictive control (NMPC) algorithm is deigned for the trajectory tracking problem of a wheeled mobile robot (WMR). The tracking problem is formulated into the standard NMPC form based on the error dynamics of the WMR. In order to improve the efficiency of the NMPC, a modified continuation/generalized minimal residual (c/GMRES) algorithm is used. A sinusoidal trajectory is tested to evaluate the proposed algorithm. The simulation results show the efficiency and effectiveness of the designed NMPC algorithm for the trajectory tracking of the WMR.

Keywords: wheeled mobile robot (WMR), trajectory tracking, nonlinear model predictive control (NMPC), modified continuation/generalized minimal residual (c/GMRES),

INTRODUCTION

Recently, wheeled mobile robots (WMRs) have attracted increasing attention in many areas from industrial applications to entertainments, e.g., the factory transportation, the floor cleaning, the space exploration and the robot soccer [1]. In above applications, the WMR is usually required to track a specific trajectory, which can be represented by several points and orientations, or even by the analytical expression. What's more, some cases require that the speed of the mobile robot should be either constant or any values within the specific range. Hence a high-performance tracking controller satisfying various requirements is desired.

The trajectory tracking, being a fundamental motion control problem for WMRs, has been intensively studied in robotic community in past few decades. One of the early publication [2] develops a local tracking controller. But the velocity and acceleration are limited to guarantee the asymptotic stability. The global tracking is explored by using the nonlinear control technique such as the feedback linearization [3], the backstepping technique [4] and the sliding mode control [5]. However, the performance of above controllers may degrade when the constraints on either states or inputs have to be considered. One general solution is to omit the constraints by the mean of saturation [6]. Recently, the model predictive control (MPC) has become popular since it is able to explicitly handle various constraints [7]. The control problem formulation for MPC is highly feasible so that it can be used for nonlinear or time-varying systems. Due to aforementioned features, MPC is an attractive method for WMR tracking control problems. In [8], a trajectory controller based on MPC technique is deigned for the linearized WMR model. The nonlinear MPC (NMPC) technique is also applied to WMRs. Gu *et al.* [9] design a NMPC algorithm based on the error dynamics of the WMR. [10] investigates the trajectory-tracking control of the WMR, in which the NMPC scheme is incorporated with the neural-dynamics optimization to improve the algorithm efficiency.

Although the NMPC scheme shows attractive features, the high computational load of the NMPC scheme limits its real-world applications. The optimization problem arisen from the NMPC scheme is required to be solved within the sampling period so that the MPC can be successfully applied in some areas in which the sampling period is sufficiently large, e.g., the chemical industry [11]. However, the computational time requirement for the system with small sampling period, e.g., robotic systems including WMRs, cannot be ignored. Consequently, different numerical algorithms, such as offline pre-computations , delay compensation , computing division and iterating with problem changes, have been developed to solve the NMPC problem efficiently [12].

In this brief, we investigate the trajectory tracking problem of the WMR.The tracking problem is formulated in the context of NMPC based on the error dynamics of the WMR. The WMR kinematic model is applied to design desired trajectory so that the potential singularity can be avoided in following computations. In addition, Ohtsuka's continuation/generalized minimal residual (c/GMRES) algorithm [13] is used to solve the NMPC problem efficiently. But the c/GMRES method can only handle the equality constraints. To overcome this limitation, the modified c/GMRES method is introduced, in which the barrier method is used to transform the original problem into an unconstrained one. The modified c/GMRES algorithm can preserve high efficiency of the c/GMRES method while handling the inequality constraints. The simulations are conducted to evaluate the modified c/GMRES algorithm, which shows the efficiency and the effectiveness of the designed algorithm.

PROBLEM FORMULATION

In this brief, we consider a WMR in Fig. 1 described by the following kinematic model

$$\dot{X} = \begin{bmatrix} \dot{x} \\ \dot{y} \\ \dot{\theta} \end{bmatrix} = \begin{bmatrix} v\cos\theta \\ v\sin\theta \\ \omega \end{bmatrix} = \begin{bmatrix} \cos\theta & 0 \\ \sin\theta & 0 \\ 0 & 1 \end{bmatrix} U, \qquad (1)$$

where $X = [x, y, \theta]^{\mathrm{T}}$ indicates the WMR's position and posture and $U = [v, w]^{\mathrm{T}}$ denotes the linear and angular velocities of the WMR. X and U are the vectors for the state and control signal.

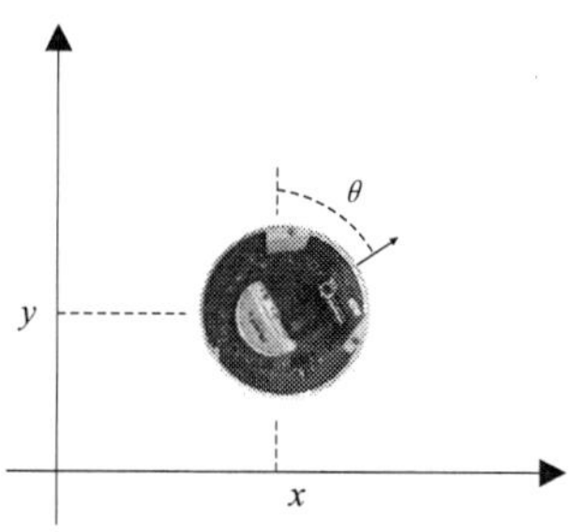

Figure 1: The WMR coordinate system

Let $X_r = [x_r, y_r, \theta_r]^{\mathrm{T}}$ be the state vector for the reference trajectory and $U_r = [v_r, w_r]^{\mathrm{T}}$ be the reference control signal vector. The reference trajectory should also satisfy the kinematic model in (1). To track the reference trajectory, an error state $X_e = [x_e, y_e, \theta_e]^{\mathrm{T}}$, which is presented in local coordinate with the origin X, can be defined as follows [2]:

$$X_e = \begin{bmatrix} x_e \\ y_e \\ \theta_e \end{bmatrix} = \begin{bmatrix} \cos\theta & \sin\theta & 0 \\ -\sin\theta & \cos\theta & 0 \\ 0 & 0 & 1 \end{bmatrix} \begin{bmatrix} x_r - x \\ y_r - y \\ \theta_r - \theta \end{bmatrix}. \qquad (2)$$

According to (1) - (2), we can derive the following error dynamic model of the WMR [9]:

$$\begin{cases} \dot{x}_e = \omega y_e - v + v_r \cos\theta_e \\ \dot{y}_e = -\omega x_e + v_r \sin\theta_e \\ \dot{\theta}_e = \omega_r - \omega \end{cases}. \qquad (3)$$

The control input signals are redefined as follows:

$$U_e = \begin{bmatrix} u_1 \\ u_2 \end{bmatrix} = \begin{bmatrix} v_r\cos\theta_e - v \\ \omega_r - \omega \end{bmatrix}.$$

For the convenience, we consider the discretized version of (3) using Euler forward approximation in the following part

$$X_e(i+1) = f(X_e(i), U_e(i))\Delta t + X_e(i),$$

where Δt is the sampling period. Here a cost function is used as a tracking performance index from kth time instant to $(k+N)$th time instant, which is defined as follows:

$$\begin{aligned} J(X_e(k), U_e(k)) = g(X_e(k+N)) + \\ \textstyle\sum_{i=0}^{N-1} L(X_e(k+i), U_e(k+i))\Delta t \end{aligned} \qquad (4)$$

where

$$\begin{aligned} L(X_e(k), U_e(k)) &= \tfrac{1}{2}(X_e(k)^{\mathrm{T}} Q X_e(k) + U_e^{\mathrm{T}} R U_e(k)), \\ g(X_e(k)) &= \tfrac{1}{2} X_e(k)^{\mathrm{T}} Q_t X_e(k). \end{aligned}$$

Here, matrices Q, Q_t, R are the positive definite weighting matrices; N is the prediction horizon. To minimize the cost function in (4) at kth time instant, the NMPC problem is formulated as follows:

$$\begin{array}{ll} \min & J(X_e(k), U_e(k)) \\ \text{s.t.} & X_e(j+1) = f(X_e(j), U_e(j))\Delta t + X_e(j) \\ & X_e(k) = X_0 \\ & h(U_e(j)) = \|U_e(j)\| - U_{max} \le 0 \\ & j = k, k+1, \ldots, k+N-1 \end{array} \qquad (5)$$

where X_0 is the state at time k and $U_{max} = [v_{\max}\ \omega_{\max}]$ is the limit of the control signal.

NMPC IMPLEMENTATION

In this section, we investigate the application of the Ohtsuka's c/GMRES method to the nonlinear programming (5). But the c/GMRES method can only handle the less general nonlinear programming subject to equality constraints, while the inequality constraints exist in the WMR trajectory tracking problem. To overcome this limitation, as described in [14], different numerical algorithms are developed, such as: the exterior penalty method, the barrier method, the auxiliary variable method and the semi-smooth transformation method. In this brief, the barrier method is used to handle the inequality constraints in (5).

The barrier method is to add an barrier function in the cost function to approximate the original problem so that the inequality constraints in (5) can be ignored. The new cost function is defined as follows:

$$\begin{aligned} \bar{J}(X_e(k), U_e(k)) = J(X_e(k), U_e(k)) - \\ \kappa \textstyle\sum_{i=0}^{N-1} \sum_{j=0} \log(-h_j(U_e(k+i))), \end{aligned}$$

where $\kappa > 0$ is the barrier parameter. Then we have the approximation of problem (5):

$$\begin{array}{ll} \min & \bar{J}(X_e(k), U_e(k)) \\ \text{s.t.} & X_e(j+1) = f(X_e(j), U_e(j))\Delta t + X_e(j) \\ & X_e(k) = X_0 \\ & j = k, k+1, \ldots, k+N-1 \end{array} \tag{6}$$

In Ohtsuka's c/GMRES method, the first order necessary conditions are obtained according to the Pontryagins minimum principle (PMP). Let H be the Hamiltonian with

$$H(X_e, U_e, \mu,) = L(X_e, U_e) + \lambda^{\mathrm{T}} f(X_e, U_e),$$

where λ is the costate. Then we have the following optimal conditions:

$$\begin{array}{l} X_e^*(j+1) = X_e^*(j) + f(X_e^*(j), U_e^*(j))\Delta t \\ \lambda^*(j) = \lambda^*(j+1) + H_{X_e}^{\mathrm{T}}(X_e^*(j), \lambda^*(j+1), U_e^*(j))\Delta t \\ \lambda^*(N) = g_{X_e}^{\mathrm{T}}(X_e(k+N)) \\ X_e(k) = X_0 \\ H_{U_e}(x^*(j), \lambda^*(j+1), U_e^*(j)) = 0 \\ j = k, k+1, \ldots, k+N-1 \end{array} \tag{7}$$

Let $U(k) = [U_e(k)^{\mathrm{T}}, U_e(k+1)^{\mathrm{T}}, \ldots, U_e(k+N-1)^{\mathrm{T}}]^{\mathrm{T}}$, then first order necessary conditions can be represented as

$$F(U^*(k), Xe(k)) = \begin{bmatrix} H_{U_e}(x^*(k), \lambda^*(k+1), U_e^*(k)) = 0 \\ H_{U_e}(x^*(k+1), \lambda^*(k+2), U_e^*(k+1)) = 0 \\ \vdots \\ H_{U_e}(x^*(k+N-1), \lambda^*(k+N), U_e^*(k+N-1)) = 0 \end{bmatrix} \tag{8}$$

By solving (8), we can get the solution for (6).

Solving (8) by using the iterative method is very expensive for calculating Jacobians, Hessian, and inverses . Instead, Ohtsuka's c/GMRES method uses the continuation method to consider the $F(U(t), Xe(t))$ as a dynamics system

$$\dot{F}(U(t), Xe(t), k) = A_s F(U(t), Xe(t), t), \tag{9}$$

where A_s is the stable matrix that stabilizes $F(U(t), Xe(t), t)$ at the origin. Here, we indicate the time independence of the system (9) by representing the state as $F(U(t), Xe(t), t)$. By calculating the derivative of $F(U(t), Xe(t), t)$ in (9), we have

$$F_U \dot{U} = A_s F - F_{X_e} \dot{X}_e - F_k k. \tag{10}$$

If the initial $U(0)$ satisfies $F(U(0), X_e(0), 0) = 0$, then we can get the solution $U_e(t)$ for (6) by integrating (10).

However, the operations, such as Jacobians and inverse, in (10) are still computationally demanding, the forward difference is used to approximate the products of the Jacobians, which is given as follows:

$$\begin{array}{l} F_U(U, X_e, t)w_u + F_{X_e}(U, X_e, t)w_x + F_t(U, X_e, t)w_t \\ := \delta^{-1}(F(U + \delta w_u, X_e + \delta w_x, t + \delta w_t) - F(U, X_e, t) \\ \simeq D_\delta F(U, X_e, t : w_u, w_x, w_t) \end{array} .$$

Here δ is a positive number. This forward difference based GMRES algorithm is called FDGMRES. The details of FDGMRES can be found in [13]. For convenience, FDGMRES is considered as a function with

$$\dot{U} = \text{FDGMRES}(U, X_e, \dot{X}_e, \hat{\dot{U}}, \delta, k_{\max}).$$

Then the modified c/GMRES algorithm is described in Algorithm 1.

1 Initialize $t = 0, k = 0$ and $l = 0$, given initial state $X_e(0) = X_0$, $\kappa(0), \alpha$ and U_0 with $F(U^l(0), X_e(0), 0) = 0$.
2 Let $l = l + 1$. If $l < l_{\max}$, go to Step 3. Otherwise, go to Step 4.
3 Let $\kappa(l) = \alpha\kappa(l-1)$ and find $U^l(0)$ with $F(U^l(0), X_e(0), 0) = 0$.
4 For $s \in [t, t + \Delta t)$, take the first of the control sequence $U(k)$ as the input to the real system.
5 At next sampling instant $t + \Delta t$, the system state $X_e(k+1) = X_e(t + \Delta t)$ is fed back. Compute the state different $\Delta X_e(k) = X_e(k+1) - X_e(k)$.
6 $\dot{U}(k)$= FDGMRES$(U(k), X_e(k), \Delta X_e(k)/\Delta t, t, \hat{\dot{U}}, \delta, k_{\max})$
7 Set $U(k+1) = U(k) + \dot{U}(k)\Delta t$
8 Update $t = t + \Delta t, k = k + 1$.
9 Go to step 4

Algorithm 1: Modified c/GMRES algorithm [15]

It should be noted that, in the modified c/GMRES algorithm, although the PMP system in (7) are solved several times with the decrease of κ at the initialization to obtain a good approximation of $U(0)$, the aforementioned progress can be done off-line. Hence the modified c/GMRES algorithm preserves the efficiency of Ohtsuka's c/GMRES algorithm.

SIMULATIONS

The reference trajectory is a sinusoid, which is defined as follows:

$$x_r = 0.5t,\ y_t = 0.5\sin(0.5t), t \in [0, 14].$$

The maximum linear and angular velocities are 0.5m/s and $\pi/2$rad/s, respectively. For the NMPC, the sampling period $\Delta t = 0.1s$, the weighting matrices $Q = \text{diag}(300, 300, 100)$, $R = \text{diag}(5, 8), Q_t = \text{diag}(0.5, 0.5, 0.1)$ and the prediction horizon $N = 7$. For the modified c/GMRES algorithm, $A_s = (1/\Delta t)I, k_{\max} = 4, \delta = 10^{-5}, \kappa(0) = 10, l = 6, \alpha = 0.1$.

The simulations are conducted to evaluate the modified c/GMRES algorithm. The initial position and posture of the WMR are $(0.2, 0.2, 0)$. The tracking results of a sinusoid are shown in Fig. 2. The blue solid line and the red dash line are the simulated WMR trajectory and the reference trajectory, respectively. The linear and angular velocities with respect to time are given in Fig. 3. From the simulation results, it is shown that the WMR can track the given trajectory simultaneously and all control signals satisfy their constraints. In addition, the average computational time per NMPC step for solving (8) by using the MATLAB function $fslove$ is 0.6351 seconds, while it is 0.0044 seconds by using the modified c/GMRES method, which is much lower than the sampling period Δt. Hence it is possible for the real-time implementation of the modified c/GMRES algorithm.

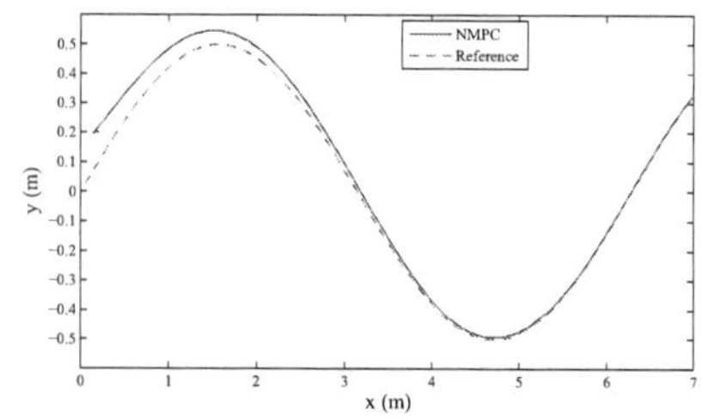

Figure 2: The trajectories

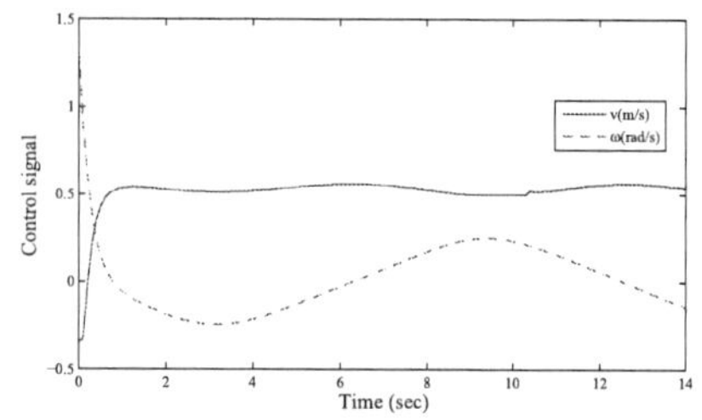

Figure 3: The NMPC control signals

CONCLUSION

In this brief, the NMPC for trajectory tracking control of the WMR is investigated. The computational time per NMPC step is significantly decreased by using the modified c/GMRES algorithm. The sinusoidal trajectory tracking results show the efficiency and effectiveness of the modified c/GMRES algorithm for the trajectory tracking problem of the WMR. The future work will focus on the experimental validation of the c/GMRES algorithm.

REFERENCES

[1] Volos, C. K., Kyprianidis, I. M., and Stouboulos, I. N. 2013, "Experimental investigation on coverage performance of a chaotic autonomous mobile robot," *Robotics and Autonomous Systems*, vol. 61, no. 12, pp. 1314–1322.

[2] Kanayama, Y., Kimura, Y., Miyazaki, F., and Noguchi, T. 1990, "A stable tracking control method for an autonomous mobile robot," *1990 IEEE International Conference on Robotics and Automation (ICRA 1990)*, Tsukuba, Japan, pp.384–389.

[3] Novel, B.d'Andréa, Campion, G., and Bastin, G. 1995, "Control of nonholonomic wheeled mobile robots by state feedback linearization," *The International Journal of Robotics Research*, vol. 14, no. 6, pp. 543–559.

[4] Jiang, Z. and Nijmeijer, H. 1997, "Tracking control of mobile robots: A case study in backstepping," *Automatica*, vol. 33, no. 7, pp. 1393–1399.

[5] Yang, J.-M. and Kim, J.-H. 1999, "Sliding mode control for trajectory tracking of nonholonomic wheeled mobile robots," *IEEE Transactions on Robotics and Automation*, vol. 15, no. 3, pp. 578–587.

[6] Oriolo, G., Luca, A. D., and Vendittelli, M. 2002, "WMR control via dynamic feedback linearization: Design, implementation, and experimental validation," *IEEE Transactions on Control Systems Technology*, vol. 10, no. 6, pp. 835–852.

[7] Mayne, D. Q., Rawlings, J. B., Rao, C. V., and Scokaert, P. O. 2000, "Constrained model predictive control: Stability and optimality," *Automatica*, vol. 36, no. 6, pp. 789–814.

[8] Kuhne, F., Lages, W. F., and Silva Jr, J. G.da 2004 "Model predictive control of a mobile robot using linearization," *Mechatronics and Robotics*, pp. 525–530.

[9] Gu, D. and Hu, H., 2006, "Receding horizon tracking control of wheeled mobile robots," *IEEE Transactions on Control Systems Technology*, vol. 14, no. 4, pp. 743–749.

[10] Li, Z., Deng, J., Lu, R., Xu, Y., Bai, J., and Su, C. 2016, "Trajectory-tracking control of mobile robot systems incorporating neural-dynamic optimized model predictive approach," *IEEE Transactions on Systems, Man, and Cybernetics: Systems*, vol. 46, no. 6, pp. 740–749.

[11] Qin, S. J. and Badgwell, T. A. 2000, "An overview of nonlinear model predictive control applications," in *Nonlinear model predictive control*, Springer, pp. 369–392.

[12] Grne, L. and Pannek, J. 2013, *Nonlinear Model Predictive Control: Theory and Algorithms*,Springer.

[13] Ohtsuka, T. 2004, "A continuation/GMRES method for fast computation of nonlinear receding horizon control," *Automatica*, vol. 40, no. 4, pp. 563–574.

[14] Huang, M., Nakada, H., Butts, K., and Kolmanovsky, I. 2015, "Nonlinear model predictive control of a diesel engine air path: A comparison of constraint handling and computational strategies," *5th IFAC Conference on Nonlinear Model Predictive Control (NMPC 2015)*, Seville, Spain, pp. 372–379.

[15] Shen, C., Buckham, B., and Shi, Y. 2016 "Modified C/GMRES algorithm for fast nonlinear model predictive tracking control of AUVs." doi:10.1109/TCST.2016.2628803.

TANGENTIAL IMPACT FOR MULTIBODY SYSTEMS WITH NUMERICAL APPLICATION ON (P-R) ROBOT

Hassan M. Alkomy
Engineering Mathematics and Physics (Mechanics)
Alexandria University
Alexandria, Egypt
h.m.alkomy@alexu.edu.eg

Khaled T. Mohamed
Mechanical Engineering
Alexandria University
Alexandria, Egypt
ktawfik64@alexu.edu.eg

Ahmed S. Ashour
Engineering Mathematics and Physics (Mechanics)
Alexandria University
Alexandria, Egypt
ahmed.ashour@alexu.edu.eg

Hesham A. Elkaranshawy
Engineering Mathematics and Physics (Mechanics)
Alexandria University
Alexandria, Egypt
hesham_elk@alexu.edu.eg

ABSTRACT

The impact engineering is an important topic in modern life. Due to the impossibility of smooth contact, the frictional impacts gain much practical importance in engineering. This paper treats some topics related to the frictional impacts in multibody systems. The impact model of rough collision for a general multibody system is derived and the impact classes is introduced using Routh method. The tangential impact, as an important class of impacts, is generalized to adapt the multibody systems and it is illustrated graphically and analytically in details. Finally, a numerical example on the tangential impact is analyzed. This example gives the solution of the famous dynamical problem, Painlevé paradox in a prismatic-revolute (P-R) robot.

KEYWORDS: Contact and Friction, Impact Dynamics, Multibody Dynamics, Painlevé Paradox

INTRODUCTION

The field of impact engineering is one of the vital research areas in applied mechanics as it is related to both; direct applications and explanation of other phenomena. The direct applications include collision problems in vehicle crashing and military purposes, while the explanation of other phenomena is such as solving Painlevé paradox. This paper concentrates on the second part only. Painlevé paradox is a dynamical problem that was, firstly, mentioned by Paul Painlevé [1]. It occurs in multibody systems in practical conditions of operation [2-8]. This paradox may happen when a multibody system slides on a rough surface under some conditions. The paradox is that the equations of motion cannot give the correct solution when the manipulator enters the paradox zone [2-12]. In order to solve this paradox some researchers suggested a method considering the local compliance at the contact point [9], while others presented a solution using the concept of tangential impact [2-4, 11]. The model of rough collision is presented in numerous articles and books such as [13-17]. In order to understand the impact process, Routh suggested a graphical method which indicates five modes of impact according to the impact lines [18]. The tangential impact is a special class of frictional impacts with zero normal velocity. Some researchers investigated the tangential impact in case of a rigid body and found valuable results [13, 14]. Other researchers used the concept of tangential impact in multibody systems without indicating how they generalized the tangential impact equations to adapt multibody systems [2-4]. Regarding solving Painlevé paradox in multibody systems, a Chinese research group performed numerical and experimental researches on a two-link robot with revolute joints only [3, 4]. Although they solved the paradox, they did not give details about the impact process. Another Egyptian research team conducted parametric studies on two different robots and suggested methods to escape the paradox zone [5-8]. Even though they presented a complete kinematical solution for the paradox, the details of the impact process were not clarified [2]. In this paper, the impact model for rough collision in multibody systems is introduced briefly, and then the impact lines are defined. After that, Routh method for multibody systems is presented and the impact modes are illustrated. The impact classes are introduced for a general multibody system, while a special attention on the tangential impact is paid. The analytical model and equations of the tangential impact for a general multibody system is derived. A numerical example of solving the Painlevé paradox for a (P-R) robot is introduced with its impact details. This paper can be considered as an illustrating paper that clarifies, to the researchers, the generalization method of tangential impact to adapt the multibody systems with a numerical example on a practical problem.

IMPACT MODEL FOR ROUGH COLLISION

In this paper, we build on the equations derived in our previous paper [2], so, the final equations are presented only. For more details about the derivation, please go through it. The equation of motion of a general manipulator slides on a rough surface is [2]

$$\boldsymbol{\Phi}(\boldsymbol{q})\,\ddot{\boldsymbol{q}} + \boldsymbol{u}(\boldsymbol{q},\dot{\boldsymbol{q}}) = \boldsymbol{w} + \boldsymbol{J}^T \begin{bmatrix} F_n \\ F_t \end{bmatrix} \tag{1}$$

where, $\boldsymbol{q}$ is the vector of the joint coordinates, $\boldsymbol{\Phi}$ is the inertia matrix, $\boldsymbol{u}$ is the vector of Coriolis and centripetal forces, $\boldsymbol{w}$ is the generalized active forces, F_n, F_t are the normal and tangential components of the contact forces, respectively, and $\boldsymbol{J}$ is the Jacobian matrix that can be defined as

$$\boldsymbol{v} = \boldsymbol{J}\dot{\boldsymbol{q}} = \begin{bmatrix} \boldsymbol{\alpha}^T \\ \boldsymbol{\beta}^T \end{bmatrix} \dot{\boldsymbol{q}}, \quad \boldsymbol{\alpha} \in \boldsymbol{R}^{nx1}, \ \boldsymbol{\beta} \in \boldsymbol{R}^{nx1} \tag{2}$$

where, $\boldsymbol{v} = \begin{bmatrix} v_n \\ v_t \end{bmatrix}$ is the vector of the normal and tangential components of the contact point velocity. Integrating Eq. (1) over the collision time period from t to $t+\Delta t$ and results in [2]

$$\Phi(q)\Delta\dot{q} = J^T I \quad (3)$$

where, I is the impulse of the contact forces. From Eq. (2)

$$\Delta v = J\Delta\dot{q} \quad (4)$$

From Equations (3, 4) it is obvious that impact equation is:

$$\frac{dv}{dI} = J\Phi^{-1}J^T = D = \begin{bmatrix} D_{11} & D_{12} \\ D_{21} & D_{22} \end{bmatrix} \quad (5)$$

where, the Jacobian inertia matrix, D, depends only on the system configuration [2]. It is clear that, $D_{11} = \alpha^T\Phi^{-1}\alpha$, $D_{22} = \beta^T\Phi^{-1}\beta$, $D_{12} = D_{21} = \beta^T\Phi^{-1}\alpha$.

In sliding mode, the combination of the impact equation with Coulomb's friction law results in [2]

$$dv = \begin{bmatrix} dv_n^s \\ dv_t \end{bmatrix} = D\begin{bmatrix} dI_n \\ dI_t^s \end{bmatrix} = \begin{bmatrix} D_{11} & D_{12} \\ D_{21} & D_{22} \end{bmatrix}\begin{bmatrix} 1 \\ -\mu\sigma \end{bmatrix} dI_n \quad (6)$$

where, the subscripts (t) refers to the tangential components. The subscript (n) refers to the normal components and the superscript (s) refers to the sliding mode.

In non-sliding (sticking) mode, the impact equation is

$$dv = \begin{bmatrix} dv_n^{ns} \\ dv_t \end{bmatrix} = \begin{bmatrix} D_{11} & D_{12} \\ D_{21} & D_{22} \end{bmatrix}\begin{bmatrix} dI_n \\ -\frac{D_{21}}{D_{22}} dI_n \end{bmatrix} \quad (7)$$

where, the superscript (ns) refers to the non-sliding (sticking).

IMPACT PROCESS

The impact process, according to the collision assumptions, can be partitioned into two periods: compression and restitution. In order to understand and classify the impact process well, *impact lines* must be defined. [13, 14]. All of these lines can be derived from integrating the scalar form of the impact equation, Eq. (5). This integrated scalar form can be shown as follows

$$v_n = v_{no} + D_{11}I_n + D_{12}I_t \quad (8)$$

$$v_t = v_{to} + D_{21}I_n + D_{22}I_t \quad (9)$$

where, v_{no}, v_{to} are the normal and tangential components of the initial velocity. These lines can be defined as follows.

The maximum compression line is the line which defines the end of the compression period and the start of the restitution. Its condition is v_n = zero. Substitution in Eq. (8), the equation of the maximum compression line appeared as:

$$v_{no} + D_{11}I_n + D_{12}I_t = 0 \quad (10)$$

At this moment the restitution period starts and it will end, according to Poisson's hypothesis, when the normal impulse value reaches a value of (1+ e) times of its value at the end of the compression period as follows

$$I_n^f = (1+e)I_n^c \quad (11)$$

where, e is the coefficient of restitution, I_n^f is the final value, at the end of the impact period, of the normal impulse and I_n^c is its value at the end of the compression period, at the maximum compression line.

The sticking line defines the condition of sticking. Its condition is v_t = zero. Hence, Eq. (9) can be written as

$$v_{to} + D_{21}I_n + D_{22}I_t = 0 \quad (12)$$

which represents the equation of the sticking line.

The limiting friction line is the line which satisfies the Coulomb's friction law with a slope equals $1/\mu$

$$F_t^s = -\mu\sigma F_n$$

$$\frac{dI_t^s}{dt} = -\mu\sigma\frac{dI_n}{dt} \quad or \quad dI_t^s = -\mu\sigma dI_n \quad (13)$$

where, σ is the sliding direction indicator, $\sigma = \frac{v_t}{|v_t|}$.

Routh incremental method is an effective way to illustrate what happens during the impact period [13, 14, 18]. Figure 1 shows the different cases of impact process according to Routh incremental method and Poisson's hypotheses. The impulse space (I_t, I_n) is constructed and the maximum compression line and the sticking line are plotted. Initial sliding motion is assumed. As the impact process starts on the limiting friction line with initial values of the impulse components (I_{to}, I_{no}) equal zero, I_n increases monotonically. If the impact ends, according to Poisson's hypothesis, Eq. (11), without reaching the sticking line, the contact point will slide during the whole impact period as in Fig. 1.a. If the contact point reached the sticking line before the impact ending, the initial sliding motion will stop and there will be two possible modes. The contact point will either stick until the impact ends as in Fig. 1.b, or will reverse sliding until the impact ends as in Fig. 1.c. In these cases, the relation between the critical coefficient, μ_s, and the limiting friction determine which mode will occur. The critical coefficient of friction is the coefficient of friction that is required to prevent sliding. It can be defined from Eq. (12) as $\mu_s = \left|-\frac{D_{21}}{D_{22}}\right|$. If the limiting friction is more than the critical coefficient of friction $\mu > \mu_s$, the contact point will stick until the impact terminates. Otherwise, the reverse motion will occur.

IMPACT MODES

According to Routh graphical method, there are five modes of contact that depend on the locations and the slopes of three impact lines [13, 14, 18]. The case of $D_{12} = D_{21} > 0, v_{to} < 0$ and $v_{no} < 0$ is the clearest example to explain the contact modes, Fig. 1.d. In region I, when the impact termination occurs before sticking, the motion will be *continuous sliding* during the entire impact period. When the instantaneous sticking occurs in the restitution period as in region II, the mode is either *restitution-sticking* or *restitution-reverse sliding* according to the values of μ and μ_s as mentioned previously. In cases when the instantaneous sticking occurs in the compression period, as in region III, the mode is either *compression-sticking* or *compression-reverse sliding.*

An analytical analysis can be conducted to derive the conditions that define these regions instead of the previous graphical analysis. This can be deduced from Fig. 1.d after defining some factors.

As shown in Fig. 1.d, I_n^{st} and I_n^c are the values of the normal impulses at the intersection points of the limiting friction line with the sticking line (point s) and the maximum compression line (point c), respectively. According to Poisson's hypothesis, the termination occurs when the final normal impulse is *(1+e)* times of the normal impulse at the end of the compression period. Accordingly, the analytical conditions can be tabulated as in Table 1.

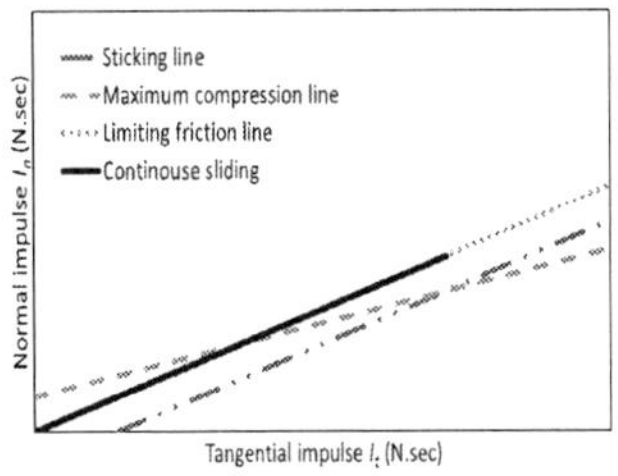

Fig. 1.a: Continuous sliding

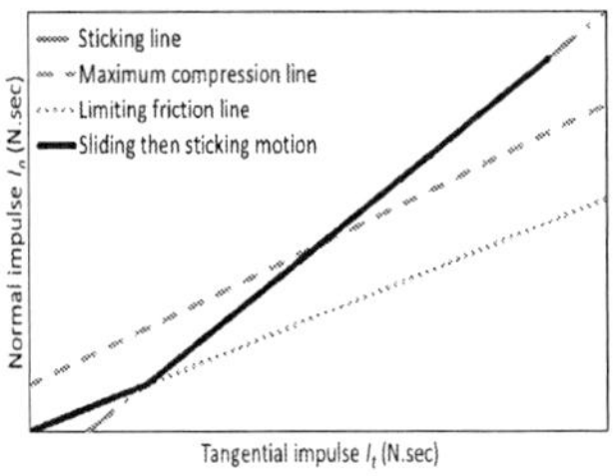

Fig. 1.b: Sticking ($\mu > \mu_s$)

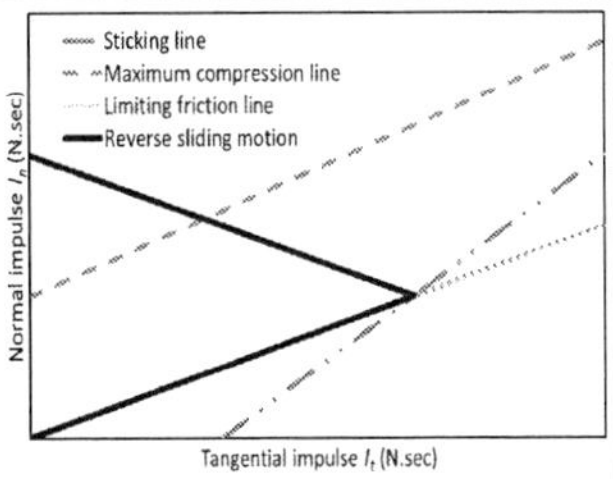

Fig. 1.c: Reverse sliding ($\mu < \mu_s$)

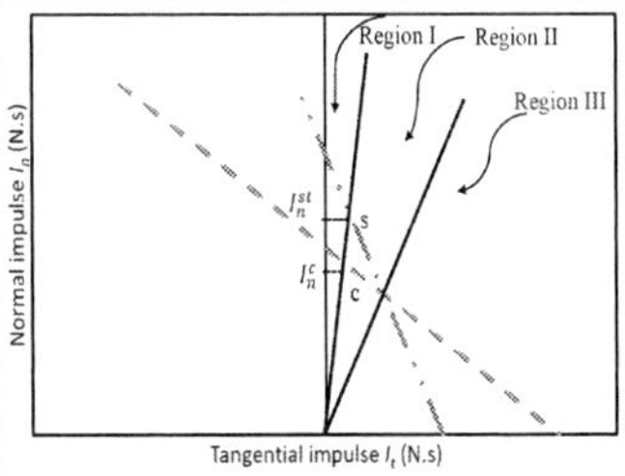

Fig. 1.d: Impact regions

Table 1: Contact modes of oblique impact

	$\mu > \mu_s$	$\mu < \mu_s$
Region I $I_n^{st} > (1+e)\ I_n^c$	Sliding	
Region II $I_n^c < I_n^{st} < (1+e)\ I_n^c$	Restitution-Sticking	Restitution-Reverse Sliding
Region III $I_n^{st} < I_n^c$	Compression - Sticking	Compression-Reverse Sliding

IMPACT CLASSES

From the definition of the Jacobian inertia matrix $\boldsymbol{D}$ in Eq. (5) only D_{12} and D_{21}, which are already equals to each other, can be positive or negative, while D_{12} and D_{21} are always positive [3]. The initial sliding (tangential) velocity v_t may be positive or negative according to the initial sliding direction. A negative initial normal velocity v_n is considered. This class of impact is called meeting impact. The cases of each impact class are defined, basically, according to the location and the slope of the maximum compression and the sticking lines. Figure 2 shows, in the impulse space, all possible cases of meeting impacts. When $v_{to} = 0$, the impact is called *direct impact*, otherwise it is called *oblique impact*. A special class is the *tangential impact*, which occurs when $v_{no} = 0$ and is illustrated in Fig. 3. In tangential impacts, although there are nine cases, only two of them are practically approved. These two cases are illustrated Figs. 3.a and 3.i. In these cases when the μ is high the impact is applicable and the impact mode will be compression-sticking. When μ is low, the impact is impossible because the compression will not exist at all. The other seven cases are not applicable as they do not agree with the collision assumptions. Figs. (2, 3) are in the impulse space (I_t, I_n). Dashed lines represent maximum compression lines and dash-double dots lines are for the sticking lines.

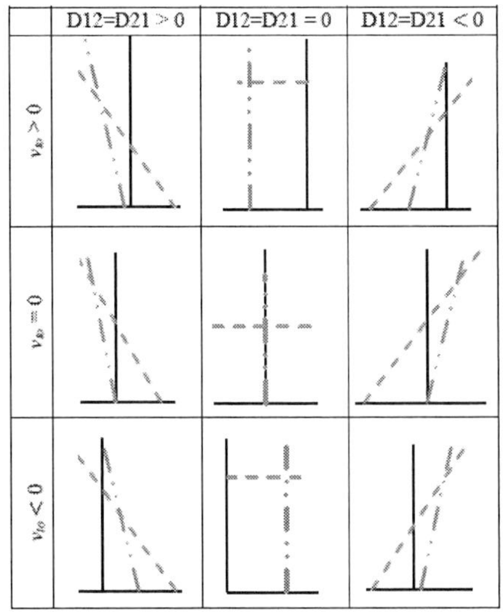

Figure 2: All meeting impact cases

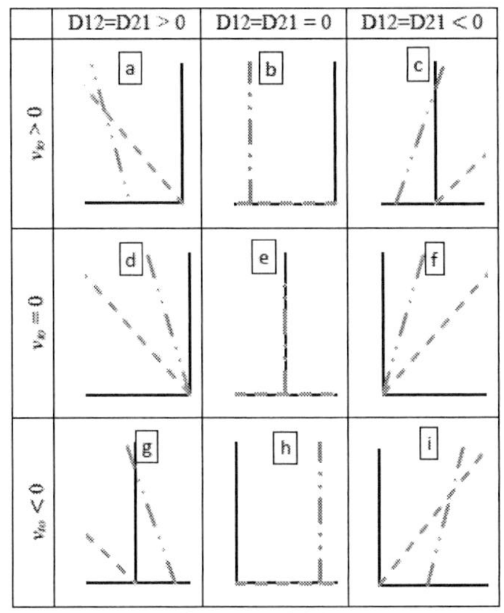

Figure 3: All tangential impact cases

THE MODEL OF THE TANGENTIAL IMPACT

According to Fig. 3, the tangential impact can be partitioned into three phases: Initial sliding in the compression period the tangential velocity vanishes, then sticking until the end of the compression period. After that, sticking in the restitution period until the end if the impact process. Utilizing Eqs. (6, 7, 11, 13), the equations that rule the tangential impact can be derived as follows. The initial conditions of the impact process are $v_{to} = v_{no} = I_{to} = I_{no} = 0$. The final conditions of each phase are the initial of the following one.

Initial sliding phase

It is worth to notice that the final condition of this phase are $v_t = 0$, $v_n = v_n^s$, $I_n = I_n^s$, $I_t = I_t^s$.

$$v_n^s = (D_{11} - \mu\sigma D_{12}) I_n^s \qquad (14)$$

$$I_n^s = \frac{-v_{to}}{(D_{21} - \mu\sigma D_{22})} \qquad (15)$$

$$I_t^s = -\mu\sigma I_n^s \qquad (16)$$

Sticking in the compression phase

It is worth to notice that the final condition of this phase are $v_t = 0$, $v_n = 0$, $I_n = I_n^c$, $I_t = I_t^c$.

$$I_n^c = I_n^s - \frac{v_n^s}{D_{11} - \frac{D_{12}^2}{D_{22}}} \tag{17}$$

$$I_t^c = I_t^s - \frac{D_{21}}{D_{22}}(I_n^c - I_n^s) \tag{18}$$

Sticking in the restitution phase

It is worth to notice that the final condition of this phase are $v_t = 0$, $v_n = v_n^f$, $I_n = I_n^f = (1+e)I_n^c$, $I_t = I_t^f$

$$I_t^f = I_t^c - e\frac{D_{21}}{D_{22}} I_n^c \tag{19}$$

$$v_n^f = e\left(D_{11} - \frac{D_{12}^2}{D_{22}}\right) I_n^c \tag{20}$$

SOLVING PAINLEVÉ PARADOX

The main objective of this section is to provide a detailed, graphical and analytical, solution of Painlevé paradox using the concept of the tangential impact. Consider a two-link prismatic-revolute, (P-R), robot as shown in Fig. 4.a. The robot's configuration parameters are $m_1 = m_2 = 0.12$ kg and $l_1 = l_2 = 0.21$ m, $\mu = 1$, $\sigma = 1$, $e = 0.7$ and the total height $H = 0.3775$ m. According to [2, 8], the robot enters the paradox zone at angle $\theta = 17^o$. The robot begins its sliding from the rest when $\theta = 0$ and considering that $F = 0.23S$ and $\tau = 0.15\theta$. Using Fehlberg 4-5th order Runge-Kutta method in solving Eq. (1), the initial conditions of the tangential impact can be determined. These are $\theta = 17^o$, $v_{to} = 1.50799$ m/sec and $v_{no} = 0$. The critical coefficient of friction $\mu_s = 0.2293$.

Using Routh graphical method, the impact process is plotted in Fig. 4.b. Using Eqs. (14-20), the final velocity and the velocity profiles can be obtained analytically. The final conditions are $\theta = 17^o$, $v_{tf} = 0$ and $v_{nf} = 0.2293$ m/sec, while the velocity profiles are plotted in Figs. 4.(c, d).

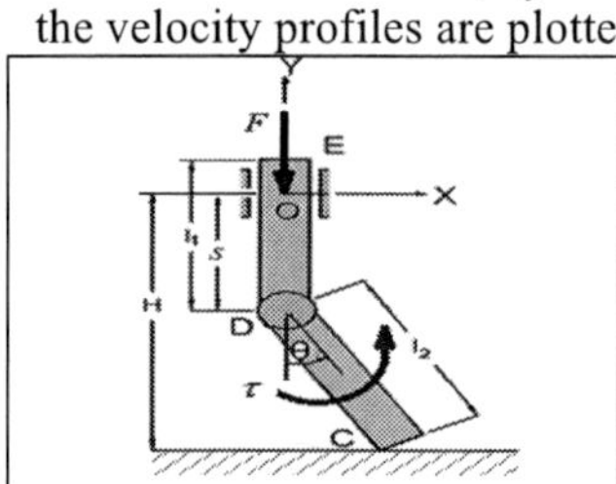

Fig. 4.a: (P-R) robot

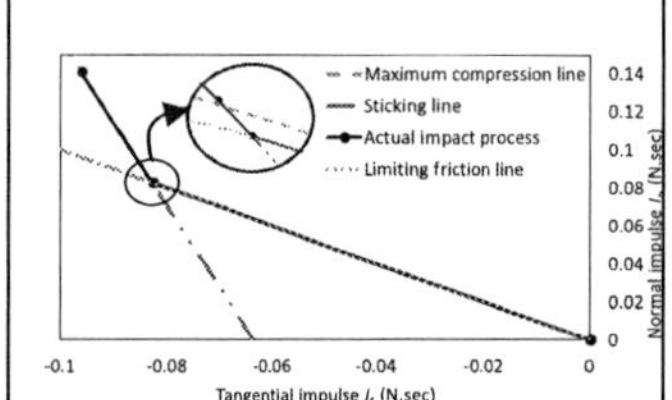

Fig. 4.b: The tangential impact

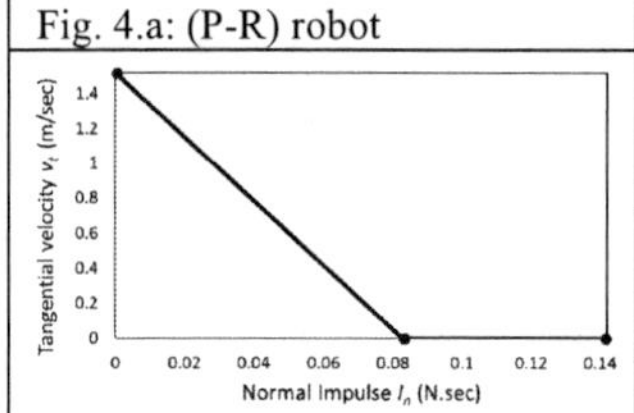

Fig. 4.c: Tangential velocity profile

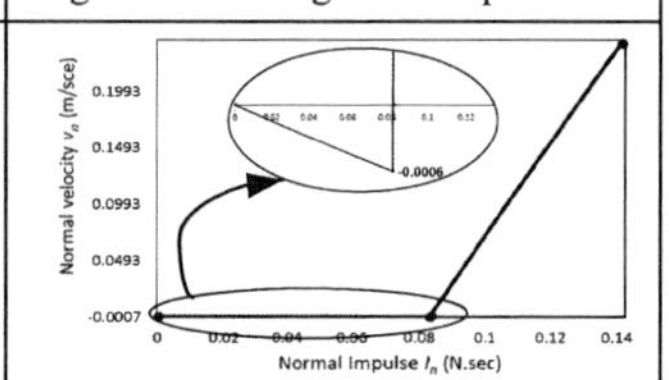

Fig. 4.d: Normal velocity profile

CONCLUSION

The aim of this paper can be portioned into two parts; investigating the tangential impact in multibody systems and using that concept to solve the Painlevé paradox. The tangential impact has been explained graphically and analytically for a general multibody system. The Painlevé paradox has been solved, in details, for a (P-R) robot and the results have been plotted and discussed.

REFERENCES

1. Painlevé, P. 1895, "Sur les lois du frottement de glissement", *Comptes Rendus Academie Science*, Vol(41)
2. Elkaranshawy, H. A., Mohamed, K. T., Ashour, A. S. and Alkomy, H. M. 2017, "Solving Painlevé paradox: (P–R) sliding robot case", *Nonlinear Dynamics*, doi: 10.1007/s11071-017-3339-y
3. Liu, C., Zhao, Z. and Chen, B. 2007, "The bouncing motion appearing in a robotic system with unilateral constraint", *Nonlinear Dynamics*, Vol(49), pp. 217-232
4. Zhao, Z., Liu, C., Ma, W. and Chen, B. 2008, "Experimental Investigation of the Painlevé Paradox in a Robotic System", *Journal of Applied Mechanics*, Vol(75), pp. 1-25
5. Elkaranshawy, H. 2011, "Using self-motion in redundant manipulators to cope with Painlevé paradox", *Proceedings of the IASTED, Pittsburgh, USA*, pp. 361-367
6. Elkaranshawy, H. and Ghazy, M. 2012, "Effect of the Link-Length Ratio on Painlevé Paradox for a Sliding Two-Link Robot"", *15th International Conference on Applied Mechanics and Mechanical Engineering, Cairo, Egypt, RC*, pp. 76-87
7. Ghazy, M. and Elkaranshawy, H. 2012, "A method to escape Painlevé paradox in a two-link robotic manipulator", *2012 International Conference on Engineering and Technology (ICET)*, New Cairo City, Egypt pp. 1-5
8. Alkomy, H. M., Elkaranshawy, H. A., Ashour, A. S. and Mohamed, K. T. 2015, "Effect of Robot Configuration Parameters, Masses and Friction on Painlevé Paradox for a Sliding Two-Link (PR) Robot", *International Journal of Mechanical, Aerospace, Industrial, Mechatronic and Manufacturing Engineering*, Vol(9), pp. 1670-1675
9. Shen, Y. and Stronge, W. 2011, "Painlevé paradox during oblique impact with friction", *European Journal of Mechanics-A/Solids*, Vol(30), pp. 457-467
10. Payr, M. and Glocker, C. 2005, "Oblique Frictional Impact of a Bar: Analysis and Comparison of Different Impact Laws", *Nonlinear Dynamics*, Vol(41), pp. 361-383
11. Hall, B. and Champneys, A. 2009, "Why does chalk squeak?", *Master Thesis*, University of Bristol, UK
12. Génot, F. and Brogliato, B. 1999, "New results on Painlevé paradoxes", *European Journal of Mechanics - A/Solids*, Vol(18), pp.653-677
13. Mason, M. T. and Wang, M. Y. Year, "On the inconsistency of rigid-body frictional planar mechanics", *Proceedings of the IEEE International Conference on Robotics and Automation*, Philadelphia, PA, USA, pp. 524-528
14. Wang, Y. and Mason, M. T. 1992, "Two-Dimensional Rigid-Body Collisions With Friction", *Journal of Applied Mechanics*, Vol(59), pp. 635-642
15. Stronge, W. J. 2013, "Smooth dynamics of oblique impact with friction", *International Journal of Impact Engineering*, Vol(51), pp. 36-49
16. Stewart, D. E. 2000, "Rigid-Body Dynamics with Friction and Impact", *SIAM Rev.*, Vol(42), pp. 3-39
17. Elkaranshawy, H. 2007, "Rough collision in three-dimensional rigid multi-body systems", *Proceedings of the Institution of Mechanical Engineers*, Vol(221), pp. 541-550
18. Routh, E. J. 1905, "An elementary treatise on the dynamics of a system of rigid bodies with numerous examples", *Macmillan*, New York, USA

Fluid Mechanics

Proceedings of the 26th CANCAM
Victoria, British Columbia, Canada, May 29 – June 1, 2017

NUMERICAL STUDY OF THE EFFECT OF REACTANT INTRUSION INTO THE GAS DIFFUSION LAYER ON THE LIMITING CURRENT OF PEMFCS

Sadegh Hasanpour
Mechanical Engineering
University of Victoria
Victoria, BC, Canada
sadeghhasanpour@uvic.ca

Mohsen Akbari
Mechanical Engineering
University of Victoria
Victoria, BC, Canada
makbari@uvic.ca

Ned Djilali
Mechanical Engineering
University of Victoria
Victoria, BC, Canada
ndjilali@uvic.ca

ABSTRACT

In this study, the effect of the intrusion of the reactant (hydrogen) into the gas diffusion layer (GDL) of polymer electrolyte membrane fuel cells (PEMFCs) on the limiting current of the PEMFC is investigated. A 2D model of channel and GDL are used in Comsol Multiphysics to simulate the flow in the channel and the adjacent layer (GDL). The governing equation in the porous media is Brinkman equation, since it considers all the effective parameters of the GDL. Two non-dimensional groups that have significant effects are Darcy ($\frac{k}{H^2}$) and Reynolds ($\frac{\rho v H}{\mu}$). Based on the non-dimensional groups, the design parameters for the intrusion of the flow inside the GDL are the in-plane, through-plane permeability and the inlet velocity. These parameters are changed and the intrusion of the reactant into the GDL is studied and used to estimate the limiting current of the fuel cells. The results indicated that the higher in-plane permeability (10^{-12} m^2 to 10^{-8} m^2) significantly improves the limiting current (1.4 A/cm^2 to 3.3 A/cm^2). This analysis indicates that in-plane permeability is the most effective parameters and the least effective parameters are the through-plane permeability and inlet velocity for the improvement of the PEMFCs limiting current density.

KEYWORDS: Porous Media, PEMFC, Gas Diffusion Layer, Permeability

INTRODUCTION

A polymer electrolyte membrane fuel cell (PEMFC) is a device that can change the chemical energy of fuel (hydrogen) into the electricity in a clean way. This device consists of flow field plate, gas diffusion layers (GDLs), catalyst layers (CLs), and a membrane. In this device, hydrogen comes from the anodic part and passes through the porous layer (GDL) to reach the reaction zone (CL). In the CL, hydrogen breaks into two protons (H+) and two electrons. The protons pass through the membrane and reaches the other side (cathode) and the electrons pass through an external load and produce electricity. At the cathodic part, oxygen, similar to hydrogen, passes through the GDL and reaches to the CL. Oxygen reacts with protons (H+) and electrons and produces water as a by-product.

In this device, GDL plays a vital role for providing the following functionalities: (1) facilitating reactants to reach to the catalyst layer, (2) being a conductive layer to pass electrons into an external load, (3) facilitating the removal of the by-product, water, from the PEMFCs, and (4) supporting the membrane mechanically [1]. These tasks are required to employ a highly porous material, with high electrical conductivity. GDLs are available commercially for a different range of porosity (60~85%) and thickness (150-500 μm). Although the thickness of this layer seems small, it is much thicker than CL (~15 μm) and thicker than membrane (~150 μm). This layer is in contact with the bipolar plate and catalyst layer. The bipolar plate is made in a way to distribute the reactants in the GDLs homogenously. Many small channels are made to keep the reactants constantly be provided to the PEMFCs. Different configurations of channels have been used to provide sufficient reactants into the membrane. Four basic channels are parallel, serpentine, parallel serpentine and interdigitated [2]. The most widely used flow channel is serpentine design [3] with a rectangular cross-section [4].

The means of transportation in the porous media of fuel cells are mainly based on diffusion (i.e. gradient in concentration causes the transportation of a species) and convection (i.e. the transportation of species under the bulk movement of a fluid). In the channel, the dominant mean of transportation is convection; however, in the porous structure, the dominant transportation is the diffusion [1]. The flow in the channel is categorized as a laminar flow ($Re < 2000$) and very rarely, a turbulent flow occurs in the fuel cell operation [6]. Channels are in contact with GDLs, which means the cross flow passes over a porous medium. The flow movement in the GDL is a function of the transport properties of the GDL. As a results, the transport properties of the GDL such as porosity, porosity distribution [5], permeability [6, 7] and effective diffusivity [8] have been investigated widely. The key characteristic of the porous medium for the convention is permeability. The GDL permeability is not isotropic and it has different permeability in the in-plane (across the thickness) and the through-plane (along the thickness) directions. Different studies have been investigated this property numerically and experimentally [6,

7 and 9]. In general, the in-plane permeability is 2 to 5 times higher than the through-plane permeability.

The limiting current density occurs when the consumption of reactants will be higher than the concentration of the reactant at the CL. Although enough reactants exist in the channel to fuel the PEMFC, it cannot reach the CL. The reason is that the limiting current is a function of the concentration at the channel and properties of the GDL. These properties are the effective diffusion and the thickness of the layer, Eq. 1:

$$j_L = nFD^{eff}\frac{C_R^0}{\delta} \qquad (1)$$

where j_L is the limiting current density, D^{eff} is the effective diffusion and δ is the thickness of the diffusion layer, C_R^0 is the concentration of the reactant at the channel, n is the number of mole of electrons transferred (for the anode is 2 and for the cathode is 4) and F is Faraday's constant.

To increase the limiting current different improvements have been done, constant distribution of reactant across the surface of the GDL, increasing the effective diffusion of GDLs and thinner GDL are the main remedies. The question is what the effective diffusion thickness is. Most studies assumed this value as the GDL thickness. This assumption is based on that at the interface of the channel and the GDL, the fluid velocity is zero and the transportation changes to the diffusion. However, if the penetration of the reactant is higher, the changing point from the convection to the diffusion is deeper in the GDL and the effective thickness is lower, and consequently, limiting current increases. The parameters that can change the intrusion of the fluid into the GDL are in-plane, through –plane permeability and inlet velocity.

My study is to investigate the effect of these parameters, permeability and inlet velocity, on the penetration of the reactant (hydrogen) into the GDLs. The in-plane, through-plane permeability and the inlet velocity are varied to obtain the depth of intrusion of hydrogen into the GDL. This analysis will facilitate to obtain the effective parameters to find the effective diffusion thickness, furthermore, the effect of the intrusion on the limiting current density of the fuel cells will be investigated.

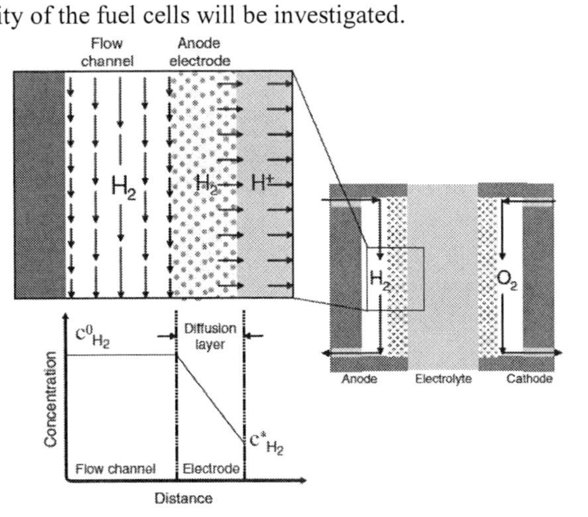

Figure 1 - Different concentration of reactant exists in the GDL. The arrow shows the velocity of hydrogen in the channel, $C^0_{H_2}$is the concentration of hydrogen at the channel and $C^*_{H_2}$is the concentration of hydrogen at the CL [2].

METHODOLOGY

To investigate the effect of these parameters, the governing equations for the flow in the channel and GDL are employed. The flow inside a channel is governed by the Navier-Stokes (N-S) equation. The challenging part is the modelling of the flow in the porous GDL. The formulation of the governing equation in the porous media requires in particular prescription of the average pore size of the effective transport properties.

The representative elementary volume (REV) is the average value in the centroid of the sample. The length scale of the REV should be much larger than the pore size of the medium and at the same time small enough in comparison to the length scale of the simulation [10]. The pore size distribution of the GDL is between 30 to $40\ \mu m$ [1], and the channel length for the 2D analysis is $10\ cm$. These length scales indicate that the REV approach is applicable for the porous GDL. The continuity equation in the porous medium can then be expressed by accounting for porosity in the time dependent term, i.e. ($\epsilon\frac{\partial\rho_f}{\partial t} + \nabla.(\rho_f v) = 0$). For the momentum equations, the Brinkman equation is employed for the porous regions with the specification of all effective parameters in the GDL [10].

SCHEMATIC OF 2D GEOMETRY

The 2D schematic of the structure is shown in Figure 2. The figure shows one cross-section along the length of the channel. The channel width is equal to 1 *mm*, and the GDL thickness is $300\ \mu m$. The governing equation in the channel section is the N-S equation and continuity:

$$\rho(u.\nabla)u = \rho g - \nabla P + \mu\nabla^2 u \qquad (2)$$

$$\rho\nabla.(u) = 0$$

where ρ and μ are the density and the viscosity of the fluid. For the porous region adjacent to the channel, the governing equation are the Brinkman and continuity equations:

$$\frac{1}{\epsilon_p}\rho(u.\nabla)\frac{u}{\epsilon_p} = -\nabla P + \frac{\mu}{\epsilon_p}\nabla^2 u - \frac{\mu u}{k} + \beta_f|u|u + \frac{Q_{br}}{\epsilon_p^2}u \qquad (3)$$

$$\rho\nabla.(u) = Q_{br}$$

where ϵ_p is the porosity of the medium, k is the permeability, β_f is the Forchheimer coefficient and Q_{br} is the mass source or sink in the medium. Experimental studies have shown that the effect of Forchheimer drag is negligible for GDLs [3]. Since the reactions occurs in the membrane, Q_{br} exists in the GDL sections. The rate of consumption of the reactant in the anode is $0.125\ kg/m^3$ per 10 cm of channel length [2], and corresponds to a current density of 2 A/cm^2.

The inlet condition is defined as a normal inflow, with a uniform velocity. And two pressure outlet conditions are assigned separately for the channel and GDL section. All other surfaces are defined as walls. The non-dimensional groups for the Brinkman equations are as follow:

$$u^* = \frac{u}{U},\ P^* = \frac{PH}{\mu U}, \nabla^* = H\nabla, \nabla^{2*} = H^2\nabla^{2*} \quad (4)$$

where U is the inlet velocity, μ is the viscosity of the fluid, P is the outlet pressure (atmospheric pressure), and H is the thickness of the GDL [14]. Using these groups the non-dimensional form of the Brinkman equation is:

$$\frac{Re}{\epsilon_P^2}(u^*.\nabla^*)u^* = -\nabla^* P^* + \frac{\nabla^{2*}u^*}{\epsilon_P} - \frac{u^*}{Da} + \frac{Q_{br}H}{\rho U \epsilon_P}u^* \quad (5)$$

where $Re = \frac{\rho v H}{\mu}$ and $D = \frac{k}{H^2}$. In this analysis the porosity and Q_{br} are constant and assigned to 0.8 and 0.125 $\frac{kg}{m^3.s}$ and the effect of the last term in the analysis is negligible ($\frac{Q_{br}H}{\rho U \epsilon_P} \rightarrow 10^{-4}$). The two important non-dimensional groups are Reynolds and Darcy numbers. Their effect is investigated by varying the inlet velocity and the permeability.

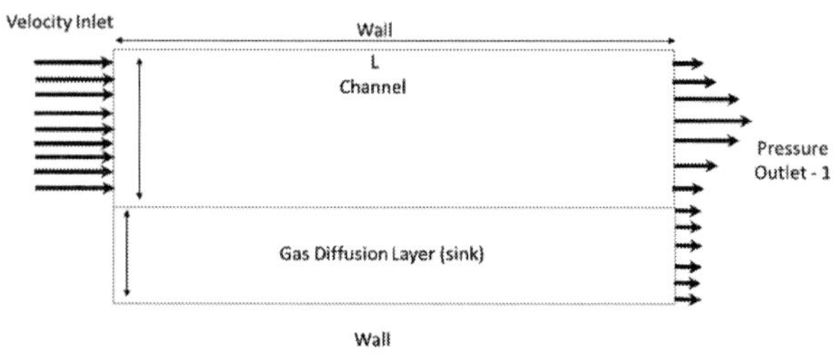

Figure 2: Schematic of the simulation domain. The channel is adjacent to the GDL, which is bounded by a solid wall on top and the flow inlet and outlets.

The inlet velocity varies between 3 and 30 m/s, and hydrogen density is specified using the ideal gas for the operating condition of $80\ ^\circ C$ and atmospheric pressure. The range of *Re* number varies between 20 and 200. This values assures the channel flow is in the laminar regime. The Darcy number depends on the permeability of the medium. With permeability values varied between 10^{-12} and 10^{-6} m^2, D is between 10^{-5} and 11. This analysis shows that for the lower permeability value, the dominant term is the permeability term ($\frac{u^*}{Da}$) and the term containing *Re* is of second order ($\frac{Re}{\epsilon_P^2}(u^*.\nabla^*)u^*$).

RESULTS AND DISCUSSION

The simulation is performed on the basic parameters with the inlet velocity of 3 m/s, the permeability of GDL is assigned to 10^{-11} m^2 for the in-plane direction and 10^{-12} m^2 for the through-plane direction. These values are chosen based on the experimental results available for commercially available GDLs [3]. The operating condition are atmospheric pressure and $80\ ^\circ C$ temperature. In typical fuel cells with serpentine flow channels, the aspect ratio of each channel is large ($\frac{L}{H+h} \approx 77$). To ensure fully developed flow conditions, the simulations are performed for in a domain with a 10 cm channel length.

The in-plane permeability values are varied between 10^{-12} and 10^{-6} m^2 and the through-plane permeability is kept at 10^{-12} m^2. In the second simulation, the through-plane permeability varies between 10^{-12} and 10^{-6} m^2 and the in-plane permeability is kept at 10^{-11} m^2. In the last part, the inlet velocity varies between 3 and 30 m/s.

Figure 3 shows the resulting velocity isocontours for different in-plane permeabilities in a cross-section far from the entrance. X-axis is the thickness of the channel and Y-axis is the length of the channel. These results show as expected that the intrusion of the reactant inside the porous structure is higher for the more permeable GDLs. To quantify the depth of intrusion we locate the depth at which the velocity magnitude is 1% of the inlet velocity. The penetration of reactants varies between 0 to 260 μm for the lowest and highest in-plane permeability, respectively. The zero value corresponds to the case when penetration is negligible (the of the 1% velocity is in this case within the channel boundary layer). The higher permeability GDL allows transport of more reactant to the vicinity of the reaction zone (CL).

The variation of the through-plane permeability has almost no effect on the depth of intrusion, and a negligible effect on convection, primarily because of the low normal velocities at the interface.

To evaluate the limiting current, Eq.1 is employed. Because this analysis performed on the anodic side, the number of electron per mole of reactant is 2 (n=2). Using Faraday constant (96485 C/mol), the value of D^{eff} is calculated based on the Bruggrmann equation ($D_{ij}^{eff} = \epsilon^{1.5} D_{ij}$) [6], in which the binary diffusion of hydrogen into the air D_{ij} depends on temperature. For and operating temperature of $80\ ^\circ C$, $D_{ij} = 1.02\ cm^2/s$), and D^{eff} is 0.654 cm^2/s. The concentration of the reactant is calculated based on the ideal gas law as:

$$C_R^0 = \frac{n_R^0}{V} = \frac{P_R^0}{RT} \quad (6)$$

For atmospheric pressure and R=8.314 J/mol.K. $C_R^0 = 34.5\ mol/m^3$. Effective diffusion thicknesses are calculated based on the variation of the design parameters. The effect of in-plane permeability on the limiting current is shown in Figure 4. This result shows that the limiting current is highly related to the in-plane permeability. By increasing the in-plane permeability, the limiting current increases significantly. The rate of change is much higher for permeabilities that are higher than 10^{-8} m^2. However, limiting current remained unchanged for different values of through-plane permeability (from 10^{-12} to 10^{-6} m^2) and also the changes are very small with change of inlet velocity, with intrusion of hydrogen into the GDL changing by up to 2 μm (for 30 m/s). The effect of such a small intrusion on the limiting current is negligible.

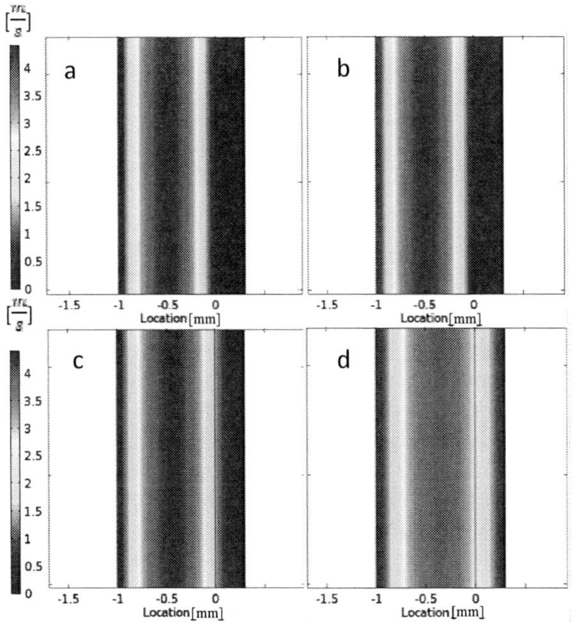

Figure 3 - Velocity magnitude for in-plane permeability of (a) $k = 10^{-12}\ m^2$, (b) $k = 10^{-10}\ m^2$, (c) $k = 10^{-8}\ m^2$ and (d) $k = 10^{-6}\ m^2$.

.

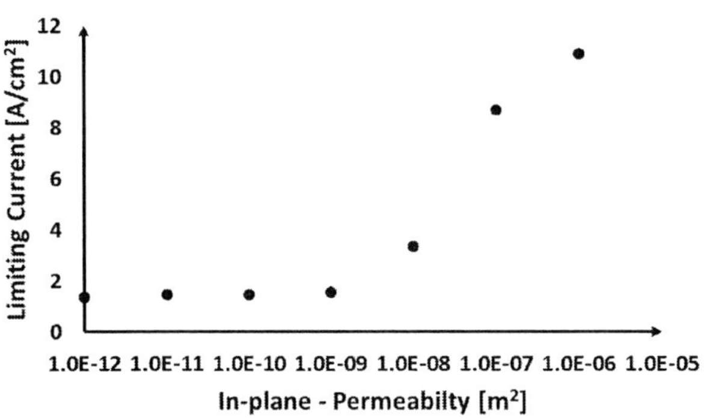

Figure 4- Variation of limiting current with in-plane permeability of the GDL.

CONCLUSION

A 2D model of the channel and the porous gas diffusion electrode of a fule cell was used to simulate the flow by solving numerically the governing equations in these two regions (channel and GDLs). The non-dimensional groups in the governing equation were Reynolds and Darcy numbers. Based on these groups, the parametric study was carried out. The effective parameters on the intrusion of the cross flow over the porous GDL was studied and the effect on the limiting current was analyzed. The parameters varied in the study are the in-plane permeability, through-plane permeability and inlet flow velocity. The results show that the intrusion depth of the reactants into the porous media is highly and primarily dependent on the in-plane permeability, whereas the other two parameters (inlet velocity and also the through-plane permeability) have a minimal effect. The limiting current density of the fuel cell depends on the penetration depth of reactant into the porous electrode. The results show that the higher permeability medium can potentially improve the performance of the fuel cell in the high current density regime. The key design parameter in this respect is the in-plane permeability.

REFERENCES

1. Mathias, M., Roth, J., Fleming, J., and Lehnert, W. "Diffusion media materials and characterisation," *Handbook of Fuel Cells*, 2003.
2. O'Hayre, R. P., Cha, S., Colella, W. and Prinz, F. B. 2006 "Fuel Cell Fundamentals. John Wiley & Sons New York.
3. Shi, Z. and Wang, X. 2008"A numerical study of flow crossover between adjacent flow channels in a proton exchange membrane fuel cell with serpentine flow field," *J. Power Sources*, vol. 185, pp. 985-992.
4. Park, J. and Li, X. 2007 "An experimental and numerical investigation on the cross flow through gas diffusion layer in a PEM fuel cell with a serpentine flow channel," *J. Power Sources*, vol. 163, pp. 853-863.
5. Fishman, Z., Hinebaugh, J. and Bazylak, A., 2010 "Microscale tomography investigations of heterogeneous porosity distributions of PEMFC GDLs," *J. Electrochem. Soc.*, vol. 157, pp. B1643-B1650.
6. Gostick, T., Fowler, M. W., Pritzker, M. D., Ioannidis, M. A. and Behra, L. M. "In-plane and through-plane gas permeability of carbon fiber electrode backing layers," J. Power Sources, vol. 162, pp. 228-238, 2006.
7. Pharoah, J. 2005,"On the permeability of gas diffusion media used in PEM fuel cells," *J. Power Sources*, vol. 144, pp. 77-82.
8. M. Ismail, D. Ingham, K. Hughes, L. Ma and M. Pourkashanian, "Effective diffusivity of polymer electrolyte fuel cell gas diffusion layers: An overview and numerical study," Int J Hydrogen Energy, vol. 40, pp. 10994-11010, 2015.
9. J. Feser, A. Prasad and S. Advani, "Experimental characterization of in-plane permeability of gas diffusion layers," J. Power Sources, vol. 162, pp. 1226-1231, 2006.
10. Nield, D. A. and Bejan, A. 2006, "Convection in Porous Media". Springer Science & Business Media.

Proceedings of the 26th CANCAM
Victoria, British Columbia, Canada, May 29 – June 1, 2017

DIFFUSION ANGLE EFFECTS ON CAVITATING FLOW OF LIQUEFIED NATURAL GAS INSIDE A LAVAL NOZZLE

Saeed Rahbarimanesh
School of Engineering
University of British Columbia
Kelowna, BC, Canada
saeed.rahbarimanesh@alumni.ubc.ca

Joshua Brinkerhoff
School of Engineering
University of British Columbia
Kelowna, BC, Canada
joshua.brinkerhoff@ubc.ca

ABSTRACT

Achieving high efficiency in turbomachinery equipment, including liquefied natural gas (LNG) fuel pumps, is highly dependent on accurately predicting cavitation. Under cryogenic conditions, thermodynamic effects are increasingly important due to the low liquid-to-vapor density ratios. A numerical model is presented that examines thermodynamic effects on cavitation of LNG through a converging-diverging Laval nozzle. To investigate how the interactions of shear-layer instabilities and vapor structures affect the cloud/sheet cavitation patterns in the resulting flow, numerical simulations are conducted for nozzle diffusion angles of 2.5°-15° at a fixed cavitation number of 1.05. Temperature variations near the liquid/vapor interface change the cavitating behavior of LNG such that growth/collapse of vapor spots is delayed. This behavior is altered at larger expansion angles where increasing flow recirculation generates larger vapor cavities.

KEYWORDS: Cryogenic cavitation, LNG, thermodynamic effects, Laval nozzle, diffusion angle

INTRODUCTION

Due to environmental and economic factors, there is a recent interest in liquefied natural gas (LNG) as an alternative fuel for commercial, marine, and mining vehicles [1]. Because of the low boiling temperature of LNG, most of the LNG-based turbomachinery will operate near the saturation point. Unsteady cavitation occurs when the local pressure and/or temperature exceed the saturation conditions and the liquid changes phase to vapor. In addition to degrading the efficiency of the machinery, cavitation leads to cyclic stresses and erosion of components [2]. The expense of experimental facilities and the complexity of measuring cavitation in cryogenic conditions limit the research available compared to cavitation studies in thermal-insensitive fluids (e.g. water). Simoneau and Hendricks [3] conducted a series of experiments to investigate liquid nitrogen and hydrogen (LN_2 and LH_2) cavitation patterns through nozzle, hydrofoil, and ogive geometries under a range of inlet velocities and temperatures. Tani and Nagashima [4] performed experimental and numerical observations of cavitating water and liquid nitrogen in a converging-diverging Laval nozzle.

The complexity of experimental studies motivates computational approaches to modeling cryogenic cavitation. Computational fluid dynamics (CFD) models of cavitation in cryogenic fluids are divided into three categories: interfacial dynamics methods (IDM), homogeneous equilibrium methods (HEM), and transport equation-based models (TEM). In IDM, liquid and vapor phases are modeled individually with an interface capturing approach, such as the marker and cell (MAC) method [5]. In HEM, the fluid is treated as a single phase and the liquid and vapor phases are modeled by solving an equation of state (EOS) for the vapor volume fraction. In TEM, a transport equation is solved for the vapor mass fraction with source terms for condensation and evaporation. Sun et al. [6] used HEM to model cavitating liquid nitrogen around a hydrofoil under isothermal and cryogenic conditions. By employing a modified Zwart cavitation model [7], which considers thermodynamic effects, they concluded that cryogenic condition causes shorter attached cavitation region around the hydrofoil. Cao et al. [8] used a TEM approach coupled with the energy equation to investigate the effects of temperature and pressure on cavitation mechanisms of LN_2 and LH_2 around hydrofoil and ogive geometries.

The prevalence of isothermal conditions in most HEM approaches imparts a barotropic form, $\rho_m = f(p)$, to the EOS used to estimate the vapor mass fraction. As a result, barotropic vorticity generation cannot be predicted with HEM methods. Due to this limitation, TEM has become more popular in recent studies. However, TEM requires several empirical constants in the condensation and evaporation source terms in the vapor transport equation. Therefore, the present paper describes a HEM-based cavitation model that corrects the density field predicted by the barotropic EOS with accurately updated thermophysical property values, obtained using JANAF-based polynomials, in order to capture unsteady cavitation in cryogenic fluids. The developed model is used to simulate the cavitation patterns of LNG flow inside a contracting/expanding nozzle.

1. NUMERICAL MODEL

1.1 MASS AND MOMENTUM CONSERVATION

The present model is developed in OpenFOAM® based on the unsteady cavitation solver *cavitatingFoam* [9]. A brief description of *cavitatingFoam* is given below, followed by modifications to account for cryogenic fluids. *cavitatingFoam* employs HEM to capture phase-change in isothermal compressible cavitating flows. The fluid is modeled as a single phase, and the variable density field is computed via a barotropic equation of state correlating mixture density ρ_m to pressure p by a compressibility function ψ [10]:

$$\frac{D\rho_m}{Dt} = \psi_m \frac{Dp}{Dt} \tag{1}$$

The compressibility is modeled as a linear combination of the vapor and liquid compressibility,

$$\psi_m = \alpha\psi_v + (1-\alpha)\psi_l \tag{2}$$

where the v and l subscripts denote the vapor and liquid phases, respectively. The vapor volume fraction, α, is defined as

$$\alpha = \frac{\rho_m - \rho_{l,sat}}{\rho_{v,sat} - \rho_{l,sat}} \tag{3}$$

where $\alpha = 0$ corresponds to fully liquid and $\alpha = 1$ refers to fully vapor. The vapor phase density at saturation is given by

$$\rho_{v,sat} = \psi_v\, p_{sat} \tag{4}$$

and the corresponding equations of state for vapor and liquid phases, respectively, are

$$\rho_v = \psi_v\, p \tag{5}$$

$$\rho_l = \rho_l^0 + \psi_l\, p \tag{6}$$

in which

$$\rho_l^0 = \rho_{l,sat} - \psi_l\, p_{sat} \tag{7}$$

is the density of liquid at the initial temperature. Combining equations (1)-(7) yields an expression for the mixture density [11]

$$\rho_m = (1-\alpha)\rho_l^0 + (\alpha\psi_v + (1-\alpha)\psi_l)p_{sat} + \psi_l\alpha(p - p_{sat}). \tag{8}$$

The density is convected via the continuity equation

$$\frac{\partial\rho_m}{\partial t} + \nabla\cdot(\rho_m \boldsymbol{u}) = 0 \tag{9}$$

where $\boldsymbol{u} = (u, v, w)$ is the velocity field obtained from the momentum equation

$$\frac{\partial(\rho_m \boldsymbol{u})}{\partial t} + \nabla\cdot(\rho_m \boldsymbol{uu}) = -\nabla p + \nabla\cdot[\mu_m(\nabla\boldsymbol{u} + (\nabla\boldsymbol{u})^T)] \tag{10}$$

where the mixture dynamic viscosity is defined as

$$\mu_m = \alpha\mu_v + (1-\alpha)\mu_l\,. \tag{11}$$

Continuity and momentum equations are solved sequentially using a merged PISO-SIMPLE (PIMPLE) algorithm [12]. In the corrector step, the velocity field is corrected by solving a transformed pressure equation by use of equation (9):

$$\frac{\partial(\psi_m p)}{\partial t} - \left(\rho_l^0 + (\psi_l - \psi_v)\,p_{sat}\right)\frac{\partial\alpha}{\partial t} - p_{sat}\frac{\partial\psi_m}{\partial t} + \nabla\cdot(\rho_m\boldsymbol{u}) = 0 \tag{12}$$

After continuity has been reached, α, ψ_m, μ_m, and ρ_m are updated.

1.2 ENERGY CONSERVATION

Temperature effects on the density field cannot be captured using the barotropic EOS currently employed in *cavitatingFoam*. Therefore, to resolve this issue in the case of cryogenic cavitation, modifications to the *cavitatingFoam* model are implemented to correct the ρ_m, μ_m, ψ_m, p, and α fields to account for temperature effects. The temperature field is obtained by solving an enthalpy-based energy equation for the mixture as follows [13]:

$$\frac{\partial\rho_m h}{\partial t} + \nabla\cdot(\rho_m \boldsymbol{u}h) + \frac{\partial\rho_m K}{\partial t} + \nabla\cdot(\rho_m\boldsymbol{u}K) - \frac{\partial p}{\partial t} = -\nabla\cdot\boldsymbol{q} + \nabla\cdot(\boldsymbol{\tau}\cdot\boldsymbol{u}) \tag{13}$$

The LHS terms represent the transport of total energy (enthalpy plus kinetic energy). The RHS terms are, respectively, conductive heat flux and viscous work terms. The stress term $\nabla\cdot(\boldsymbol{\tau}\cdot\boldsymbol{u})$ uses Stokes' theorem for compressible Newtonian fluids,

$$\boldsymbol{\tau} = \mu_m(\nabla\boldsymbol{u} + (\nabla\boldsymbol{u})^T) - \frac{2}{3}\mu_m\nabla\cdot\boldsymbol{u}\,. \tag{14}$$

By substitution of this equation in equation (13), the system of governing equations is closed.

1.3 THERMOPHYSICAL PROPERTIES

Equations (10) and (13) require thermophysical properties such as specific heat and dynamic viscosity at different temperature and pressures. A series of numerical simulations were performed to examine the accuracy of different thermophysical models available in OpenFOAM for cryogenic fluids. Reasonable agreements against available experimental data are obtained for Sutherland, and JANAF models for viscosity μ_m and specific heat c_p, respectively. The Sutherland model calculates the liquid and vapor dynamic viscosity as follows:

$$\mu_l(T) = \frac{(2.48\times10^{-6})\sqrt{T}}{1\text{-}\frac{84.87}{T}} \qquad \mu_v(T) = \frac{(8.02\times10^{-7})\sqrt{T}}{1+\frac{90.04}{T}} \tag{15}$$

where coefficients are found from NIST property data for methane at 6.89 bar and temperature range of 95 K < T < 180 K [14]. Within the pressure range considered in this study (4-8 bar), the pressure effect on dynamic viscosity is less than 0.5%; thus, it is assumed for this work that the viscosity only varies with temperature.

The JANAF thermophysical model calculates specific heat (c_p) as a polynomial function of temperature [14]. At a given pressure, two polynomials predict c_p for the vapor and liquid phases, respectively. The main downside of the JANAF model is its temperature-based derivation. In the present study it is extended to account for both temperature and pressure effects on cryogenic cavitation. The effect of pressure on c_p is obtained by using separate sets of JANAF polynomials at a high and low pressure, and then performing linear interpolation for intermediate pressures. Equations (16) and (17) present the JANAF polynomials for LNG at 6.89 bar and 4.13 bar, respectively:

6.89 bar:
$$\begin{aligned} c_{p\_v}(T) &= \left(2.15\times10^{-7}\right)T^4 - \left(1.44\times10^{-4}\right)T^3 + \left(3.65\times10^{-3}\right)T^2 - 4.12\,T + 177.5 \\ c_{p\_l}(T) &= \left(5.13\times10^{-8}\right)T^4 - \left(2.18\times10^{-5}\right)T^3 + \left(3.58\times10^{-3}\right)T^2 - 0.26\,T + 10.4 \end{aligned} \tag{16}$$

4.13 bar:
$$\begin{aligned} c_{p\_v}(T) &= \left(6.96\times10^{-8}\right)T^4 - \left(4.6\times10^{-5}\right)T^3 + \left(1.14\times10^{-3}\right)T^2 - 1.27\,T + 55.7 \\ c_{p\_l}(T) &= \left(3.75\times10^{-8}\right)T^4 - \left(1.54\times10^{-5}\right)T^3 + \left(2.45\times10^{-3}\right)T^2 - 0.173\,T + 7.86 \end{aligned} \tag{17}$$

Note that the polynomial coefficients are found from a fitting function correlating specific heat at the given pressure to the operating temperature range of 95 K < T < 180 K [14]. This generic model of JANAF for $c_p(p,T)$ is developed by using available interpolation/extrapolation libraries in OpenFOAM.

Since saturation vapor pressure of LNG varies with temperature, a temperature-variant equation based on the Prydz-Goodwin theory [15] and methane look-up tables [14] is used in this study to precisely correlate LNG saturation pressure to temperature; thus, making the saturation properties of LNG dependent on temperature:

$$\mathrm{Log}\left(p_{v,\ sat}(T)\right) = \left[3.98 - \left(\frac{443.03}{T - 0.49}\right)\right] \tag{18}$$

In order to couple the fluid properties obtained from the thermophysical model to those employed in the conservation equations, the thermophysical properties are rewritten in terms of thermophysical classes to link them to the solved enthalpy field. By doing so, the α and ρ_m fields are corrected to incorporate the effects of cryogenic conditions on cavitation. It also allows the boundary and initial conditions for the enthalpy equation to be imposed in terms of temperature, which simplifies comparison to experiments.

1.4 COUPLING OF MASS, MOMENTUM AND ENERGY EQUATION

The temperature field which is found from the solution of equation (13) is outside the predictor-corrector loops of the PIMPLE algorithm. As a result, the solved vapor volume fraction field is decoupled from solution of the energy equation. In order for flow parameters to be updated and coupled to the computed temperature field, following modifications are made to the PIMPLE loop to guarantee the dependency of cavitation field on temperature and pressure:

- Equations (13) and (18) are added to the inner PIMPLE loop.
- ρ_m is corrected using equation (18) to account for temperature effects.
- Continuity equation is solved with the corrected ρ_m field.
- The compressibility model (2), barotropic EOS, and pressure equation (12) are solved for corrected p and α fields.
- Momentum (10), and energy equations (13) are updated.
- Thermophysical model and flow parameters are updated for the next outer loop.

2. SIMULATION OF LNG FLOW IN LAVAL NOZZLE

Following the experimental study of Nagashima and Tani [4], cavitating flow of LNG inside a 2D Laval nozzle (Fig. 1) with throat length of 1.054 cm and an expansion angle of 2.5° -15° is studied numerically using the developed solver. A total pressure of 6.89 bar and 4.13 bar are specified (pressure ratio = 1.67) at the inlet and outlet boundaries, respectively. The Reynolds number based on the throat length and throat velocity is 8720, indicating a laminar flow regime. Initial flow field contains only liquid methane with inlet density of 390.92 kg/m^3 at an initial and inlet temperature of 132 K.

A computational domain of 32A×12A (A: throat cross-section area) is meshed for the nozzle using 721,600 non-uniform structured 2D elements. Near-wall regions are meshed with 12 prismatic layers with smallest element thickness of 0.0004 m with wall-normal grading rate of 1.176. A buffer zone with the size of 20 × 12 cm^2 follows the nozzle diverging section to damp any outlet boundary effects.

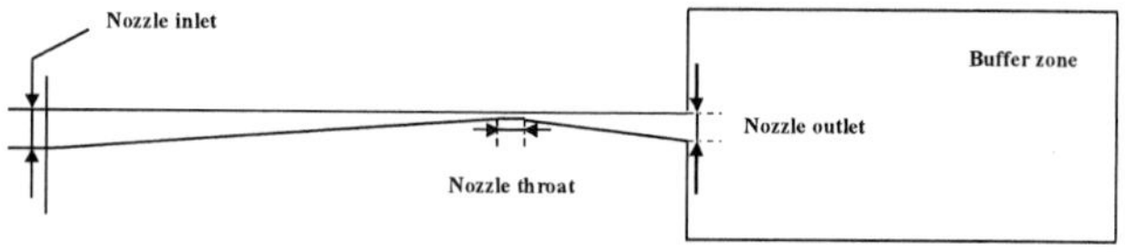

Figure 1: Computational domain for Laval nozzle

Spatial discretization is via a second-order total variation diminishing (TVD) scheme. Second-order backward discretization is employed for temporal terms. The maximum Courant number is 0.3 with $\Delta t = 1.47\times10^{-6}$ s. The maximum acoustic Courant number was set to 10 to stabilize compressible pressure waves, and to reduce artificial diffusion between phases. All simulations were run in parallel on 12 processors of a 3-GHz Intel-Xeon machine using the message passing interface (MPI).

3. RESULTS AND DISCUSSION

To validate the developed numerical model and to examine the reliability of the results for different inflow conditions, a series of validating simulations were performed and compared against the experimental study of Simoneau and Hendricks [3]. Liquid nitrogen (LN_2) flow inside the Laval nozzle was simulated for five reduced pressures between 0.2 and 2.0 at a reduced temperature of 0.75, where the reduced pressure/temperature is defined as the ratio of stagnation (inlet) to critical pressure/temperature. Comparison with the experimental data in Fig. 2 is based on the dimensionless pressure and mass flow rate at the nozzle throat. The mass flow rate (Fig. 2(a)) increases with increasing the total pressure while an opposite behavior is observed for the throat pressure (Fig. 2(b)). The numerical results achieve very good agreement with the experiments, and the observed differences are likely due to three-dimensional effects in the experiment that cannot be captured in the 2D simulations.

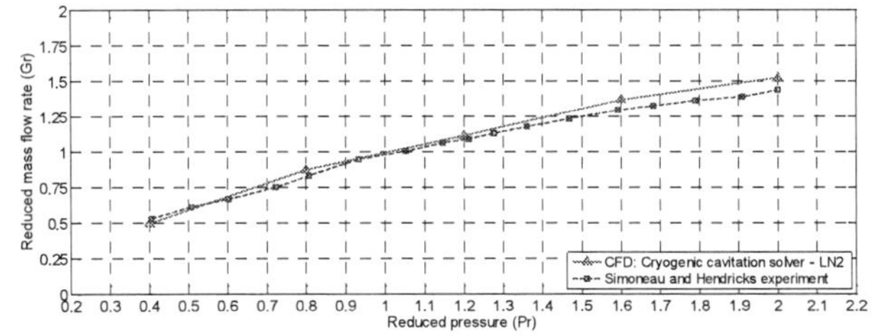

(a)

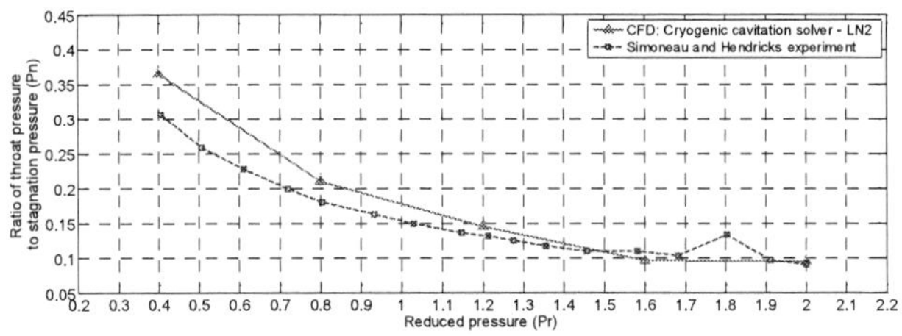

(b)

Figure 2: Comparison of the averaged numerical results for cavitating flow of LN_2 in the Laval nozzle against the Simoneau and Hendricks's experiment [3]

Temporal evolution of the vapor phase fraction field for the diffusion angles (DA) of 7.5° at t = 0.54 s and 15° at t = 0.33 s is depicted in Fig. 3(a). The local pressure at the intersection between the throat and diverging part of the nozzle is reduced such that phase-change of LNG is first observed on the lower wall of the diverging section. In the case of DA = 15°, this process occurs faster because of stronger separation of the flow on the lower wall. As the flow continues through the diverging portion of the nozzle, the attached sheet cavity region on the lower wall is followed by the formation of unsteady cavitation clouds that are detached from the lower wall. Due to the interaction with shear layers in the flow—as illustrated in the vorticity field in Fig. 3(c)—these cavitating clouds possess high vaporization rates such that large detached cavities are formed downstream of the nozzle. Detachment is mainly because of flow separation and recirculating flows near the lower wall, termed re-entrant jets, which cause separation of the sheet cavities [16]. Increasing the expansion angle of the nozzle strengthens the re-entrant jets near the lower wall such that the larger detached vortex structures occur after the throat [4]. These large vortex structures provide necessary conditions for generating stronger cavity spots in the nozzle with DA = 15°.

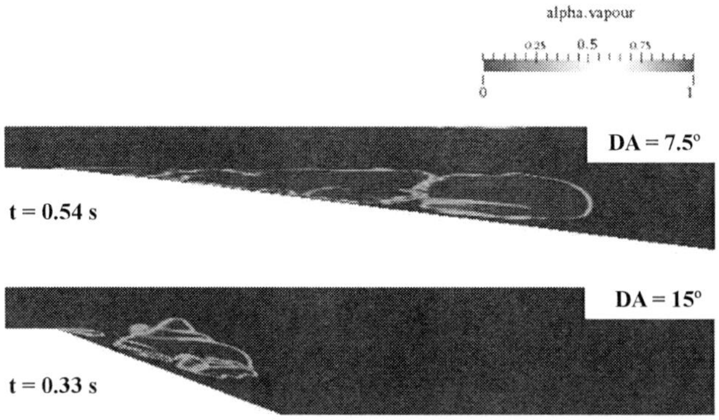

(a)

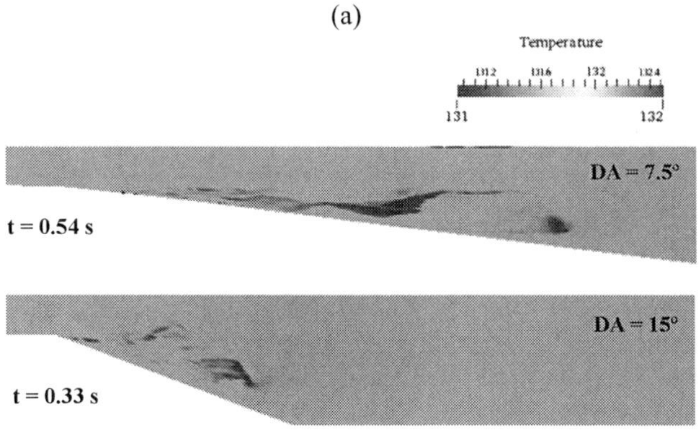

(b)

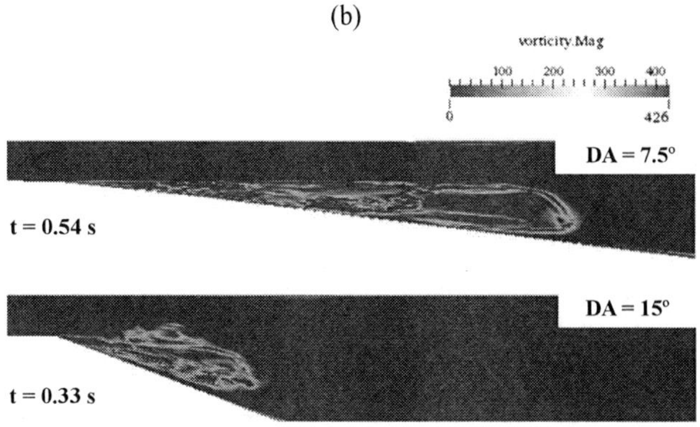

(c)

Figure 3: Unsteady evolution of the cavitating LNG flow inside the diverging part of the nozzle with diffusion angle DA = 7.5° and 15° (zoomed-in view of vapor cavity): a) vapor phase fraction, b) temperature (K), and c) vorticity magnitude (1/s) fields.

Another important observation in Fig. 3(a) and (c) for both case studies is that the presence of growing unsteadiness within cavitating clouds, which is due to vortex core instabilities and effects of micro-bubble accumulations at the vortex core, causes the generated vortical structures to be elongated and distorted [17]. The strength of these vortical structures is enhanced by collapse/formation of surrounding cavities leading to sudden changes in the overall vapor volume generated in the domain [18].

The corresponding temperature field is provided in Fig. 3(b). A slight temperature variation of ±2 K is observed in both cases. This small variation, however, affects the cavitation development significantly. When vapor spots are generated and grow, the latent heat of evaporation is absorbed from the surrounding liquid, thus, a temperature drop around the vapor area occurs. Because of the direct connection between temperature and saturation pressure, the drop in temperature corresponds to a decrease in saturation pressure [6]. As a result, the inception and growth rate of the vapor regions are delayed. It is seen that the diffusion angle of the nozzle triggers this behavior of cavity patterns by influencing the shear-layer instability process such that the longer cavities in the larger angle case are more strongly affected by temperature variations.

CONCLUSIONS

Cavitating flow of LNG through a 2D Laval nozzle is studied numerically using the open-source CFD package OpenFOAM. In order to address the thermodynamic effects on cavitation patterns, a new solver based on the existing *cavitatingFoam* solver is developed, and works in conjunction with a modified thermophysical model that accurately captures the thermal properties of cryogenic fluids (LNG and LN_2). The developed solver employs an enthalpy-based energy conservation equation to yield the transient temperature-dependent fields. In addition to implementing a procedure to update the thermophysical properties at local temperatures, a fitting function correlating the saturation pressure to temperature is adapted to make the available barotropic equation of state for the saturation properties dependent on temperature. An extended PIMPLE algorithm iteratively couples the enthalpy, pressure, velocity, and vapor fraction fields.

Cavitation patterns in the form of recirculating vortex and sheet cavities are successfully captured from the developed solver. It is observed that in the case of LNG temperature changes alter the overall behavior of the cavitation such that cavity inception/collapse is delayed. The onset of cavitation is linked to coherent vortices that are formed following the roll-up of shear layers past the nozzle throat. The diffusion angle affects the cavity patterns by influencing the shear-layer instability process such that larger cavities are observed at larger expansion angles. This behavior also enhances the vapor production rate through decreasing the local temperature field in the liquid regions surrounding a cavity.

ACKNOWLEDGMENTS

The present work was sponsored by the Natural Sciences and Engineering Research Council of Canada (NSERC) with cooperation of Westport Fuel Systems©, and WestGrid®.

REFERENCES

1. Arteconi, A., and Polonara, F. 2013, "LNG as Vehicle Fuel and the Problem of Supply: The Italian Case Study", *Energy Policy*, 62, pp. 503-512.
2. Leroux, J., Coutier, O., and Astolfi, J. 2005, "A Joint Experimental and Numerical Study of Mechanisms Associated to Instability of Partial Cavitation on Two-dimensional Hydrofoil", *Physics of Fluids*, 5(17), pp. 52-101.
3. Simoneau, R. J., and Hendricks, R. C. 1979, "Two-Phase Choked Flow of Cryogenic Fluids in Converging-diverging Nozzles", *NASA Report*, TP-1484.
4. Nagashima, T., and Tani, N. 2003, "Cryogenic Cavitating Flow in 2D Laval Nozzle", *Journal of Thermal Science*, 2(12), pp. 157-161.
5. Harlow, F. H., and Welch, J. E. 1965, "Numerical Calculation of Time-Dependent Viscous Incompressible Flow of Fluid with Free Surface", *Physics of Fluids*, 8(12), pp. 2182-2189.
6. Sun, T., Ma, X., Wei, Y., and Wang, C. 2015, "Computational Modeling of Cavitating Flows in Liquid Nitrogen by an Extended Transport-based Cavitation Model", *Science China (Technological Sciences)*, 2(59), pp. 337-346.
7. Zwart, P. J., Gerber, A. G., and Belamri, T. 2004, "A Two-Phase Flow Model for Predicting Cavitation Dynamics", *5th International Conference on Multiphase Flow*, Yokohama, Japan.
8. Cao, X., Zhang, X., Qiu, L., and Gan, Z. 2009, "Validation of Full Cavitation Model in Cryogenic Fluids", *Chinese Science Bulletin*, 10(54), pp. 1633-1640.
9. https://github.com/OpenFOAM/OpenFOAMdev/tree/master/applications/solvers/multiphase/cavitatingFoam
10. https://github.com/OpenFOAM/OpenFOAM-3.0.x/blob/master/src/thermophysicalModels/barotropicCompressibilityModel/Wallis/Wallis.H
11. Kärrholm, F. P., Weller, H., and Nordin, N. 2007, "Modeling Injector Flow Including Cavitation Effects for Diesel Applications", *ASME Fluids Engineering Conference*, San Diego, USA, pp. 465-474.
12. http://cfd.direct/openfoam/user-guide/fvsolution/
13. http://cfd.direct/openfoam/energy-equation/
14. http://webbook.nist.gov/cgi/cbook.cgi?ID=74-82-8
15. Prydz, R., and Goodwin, R. D. 1972, "Experimental Melting and Vapor Pressures of Methane", *Journal of Chemical Thermodynamics*, 1(4), pp. 127-133.
16. Sarraf, C., Bouziad, Y. A., Djeridi, H., et al. 2006, "Effect of Cavitation on the Structure of the Boundary Layer in the Wake of a Partial Cavity", *6th International Symposium on Cavitation CAV2006*, Wageningen, Netherlands.
17. Callenaere, M., Franc, J. P., and Michel, J. M. 1998, "Influence of Cavity Thickness and Pressure Gradient on the Unsteady Behavior of Partial Cavities", *3rd International Symposium on Cavitation*, Grenoble, pp. 209-214.
18. Gopalan, S., and Katz, J. 2000, "Flow Structure and Modeling Issues in the Closure Region of Attached Cavitation", *Physics of Fluids*, 4(12), pp. 895-911.

EXPERIMENTAL MEASUREMENT OF FLOW DISTRIBUTION TO ENHANCE AND CONTROL PIPE LINING

Michael McKinnon
Mechanical and Mechatronics Engineering
University of Waterloo
Waterloo, Ontario, Canada
mmckinno@uwaterloo.ca

David A. Johnson
Mechanical and Mechatronics Engineering
University of Waterloo
Waterloo, Ontario, Canada
david.johnson@uwaterloo.ca

ABSTRACT

Coating techniques are used to effectively distribute a lining on the interior walls of water pipes. Envirologics uses a system to clean and apply a coating on the walls of the pipe. First the pipe needs to be cleaned of any corrosion, tuberculation, and bitumen, using their airborne abrasive cleaning method. The pipe is then coated with a lining material. This process uses a distributive body, referred to as a DB, to apply a coating material to the walls while moving through the pipe by the use of vacuum suction. The design of the DB was modified to apply a more even coating on the walls of the pipe. The design was then compared using computational fluid dynamics (CFD) and experimental testing to the original DB design. Testing of the DBs was performed using a clear pipe attached to a vacuum pump while taking pitot tube measurements to determine the flow characteristics. This project was in affiliation with Envirologics Engineering Inc.

KEYWORDS: water pipe, pipe lining, tuberculation, airborne abrasive cleaning, distributed body

INTRODUCTION

Water pipe corrosion is a concern in most city infrastructures. The underground pipes used to provide a supply of water to residences and businesses can develop corrosion, deteriorating the pipe, and tuberculation, creating extrusions on the walls of the pipe that block flow. This leads to decreased flow capacity necessary to meet firefighting requirements as well as reduced water quality. The current method of dealing with this issue is to excavate, remove, and replace the water pipe. This is an expensive process that can lead to extended water outages.

Envirologics Engineering Inc. has developed a method to clean and prevent future corrosion while keeping the original pipes in the ground. The process involves an airborne abrasive cleaning method followed by lining the walls of the pipe with a coating material. This method performed by their Tomahawk System essentially uses abrasive material moved through the water pipe by a vacuum truck, shown in Fig. 1 [1]. The abrasive material bombards the walls of the pipe, removing the corrosion and tuberculation, as well as any tar or bitumen in the pipe.

Figure 1: Envirologic's Tomahawk System and vacuum truck setup [1]

Once the pipe has been cleaned, a distributed body (DB) is used to distribute a lining on the walls of the pipe that will prevent any leakage and future deterioration of the pipe. The lining process is effective in coating the entire wall of the pipe and covering any pipe joints [2].

Collaborative research with the University of Waterloo and Envirologics to test their Tomahawk System on watermain pipes in Waterloo, Ontario was undertaken to optimize and test base case DB designs to provide a more effective and even lining distribution.

FIELD DEMONSTRATIONS

Envirologics worked with the University of Waterloo and City of Waterloo to demonstrate their Tomahawk system in two locations in Waterloo, Ontario. The first demonstration was completed for a 130 m pipe on Hazel Street and the second for two sections of 50 m pipes on Albert Street. The pipe diameters for each test were 6 inches in diameter.

In each case, the University of Waterloo conducted airflow measurements for comparison before and after cleaning. Fig. 2 shows a visual result of the cleaning and lining for the Albert Street demonstration. It can be seen that the airborne abrasive cleaning method was successful in removing the tuberculation, corrosion, and bitumen from the pipe.

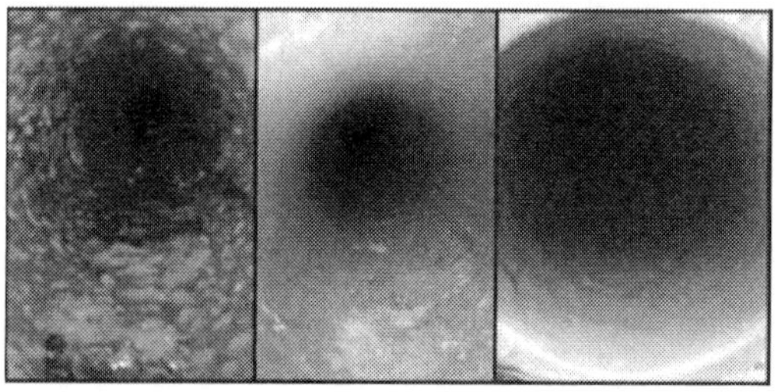

Figure 2: Alberta Street pipe before cleaning showing tuberculation (left), after cleaning (middle), and after lining (right) [3]

DISTRIBUTED BODY DESIGN

In the lining process of the Tomahawk System, a DB was designed to apply coating material on the walls of the pipe as a result of the airflow moving around it. Envirologics has several patented designs for the DB, shown in Fig. 3 to distribute the coating material.

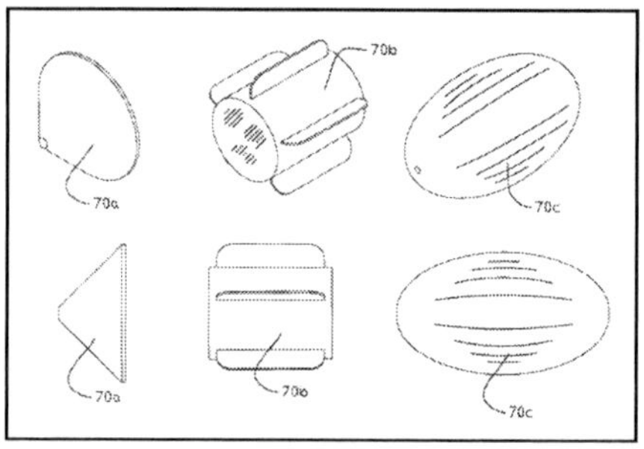

Figure 3: Three differing designs for the base case DB [4]

The DB moves through the pipe as a result of the pressure difference created by the vacuum truck situated on the outlet of the pipe showed previously in Fig. 1. The DB is attached to a cable running from the inlet which is slowly released to allow the DB to move down the pipe. Coating material is pooled in front of the DB and pushed along as the DB moves, as shown in Fig. 4 in which the DB is represented by an orb. The figure shows the material pooled in front of the DB orb as it moves in the direction of the arrow and the material is distributed along the walls of the pipe behind the DB. As the airflow moves over the DB, it distributes the coating material on the walls of the pipe.

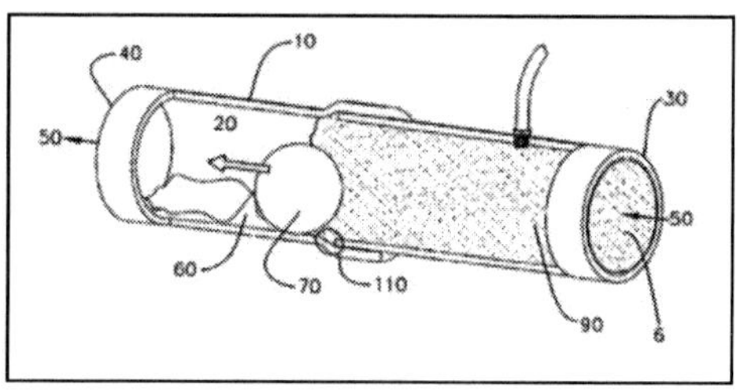

Figure 4: A cut-away perspective of a pipe having the DB (shown as an orb) to displace and distribute liquid lining along the wall [4]

This design has proven effective in distributing the coating material on the walls of the pipe however a thicker coating was noticed to form at the bottom of the pipe compared to the top. This is primarily due to gravitational forces making it challenging to apply more material to the top of the pipe. Therefore, research has been undertaken at the University of Waterloo to understand the coating process and potentially redesign the DB to solve this issue.

Several designs were proposed and one was selected as the best solution to problem. This design, named the HF DB, was designed and the flow was simulated using computational fluid dynamics. It was then manufactured and experimentally tested to determine the air flow distribution and verify the improvement in design.

The HF DB was designed to direct a portion of the airflow at the pool of coating material in front of the DB to move more of the material to the top of the pipe. This is achieved through a curved nozzle at the bottom front of the DB, labelled the mouth. Air flow behind the DB enters a channel which is contracted to the nozzle in the front. The HF design can be seen in Fig. 5.

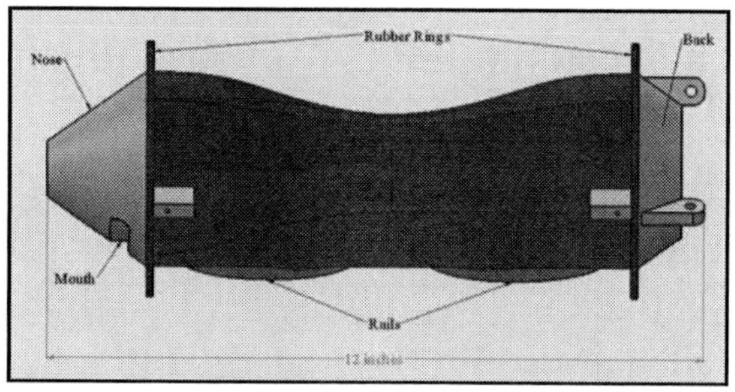

Figure 5: Sectional view of CAD model for HF DB

The design uses rubber rings located near the back and front of the DB to push the coating material and distribute the material on the walls of the pipe as a result of a pressure drop which is created as the air flows in the gap between the rubber ring and the pipe wall. The DB runs on four metal rails to keep it centered within the pipe. The total length of the DB is approximately 30 cm in length and designed for a 6 inch watermain pipe. Additionally, the back of the DB allows it to be attached to the same cable as the base case DB.

COMPUTATIONAL FLUID DYNAMIC SIMULATION

The HF DB and an additional Base Case (BC) DB design with a front shape similar to 70a in Fig. 3 were simulated using ANSYS version 16.2 for comparison and to verify the purpose of the HF DB. Each DB was simulated in a 6 inch pipe with an approximate flow rate that would be seen in operation. The velocity vectors for the HF DB can be seen in Fig. 6 from the results of the CFD simulation.

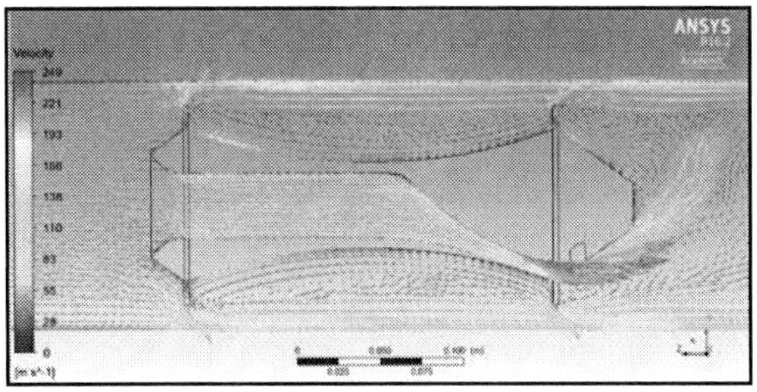

Figure 6: ANSYS simulated velocity vectors for the HF DB

The CFD simulations showed an increased velocity out of the mouth of the HF DB compared to around the rubber rings. The air flow out of the mouth of the DB is directed upwards, following the shape of the nose and dissipated much quicker than the flow over the rubber rings near the walls of the pipe. This has potential to direct some of the coating material to the top of the pipe providing the pool of material in front of the DB is large enough, but could potentially fail to reach the pooled material to carry it to the top of the pipe. The flow over the rubber rings of the DB is similar for both DBs where there is an expected increased velocity and pressure drop for the HF DB. Recirculation areas occur surrounding the body of the DB in a 3-dimensional view between the rubber rings. However, due to the increased flow out of the mouth of the HF DB, there appears to be a larger recirculation area above the nose compared to below. Assuming the material is carried from the pool in front of the DB to the recirculation area, this large recirculation area should help to apply coating on the upper surface of the pipe.

Once the performance of the HF DB was validated for the intended purpose, the design was manufactured. The nose, back, and body pieces of the DB were 3D printed using PC-ABS material from a fused deposition modelling machine with standard resolution. The nose and back were printed with 100% fill for structural purposes while the body pieces were printed with 30% sparse fill. The rubber rings and metal rails were added and the entire DB was assembled using standard machine screws for ease of disassembly.

EXPERIMENTAL TESTING

In order to provide further validation of the effectiveness of the HF DB and compare to the BC DB, experimental testing was conducted to determine the air velocities at certain points of interest. The test setup used a vacuum pump and tank system to create and maintain a desired airspeed. The vacuum pump was driven by a variable frequency drive motor to adjust the pump speed and control the velocity through the pipe. The pump has an operating range of 400 to 1750 RPM. A 6 inch clear acrylic pipe was used for the test section, which was attached to a straight pipe section leading into the vacuum tank. The DB was secured inside of the acrylic pipe with sufficient distance away from the pipe inlet and connection to the straight pipe section to minimize any additional disturbances upstream and downstream of the test section. A pitot tube connected to a digital handheld manometer was placed upstream of the DB. The position of the DB was variable as well as the height of the pitot tube. The test setup with the DB and pitot tube inside the acrylic pipe can be seen in Fig. 7.

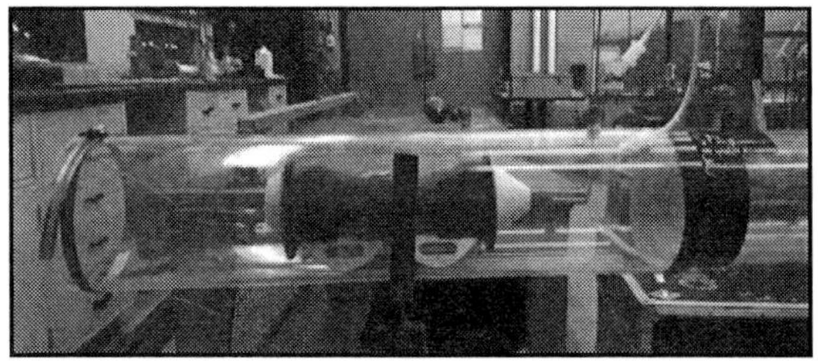

Figure 7: Side view of test setup

The vacuum pump was set to create a freestream velocity inside the pipe of approximately 6 m/s. This was based on the capabilities of the pump and setup. For both the BC and HF DBs, various measurements spanning the height of the pipe were taken at distances of 0, 1.25 cm, and 2.5 cm downstream of the DB nose. The results of the measurements are shown in Fig. 8, where y/D represents the measurement height inside the pipe with respect to the pipe diameter, u/U represents the measured velocity with respect to the free stream velocity, and x represents the distance of the measurement from the nose of the DB.

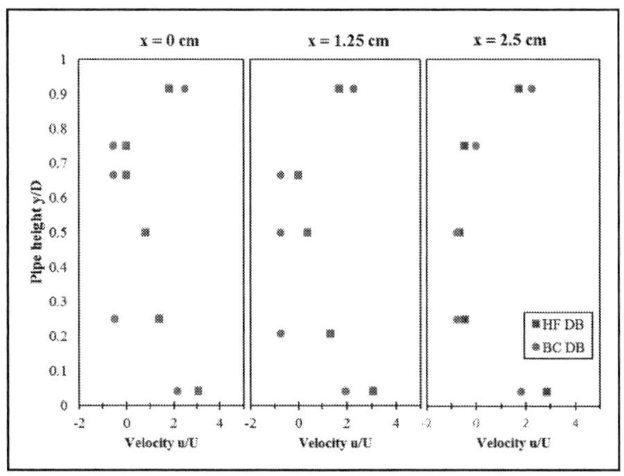

Figure 8: Velocity profile comparison at distance x downstream of nose of DB

Additional points of interests included in front of the mouth of the DB, between the gap of the front rubber ring and the pipe wall, as well as the recirculation area above the body of the DB between the rubber rings. The measurements of these additional points are shown in Table 1.

Table 1: Single point velocities of DB

Measurement Location	y/D	HF u/U	BC u/U
Mouth	0.25	3.66	0.47
Lower Rubber gap	0.08	3.15	-
Upper Rubber gap	0.92	2.31	2.77
Recirculation	0.83	1.06	0.61

From the measurements results in Fig. 8, the flow around the mouth of the HF DB, y/D = 0.25, shows an increase in air speed of approximately 400% and 270% at 0 and 1.25 cm from the nose respectively. In all distances from the nose, the airspeed at the top of the pipe was lower in the HF DB compared to the BC DB whereas the bottom of the pipe showed a higher air speed.

The maximum airspeed was measured directly in the mouth of the HF DB, as see in Table 1. The value in the gap of the rubber ring and upper pipe wall of the pipe showed similar airspeeds for both DBs with the BC DB having a slightly higher speed. The difference in airspeed in front of the rubber ring gap for the upper pipe wall and the lower wall at the DB nose was approximately 40% increase and 30% decrease in the measurement of the HF DB with respect to the BC DB.

On the HF DB, there are possible signs of recirculation through the measurement of zero and negative flow as predicted in the CFD analysis, however backwards flow is difficult to determine using a pitot tube. Compared to the flow at walls of the pipe, the airspeed out of the mouth of the HF DB seemed to decelerate and dissipate much quicker. This was predicted in the simulation.

Overall, there seems to be an improved performance of the HF DB from the original BC DB design. The increased flow at the bottom of the pipe shows promising results in order to distribute more material to the top of the pipe.

CONCLUSIONS

The process of cleaning and lining water main pipes through Envirologics' Tomahawk System was verified and optimized through the work of the University of Waterloo. The work consisted of field demonstrations in Waterloo, Ontario for watermains on the streets of Hazel Street and Albert Street. The results were successful in removing the tuberculation, corrosion, and bitumen from the pipe using the airborne abrasive cleaning method. A coating of the lining material was successfully applied to the walls of the pipe.

It was determined that the original BC DB produced a thicker coating of lining material at the bottom of the pipe compared to the top. In order to achieve a more even coating thickness distribution, a new design, the HF DB, was devised. The design was simulated using CFD, which showed an increased airspeed at the bottom of the pipe as intended. Additionally, it was observed that the airflow coming out of the nozzle mouth was directed upwards, following the shape of the nose. This could either be beneficial, providing the material pool in front of the DB is above the mouth, or else it may not capture the material to direct it to the top of the pipe.

The HF DB was manufactured and tested to determine the air speed at various points downstream of the DB. The BC DB was also tested for comparison. From the experiment, there was an increased airspeed at the bottom of the pipe and out of the mouth of the HF DB compared to the BC DB at a distance of 0 and 1.25 cm downstream of the nose. At 2.5 cm downstream of the nose, the two DBs seemed to have similar performance. At the nose of the DB, the gap between the rubber ring and the pipe produced a higher airspeed at the bottom of the pipe for the HF DB compared to the BC DB with a 40% increase, whereas the BC DB had a higher airspeed at the top gap with a 30% difference.

The results of the experimental test was promising in that it showed an increased airspeed towards the bottom of the pipe that will potentially distribute more lining material to the top wall of the pipe. The results were also fairly consistent with the CFD simulation.

To completely test the performance of the HF DB, the DBs will be compared using Envirologics' testing setup to achieve higher flow rates. In addition, the coating material will be added to determine the performance in a wet test run. Other designs may be devised in order to direct the airflow which will also be validated, optimizing the DB lining distribution.

ACKNOWLEDGMENTS

The authors acknowledge the help and support of the Wind Energy Group, including Farid Samara, Tyler Gallant, Alison Zilstra, and Michael Lenover, for their work on the project. This project was in affiliation with Envirologics Engineering Inc.

REFERENCES

1. Verwey, K. 2015, "Waterless pipeline cleaning process saves time, money and resources", *Environmental Science & Engineering Magazine*, Issue: May/June.
2. Envirologics Engineering Inc. 2017, http://www.tomahawksystem.ca/lining
3. Thorogood, B., 2015, "Trenchless and Dry Watermain Rehabilitation using Tomahawk in Victoria and Vancouver BC", *Trenchless Technology Roadshow*, Richmond, BC.
4. Envirologics Engineering Inc. 2016, "Method to Line Pipe Homogeneously", CA 2902708.

THREE DIMENSIONAL UNSTEADY CONFINED VISCOUS FLOWS WITH TIME-VARIABLE INFLOW VELOCITY AT LOW REYNOLDS NUMBERS

Abdurrazag Khaled
Department of Mechanical Engineering
McGill University
Montreal, Quebec, Canada
Abdurrazag.Khaled@mail.mcgill.ca

Dan Mateescu
Department of Mechanical Engineering
McGill University
Montreal, Quebec, Canada
Dan.Mateescu@mcgill.ca

ABSTRACT

This paper presents the analysis of the three-dimensional unsteady confined viscous flows past a backward-facing step generated by the variation in time of the inflow velocities. The present solution of the Navier-Stokes equations is obtained by the authors using a time spectral method in which the fluid velocity components and pressure are expressed by truncated Fourier series expansions.

KEYWORDS: Unsteady viscous flows, low Reynolds numbers, time-variable inflow velocity

INTRODUCTION

The unsteady flows in many engineering applications involve time-variable inflow velocity, which are studied for the fluid-structure interaction and flow-induced vibration problems. In these studies, the unsteady flow solutions should have a very good computational efficiency.

For the unsteady flows at low Reynolds numbers past airfoils and wings, or in confined 2D and 3D configurations, the second author and his graduate students developed time-accurate numerical methods for solving the Navier-Stokes equations [1-4], obtaining solutions in good agreement with the experimental results.

To increase substantially the computational efficiency, the authors developed a time spectral method using truncated Fourier series for the analysis of the confined unsteady two-dimensional viscous flows [5].

The aim of this paper is to present solutions of the Navier-Stokes equations using a time spectral method based on truncated Fourier series for the analysis of the unsteady viscous flows in confined three-dimensional configurations at low Reynolds numbers and low reduced frequencies. The time spectral method is applied to obtain solutions of the unsteady flows with separation regions past the benchmark configurations with a downstream-facing step, which is generated by harmonic variation in time of the inflow velocity.

PROBLEM FORMULATION

Consider the flow in a rectangular duct of constant width *2bH* with an upstream portion of height *Hh* and length Hl_0 and a sudden step followed by a downstream portion of height *H* and length Hl_1 (Figure 1). The downstream-facing step of height (1-*h*)*H* is considered for *h*=0.5 in the numerical results presented further. This flow is referred to the Cartesian coordinates *Hx*, *Hy* and *Hz* centered at the step corner in the plane of symmetry of the duct, where *x*, *y* and *z* are nondimensional coordinates.

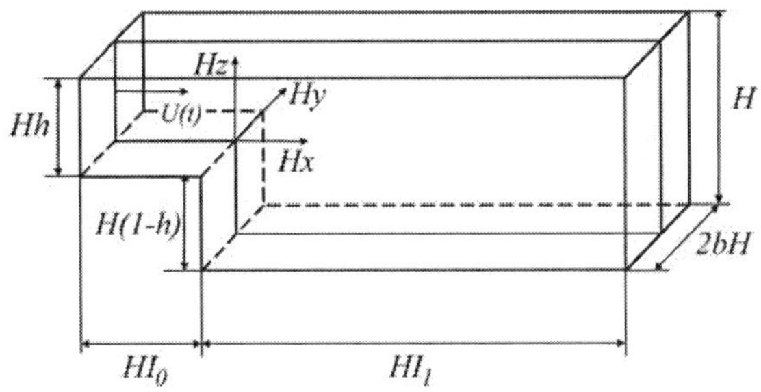

Figure 1: Geometry of the duct of a rectangular cross-section with a downstream-facing step.

At the channel inlet ($Hx = -Hl_0$) there is a fully developed laminar flow defined by the axial velocity $U(y,z,t)$ which is variable in time according to the equation

$$u^*(-l_0, y, z, t) = U_0(1 - a\sin\omega t)F(z) \quad , \tag{1}$$

$$F(z) = \frac{16h^2}{\pi^3} C \sum_{K=0}^{\infty} \frac{(-1)^k}{(2k+1)}\left[1 - \frac{\cosh(2k+1)(\pi y/h)}{\cosh(2k+1)(\pi b/h)}\right] + \cos\left[(2k+1)\frac{\pi}{2}\frac{2z-h}{h}\right] \tag{2}$$

where C at the nondimensional time $t = U_0 t^*/H$ are expressed in the form

$$C = (3/h^2)\left[1 - \frac{192}{\pi^5}\frac{h}{b}\sum_{k=0}^{\infty}\frac{1}{(2k+1)^5}\tanh(2k+1)\frac{\pi b}{h}\right]^{-1} \tag{3}$$

in which U_0 is the time-averaged mean flow velocity, and ω is the reduced frequency of oscillation, defined as $\omega = \omega^* H/U_0$ ($\omega^* = 2\pi f$, is the radian frequency of the oscillations), and *a* is the nondimensional amplitude of the inflow velocity oscillations; the mean flow velocity is $U(t) = U_0(1 - a\sin\omega t)$.

The time dependent Navier-Stokes and continuity equations for unsteady incompressible flows can be expressed in nondimensional conservation law form as

$$\frac{\partial \mathbf{V}}{\partial t}+\mathbf{Q}(\mathbf{V},p)=\mathbf{0}, \qquad \nabla \bullet \mathbf{V}=\frac{\partial u}{\partial x}+\frac{\partial v}{\partial y}+\frac{\partial w}{\partial z}, \tag{4}$$

$$\mathbf{Q}(\mathbf{V},p)=\{Q_u(u,v,w,p),Q_v(u,v,w,p),Q_w(u,v,w,p)\}^{\mathrm{T}}, \tag{5}$$

where $\mathbf{V}=\{u,v,w\}^{\mathrm{T}}$ represents the dimensionless fluid velocity vector (nondimensionalized with respect to the time average mean flow velocity U_0), and $\mathbf{Q}(\mathbf{V},p)$ is expressed in 3D Cartesian coordinates in the forms

$$Q_u=\frac{\partial(uu)}{\partial x}+\frac{\partial(vu)}{\partial y}+\frac{\partial(wu)}{\partial z}+\frac{\partial p}{\partial x}-\frac{1}{\mathrm{Re}}\left(\frac{\partial^2 u}{\partial x^2}+\frac{\partial^2 u}{\partial y^2}+\frac{\partial^2 u}{\partial z^2}\right),$$

$$Q_v=\frac{\partial(uv)}{\partial x}+\frac{\partial(vv)}{\partial y}+\frac{\partial(wv)}{\partial z}+\frac{\partial p}{\partial y}-\frac{1}{\mathrm{Re}}\left(\frac{\partial^2 v}{\partial x^2}+\frac{\partial^2 v}{\partial y^2}+\frac{\partial^2 v}{\partial z^2}\right), \tag{6}$$

$$Q_w=\frac{\partial(uw)}{\partial x}+\frac{\partial(vw)}{\partial y}+\frac{\partial(ww)}{\partial z}+\frac{\partial p}{\partial z}-\frac{1}{\mathrm{Re}}\left(\frac{\partial^2 w}{\partial x^2}+\frac{\partial^2 w}{\partial y^2}+\frac{\partial^2 w}{\partial z^2}\right),$$

in which $\mathrm{Re}=HU_0/\nu$ represent the time-average mean Reynolds number based on the downstream channel height, and u, v, w and p represent the dimensionless velocity components and pressure (with respect to U_0 and ρU_0^2, respectively, where ρ and ν are the fluid density and kinematic viscosity).

TIME SPECTRAL METHOD

This unsteady confined flow is a periodic flow with the basic reduced frequency $\omega=2\pi f\,H/U_\infty$, the fluid velocity and pressure can be expressed by truncated Fourier series expansions, where the reduced amplitudes $\hat{u}_n(x,y,z)$, $\hat{v}_n(x,y,z)$, $\hat{w}_n(x,y,z)$ and $\hat{p}_n(x,y,z)$ associated with the harmonic frequency $n\omega$ are complex functions ($i=\sqrt{-1}$)

$$\mathbf{V}(t,x,y,z)=\sum_{n=0}^{N}\hat{\mathbf{V}}_n(x,y,z)e^{in\omega t}, \quad \hat{\mathbf{V}}_n=\{\hat{u}_n,\hat{v}_n,\hat{w}_n\}^{\mathrm{T}} \tag{7}$$

$$u(t,x,y,z)=\sum_{n=0}^{N}\hat{u}_n(x,y,z)e^{in\omega t}, \quad \hat{u}_n(x,y,z)=\hat{u}_{nR}(x,y,z)+i\,\hat{u}_{nI}(x,y,z) \tag{8}$$

$$v(t,x,y,z)=\sum_{n=0}^{N}\hat{v}_n(x,y,z)e^{in\omega t}, \quad \hat{v}_n(x,y,z)=\hat{v}_{nR}(x,y,z)+i\,\hat{v}_{nI}(x,y,z) \tag{9}$$

$$w(t,x,y,z)=\sum_{n=0}^{N}\hat{w}_n(x,y,z)e^{in\omega t}. \quad \hat{w}_n(x,y,z)=\hat{w}_{nR}(x,y,z)+i\,\hat{w}_{nI}(x,y,z) \tag{10}$$

$$p(t,x,y,z)=\sum_{n=0}^{N}\hat{p}_n(x,y,z)e^{in\omega t} \quad \hat{p}_n(x,y,z)=\hat{p}_{nR}(x,y,z)+i\,\hat{p}_{nI}(x,y,z) \tag{11}$$

Based on these truncated Fourier series expansion, the Navier-Stokes and continuity equations (4) and applying a process of identification of all terms in the Navier-Stokes and continuity equations which are associated to each harmonic frequency $n\omega$ (and discarding all the harmonic terms higher than N), one can obtain the following truncated set of equations for each $n\in\{0,1,2,...,N\}$:

$$i\,n\omega\,\hat{u}_n+G_{Lu}(\hat{u}_n,\hat{v}_n,\hat{w},\hat{p}_n)+G_{Nu}(\hat{u},\hat{v},\hat{w})_{n,0}+G_{Nu}(\hat{u},\hat{v},\hat{w})_{0,n}=-\sum_{k=1}^{n-1}G_{Nu}(\hat{u},\hat{v},\hat{w})_{n-1-k,k} \tag{12}$$

$$i\,n\omega\,\hat{v}_n+G_{Lv}(\hat{u}_n,\hat{v}_n,\hat{w},\hat{p}_n)+G_{Nv}(\hat{u},\hat{v},\hat{w})_{n,0}+G_{Nv}(\hat{u},\hat{v},\hat{w})_{0,n}=-\sum_{k=1}^{n-1}G_{Nv}(\hat{u},\hat{v},\hat{w})_{n-1-k,k} \tag{13}$$

$$i\,n\omega\,\hat{w}_n+G_{Lw}(\hat{u}_n,\hat{v}_n,\hat{w},\hat{p}_n)+G_{Nw}(\hat{u},\hat{v},\hat{w})_{n,0}+G_{Nw}(\hat{u},\hat{v},\hat{w})_{0,n}=-\sum_{k=1}^{n-1}G_{Nw}(\hat{u},\hat{v},\hat{w})_{n-1-k,k} \tag{14}$$

$$\frac{\partial\hat{u}_n}{\partial x}+\frac{\partial\hat{v}_n}{\partial y}+\frac{\partial\hat{w}_n}{\partial z}=0 \tag{15}$$

where $G_{Lu}(\hat{u},\hat{v},\hat{w})_{n,m}$, $G_{Lv}(\hat{u},\hat{v},\hat{w})_{n,m}$, $G_{Lw}(\hat{u},\hat{v},\hat{w})_{n,m}$ and $G_{Nu}(\hat{u},\hat{v},\hat{w})_{n,m}$, $G_{Nv}(\hat{u},\hat{v},\hat{w})_{n,m}$, $G_{Nw}(\hat{u},\hat{v},\hat{w})_{n,m}$ are the linear and nonlinear derivation terms.

By separating the real and imaginary parts of these equations one obtains

$$-n\omega\hat{u}_{nI}+G_{Lu}(\hat{u}_{nR},\hat{v}_{nR},\hat{w}_{nR},\hat{p}_{nR})+\mathrm{Re}\{G_{Nu}(\hat{u},\hat{v},\hat{w})_{n,0}\}+\mathrm{Re}\{G_{Nu}(\hat{u},\hat{v},\hat{w})_{0,n}\}=-\sum_{k=1}^{n-1}\mathrm{Re}\{G_{Nu}(\hat{u},\hat{v},\hat{w})_{n-1-k,k}\} \tag{16a}$$

$$-n\omega\hat{v}_{nI}+G_{Lv}(\hat{u}_{nR},\hat{v}_{nR},\hat{w}_{nR},\hat{p}_{nR})+\mathrm{Re}\{G_{Nv}(\hat{u},\hat{v},\hat{w})_{n,0}\}+\mathrm{Re}\{G_{Nv}(\hat{u},\hat{v},\hat{w})_{0,n}\}=-\sum_{k=1}^{n-1}\mathrm{Re}\{G_{Nv}(\hat{u},\hat{v},\hat{w})_{n-1-k,k}\} \tag{16b}$$

$$-n\omega\hat{w}_{nI}+G_{Lw}(\hat{u}_{nR},\hat{v}_{nR},\hat{w}_{nR},\hat{p}_{nR})+\mathrm{Re}\{G_{Nw}(\hat{u},\hat{v},\hat{w})_{n,0}\}+\mathrm{Re}\{G_{Nw}(\hat{u},\hat{v},\hat{w})_{0,n}\}=-\sum_{k=1}^{n-1}\mathrm{Re}\{G_{Nw}(\hat{u},\hat{v},\hat{w})_{n-1-k,k}\} \tag{16c}$$

$$\frac{\partial\hat{u}_{nR}}{\partial x}+\frac{\partial\hat{v}_{nR}}{\partial y}+\frac{\partial\hat{w}_{nR}}{\partial z}=0 \tag{16d}$$

$$n\omega\hat{u}_{nR}+G_{Lu}(\hat{u}_{nI},\hat{v}_{nI},\hat{w}_{nI},\hat{p}_{nI})+\mathrm{Im}\{G_{Nu}(\hat{u},\hat{v},\hat{w})_{n,0}\}+\mathrm{Im}\{G_{Nu}(\hat{u},\hat{v},\hat{w})_{0,n}\}=-\sum_{k=1}^{n-1}\mathrm{Im}\{G_{Nu}(\hat{u},\hat{v},\hat{w})_{n-1-k,k}\} \tag{17a}$$

$$n\omega\hat{v}_{nR}+G_{Lv}(\hat{u}_{nI},\hat{v}_{nI},\hat{w}_{nI},\hat{p}_{nI})+\mathrm{Im}\{G_{Nv}(\hat{u},\hat{v},\hat{w})_{n,0}\}+\mathrm{Im}\{G_{Nv}(\hat{u},\hat{v},\hat{w})_{0,n}\}=-\sum_{k=1}^{n-1}\mathrm{Im}\{G_{Nv}(\hat{u},\hat{v},\hat{w})_{n-1-k,k}\} \tag{17b}$$

$$n\omega\hat{w}_{nR}+G_{Lw}(\hat{u}_{nI},\hat{v}_{nI},\hat{w}_{nI},\hat{p}_{nI})+\mathrm{Im}\{G_{Nw}(\hat{u},\hat{v},\hat{w})_{n,0}\}\;\mathrm{Im}\{G_{Nw}(\hat{u},\hat{v},\hat{w})_{0,n}\}=-\sum_{k=1}^{n-1}\mathrm{Im}\{G_{Nw}(\hat{u},\hat{v},\hat{w})_{n-1-k,k}\} \tag{17c}$$

$$\frac{\partial\hat{u}_{nI}}{\partial x}+\frac{\partial\hat{v}_{nI}}{\partial y}+\frac{\partial\hat{w}_{nI}}{\partial z}=0 \tag{17d}$$

The equation systems (16) and (17) are then solved sequentially, advancing from the steady flow component n=0 to the highest component considered (n=N). This is done, for each harmonic mode n, using an iterative pseudo-time relaxation procedure with artificial compressibility developed in our previous papers [1, 2].

Inflow boundary conditions for the harmonic flow components. The fluid velocity at the inlet is defined by equation (1), resulting thus the following inflow boundary conditions for the harmonic flow components:

$u_{0R} = U_0 F(z)$, $v_{0R} = 0$, $w_{0R} = 0$, $u_{0I} = 0$, $v_{0I} = 0$, $w_{0I} = 0$ for the steady harmonic flow component, n=0;

$u_{1R} = 0$, $v_{1R} = 0$, $w_{1R} = 0$, $u_{1I} = -aU_0 F(z)$, $v_{1I} = 0$, $w_{1I} = 0$ for the first harmonic flow component, n=*1*; and

$u_{nR} = 0$ $v_{nR} = 0$ $w_{nR} = 0$ $\quad u_{nI} = 0$ $v_{nI} = 0$ $w_{nI} = 0$,
for the higher harmonic flow components, $n \geq 2$.

Fluid velocity and pressure. When each harmonic flow component n is solved, the flow velocity components and pressure are also obtained.

RESULTS FOR CONFINED FLOW WITH VARIABLE INFLOW VELOCITY

The solutions obtained with this method are presented in the following. Computations have been performed for the step height *Hh=H/2*, and at various low Reynolds numbers, $\mathrm{Re} = HU_0/\nu$, amplitudes of the inflow velocity, *a*, and reduced frequencies of oscillation, ω. The numerical results were obtained using a stretched staggered grid with 300x48x68 grid points for each variable (that is, 599x95x135 grid points in total). The pseudo time computations have been performed using δ=0.3 and Δτ=0.05 and convergence was reached when the RMS residuals were less than 10^{-5}.

Profiles of the velocity components u, v, w and pressure p are calculated at different downstream locations by using the present time spectral method with two harmonic flow components *N*=1 (denoted by H1), with three components *N*=2 (denoted by H2) and with four components *N*=3 (denoted by H3), and compared with the time accurate results (denoted by TA). A very good agreement can be observed between these solutions, as shown in Figure 2.

These comparisons indicated that the time spectral method with $N = 3$ (four harmonic flow components) required less than one quarter of the computational time needed by the time accurate method, and about one seventh of the time for two harmonic flow components, $N = 1$.

Two flow separation regions are formed on the lower and upper walls. The variations during the oscillatory cycle, *t/T*, of the upper and lower wall separation and reattachment locations are shown in Figures 3 and 4.

The influence of the inflow velocity amplitude, *a*, and the effect of the duct width are shown in Figure 3. One can notice that for some larger values of the inflow velocity amplitude, such as *a*=0.075 or 0.1, the upper wall flow separation is not present during the entire oscillatory cycle (appearing and then disappearing). The upper flow separation is much reduced for the narrow duct (2*b*=16) than for the wider duct (2*b*=26.6) as one can see from the same figure.

The variation of the flow separation along the duct width is shown in Figure 4.

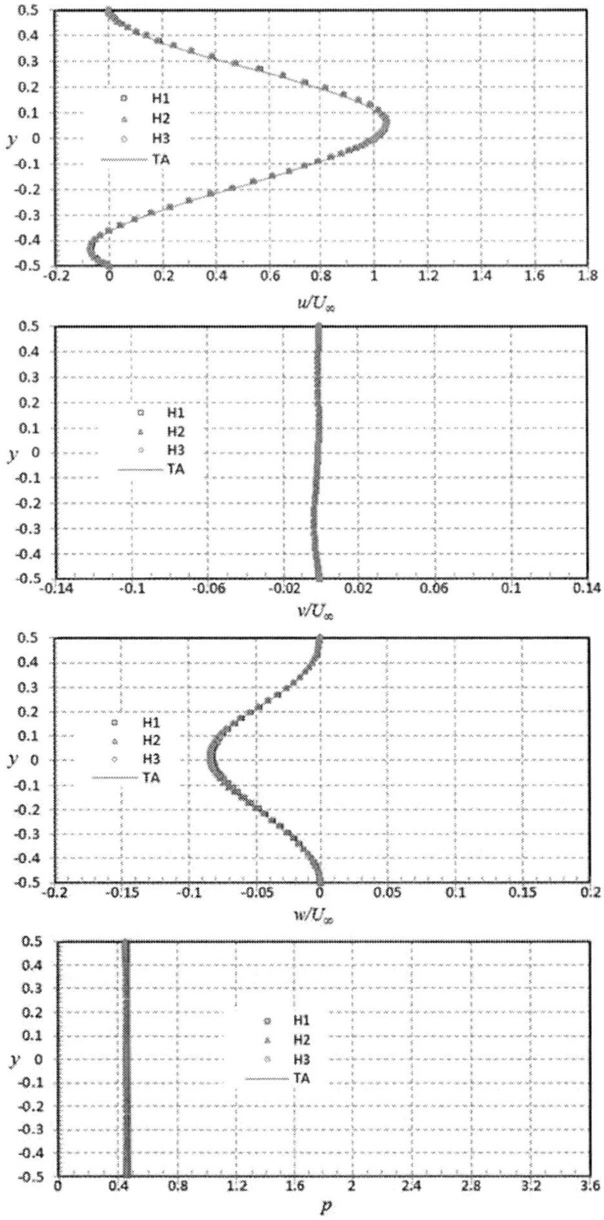

Figure 2: Profiles of the velocity components u, v, w and pressure p at *x*=5 in the symmetry plan (*y*=0) for 2*b*=26.6, Re=600, inflow velocity amplitude a =0.05, reduced frequency ω=0.05, at t/T=0.25.

CONCLUSIONS

The time spectral method presented in this paper is very efficient computationally, requiring for four harmonic flow components less than one quarter of the time needed by the time accurate method, and one seventh for two components.

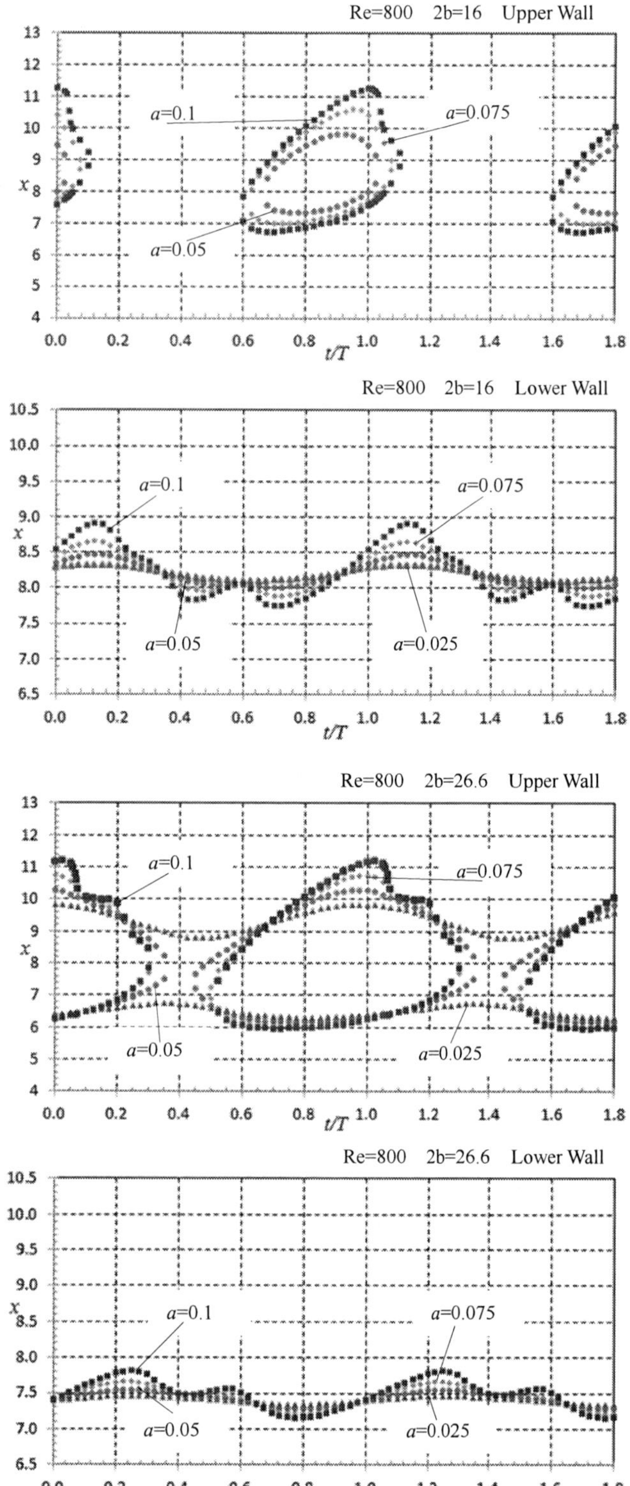

Figure 3: Influence of the inflow velocity amplitude, a : Variation during the oscillatory cycle, t/T, of the upper and lower wall separation and reattachment locations at Re=800, ω =0.05 and y=0, for two duct widths, $2b$=16 and 26.6.

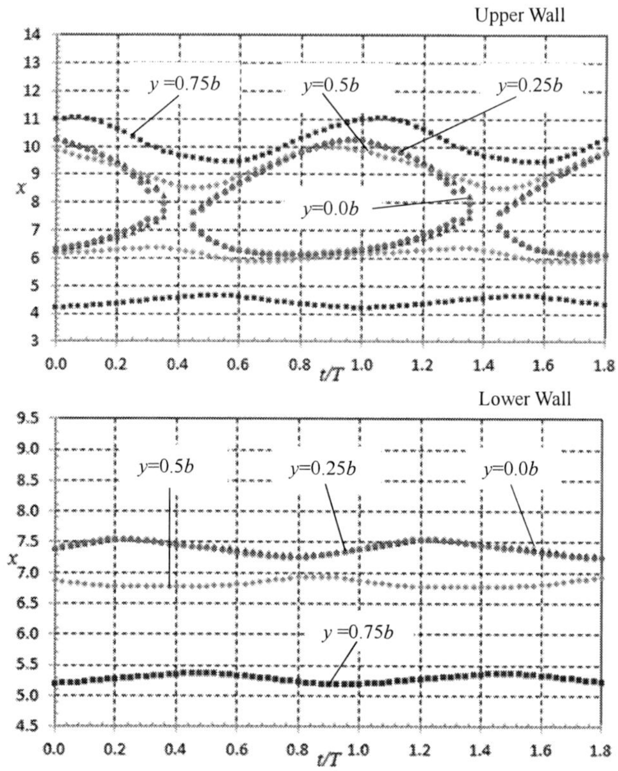

Figure 4: Spanwise variation of the unsteady flow separation: Variation during the oscillatory cycle, t/T, of the upper and lower separation and reattachment locations, for Re=800, a=0.05, ω =0.05, $2b$=26.6, at several locations along the duct span: y=0, 0.25b, 0.5b, and 0.75b.

REFERENCES

1. Mateescu, D., Munoz, M., and Scholz, O. 2010, "Analysis of unsteady confined viscous flows with variable inflow velocity and oscillating walls." *ASME J. of Fluids Engineering*, Vol. 132, 4, 41105, pp. 1-9.
2. Mateescu, D., and Abdo, M. 2010, "Analysis of flows past airfoils at very low Reynolds numbers." *Journal of Aerospace Engineering*, Vol. 224, pp. 757-775.
3. Mateescu, D., Munoz, M., and Scholz, O. 2011, "Unsteady flows past oscillating airfoils at very low Reynolds numbers." *29th AIAA Applied Aerodynamic Conference*. Honolulu, Hawaii, AIAA Paper 2011-3657, pp. 1-15.
4. Mateescu, D., and Panahi, A. 2013, "Analysis of three dimensional confined flows with oscillating walls and variable inflow velocity." *31thAIAA Applied Aerodynamics Conference, 51th AIAA Aerospace Science Meeting*, Dallas, Texas, AIAA Paper 2013-0063, pp.1-18.
5. Mateescu, D., and Khaled, A. 2016, "Time spectral method for unsteady confined viscous flows with variable inflow velocity at low Reynolds numbers." *AIAA Applied Aerodynamics Conf., Aviation and Aeronautics Forum*, San Diego, California, AIAA Paper 2016-1065, pp.1-22.

AEROACOUSTIC MODELING FOR FLOW PAST A BLUFF BODY AT TWO DISTINCT REYNOLDS NUMBERS

Sultan Ibrahim Alqash
Department of Mechanical and Industrial Engineering
University of Toronto
Toronto, ON, Canada
Sultan.alqash@mail.utoronto.ca

Prof. Kamran Behdinan
Department of Mechanical and Industrial Engineering
University of Toronto
Toronto, ON, Canada
behdinan@mie.utoronto.ca

ABSTRACT

Landing gears (LG) are primarily designed to support all the loads of an aircraft during landing, taxiing, and taking off. It consists of many components with different lengths and sizes. To understand the physics and analyze both the flow structure and acoustic propagation, a two-dimensional (2D) case of flow past a circular cylinder was simulated using ANSYS Fluent. Two different Reynolds numbers, Re (150 and 90,000) were examined. For low Re, two distinct numerical conditions steady and unsteady flow were simulated and compared to examine the effect of the time dependency on the associated acoustic field. For high Re, the acoustic field was computed using the built-in Ffowcs William and Hawkings (FW-H) acoustic analogy solver in Fluent. The results show the importance of considering the time variable to extract the corresponding flow data. The far-field noise prediction highly depends on the location of the near-filed data and its associated integral source terms (surface/volume).

KEYWORDS: Reynolds number, acoustic analogy, near-field, far-field, pressure fluctuation

INTRODUCTION

The aircraft noise was recognized as an environmental problem in the early 1950s, when the turbo-jet powered aircraft was introduced to the markets. Since then, the aircraft noise issue has been dominated by jet noise, known as propulsive noise. In the early 1970s, aircraft noise was successively reduced after the introduction of the first twin cycle bypass turbofan engines [1]. Due to the noise reduction of the propulsive noise that originates from jet-engines, the airframe itself becomes the major contributor to the overall noise. As a result of decades of research and development into noise reduction, by the year 2000 the aircraft noise level had become 20-30 dB quieter compared to the first generation of turbofan powered aircraft. The US and European communities have set a target to minimize the noise level by 50% near airport areas by 2020 [2]. By 2050, the overall noise level emission of flying aircraft should be further reduced by 65% [3]. When the aircraft approaches the airport, the engines are almost at low speed which reduces its noise contribution. That yields to put a high portion of the generated noise on the airframe itself. Among the airframe components, the landing gears have been identified as a main source of noise during the approach phase, when they are fully deployed prior to landing for both aircraft types: long range and short range. It has a significant noise level due to its complex structural design that consists of many components with different lengths and sizes.

For a commercial aircraft, the noise generated from the airframe has received much attention in recent years. This has led the Federal Aviation Administration (FAA) to impose strict regulations on the aircraft noise level [4]. All these demands created pressure on industries and academic communities to develop improved techniques for noise reduction and better methods for its prediction. To ease inspection and maintenance, the aerodynamic design is not refined. As a result, many components such as, hydraulic cables, electric wiring, torque links, front and rear braces are exposed to the air flow. Therefore, the flow separation over the LG components constitutes the main noise source mechanism through unsteady wake flow and large-scale vortex instability and deformation [5]. In the literature, there are three distinct approaches to study flow field and associated acoustic field around the LG within certain limitations. Those approaches are experimental, empirical, and numerical. The main difference between flow around landing gears and that around simple geometries, like single or tandem circular cylinders, is the three-dimensionality flow characteristic. For that reason, the experimental studies for the isolated or installed LG system are recommended but too expensive. Thus, the numerical approach becomes an alternative candidate to resolve this issue. Although current computational approaches are infeasible to handle all the turbulence scales that inevitably exist in cases of complex geometry such as a LG, there are some computational fluid dynamic (CFD) approaches that have been utilized to resolve this issue. The most popular methods are Direct Numerical Simulations (DNS), Large-Eddy Simulations (LES), and Reynolds Averaged Navier-Stokes (RANS). Recently, there is a common CFD method known as Detached-Eddy Simulations (DES). The DES approach is a combination of LES and RANS solvers, the LES takes care of the large-scale separated flow regions, while RANS handles the modeling of the turbulence boundary layer [6].

Any aeroacoustics problem consists of two fields as its name implies fluid dynamics and acoustics, respectively. It is a two-step process: 1) sound generation based on the study of the aerodynamic flow which generates sound/noise sources and then, 2) sound propagation based on the investigation of the acoustic field spreading in a medium at rest that generated from step 1. The Lighthill's acoustic analogy is considered as the first pioneering work in sound generated aerodynamically and is named after Jims Lighthill [7] in 1952. In 1954, he assumed that the turbulent fluctuations of the stream are equivalently represented by a distribution of quadrupole sources in the same volume of the flow. In 1955, Curle [8] modified the Lighthill's analogy to include the effect of solid boundaries. In 1969, Ffowcs William and Hawkings [9] considered both the influence of the solid surfaces and the turbulence in arbitrary motion and developed the FW-H analogy. The FW-H analogy is considered as the generalized Lighthill's acoustic analogy which makes it more common and applicable for a wide range of applications. Van Mierlo et al. [10] studied the influence of the main landing gear (MLG) inclination angle on the noise level using ANSYS Fluent. However, they did not include the contribution of the quadrupole sources corresponds to the volume integral due to its high computational cost. Thus, the selection of the surface always coincides with the wall of the bluff body. The same approach was used by Long et al. [11] to study the far-field noise of a simplified nose landing gear (NLG). They noticed that the directivity of the noise is a dipole like source. Another method based on CFD and the FW-H equation with penetrable surface has advanced considerably. For instance, Lockard et al. [12] compared the results and noticed that the noise predicted based on solid surface agreed well with the near-field CFD results, and that noise predicted based on the penetrable surface was highly suspected to be corrupted by the pseudo-sound created at the closing FW-H surface in the wake region. Also, Spalart et al. [13] and Sanders et al. [14] argued that by calculating the noise based only on a solid surface would lead to inaccurate results even at a Mach number as low as 0.115. The results revealed a typical discrepancy of 3 dB compared to the permeable surface.

In this study, since the LG system is too complex and contains multiple bluff bodies, 2D flow past a circular cylinder with two different Re numbers were simulated. This simplified case will pave the road to understand the underlying physics of both flow structure (near-field) as well as its associated noise propagation (far-field). Different turbulent models were examined according with the flow anticipated behavior (laminar/turbulent). The FW-H solver in Fluent was utilized to study the far-field noise prediction.

ACOUSTIC ANALOGIES FORMULATION

First, Lighthill formulated a wave equation of acoustic analogy where the source derived by comparing the exact equations of motion of a fluid with the equations of sound propagation in a medium at rest as follows [7],

$$\frac{1}{c_o^2}\frac{\partial^2 \rho'}{\partial t^2} - \nabla^2 \rho' = \frac{\partial^2 T_{ij}}{\partial x_i \partial x_j} \tag{1}$$

$$T_{ij} = \rho v_i v_j + P_{ij} - c_o^2 \rho' \delta_{ij} \tag{2}$$

where $(\rho' = \rho - \rho_o)$ indicates density perturbation, T_{ij} is the Lighthill's stress tensor, $P_{ij} = p'\delta_{ij} - \tau_{ij}$ refers to the compressive stress tensor with p' being the fluctuating pressure and τ_{ij} being the total stress tensor equal to $\sigma_{ij} - p\delta_{ij}$ with σ_{ij} refers to the residual stress tensor. Note that v_i and v_j indicate the velocity components in x_i and x_j directions, respectively, and c_o is the speed of sound.

Second, Curle [8] expanded the Lighthill's analogy considering a rigid surface; after that, Ffowcs Williams and Hawkings further generalized the theory considering a rigid object in arbitrary motion as follows [9],

$$\left(\frac{\partial^2}{\partial t^2} - c_o^2 \frac{\partial^2}{\partial x_i^2}\right)(H(f)\rho') = \frac{\partial^2}{\partial x_i \partial x_j}\left(T_{ij}H(f)\right) - \frac{\partial}{\partial x_i}\left(F_i\delta(f)\right)$$

$$+ \frac{\partial}{\partial t}\left(Q\delta(f)\right) \tag{3}$$

$$F_i = \left(P_{ij} + \rho v_i(v_j - u_j)\right)\frac{\partial f}{\partial x_j} \tag{4}$$

$$Q = (\rho_o u_i + \rho(v_i - u_i)) + \frac{\partial f}{\partial x_i} \tag{5}$$

where f represents the body surface as function of $f(x,t) = 0$; in which $f < 0$ and $f > 0$ refer to inside and outside of rigid body, respectively. The right-hand side of eqn. (3) is the source terms; the first term shown explicitly in eqn. (2) physically means unsteadiness inside the fluid expressed in quadruple source as Reynolds stress which is called turbulent-induced term, the second term is dipole source due to local fluctuating stresses exerted by the surface on the fluid and third term is monopole source generated by mass-flux fluctuating around the surface. The last two terms are represented in equations (4) and (5) which are called loading noise and thickness noise, respectively.

The differential form of FW-H shown in eqn. (3) can be written in terms of fluctuating pressure based on the following relationship $(p' = c_o^2 \rho')$ which shows the isentropic relation. There are three important functions used in the above equations namely Kronecker delta δ_{ij}, Heaviside $H(f)$, and the derivative of the Heaviside function $H'(f) = \delta(f)$ is the Dirac delta. The turbulent term associated with the quadrupole source is in volume integration. Hence, the decision of including the contribution of the volumes in fluid around rigid body to integrate and evaluate the integration kernel is difficult. Thus, it requires a better algorithm/numerical technique to handle the computational cost and memory management. This approach is still under investigation and deserves more developments.

NUMERICAL SET UP

A two-dimensional case of flow past a circular cylinder was simulated using ANSYS Fluent. First, the flow domain was designed using ANSYS Workbench. The computational domain was extended 5 times and 20 times the cylinder of unity diameter (*D*) upstream and downstream, respectively. The vertical dimension is 5*D* equally in both Y-directions from the cylinder center. Then, the domain was meshed using an unstructured method of a triangular. However, close to the cylinder wall the structured mesh with rectangular shape elements were utilized to account for the boundary layer (BL) effect. The total number of cells is 22370. After that, the boundary conditions (B.C's) were assigned within corresponding input values. All the numerical modeling process is shown in Figure 1(a), (b), and (c) starts from the domain, mesh, and B.C's, respectively, including the values of B.C's as shown in Table 1.

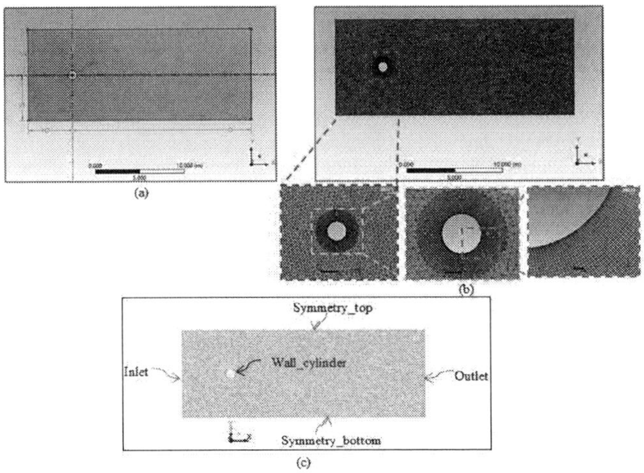

Figure 1: Numerical Modeling: (a) Computational domain, (b) Mesh generation, and (c) Boundary conditions

Table 1: Boundary conditions and corresponding values

B.C.	Value	Remarks
Inlet	V = 1 [m/s]	These B.C.'s are common for incompressible flows
Outlet	Gauge Pressure = 0 [Pa]	
Wall_cylinder	Stationary wall, V_{wall} = 0 [m/s]	With no-slip shear condition
Symmetry_top	No values needed to be entered	To ensure that flow does not form B.L. on these B.C.'s
Symmetry_bottom		

Two different Reynolds numbers were examined Re = 150 and 90,000. In case of low Re, two distinct numerical conditions steady and unsteady flow were simulated and compared to examine the effect of the time dependency. For both flow conditions the viscous laminar model was utilized. In case of high Re, the acoustic field was computed using the built-in FW-H solver available in Fluent. For this case, the same computational domain is utilized, but the DES with a Spalart-Allmaras (one-equation) model is used to capture the turbulent flow.

RESULTS AND DISCUSSION

The purpose of steady-state simulation is to investigate the mean flow patterns around the bluff-body. The results showed that only the mean velocity and pressure can be captured as shown in Figure 2(a) and (b), respectively. This is evidence that the acoustic field cannot be extracted, but the results could be utilized as an initial value for simulating unsteady case. On the other hand, the unsteady flow simulation includes the transient behavior which then clearly depicts the flow motion as shown in Figure 2(c), (d), (e) and (f). The flow instability downstream known as Karman Vortex Street was successfully captured.

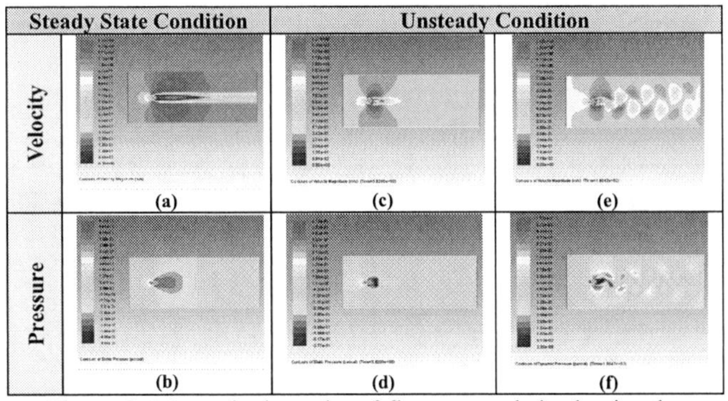

Figure 2: Numerical results of flow around single circular cylinder with Re = 150

Because the flow depends on time, it is important to calculate the time step. For example, in the region behind the circular cylinder a vortex shedding will be created. From the experiments [15], the Strouhal number, St = *f D*/v is 0.172, then the period, T=1/*f* is found to equal 5.8139 seconds if the frequency, *f* is calculated. Assuming the number of time steps is n = 20, then ΔT = T/n = 0.291 seconds. Actually, this is the first vortex shedding that is captured. Note that the lift coefficient, C_L has a sinusoidal pattern which is an indication of the sustained vortex shedding that takes place in the wake region. From Fluent the drag force, F_d, is found to equal 0.6961 N, then the drag coefficient, C_d was calculated using this relation: $C_d = \frac{2 F_d}{\rho V^2 A}$, where *A* is the projected area normal to the flow direction (i.e. for a 2D cylinder is equal to the diameter), and it turned out to be equal to 1.392 which differs by about 7.2% from the experimental value [15].

The near-field data with its corresponding far-field acoustic prediction is captured as seen in Figure 3. The wall of the circular cylinder is selected to extract all the relevant acoustic source date such as velocity and pressure. In case of 2D, the source correlation length is assigned to be about 5 *D* since the Fluent assumes that the sound sources are perfectly correlated over the specified correlation length and zero outside. The acoustic pressure signals at the two receiver locations that were detected. Also, the spectral of the receiver signals are plotted using the Fast Fourier Transform (FFT). Then, the sound pressure level (dB) versus the frequency (Hz) is calculated as shown in Figure 3(e) and (f). Due to the turbulence behavior, different size of eddies generated, but the small-scale eddies are not captured within this number of cells. The results prove that the noise generated is a broadband as detected by receivers 1 and 2. The pressure is plotted at the different zone level downstream and revealed asymmetric behavior. This indicates that a higher frequency creates a more rapid variation in air pressure and results in a higher pitch. The pressure distribution around the cylinder was computed for both cases and validated with analytical as well as experimental data [16] and showed good agreement as shown in Figure 3(c). The pressure coefficient, $C_p = 1 - 4\sin^2\theta$, with respect to different azimuthal angles around the cylinder ($0^o \leq \theta \leq 180^o$) depicts the symmetrical pattern of the flow. However, in the real-world problem the viscosity affects the streamline of the flow and initiates the separation which causes flow instability based on the flow velocity.

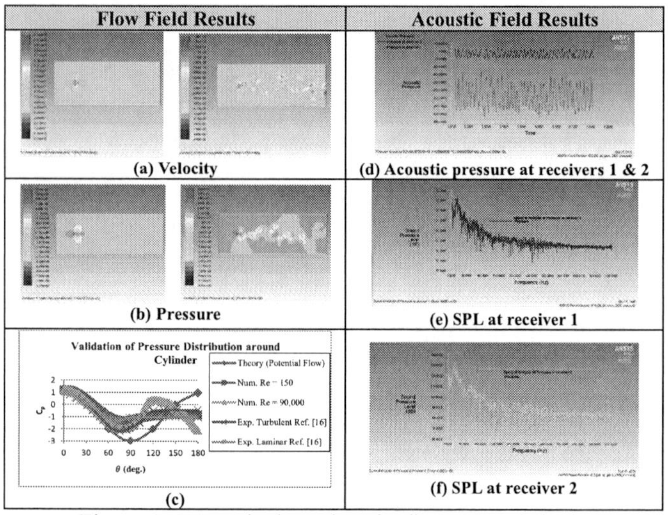

Figure 3: Numerical results of a flow around single circular cylinder with re = 90,000

CONCLUSIONS

In order to understand the physics of both flow and acoustic fields of landing gear systems, 2D flow past a bluff body (circular cylinder) was simulated using ANSYS Fluent. Two distinct Re numbers 150 and 90,000 were examined. The time dependency consideration affects the extraction of near-field flow data as well as the far-field noise prediction. The FW-H solver in Fluent does not include the contribution of the quadrupole sources (volume integration). Thus, type and location of the surface data are crucial in capturing the far-field noise level. Further investigations are needed to adjust for all source terms. An efficient algorithm/numerical technique in terms of computational cost and memory management are highly required to tackle such complex case like a LG.

ACKNOWLEDGMENTS

The author is grateful to Prof. Behdinan for supporting and supervising this research. The computing facility provided by his lab (University of Toronto) is gratefully acknowledged. I acknowledge Saudi Arabia government (Taibah University) for granting a scholarship.

REFERENCES

[1] Morgan, H. G. & Hardin, J. C. 1975, "Airframe Noise: The Next Aircraft Noise Barrier", *Journal of Aircraft*, 12 (7), pp. 622–624.

[2] Argüelles, P., Bischoff, M., Busquin, P., Droste, B.A.C., Evans, S.R., Kröll, W., Lagardere, J.L., Lina, A., Lumsden, J., Ranque, D. and Rasmussen, S. 2001, "European Aeronautics: A vision for 2020", *Advisory Council for Aeronautics Research in Europe*, Report.

[3] Flightpath, A.C.A.R.E., 2011, "2050-Europe's Vision for Aviation", *Advisory Council for Aeronautics Research in Europe.*

[4] FAA 2009 Airport noise and access restrictions, *In FAA Airport Compliance Manual*, 5190.6B.

[5] Lazos, B. 2002, "Surface topology on the wheels of a generic four-wheel landing gear", *AIAA journal*, 40(12), pp. 2402-2411.

[6] Hedges, L. S., Travin, A. K. & Spalart, P. R. 2002, "Detached-Eddy Simulations Over a Simplified Landing Gear", *Journal of Fluids Engineering*, 124 (2), pp. 413-423.

[7] Lighthill, M.J. 1952, "On sound generated aerodynamically: I. General theory", *In Proceedings of the Royal Society of London A: Mathematical, Physical and Engineering Sciences*, 211(1107), pp. 564-587.

[8] Curle, N. 1955, "The influence of solid boundaries upon aerodynamic sound", *In Proceedings of the Royal Society of London A: Mathematical, Physical and Engineering Sciences*, 231(1187), pp. 505-514.

[9] Williams, J.F. and Hawkings, D.L. 1969, "Sound generation by turbulence and surfaces in arbitrary motion", *Philosophical Transactions of the Royal Society of London A: Mathematical, Physical and Engineering Sciences*, 264(1151), pp. 321-342.

[10] Van Mierlo, K.J., Takeda, K. and Peers, E. 2010, "Computational analysis of the effect of bogie inclination angle on landing gear noise", *In 16th AIAA/CEAS Aeroacoustics Conference, AIAA Paper*, pp. 7-9.

[11] Long, S.L., Nie, H. and Xu, X. 2012, "Aeroacoustic study on a simplified nose landing gear", *In Applied Mechanics and Materials*, pp. 18-23.

[12] Lockard, D.P., Khorrami, M.R. and Li, F. 2004, "High resolution calculation of a simplified landing gear", *AIAA paper*, pp. 2887.

[13] Spalart, P.R., Shur, M.L., Strelets, M.K. and Travin, A.K. 2011, "Initial noise predictions for rudimentary landing gear", *Journal of Sound and Vibration*, 330(17), pp. 4180-4195.

[14] Sanders, L., Manoha, E., Ben Khelil, S. and Francois, C. 2011, "LAGOON: CFD/CAA coupling for landing gear noise and comparison with experimental database", *American Institute of Aeronautics and Astronautics.*

[15] Shames, I. H. 1982, *Mechanics of Fluid*, 3rd Ed., Ch.13: pp. 669-675, McGraw-Hill, New York.

[16] Nakayama, Y. and Boucher, R. 1998. *Introduction to fluid mechanics*, Butterworth-Heinemann.

Computational Solid Mechanics

Proceedings of the 26th CANCAM
Victoria, British Columbia, Canada, May 28 – June 1, 2017

EFFECTIVE CONVERGENCE CHECKS FOR FINITE ELEMENT ANALYSIS OF STRESS CONCENTRATIONS

G. B. Sinclair
Department of Mechanical Engineering
Louisiana State University
Baton Rouge, Louisiana, U.S.A.
sinclair@me.lsu.edu

J. R. Beisheim
Development Department
ANSYS Inc.
Canonsburg, Pennsylvania, U.S.A.
jeff.beisheim@ansys.com

ABSTRACT

Control of discretization error in the finite element analysis of stress concentrations is essential for results to be reliably used in practice. Here we offer some convergence checks and error estimates that can provide such control. These checks are applied to seven 2D and twelve 3D stress concentration problems that have exact solutions. A total of 148 stress error estimates are so obtained. Estimates agree with actual errors for 89%, and are somewhat larger than actual errors and hence conservative for the remaining 11%.

KEYWORDS: Stress concentrations, finite elements, verification

1. INTRODUCTION

The accurate determination of stress concentration factors is important in engineering failure and reliability analysis. Finite element analysis (FEA) has become the method of choice in making such determinations when new configurations with unknown stress concentrations are encountered in practice. Inherent in such FEA is discretization error: discretization error has to be controlled for stress concentration results from FEA to be useful. The convergence checks described here seek to provide an effective means of assessing and thereby controlling discretization error.

One example of the potential usefulness of effective convergence checks in the FEA of stresses comes from the gas turbine industry. This example involves the attachment of blades to disks in gas turbines. On occasion, these attachments fail. Large edge-of-contact stresses are seen as a primary underlying reason for such failures, and consequently an accurate determination of these stresses is of value to the industry.

To this end, [1] provides an FEA of a dovetail attachment. The first mesh used in [1] (m = 1) has element extents in the contact region that are 1/50 of the contact length, and is thus a more refined mesh than earlier FEA treatments. The second mesh (m = 2) has element extents that are 1/100. The normalized maximum contact pressures, $\bar{\sigma}_{\max}$, resulting from these two meshes are $\bar{\sigma}_{\max}$ = 8.26, 8.63 for m = 1,2, respectively. With these results if one were to adopt the common practice of estimating error by taking the quotient of the change in stress divided by its best value, one would have

$$\hat{\varepsilon}_2 = 4.3\% \quad (1)$$

as an estimate of the percentage error for the stress for m = 2. In fact, in [1], four further refined meshes are used, ultimately resulting in $\bar{\sigma}_{\max}$ = 12.0 for m = 6. This value is analytically confirmed in [2] which has $\bar{\sigma}_{\max}$ = 12.1 and hence an error of but 0.8% in the FEA result for m = 6. This analytical determination also has that the actual error in the FEA stress for m = 2, ε_2, is

$$\varepsilon_2 = 28.7\%. \quad (2)$$

Clearly the simple approach leading to the error estimate in (1) is seriously awry. The intent here then is to furnish convergence checks and companion error estimates that avoid such erroneous underestimation of errors.

The approach adopted to this end is an outgrowth of developments in computational fluid dynamics by de Vahl Davis and Roache: an excellent account of this research is given in Roache [3]. In applying this work to stress analysis in Sinclair et al. [4], we sought to promote conservative estimation of errors in stresses.

In what follows, we first summarize the convergence checks and error estimates of [4] in Sect. 2, then describe their application to 2D and 3D stress concentrations in Sects. 3 and 4, respectively. The test problems treated in Sects. 3 and 4 have exact solutions, thereby affording an accurate assessment of the effectiveness of the convergence checks and error estimates. We close with some concluding remarks in Sect. 5.

2. CONVERGENCE CHECKS AND ERROR ESTIMATES

The mesh refinement used for the convergence checks takes any of the linear element sizes in mesh m, h_m, to follow from the corresponding size in the preceding mesh by being reduced by a constant scale factor λ. That is

$$h_m = h_{m\text{-}1}/\lambda, \quad (3)$$

for all $m \geq 2$. Complete compliance with (3) is readily achieved with meshes that are structured in their entirety. In practice, local structured meshes containing the stress riser of concern are often used with global unstructured meshes. For such meshes, compliance with (3) is readily achieved for their structured portions. For their global meshes, we just require that the number of elements increase in accord with $N_m = \lambda^2 N_{m-1}$ and $N_m = \lambda^3 N_{m-1}$ for 2D and 3D analysis respectively, where N_m is the number of global elements in mesh m. Thus (3) is complied with on average in the global portion. Here we take $\lambda = 2$. This is primarily because this choice is readily implemented. However, adaptation of what follows for $\lambda \neq 2$ is straightforward.

For a given configuration, when FEA is performed on a sequence of meshes so refined, we denote the stress of interest determined by the FEA on mesh m by σ_m, and the stress increment attending mesh refinement from m-1 to m by $\Delta\sigma_m = \sigma_m - \sigma_{m-1}$. With monotonic convergence, discretization errors in FEA are of the same sign and decreasing in magnitude with mesh refinement. To parallel such convergence, we require the stress increments to do likewise. That is

$$\Delta\sigma_{m-1}\,\Delta\sigma_m > 0,\ |\Delta\sigma_{m-1}| > |\Delta\sigma_m|. \quad (4)$$

We do not attempt to estimate errors until stress increments comply with the *convergence checks* of (4).

When (4) holds for $\lambda = 2$, we obtain an estimate the effective convergence rate attending mesh refinement from m-1 to m, $\hat{c}_m$, from

$$\hat{c}_m = \frac{\ln\left(\Delta\sigma_{m-1}/\Delta\sigma_m\right)}{\ln 2}. \quad (5)$$

As a result of complying with (4), $\hat{c}_m$ of (5) is real and positive. Then a companion absolute percentage *error estimate*, $\hat{\varepsilon}_m$, is given by

$$\hat{\varepsilon}_m = \frac{|\Delta\sigma_m|}{|\sigma_m|\left(2^{\hat{c}_m} - 1\right)} \mathrm{x}100. \quad (6)$$

Derivations of (5) and (6) are furnished in [4].

The error estimate in (6) is being increased to try and compensate for slow convergence if $\hat{c}_m < 1$. Hence to promote conservative error estimation we use (6) to obtain an error estimate whenever $\hat{c}_m < 1$. Conversely, when $\hat{c}_m > 1$ the estimate in (6) is being reduced. While we would like to take advantage of this faster convergence, we need to be careful if we are to promote conservative error estimation. In [4], it is shown that error estimates are conservative whenever actual effective convergence rates are increasing. To reflect this situation, we proceed as follows. We use (6) to obtain error estimates if one of the following situations applies:

$$\begin{aligned}&(\mathrm{i})\,\hat{c}_m \le 1,\\ &(\mathrm{ii})\,\hat{c}_m > 1, m \ge 4, \hat{c}_m \ge \hat{c}_{m-1},\\ &(\mathrm{iii})\,\hat{c}_m > 1, m = 3, \hat{c}_4 \ge \hat{c}_3.\end{aligned} \quad (7)$$

Otherwise we simply estimate percentage errors with $\hat{\varepsilon}_m = \left(|\Delta\sigma_m| / |\sigma_m|\right) \mathrm{x}100$. This is tantamount to setting $\hat{c}_m = 1$ in (6) on occasion when $\hat{c}_m > 1$. With the convergence checks of (4) and the error estimate of (5), (6) and (7), the erroneous estimate of (1) for the dovetail attachment mentioned in the Introduction is removed and reasonable error estimates obtained (see [4] for specifics).

With a percentage error estimate so obtained, we judge the FEA to have converged on mesh m to one of the three following levels of accuracy:

$$\begin{aligned}&1 < \hat{\varepsilon}_m \le 5\% \text{ - satisfactory},\\ &1/5 < \hat{\varepsilon}_m \le 1\% \text{ - good},\\ &0 \le \hat{\varepsilon}_m \le 1/5\% \text{ - excellent}.\end{aligned} \quad (8)$$

We have found assigning these error ranges to these three accuracy levels to be reasonable for peak stresses in practice, but certainly other ranges could be used. Consistent with (8), we view errors that are larger than 5% to be unsatisfactory.

3. PERFORMANCE ON 2D TEST PROBLEMS

For 2D test problems, we have an elastic plate with hyperbolic notches on opposite edges under a pair of tensile forces, F (Fig. 1). As notch acuity is increased, increasing stress concentrations at notch roots can be produced to challenge FEA. Here we consider four notch acuities that result in normalized maximum stresses, $\bar{\sigma}_{\max}$, at the notch root ranging in value from 2.7 to 90.

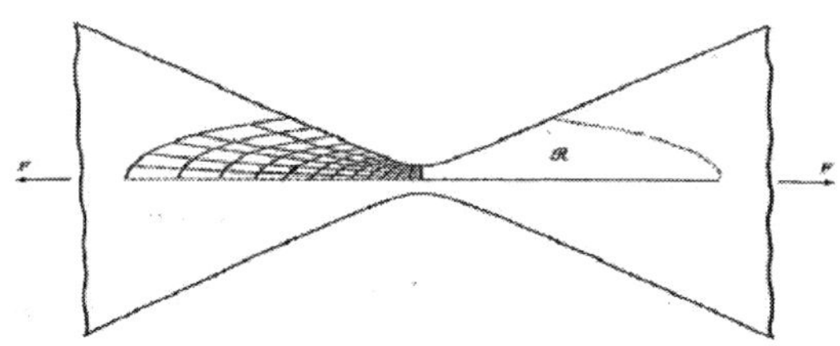

Figure 1. Notched plate under tension: initial finite element mesh

For an infinite elastic plate, a complete exact solution for this notch configuration is derived in Neuber [5]. By drawing on this solution, tractions can be imposed at a finite station to realize problems amenable to FEA (see [4] for details).

For the FEA of these four test problems, throughout we start with coarse meshes then successively refine these meshes by halving element extents. Our initial mesh sequence starts with a coarse mesh (m = 1) of four-node quadrilateral (4Q) elements. This coarse mesh takes advantage of the symmetry of the test problem configuration to confine attention to a quadrant of the elastic plate involved, $\mathscr{R}$ (Fig. 1). It is structured by having nodes placed at equal increments in elliptical cylindrical coordinates and is comprised of 72 elements (Fig. 1, mirror image). For the lower $\bar{\sigma}_{\max}$, seven further meshes (m = 2-8) are produced by halving the increments in elliptical cylindrical coordinates. For the higher $\bar{\sigma}_{\max}$, one further mesh (m = 9) is run so as to be more likely to realize satisfactory stress values. However, the resolution of the mesh sequences employed is not a key issue here: it is every bit as important for our convergence checks to inform finite element engineers when their FEA results in unsatisfactory stress results as when it results in satisfactory or better.

For the first two test problems with lower stress concentrations, we run 4Q elements with free meshes. With some care in implementation, these free meshes locally comply with (3) and have the same numbers of elements as their structured counterparts. To assess the performance of the convergence checks with higher-order elements, a parallel sequence of meshes is run with eight-node quadrilateral (8Q) elements. All told we run 94 meshes on the four 2D test problems (see [4] for details).

Table 1 displays an example of the results so obtained. The stresses in Table 1 are from structured 8Q meshes. The convergence checks of (4) are complied with throughout the mesh sequence, enabling the calculation of estimated effective convergence rates, $\hat{c}_m$ of (5). These $\hat{c}_m$ decrease on meshes m = 3-5, thus (ii), (iii) of (7) applies and $\hat{c}_m$ is reduced to 1 in estimating errors, $\hat{\varepsilon}_m$, with (6). Error estimates are then conservative relative to actual true errors, ε_m: without (ii), (iii) of (7), error estimates would be nonconservative for m = 3 and 4. Even so, error estimates classified in accord with (8) are at the same levels as true errors for three meshes (m = 4-6), one level worse for the other mesh (m = 3).

Table 1. Convergence checks and error estimates for $\overline{\sigma}_{\max} = 2.7310$

m	σ_m	$\Delta\sigma_m$	$\hat{c}_m$	$\hat{\varepsilon}_m$	ε_m
1	2.5135				
2	2.6714	0.1579			
3	2.7130	0.0416	1.92	1.50	0.66
4	2.7254	0.0124	1.75	0.45	0.21
5	2.7294	0.0040	1.63	0.15	0.06
6	2.7306	0.0012	1.74	0.02	0.02

Further results for the present hyperbolic notch problems are given in [4], which also reports results for test problems for three plates with elliptical holes. An additional 35 finite element meshes are run on these elliptical hole problems. For all seven 2D test problems, 82 error estimates are obtained. In terms of the levels of (8), estimated errors are at the same level as actual errors on 74 occasions, one level worse, hence conservative, for the other 8.

These error estimates do not require c_m to have attained their asymptotic theoretical values (here 1 for 4Q elements, 2 for 8Q). For 51% of the error estimates found for the test problems, c_m are not within 10% of their theoretical values yet error assessment is largely accurate and otherwise conservative. For 35%, c_m are not within 20% of their theoretical values yet error assessment is effective.

Error estimation using the convergence checks of (4) and the companion error estimates of (5), (6) and (7) is uniformly conservative for all 2D test problems. That is,

$$\hat{\varepsilon}_m \geq \varepsilon_m \tag{9}$$

for all 82 error estimates obtained. This would not be the case without the convergence checks of (4): absent (4) there are 23 instances of nonconservative error estimates. It would also not be the case without the adoption of (7): absent (7) there are 24 further instances of nonconservative error estimates.

4. PERFORMANCE ON 3D TEST PROBLEMS

For 3D test problems, we have a solid elastic ellipsoid weakened by a hyperbolic notch encircling its equator and simultaneously being stretched by forces F and twisted by torques T (Fig. 2). Again as notch acuity is increased, increasing stress concentrations at notch roots can be produced to challenge FEA. Here we consider four notch acuities with three normalized peak stresses for each for a total of twelve 3D problems (see [6] for details).

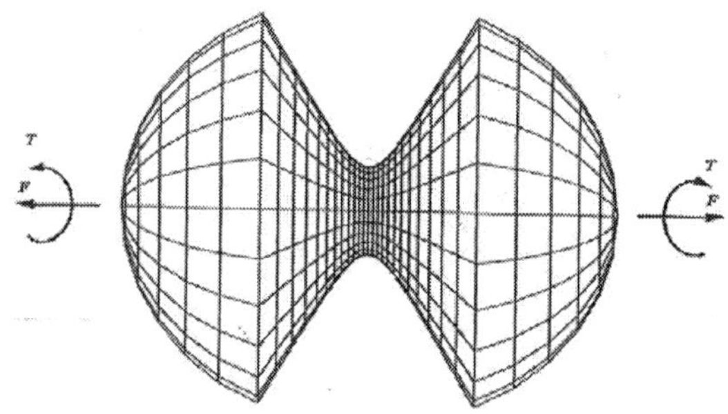

Figure 2. Notched ellipsoid under tension and torsion: initial finite element mesh

For an infinite elastic solid, complete exact solutions for both F and T are derived in Neuber [5]. Again by drawing on these solutions, tractions can be imposed at a finite station to realize problems amenable to FEA (see [6] for details).

For the FEA for these twelve test problems, we continue to start with coarse meshes, then successively refine meshes by halving element extents. Our initial mesh sequence starts with a coarse mesh (m = 1) of eight-node hexahedral (8H) elements. It is structured by having nodes placed at equal increments in oblate spheroidal coordinates and is comprised of 3456 elements (Fig. 2). Four further meshes (m = 2-5) are produced by halving increments in these coordinates. To assess the performance of the convergence checks with higher-order elements, a parallel sequence of meshes is run with twenty-node hexahedral (20H) elements. All told we run 40 meshes on the four 3D test configurations (see [6] for details).

Table 2 displays an example of the results so obtained. The stresses in Table 2 are from structured 8H meshes. The convergence checks of (4) are complied with throughout the mesh sequence, enabling the calculation of estimated effective convergence rates, $\hat{c}_m$ of (5). These $\hat{c}_m$ are less than 1, reflecting the slow initial rates of convergence for this demanding problem. Such $\hat{c}_m$ significantly increase error estimates $\hat{\varepsilon}_m$ from (6). While these $\hat{\varepsilon}_m$ then markedly overestimate true errors ε_m initially, this is not a significant drawback since both are at unacceptable error levels. For $m = 5$, the convergence rate has improved and both $\hat{\varepsilon}_m$ and ε_m attain a satisfactory level of accuracy. Absent $\hat{c}_m < 1$ in (6), a nonconservative error estimate would result for $m = 3$.

Table 2. Convergence checks and error estimates for $\bar{\sigma}_{\max} = 7.2302$

m	σ_m	$\Delta\sigma_m$	$\hat{c}_m$	$\hat{\varepsilon}_m$	ε_m
1	3.7583				
2	5.0663	1.3080			
3	6.0348	0.9685	0.43	46	17
4	6.6114	0.5766	0.75	13	8.6
5	6.9163	0.3049	0.92	4.9	4.3

Further results for the present notched ellipsoidal problems are given in [6]. For all twelve 3D test problems, 66 error estimates are obtained. In terms of the levels of (8), estimated errors are at the same level as actual errors on 57 occasions, one level worse, hence conservative, for the other 9.

Again these error estimates do not require c_m to have attained their asymptotic theoretical values (here 1 for 8H elements, 2 for 20H). For 51% of the error estimates c_m are not within 10% of their theoretical values, for 31% not within 20%, yet error assessment is effective.

Error estimation using the convergence checks of (4) and the companion error estimates of (5), (6) and (7) is uniformly conservative for all 3D test problems. That is, (9) holds for all 66 error estimates obtained. This would not be the case without the convergence checks of (4): absent (4) there are 12 instances of nonconservative error estimates. It would also not be the case without the adoption of (7): absent (7) there are 17 further instances of nonconservative error estimates.

5. CONCLUDING REMARKS

Here we offer some convergence checks and a companion a posteriori error estimate that enable engineers to control discretization errors when using finite elements to analyze stress concentrations. The checks employ a sequence of at least three finite element meshes, with meshes being refined by reducing element extents by a constant factor. In applying these convergence checks, we have specifically chosen to start with a coarse mesh and refine it by halving element extents. Alternatively, one could start with a baseline mesh that is judged to be of sufficient resolution, then successively coarsen this mesh.

The specifics of the convergence checks are given in Sect. 2. Essentially they are an adaptation to stress analysis of checks developed in computational fluid dynamics (see [3], [4] for details). By considering changes in effective convergence rates, the present checks identify circumstances under which accompanying error estimates can be expected to be conservative. This attribute occurs even when effective convergence rates have not attained asymptotic theoretical values. Implementation of these checks once finite element stresses have been obtained is straightforward.

To gauge the effectiveness of the checks we apply them to seven 2D and twelve 3D test problems that have exact values for the peak stresses involved. These peak stresses span a range of values that is greater than that normally encountered in practice. They are subjected to FEA with first order (4Q, 8H) and second order (8Q, 20H) elements. A total of 169 finite element meshes are run on the test problems. The convergence checks then provide 148 error estimates. When these error assessments are used to classify accuracy levels as being unsatisfactory, satisfactory, good, or excellent in accordance with (8) of Sect. 2, errors are at the same levels as true errors for 131 of these assessments, one level worse than true errors for the other 17. Thus error estimation that is typically reasonably correct (89%), and otherwise somewhat conservative (11%).

REFERENCES

1. Sinclair, G.B., Cormier, N.G., Griffin, J.H., and Meda, G., 2002, "Contact stresses in dovetail attachments: finite element modeling," *J. Eng. Gas Turb. Power,* **124**, pp. 182-189.
2. Sinclair, G.B., 2013, "Edge-of-contact stresses in blade attachments in gas turbines," *Proc. Sur. Effects Contact Mech. XI*, WIT Press, Wessex, England, pp. 87-108.
3. Roache, P.J., 2009, *Fundamentals of Verification and Validation*, Hermosa Pub., Socorro, New Mexico.
4. Sinclair, G.B., Beisheim, J.R., and Roache, P.J., 2016, "Effective convergence checks for verifying finite element stresses at two-dimensional stress concentrations," *J. Verif. Valid. Uncert. Quant.*, **1**, pp. 041003-1-8.
5. Neuber, H., 1946, *Theory of Notch Stresses*, Edwards Inc., Ann Arbor, Michigan.
6. Beisheim, J.R. and Sinclair, G.B., 2017, "Effective convergence checks for verifying finite element stresses at three-dimensional stress concentrations," Report No. ME-MS1-17, Dept. of Mech. Eng., LSU.

TUNED TEST PROBLEMS FOR VERIFYING FINITE ELEMENT ANALYSIS OF STRESS CONCENTRATIONS

B. Zhang
Department of Mechanical Engineering
Louisiana State University
Baton Rouge, Louisiana, U.S.A.
bzhan13@lsu.edu

Y. Zhang
Department of Mechanical Engineering
Louisiana State University
Baton Rouge, Louisiana, U.S.A.
yzhan84@tigers.lsu.edu

G.B. Sinclair
Department of Mechanical Engineering
Louisiana State University
Baton Rouge, Louisiana, U.S.A.
sinclair@me.lsu.edu

ABSTRACT

When stress concentration factors are not available in handbooks, finite element analysis has become the preeminent method for determining their values. For such determinations, there is a need to know if they have sufficient accuracy. Tuned Test Problems (TTP) can provide a way of assessing the accuracy of stress concentration factors found with finite elements. Here we offer a means of constructing such TTP for stress concentrations at features that have a local constant radius of curvature. These problems are tuned to their originating applications by sharing exactly the same global geometries and having slightly higher peak stresses. They also have exact solutions, thereby enabling a precise determination of the errors incurred in their finite element analysis. Demonstrations of the construction of such TTP are provided for U notches and shoulder fillets.

KEYWORDS: Stress concentrations, finite elements, verification

1. INTRODUCTION

Quite often in practice configurations are encountered that do not have their stress concentration factors quantified in handbooks. For example, the single U notch in an elastic plate under a tensile force F of Fig. 1: when the line of action of F bisects the net section width, stress concentration factors are given in Peterson [1]; when, instead, it bisects the entire width of the plate as in Fig. 1, there are no factors in [1]. Today, finite element analysis (FEA) is the most likely recourse for stress analysts seeking to determine the peak stresses occurring in configurations like that of Fig. 1.

For such FEA to be reliably used in practice, discretization error must be controlled. Test problems with known exact elasticity solutions can afford a means of assessing and thus controlling discretization error. While there are a significant number of exact elasticity solutions available in the literature, seldom do they conform to the geometry of interest. For example, the single U notch of Fig. 1 does not have any exact solutions reported in [1] or the references therein (the limited stress concentration factors given in [1] are from photoelastic studies).

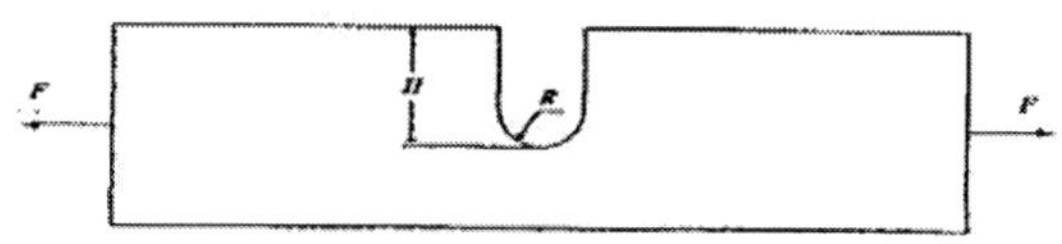

Figure 1. Plate with U notch under tension

Building on the approach in [2], here we furnish a way of constructing problems with exact solutions for geometries like that of Fig. 1, as well as procedures for tuning these problems so that the stress concentrations present challenge their FEA at least as much as the initiating configuration. The specifics of such a construction process are described next in Sect. 2. Thereafter it is demonstrated for a 2D U notch configuration in Sect. 3, and a 3D shoulder fillet configuration in Sect. 4. We close with some concluding remarks in Sect. 5.

2. CONSTRUCTION OF TUNED TEST PROBLEMS

In local (r,θ) coordinates, the stress fields used to construct TTP are:

$$\begin{aligned}\sigma_r &= \alpha\left(1-\frac{R^2}{r^2}\right)-\beta\left(1+3\frac{R^4}{r^4}-4\frac{R^2}{r^2}\right)\cos 2\theta+\frac{\gamma R}{r}\left(1-\frac{R^2}{r^2}\right)\cos\theta,\\ \sigma_\theta &= \alpha\left(1+\frac{R^2}{r^2}\right)+\beta\left(1+3\frac{R^4}{r^4}\right)\cos 2\theta+\frac{\gamma R}{r}\left(1+\frac{R^2}{r^2}\right)\cos\theta, \quad (1)\\ \tau_{r\theta} &= \beta\left(1-3\frac{R^4}{r^4}+2\frac{R^2}{r^2}\right)\sin 2\theta+\frac{\gamma R}{r}\left(1-\frac{R^2}{r^2}\right)\sin\theta,\end{aligned}$$

wherein R is the local radius of curvature at the stress concentration, and α, β and γ are constant stresses. In (1), the α terms are from Lamé's stress field for thick wall pressure vessels ([3], p. 71), the β terms are from Kirsch's field for stress at a hole ([3], p. 91), and the γ terms are from an adaptation of Williams' stress field for singularity asymptotics ([3], p. 142). The fields in (1) satisfy the stress equations of equilibrium and compatibility for 2D elasticity. In addition, they satisfy stress-free conditions on $r = R$; that is

$$\sigma_r = \tau_{r\theta} = 0, \tag{2}$$

on $r = R$.

With the fields of (1) we have, in effect, four adjustable parameters to use to construct TTP: the three constants therein as well as the location of $\theta = 0$ where the hoop stress, σ_θ, is a maximum. To set these parameters for a given application, we first perform an FEA of the application, then choose the location of $\theta = 0$ so that it matches the maximum stress location found with the FEA. Thereafter we choose α, β and γ so that the $\sigma_\theta, \partial\sigma_\theta/\partial r$ and $\partial^2\sigma_\theta/\partial\theta^2$ at $r = R$, $\theta = 0$ exceed the corresponding values found with the FEA by about 10%. With this choice the intent is to provide a TTP that is somewhat more challenging to analyze with finite elements than the originating application, but not severely more difficult.

To complete the construction of a TTP, we superimpose the fields of (1), now with α, β and γ set, throughout the underlying application's geometry. This results in stresses complying with (2) on that part of the application with $r = R$. Elsewhere we simply take displacement boundary conditions to apply with the displacements so prescribed being whatever the superposition realizes thereon. To this end, we use the displacements attending (1) that have:

$$\begin{aligned} u_r &= \frac{1}{4\mu}\left[\alpha\left((\kappa-1)r+\frac{2R^2}{r}\right)-2\beta\left(r+(\kappa+1)\frac{R^2}{r}-\frac{R^4}{r^3}\right)\cos 2\theta \right. \\ &\left. +\gamma R\left(\left((\kappa-1)\ln\frac{r}{R}-1+\frac{R^2}{r^2}\right)\cos\theta+(\kappa+1)\theta\sin\theta\right)\right], \\ u_\theta &= \frac{1}{4\mu}\left[2\beta\left(r+(\kappa-1)\frac{R^2}{r}+\frac{R^4}{r^3}\right)\sin 2\theta \right. \\ &\left. -\gamma R\left(\left((\kappa-1)\ln\frac{r}{R}+1-\frac{R^2}{r^2}\right)\sin\theta-(\kappa+1)\theta\cos\theta\right)\right], \end{aligned} \tag{3}$$

wherein μ is the shear modulus and $\kappa = 3-4\nu, (3-\nu)/(1+\nu)$ for plain strain, plane stress, respectively, where ν is Poisson's ratio. Taken together, (1) and (3) comply with the stress-displacement relations of 2D elasticity. In prescribing displacements using (3), displacements have to be resolved into the components appropriate for whatever coordinate system is being used to specify the geometry of the application away from $r = R$.

Thus we have a test problem for a given application. It is a "test" problem because it has a known solution, namely the stresses in (1) for the chosen values of α, β and γ in general, and the attendant peak stress in particular. It is a test "problem" because we know the governing field equations applicable in the interior of the application's geometry as well as the boundary conditions on its exterior. Furthermore, as a consequence of sharing exactly the same geometry as the application, this test problem can be analyzed with exactly the same finite element meshes, thereby affording an assessment of the accuracy of these meshes.

3. APPLICATION: U NOTCH UNDER TENSION

The specific U notch configurations treated all have an overall plate aspect ratio of 4:1 and a notch depth, H, penetrating 60% of the plate width (Fig. 1). In addition, they have reducing notch root radii, R, such that

$$R/H = 1/3, 1/12, 1/48, 1/196. \tag{4}$$

These reductions tend to approach a doubling of stress concentration factors. Such geometries are loaded with the line of action of F through the center of the plate (as in Fig. 1) to generate new results, and with this line of action through the center of the net section to regenerate old results for comparison with [1]. All told, 8 U notch configurations are analyzed.

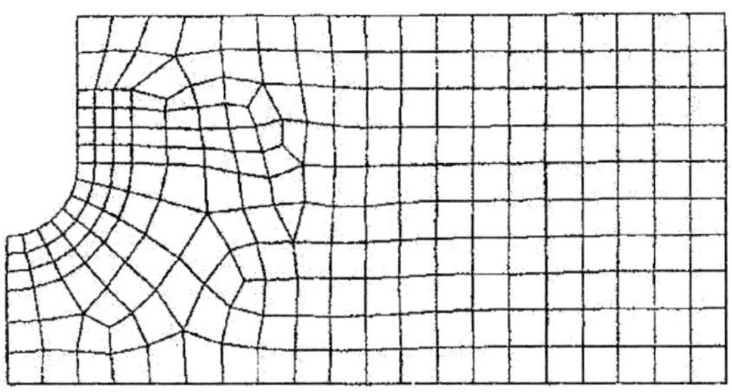

Figure 2. Initial finite element mesh

Taking advantage of the symmetry present, the initial coarse mesh of four-node quadrilateral (4Q) elements has an inner structured mesh, an outer free mesh, and a total of 228 elements (Fig. 2). Mesh refinement proceeds by taking

$$\begin{aligned} h_m &= h_{m-1}/2 \text{...structured mesh}, \\ N_m &= 4N_{m-1} \text{...free mesh}, \end{aligned} \tag{5}$$

where h_m is an element extent in structured mesh m and N_m is the number of elements in free mesh m. Five further meshes of 4Q elements are so produced (m = 2-6). To enable comparison with a second order element, a parallel sequence of five meshes (m = 1-5) of eight-node quadrilateral (8Q) elements is also employed. In sum, 88 finite element meshes are used.

For R/H = 1/3 when the plate is loaded as in Fig. 1, we obtain results for the maximum hoop stress, $\sigma_\theta^{\max}$, that are consistent to four figures from the last two meshes with 8Q elements. Specifically we find

$$\sigma_m = 19.97, \tag{6}$$

for m = 5, 6 with 8Q elements, where $\sigma_m = \sigma_\theta^{\max}/\sigma$, $\sigma = F/A$ and A is the total cross-sectional area of the plate. Increasing this peak value by about 10%, we take

$$\sigma_e = 21.94\dot{0}, \tag{7}$$

where $\sigma_e = \sigma_\theta^{\max}/\sigma$, and $\sigma_\theta^{\max}$ is now the exact value for the TTP. This occurs on $\theta = 0$ which is aligned with the symmetry axis in this application. We also increase the slope present in the FEA by about 8% in the TTP, and the curvature by about 10%. These choices have the following exact values for the constant stresses in (1), (3): α = -2.55σ, $\beta = 1.26\sigma$, and $\gamma = 11.00\sigma$. As illustrated in Fig. 3, the local stress distributions for the FEA of the original application and the exact solution for the TTP are quite similar, with the TTP having both a somewhat higher maximum value as well as higher stress gradients as this value is approached (indicated by the slightly inclined lighter lines as $r/R \to 1$ in Fig. 3(a), and the greater reduction of the stresses for the TTP in Fig. 3(b)).

We can now use this TTP to assess the error present in its FEA using exactly the same meshes as for the original application. Results for the 4Q mesh sequence are set out in Table 1. Therein the exact percentage error, ε_m, and its corresponding effective convergence rate, c_m, are defined by

$$\varepsilon_m = \frac{\sigma_e - \sigma_m}{\sigma_e} \text{ x } 100(\%), \qquad c_m = \frac{\ln(\varepsilon_{m-1}/\varepsilon_m)}{\ln 2}. \tag{8}$$

Table 1. Errors and convergence rates with 4Q elements when σ_e = 21.94

m	σ_m	$\varepsilon_m(\%)$	c_m
1	18.89	13.9	
2	20.25	7.70	0.85
3	21.04	4.10	0.91
4	21.47	2.14	0.94
5	21.70	1.09	0.97
6	21.82	0.55	1.00

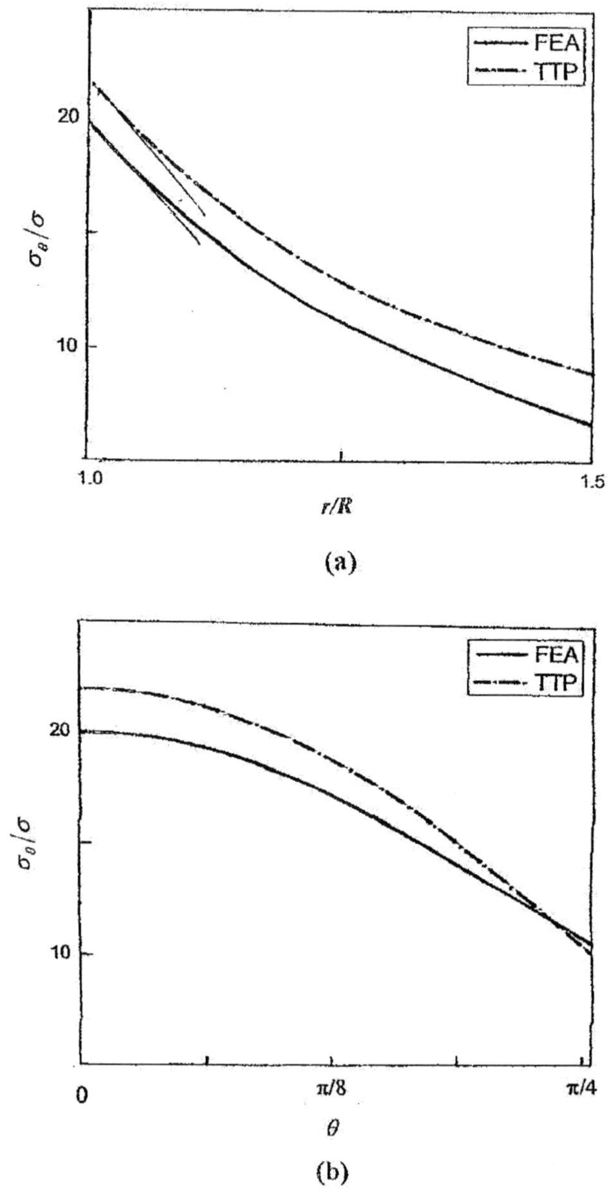

Figure 3. Hoop stress comparison: (a) radial direction, (b) theta direction

Apparent in Table 1 is that the FEA is monotonically converging and doing so with effective convergence rates that approach then become their asymptotic theoretical value of 1 for the present problem with 4Q elements. Viewing errors that are less than 5% as satisfactory and less than 1% as good, this FEA has converged to a satisfactory level with the third mesh (m = 3), a good level with the sixth (m = 6).

Results for the 8Q mesh sequence are set out in Table 2. Again the FEA is monotonically converging, but now with c_m approaching their asymptotic value of 2 for the present problem with 8Q elements. As a result of this more rapid convergence, satisfactory accuracy is achieved on $m = 1$, good accuracy on $m = 2$, and excellent accuracy $(\varepsilon_m \le 1/5\%)$ on $m = 3$.

Table 2. Errors and convergence rates with 8Q elements when $\sigma_e = 21.940$

m	σ_m	$\varepsilon_m(\%)$	c_m
1	21.554	1.759	
2	21.822	0.538	1.71
3	21.904	0.164	1.71
4	21.930	0.046	1.85

For loading as in Fig. 1, peak stresses do approach successively doubling as *R*/*H* decreases in accord with (4). As a result, companion TTP tend to identify somewhat more refined meshes for a given level of accuracy. For example, for *R*/*H* = 1/196, the smallest value in (4), $m = 4$ and 6 for satisfactory and good levels of accuracy with 4Q elements, and $m = 1$, 2 and 4 for satisfactory, good and excellent levels with 8Q elements (see [4] for details). For loading as in [1], TTP continue to identify meshes with excellent levels of accuracy. When these meshes are employed on their originating applications, they furnish peak stresses that are in agreement with corresponding experimental results reported in [1].

4. APPLICATION: SHOULDER FILLET UNDER BENDING

For a second application, we consider the stress concentration occurring in a shaft with a shoulder fillet under a bending moment *M* (Fig. 4). Some stress concentration factors for this configuration are given in [1] (based on photoelastic studies). On occasion, however, fillet radii, *R*, are encountered in practice that are smaller than the radii included [1].

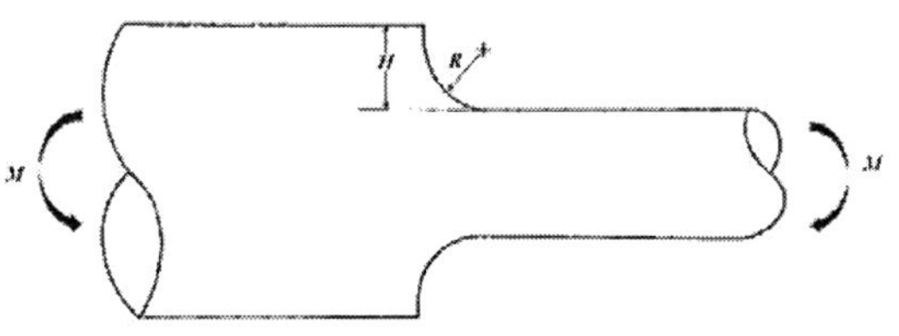

Figure 4. Shaft with shoulder fillet under bending

The specific shoulder fillet configurations treated all have an overall length to outer diameter ratio of 1.87:1, and a shoulder depth *H* that reduces the outer diameter by 4.8%. For the largest shoulder fillet radius treated, these choices enable direct comparison with experimental values reported in [1]. In addition to this radius, we treat two further radii by successively dividing by a factor of four. Thus we have

$$R/H = 10/11, 10/44, 10/176. \tag{9}$$

As previously, such reductions tend to approach a doubling of stress concentration factors.

Taking advantage of the antisymmetry present to confine discretization to the upper half of the shaft, Fig. 5 shows an initial finite element mesh for the central plane with the smallest shoulder fillet radius, $R/H = 10/176$. This mesh has an inner structured mesh and an outer free mesh. The extent of the inner structured mesh is an order of magnitude greater than *R*. The expectation therefore is that stress gradients become minor near its boundaries, thereby enabling rapid transition to a far coarser outer mesh with consequent computational savings. Even with this coarse meshing away from the stress concentrator, for this 3D problem the initial mesh has 21,000 elements.

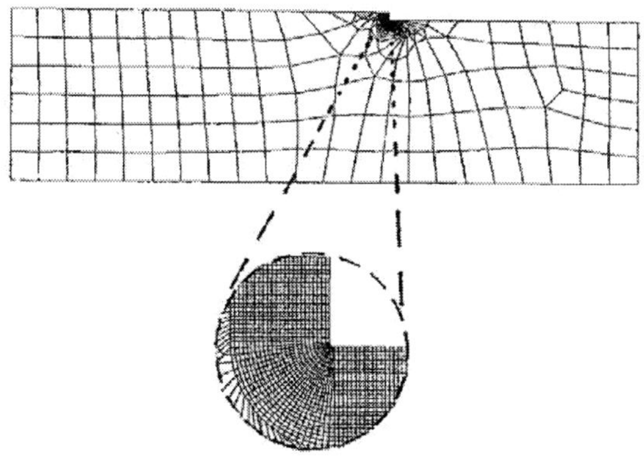

Figure 5. Initial finite element mesh

Structured meshes are refined in accordance with the first of (5) (i.e., halving h_m). Such refinement results in an eightfold increase in elements since the problem is 3D, and accordingly the second of (5) is replaced with $N_m = 8N_{m-1}$. Consequently only a sequence of three 3D meshes ($m = 1$-3) is possible with our current computational capability. These are run with eight-node hexahedral (8H) elements, and twenty-node hexahedral (20H) elements.

For a companion TTP it is probably reasonable to confine attention to just the central plane because it has the greatest stress and stress gradients within plane, with no stress gradient out of plane because of symmetry. Then we can use the fields of (1), (3). For this now 2D analysis, we run six meshes of 4Q and 8Q elements refined in accordance with (5), starting with the mesh of Fig. 5. We do this to facilitate the construction of the TTP.

To begin to tune our test problem, we first need to determine the location of the maximum stress. We find the same location is identified on $m = 5,6$ with 8Q elements to 5 significant figures. This results in us setting the line $\theta = 0$ at an angle of 25.66° to the vertical. These same meshes also both have $\sigma_m = 5.40$, where σ_m continues to be $\sigma_\theta^{\max}/\sigma$ but now σ is the maximum nominal stress in the shaft's smaller diameter. Accordingly we increase this by about 10% and take $\sigma_e = 6.\dot{0}$. We also increase the gradient and curvature for m = 6 by about 10%. These choices have the following exact values for the constant stresses in (1), (3): $\alpha = 2.6\sigma, \beta = 0.6\sigma, \gamma = -0.8\sigma$. This results in local stress distributions from FEA and the TTP that look similar to that in Fig. 3(a) for radial variations, and similar to that in Fig. 3(b) with a reflection about $\theta = 0$ for theta variations.

Results for this TTP for 8Q elements are set out in Table 3. These are converging with c_m approaching and even slightly exceeding its asymptotic value of 2. Satisfactory accuracy is achieved on m = 3, good on m = 4, and excellent on m = 5. This supports the expectation that results for the application with our third 3D mesh and 20H elements should at least be satisfactory.

Table 3. Errors and convergence rates with 8Q elements when $\sigma_e = 6.0000$

m	σ_m	$\varepsilon_m(\%)$	c_m
1	5.1279	14.5	
2	5.6300	6.17	1.24
3	5.8744	2.09	1.56
4	5.9633	0.61	1.77
5	5.9910	0.15	2.03

For the other larger radii of (9), TTP results for m = 3 with 8Q elements are good when R/H = 10/44, excellent when R/H = 10/11 (see [4] for details). Accordingly results for the 3D application with 20 H elements can be expected to be at corresponding levels. The result for R/H = 11/44 agrees with that obtained experimentally that is reported in [1].

5. CONCLUDING REMARKS

Here we have provided a means of constructing Tuned Test Problems for application configurations that have a stress concentration occurring within a traction free boundary that has a constant local radius of curvature. These problems are "test" in that they have exact closed-form solutions. These problems are "tuned" in that they share exactly the same global geometry as the originating application configuration, as well as in that they have slightly higher peak stresses and stress gradients. As a result they can be analyzed with exactly the same finite element meshes as used on the application and provide an assessment of the discretization error present with these meshes. This in turn can enable finite element engineers to select appropriate meshes for such application configurations.

The approach is demonstrated on 2D U notch and 3D shoulder fillet configurations. The TTP constructed for these applications identify elements and meshes that have satisfactory, good, and excellent levels of accuracy. Whenever available, these levels of accuracy are confirmed by experimental results reported in [1].

REFERENCES

1. Pilkey, W.D., and Pilkey, D.F., 2008, *Peterson's Stress Concentration Factors*, John Wiley, Hoboken, New Jersey.

2. Sinclair, G.B., Anaya-Dufresne, M., Meda, G., and Okajima, M., 1997, "Tuned test problems for numerical methods in engineering," *I. J. Num. Meth. Eng.*, **40**, pp. 4193-4209.

3. Timoshenko, S.P., and Goodier, J.N., 1970, *Theory of Elasticity*, McGraw-Hill, New York.

4. Zhang, B., Zhang, Y., and Sinclair, G.B., 2017, "Verification of finite element analysis for the stress concentrations occurring at U-shaped notches and shoulder fillets using tuned test problems," Report No. ME-MS2-17, Dept. of Mech. Eng., LSU.

FATIGUE ANALYSIS OF THE LONGITUDINALLY WELDED LARGE-DIAMETER OIL PIPELINES

Evgeny Anisimov
Department of Mechanical and Manufacturing Engineering
University of Calgary
Calgary, Alberta, Canada
evgeny.anisimov@ucalgary.ca

Meera Singh
Department of Mechanical and Manufacturing Engineering
University of Calgary
Calgary, Alberta, Canada
meera.singh@ucalgary.ca

Les Sudak
Department of Mechanical and Manufacturing Engineering
University of Calgary
Calgary, Alberta, Canada
lsudak@ucalgary.ca

ABSTRACT

In the energy industry, there is an increasing demand for liquid pipelines with higher transmission capabilities. This leads to the implementation of pipelines with larger diameters and increased diameter to thickness ratios. Aside from the availability of suitable manufacturing and testing techniques to satisfy this need, there is an issue with clarity within available standard design codes related to assessment of structural stresses due to weld discontinuity, which is the major stress raiser in large diameter pipelines. This study highlights the challenges associated with modeling of the weld, its effects on calculation of stresses using finite element method (FEM) and fatigue life assessment.

KEYWORDS: Pipeline, Weld, Stress Concentration, Fatigue Analysis, ABAQUS

INTRODUCTION

Larger pipes normally have outside diameters (O.D.) of 711-2174 mm, wall-thicknesses (w.t.) of 7.1-52 mm, and come in standard lengths of 6-24 m [1] [2]. In Canada, such pipes usually buried 0.6-1.2 m below the ground [3]. They are normally manufactured by cold forming flat plate using the UOE process. That is, plate is first formed into a U-shape, and then pressed into an O-shape between two semicircular dies, and finally expanded using an internal mandrel to improve its roundness. The longitudinal seam is welded using the double submerged arc welding (DSAW) process.

Pipelines, particularly liquid lines, are subjected to repeated thermal and pressure loads in service. This cyclic loading can result in fatigue failures at loads much lower than those observed in static failures. Geometric discontinuities within the loaded pipe result in a magnification of the local stress. These stress raisers further promote fatigue failure as they act as perfect sites for cracks to initiate. There are several such stress raisers in large diameter pipe that arise from the UOE manufacturing process. Specifically, stress raisers arise due to peaking and radial misalignment, both shown in Figure 1. Furthermore, the DSAW welding process used to create the longitudinal seam of the potentially misaligned pipe introduces a large stress concentration due to the weld profile, shown in Figure 2. In order to predict the fatigue life and therefore size large diameter pipe, all stress raisers must be accounted for.

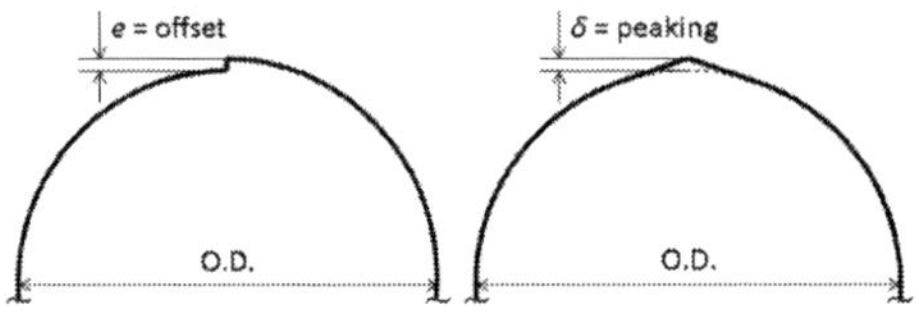

Figure 1: Misalignments of a welded pipe

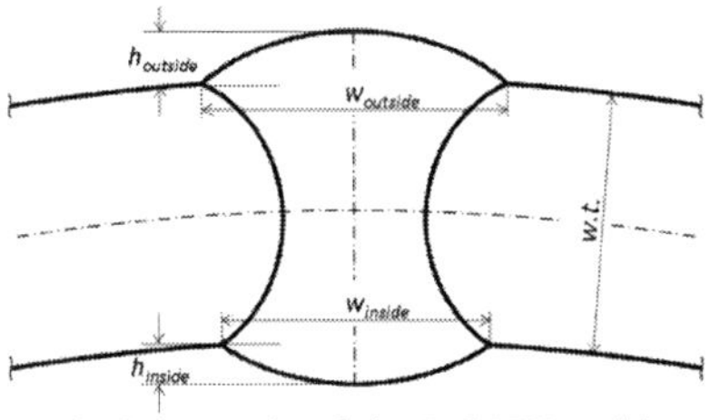

Figure 2: Schematic of the DSAW weld profile

In Canada, pipelines are generally designed with reference to codes such as: API 5L [1], CSA Z245.1 [2], CSA Z662-15 [3], BS 7608 [4]. Although the codes are critical in assuring safe operation of pipelines, the application of the code requirements often is not straightforward. In performing stress analysis and fatigue life predictions of large diameter pipes, engineers generally attempt to follow the codes and incorporate conservatism in their design to avoid failure. However, this often results in a design wall thickness (w.t.) higher than is likely required and is accompanied by a huge cost increase to the pipeline.

The global objective of this project is to provide a systematic fatigue life assessment methodology that can be used to analyze and design pipelines that will be safe from a fatigue perspective and also cost efficient. In this work, some difficulties associated with interpreting aspects of the codes, modeling misalignments associated with the UOE process, and modeling the weld seam profile are highlighted. The work discusses these difficulties in the context of a stress analysis performed on a standard large diameter pipe using the finite element method (FEM) and includes the effects of manufacturing tolerances and typical mechanical loading. The FEM results are used to make simple fatigue life predictions. The results are used to guide best practice in large diameter pipe design.

PIPE GEOMETRY, MATERIAL PROPERTIES, AND LOADS CONSIDERED

Standard 914 mm (36 in) O.D. pipe of three different wall thicknesses were chosen for analysis. The pipeline geometries considered in this study are shown in Table 1.

Table 1: Pipeline geometries used in this study

O.D., mm	914	914	914
w.t., mm	17	20	22

It is assumed that the pipe is buried 1000 mm below the ground. To account for the external pressure exerted on the pipe by the soil, a Mohr-Coulomb plasticity model was employed for the soil. The properties of the API X56 [1] pipe and the soil are shown in Table 2.

Table 2: Properties of materials used in this study

Elasticity			Mohr-Coulomb Plasticity			
ρ, g/cm^3	E, MPa	Yield Stress, MPa	Friction angle, °	Dilation angle, °	Cohesion YS, MPa	Abs plastic strain
			Pipe			
7.8	207000	390-545	-	-	-	-
			Soil			
1.5	10	0.3	0	0	0.03	0

For this initial stress analysis, the pipe was loaded by the external soil and the internal operating pressure of 9.93 MPa (1440 psi).

FEM MODEL

Initially, two geometric orders of plane strain elements were tested for convergence using a simplified two-dimensional pipe model free of stress raisers. The results of the study suggested that 8-node biquadratic plane strain quadrilateral, reduced integration (CPE8R) elements [5] provided accurate results with less computational effort and were therefore chosen for further analysis.

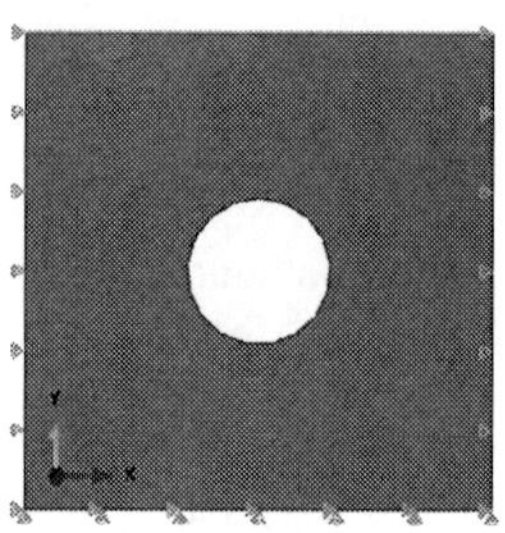

Figure 3: Boundary conditions for the soil box

The actual model consisted of a cross-sectional ring of a pipe contained within a soil box (Figure 3). The sides and bottom of the soil box were constrained in the horizontal direction, and the bottom was also constrained in the vertical direction.

Radial and peaking misalignments were modelled based on maximum values tolerable as stated in Canadian and American standards [1]- [3] (Table 3). The actual values used in the five FEM's analyzed (variants A-F) are tabulated in Table 4. Although both codes suggest the same values which are used in the models, two models (variant A and B) were generated to evaluate the effect of peaking and offset on the stress field.

Table 3: Acceptable misalignments

Offset e, mm	Peaking δ, mm	Reference
$0.1w.t.$ to a max of 1.5	3.2	[1] [2]
1.6	-	[3]

Table 4: Geometrical parameters used in this study

Variant	O.D., mm	w.t., mm	Offset e, mm	Peaking δ, mm	Weld Toe Radius r, mm
A	914	17	-	-	5
B	914	17	1.0	2.0	5
C	914	17	1.5	3.2	5
D	914	20	1.5	3.2	5
E	914	22	1.5	3.2	5
F	914	22	1.5	3.2	1

In modeling the DSAW weld seam, it is critical to model the weld profile as closely representing the real weld as possible. Inaccuracies in modeling the geometry of a weld toe, a stress riser, leads to inaccurate stress calculations. In the British Standard [4], there is a brief discussion of the importance of the weld width to the w.t. ratio $\frac{w}{w.t.}$, and it suggests ratio to use in analysis of 0.5. However, this may not properly reflect the real situation with the large pipes since the relationship between w and w.t. may not be linear. Although there is no information regarding the weld width in [1]- [3], there is more information regarding the weld bead height (see Figure 2). The weld reinforcements tolerable by standards for the w.t. used in this study are summarized in Table 5.

Table 5: Weld Bead Dimensions

Width w, mm	$h_{outside}$, mm	h_{inside}, mm	Reference
-	4.5	3.5	[1]
-	4	3.5	[2]
-	4	-	[3]
0.5 w.t	-	-	[4]

Additional information needed for the FEM modeling is shown in Figure 4 and is not present in the discussed codes. Geometry of a pipe with a misaligned weld was modeled as shown in Figure 4(a) by arrow 1.

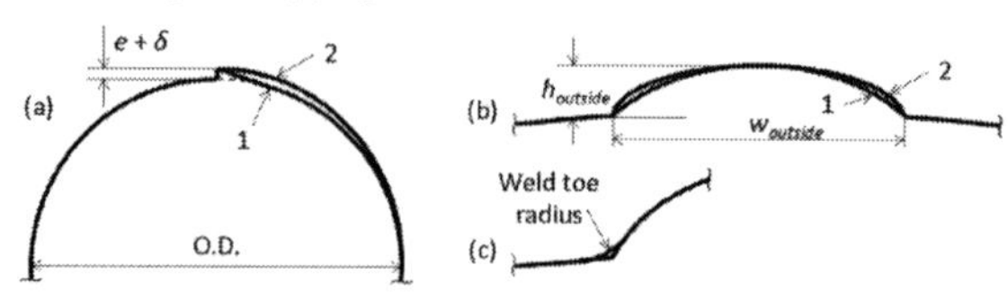

Figure 4: Additional information needed for the FEM model, (a) misaligned geometry, (b) weld profile, (c) weld root radius

Abating edges of the pipe wall were shifted outward and kept tangential to the true round shape of pipe. The geometrical center of weld was placed between the abating edges and rotated to allow for the weld toes to coincide with the plate surfaces.

The DSAW profiles of 30 different welds on pipes with O.D. of 846-1422 mm and w.t. of 8.74-22 mm were taken from the available literature such as [6] [7] and the traces of inside and outside weld beads were obtained for further analysis of the weld profile shown in Figure 4(b). The lengths of weld connections from toe to toe, the weld width, w, (Figure 2), and the height of the weld beads, h, were measured on each side of the weld. The DSAW process generates weld profiles with similar dimensions, shown in Table 6, for a number of pipe wall thicknesses.

Table 6: Weld Bead Dimensions Obtained In This Study

Weld Bead	Weld Width w, mm	Weld Height h, mm
Outside	18.98 ±4.71	2.02 ±0.40
Inside	17.29 ±4.58	1.84 ±0.44

The traces were normalized to the same arbitrary width, shown in Figure 5, as they show similar variations in geometrical appearance of the outside and inside weld beads.

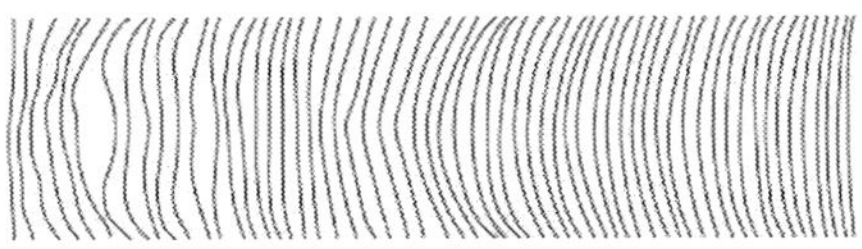

Figure 5: Traces of the DSAW profiles

The polynomial of 4th order was obtained for the weld height (y) as a function of the weld width (x), Eq. (1), with $R^2 = 0.999$ based on 16 sections of 60 traces.

$$y = -9 \cdot 10^{-5}x^4 + 0.0035x^3 - 0.0629x^2 + 0.5671x \quad (1)$$

The sought generalized weld profile is shown in Figure 6.

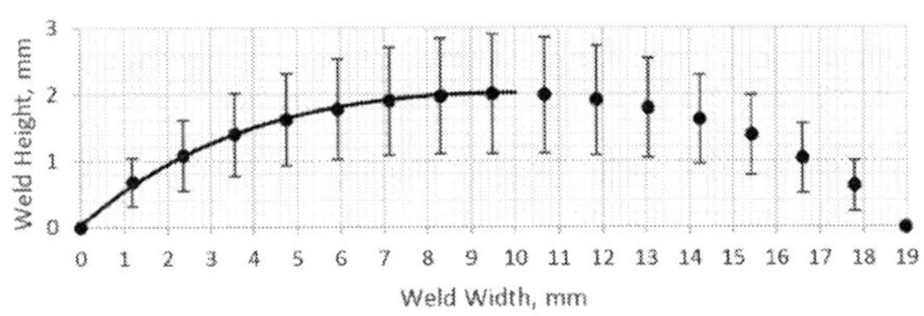

Figure 6: Weld profile (points) experimental and (line) approximated

Although not mentioned in [1]- [3], the British standard [4], specifies that the weld root radius, r, should be taken as: $r > 0.25 \times w.t.$. In this work, a 5 mm weld root radius is used to model the weld profile, Figure 4(c). Additionally, one model with a sharper radius (variant F in Table 4), was analyzed to evaluate the effect of the radius on the stress field.

The partitioning of a weld region was done in a way that enables further refinement of a mesh at four weld toes for a more accurate approximation of the stress in that critical regions. Mesh size of 0.01 was chosen for the weld toes based on a convergence study.

The standard practice for calculating the stresses at critical locations, hot-spots, makes use of a quadratic surface stress extrapolation (SSE) from the points 4 mm, 8 mm, and 12 mm away from the hot-spot, and is performed by using Eq. (2) [4].

$$\sigma_H = 3\sigma_{4mm} - 3\sigma_{8mm} + 8\sigma_{12mm} \quad (2)$$

This approach was implemented in this study to calculate the SSE Hoop stresses at the weld toe.

FEM RESULTS

The SSE hoop stress due to internal pressure of 9.93 MPa and soil pressure at the depth of 1000 mm below the ground was calculated first for pipe with w.t. of 17 mm that did not contain a weld misalignment (variant A in Table 4). The local hoop stress distribution is shown in Figure 7 and indicates that the maximum hoop stress occurs at the inner weld toe (red color), where the SSE hoop stress was calculated to be 301.2 MPa.

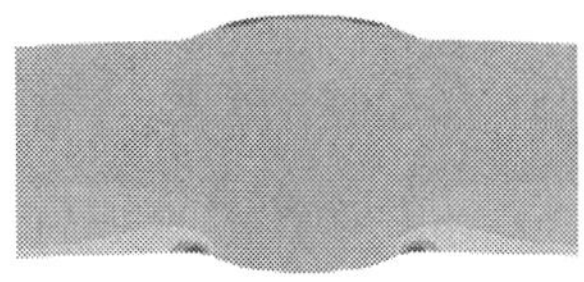

Figure 7: Hoop Stress due to internal pressure and soil

Weld misalignment applied to the pipe with w.t. of 17 mm was considered in a separate model (variant B in Table 4). The resulting local stress distribution is shown in Figure 8. The weld region experiences secondary bending due to the misalignment [4], which causes the sharp 44% increase in the SSE hoop stress to 431.6 MPa at the inner weld toe.

Figure 8: Hoop Stress due to misalignment B

Results of the analysis whereby the misalignments were increased to their maximum values (variant C) are shown in Figure 9. As can be seen, the SSE hoop stress increased to 500.8 MPa, which is 66% larger than that observed in a pipe without weld.

Figure 9: Hoop Stress due to maximum misalignment C

Similar models were analyzed for the other two w.t. with misalignment values shown in Table 4. The SSE hoop stresses for the three w.t. are summarized in Table 7, showing the exponential decrease with w.t. It should be noted that as expected, there is an increase in the SSE hoop stress (of 12 %)

when a sharper weld toe radius considered (compare E to F) for a given wall thickness. Furthermore, using the sharper weld root radius raises the SSE stress above that obtained for a thinner pipe (variant D) with a larger radius. This indicates a need for the standardization of weld root radii to be used in analysis. These stresses will be further used in fatigue analysis.

Table 7: The hoop stress at the weld toe due to misalignment

Variant	A	B	C	D	E	F
Hoop Stress, MPa	301.2	431.6	500.8	381.4	361.1	407.1

FATIGUE ANALYSIS

In this study authors use 50 years' service life with two different constant pressure cycles to simplify the actual loading, one full pressure range each two days (upset) with $n_U = 50 \times \frac{365}{2} = 9125$ cycles to simulate a full pressure drop in the line and 75% to 100% of the full pressure range each day (batching) with $n_B = 50 \times 365 = 18250$ cycles to account for a steady service condition.

The type of weld corresponding to longitudinal DSAW seam weld of a large diameter pipe is a double-sided full penetration butt weld between co-planar plates [4]. In [5], the number of cycles to failure, N, due to cyclic loading for each weld design class be simply done by using Eq. (3).

$$logN = logC_0 - d \cdot SD - m \cdot logS_r \tag{3}$$

where N is a number of cycles to failure, C_0 is a constant, SD is a standard deviation for $logN$, d is a factor for a nominal probability of failure (2.3 %), m is a slope of S-N curve, and S_r is applied stress range. In case of the DSAW connection, the number of cycles N to failure at a stress range S_r for a basic S-N curve is given by Eq. (4).

$$N = 10^{log(3.988\times10^{12})-2\cdot0.2095-3\cdot\log(S_r)} \tag{4}$$

Finally, Miner's rule was applied to assess the damage accumulation D during the intended service life. Therefore, considering the total number of cycles to failure, N_U and N_B, obtained from Eq. (4) using the stress range S_r for each of the two types of the pressure cycles, n_U and n_B, the Miner's damage rule is given by Eq. (5) [4].

$$D = \frac{n_U}{N_U} + \frac{n_B}{N_B} \tag{5}$$

The results are summarized in Table 8.

Table 8: Damage accumulation due to Hoop stress

Variant	w.t., mm	Hoop Stress, MPa	N_U cycles	N_B cycles	D
A	17	301.2	55615	3559346	0.17
B	17	431.6	18902	1209737	0.50
C	17	500.8	12099	774359	0.78
D	20	381.4	27391	1753045	0.34
E	22	361.1	32275	2065631	0.29
F	22	407.1	22524	1441557	0.42

The results of the simple fatigue analysis performed indicate that any one of the pipe sizes or misalignments considered would be safe for a 50 year service life. It should be noted, however, that the method of fatigue analysis employed assumes that the material behaves elastically during the service life.

Although for the most part, the SSE stress does remain elastic, the actual local stress at the weld toe generally exceeds the elastic limit of the material. The fatigue analysis method employed may yield erroneous results. This indicates that further investigation into a more appropriate fatigue analysis method, whether it be strain based, or fracture mechanics based, may support more realistic results.

CONCLUSIONS

The results of the static analysis on various pipe geometries highlight the need for a standardized method of modelling aspects of the pipe geometry such as the misalignment, weld profile, and weld root radius. Differences in the method employed in modelling these parameters has an effect on the stress field and therefore the ultimate selection of the pipe wall thickness. Additionally, the fatigue analysis results suggest a need careful consideration into the life assessment method employed, instead of a basic stress based method, as used in this analysis.

ACKNOWLEDGMENTS

The authors would like to acknowledge the critical contributions made to this work by Stress Engineering Services Canada.

REFERENCES

1. API 5L:2012: Specification for Line Pipe, American Petroleum Institute, 2012.
2. CSA Z245.1-14: Steel pipe, Toronto: CSA Group, 2014.
3. CSA Z662-15: Oil and gas pipeline systems, Toronto: CSA Group, 2016.
4. BS 7608:2014+A1:2015: Guide to fatigue design and assessment of steel products, BSI, 2014.
5. "Abaqus Analysis User's Manual Volume IV: Elements," Dassault Systèmes, Providence, 2012.
6. X. Chen, H. Lu, G. Chen and X. Wang, "A comparison between fracture toughness at different locations of longitudinal submerged arc welded and spiral submerged arc welded joints of API X80 pipeline steels," *Engineering Fracture Mechanics,* vol. 148, no. 8, pp. 110-121, 2015.
7. R. Li, X. Zuo, Y. Hu, Z. Wang and D. Hu, "Microstructure and properties of pipeline steel with a ferrite/martensite dual-phase microstructure," *Materials Characterization,* vol. 62, no. 8, p. 801–806, 2011.

SHEAR FEASIBILITY: NEW THEORETICAL UNDERSTANDING REGARDING THE GROWTH OF NANOVOID

Yi Cui
Department of Mechanical Engineering
University of Alberta
Edmonton, Alberta, Canada
cui7@ualberta.ca

Zengtao Chen
Department of Mechanical Engineering
University of Alberta
Edmonton, Alberta, Canada
zengtao.chen@ualberta.ca

ABSTRACT

Recently, Bulatov et al. [V. V. Bulatov, W. G. Wolfer, M. Kumar, Scr. Mater. 63 (2010) 144] asserted the impossibility of void growth via shear loop by commenting on a series of work including both the theoretical one [V.A. Lubarda, M.S. Scheider, D.H. Kalantar, B.A. Remington, M.A. Meyers, Acta Materialia 52 (2004) 1397] and the atomistic simulations [S. Traiviratana, E.M. Bringa, D.J. Benson, M.A. Meyers, Acta Materialia 56 (2008) 3874]. Our key argument is that the possible surface cut to form a shear dislocation loop is not necessary its loop plane. Further, the integral to evaluate local mass transport is not invariant with respect to arbitrary surface cut. We clarify the feasibility of nanovoid growth via shear loop from viewpoints of both continuum mechanics and molecular dynamics simulations.

KEYWORDS: Void growth, Shear dislocation, Mass transport

INTRODUCTION

The shear dislocation loop, by definition, has its Burgers vector inside the loop plane [1]. In contrast, Burgers vector of a prismatic dislocation loop is not in the loop plane [1]. The mass transport associated with a shear dislocation loop near a void is a nontrivial issue. Understanding of this issue can be achieved from two perspectives. First, in order to create the dislocation loop [2], the amount of mass that needs to be removed from or inserted into the bulk material can be examined. Dislocation loops/curves are prone to form and emit from void surface under loading. The creation of dislocation by removing or inserting material also introduces some displacement to the surrounding medium. This displacement causes another part of mass transport. By using a point force as a test probe, this displacement can be determined through the work done on certain surface bounding the dislocation loop [2]. For a shear loop, the second part of mass transport can be found in the work of Ohr [3]. This deformation induced by the shear loop, while is negligible for a microvoid, could be influential for a nanovoid.

TWO PARTS OF VOID GROWTH

We start with a brief review of the first kind of mass transport in the literature. Bulatov et al. [4] has employed the mass conservation to formulate the mass transport, based on the mass removed from or inserted into the bulk material to create a dislocation loop. However, what might have been overlooked is the definition of this bounding surface A with mass transport on it (as shown in Fig.1). Although Fig.1 is originally generated to represent a general dislocation loop [2], what showed in the figure is closer to, if not exactly a shear loop since its Burgers vector **b** appears in the loop plane. Fig. 1 comes from the same reference [2] cited by Bulatov et al. [4]. To assert the "shear impossibility", they formulated the following equation [4]

$$\Delta V = \int_{Surface} \mathbf{b} \cdot \mathbf{n} \mathrm{d}A \quad (1)$$

where **n** is the unit normal of the infinitesimal area $\mathrm{d}A$ of an arbitrary, curved surface bounding the shear loop, ΔV is the mass transported. However, the original equation for material removed [2] is expressed in terms of a differential volume element

$$\delta V = \mathbf{b} \cdot \mathbf{n} \mathrm{d}A \quad (2)$$

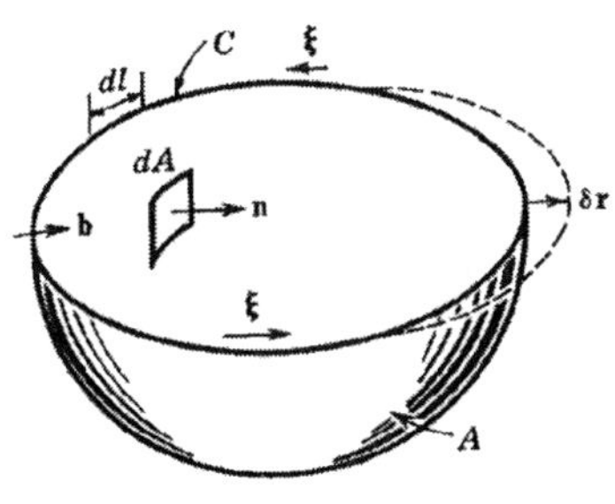

Figure 1: Closed dislocation loop C and its bounding surface A unaltered from [2]. The close loop is denoted as C, the normal of surface A is n, the segment vector of loop is **d**l=ξdl and the motion of dislocation loop is δ**r.**

Fig. 1 is the schematic corresponding to Eq. (2) [2]. As defined in [2], "any surface A bounded by loop C could be used for the operation". Indeed, Fig. 4-2 in the text book [2] clearly shows that a pure edge dislocation could be generated by different possible cuts. This idea can be easily extended to a shear loop. One side of the surface cut A is displaced by **b** relatively and the material on both side of the cut are rejoined afterwards [2, 5]. The inner product **b·n** therefore highly depends on the shape of surface A. In the comments of Bulatov et al. [4], the flat area encircled by the loop C was chosen as this surface A. It is only under this premise that their asserted condition **b·n**=0 [4] is valid and so is the "shear impossibility". Clearly, Burgers vector **b** and **n** are not perpendicular to each other everywhere at least in Fig. 1. As a result, Eq. (2) should be nonzero on most part of surface *A*. This indicates that Eq. (1) cannot be used to claim "shear impossibility". Further, the integral form in Eq. (1) is not suitable to evaluate the local mass transport. Even if Eq. (1) does equal zero when symmetry in Fig. 1 is employed, it only implies that the removed and inserted material volumes are the same. However, local mass transport is still true for specific locations on the surface as described by Eq. (2). Indeed, the integral to evaluate local mass transport should be operated on a small surface element. Neither Stokes integral theorem nor Gauss integral theorem is applicable on either the small surface element or the whole cutting surface. The reason is that the Burgers vector is neither a curl nor a continuously differentiable vector field defined inside a neighborhood volume. Therefore, these integrals are not invariant with respect to arbitrary surface cut. As long as the surface A is not presumed as the flat one encircled by loop C, there would be local mass transport. To determine the exact shape of the surface *A*, other information is required. Such information could, for example, come from molecular dynamics simulations. The mere overall mass conservation of Eq. (1), with a presumption of a flat surface *A* appears to be inadequate to disprove the void growth via shear loops/curves.

Next, we distinguish the mass transport caused by the formation of a shear loop and that by the motion of a shear loop. The mass transport caused by the motion of a dislocation loop is given as [2]

$$\delta V = \oint_C \mathbf{b} \cdot (\delta \mathbf{r} \times d\mathbf{l}) \tag{3}$$

Notably, here the integral is operated on the loop curve. If the motion is a pure slip, Eq. (3) is indeed zero for a shear dislocation loop since the three vectors **b**, δ**r** and d**l** are in the same plane. The motion other than a pure slip, as long as the projection area of the loop in the direction of Burgers vector is nonzero, should result in mass transport [2].

Second to be considered is the elastic void growth caused by the presence of a shear loop. This part is true even if the bounding surface *A* is taken as the flat one encircled by the loop as shown in Fig. 1. The presence of a shear loop introduces a strong lattice structure discontinuity to the surrounding material. Although the classical elasticity may fail at the dislocation core, modifications [6] to this approach can still lead to a solution of physical significance. A simple example of displacement field induced by an edge dislocation can be found in [7]. Similarly, but more sophisticatedly, the displacement field due to the presence of a shear loop could be obtained by the Green's function method [3]. Through integration treatment, the analytical solution $\mathbf{u}^{\text{loop}}$ of a shear loop as shown in Fig. 2 can be obtained as [3]

$$u_1^{loop}(x_1,x_2,x_3) = -\frac{b\zeta}{4(1-\nu)}\left[\frac{2(1-\nu)}{|\zeta|}I_0^0 + \frac{(x_1^2-x_2^2)}{\rho^3R^2}I_1^0 - \frac{x_1^2}{\rho^2R^2}I_0^1\right] \tag{4}$$

$$u_2^{loop}(x_1,x_2,x_3) = -\frac{bx_1x_2\zeta}{4(1-\nu)\rho^3R^2}\left(2I_1^0 - \rho I_0^1\right) \tag{5}$$

$$u_3^{loop}(x_1,x_2,x_3) = -\frac{bx_1}{4(1-\nu)\rho R}\left[(1-2\nu)I_1^0 + |\zeta| I_1^1\right] \tag{6}$$

where $\rho = \sqrt{x_1^2 + x_2^2}/R$, $\zeta = x_3/R$, R is the radius of shear loop, b is the magnitude of Burgers vector, υ is Poisson's ratio and the function I_m^n is defined as [3]

$$I_m^n(\rho,|\zeta|) = \int_0^\infty t^n J_1(t) J_m(\rho t)\exp(-|\zeta| t)dt \tag{7}$$

where J_m is the mth order Bessel function of the first kind. One can refer to Figure 2 in Ohr [3] for details. This elastic deformation is negligible for a microvoid since its b/R is extremely small. Nonetheless, for a small nanovoid adjacent to multiple emitted shear loops/curves, the elastic deformation induced by shear loops/curves could be a contributor. Such situation can be found, for instance, in the neutron-irradiated metals. Both nanovoids and dislocations can be formed in the metal due to irradiation. In the work of Shimomura and Mukouda [8], nanovoids of about 3 nm in diameter were observed inside the irradiated copper. In Figure 2 of Ohr [3], the influence of the shear loop extends to a relatively large area compared with its radius. To a rather nearby nanovoid, the void surface would be distorted by the shear loop and therefore develop a growth or shrinkage. Hence, an important focus of this report is to extend the elastic solution [3] to the case with a nearby nanovoid. The solution of the elasticity problem can be obtained by the superposition of the displacement field $\mathbf{u}^{\text{loop}}$ caused by the shear loop and the image field $\mathbf{u}^{\text{void}}$ due to presence of the void. This method has been adopted by Wolfer and Drugan [9] to deal with the interaction between a prismatic loop and a void. The lack of symmetry of the elasticity problem here makes it difficult to pursue any analytical solution. As an alternative, the image field is solved by the finite element method to satisfy the traction-free boundary condition on the void surface. Once $\mathbf{u}^{\text{void}}$ is solved, the total displacement field could be expressed as

$$\mathbf{u} = \mathbf{u}^{loop} + \mathbf{u}^{void} \tag{8}$$

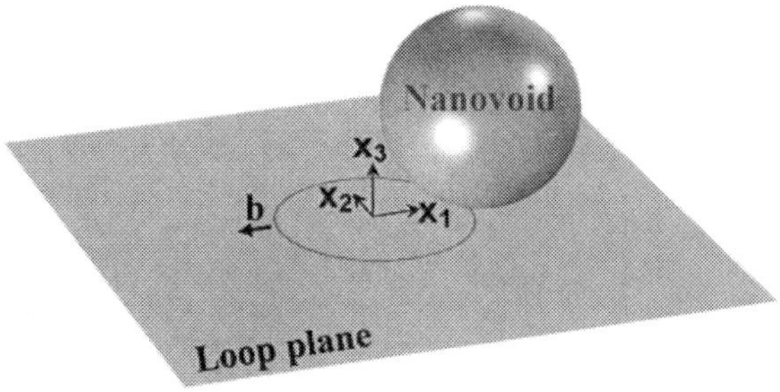

Figure 2: Schematic of the corresponding geometry. The center of the circular shear loop with a radius of R coincides with the coordinate origin. The spherical void of a same radius R is centered at the coordinates (1.1R, 0, 1.05R). The ratio b/R is set as 0.1.

The corresponding strain field can be easily obtained through the displacement field caused by the shear loop. Next, the stress tensor $\boldsymbol{\sigma}^{loop}$ caused by the shear loop can be found through applying Hooke's law. Consequently, the boundary loading condition of the image field can be determined. The combined stress, caused by both the shear loop and the image field due to the presence of the nanovoid, must satisfy the traction-free boundary conditions on the void surface:

$$\boldsymbol{\sigma}^{void} \cdot \mathbf{n} + \boldsymbol{\sigma}^{loop} \cdot \mathbf{n} = 0 \tag{9}$$

where **n** is the normal vector on the void surface. The geometry configuration is shown in Fig. 2. Finite element calculation has been performed by modeling this geometry after a mesh convergence test. The Young's modulus and Poisson's ratio are set to be 120 GPa and 0.33, respectively. For brevity, only the magnitude of total displacement **u** is displayed in Fig. 3.

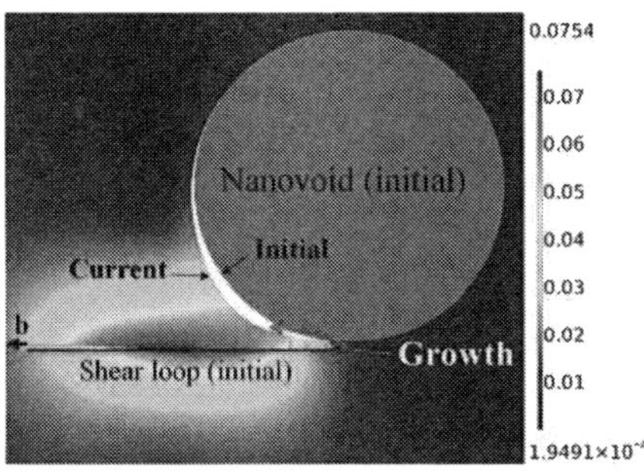

Figure 3: Deformed slice plot of the magnitude of total displacement **u** with the cutting plane at x_2=0. Growth of the nanovoid towards the shear loop can be clearly seen (Deformation has been magnified by a scale factor of 2). The color bar is in the unit of R.

The white crescent shows the difference between the current profile and the initial profile of the nanovoid, which is displayed as a grey circle. It is worth noting that the magnitude of displacement of a single shear loop without a nearby void should be symmetric with respect to the plane x_3=0. Yet, the presence of the nearby void affects the displacement field and breaks this symmetry. The volume increase of nanovoid can be calculated by

$$\Delta V_{void} = \int_{S_{void}} (\mathbf{u} \cdot \mathbf{n}) \mathrm{d}S \tag{10}$$

The integral in Eq. (10) is directly defined on the void surface and, so is the unit normal vector **n**. The volume increase calculated by Eq. (10) is 0.062 in the unit of R^3. Hence, the ratio of increase $\Delta V_{void}/V_{void}$ is found to be 0.015. Since the solution is only for a complete shear loop [3], the modeled shear loop is assumed to be close but not attached to the void surface. Compared with the image field, the displacement field by the shear loop contributes more to the void growth.

CONCLUSIONS

The two parts of possible mass transport induced by a shear loop have now been obtained. Once the shear loop forms, its contribution to the nearby nanovoid growth would consist of these two parts if it moves by pure slip. As mentioned, the bounding surface A can take any arbitrary shape. This shape is crucial to the mass transport. The exact shape of this surface A is out of the scope of a pure theoretical understanding. The first part of mass transport to the void growth is irreversible. The second part depends on the distance between the shear loop and the void surface. Evidence and explanation for void growth via shear dislocation curve/loop can also be found in Refs. [12, 13]. Nevertheless, it should be emphasized that the formed pattern of dislocation structure is subject to the simulation conditions such as loading pattern or lattice orientation.

ACKNOWLEDGMENTS

Financial support for this work has been provided by the Natural Sciences and Engineering Research Council of Canada (NSERC), and is gratefully acknowledged.

REFERENCES

1. Hull, D. and Bacon, D.J. 1984, *Introduction to Dislocations*, Butterworth-Heinemann, London.
2. Hirth, J.P., Lothe J. 1982, *Theory of Dislocations*, second ed.,Krieger, Malabar, FL, pp. 96–97.
3. Ohr, S.M. 1972, "Displacement field around a shear dislocation loop" *Philosophical Magazine*, Vol(26), pp. 1307-1312.
4. Bulatov, V.V., Wolfer W.G., Kumar, M. 2010, "Shear impossibility: comments on "void growth by dislocation emission" and "void growth in metals: Atomistic calculations"" *Scripta Materialia*, Vol(63), pp. 144-147.
5. Lehto, N., Öberg, S. 1998, "Effects of Dislocation Interactions: Application to the Period-Doubled Core of

the 90° Partial in Silicon" *Physical Review Letters,* Vol(80), pp. 5568-5571.

6. Lazar, M. 2013, "The fundamentals of non-singular dislocations in the theory of gradient elasticity: Dislocation loops and straight dislocations". *International Journal of Solids and Structures*, Vol(50), pp. 352-362

7. Mura, T. 1987, *Micromechanics of Defects in Solids, second ed.* Martinus Nijhoff, Dordrecht, The Netherlands, pp. 18-19.

8. Shimomura, Y., Mukouda, I. 2000, "Development of vacancy clusters in neutron-irradiated copper at high temperature", *Journal of Nuclear Materials*, Vol(283-287), pp. 249-254

9. Wolfer, W.G., Drugan, W.J. 1988, "Elastic interaction energy between a prismatic dislocation loop and a spherical cavity", *Philosophical Magazine A*, Vol(57), pp. 923-937.

10. Cui, Y., Chen Z., 2016, "Material transport via the emission of shear loops during void growth: A molecular dynamics study", *Journal of Applied Physics*, Vol(119), pp. 225102.

11. Begau, C., Hua, J., and Hartmaier, A. 2012, "A novel approach to study dislocation density tensors and lattice rotation patterns in atomistic simulations", *Journal of the Mechanics and Physics of Solids*, Vol(60), pp. 711-722.

12. Lubarda, V.A., Scheider, M.S., Kalantar, D.H., Remington, B.A., and Meyers, M.A. 2004, "Void growth by dislocation emission", *Acta Materialia,* Vol(52), pp. 1397-1408.

13. Traiviratana, S., Bringa, E.M., Benson, D.J., Meyers, M.A. 2008, "Void growth in metals: Atomistic calculations", *Acta Materialia*, Vol(56), pp. 3874-3886

ON THE FINITE ELEMENT MODELLING OF OCTET-TRUSS LATTICE STRUCTURES

Mohamed Abdelhamid
Department of Mechanical Engineering
York University
Toronto, Ontario, Canada
mahamid@yorku.ca

Aleksander Czekanski
Department of Mechanical Engineering
York University
Toronto, Ontario, Canada
alex.czekanski@lassonde.yorku.ca

ABSTRACT

Inspiration for cellular structures can be found in natural cellular materials like sponge, trabecular bone and wood. Continuum analytical models of the octet-truss lattice have been previously developed to obtain the stiffness tensor that describes its macroscopic behavior. Finite element modelling using 3D solid and 1D beam/truss elements is performed for different lattice geometries to assess the validity of boundary conditions and the choice of mesh elements.

KEYWORDS: Octet-truss, lattice structure, structural effects, stiffness tensor, finite element analysis

INTRODUCTION

Architectural structures refer to materials where a unit cell is repeated in three orthogonal directions to form a lattice structure. Such hierarchically-designed materials have proven to be an effective load-carrying yet lightweight material, among other applications. Inspiration can be found in natural cellular materials like honeycomb, trabecular bone, butterfly wings and sponge [1]. The mechanical properties of such architected materials on the macroscale depend on three parameters: the constituent material properties, the unit cell configuration and the relative density, $\overline{\rho}$, where $\overline{\rho}$ is commonly defined as the solid volume within the unit cell divided by the whole volume of the unit cell and is usually expressed in decimals or percentage.

Lattice structures can deform by either stretching or bending of their individual members depending on the nodal connectivity of the unit cell. A nodal connectivity of $Z = 6$ is required for a 3D structure to be rigid, while for bend-dominated structures $6 \leq Z < 12$ and for stretch-dominated ones $Z = 12$ [2–5]. Previous studies have proven that the elastic modulus $E_{lattice}$ and strength $\sigma_{lattice}$ of lattice materials are coupled to their relative density $\overline{\rho}$ following Eq. (1) & (2) respectively:

$$E_{lattice} \;\alpha\; \overline{\rho}^{n} \tag{1}$$

$$\sigma_{lattice} \;\alpha\; \overline{\rho}^{m} \tag{2}$$

Stretch-dominated unit cells have better coupling relations between the mechanical properties and the relative density than bend-dominated ones, where the dimensionless parameters n & m are unity for stretch-dominated types and greater than unity for bend-dominated ones. Hence, the unit cell of choice in our work is the octet-truss, which is a stretch-dominated type [3,4]. Metamaterials refer to architected structures with dimensions scaling down to the nano and micro length scales. At such scales, metamaterials' mechanical properties are determined by their structure or geometry rather than their composition [4].

CONTINUUM-BASED MODEL

Continuum models were developed previously for the octet-truss lattice to describe its macroscopic behavior [2,6–10]. Nayfeh and Hefzy (1978) employed 3D coordinate transformation and volume averaging in order to obtain the macroscopic stiffness tensor [11]. Lake (1992) constructed a stiffness tensor by converting applied stresses to strains for each parallel group of members using the macroscopic compliance matrix [6]. Deshpande et al. (2001) performed experimental testing on truss core sandwich plates under tensions, compression, and shear loading conditions. The reported effective properties agreed well with the continuum model and showed a stretch-dominated behavior in most cases. It's worth noting that earlier continuum models considered only the case of cubic symmetry, i.e. the lattice angle θ equals 45°, where θ is defined as the angle between the lattice members and the horizontal midplane [3] as shown in Fig. 1.

Recently, Abdelhamid et al. developed a continuum-based model of the octet-truss lattice in order to obtain the macroscopic stiffness tensor at different lattice angles [12]. Homogenous isotropic constitutive material properties were used along with a constant cross sectional area for all members. The following steps were used to develop the model: a tetrahedron was selected as the least substructure that composes the whole unit cell as shown in blue in Fig. 1 [12], the individual members were transformed from each member's local coordinates to the global coordinates (x, y, z), and lastly homogenization was applied to relate the microscopic stresses and strains to the macroscopic stresses and strains.

Since the octet-truss lattice is of the stretch-dominated type, it's safe to assume that the only contribution to the macroscopic stiffness is from the elastic uniaxial tensile/compressive stiffness of the individual members with no contribution from the bending resistance of the members or the nodes. This is equivalent to a structure of 1D truss elements capable of sustaining only tensile/compressive loads with pin-jointed nodes [3,10,12]. The before-mentioned assumption is essential in defining the proper boundary conditions in the finite element model in the next section.

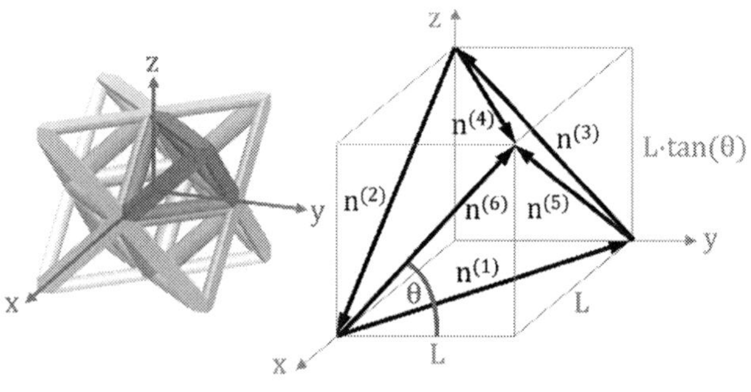

Figure 1: Octet-truss unit cell (left) and the tetrahedron substructure with its six members (right).

In order to understand how the octet-truss lattice behaves under tensile/compressive loads from different loading directions, two consecutive stiffness tensor transformations were performed, where (x, y, z) is the original coordinate system and (x', y', z') and (x'', y'', z'') are the first and second transformation coordinate systems respectively. The loading direction aligns with the x'' axis, which is defined by the tensor transformation angles φ and α as shown in Fig. 2.

In addition, the transformed elastic modulus equation was used to obtain an analytical expression that defines the angles of the highest elastic modulus' loading direction at different lattice angles [13].

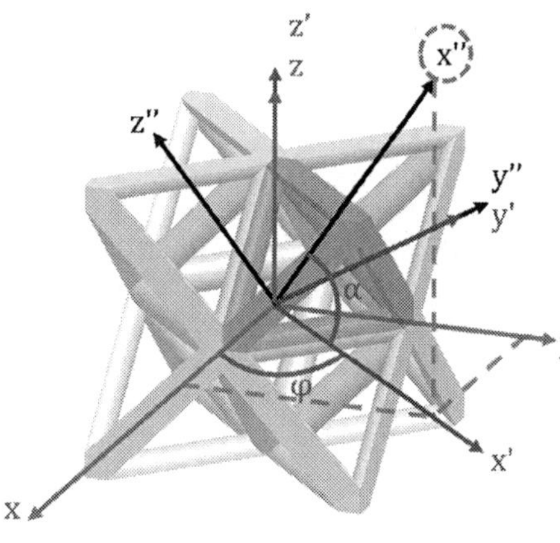

Figure 2: Stiffness tensor transformations, the loading direction is defined by the x'' axis.

FINITE ELEMENT ANALYSIS

In this work, finite element modelling of the octet-truss lattice is performed, and its accuracy is assessed by comparing the resulting elastic moduli values with those obtained using the continuum-based analytical model. The software of choice is HyperWorks, with HyperMesh as the meshing solution and OptiStruct as the solver. Due to the high-symmetry of the octet-truss lattice, only the tetrahedron substructure is considered in the finite element analyses while applying the proper symmetry conditions [2,10–12]. This results in significant computational cost savings in the case of 3D solid elements, unlike for 1D elements where the savings are less significant. The lattice members' cross-sections are modelled as circular solid in the analyses, nonetheless hollow cross-sections of any shape are also valid.

▪ FINITE ELEMENT MODEL

Three different mesh elements were employed: 3D solid, 1D beam, and 1D truss elements. The same parameter matrix varying the lattice angle and the aspect ratio was followed for every mesh element type used. Aspect ratios between 0.02 and 0.12 and lattice angles between 30 and 60 degrees were analyzed. In this context, the aspect ratio is defined as the ratio of the individual lattice member radius to its length $r/\sqrt{2}L$. The base side length of the tetrahedron (L in Fig. 1) is fixed, while the height of the tetrahedron changes with the lattice angle and the diameter of individual member is changed to obtain different aspect ratios.

Firstly, in order to assess the validity of the pin-jointed nodes assumption (i.e. ignoring the bending resistance of the nodes and members), the octet-truss lattice is modelled using 3D solid elements, which are capable of capturing member and node bending resistance in addition to the tensile/compressive and torsional stiffness of the lattice members. Secondly, 1D beam elements were employed to assess the effect of the bending of the lattice members on the macroscopic stiffness. Finally, 1D truss elements were utilized where only the tensile/compression stiffness of the lattice members is considered. It's worth noting that the torsional stiffness of the members is only significant when there exists asymmetry at the common nodes, a possible phenomenon in microfabrication [14].

A certain force F is applied on the tetrahedron substructure, and the displacement is extracted from the FEA model. The elastic modulus is then calculated as the ratio of stress to strain per Eq. (3), where the stress is the force divided by the projected area and the strain is the displacement divided by the original height.

$$E_{lattice} = \frac{F\ x\ Height}{L^2\ x\ Displacement} \tag{3}$$

▪ MESH SELECTION

For the case of 3D solid elements, a tetrahedral mesh was

generated using HyperMesh. It was manually-refined until the displacement converged, with a final average mesh density of roughly 40,000 elements per cubic millimeter. Various quality parameters need to be checked to measure how far these elements deviate from the ideal shape of a tetrahedral elements, which is an equilateral tetrahedron. The most important of these quality parameters is the Jacobian, which has an ideal value of 1.0 and is recommended to be greater than 0.5. The Jacobian value was found to be above 0.5 for more than 99.9% of the tetrahedral elements, which gives a proper indication of the mesh quality [15].

For the case of 1D elements, a single member was applied to each of the tetrahedron member since the cross-section is constant and axisymmetric. Care was given to the orientation of the half-cylinder cross-section of the beam in the case of 1D beam elements so that the bending resistance of the members is correctly simulated.

- **BOUNDARY CONDITIONS**

Based on the assumption that the only contribution to the macroscopic elastic stiffness is from the elastic uniaxial tensile/compressive stiffness of the individual members, attention should be given to how the boundary conditions are defined in each case of the FEA model. In the case of 3D solid elements, all the nodes on the upper surface were linked together through rigid elements (RBE2) with the compression force applied to the independent node. The nodes on the bottom surface were constrained from translation in z direction and free to move in the x and y directions. Precautions were taken to maintain the symmetry conditions and to prevent rigid body movement of the model. Fig. 3 shows the original vs deformed shape of the tetrahedron substructure. In the case of 1D elements, the boundary conditions are applied directly to the individual nodes. Fig. 4 shows the original vs deformed shape of a full octet-truss unit cell with 1D elements.

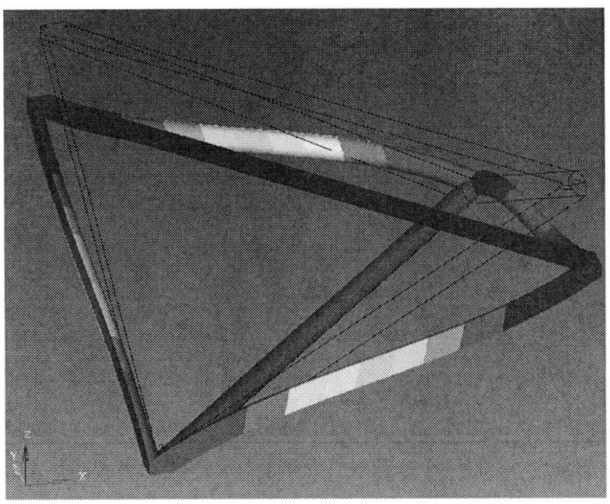

Figure 3: The deformed vs original shape of the tetrahedron model under a compressive force along the z direction. Colors represent displacement in the z direction.

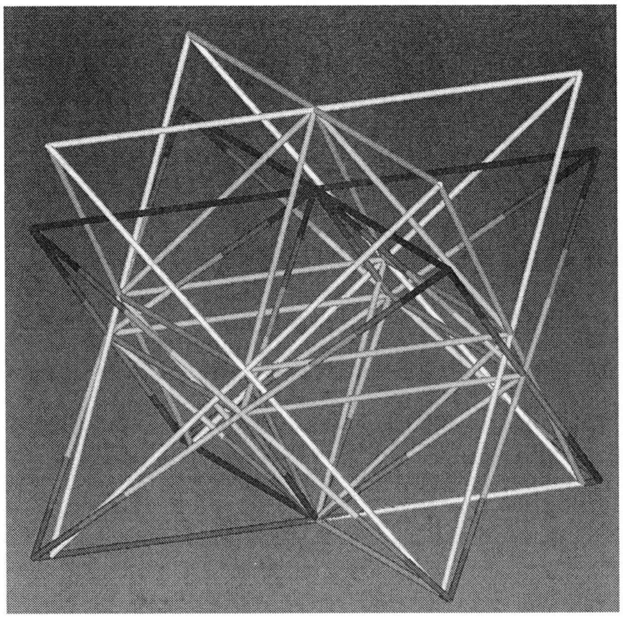

Figure 4: An octet-truss unit cell before (yellow) and after deformation (colors represent displacement values) under a compressive force.

RESULTS

Elastic moduli values were calculated per the continuum-based analytical model, and compared to those obtained from the FEA for each geometrical configuration. For the 3D solid elements, elastic moduli demonstrated errors as high as 50% for the 0.12 aspect ratio, and lower error values for smaller ones. For the case of 0.02 aspect ratio and 60° lattice angle the error was found to be around 4%. This is mainly because the nodes and member bending contribution to the overall stiffness of the structure is higher for large aspect ratios and small lattice angles (more solid material at the nodes) that for small aspect ratios and large lattice angles.

For the case of 1D beam elements, the elastic moduli values were more consistent with a maximum error of 7%. This is mainly due to the bending resistance of the tetrahedron members. As for the case of 1D truss elements, the results showed accurate results with less than 1% error. This is because this case completely aligns with the assumptions followed in developing the continuum-based analytical model.

The choice of which FEA model to follow depends on the nature of the structure and application. For space structures, the aspect ratio is usually low and the cross-section is hollow, hence using the 1D element is justified. This is also the case of mechanical metamaterials; the main difference is that the dimensions are usually at the nano and micro length scales.

For structural geometries of high aspect ratio, the 3D solid elements can produce more accurate results albeit at higher computational costs. Altair Engineering, Inc. recently included lattice structure optimization within their

HyperWorks software package. It uses 1D tapered beam elements to represent lattice structure elements, which can't be reliably used for high aspect ratios.

It's worth noting that the highest stresses are found to be around the nodes. Although the continuum-based model can be utilized to investigate the macroscopic strength of lattice structures, it only considers the failure of the members (elastic buckling or plastic yielding). Hence, a correction parameter needs to be introduced to account for the plastic yielding at the nodes.

CONCLUSIONS

In comparison with the continuum-based analytical model of the octet-truss, 3D solid elements along with 1D truss/beam elements were utilized. The 1D truss elements were found to be the most accurate with respect to the macroscopic elastic modulus of the structure since it completely aligns with the continuum-based model assumptions. However, for geometries with high aspect ratios, the analytical model tends to produce higher errors. For small aspect ratios, it is advised to use 1D beam/truss elements to represent lattice structures given the computational cost saving associated with 1D over 3D elements. Although, both 1D elements capture accurate displacement values (i.e. structural stiffness), accurate stress analysis can only be performed using 3D solid elements.

REFERENCES

1. Montemayor, L., Chernow, V., and Greer, JR. 2015, "Materials by design: Using architecture in material design to reach new property spaces", *MRS Bull*, Vol(40), pp. 1122-1129.
2. Deshpande, VS., Fleck, NA., and Ashby, MF. 2001, "Effective properties of the octet-truss lattice material", *J Mech Phys Solids*, Vol(49), pp. 1747-1769.
3. Montemayor, LC., and Greer, JR. 2015, "Mechanical Response of Hollow Metallic Nanolattices: Combining Structural and Material Size Effects", *J Appl Mech*, Vol(82), pp. 1-10.
4. Zheng, X., Lee, H., Weisgraber, TH., Shusteff, M., DeOtte, J., Duoss, EB., Kuntz, JD., Biener, MM., Ge, Q., Jackson, J a., Kucheyev, SO., Fang, NX., and Spadaccini, CM. 2014, "Ultralight, ultrastiff mechanical metamaterials.", *Science*, Vol(344), pp. 1373-1377.
5. Schaedler, TA., Jacobsen, AJ., Torrents, A., Sorensen, AE., Lian, J., Greer, JR., Valdevit, L., and Carter, WB. 2011, "Ultralight Metallic Microlattices", *Science (80-)*, Vol(334), pp. 962-965.
6. Lake, MS. 1992, "Stiffness and strength tailoring in uniform space-filling truss structures", *Nasa Tech Pap*,
7. Christensen, RM. 2000, "Mechanics of cellular and other low-density materials", *Int J Solids Struct*, Vol(37), pp. 93-104.
8. Challapalli, Adithy1. Challapalli A, Ju J. Continuum model for effective properties of orthotropic octet-truss lattice materials. In: ASME International Mechanical Engineering Congress and Exposition. Montreal, Canada, pp. 115–119. ., and Ju, J. 2014, "Continuum model for effective properties of orthotropic octet-truss lattice materials", *ASME International Mechanical Engineering Congress and Exposition*, Montreal, Canada.pp. 115-119.
9. DOYOYO, M., and WANHU, J. 2006, "Plastic failure analysis of an auxetic foam or inverted strut lattice under longitudinal and shear loads", *J Mech Phys Solids*, Vol(54), pp. 1479-1492.
10. Wan Hu, J., and Park, T. 2013, "Continuum Models for the Plastic Deformation of Octet-Truss Lattice Materials Under Multiaxial Loading", *J Eng Mater Technol*, Vol(135), pp. 21004.
11. Nayfeh, AH., and Hefzy, MS. 1978, "Continuum modeling of three-dimensional truss-like space structures", *AIAA J*, Vol(16), pp. 779-787.
12. Abdelhamid, M., and Czekanski, A. 2016, "On the Effective Properties of 3D Metamaterials", *ASME, Volume 9: Mechanics of Solids, Structures and Fluids; NDE, Diagnosis, and Prognosis*, ASME, Phoenix, USA.pp. V009T12A062.
13. Abdelhamid, M., and Czekanski, A. 2016, "Effects of the Lattice Angle and the Loading Direction on the Effective Properties of the Octet-Truss Lattice Structure", Manuscript submitted for publication.
14. Jang, D., Meza, LR., Greer, F., and Greer, JR. 2013, "Fabrication and deformation of three-dimensional hollow ceramic nanostructures.", *Nat Mater*, Vol(12), pp. 893-898.
15. Altair. 2016, *HyperWorks 14.0 Desktop User Guides*, http://www.altairhyperworks.ca/.
16. Meza, LR., Das, S., and Greer, JR. 2014, "Strong, lightweight, and recoverable three-dimensional ceramic nanolattices", *Science (80-)*, Vol(345), pp. 1322-1326.

Mechatronics and Control Systems

CONTRÔLE D'ATTITUDE D'UN SATELLITTE PAR SYSTÈME INERTIEL

Marie-France BARRIOL DANDINE
Professeur
Arts et Métiers ParisTech
France
dandinemariefrance@gmail.com

Danièle DARGAUD
Ingénieur
Cnam Le Mans
France
daniele.dargaud@wanadoo.fr

ABSTRACT

L'objet du contrôle d'attitude d'un satellite artificiel est de gérer son orientation dans l'espace. Ceci se distingue du contrôle d'orbite, ces contrôles sont souvent associés (SCAO). Ce contrôle est indispensable pour l'exécution des missions du satellite : Observations, mesures diverses, transmission.
Ceci consiste, à partir d'un point de repère de l'espace tel qu'une étoile, à détecter l'écart entre l'orientation initiale et l'orientation en cours (actuelle) puis à corriger l'orientation du satellite. Les méthodes pour effectuer ce contrôle d'attitude sont scindées en deux modes : passif et actif. Le mode actif qui est le plus précis, fait appel à des boucles d'asservissement pour corriger l'attitude du satellite, deux méthodes sont utilisées : Par action interne (échange de quantité de mouvement interne : roue à réaction, volant gyroscopique), ou par action externe (échange de quantité de mouvement avec l'extérieur) ; ces deux méthodes coexistent très fréquemment sur un même satellite.

Pour la méthode par actions internes, il n'y a pas de problème quant à l'emport de masse ; il y a seulement un problème d'énergie, ce qui est très important pour les missions de longue durée.

C'est selon ce principe des actions internes que sont effectués (en tout ou partiellement) les contrôles d'attitude de : Spot 5, Galileo (Constellation de satellites européen pour système de navigation), Rosetta, Le télescope spatial Hubble, etc

KEYWORDS: satellite, attitude control, picture processing, inertial actuator, automatic-control, microcontroller.

INTRODUCTION

L'objectif du travail présenté est de tester la faisabilité de différentes lois de commande à l'aide d'un prototype holomorphe que j'avais réalisé antérieurement, et qui fonctionne sur le principe du contrôle par actionneur inertiel.
Outre le caractère holomorphe de la maquette terrestre, des différences par rapport aux satellites réels sont à mentionner : Ainsi pour déterminer l'écart entre l'orientation souhaitée du satellite et son orientation en cours, nous utilisons une méthode de traitement d'image, avec prise par une caméra extérieure au satellite. De même les éléments de décision sont extérieurs au satellite.
Le cahier des charges est également différent, en particulier moins d'exigence pour la précision qui sera de 0,5°.

1/ Le prototype de laboratoire – Théorie – équations,

Le prototype est unidimensionnel, ceci en vue d'appliquer la méthode de contrôle, et maitriser les techniques à mettre en œuvre (sans les équations couplées du cas tridimensionnel). L'objet de la présentation est d'approfondir la question pilotage pour cette configuration de base.
Le montage est le suivant :
Un disque correspondant à l'inertie d'un microsatellite, dont l'axe de rotation vertical est monté sur pivots de sorte à avoir de très faibles frottements. Cet axe se termine par des pointes portant des billes en carbure de tungstène, leur rotation s'effectue dans des cuvettes en dural, légèrement olivées.
Pour simplifier la construction, l'actionneur inertiel (moteur DC + DT) a été excentré.
Ce disque porte la cible destinée à la mesure de sa position.

L'action sur le satellite est modélisée en appliquant l'équation de conservation de la quantité de mouvement, en excluant toutes actions perturbatrices extérieures, (ce qui fait l'objet d'une autre étude).

Les notations :

Is	moment d'inertie du satellite
Iv	moment d'inertie des parties tournantes
It	moment d'inertie total du satellite
θ	angle de rotation du satellite
β	angle de rotation des parties tournantes par rapport au satellite
ω	vitesse de rotation du moteur

Équation du mouvement satellite + actionneur :

$$I_s \frac{\partial^2 \theta}{\partial t^2} + I_v \left(\frac{\partial^2 \theta}{\partial t^2} + \frac{\partial^2 \beta}{\partial t^2}\right) = 0 \rightarrow I_t \frac{\partial^2 \theta}{\partial t^2} + I_v \frac{\partial^2 \beta}{\partial t^2} = 0 \qquad (1)$$

$$\textit{conditions initiales}\, 0 \rightarrow \theta = -\frac{I_v}{I_t} \int_0^t \omega . dt \rightarrow \theta = -\frac{I_v}{I_t} \beta \qquad (2)$$

Cette équation est celle d'un système « rigide », alors que le couplage est uniquement lié à la conservation du moment cinétique ; on devra par conséquence porter une attention particulière sur les conditions initiales (satellite en rotation, ou bien roue d'inertie en mouvement). De plus l'effet gyroscopique qui résulte de la rotation de la roue d'inertie n'est pas pris en compte sur cette maquette...
L'équation électrique du moteur DC est classique et non reprise ici,

2/ Commande de l'actionneur inertiel, Simulation Matlab Simulink de la réponse du satellite avec le calcul de l'asservissement.

Pour l'asservissement en position du satellite, il est indispensable de maitriser parfaitement la réponse du moteur DC, ce qui nécessite de l'asservir en vitesse afin de tenir compte des non-linéarités, des frottements etc. ; la vitesse de rotation est mesurée par une dynamo tachymétrique (D.T.).
L'asservissement comporte deux boucles (Fig1) :
Une boucle interne pour l'asservissement en vitesse du moteur (moteur + D.T. + correcteur tachymétrique cΩ), l'équation de ce correcteur est implantée dans le micro contrôleur à l'aide d'un programme en C. Le microcontrôleur est un 32 bits, cœur ARM adapté pour systèmes embarqué.
Une boucle externe pour l'asservissement en position du satellite (moteur asservi + Sat. + Mes. + Correcteur position CΘ), ce correcteur est implanté dans le même microcontrôleur également programmé en C.
La mesure de position (Mes.) est effectuée par imagerie, traitée par Matlab avec un modèle original que j'ai mis au point.

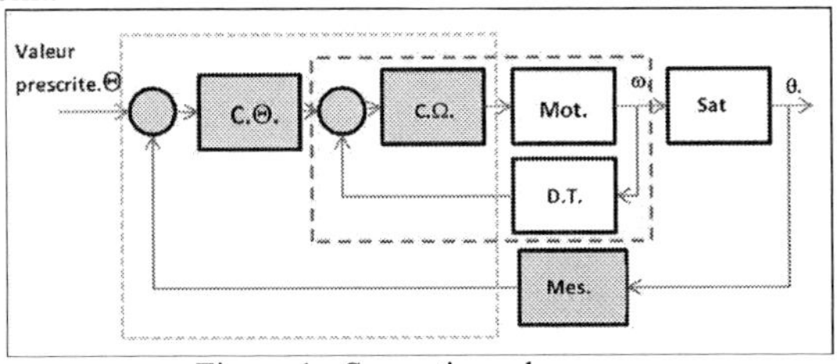

Figure 1 : Synoptique du processus.

La fonction de transfert de la boucle interne est d'abord déterminée, ceci conduit à un bloc moteur+DT+correcteur. Ce bloc est ensuite inclus dans la boucle (externe) contrôlant la position du satellite qui est caractérisé par sa fonction de transfert (entrée en vitesse de rotation du moteur, sortie en position angulaire).
Pour l'implantation sur le microcontrôleur, à ces blocs analogiques, nous associons les fonctions de transfert échantillonnées correspondantes.
Les correcteurs ont été optimisés par calcul, et simulés à l'aide de Matlab-Simulink
Les résultats ci-dessous donnés par la simulation sont probants, et démontrent bien la théorie du contrôle inertiel.

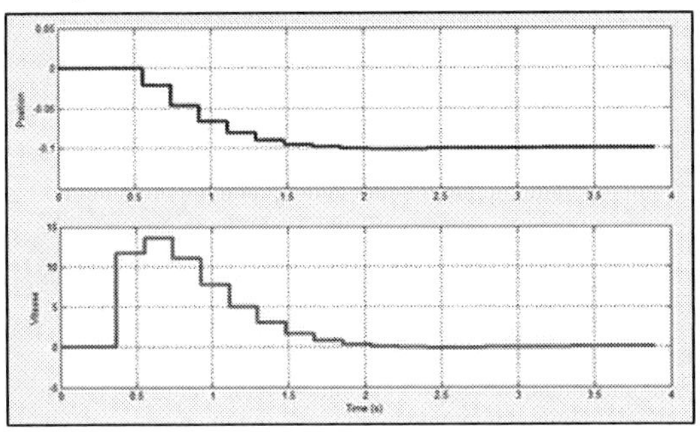
Figure 2 : Réponse pour une entrée échelon de 0,1rad

3/ Validation de la commande par microcontrôleur par des essais avec un simulateur analogique de l'ensemble moteur + DT + satellite.

Préalablement aux essais sur le prototype, et afin d'attendre la mise au point du dispositif de mesure de position par imagerie, nous avons effectué les premiers essais avec un montage analogique qui a les mêmes équations que le système réel, d'où la même réponse. Le but étant de tester le bon fonctionnement des programmes implantés dans le microprocesseur, et de qualifier le travail effectué.
Le système est simulé par un montage à base de composants électroniques analogiques, il respecte fidèlement le schéma fonctionnel, il comporte une entrée correspondant à la tension appliquée au moteur, une sortie correspondant à la tension délivrée par la dynamo tachymétrique (d'où sa vitesse), et une sortie correspondant à la position satellite, et accessible l'information sur le courant moteur.
Nous avons choisi des échelles pour la représentation de variables physiques, de sorte à utiliser toute la pleine échelle des composants,
La simulation est en temps réel et a donné les résultats (Fig3) qui sont en complète concordance avec ceux obtenus à l'aide de Matlab-Simulink (Fig2).

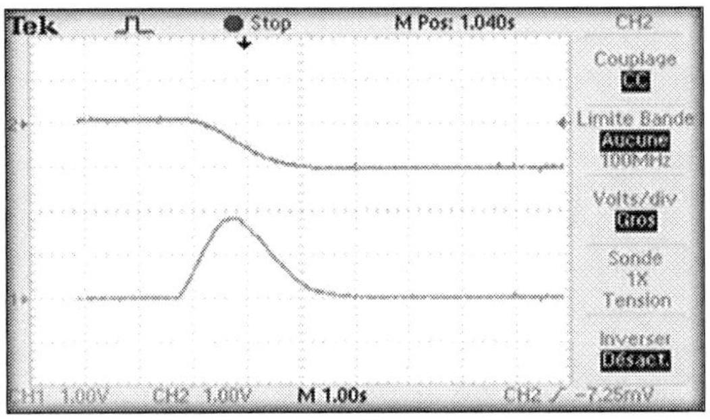

Figure 3 : Expérimentation avec simulation analogique
En orange la vitesse (tension donnée par la DT), ce qui correspond à un maximum de 225rd/s
En bleu la position sat, le décalage correspond à 0.1rd

4/ Mesure de la position angulaire du satellite par imagerie, cybercaméra et traitement d'image.

La détermination de la position angulaire est effectuée à partir de la détermination de la position instantanée d'une cible sur le satellite, l'image sur la surface sensible de la caméra (monde image) est traitée numériquement.
La correspondance nécessite une opération de calibration.

Pour cela on utilise une grille de calibration fournie par Matlab qui remplace temporairement le satellite.
(Par un calcul dans le monde image la position de la cible sera définie dans le repère lié à cette grille).

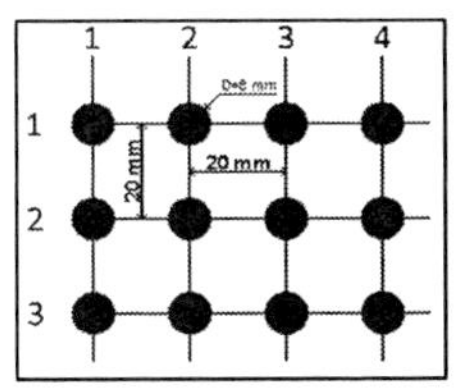

Figure 4 : Grille calib.

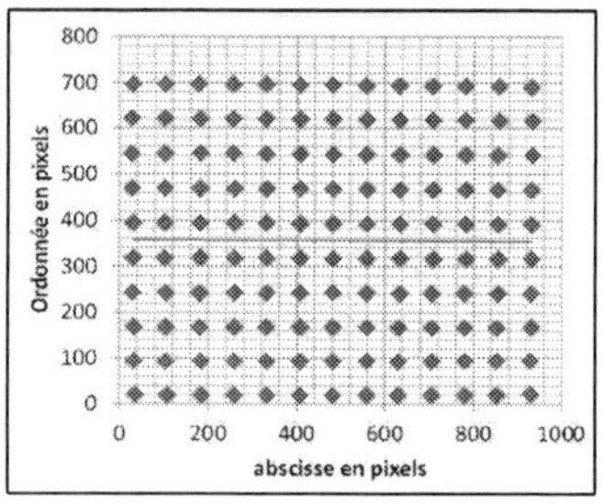

Les coordonnées en pixels des centres de chaque région sont déterminées en Pixels à l'aide d'un programme Matlab.
La grille de calibration couvre tout le champ de la caméra, et est centrée.

Figure 5 : Image, Coordonnées en Pixels des régions

L'étude numérique conjointement de l'image et de la grille de calibration a permis de constater qu'avec un montage correct de la caméra (perpendicularité de son axe optique avec le centre de la grille, et respect même approximatif de la correspondance des axes x et y), il n'y avait pas de distorsion appréciable de l'image, et que seuls les coins de l'image étaient légèrement affectés. Dans le domaine utile, il y a correspondance linéaire.

Une transposition dans le monde réel est ensuite effectuée. On obtiendra la correspondance en mm des centres des marques dans le monde réel vers les pixel dans le monde image.

Pratiquement, pour éviter les difficultés consécutives aux signaux parasites, nous avons travaillé avec des marques blanches sur fond noir, et effectué une opération de seuillage.

Un premier modèle effectué par Matlab qui aurait permis de déterminer les coordonnées de la cible dans le monde réel n'a pas été exploitable car l'erreur varie avec la position de la cible, ce qui est inadapté à notre application.

Nous avons établi notre propre modèle, présenté ci-dessous :
Comme la cible ne coïncide pas avec les marques servant à l'étalonnage, j'ai mis au point une méthode d'interpolation linéaire dans le monde image ; pour cela nous nous basons sur les trois marques de la grille les plus proches de la position de la cible, et dont nous connaissons les coordonnées des centres dans le monde de l'image et dans le monde réel, et appliquons la méthode d'interpolation linéaire

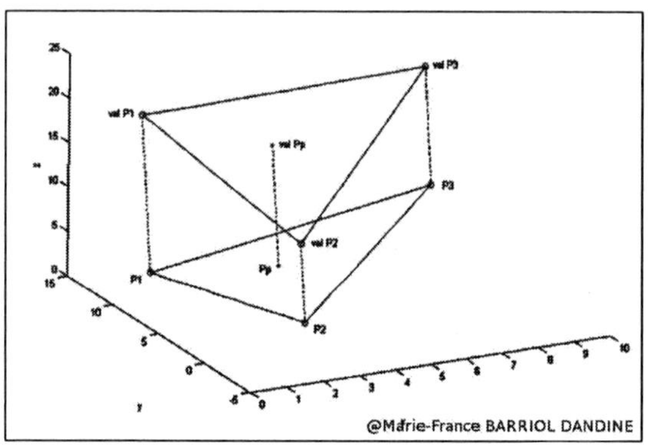

Figure 6 : Schéma méthode de linéarisation
Verticalement : Abscisse ou ordonnée dans le monde réel
Horizontalement : Axe vers avant : Ordonnée en pixels,
Axe vers la droite : Abscisse en pixels.

consistant à déterminer les équations de deux plans passant par les coordonnées géométriques des centres des trois marques de référence (Fig6) ; ces équations permettent ensuite d'obtenir les deux coordonnées géométriques du centre de la cible (qui est connue en Pixels) (Fig8).
Précision obtenue sur la position de la cible :
Abscisse : $\Delta x = \pm 0{,}14\ pixels$
Ordonnée : $\Delta y = \pm 0{,}18\ pixels$

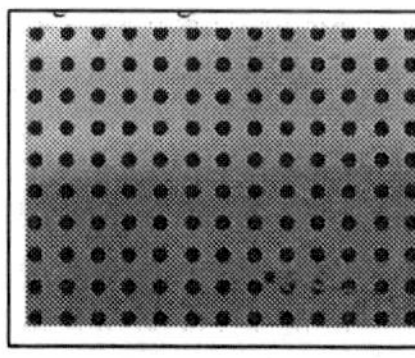

Pour valider la méthode, nous avons Mesuré les coordonnées du centre de la cible sur la grille.
Abscisse de la cible : 149,5 $\pm$ 0,5 *mm*
Ordonnée de la cible : 154 $\pm$ 0,5 *mm*

Figure 7 : Image de la grille avec la cible.
Calibrée géométriquement

	7	8		9		10
	y	x	y	x	y	x
6	394,9273	558,705757	394,839415	636,588741	470,991809	711,880404
7	471,395491	558,725396	471,350127	636,61386	547,580177	711,77321
8	547,932217	558,69071	547,909538	595,704311	601,029731	711,919228
9	624,623157	558,7288	624,612702	636,562923	624,322389	711,973867
10	701,29796	558,637978	701,133954	636,497629	700,848072	712,01433

Coordonnées du point exclu
Point utilisé pour le calcul
Coordonnée de la cible

Notre calcul conduit à :
Abscisse cible :149,504 *mm*
Ordonnée cible : 153,851 *mm*

Figure 8 : Exemple d'application de la méthode
Coordonnées en Pixels des centres des marques la grille de calibration.
et du centre de la cible

La précision obtenue par cette méthode est supérieure à celle fournie par une mesure géométrique (graduation 0,5mm).

Détermination de l'angle de rotation du satellite.

L'angle de rotation est déterminé à partir de la connaissance du centre de rotation du satellite, et des positions en deux instants considérés de la cible.

Équation de l'angle plan θ (3)

$$\theta = 2.Arc\ sin\frac{l}{2r} \quad avec \begin{cases} l\ distance\ parcourue \\ r\ le\ rayon\ de\ rotation \end{cases}$$

$$l = \sqrt{(x_t - x_0)^2 + (y_t - y_0)^2}$$

Avec (x_t, y_t) les coordonnées de la cible à l'instant t
et (x_0, y_0) les coordonnées initiales de la cible

La précision obtenue est : $\Delta depl\_ang = \pm 0{,}0025\ rad$

5/ Expérimentation.

La méthode :

Lecture de la position instantanée de la cible :

L'ordinateur donne à l'aide de Matlab l'information de la cible en Pixels, un calcul rapide avec Excel conduit avec notre méthode à la détermination des coordonnées géométriques de la cible et de la valeur du déplacement angulaire, qui sont envoyées à Matlab.
L'information précédente est envoyée au microcontrôleur par liaison RS232, Ainsi que l'information donnée par la DT, qui après filtrage est connectée sur une entrée analogique CAN.

La sortie destinée à l'alimentation du moteur est faite par un CNA.
La valeur prescrite est incluse dans le programme en C.
Des précautions ont été prises du fait que les tensions moteur et DT peuvent être positives ou négatives, tandis que les entrées/ sorties du µP sont toujours d'un même signe.

Les mesures effectuées :

Calibration de la mesure de l'angle de rotation par traitement d'image.

Pour différentes valeurs prescrites, les valeurs du déplacement angulaire déterminées par le programme ont été reportées en fonction des déplacements angulaires mesurés.

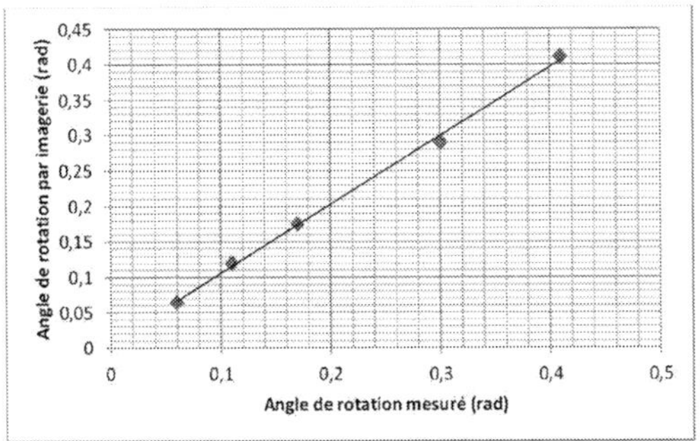

Figure 7 : Calibration de la mesure du déplacement angulaire par traitement d'image

Les mesures effectuées ont permis la validation :

Du programme d'asservissement implanté dans le microcontrôleur.
Du modèle pour la détermination des coordonnées de la cible dans le monde réel.
De l'automatisation du programme employant Matlab, Excel et le Microcontrôleur avec la boucle d'asservissement utilisant le principe d'imagerie.

Régime bouclé

Les premiers essais se sont heurtés à des perturbations liées à la raideur des fils de liaison (moteur et DT), malgré les précautions prises les effets mécaniques étaient non négligeables ; ceci a été quantitativement expliqué, et le système complet sera prochainement testé.

▪ CONCLUSION, PERSPECTIVES.

Conclusion.

En utilisant la méthode originale proposée pour l'imagerie, nous avons réussi à caractériser avec précision la position d'une cible sur le satellite, et à obtenir la valeur de sa position, d'où sa rotation au cours d'une manipulation

Nous avons établi un modèle du satellite, et du moteur qui a permis la mise en équation, ceci a été traité dans le domaine échantillonné, les correcteurs ont été calculés ; Ceci a été simulé, puis a été implanté sur microcontrôleur.

Une première série de mesures, en associant à l'asservissement numérique par microcontrôleur un simulateur, (crée en même temps que le prototype) a conduit à une très bonne concordance ; ce qui permet de valider incontestablement la méthode d'asservissement et la théorie du contrôle inertiel.

Le passage sur le prototype a fourni des résultats intéressants permettant de valider totalement la méthode de détermination par traitement d'image de l'angle de rotation. La commande de l'asservissement en vitesse a été validée ; Il reste encore quelques mesures à effectuer pour finir de valider l'asservissement de position.

Perspectives

Embarquer l'électronique ainsi que les sources d'énergie, de sorte à supprimer les problèmes dus aux fils de liaison (ce qui nécessitera une liaison sans fil avec l'ordinateur).
Tester par la suite un algorithme de correction de type logique floue.

REMERCIEMENTS.

Ce travail personnel est poursuivi en relation avec le Professeur Yves GOURINAT de l'ISAE-SUPAERO Toulouse.

REFERENCES

1. Logitech. (2012, Août) Spécifications techniques Webcam HDc310. Document.
2. Marion, A. 1997, « Acquisition & visualisation des images ». Paris: Eyrolles.
3 Mathieu, H. 2001, « Physique des semiconducteurs et des composants électroniques » 5th ed. Paris Dunod,
4 Belaroussi, R. 2010, « Traitement de l'image et de la vidéo ». Paris: Ellipses,
5 Rivoire, M. and Ferrier, J-L. 1997 « Cours d'automatique - Commande par calculateur » Tome 3,
6 Alengrin, G. 1996 « Méthode de synthèse de correcteurs numériques », Techniques de l'ingénieur - R7420.
7 Yiu, J. 2010 « The Definitive Guide to the ARM Cortex-M3 », 2nd ed., Texas Instruments, Ed. USA: Elsévier,
8 STMicroelectronics. (2011, Octobre) RM0008 - Reference manual,
9 ARM. (2010) ARMv7-M Architecture Reference Manual,
10 STMicroelectronics. (Aout, 2011) STM32F105xx - STM32F107xx,
11 Asch, G. 1999, « Les capteurs en instrumentation industrielle » Dunod,

Stepper Motor Controller Based on S-shaped Control Algorithm and Its Application

Wang Su	Xu Xi	Yi Mengyun	Zhu Fei	Wang Jiaqi	Chang Yufang
School of Electrical and Electronic Engineering Hubei University of Technology Wuhan,Hubei,,China 1643124698@qq.com	School of Electrical and Electronic Engineering Hubei University of Technology Wuhan,Hubei,,China 564588703@qq.com				

● ABSTRACT

At present, ladder speed control algorithm, segmented linear speed control algorithm and other algorithms are applied most to start the stepper motor. These algorithms have many good characteristics, such as simple algorithm, fast response and less occupancy. However, the starting process of stepper motor using these algorithms often has disadvantages of startup slowly and out of step. In this paper, an embedded controller based on microprocessor S3FN41F is proposed for stepper motor and a control algorithm of S-shaped are designed and implemented. The result shows that S-shaped control algorithm can improve the starting characteristic of stepper motor.

KEYWORDS: stepper motor,microprocessor,S-shaped algorithm, Control System, Presser

1 Moment Frequency Characteristics of stepper Motor

In the process of stepping motor operation, the relationship between the output torque and the drive frequency is called the moment frequency characteristic, and the torque decreases with the increase of the frequency[6-9]. Stepper motor torque frequency characteristic curve, as shown in Figure 1. f is the frequency at which the stepper motor starts; f_{st} is the starting frequency for just starting the stepper motor; T_d is the starting torque; T_{st} is the starting start torque.

The moment frequency characteristic of the stepping motor can be described by the kinetic model, expressed as

$$J\frac{d^2\theta}{dt^2}+\beta\frac{d\theta}{dt}+K\theta+T_Z=T_d \quad (1)$$

In the formula (1): J is the total moment of inertia of the stepper motor after some conversion; β is the motor damping coefficient ;

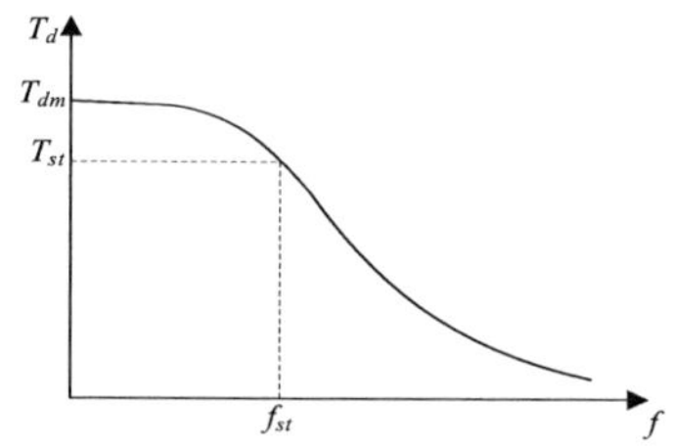

Fig.1 The character of torque-frequency of step motor

θ is the stepper motor rotation angle; K is the proportional coefficient with respect to θ as a function; T_z is the sum of the frictional resistance moment and the resistance torque independent of θ;T_d is the electromagnetic torque generated by the stepper motor[4-5].From Formula 1 can be derived

$$J\varepsilon=J\frac{d^2\theta}{dt^2}=T_d-\beta\frac{d\theta}{dt}-K\theta-T_Z \quad (2)$$

In the formula (2): $J\varepsilon$ is the inertia torque of the system.It can be seen from the formula (2) that the inertia torque $J\varepsilon$ is smaller than the maximum electromagnetic torque, In order to improve the speed of the system, the angular acceleration ε should take a larger value in the acceleration phase,but the value should not be too large, otherwise, the stepper motor is prone to out of step; At the same time, ε should not be too small, otherwise, the system is too fast. Therefore, in order to meet the system without losing the pace, fast and small impact, the angular acceleration ε should not be a big mutation.

2 Design and Application of Stepping Motor Control System

2.1 Stepping motor control system structure

As shown in Figure 2 is the stepper motor controller system block diagram. S3FN41F is a 32-bit high-performance processor with ARM Cortex-M0 as the core of Samsung

Semiconductor, with Serial Communications (SPI), Universal Serial Bus (USB), Controller Area Network (CAN) Directly with the computer, control chip, drive the chip to communicate. TMC4210 is TRINAMIC company's stepper motor control chip, the main control drive chip operation. TMC2660 is TRINAMIC company's stepper motor driver chip, equipped with memory and registers, through the SPI communication configuration parameters, while it also has the function of out-of-step protection, in the event of an out-of-synchronization through the SPI to send information to the microcontroller to adjust. The system work when the stepper motor with a dedicated switching power supply.

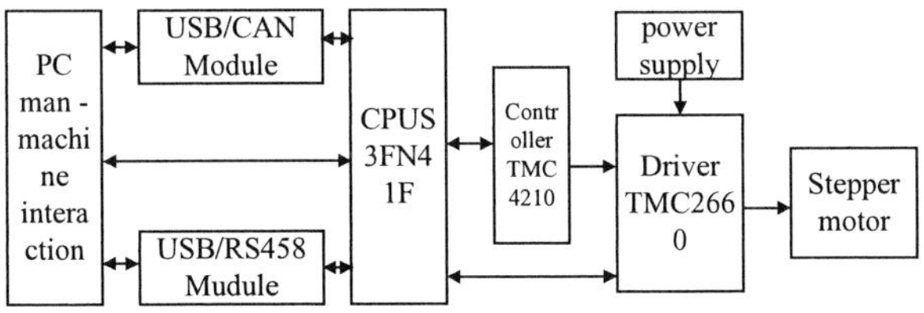

Fig.2 The principle of the stepper motor controller system

2.2 The principle of LCD screen presses

LCD screen presses is a kind of liquid crystal screen and the body of the mother press or stitch together equipment.Figure 3 shows the principle of liquid crystal screen press. To LCD TV production, for example, large quantities of LCD TV in the production of more use of dual-line work, a pipeline responsible for the assembly of LCD rear shell, by the artificial assembly of the body after the shell into the mother groove Fixed, another line is responsible for liquid crystal screen before the check and accessories, and sent to the pre-press before the scheduled position.

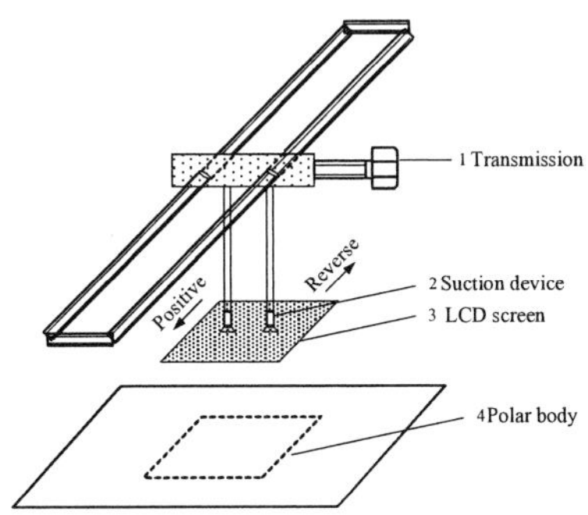

Fig.3 Lamination Device working principle

Position accuracy requirements in the process of splicing the LCD screen and the mother can be accurately positioned between to prevent the target and the mother of the dislocation between the resulting unnecessary damage; speed smoothness requirements stepper motor in the process of slow start increase or decrease, to prevent the impact of the resulting unnecessary jitter, resulting in LCD screen oscillation or damage to its damage. Some of the traditional control algorithm is clearly unable to achieve the above indicators, so the stepper motor control algorithm is very necessary.

2.3 Principle of the production line device

Figure 4 shows the LCD TV production line of the working principle.

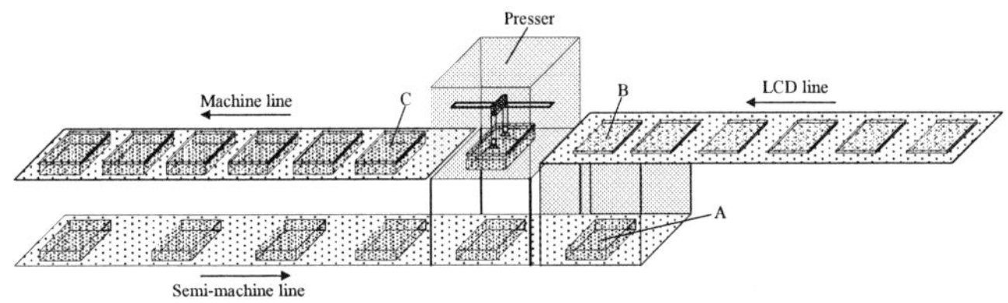

Fig.4 The principle of assembly line

In Fig.4 a general process of overall production is plotted. Which is mainly used to assemble the rear part of the rear panel of the camera body that is the body of the rear shell of the various components, the production line speed by the artificial control; LCD line is responsible for the LCD screen quality inspection and the corresponding cable accessories, LCD B is equipped with infrared sensor, LCD screen flow to the B at the LCD line automatically stop the flow, without manual control of the pipeline; the whole line is responsible for the completion of the pressure after the LCD TV flow down for the rear line engineer test performance. During the pressing process, the speed of the LCD line is large enough to fully match with the semi-machine line. The pressing process of the staff of the small demand, linear speed control, with high efficiency advantages.

3 The algorithm of Start the control

3.1 S-type control and control algorithm and its implementation

The velocity curve of the S-type acceleration control algorithm is shown in Fig.6,because the system in the acceleration and deceleration phase of the velocity curve shape S called S-type acceleration curve[10]. As shown in the figure,

S-type curve can be divided into the process of acceleration, acceleration, acceleration, uniform, deceleration, deceleration and acceleration and deceleration 7.

The acceleration curve of the S-type acceleration control algorithm is shown in Fig. 7. Since the drive pulse frequency is proportional to the rotor speed, the angular acceleration of the rotor should be proportional to the step , As shown in Figure 7, each of the lines represented by the equations are the primary function.

$$\varepsilon = df/dt = C - Df \tag{3}$$

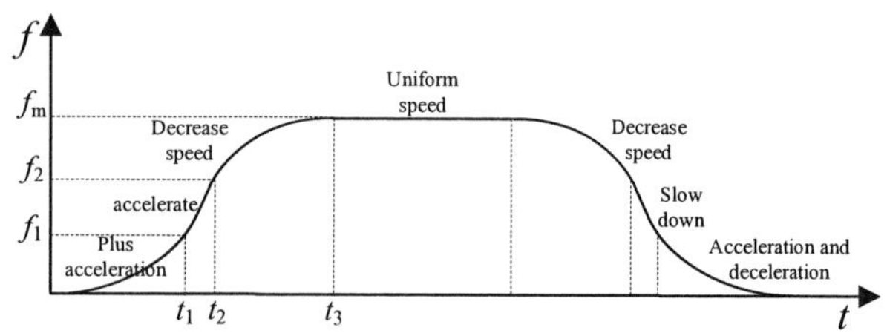

Fig.6 S-shaped acceleration and deceleration curve

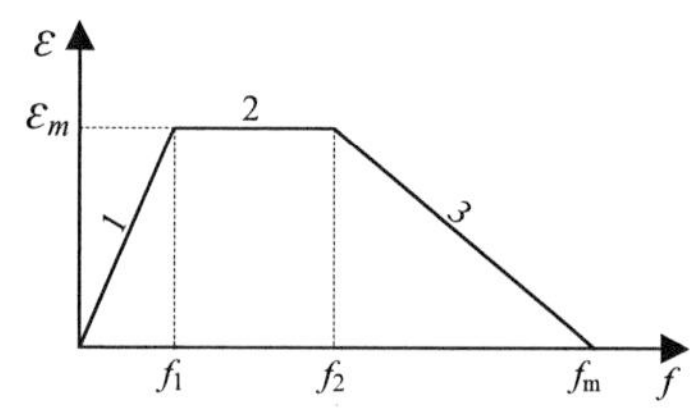

Fig.7 The acceleration curve during the process of startup

Where C and D are constants. If the initial pulse frequency is set, the formula (3) is transformed by Laplace.

$$sF(s) - f_0 + DF(s) - C/s = 0 \tag{4}$$

Solutions have to:

$$F(s) = C/[s(s+D)] + f_0/(s+D) \tag{5}$$

Laplace inverse transformation can be obtained,

$$f(t) = f_0 + E(1 - e^{-Dt}) \tag{6}$$

In the type, $E=C/D-f_0$, It can be seen from equation (6) that the frequency of the driving pulse varies with time by the coefficient[4-5].D<0,that the acceleration process of the first paragraph of the acceleration curve, the speed curve for the concave line. $D = 0$ represents the second stage of the acceleration curve during the acceleration, the velocity curve is a straight line. $D > 0$ represents the third stage of the acceleration curve during the acceleration. The velocity curve is convex. Figure 7 corresponds to the ascending (start) curve of Figure 6. Assuming t, the stepper motor receives the Nth pulse.

$$N(t) = \int_0^t f(t)dt = \int_0^t [f_0 + E(1 - e^{-Dt})]dt = (f_0 + E) - E/D(1 - e^{-Dt}) \tag{7}$$

Use the Newton iteration method to find the moment when the Nth pulse is received.

$$t_{k+1} = t_k - [(f_0 + E)t_k - E/D(1 - e^{-Dt_k}) - N] / f_0 + E(1 - e^{-Dt_k}) \tag{8}$$

In order to obtain a more accurate value, we can use Newton iterative method to evaluate, that is, (8)after repeated iterations, we can find the required value. And then the required substitution (6) can be obtained for each N corresponding $f(N)$. In the process of writing the program, according to equation (6) (7) (8) can calculate a set of acceleration curve of the pulse sequence of the pulse, so as to determine the timing of the drive pulse.

4 Realization of S - type Control Algorithm in Compressor Controller

The proposed stepper motor controller and stepper motor speed control algorithm applied to a liquid crystal TV production line, in which the production process is shown in Figure 4, the press in the transmission device using single-axis stepper motor control , The motor selection of two-phase stepper motor, rated phase current of 1A, rated pulse voltage of 5V, step angle of 1.8° . The stepper motor drives the gears on the gearbox through the drive chain. The distance between the starting position of the machine and the position of the liquid crystal display is 0.58m, the radius of the gears of the motor rotor is 0.05m, the step angle of the stepping motor is 1.8 ° , and the total number of steps is,

$$N = \frac{360^\circ S_0}{2\pi r\theta} = \frac{360 \times 0.58}{2 \times \pi \times 0.05 \times 1.8} \approx 369 \tag{9}$$

In order to better verify the S-type control algorithm to improve the start-up characteristics of the stepper motor, and the index control algorithm for the corresponding comparison. According to the overall number of steps and the corresponding known conditions to complete the index and S-type two kinds of algorithm programming. The target position, target speed, minimum speed, maximum speed and maximum acceleration can be set by CAN communication on the PC, TMC4210 and TMC2660. Where the lower left corner of the Ramp Mode can choose the operating mode,and this selection is exponential S-mode.

By adjusting the speed of different production lines, the number of LCD TV (60 ") produced in unit time is obtained under the two indexes of application index and S type, and the comparison is made. The comparison results are shown in Table 1 shows. According to the different production line speed, through a large number of experiments, the S-type control algorithm under the stepper motor out of the phenomenon and oscillation phenomenon were compared, see Table 2 shows.

Tab.1 The production number of LCD TV in both control algorithms

Serial number	Line speed *s/m*	Spwcies	
		Index acceleration/deceleration algorithm	S-type acceleration/deceleration control algorithm
1	12	164	180
2	14	144	160
3	16	129	138
4	18	116	124
5	20	106	112
6	22	97	102

Through this interface can be more convenient to complete the test parameters of the debugging, without professional engineering staff will be able to complete this work, with the characteristics of personnel adaptability.

Tab.2 The out-of-step and oscillation phenomena of S-shaped algorithm

Serial number	Line Speed *s/m*	S - type acceleration / deceleration control algorithm	
		Out of step	Shock phenomenon
1	12	When out of step	Slight concussion
2	14		
3	16	Slightly out of step	Little shock
4	18		
5	20	Almost no step out	Almost no shock
6	22		

The data in Table 1 is the number of LCD televisions stowed per hour. As can be seen from the data in Table 1, the system using the S-type control algorithm is the same as the system using the exponential control algorithm at the same line speed and time The number of television stations to suppress more, more efficient production, and the faster the line speed, the more obvious performance. The smaller the line speed in Table 2, the less the number of start and stop times of the stepping motor per unit time. As can be seen from Table 2, the stepper motor with S-type control algorithm has the advantages of less step-down and less oscillation, and unit time The less the number of stepper motor start and stop, the lower the rate of out-of-step and oscillation

5 Concluding remarks

In this paper, the exponential and S acceleration / deceleration control algorithms are applied in the stepper motor control system of liquid crystal screen presses. According to the number of LCD TV in the same line speed and the same time, the efficiency of S-type control algorithm is verified The It is concluded that the S acceleration / deceleration control algorithm has the advantages of low starting step, low oscillation and high production efficiency, and can be better applied in high speed and high precision control system.

REFERENCES

[1] 马文斌，杨延，洪运. 步进电机控制系统的设计及应用[J]. 电子技术应用，2015, 41(11): 11-13.

[2] 崔洁，杨凯，肖雅静，等. 步进电机加减速曲线的算法研究[J]. 电子工业专用设备，2013, 222: 45-49.

[3] 张晓莉，孙立斌. 步进电机精确控制算法在电力控制系统中应用[J]. 科技通报，2014, 30(2): 86-89.

[4] 王邦继，刘庆想，周磊，等. 步进电机速度曲线的设计与 FPGA 实现[J]. 微电机，2012, 45(8): 67-71.

[5] 王勇，王伟,杨文涛. 步进电机升降速曲线控制系统设计及其应用[J]. 控制工程，2008, 15(5): 576-579.

[6] 李汉. 一种步进电机最佳升降速的控制方法[J]. 电气传动，2012, 34(5): 32-34.

[7] Benetta Aranjo, Prashant Kumar Soori, Puja Talukder. Stepper motor driver for robotic applications[J]. International Power Engineering and Optimization Conference, 2012: 361-366.

[8] Divic Josip, Duric Jurica, Vrancic Koca. Microcontroller implementation of dynamically adaptable control of stepper motor with continuous second derivative of speed curve[J]. International Convention on Information and Communication Technology, Electronics and Microelectronics, 2014: 146-149.

[9] 曾康玲，杨向宇，赵世伟，等. 步进电机指数型加速曲线优化与实现[J]. 微电机，2014, 47(6): 36-40.

[10] 刘筱，吴文江，郑飂默. 柔性 S 型加减速控制算法研究[J]. 组合机床与自动化加工技术，2014, (3): 66-68,72

Positioning Control Method for Linear Switched Reluctance Machine System

Li Qiu

Abstract—**This paper reviews the recent work in the positioning control for a linear switched reluctance machine (LSRM) system. The survey is introduces some notable results and method in LSRM control area. At the same time, this paper proposed the stability analysis method for LSRM motion control system. The positioning controller design method for the closed-loop system is introduced.**

Index Terms—**Linear switched reluctance machine, positioning controller design, stability and stabilization.**

I. Introduction

As demand increases for high-speed/high-accuracy motion, various types of linear motor-driven motion systems have received significant attention for precision motion control [1]. The linear motion control system based on minimum reluctance principle is extended to a variety of industrial applications such as parts assembly, printed circuit and board drilling, etc. since it has advantages such as high speed, high precision, fast response, no dead-path error, large output, straight line execution and so on. However, it is challenged to achieve high precision positioning control because that it is difficult to obtain the precision mathematical model parameters induced by the sensitive and variations friction coefficients traction pulsation. In recent years, some work has been addressed for linear switched reluctance machines (LSRM) control system, for example, a flux feedback reluctance actuator linearization scheme has been described based on cascaded analog sensing coil and digital hall probe feedback control [2]. A systematic control design method is studied for the direct-drive linear motor to provide high-speed and high-precision performance by using fast non-singular terminal sliding mode in [3]. In [4], a new methodology is presented to determine the magnetically guided robot position in horizontal planes by using magnetic flux sensors. The adaptive robust control and optimized adaptive motion control methods are proposed in [5] and [6], respectively. In order to overcome the motor unknown parameters effects, minimize the disturbance effect and reduce the disturbance response time, an implements modified disturbance observer and compensator are proposed in [7]. In order to reduce the force ripple, a new control strategy is proposed for LSRM to control the motor instantaneous position, speed, phase current, and force in [8]. In [9], the design method and performance comparison studies of dual side mover and segmental stator LSRM are addressed. In order to evaluate the winding configurations effects, the electromagnetic performances of half-teeth-wound and full-teeth-wound switched reluctance machines which including air-gap flux density, coil, phase flux linkage, inductance, torque, volt-ampere and radial magnetic force are comprehensively comparative studied in [10]. In [11], the coordinated tracking control for the group motion control system based on three direct-drive double sided LSRM is proposed. In [12], a solution for the 4/2 SRM startup problem is provided and the proposed control algorithm respects the varying average torque bounds and explicitly enforces the maximum phase current limit. In [13], a mover position estimation method is presented for the LSRM based on the eddy-current effect, and the equivalent eddy-current resistance model is presented through the finite element analysis. Current studies mainly focuses on distributed control designs, PID parameters tuning control and stability analysis for LSRM control system.

Li Qiu is with the College of Mechatronics and Control Engineering, Shenzhen University, Shenzhen, 518060, China, and with the Department of Mechanical Engineering, University of Victoria, Victoria, BC V8W 3P6, Canada(qiuli@szu.edu.cn).

The traditional PID positioning control method for LSRM systems faces with enormous challenges because that the PID parameter setting has a strong dependence on the experience. Some work has been researched for LSRM system, for example, in [14], the voltage output ripple reduction and error minimization are discussed for the direct-drive linear switched reluctance generator wave power generation system. In [15], the closed-loop speed control performance is investigated for conventional linear flux-switching permanent-magnet motors based on primary flux-oriented control. In [16], an active suspension system utilizing a low-cost high-performance linear switched reluctance actuator with proportional-derivative control is presented. In [17], a nonlinear inductance modeling method is proposed by introducing a K factor that reflects both current and position effect on inductance. Considering the uncertainties and disturbances from the dual LSRM motion system, the online parameter estimation with adaptive control strategy is proposed to detect system parameter variations and regulate control parameters in real time. A mover position estimation method in the voltage pulse injection for the LSRM based on the eddy-current effect is addressed in [18].

In this paper, first, the positioning control problem is addressed for a second-order direct-drive LSRM system. Second, the stability conditions for LSRM motion control system are given. Third, the positioning controller gain range is obtained by adopting the matrix eigenvalue numerical analysis method.

II. Control Method for LSRM

The discrete-time state-space describe of the second-order LSRM motion control system can be denoted as

$$x(k+1) = Ax(k) + Bu(k), \tag{1}$$

where second-order system parameters A and B, system state vector $x(k) = \begin{bmatrix} s(k) \\ \dot{s}(k) \end{bmatrix}$, $s(k)$ and $\dot{s}(k)$ denote the positions and the velocity of the LSRM at time step k, $u(k)$ denotes the control input signal for LSRM. The state feedback positioning controller for the LSRM is proposed as follows

$$u(k) = Kx(k), \tag{2}$$

where K is the state feedback controller gain.

Substituting controller (2) into equation (1), the closed-loop LSRM motion control system can be obtained as follows

$$x(k+1) = (A + BK)x(k) \tag{3}$$

Definition 1: System (3) is stabilizable, if there exists a feedback control $u(k) = Kx(k)$, such that for any $x_0 \in \mathbb{R}^n$, the closed-loop system $x(k+1) = (A+BK)x(k)$, $x(0) = x_0$ is asymptotically stable, that is, we have

$$\begin{cases} \lim_{k\to\infty} \|x(k) - 0\| = \lim_{k\to\infty} \|A^k x_0\| = 0, \\ x_0 \neq 0, k = 0, 1, 2... \end{cases} \tag{4}$$

where $K \in \mathbb{R}^{m\times n}$ is a constant matrix.

Theorem 1: Under the proposed state feedback control law in (2), the closed-loop LSRM motion control system (3) is asymptotic stable if the following condition hold

$$| \sigma((H_n^{\mathrm{T}} H_n)^{-1} H_n^{\mathrm{T}}((A+BK)^{\mathrm{T}} \otimes (A+BK)^{\mathrm{T}} - I\otimes I)H_n) |< 1. \tag{5}$$

Proof: It can be obtained by combining the vectorization technique and H-representation method, the proof is omitted.

Theorem 2: Under the proposed state feedback control law in (2), the closed-loop second-order LSRM motion control system (3) is asymptotic stable if all eigenvalues λ_i, $(i = 1, 2, 3)$ locate in the unit circle, and the non-zero control parameter K satisfies

$$-1 < \lambda_i < 1, i = 1, 2, 3 \tag{6}$$

where λ_i, $(i = 1, 2, 3)$ is the three-order square matrix $\Phi = (H_n^{\mathrm{T}} H_n)^{-1} H_n^{\mathrm{T}}((A + BK)^{\mathrm{T}} \otimes (A + BK)^{\mathrm{T}} - I \otimes I)H_n$ eigenvalues. Then the controller gain K can be determined by matrix eigenvalues numerical analysis method.

Proof: It can be obtain by using matrix eigenvalues numerical analysis method, the proof is omitted.

III. Conclusions

In this paper, some positioning control methods are reviewed for the linear switched reluctance motor. The stability conditions and the positioning controller gains of closed-loop LSRM system are given by combining with stability analysis theory and vectorization method. The matrix eigenvalues numerical analysis method is adopted to obtain the controller gain range.

References

[1] K. Cho, J. Kim, S. B. Choi, and S. Oh, "A high-precision motion control based on a periodic adaptive disturbance observer in a PMLSM," *IEEE/ASME Transactions on Mechatronics*, 20(5): 2158-2171, 2015.

[2] A. Katalenic, H. Butler, and P. P. J. van den Bosch, "High-precision force control of short-stroke reluctance actuators with an air gap observer," *IEEE/ASME Transactions on Mechatronics*, 21(5): 2431-2439, 2016.

[3] J. C. Zheng, H. Wang, Z. H. Man, J. Jin, and M. Y. Fu, "Robust motion control of a linear motor positioner using fast nonsingular terminal sliding mode," *IEEE/ASME Transactions on Mechatronics*, 20(4): 1743-1752, 2015.

[4] X. D. Zhang, M. Mehrtash, and M. B. Khamesee, "Dual-axial motion control of a magnetic levitation system using hall-effect sensors," *IEEE/ASME Transactions on Mechatronics*, 21(2): 1129-1139, 2016.

[5] Z. Chen, B. Yao, and Q. F. Wang, "μ-Synthesis-based adaptive robust control of linear motor driven stages with high-frequency dynamics: a case study," *IEEE/ASME Transactions on Mechatronics*, 20(3): 1482-1490, 2015.

[6] H. H. Chiang, K.C. Hsu, and I. H. Li, "Optimized adaptive motion controlt through an SoPC implementation for linear induction motor drives," *IEEE/ASME Transactions on Mechatronics*, 20(1): 348-3600, 2015.

[7] H. W. Chow and N. C. Cheung, "Disturbance and response time improvement of submicrometer precision linear motion system by using modified disturbance compensator and internal model reference control," *IEEE Transactions on Industrial Electronics*, 60(1): 139-150, 2013.

[8] S. Masoudi, M. R. Feyzi, and M. B. B. Sharifian, "Force ripple and jerk minimisation in double sided linear switched reluctance motor used in elevator application," *IET Electric Power Applications*, 10(6): 508-516, 2016.

[9] D. Wang, X. H. Wang, and X. F. Du, " Design and comparison of a high force density dual side linear switched reluctance motor for long rail propulsion application," *IEEE Transactions on Magnetics*, DOI:10.1109/TMAG.2017.2659804, 2017.

[10] W. Hua, H. Hua, N. Y. Dai, G. S. Zhao, and M. Cheng, "Comparative study of switched reluctance machines with half-and full-teeth-wound windings," *IEEE Transactions on Industrial Electronics*, 63(3): 1414-1424, 2016.

[11] B. Zhang, J. Yuan, L Qiu, N. Cheung, and J. Pan, "Distributed coordinated motion tracking of the linear switched reluctance machines based group control system, "*IEEE Transactions on Industrial Electronics*, 63(3): 1480-1489, 2016.

[12] S. Marinkov and B. D. Jager, "Four-quadrant control of 4/2 switched reluctance machines," *IEEE Transactions on Industrial Electronics*, 63(12): 7393-7403, 2016.

[13] Q. L. Wang, H. Chen, T. Xu, R. Nie, J. W. Wang, and S. Abbas, "Position estimation of linear switched reluctance machine with iron losses based on eddy-current effect," *IEEE Transactions on Industrial Electronics*, 10(8): 772-778, 2016.

[14] J. F. Pan, Y Zou, N. Cheung, and G. Z. Cao, "On the voltage ripple reduction control of the linear switched reluctance generator for wave energy utilization," *IEEE Transactions on Power Electronics*, 29(10): 5298-5307, 2014.

[15] R. W. Cao, M. Cheng, and B. F. Zhang, "Speed control of complementary and modular linear flux-switching permanent-magnet motor," *IEEE Transactions on Industrial Electronics*, 62(7): 4056-4064, 2015.

[16] J. K. Lin, K. W. E. Cheng, Z. Zhang, N. C. Cheung, X. D. Xue, and T. W. Ng, "Active suspension system based on linear switched reluctance actuator and control schemes," *IEEE Transactions on Vehicular Technology*, 62(2): 562-572, 2013.

[17] J. F. Pan, Y. Zou, and G. Z. Cao, "Adaptive controller for the double-sided linear switched reluctance motor based on the non-linear inductance modelling," *IET Electric Power Applications*, 7(1): 1-15, 2013.

[18] Q. L. Wang, H. Chen, T. Xu, R. Nie, J. W. Wang, and S. Abbas, "Position estimation of linear switched reluctance machine with iron losses based on eddy-current effect," *IET Electric Power Applications*, 10(8): 772-778, 2016.

[19] J. F. Pan, Y. Zou, and G. Cao, "Adaptive controller for the doublesided linear switched reluctance motor based on the nonlinear inductance modelling," *IET Electric Power Applications*, 7(1): 1-15, 2013.

[20] W. Zhang, B. Chen, "On stabilizability and exact observability of stochastic systems with their applications," *Automatica*, 40: 87-94, 2004.

[21] W. Zhang, B. Chen, "H-representation and applications to generalized Lyapunov equations and lineat stochastic systems," *IEEE Transactions on Automatic Control*, 57(12): 511-518, 2006.

[22] R. Bellman, "Introduction to Matrix Analysis," Philadelphia, PA:SIAM, 1995.

TETHERED SATELLITE DEPLOYMENT BY FRACTIONAL ORDER SLIDING MODE CONTROL

Zheng H. Zhu
Dept. of Mechanical Engineering
York University
Toronto, Ontario, Canada
gzhu@yorku.ca

Junjie Kang
Dept. of Earth & Space Science & Engineering
York University
Toronto, Ontario, Canada
kangjunjie1989@gmail.com

ABSTRACT

This paper develops a fractional order sliding mode control for tethered satellite deployment with the consideration of external disturbances and unmodeled system dynamics. The proposed control law consists of two sub-sliding manifolds that are defined separately for the actuated and unactuated states. This, in turn, generates a control scheme to make all states move toward to the desired states. The proposed control law is proved stable and is capable of suppressing disturbances satisfactorily. Parametric studies are conducted to investigate the influences of fractional order and sub-sliding manifold of unactuated states on the performance of the proposed control law. Effectiveness and robustness of the proposed control law are demonstrated by computer simulations.

KEYWORDS: Tethered satellite; Tether deployment; Fractional order; Sliding mode control; Uncertainty

INTRODUCTION

A fundamental challenge in missions of tethered satellite systems (TSS) is the successful deployment of tether in a quick and stable way. It is well known that the space tether is prone to the undesirable libration in the deployment process, which may lead to tether slackness or winding around the main spacecraft during the deployment [1]. Model-based controllers have been proposed. However, they are affected adversely by nonlinear tension disturbances caused by the unmodeled system dynamics as well as the uncertainty of external disturbances. For example, one of the main nonlinear disturbances, friction in deployment/retrieval mechanisms, is difficult to model precisely. This is because its mathematic models are either too complicated for implementation in control design or hard to be validated experimentally in space environment to sufficient accuracy. The other unmodeled dynamics and environmental/orbital disturbances are the variation of gravitational field, atmospheric drag, temperature oscillation, lunisolar perturbations and other minor forces. Therefore, it is necessary to develop advanced, nonlinear and robust control schemes that are capable of tolerating the inaccuracy of system model to ensure better system performance. In the current work, a fractional order sliding model control (FOSMC) is developed for the control of tether deployment by considering uncertainties.

FRACTIONAL ORDER SLIDING MODE CONTROL

Consider a tethered space system (TSS) with a long tether connecting two satellites at its ends. The system is orbiting in a circular Keplerian orbit in a central gravitation field of earth. The tether is assumed straight, inextensible and massless and the tethered satellites are considered as lumped masses due to the large ratio of tether length over satellites' dimensions. Accordingly, the TSS can be simplified as a dumbbell model. Based upon the assumptions, the TSS motion can be decomposed into the orbital motion of the system's center of mass (CM) and the local libration motion of tether about the CM in the orbital plane, as shown in Fig. 1. It is well-known that the out-of-plane libration is coupled with the in-plane libration through the higher order term, which is very weak. The influence of the out-of-plane libration angle changes has very little effect on the pitch angle in the short period of time, such as in the tether deployment process. Thus, the out-of-plane libration is ignored. The libration motion of the system is described in a local orbital reference frame with its origin at the CM. The y-axis points toward the Earth center. The x-axis lies in the orbital plane perpendicular to the y-axis and pointing towards the moving direction of main satellite. The z-axis completes a right-hand coordinate system. By ignoring the effect of aerodynamic drag due to the short deployment time, the dynamics of tether deployment can be described by a second order nonlinear system [2] in a dimensionless form,

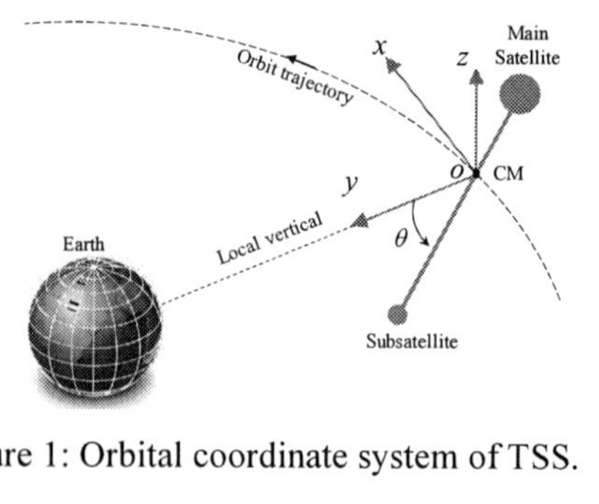

Figure 1: Orbital coordinate system of TSS.

$$l'' - l[(1+\theta')^2 - 1 + 3\cos^2\theta] = -\bar{T}$$
$$\theta'' + 2(l'/l)(1+\theta') + 1.5\sin 2\theta = 0 \quad (1)$$

with dimensionless variables

$$l = L/L_{\max},\ \bar{T} = T/(mL_{\max}\Omega^2),\ \tau = \Omega t,\ ()' = \frac{d()}{d\tau}$$

where $L_{\max}$ is the maximum or desired tether length to be deployment, $\bar{T}$ is the dimensionless tension that is the designed control input in the tension control law, τ is the dimensionless time, and the prime denotes the dimensionless time derivative. Equation (2) shows that the tension and libration is coupled by the nonlinear term $(l'/l)(1+\theta')$. It should be noted that $l \in [0,1]$ and $\bar{T} = 3$ at equilibrium state ($l = 1,\ l' = \theta = \theta' = 0$).

Since we are interested in the state at the end of tether deployment, Eq. (2) is transformed into a set of first order differential equations by introducing new state variables $x_1 = l - 1,\ x_2 = l',\ x_3 = \theta,\ x_4 = \theta'$, such that,

$$\begin{aligned} x_1' &= x_2 \\ x_2' &= f(x_1, x_2, x_3, x_4) - \bar{T} \\ x_3' &= x_4 \\ x_4' &= g(x_1, x_2, x_3, x_4) \end{aligned} \quad (2)$$

where

$$f(x_1, x_2, x_3, x_4) = (1+x_1)(1+x_4)^2 - (1+x_1) + 3(1+x_1)\cos^2 x_3$$
$$g(x_1, x_2, x_3, x_4) = -2x_2(1+x_4)/(1+x_1) - 3\cos x_3 \sin x_3$$

and the desired stable (equilibrium) states at the end of deployment are $x_{1,d} = x_{2,d} = x_{3,d} = x_{4,d} = 0$.

Now, let us introduce the uncertainty of dynamic model as Δf and the external disturbances to control input as Δu. Then, Eq. (3) is revised as,

$$\begin{aligned} x_1' &= x_2 \\ x_2' &= f(x_1, x_2, x_3, x_4) + u + d \\ x_3' &= x_4 \\ x_4' &= g(x_1, x_2, x_3, x_4) \end{aligned} \quad (3)$$

where $u = -\bar{T}$ is the control input and $d = \Delta f + \Delta u$ is the total disturbance. In reality, it is hard to (i) distinguish the disturbance of different sources and (ii) get real mission data about the disturbance. Thus, these two sources of disturbance are combined into a single total disturbance in the current work to demonstrate the ability of disturbance rejection by the proposed controller. Furthermore, the precise knowledge of upper bound of the disturbance is, in general, rather difficult to obtain. In the current work, it is assumed limited by an arbitrary positive value η, such that, $\|d\| \le \eta$. . To compensate for the total disturbance, a fractional order sliding mode control law is proposed in the following.

The dynamics of the tether deployment in Eq. (4) is a typical nonlinear coupled and underactuated system, where the libration mode is indirectly affected by the control input through nonlinear modal coupling. To ensure the convergence of all modes, define a fractional order sliding manifold that consists of two sub-manifolds (s_θ and s_l). The sub-manifolds correspond to the libration and length modes of TSS, respectively

$$s_\theta = x_3 \quad (4)$$
$$z = (c_3/c_1)s_\theta \quad (5)$$
$$s_l = c_1 D^{1-\alpha}(x_1 - z) + c_2 D^{1-\alpha} x_2 \quad (6)$$

where α is the order of fractional operator, $\alpha \in (0,1)$, $c_1 > 0$, $c_2 > 0$ and c_3 is the positive parameter to be determined. The intermediate state z in s_l, Eq. (6), interconnects two decoupled manifolds and transforms the equilibrium point ($x_1 = 0$, $x_2 = 0$) to ($x_1 = z$, $x_2 = 0$). Note that from Eq. (5) $s_\theta = 0$ is equivalent to $z = 0$. The variable $s_l = 0$ is reduced to zero when $s_\theta = 0$. Thus, all states are driven to the point of equilibrium.

Applying the fractional operator D^α to both sides of Eq. (6) and then combining with Eq. (3) yield

$$D^\alpha s_l = c_1(x_1' - z') + c_2 x_2' = c_1 x_2 + c_2 f - c_3 x_4 + c_2 u + c_2 d \quad (7)$$

Based on Eq. (7), the fractional order sliding mode control is proposed as,

$$\begin{cases} u = u_{eq} + u_{sw} \\ u_{eq} = -[c_1 x_2 + c_2 f(x) - c_3 x_4]/c_2 \\ u_{sw} = \dot{s}_l = -(k\,\mathrm{sat}(s_l/\varepsilon) + \delta s_l) \end{cases} \quad (8)$$

where

$$\mathrm{sat}(s_l/\varepsilon) = \begin{cases} \mathrm{sgn}(s_l/\varepsilon) & \text{if } |s_l/\varepsilon| \ge 1 \\ s_l/\varepsilon & \text{if } |s_l/\varepsilon| < 1 \end{cases}$$

u_{eq} and u_{sw} are the equivalent and switch control inputs, $\mathrm{sat}(s_l/\varepsilon)$ is the saturation function with $\varepsilon = 0.01$, k and δ are the positive parameters to be determined to ensure the existence and convergence of sliding mode control law, respectively.

Define a positive definite Lyapunov function candidate

$$V = s_l^2 \quad (9)$$

Applying the fractional operator D^α to both sides of Eq. (9) yields [3]

$$\begin{aligned} D^\alpha V &= s_l D^\alpha s_l + \sum_{i=1}^{\infty} \frac{\Gamma(1+\alpha)}{\Gamma(1-i+\alpha)\Gamma(1+i)} D^i s_l D^{\alpha-i} s_l \\ &\le s_l D^\alpha s_l + \left| \sum_{i=1}^{\infty} \frac{\Gamma(1+\alpha)}{\Gamma(1-i+\alpha)\Gamma(1+i)} D^i s_l D^{\alpha-i} s_l \right| \end{aligned} \quad (10)$$

Substituting $\left| \sum_{i=1}^{\infty} \frac{\Gamma(1+\alpha)}{\Gamma(1-i+\alpha)\Gamma(1+i)} D^i s D^{\alpha-i} s \right| \le \gamma |s|$ (γ is a positive constant) [4] and Eqs. (3), (7)-(8) into Eq. (10) yields

$$\begin{aligned} D^{\alpha}V &\le s_l[c_1 x_2 + c_2 f - c_3 x_4 + c_2 u + c_2 d] + \gamma|s_l| \\ &= -c_2 s_l[ksgn(s_l) + \delta s_l - d] + \gamma|s_l| \\ &\le c_2(\eta + \gamma/c_2 - k)|s_l| \end{aligned} \tag{11}$$

By letting $k > \eta + \gamma/c_2$ leads to $D^{\alpha}V < 0$. Thus, the system will reach and stay on the sliding manifold $s_l = 0$. Once the sliding mode is satisfied, i.e., $D^{\alpha}s_l = c_1\dot{x}_1 + c_2\dot{x}_2 - c_3\dot{x}_3 = 0$, one can derive the following relationship

$$\dot{x}_2 = -(c_1/c_2)x_2 + (c_3/c_2)x_4 \tag{12}$$

Substituting Eq. (12) into Eq. (2) and linearizing the resulting equation in the neighborhood of the equilibrium point ($x_{1,d} = x_{2,d} = x_{3,d} = x_{4,d} = 0$) lead to

$$\begin{Bmatrix} x_1' \\ x_2' \\ x_3' \\ x_4' \end{Bmatrix} = \begin{bmatrix} 0 & 1 & 0 & 0 \\ 0 & -c_1/c_2 & 0 & c_3/c_2 \\ 0 & 0 & 0 & 1 \\ 0 & -2 & -3 & 0 \end{bmatrix} \begin{Bmatrix} x_1 \\ x_2 \\ x_3 \\ x_4 \end{Bmatrix} \tag{13}$$

The eigenvalue equation of Eq. (13) is

$$\lambda\{c_2\lambda^3 + c_1\lambda^2 + (3c_2 + c_3)\lambda + 3c_1\} = 0 \tag{14}$$

Obviously, the roots of Eq. (14) are $\lambda_1 = 0$ and $\mathrm{Re}(\lambda_{2,3,4}) < 0$ because $c_{1,2,3} > 0$. Accordingly, the state variables (x_2, x_3 and x_4) are asymptotically stable. For the state variable x_1 with zero eigenvalue $\lambda_1 = 0$, the state variable x_1 converges to zero and becomes stable asymptotically based on Eq. (12). Thus, the proposed fractional order sliding mode control law is asymptotically stable.

SIMULATION RESULTS AND DISCUSSION

The proposed fractional order sliding mode (FOSM) tension controller is examined by computer simulations and its effectiveness is compared with three existing control methods, such as, the standard integer order sliding mode (SM) tension controller, the PD tension controller [2], and the fractional order PD (FOPD) tension control strategy [5]. For comparison purposes, the cases with the same system parameters considered in [2] and [5] are used as a benchmark, such that, the TSS is orbiting in a circular orbit with the orbit rate of 1.18x10^{-3} rad/s. The initial conditions of dimensionless state variables are assumed as $x_1(0) = -0.99$, $x_2(0) = 0.5$, $x_3(0) = 0$, $x_4(0) = 0$. The parameters of sliding manifold are chosen as $c_1 = 1$, $c_2 = 1$, $c_3 = 0.8$ to achieve a fast and stable deployment. It should be noted that all three parameters will affect the results generally. However, by the definitions of two sub-sliding manifolds, one can easy find the c_1 and c_2 mainly act on the tether length and the c_3 acts on the sliding manifold s_θ and appears in the first sub-sliding manifold s_l as the intermediate variable z. To focus on the influence of the sliding manifold s_θ on the tether deployment, the c_1 and c_2 are kept unchanged within the stable domain and only the variation of c_3 is investigated. The gains of $k = 0.4$ and $\delta = 0.2$ are used in the switching control law for both SM and FOSM control laws. Furthermore, the fractional order in all fractional order controllers is defined as $\alpha = 0.5$ except in Section 4.1, where the influence of fractional order is studied parametrically. The total disturbance resulting from the model uncertainty and environmental disturbance is introduced as $d(\tau) = 0.2\sin(\pi\tau)$, in order to demonstrate the ability of suppressing disturbance of different control laws.

Influence of fractional order on controller

First of all, the influence of fractional order on the performance of the proposed FOSM tension controller is investigated by varying the fractional order, when subjected to disturbance mentioned above. A series of fractional orders α = 0.1, 0.3, 0.5, 0.7 and 0.9 are used with the gains of $k = 0.4$ and $\delta = 0.2$ in the switching control law.

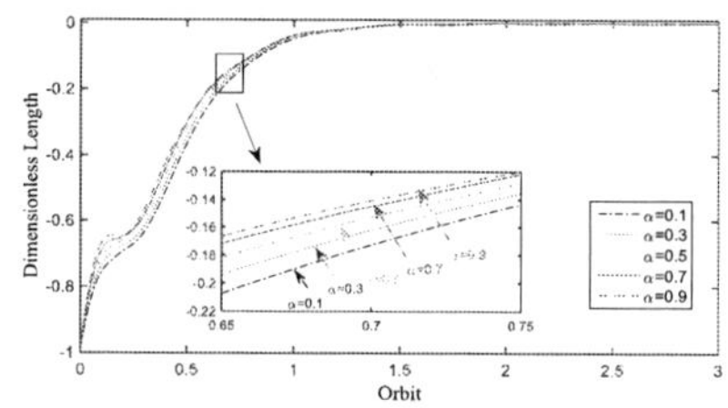

Fig. 2. The tether length vs different fractional orders.

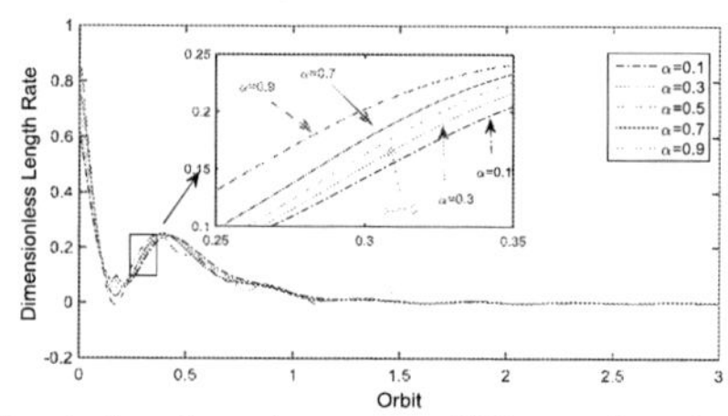

Fig. 3. The tether length rate vs different fractional orders.

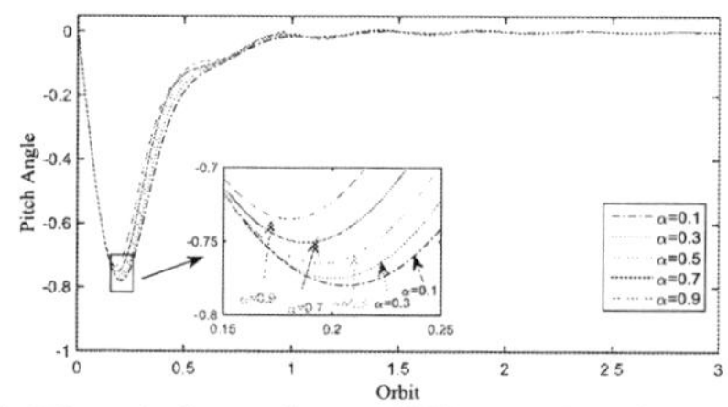

Fig. 4. The pitch angle vs different fractional orders.

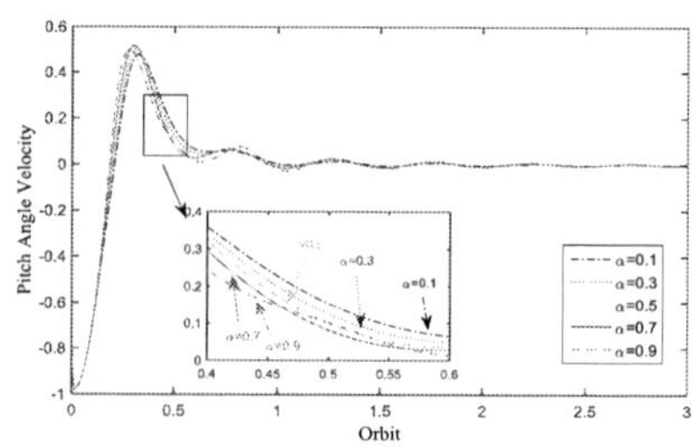

Fig. 5. The pitch angular velocity vs different fractional orders.

Influence of sliding sub-manifold s_θ

The control on the un-actuated state (x_3 and x_4) is achieved by the sub-manifold s_θ in the current underactuated system, where the tether is deployed without thrust at the subsatellite. The influence of the sliding sub-manifold is investigated by varying the parameter c_3 from 0.7 to 1.5. The other parameters of sliding surface, the disturbance and the gains in the switch control law are kept the same as the previous section. Finally, the fractional orders $\alpha = 0.5$ is used in the current case.

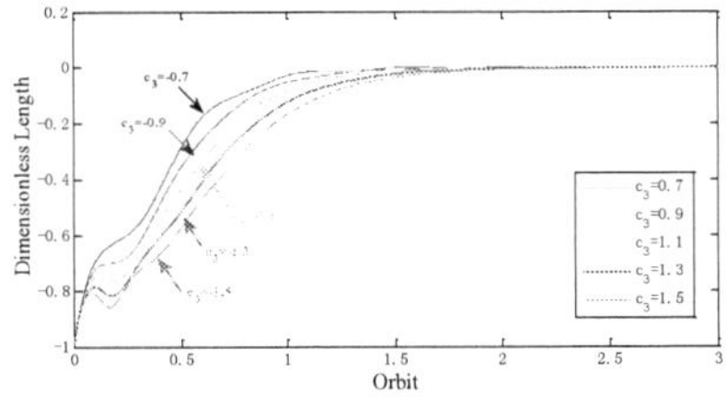

Fig. 6. The tether length vs different parameters c_3

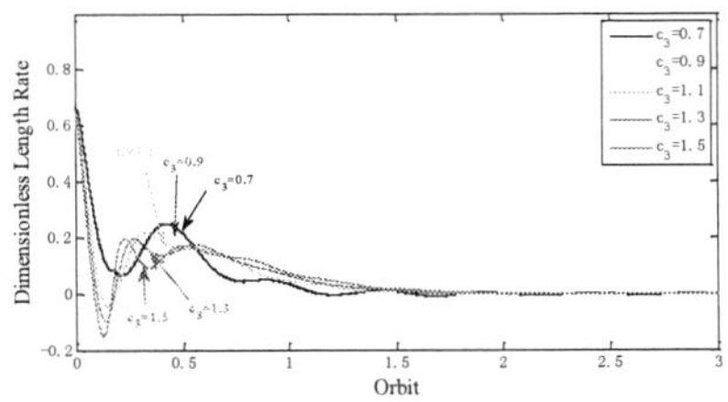

Fig. 7. The tether length rate vs different parameters c_3

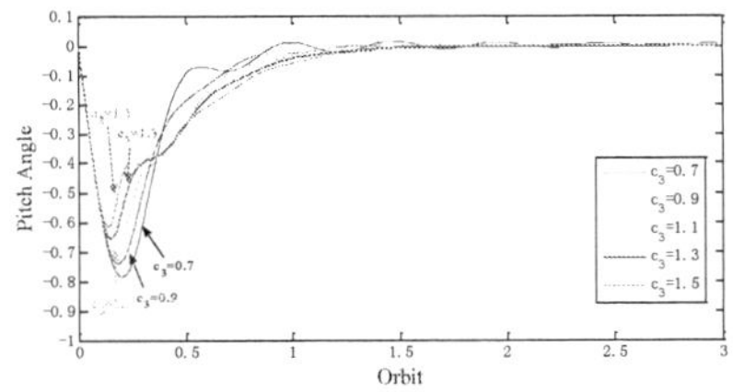

Fig. 8. The pitch angle vs different parameter c_3

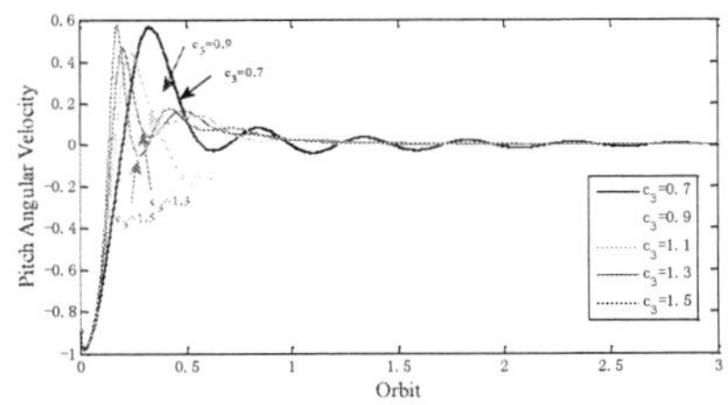

Fig. 9. The pitch angular velocity vs different parameter c_3

Comparisons with other control methods

The effectiveness of the proposed FOSM control law is demonstrated by comparing the control performance of three different control laws (PD, FOPD, SM). The parameters of the PD tension control law are the same as those in [2], and the FOPD tension control strategy in [5].

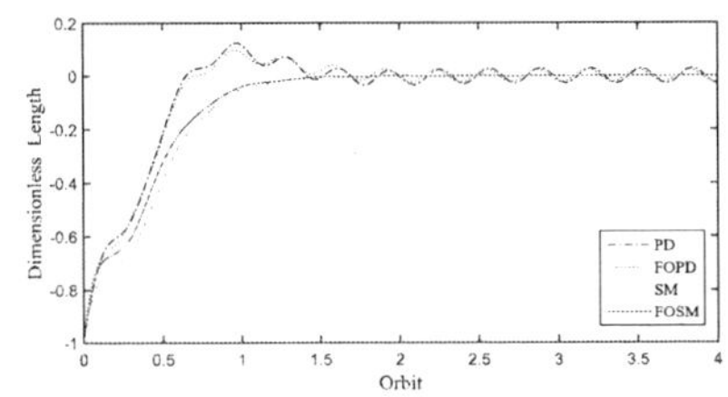

Fig. 10. Comparison of tether length with different control methods

To further demonstrate the effectiveness of the proposed fractional order sliding mode control law in suppressing the disturbance in a wider frequency range, an external disturbance that combining low and high frequencies is inputted into the TSS,

$$d_1(\tau) = 0.2\left[\sin(\pi\tau) + \sin(100\pi\tau)\right] \tag{15}$$

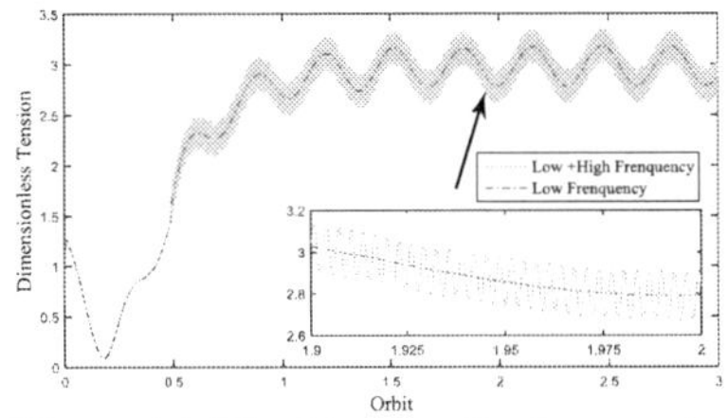

Figure 14. The history of tether tension

CONCLUSIONS

The proposed fractional order sliding mode control law demonstrates excellent disturbance rejection capability in a wide frequency range studied. The simulation results show that the fractional order sliding mode control law is more competitive on these two performance indexes, the settling time and the maximum value of pitch angle subject to unwanted disturbance in the control input, compared with the existing methods. Finally, the control parameter c_3 for the sliding sub-manifold of unactuated states has a positive effect in constraining the tether libration.

ACKNOWLEDGMENTS

This work is supported by Natural Sciences and Engineering Research Council of Canada.

REFERENCES

1. Carroll, J.A., 1985. Guidebook for analysis of tether applications. National Aeronautics and Space Administration, Springfield, VA.

2. Pradeep, S., 1997. A new tension control law for deployment of tethered satellites. Mechanics Research Communications 24, 247-254.

3. Fugere, J., Gaboury, S., Tremblay, R., 2012. Leibniz rules and integral analogues for fractional derivatives via a new transformation formula. Bulletin of Mathematical Analysis & Applications 4(2), 72-82.

4. Aghababa, M.P., 2014. A Lyapunov-based control scheme for robust stabilization of fractional chaotic systems. Nonlinear Dyn 78, 2129-40.

5. Sun, G., Zhu, Z.H., 2014. Fractional-order tension control law for deployment of space tether system. Journal of Guidance, Control and Dynamics 37, 2062-2066.

Proceedings of the 26th CANCAM
Victoria, British Columbia, Canada, May 28-June 1, 2017

DISCRETE-TIME PIECEWISE AFFINE SYSTEM IDENTIFICATION FOR MPPT OPERATION OF A WIND TURBINE

Peyman Sindareh-Esfahani
Mechanical and Manufacturing Engineering Department
University of Calgary
Calgary, Alberta, Canada
peyman.sindarehesfah@ucalgary.ca

Jeffery Kurt Pieper
Mechanical and Manufacturing Engineering Department
University of Calgary
Calgary, Alberta, Canada
pieper@ucalgary.ca

ABSTRACT

In this paper, a discrete-time PieceWise Affine (PWA) model of a wind turbine during Maximum Power Point Tracking (MPPT) region is identified. A clustering based identification method is utilized to create PWA maps for nonlinear aerodynamic torque and thrust force functions. These identified maps are replaced with the nonlinear static functions of the dynamic model of the wind turbine. Characteristics of a 5MW wind turbine are considered and the resulting model, which consists of 20 sub-regions are compared with the nonlinear dynamic model. Simulation results represent the effectiveness of the PWA model since the response of the identified PWA model is close to the nonlinear one. The simulation results show less than 0.1% root mean square error between two models.

KEYWORDS: Piecewise affine (PWA) models, wind turbine, clustering based identification, system identification, nonlinear systems

INTRODUCTION

Wind energy systems have gained tremendous attention during past decades as an abundant and affordable renewable energy resource. To capture more energy from the wind flow, dimensionally bigger wind turbines are needed. As wind turbines become larger, more demand on design, quality of materials and also new control strategies are required. Moreover, In order to decrease the cost to power ratio of wind turbines, advanced control methods play an important role.

Model Predictive Control (MPC) is an advanced control approach which offers attractive solutions for regulation and tracking problems. Nowadays, MPC is a preferred control strategy for many industrial processes [1], [2]. The practical use of MPC is due to the fact that processes need to be operated under tight performance specifications. Besides, more constraints need to be satisfied especially in sensitive processes like a wind turbine. An accurate model is essential in order to design an efficient model based controller.

Hybrid models have gained attention over the past two decades since they provide a powerful means of analysis and design for realistic processes. Significant activity in this field to analysis the behavior of hybrid systems has been done [3]. This category of systems can be modeled applying different strategies including mixed logical dynamical systems [4], switched systems [5], and piecewise affine (PWA) or linear systems [6]. PWA systems are multi-model and propose a useful and efficient framework for analyzing complex nonlinear dynamic models. Many common nonlinear elements including relays, saturations, and switches can be modeled by means of PWA systems. In [7], a PWA model of a pitch-controlled wind turbine is identified in the vicinity of the optimal wind turbine static characteristics. Same work had been done in [8] for the optimal active power control of a wind farm.

In this work, regardless of the optimal characteristic of the wind turbine, a PWA model is developed for wind turbines. A PWA map for nonlinear functions of the wind turbine model including aerodynamic torque and thrust force is constructed. A set of 1000 data points which randomly choose is used for identification purpose. This model is identified based on clustering based PWA identification technique. The wind speed and rotor speed are considered as the inputs of nonlinear aerodynamic torque and thrust force functions. By generating a wind speed profile and acquiring the characteristics of a 5MW wind turbine, both models are simulated in SIMULINK. A comparison between the nonlinear model of the 5MW wind turbine and PWA model is studied to show a good approximation of the PWA model.

The rest of the paper is consisting of an introduction of PWA systems and the clustering based identification technique. After that, the nonlinear dynamic model of a wind turbine introduces and PWA maps of the nonlinear functions identify. At last, simulation results bring and the paper conclude.

DISCRETE-TIME PIECEWISE AFFINE MODEL

Let consider the following discrete-time piecewise affine system,

$$x(k+1) = A_i x(k) + B_i u(k) + B_{w,i} w(k) + b_i$$

$$z(k) = \begin{bmatrix} C_i x(k) \\ D_i u(k) \end{bmatrix} \tag{1}$$

$$for \quad x(k) \in X_i, \quad i \in \wp$$

where $x(k) \in \mathbb{R}^n$, $u(k) \in \mathbb{R}^m$, $w(k) \in \mathbb{R}^r$, and $z(k) \in \mathbb{R}^z$ are the state, controlled input, exogenous input, and controlled output, respectively. b_i is the affine term in each sub-region. In Eq. (1), the set X_i is defined as a partition of the state space into a number of closed polyhedral subspaces,

$$X_i = \{(x,u) \mid H_i \begin{bmatrix} x \\ u \end{bmatrix} \le K_i\}_{i\in\wp} \tag{2}$$

where H_i, K_i are appropriate dimensional matrices. $\wp$ is the index set of these sub-regions which its values are the elements of the set $\{1,2,...,s\}$. s is the number of PWA system sub-regions.

CLUSTERING BASED IDENTIFICATION OF A PIECEWISE AFFINE SYSTEM

In this section, clustering based identification is illustrated. This identification technique is used for nonlinear static functions although the main process is utilized for the identification of a class of piecewise affine systems called a piecewise affine autoregressive exogenous (PWARX) model [8]. A static nonlinear function can be described as,

$$y_k = f(u_k) + e_k \tag{3}$$

This function can be approximated with the following PWA map,

$$f(u_k) = \begin{cases} \theta_1^T [u_k \ 1]^T & if \ u_k \in X_1 \\ \vdots & \\ \theta_s^T [u_k \ 1]^T & if \ u_k \in X_s \end{cases} \tag{4}$$

where u_k is the static function input, θ_i is the affine function parameters, $f(u_k)$ is the output and closed polyhedral subspaces are described by X_i. Consider that N sets of data (y_k, u_k) are collected for identification purpose. The considered identification problem consists in finding the PWA map that best matches the given data according to a specified fitting criterion [10]. Clustering based identification mainly consisted of four estimation steps of the following [11].

1) Building local data
2) Clustering Algorithm
3) Sub-model parameter estimation
4) Region estimation

In the first step, a local dataset collects a point (y_k, u_k) and its nearest c - 1 points. Based on the proposed PWA map identification method in [9], this local data set can be a dataset which used for identification propose if the N points of dataset chose randomly so that covered the entire possible response of the nonlinear function. In the second step, the dataset is classified into several clusters by using a data clustering technique like the well-known K-means methods used in [12]. Once the dataset is properly clustered, we can apply the standard Least Squares (LS) method to identify the parameters of each sub-region [13], [14]. Moreover, since the polyhedral regions on the identification space are characterized by boundary hyperplanes between adjacent clusters, we need to estimate these hyperplanes. The well-known Support Vector Classifier (SVC) [15] are typically used for this purpose.

WIND TURBINE SYSTEM OPERATION

Three regions can be identified for the operation of a wind turbine. For the wind speeds below the rated speed, the turbine captures the maximum available power in the wind flow. This region is specified by Region I. In this region, the pitch angle is fixed and the optimal rotor speed is tracked via generator torque control. In Region III, the turbine should work at the constant rated power. In this region, there is more wind power available than the captured rated power. Therefore, the turbine works at lower efficiency. In this region, the generator torque is constant at its rated value and pitch control limits the rotor speed for less fluctuation in output power. Besides, the generator torque can vary so that the multiply of rotor speed and generator torque remains constant. Both constant torque and constant power strategies can be used in Region III. Region II is a transition region between Regions I and III, and authority over rotor speed is transferred between the generator torque and the blade pitch. In some cases, this region may be ignored by connecting the maximum power region to the constant power region [16]. The purpose of this paper is bounded to Region I. Fig. 1. shows the operation regions of a wind turbine.

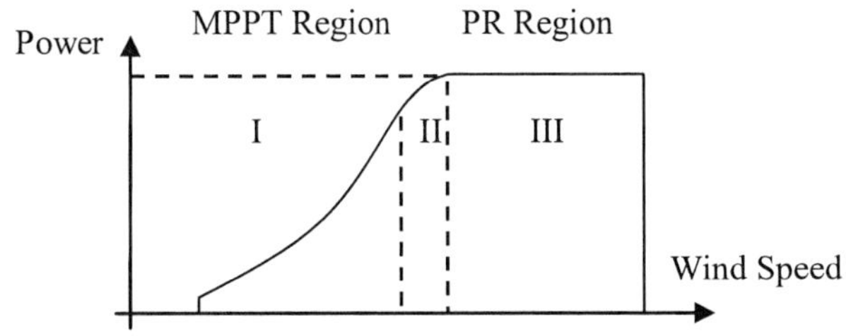

Figure 1: Ideal power versus wind speed curve.

NONLINEAR MODEL OF A WIND TURBINE SYSTEM AND PROBLEM DEFINITION

Two main dynamical equations express the aerodynamic and mechanical modeling of a wind turbine at MPPT operation region. Eq. (1) illustrates the rotor speed dynamics and Eq. (2) depicts elastic tower fore–aft motion [17].

$$(J_R + J_G)\dot{\Omega} + C_t\Omega + T_g - F_a(\Omega, V_w) = 0 \tag{5}$$

$$M_T\ddot{d} + C_T\dot{d} + K_T d - F_b(\Omega, V_w) = 0 \tag{6}$$

In Eq. (1) V_w is the wind speed and Ω is the rotor speed and T_g is the electrical torque induced in the generator. The term $C_t\Omega$ is the resistance torque in the gearbox proportional to rotor speed. J_R and J_G are the rotor and generator moments of inertial, respectively. In Eq. (2), the term d is tower-top fore-aft displacement. M_T, C_T, and K_T are the equivalent mass, damping coefficient, and stiffness of the wind turbine tower, respectively. F_a is the aerodynamic torque induced in the blades and F_b is the drag force on the turbine due to the wind flow. F_a and F_b are given by,

$$F_a = \frac{1}{2}\rho\pi R^3 \frac{C_p(\lambda,\beta)}{\lambda}(V_w - \dot{d})^2 \qquad (7)$$

$$F_b = \frac{1}{2}\rho\pi R^2 C_f (V_w - \dot{d})^2 \qquad (8)$$

Where λ is the so-called tip-speed ratio given by,

$$\lambda = \frac{R\Omega}{V_w - \dot{d}} \qquad (9)$$

where ρ is the air density and R is the length of blades. C_p is the power coefficient and C_f is the drag coefficient.

PIECEWISE AFFINE MODEL IDENTIFICATION OF A WIND TURBINE

To the rest of the paper, a 5 MW wind turbine is considered for the simulation study. The goal is to identify a very good approximation of nonlinear static functions (F_a and F_b) based on clustering based identification method introduced in the previous section. In other word, a PWA map is identified to approximate the static nonlinear function. Fig. 2. shows the behavior of the nonlinear functions F_a and F_b with respect to wind speed and rotor speed. A set of 1000 data points is created to identify the PWA maps. For clustering purpose, the well-known K-means is adopted and in order to identify boundary hyperplanes, a perceptron based multi-class linear classifier using the Kesler's construction is utilized. 20 regions are considered to identify a good approximation for nonlinear functions. Identified PWA maps for aerodynamic torque (F_a) and thrust force (F_b) are used to substitute nonlinear static functions in the nonlinear wind turbine model (5), (6). This conclude a PWA model of the form of Eq. (1) with,

$$x = [\Omega, d, \dot{d}],\quad u = [T_g],\quad w = [V_w] \qquad (10)$$

Here, the electrical torque T_g is considered as the control input of the system and V_w as the exogenous input. Moreover, in should be noted that the inequality in Eq. (2) is valid for one polytope since there are no polytopes that overlap. On the other words, we have, $\cup_1^s X_i = X$.

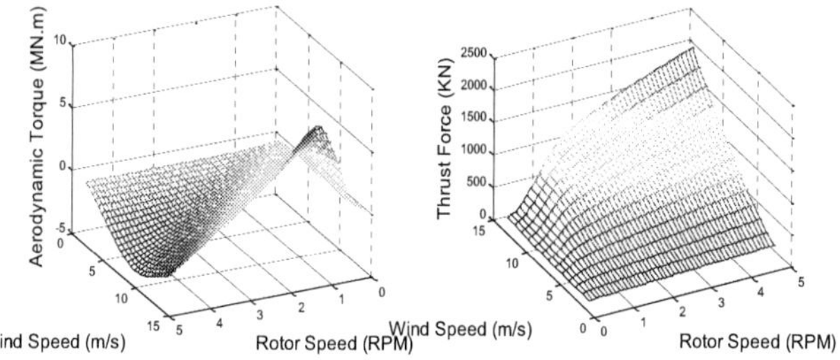

Figure 2: Aerodynamic torque and thrust force vs. wind speed and rotor speed

SIMULATION RESULTS AND MODEL VERIFICATION

The PWA model is implemented in SIMULINK. A wind profile with the reference wind speed of 8 (m/s) is generated by TurbSim software shown in Fig. 3. Figs. 4. shows the simulation results for the nonlinear parts of the wind turbine model. The simulation results depict that the PWA maps can properly approximate the nonlinear functions of the wind turbine model. Figs. 5. and 6. depict the states of the system which are rotor speed and tower displacement of the nonlinear wind turbine model compare to PWA wind turbine model. These figures show good fit between two models with the root mean square error of 0.02% for rotor speed and 0.1% for tower displacement.

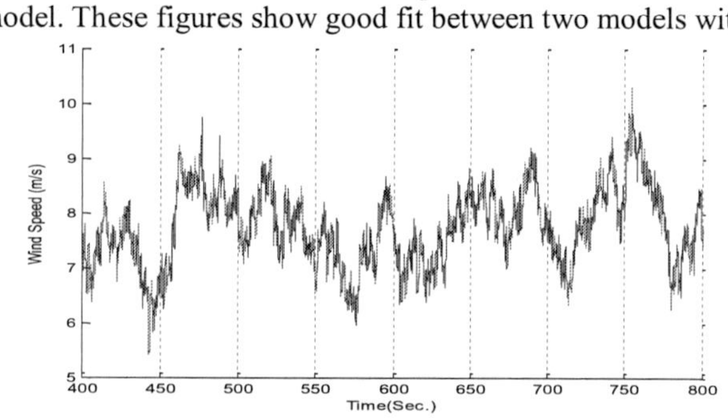

Figure 3: Turbulent wind speed profile of 8 m/s mean value

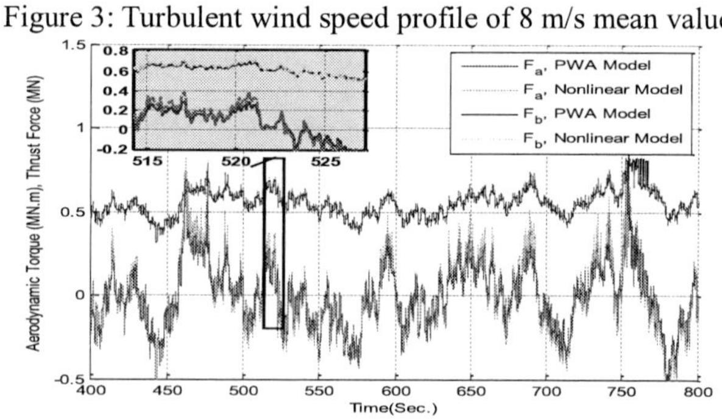

Figure 4: F_a and F_b for open loop simulation

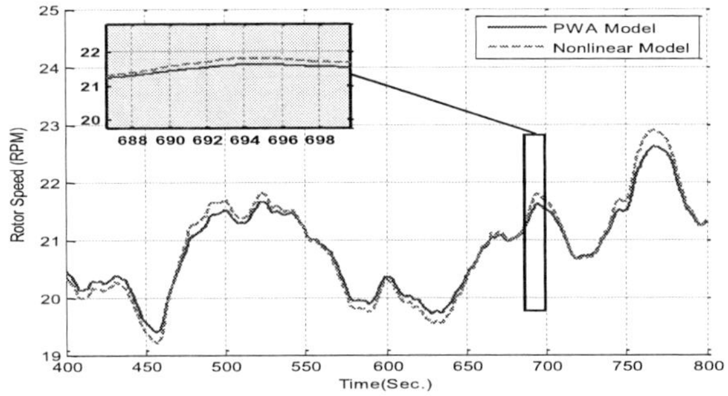

Figure 5: Rotor speed for open loop simulation

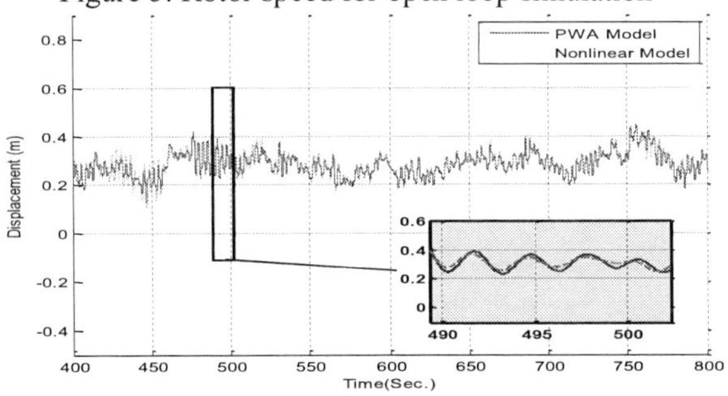

Figure 6: Top-tower displacement for open loop simulation

CONCLUSIONS

A discrete-time PWA system model of a wind turbine in MPPT region was presented in this paper. Clustering based identification method for PWA systems applied to the nonlinear functions of a wind turbine model including aerodynamic torque and thrust force. A set of 1000 data points is generated with 20 sub-regions for the identification process. According to the simulation results, a precise PWA model is obtained throughout the clustering based identification method. Obtained PWA map can be used for designing model-based controllers for the wind turbine.

REFERENCES

1. Akter, M.P., Mekhilef, S., Tan, N.M.L. and Akagi, H., 2016. "Modified model predictive control of a bidirectional ac–dc converter based on Lyapunov function for energy storage systems". *IEEE Transactions on Industrial Electronics*, *63*(2), pp.704-715.
2. Sindareh-Esfahani, P. and Tabatabaei, S.S. Pieper, J.K., 2017. Model Predictive Control of a Heat Recovery Steam Generator during Cold Start-up Operation Using Piecewise Linear Models. *Applied Thermal Engineering*. http://dx.doi.org/10.1016/j.applthermaleng.2017.03.041
3. Cervantes, I., Femat, R. and Leyva-Ramos, J., 2007. "Study of a class of hybrid-time systems". Chaos, Solitons & Fractals, 32(3), pp.1081-1095.
4. Bemporad, A., 2004. "Efficient conversion of mixed logical dynamical systems into an equivalent piecewise affine form". IEEE Transactions on Automatic Control, 49(5), pp.832-838.
5. Skafidas, E., Evans, R.J., Savkin, A.V. and Petersen, I.R., 1999. "Stability results for switched controller systems". Automatica, 35(4), pp.553-564.
6. Ferrari-Trecate, G., Cuzzola, F.A., Mignone, D. and Morari, M., 2002. "Analysis of discrete-time piecewise affine and hybrid systems". Automatica, 38(12), pp.2139-2146.
7. Vašak, M., Hure, N. and Perić, N., 2011, May. "Identification of a discrete-time piecewise affine model of a pitch-controlled wind turbine". In *MIPRO, 2011 Proceedings of the 34th International Convention* (pp. 744-749). IEEE.
8. Zhao, H., Wu, Q., Guo, Q., Sun, H. and Xue, Y., 2015. Distributed Model Predictive Control of a Wind Farm for Optimal Active Power ControlPart II: Implementation With Clustering-Based Piece-Wise Affine Wind Turbine Model. *IEEE Transactions on Sustainable Energy*, *6*(3), pp.840-849.
9. Nakada, H., Takaba, K. and Katayama, T., 2005. "Identification of piecewise affine systems based on statistical clustering technique". *Automatica*, *41*(5), pp.905-913.
10. S. Paoletti., 2004 "Identification of piecewise affine models," Ph.D. dissertation, Dept. Inf. Eng., Univ. Siena, Siena, Italy,.
11. Paoletti, S., Juloski, A.L., Ferrari-Trecate, G. and Vidal, R., 2007. "Identification of hybrid systems a tutorial. *European journal of control"*, *13*(2), pp.242-260.
12. Ferrari-Trecate, G., Muselli, M., Liberati, D. and Morari, M., 2003. "A clustering technique for the identification of piecewise affine systems". *Automatica*, *39*(2), pp.205-217.
13. Sindareh-Esfahani, P., Ghaffari, A. and Ahmadi, P., 2014. Thermodynamic modeling based optimization for thermal systems in heat recovery steam generator during cold start-up operation. *Applied Thermal Engineering*, *69*(1), pp.286-296.
14. Sindareh-Esfahani, P., Habibi-Siyahposh, E., Saffar-Avval, M., Ghaffari, A. and Bakhtiari-Nejad, F., 2014. Cold start-up condition model for heat recovery steam generators. *Applied Thermal Engineering*, *65*(1), pp.502-512.
15. Söderström, T., & Stoica, P. 1989. "*System identification"*. Englewood, Cliffs, NJ: Prentice-Hall
16. Abdullah, M.A., Yatim, A.H.M., Tan, C.W. and Saidur, R., 2012. "A review of maximum power point tracking algorithms for wind energy systems". *Renewable and sustainable energy reviews*, *16*(5), pp.3220-3227.
17. Jafarnejadsani, H., Pieper, J. and Ehlers, J., 2013. "Adaptive control of a variable-speed variable-pitch wind turbine using radial-basis function neural network". *IEEE Transactions on Control Systems Technology*, *21*(6), pp.2264-2272.

Micro-fluidics

ENDOSCOPIC ROBOTIC TOOL SYSTEM FOR PROJECT NEUROARM

Don Peterson
Graduate Programme in Mechanical Engineering
The University of Calgary
degpeter@ucalgary.ca

Kourosh Zareinia, Garnette Sutherland
Department of Clinical Neurosciences and Hotchkiss Brain Institute
The University of Calgary
neuroarm@ucalgary.ca

Salvatore Federico
Department of Mechanical & Manufacturing Engineering
The University of Calgary
salvatore.federico@ucalgary.ca

ABSTRACT

Technology and surgical advances are facilitating the development of new modalities to improve neurosurgical patient outcome. The development of an endoscopic robotic tool system is described which will facilitate autonomous and surgeon driven neoplasm assessment and intervention within an MRI environment.

KEYWORDS: neurosurgery, neuroArm, robotic tool, neuro endoscopy, brain tumor

INTRODUCTION

Advances in imaging technology, surgical simulation, sensor technology and robotics are enabling surgical procedures to be performed with increased safety, higher precision, improved outcomes, and reduced costs. These advances allow for more complete removal of pathological tissue, preservation of eloquent tissue, fewer post-operative complications, shorter hospital stays and reduced follow-up [1].

This paper discusses the development of an endoscopic robotic tool system designed to leverage these technological and surgical advances in order to improve neurosurgical patient care.

PROJECT NEUROARM

Project neuroArm includes a seven degree of freedom (DOF) image guided, MRI compatible / safe robot (Figure 1), and its associated technologies developed at the University of Calgary [2-4]. The surgical suite, shown in Figure 2, consists of the neuroArm robot, a moveable 3.0 Tesla intraoperative (iMRI) imaging system [5, 6], and a separate control room. The control room, shown in Figure 3, includes haptic hand controllers, three-dimensional (3D) MRI display with tool overlay, stereoscopic view of the operating field, and a virtual image of the manipulators relative to the patient [7].

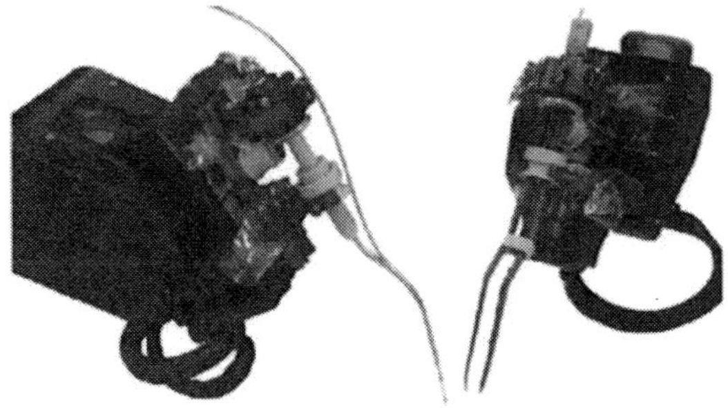

Figure 1 – Project neuroArm robot manipulators

NeuroArm replicates the micro-neurosurgical environment and has been used for more than 70 cases including, but not limited to, low and high grade glioma, meningioma, cavernous angioma, and acoustic Schwannoma. NeuroArm augments the surgical environment with sub-millimeter precision and accuracy, physiological tremor removal via low-pass filtering, haptic force feedback and motion and force scaling.

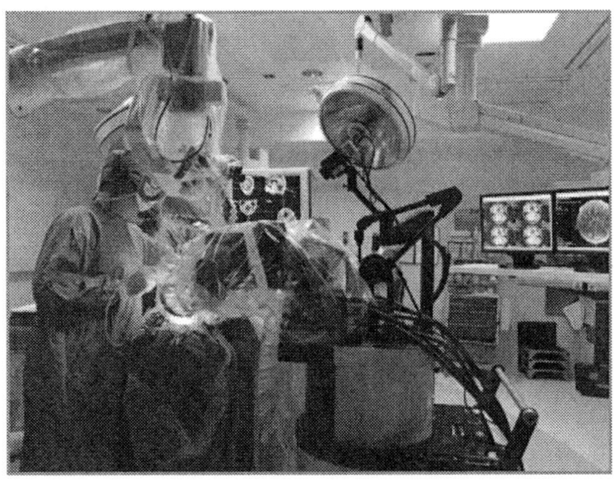

Figure 2 – Project neuroArm surgical suite

Research is currently underway to quantify surgical tool forces throughout various procedures in order to enhance robot hardware and software design, and for training purposes [8].

Figure 3 – Project neuroArm control room

ENDOSCOPIC ROBOTIC TOOL SYSTEM SCOPE

The endoscopic robotic tool system adds to Project neuroArm by providing non-line-of-sight access to the desired surgical location. The system is designed to be mounted on the manipulator(s) of the neuroArm robot or to be used stand-alone, similarly to existing manual neuro-endoscopes.

The system is designed to address a variety of pathologies - pineal cysts and neoplasms; third ventricle lesions (i.e. colloid cysts); fourth ventricle lesions (i.e. medulloblastoma); lateral ventricle meningioma, ependymoma, and astrocytoma; and skull base neoplasms (in particular meningioma and pituitary adenoma) [9]. Additionally, the system is designed to augment neuroArm in accessing difficult-to-reach pathologies [10], to serve as a platform for emerging in-situ imaging and tissue characterization technologies [11-14], and to provide localized therapeutic delivery in an MRI environment.

MECHANICAL DESIGN

The endoscopic robotic tool system is configurable in a timely and cost-effective manner to provide a patient specific solution with respect to tool geometry, DOF, actuation and control, working channel size and distribution, tool and sensor suite, and integration with the neuroArm robot [15]. The design of a 2 DOF system with 5 working channels, is shown in Figure 4.

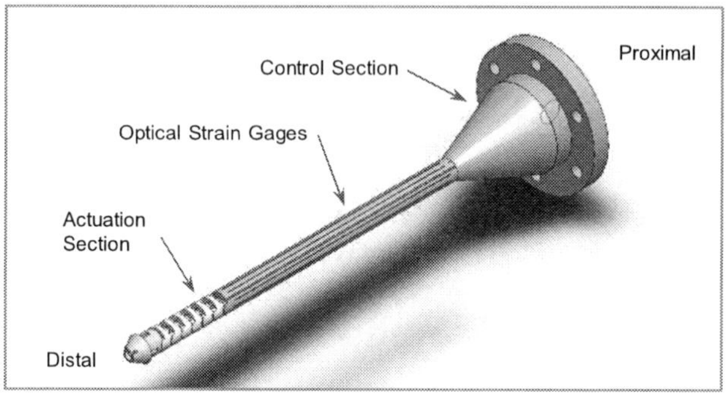

Figure 4 – Five working channel, 2 DOF endoscopic robotic tool design

The tools are designed to minimize the use of metal parts to limit image interference and tissue heating concerns [16]. Articulated joints employ simple 2D - 1 DOF and 3D - 2 DOF motions most suitable to the procedure and to patient anatomy [17]. Tool designs are composed of discrete actuating components; however, the stiffness of continuous elements permits the analysis of overall deformation as a continuum robot [18]. Figure 5 shows details of the distal end of the tool design of Figure 4.

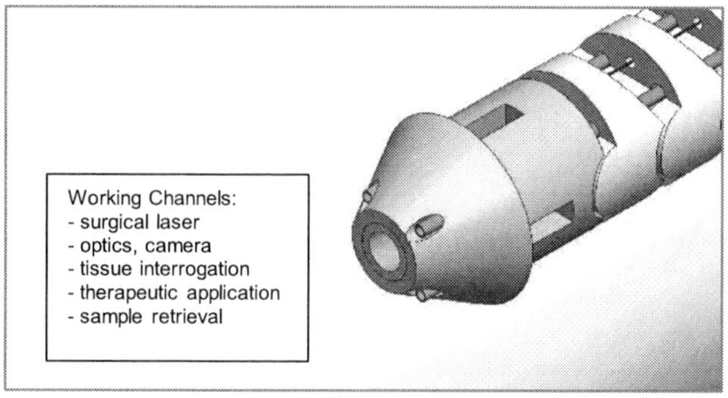

Figure 5 – Working channel arrangement

Figure 6 shows a prototype of robotic tool based on the design of Figures 4 and 5. Each robotic tool consists of a distal section which is configured to the patient and procedure planned. This section is intended for single use. The proximal section contains drive and control components which are common to all patients and are reused after inspection and function checks.

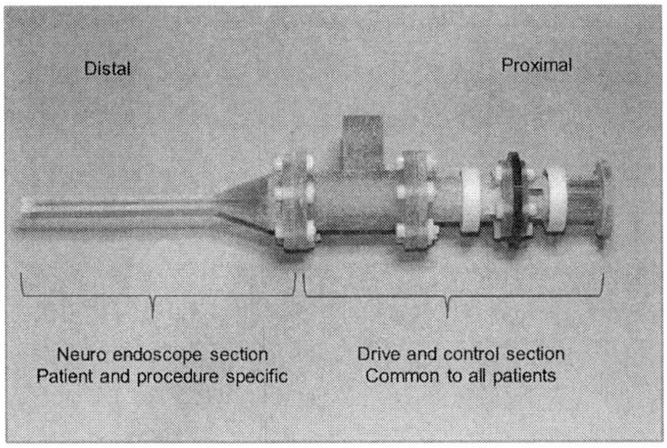

Figure 6 – Prototype of the robotic tool

CONTROL PHILOSOPHY

The control system will utilize autonomous or surgeon control of the robotic tool at different times during a procedure. Autonomous control is generally employed to provide tool tracking through complex anatomies and/or narrow surgical corridors as well as to minimize tissue shear stress by minimizing friction. Tools employ closed loop, MRI compatible control elements [19] combined with preoperative and iMRI, ifMRI, Diffusion Tensor Imaging (DTI) or radiographic images. Initial control system testing will utilize a material mimicking neural tissue [20].

SIMULATION

Brain tissue is a very soft, incompressible, nonlinear viscoelastic material with a linear viscoelastic strain limit of 0.1-0.3%. Tissue response depends on load history and strain rate, with increasing stiffness observed with increasing strain rate. Failure occurs at strains of 25-100%, depending on loading type [21].

Numerical simulation is being implemented to support various phases of neurosurgery. Brain shift between initial assessment and surgery must be known in order to update surgical planning and to ensure a safe and efficient procedure [22]. Numerical models will be overlaid with imaging results from the entire brain or from a reduced volume in the immediate vicinity of the tool tip.

The robotic tool is being designed with higher compliance than typical neurosurgical instruments. Thus, knowledge of estimated and measured loads on the tool and surrounding tissue is important. These data will be used to support surgical planning, to minimize tissue damage associated with excessive strain and ischemia, to enhance surgical training and as feedback to future tool improvements [23].

CONCLUSIONS

This study proposes the development of an endoscopic robotic tool system which will facilitate improved outcomes for neurosurgical patients. The system includes patient-specific endoscopic tool design, an MRI-compatible navigation system, and integration of interoperative imaging and computational model results synchronized with surgical workflow. Combined with other elements of Project neuroArm, a complete neurosurgical solution for micro neurosurgery and non-line-of-sight neuroendoscopic procedures will be available.

ACKNOWLEDGMENTS

The authors wish to express their appreciation to the staff of the Hotchkiss Brain Institute and to the Department of Mechanical and Manufacturing Engineering for their ongoing support of this work.

REFERENCES

1. Weissman, J. and Zinner, M. 2013, "Comparative Effectiveness Research on Robotic Surgery", *JAMA*, 309(7), pp. 721-722.
2. Pandya, S. et. al. 2009, "Advancing neurosurgery with image-guided robotics", *J Neurosurg*, 111(12), pp. 1141-1149.
3. Sutherland, G. R. et. al. 2013, "Merging machines with microsurgery: clinical experience with neuroArm", *J Neurosurg*, 118(3), pp. 521-529.
4. Sutherland, G. R. et. al. 2013, "The Evolution of neuroArm", *Neurosurgery*, 72(51), pp. A27-A32.
5. Lang, M. J., Greer, A. D., and Sutherland, G. R., 2011, "Intra-operative MRI at 3.0 Tesla: A Moveable Magnet", *Intraoperative Imaging*, 109, pp. 151-156.
6. Lang, M. J., Kelly, J. J., and Sutherland, G. R., 2011, "A Moveable 3-Tesla Interoperative Magnetic Resonance Imaging System", *Operative Neurosurgery 1*, 68(3), pp. 168- 179.
7. Sutherland, G. R. et. al. 2015, "Robotics in the neurosurgical treatment of glioma", *Surgical Neurology International*, 2015(6), pp. S1-8.
8. Marcus, H. J., et. al. 2014, "Force exerted during microsurgery: a cadaver study", *The International Journal of Medical Robotics and Computer Assisted Surgery*, 2014(10), pp. 251-256.
9. Yasargil, M. G., 1996, *Microneurosurgery Vol. IVB Microneurosurgery of CNS Tumors*, Theime, New York.
10. Draf, W. et. al., ed., 2015, *Endonasal Endoscopic Surgery of Skull Base Tumors: An Interdisciplinary Approach*, Thieme, New York.

11. Wilson, R. H., et. al. 2015, "Review of short-wave infrared spectroscopy and imaging methods for biological tissue characterization", *Journal of Biomedical Optics*, 20(3), 03090901 pp. 1-10.

12. Beekman, D., Bijadi, S., and Kowalewski, T. 2014, "Real-Time Tissue Differentiation Using Fiber Optic Sensing in Laser Catheters", *Journal of Medical Devices*, 2014(8), 030934 pp. 1-2.

13. Drexler, W., et. al. 2014, "Optical coherence tomography today: speed, contrast, and multimodality", *Journal of Biomedical Optics*, 19(7), 071412 pp. 1-34.

14. Natalin, R. A. and Landman, Jaime. 2008, "Where Next the Endoscope?", *Medscape Multispeciality, article 712248_1*, pp. 1-11.

15. Bass, B. L. and Garbey, M. 2014, "A road map for computational surgery: challenges and opportunities", *Journal of Computational Surgery*, 1(2), pp. 1-18.

16. Clogenson, H. C. M., Dankelman, J. and van den Dobbelsteen, J. J. 2014, "Steerable Guidewire for Magnetic Resonance Guided Endovascular Interventions", Journal of Medical Devices, 8(6), 021002-1, pp. 1-7.

17. Jelinek, F. et. al. 2015, "Classification of Joints Used in Steerable Instruments for Minimally Invasive Surgery – A Review of the State of the Art", *Journal of Medical Devices*, 9(3), 010801, p. 1-11.

18. Kato, T. et. al. 2014, "Extended Kinematic Mapping of Tendon-Driven Continuum Robot for Neuroendoscopy", *2014 IEEE / RSJ International Conference on Intelligent Robots and Systems (IROS 2014),* Chicago, IL, USA, pp. 1997-2002.

19. El Bannan, K., Chronik, B., and Salisbury, S. 2015, "Development of an MRI-Compatible, Compact, Rotary-Linear Piezoworm Actuator", *Journal of Medical Devices*, 9(3), 014501, pp. 1-7.

20. Niu, C. J. et. al. 2015, "Polyacrylamide gel substrates that simulate the mechanical stiffness of normal and malignant neuronal tissues increase protoporphyin IX synthesis in glioma cells", Journal of Biomedical Optics, 20(9), 098002, pp. 1-7.

21. Miller, K. ed. 2011, *Biomechanics of the Brain*, Springer, New York.

22. Miller, K. 1999, "Constitutive model of brain tissue suitable for finite element analysis of surgical procedures", *Journal of Biomechanics*, 32, pp. 531-537.

23. Wittek, A. et. al. 2006, "Subject-specific non-linear biomechanical model of needle insertion into brain", *Computer Methods in Biomechanics and Biomedical Engineering*, 11(2), pp. 135-146.

Numerical Simulation of Magnetophoretic Based Particle Capturing Using Continuous Flow Microfluidic Device

Amir Seyfoori
Advanced Magnetic materisla Research Center, College of Engineering, University of Tehran, Tehran, Iran.
Department of Mechanical Engineering, Laboratory for Innovation in MicroEngineering (LiME), University of Victoria, 3800 Finnerty Rd., Victoria, BC, Canada, V8P 2C5
Klm.1985@yahoo.com

Seyyed Ali Seyyed Ebrahimi
Advanced Magnetic materisla Research Center, College of Engineering, University of Tehran, Tehran, Iran.
Saseyyed@ut.ac.ir

Mohamadmahdi Samandari
School of Mechanical Engineering, College of Engineering, University of Tehran, Tehran, Iran.
mm_samandari@ut.ac.ir

Mohsen Akbari*
Department of Mechanical Engineering, Laboratory for Innovation in MicroEngineering (LiME), University of Victoria, 3800 Finnerty Rd., Victoria, BC, Canada, V8P 2C5
makbari@uvic.ca

ABSTRACT

Capturing and isolation of magnetic microparticles in microfluidic systems has been widely used for diagnostic applications including the detection of HIV, malaria, and cancer. One of the main challenges with the conventional magnetophoretic systems is that common permanent magnet which are held out of channel suffer from low magnetic gradient generation to trap particles in a defined position in microchannel. In order to modulate magnetic field for further cell capturing efficiency, localized magnetic field generation within the fluid stream is needed and to meet this target micromagnet patterns are considered as a proper solution. In our study, micromagnets patterns in microfluidic channel were modeled in a channel containing microtraps (microwells) and simulated using COMSOL Multiphysics software. The results of the simulation showed that the presence of transverse micromagnet can have a profound effect on the directional magnetophoresis of the particles toward the traps. Moreover, the extract results showed that positive and negative magnetic force applied evenly on microparticles towards the both sides of the channel which is decreased with increasing the height. Totally, the presented model can strongly localize the uniform magnetic field to guide and distribute the microparticles in such a way to which can further be used for capturing and incubating the rare bio-microparticles like Circulating tumor cells.

KEYWORDS: magnetophoretic, microfluidic, micromagnet, FEM simulation

INTRODUCTION

In the context of particle sorting and isolating, magnetophoresis is a powerful concept in field flow fractionation by applying the external magnetic field which is perpendicular to the flow direction and in this way particles with intrinsic or adventitious susceptibility can have a motion through the out of flow streamlines. Magnetophoresis in microfluidic platforms have two main advantages of high-throughput and high efficiency in comparison to conventional macroscopic magnetic based particle sorting [1,2]. During magnetophoresis and while passing through the microfluidic device, several key forces have contribution. Among these forces, magnetic and drag forces have significant role on the final particle direction in the channel. Magnetic force is the main force, which is applied on magnetic particles and mainly depends on the volume of the particle and the magnetic field. On the other hand, the drag force depends on the particle diameter and the relative velocity of the particle and fluid flow [3]. Indeed, the final trajectory of a particle in microfluidic channel using magnetophoresis can be calculated by linking the Newton's second law which makes a balance between these two main forces and the other forces which are mentioned in the equation 1.

$$M_p\,(dU_p/dt) = F_m + F_d + F_g + F_b + F_l \qquad (1)$$

Where, U_P is the particle velocity and F_m, F_d, F_g, F_b, F_l are magnetic, drag, gravity, brownian, and lift forces, respectively.

Despite tremendous progress in the fields of experimental and analytical researches, 3D FEM simulations to show the effect of magnetic fields in the microfluidic devices are the missing piece of the Magnetophoretic microfluidic research chain. To address this challenge, the present study analyze this problem using a Multiphysics CFD-magnetic approach. Furthermore, in this work, a novel microfluidic design is developed and the proof of the concept is provided using the mentioned FEM simulation.

2. THEORY AND FEM SIMULATION

Device Model

In this study, COMSOL Multiphysics (version 5.2) was used to simulate the magnetic field and consequently magnetic force produced by the permanent magnet adjacent the channel.

Following assumptions are used to simplify the microfluidic model and computing the magnetic gradient and forces applied on magnetic particles.

1- Magnetic field is uniform in X direction.

2- Flow is laminar.

As mentioned before, two important forces are applied on the magnetic particles in the microchannel with respect to the two above assumptions [4].

First, magnetic force under the gradient magnetic field which is expressed as:

$$\mathbf{F}_m = \mu_0 \int (M.\nabla)H dV \tag{2}$$

Where, μ_0 is the vacuum permeability; **H** is the magnetic flux intensity; **M** is the field-dependent particle magnetization and V is the particle volume.

By doing integrand over the magnetic particle volume, we have:

$$\mathbf{F}_m = V_P M_P G \tag{3}$$

Where V_P is the volume of the magnetic particle and **G** is the magnetic field gradient in the direction of **H**. So regarding to the substitution of the **H** with B the final magnetic force applied on the magnetic particles is:

$$\mathbf{F}_m = \frac{V_P \Delta\chi_P}{2\mu_0} \nabla B^2 \tag{4}$$

Where, $\Delta\chi_p$ is the susceptibility difference between particle and fluid.

Since this simulation will be used for circulating tumor cell (CTC), capturing and magnetic nanoparticles are attached to the surface of the single CTC, so the effective magnetic susceptibility is introduced with respect to the whole cell-magnetic particle complex.

$$\Delta\chi_{complex} = N\,(Rp^3/ Rc^3)\,\Delta\chi_p \tag{5}$$

Where, $\Delta\chi_{complex}$ is the relative susceptibility for cell-particle complex and fluid, N is the magnetic particle numbers, R_P and R_C are radius of particle and cell respectively.

In the fluid flow, another significant force which is sensed by the magnetic particles is drag force which is determined by the stocke's equation:

$$F_{drag} = 6\pi\eta\, Rc\, \Delta v \tag{6}$$

Where η is the medium viscosity and Δv is difference between particle and medium velocity.

By calculating the effect of both magnetic and drag force on the particles in the fluid, the final relative velocity which determine the direction of the particle trajectory is:

$$\Delta v = (Rc^2 \Delta\chi_{complex} / 9\mu_0\eta)\, \nabla B^2 \tag{7}$$

In this study due to calculate the effective magnetic force applied on magnetic microparticles with the saturation magnetization of 50 emu/g and $\Delta\chi$ of 5, a microchannel with the narrow and wide width of 70 and 200 μm and height of 100 μm was modeled, Fig.1.

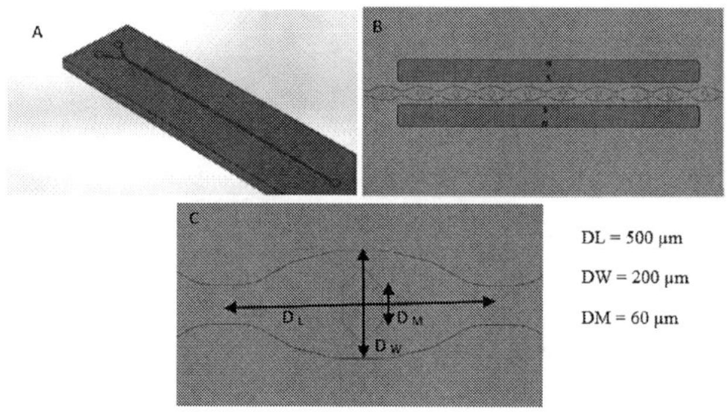

Fig.1: A) microfluidic design for magnetic particle capturing. B, C) magnified design scheme with dimension.

Transverse micromagnets from Ni material with the height of 500 nm were positioned in the channels to produce the gradient magnetic field at the broader zone of the channel capture the magnetic particles where the fluid velocity is very low in comparison to the narrow part of the microchannel.

As shown in Fig.1 two block permanent NdFeB magnet with the B_r of 0.2 T at the central zone of the magnetic flux was used to produce the uniform magnetic field.

RESULTS AND DISCUSSIONS

Fig. 2 shows the magnetic field results of the simulations in a single microchamber along with the computational mesh structure. The main challenge of this work was different length scale for meshing between micromagnets thickness and permanent magnets. As a result a model with approximately six million degrees of freedom was solve in the magnetic field simulation stage. The magnetic flux density map of one chain of the whole channel in the presence of the micromagnet was extracted at two different heights of 10 and 50 µm above the micromagnet (Fig. B1 and B2). It is clear that at 10 µm height adjacent the micromagnet the magnitude of the magnetic flux is greater than that of 50 µm one and in both of them the flux density is localized at the sharp boarder of the micromagnet. Moreover, at the central zone of the micromagnet the flux density concentration is dropped and hence the total gradient and magnetic force which is suitable for particle capturing is attained. According to Fig. 2C$_1$, the variation of the magnetic flux density in X, Y and Z direction shows that created magnetic flux density is dependent on the height of the channel so that a sharp flux variation at X (Fig 2C$_2$) and Z (Fig 2C$_3$) direction. This depiction can predict the direction of the magnetic particles attraction by the micromagnets to the side-down of the microchannel with the capability to be captured in the trap sites.

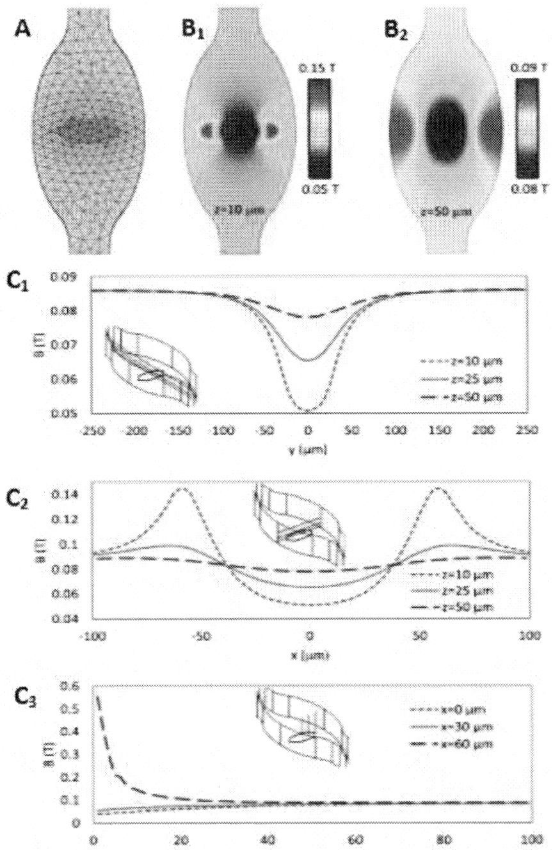

Fig.2: A) Mesh used for the simulation of the particle trapping device, B) Magnetic flux density map above the micromagnet with two heights, B$_1$)10µm and B$_2$) 50µm. C$_1$ &C$_2$) variation of the magnetic flux density at length and width of the microchannel for 3 different Z measurements. C$_3$) variation of the magnetic flux density at height of the microchannel for 3 different X measurements.

The specific geometry of this design allow the magnetic particles to be captured in a region of channel with the greater width. This phenomenon will be facilitated due to low fluid velocity at this section of the channel and subsequently more time for particle retention which case efficient capturing (Fig. 3A). Magnetic force map which is generated by the gradient magnetic field of the micromagnet is depicted in Fig.3B. It is obvious that magnetic force distribution is localized at lower heights adjacent to the micromagnet surface and by increasing the height, localization and magnitude of the magnetic force is decreased. Moreover, it is illustrated that magnetic force at both X and Z directions has attractive and repulsive nature at sides of the micromagnet which causes particle movement towards the trap zones of the channel. Results demonstrated that by applying uniform magnetic field in the presence of the designed micromagnets, particles can be divided to both channel sides and hence particle trapping can be done more effectively in comparison to the previous straight channels with constant width from inlet to outlet.

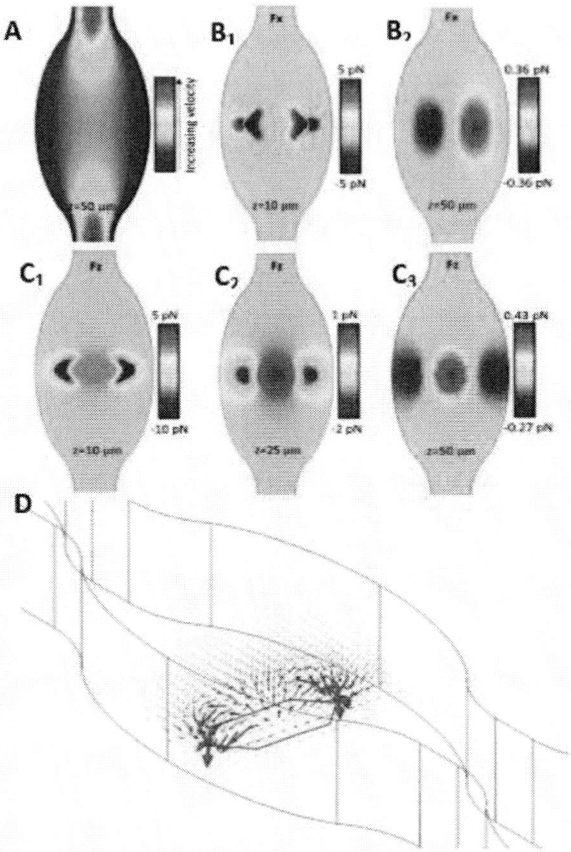

Fig. 3: A) Velocity magnitude of the liquid in the microfluidic channel (surface plot). B) magnetic force distribution map at X direction above micromagnet B$_1$) 10µm and B$_2$) 50µm. C$_1$-C$_3$) magnetic force distribution map at Z direction above micromagnet for Z = 10, 25, 50 µm respectively. D) Magnetic force direction adjacent the micromagnet applied on magnetic particle.

CONCLUSIONS

A novel 3D model of a magnetophoresis based microfluidic was generated and simulated by COMSOL Multiphysics software to calculate the magnetic flux density and forces applied on the magnetic particles moving in the microchannel. The new designed microchannel has generally two fluid velocity zones which is beneficial for more particles retention and eventually better particle trapping with more efficiency.

ACKNOWLEDGMENTS

M.A. would like to thank the Natural Sciences and Engineering Research Council of Canada (NSERC) and Canadian Foundation for Innovation for supporting this work

REFERENCES

1. Zhou R., Wang C., 2016, "Microfluidic separation of magnetic particles with soft magneticmicrostructures", Microfluid Nanofluid, 20:48, pp. 388-398.

2. Hoshino K., Huang Y., Lane N., Huebschman M., 2011, "Microchip-based immunomagnetic detection of circulating tumor cells", Lab Chip, 11, pp. 3449–3457.

3. Ying Wong Q., Liul N., Koh C., Li H., 2016, "Isolation of magnetically tagged cancer cells throughan integrated magnetofluidic device, Microfluid Nanofluid, 20:139.

4. Hejazian M., Liband W., Nguyen N., 2015, "Lab on a chip for continuous-flow magnetic cell separation", Lab Chip, 15 (4), pp. 959-970.

EXPERIMENTAL INVESTIGATION OF THE EFFECT OF VISCOSITY ON DEAN FLOW VELOCITY

Pouriya Bayat
Mechanical Engineering
York University
Toronto, Ontario, Canada
pouriya@yorku.ca

Pouya Rezai
Mechanical Engineering
York University
Toronto, Ontario, Canada
prezai@yorku.ca

ABSTRACT

Spiral microfluidic devices have been under investigation for a variety of applications such as microparticle separation [1] and focusing [2] as well as micro-mixing [3]. The radial pressure gradient that fluid experiences in curved microchannels results in two counter rotating vortices. The presence of these vortices can be employed for fluid and particle manipulation. The resultant competition between the drag force along these vortices and the inertial forces on microparticles has been used to sort and separate microparticles in curved microchannels. Despite the significant need to characterize the fluid behavior, this phenomenon has not been systematically investigated compared to the efforts that have been made to characterize the microparticles. In this study, we investigate the effect of fluid viscosity on Dean velocity in curved microchannels using a simple but practical methodology. Our method is based on quantitative characterization of radial position exchange of co-flows of liquids.

KEYWORDS: Dean Velocity, Spiral microfluidics, Dean flow

INTRODUCTION

It is of great importance to be able to characterize the flow behavior accurately in curved microchannels to design microfluidic devices for optimized performance. For example, multiple forces are exerted on microparticles moving in a curved microchannel and precise formulas are needed for calculation of these forces in order to design efficient microparticle sorters and separators. A major force that is exerted over the microparticles in curved microchannels is called Dean drag force that is caused due to the presence of Dean vortices and can be assumed to follow the Stokes law:

$$F_d = 6\pi r \mu v \qquad (1)$$

where r, μ and v represent radius of microparticles, dynamic viscosity of the fluid and microparticle velocity, respectively. Assuming the microparticle velocity is equal to the fluid Dean velocity (V_{De}), an accurate formula for Dean velocity is needed to calculate the Dean drag force exerted on microparticles in a curved microchannel.

Three types of approaches can be taken to characterize the Dean flow in curved micro- and macro-channels which are the analytical[4], [5], experimental [6], [7] and numerical [8], [9] approaches. However, to the best of our knowledge, none of the aforementioned attempts has been successful to provide a comprehensive solution for Dean flow problem in curved channels. However, the most widely used formula for calculation of the average Dean velocity was proposed by Ookawara et al. [8] by performing numerical simulation in a rectangular curved microchannel. It was shown that the velocity over the Dean vortices (V_{De}) vs. Dean number follows a power function.

$$V_{De} = 1.8\times10^{-4}\ De^{1.63} \quad \frac{m}{s} \qquad (2)$$

$$De = Re\sqrt{\frac{D_{eq}}{2R}} \qquad (3)$$

where De is the dimensionless Dean number, Re is the Reynolds number and D_{eq} and R represent hydraulic diameter and radius of curvature of the channel.

In this study, we utilized a combination of an experimental and numerical approach to investigate the effect of viscosity on Dean velocity in a curved microchannel by optically investigating the lateral displacement of fluids in the channel. With this study, we could show that the current widely used formulas for V_{De} is not capable of capturing the effect of viscosity and a more comprehensive correlation is required to characterize this phenomenon.

MATERIALS AND METHODS

In order to observe the flow behavior in microfluidic environments, we utilized a curved microchannel with R=0.5 cm radius of curvature and 150μm×150μm cross-section with two inlets as shown in Fig.1. Tap water and dyed water (with methylene blue) were pumped into the microchannel at different axial velocities (V_x) from the two inlets using a Legato 110 (KD Scientific, USA) syringe pump. A mixture of water and glycerol was used for increasing the kinematic viscosity of the fluid[10]. The fluids then passed through a 330° section of the channel and a video along the channel was recorded using a Point Grey camera (BC, Canada) connected to a Bioimager microscope (ON, Canada).

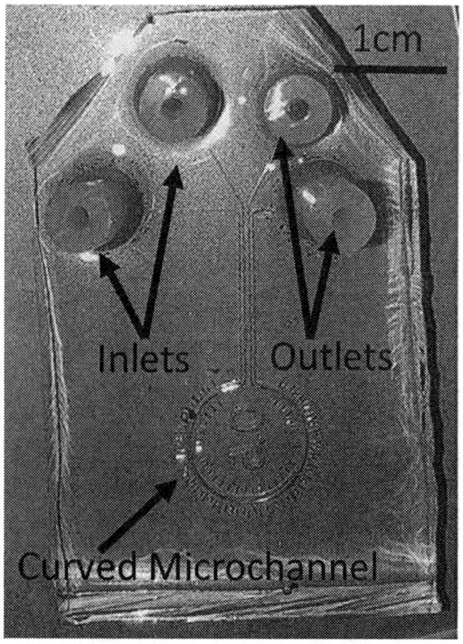

Figure 1: Microfluidic curved microchannel device with R=0.5 cm radius of curvature that contained two inlets and two outlets. Water and methylene blue solutions were injected into the device from the two inlets at various flow rates and the Dean velocity-based displacement of the fluids along the channel width was imaged under a microscope.

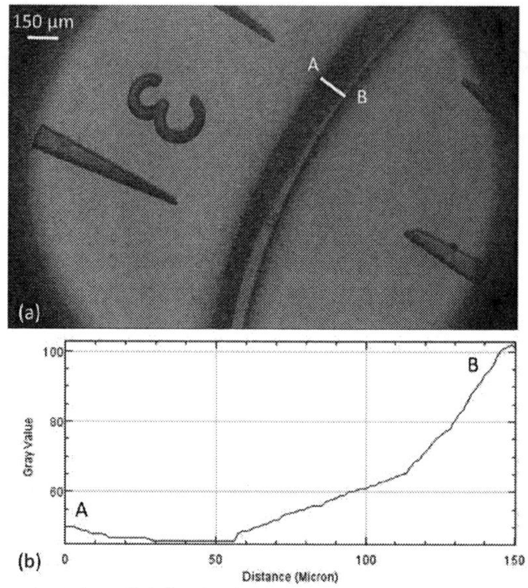

Figure 2: Microfluidic device shown under the microscope. Dyed water and glycerol mixtures at Q=0.8 ml/min were pumped into the device and the process of fluid switching started to occur immediately as shown in (a). (b) Gray intensity plot of the cross-section line AB shown in (a).

The recorded videos were then imported into ImageJ software (National Institutes of Health, Bethesda, Maryland, USA) and gray intensity values were plotted at every 10° angle over the width of the channel (e.g. line AB in Fig. 2). These data were then plugged into the following formula to calculate the switching index (SI) at each angle line AB.

$$\sigma = \sqrt{\frac{1}{N}\sum_{i=1}^{N}(c_i - \bar{c})^2} \quad (4)$$

$$SI = \sigma/\sigma_{max} \quad (5)$$

where c_i, $\bar{c}$ and N are intensity values, average intensity and number of the points on a cross section, respectively.
Finally the SI values were plotted along the length of the channel as depicted in Fig. 3. This experiment was done at various flow rates of 0.2 ml/min to 1 ml/min. The peak values of SI in these curves show the location at which the fluid exchange their radial position along the channel. We measured the switching angle (θ_s) and converted it to the switching length (L_s) to find the time required for a switch to happen.

$$t = L_s/V_x \quad (6)$$

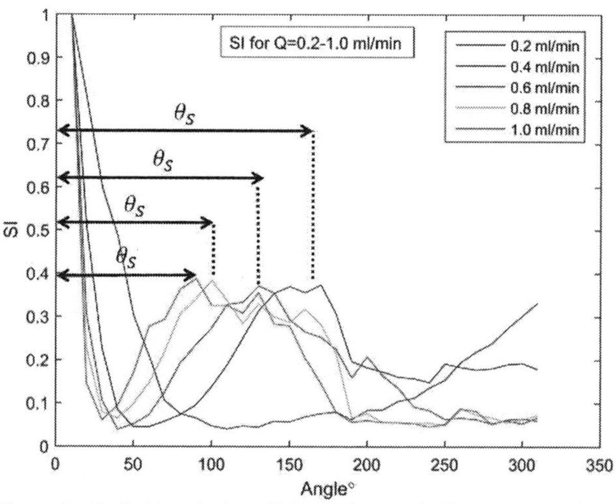

Figure3: Switching index (SI) vs. the angle for water and MB flow in the microfluidic device. Switching does not happen at low flow rates (<0.2 ml/min). Increasing the flow rate resulted in lower switching angles or faster switch.

Travelled distance of the fluid elements over the cross section that is a representative of the average length of a Dean vortex was also calculated using the Ookwara's estimation [8] (0.75 D_{eq}, where D_{eq} is the hydraulic diameter of the channel). Dividing the average travelled distance by the time that is required for one switch results in the average V_{De}.

$$V_{De} = L_t/t \quad (7)$$

The average V_{De} is then reported for different viscosities of the fluids.

RESULTS

We used our microfluidic device to investigate the effect of fluid viscosity on V_{De}. V_{De} is plotted vs. Dean number in Fig. 4a for pure water and 15% water/glycerol mixture. We could not increase the percentage of the glycerol further due to high pressure drops in the device that could not be handled by our syringe pump. Numerical results that matched our experimental data were used to investigate the effect of higher viscosities on Dean Velocity. As seen in Fig. 4a, in both of these cases we see an increase in V_{De} as the Dean number increases. This is due to the fact that inlet velocity should increase to increase the Dean number and the inlet velocity directly affects the centrifugal force ($\propto \frac{V_x^2}{r}$) thus increasing the V_{De}. However, V_{De} was higher for water/glycerol mixture with higher kinematic viscosity. In other words, two flows with the same Dean number but different kinematic viscosities did not possess equal V_{De}. This cannot be captured by presently available formulas in the literature for V_{De} (Eq. 2). In Fig. 4b, our validated numerical results for higher amounts of glycerol in water (i.e. higher kinematic viscosities) are shown. The water/glycerol viscosity was calculated from the work of Cheng [10] and was inputted in COMSOL Multiphysics to simulate the fluid flow. It is seen that the direct effect of viscosity on V_{De} continues as the viscosity increases. Also, a power function behavior for V_{De} as a function of De number that was reported by Ookawara [8] is seen here too. However, one single power function could not be fitted over all data points.

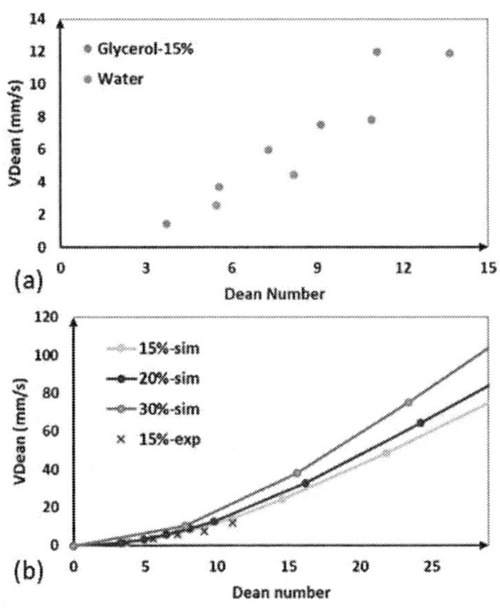

Figure 4: V_{De} is plotted vs. Dean number in (a). Higher Dean number contributes to strengthening the Dean vortices and increases the V_{De}. Numerical and experimental results for 15% water/glycerol mixture are shown in (b).

Ookawara has shown that V_{De} vs. Dean number follows a power function with the power of 1.63. We also showed in Fig 4 that kinematic viscosity directly affects the V_{De}. Combining these two, we plotted V_{De} as a function of $\upsilon De^{1.63}$ (υ being the kinematic viscosity) in Fig. 5. Interestingly, a linear function could be fitted over all the data points (R^2=0.999).

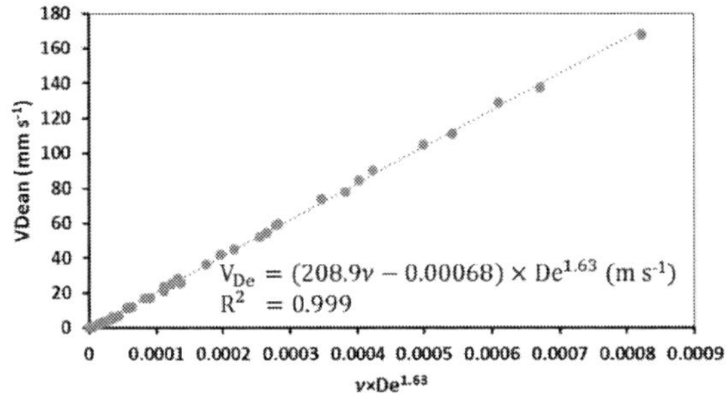

Figure 5: V_{De} vs. $\upsilon De^{1.63}$ is plotted and a linear function was fitted over the results.

This formula can be simplified to arrive at a more comprehensive correlation that captures the effect of kinematic viscosity on V_{De}.

$$V_{De} \sim 208.9 \ \upsilon \ De^{1.63} \quad \frac{m}{s} \tag{8}$$

Our correlation can be used in the range of 0<De<30 that is a typical range in microfluidics.

CONCLUSIONS

In this study, we investigated the effect of viscosity on Dean velocity and examined the capability of the currently used formulas in the literature for capturing this effect. We utilized an experimental approach along with a numerical scheme to arrive at a correlation that sufficiently captures the effect of viscosity on V_{De} as opposed to the aforementioned equations.

REFERENCES

[1] G. Guan, L. Wu, A. A. Bhagat, Z. Li, P. C. Y. Chen, S. Chao, C. J. Ong, and J. Han, "Spiral microchannel with rectangular and trapezoidal cross-sections for size based particle separation.," *Sci. Rep.*, vol. 3, p. 1475, 2013.

[2] N. Xiang, H. Yi, K. Chen, D. Sun, D. Jiang, and Q. Dai, "High-throughput inertial particle focusing in a curved microchannel: Insights into the flow-rate regulation mechanism and process model," vol. 044116, pp. 1–10, 2013.

[3] A. P. Sudarsan and V. M. Ugaz, "Multivortex micromixing.," *Proc. Natl. Acad. Sci. U. S. A.*, vol. 103, no. 19, pp. 7228–7233, 2006.

[4] M. Norouzi and N. Biglari, "An analytical solution for Dean flow in curved ducts with rectangular cross

section An analytical solution for Dean flow in curved ducts with rectangular cross section," vol. 053602, no. May 2016, pp. 0–15, 2013.

[5] S. Yanase, N. Goto, and K. Yamamoto, "Dual solutions of the flow through a curved tube," vol. 191, 1873.

[6] P. M. Ligrani and R. D. Niver, "Flow visualization of Dean vortices in a curved channel with 40 to 1 aspect ratio.," *Phys. Fluids*, vol. 31, no. 12 , Dec. 1988, pp. 3605–3617, 1988.

[7] B. Bara, K. Nandakumar, and J. H. Masliyah, "An experimental and numerical study of the Dean problem: flow development towards two-dimensional multiple solutions," *J. Fluid Mech.*, vol. 244, pp. 339–376, 1992.

[8] S. Ookawara, R. Higashi, D. Street, and K. Ogawa, "Feasibility study on concentration of slurry and classification of contained particles by microchannel," *Chem. Eng. J.*, vol. 101, no. 1–3, pp. 171–178, 2004.

[9] T. T. Chandratilleke and A. Nursubyakto, "Numerical prediction of secondary flow and convective heat transfer in externally heated curved rectangular ducts," *Int. J. Therm. Sci.*, vol. 42, no. 2, pp. 187–198, 2003.

[10] N.-S. Cheng, "Formula for the Viscosity of a Glycerol−Water Mixture," *Ind. Eng. Chem. Res.*, vol. 47, no. 9, pp. 3285–3288, May 2008.

2D CONTINUUM MODEL FOR SIMULATING ANGIOGENESIS STUDIES IN MICROFLUIDIC SYSTEMS

Nikola Kuzmic
Dept. Mechanical & Industrial Engineering
University of Toronto
Toronto, ON, Canada
nikola.kuzmic@mail.utoronto.ca

Edmond W. K. Young
Dept. Mechanical & Industrial Engineering
University of Toronto
Toronto, ON, Canada
eyoung@mie.utoronto.ca

ABSTRACT

Tumour-induced angiogenesis is a process which involves growth of blood vessels from existing vasculature for purpose of increasing nutrient supply to undernourished parts of the tumour tissue. Being able to predict the growth and behavior of blood vessel networks under different stimuli is crucial to understanding the growth of tumours and ultimately designing better cancer treatment plans. In this paper, we present a 2D simulation of angiogenesis based on continuum models proposed by Orme and Chaplain [1] and Anderson and Chaplain [2]. The system is nondimensionalized and simulations are performed using FEniCS [3]. Results reveal cell migration on a gradient of surface adhesion sites, process known as haptotaxis, as well as propagation in the direction of a vascular endothelial growth factor (VEGF), which is referred to as chemotaxis. The model allows for studying the effects of various pro- and anti-angiogenic factors. The novelty of this study is integration of mathematical modeling with microfluidic devices to aid biologists and chemists in drug discovery.

KEYWORDS: mathematical modeling, continuum models, angiogenesis, chemotaxis, haptotaxis, microfluidics

INTRODUCTION

Angiogenesis is a biological process that involves the growth of new blood vessels from existing vasculature. In certain instances such as embryo development and wound healing, angiogenesis is a normal and necessary process. However, in the case of cancer, angiogenesis is potentially lethal as it enables tumour development and metastatic spread. As avascular tumours grow, they reach a certain size where nutrient diffusion from existing blood vessels becomes insufficient due to the nutrients being consumed on the outer regions of the tumour. In turn, the central region of the tissue becomes undernourished and induces signals to existing blood vessels to proliferate towards it. This is the initiation of angiogenesis. Understanding angiogenesis enables one to control tumour size, induce starvation, and assess the impact of new drugs. Controlling angiogenesis is considered a major part of regulating tumour growth, which in some cases may assist in putting the cancer into remission. Angiogenesis is an active area of research which has the potential to play a significant role in advancing cancer therapy [4]. Angiogenesis involves several cellular processes, including chemotaxis and haptotaxis. Induced signals by undernourished cells is communicated through secretion of vascular endothelial growth factor (VEGF). Endothelial cells (ECs) from the parent blood vessels respond to the VEGF gradient via cell-surface receptors for VEGF. Once the basement membrane is breached, cells begin proliferating towards the undernourished region. Another well-established angiogenesis process is migration of parent vessel ECs along an adhesive chemical gradient, such as fibronectin (FN) [1]. As ECs move through the extracellular matrix (ECM) they bind to FN macromolecules via cell-surface receptors. While these angiogenesis processes have been modeled before, the models have yet to be applied to microfluidic systems, which are increasingly being used for experimental angiogenesis research. Our main goal is to study the effects of VEGF and FN (and other chemicals) on angiogenesis, specifically in a microfluidic context.

METHODOLOGY

Our aim is to study VEGF and FN effects on angiogenesis *in silico*, with a microfluidic device concept as shown in Fig. 1. This device concept is based on a previous design by Bischel et al. [5].

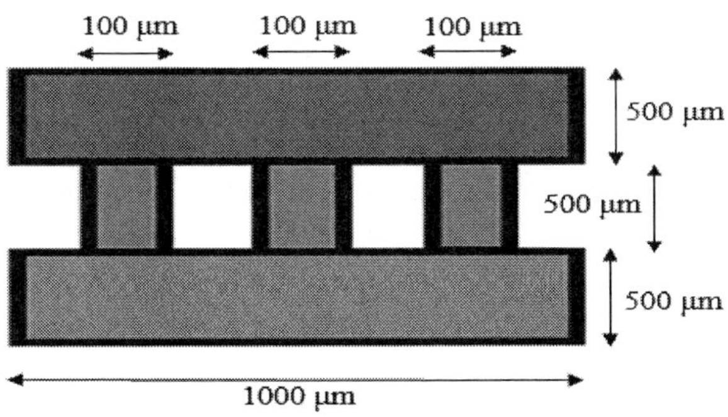

Figure 1: Microfluidic device concept (not to scale)

This would allow us to apply constant FN concentration and ECs on the top side and constant concentration of VEGF on the bottom side of the diffusion ports (blue regions).
In our analysis, we simplify the diffusion port geometry to a

uniform square. The model utilized in our study is based on the continuum model proposed by Anderson and Chaplain [2] and is summarized in Eq. (1):

$$\frac{\partial n}{\partial t} = D_n \nabla^2 n - \chi_0 \nabla \cdot (n \nabla c) - \rho_0 \nabla \cdot (n \nabla f) \qquad (1)$$

where n is number of ECs per unit area, D_n is EC random motility coefficient, χ_0 is chemotactic coefficient, c is VEGF concentration, ρ_0 is haptotactic coefficient and f is FN concentration. The model is derived from the conservation equation for ECs. The first term on the right side, indicates random motility of ECs, while the second and third terms indicate chemotactic and haptotactic responses, respectively. The nature of the model implies migration of cells towards a positive VEGF and FN gradient, but down a gradient of EC. Eq. (2) below presents the model in dimensionless form.

$$\frac{\partial n'}{\partial t'} = D \nabla^2 n' - \chi \nabla \cdot (n' \nabla c') - \rho \nabla \cdot (n' \nabla f') \qquad (2)$$

where

$$n' = \frac{n}{n_0}, \quad c' = \frac{c}{c_0}, \quad f' = \frac{f}{f_0}, \quad t' = \frac{t}{\tau}$$

$$D = \frac{D_n}{D_c}, \quad \chi = \frac{\chi_0 c_0}{D_c}, \quad \text{and} \quad \rho = \frac{\rho_0 f_0}{D_c}$$

(n_0, c_0, f_0 and τ are appropriate reference variables).

Model parameters were obtained based on experiments performed by Stokes et al. [6] and Rupnick et al. [7]. Anderson and Chaplain [2] utilize the values summarized in Table 1 and $\tau = 1.5$ *days* was used as the reference time to nondimensionalize the model.

Table 1 – Model Parameters

Variable Name	Value
EC random motility coefficient (D_n)	10^{-10} [$cm^2 s^{-1}$]
VEGF diffusion coefficient (D_c)	$2.9 \cdot 10^{-7}$ [$cm^2 s^{-1}$]
Chemotactic coefficient (χ_0)	2600 [$cm^2 s^{-1}$]
VEGF concentration (c_0)	approx. 10^{-10} [M]

Hence, $D = 0.00035$ and $\chi = 0.38$ are obtained.

Due to difficulty of estimating parameters regarding haptotactic coefficient (ρ), following the findings presented by Terranova et al. [8] it was assumed that $\chi > \rho$, and ρ was taken to be 0.34.

Simulations were performed in FEniCS [3]. Eq. (2) is casted into the variational form and is presented below (after dropping the apostrophes for clarity):

$$\int \left\{ \frac{n - n^{n-1}}{dt} v + D(\nabla n \cdot \nabla v) + \chi \nabla v \cdot (n \nabla p) + \rho \nabla v \cdot (n \nabla c) \right\} d\Omega = 0 \qquad (3)$$

where v is the test function. The system was solved using Newton solver with a time-step of $dt = 0.05$, on a 30 x 30 unit-square grid. Due to VEGF and FN diffusion coefficients being significantly larger than the EC random motility coefficient, VEGF and FN distributions were assumed to be in steady-state. A zero Neumann boundary condition was applied on all four sides of the domain for the EC (n) field.

Anderson and Chaplain [2] assumed approximately circular tumour at (1, 0.5) to act as the source of VEGF, which is illustrated with Eqs. (4) and (5) below:

$$c(x, y, t) = \frac{(\nu - r)^2}{\nu - 0.1} \qquad (4)$$

$$r = \sqrt{(x-1)^2 + \left(y - \frac{1}{2}\right)^2} \qquad (5)$$

where $\nu = 1.728$.
Steady-state VEGF distribution is illustrated in Fig. 2.

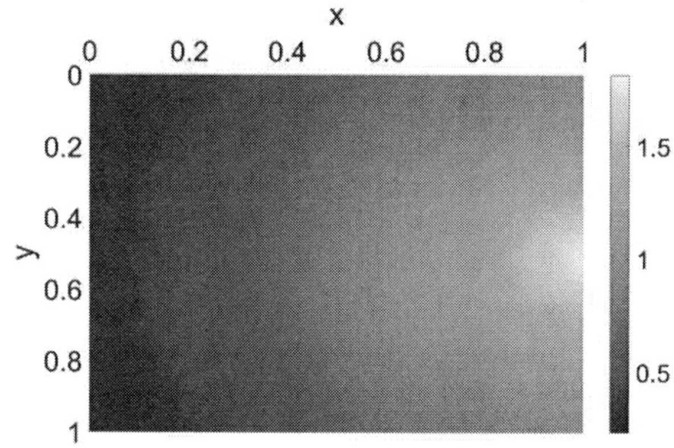

Figure 2: Steady-state VEGF distribution

FN distribution used in [2] is represented with Eq. (6) and illustrated in Fig. 3.

$$f(x, y, t) = 0.75 e^{-\frac{x^2}{0.45}} \qquad (6)$$

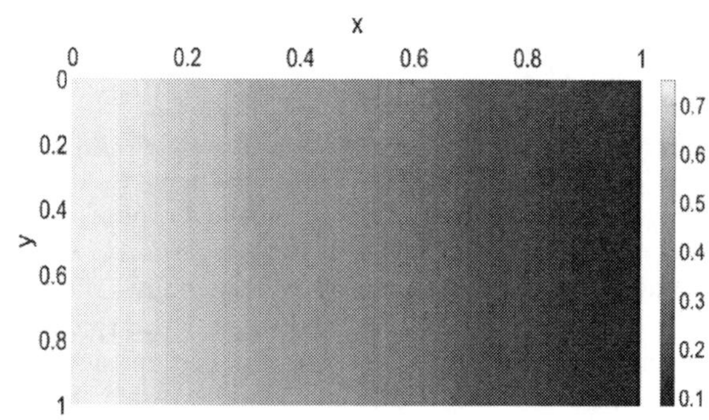

Figure 3: Steady-state FN distribution

It was also of interest to investigate the influence of an increasing FN distribution and we simply used:

$$f(x, y, t) = 0.75x \qquad (7)$$

to study the effect of FN on angiogenesis, as illustrated in Fig. 4.

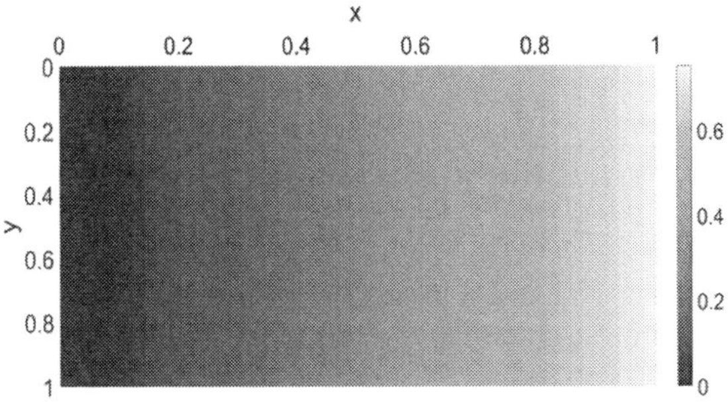

Figure 4: Linear steady-state FN distribution

Initial EC condition applied in [2] is particularly interesting to us because it allows us to incorporate what would usually be found in a typical experiment. Here we apply four clusters of ECs that are described by Eq. (8) and are illustrated in Fig. 5.

$$n(x,y,0) = e^{-\frac{x^2}{0.001}}\sin^2(4\pi y) \tag{8}$$

Figure 5: Initial EC distribution
(*red indicates high and blue indicates low EC density*)

RESULTS AND DISCUSSION

Based on the above model, we numerically solved for the EC density, n, over time. It was of interest to examine how ECs would behave under the influence of VEGF alone, and under the combined influence of VEGF and FN. Figs. 6 and 7 illustrate EC migration when the cells are only exposed to the VEGF distribution illustrated in Fig. 2, at t = 1.125 days and t = 2.25 days, respectively.

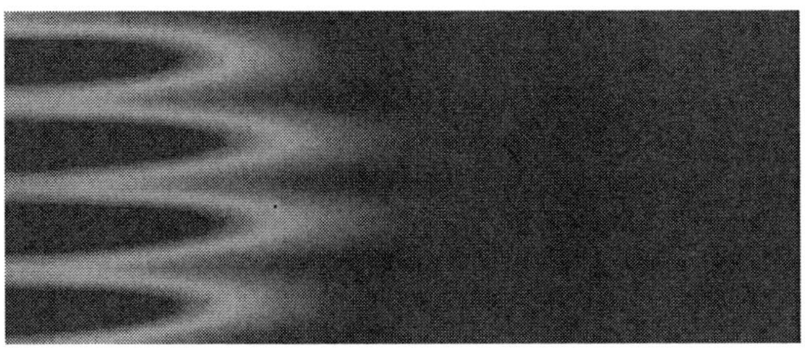

Figure 6: EC migration due to VEGF at t = 1.125 days

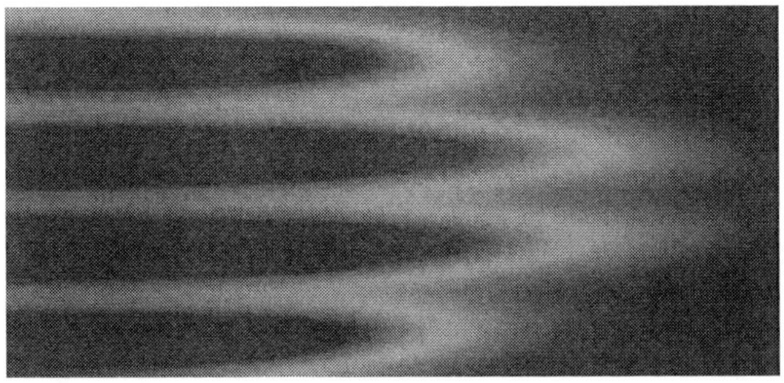

Figure 7: EC migration due to VEGF at t = 2.25 days

From the results, it is apparent that only chemotaxis is exhibited by the cells due to the tumour-induced VEGF gradient. EC clusters only merge together at the very right, due to the circular VEGF distribution.

When ECs are exposed to VEGF and the FN distribution presented in Fig. 3, cell clusters exhibit chemotaxis and slight haptotaxis. Results are summarized in Figs. 8 and 9, at t = 1.125 days and t = 2.25 days, respectively.

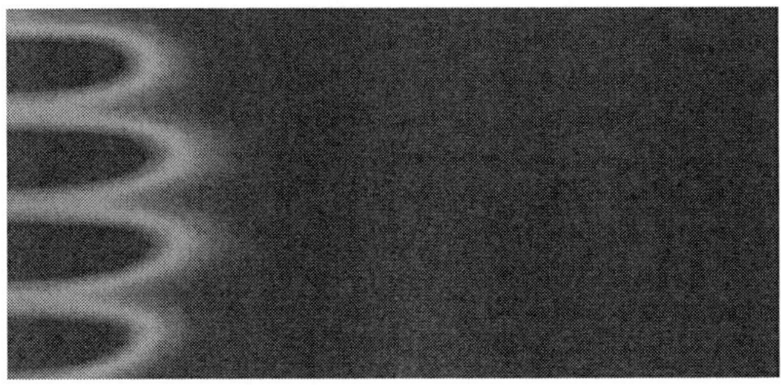

Figure 8: EC migration due to VEGF and the negative FN gradient at t = 1.125 days

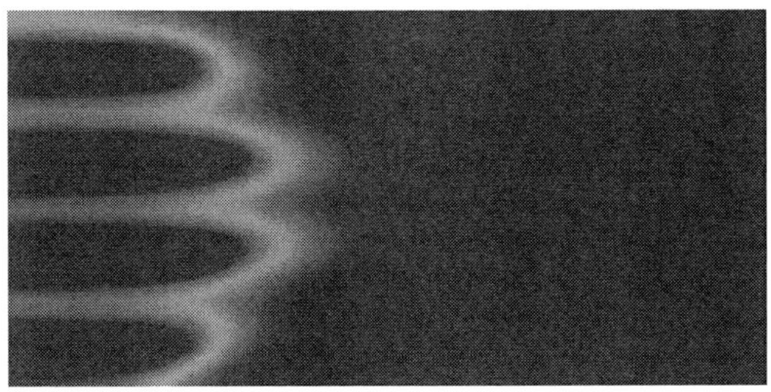

Figure 9: EC migration due to VEGF and the negative FN gradient at t = 2.25 days

It is important to note that FN gradient is negative for the above results, whereas VEGF is increasing in the direction of tumour. Here, ECs exhibit chemotaxis and only a slight haptotaxis, as a small merging of clusters can be observed from Fig. 9. Comparing Figs. 6 and 8, as well as Figs. 7 and 9, we see that ECs migrate more when only exposed to

VEGF. The reason is the fact that negative FN gradient 'slows' down cells due to the nature of haptotactic term of the mathematical model.

Last combination that we investigated was that of VEGF and FN distribution illustrated in Fig. 4. Results are illustrated in Figs. 10 and 11.

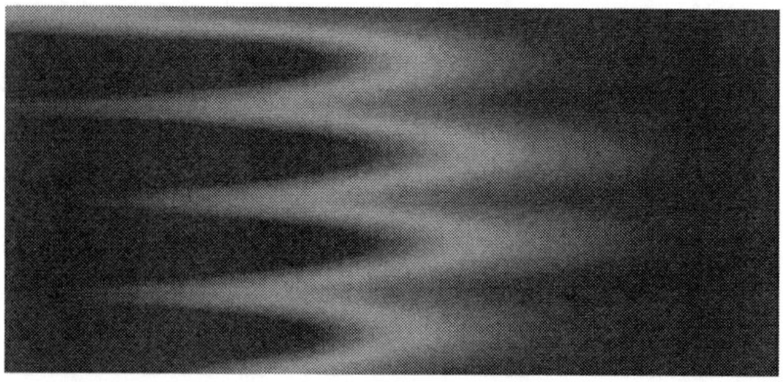

Figure 10: EC migration due to VEGF and the positive FN gradient at t = 1.125 days

Figure 11: EC migration due to VEGF and the negative FN gradient at t = 2.25 days

As expected, chemotaxis is exhibited by the cells due to VEGF gradient and pronounced haptotaxis occurs due to a positive FN gradient. Strong sprout merging is observed because both VEGF and FN are positive and induce ECs to migrate at higher rates than in either of the two previous cases.

CONCLUSIONS

We have presented 2D simulations of angiogenesis inside a microfluidic channel, based on the continuum approach, that would enable investigating the influence of VEGF and FN on angiogenesis. The results show that ECs move in the direction of positive VEGF gradient and that they merge together due to FN gradient, which are two processes known as chemotaxis and haptotaxis, respectively. In addition, negative FN gradient impedes chemotaxis whereas positive FN gradient promotes it. Developing accurate mathematical models will allow us to gain insight into various modeling parameters that govern angiogenesis, and have the potential to advance our understanding of mechanisms that promote and inhibit angiogenesis. The above approach is novel and significant because it integrates microfluidic systems and mathematical modeling for the purposes of advancing antiangiogenesis drug discovery and development, with the ultimate goal of improving quality of cancer treatment.

ACKNOWLEDGMENTS

The authors would like to acknowledge financial support from The Natural Sciences and Engineering Research Council of Canada (NSERC).

REFERENCES

1. Orme, M. E., Chaplain, M. A. (1997), Two-dimensional models of tumor angiogenesis and anti-angiogenesis strategies, IMA Journal of Math. App. In Medicine & Biology, Vol (14), 189-205.
2. Anderson, A. R. A., & Chaplain, M. A. J. (1998). Continuous and Discrete Mathematical Models of Tumor-induced Angiogenesis. Math. Biology, 60, 857-900.
3. Alnaes, M. S., Blechta, J., Hake, J., Johansson, A., Kehlet, B., Logg, A., Richardson, C., Ring, J., Rognes, M.E., Wells, G. N., The FEniCS Project Version 1.5, Archive of Numerical Software, Vol (3).
4. Carmeliet, P. (2005). Angiogenesis in life, disease and medicine, Nature, Vol (438).
5. Bischel, L. L., Young, E. W. K., Mader, B. R., & Beebe, D. J. (2013). Biomaterials Tubeless micro fl uidic angiogenesis assay with three-dimensional endothelial-lined microvessels. Biomaterials, 34(5), 1471–1477. http://doi.org/10.1016/j.biomaterials.2012.11.005
6. Stokes, C. L., M. A. Rupnick, S. K. Williams and D. A. Lauffenburger (1990). Chemotaxis of human microvessel endothelial cells in response to acidic fibroblast growth factor. *Lab. Invest.* **63**, 657–668.
7. Rupnick, M. A., C. L. Stokes, S. K. Williams and D. A. Lauffenburger (1988). Quantitative analysis of human microvessel endothelial cells using a linear under-agarose assay. *Lab. Invest.* **59**, 363–372.
8. Terranova, V. P., R. Diflorio, R. M. Lyall, S. Hic, R. Friesel and T. Maciag (1985). Human endothelial cells are chemotactic to endothelial cell growth factor and heparin. *J. Cell Biol.* **101**, 2330–2334.

ON-DEMAND STIMULATION AND MOVEMENT ANALYSIS OF ZEBRAFISH LARVAE INSIDE MICROCHANNELS USING ELECTRIC AND FLUIDIC STIMULI

Amir Reza Peimani[1], Georg Zoidl[2], and Pouya Rezai[1*]
[1]Department of Mechanical Engineering, York University, Toronto, ON, CANADA
[2]Department of Biology, York University, Toronto, ON, CANADA
*Corresponding Author: BRG 433B, 4700 Keele St, Toronto, ON, M3J 1P3, Canada; Tel: 416-736-2100 ext. 44703; Email: prezai@yorku.ca

ABSTRACT

The zebrafish larva is a favorable model organism for systematic analysis of the sensory-motor transformations and activities due to its unique merits such as genomic homology to humans, small size, optical transparency and rapid embryonic development. Investigation of zebrafish motor and neural activities requires proper manipulation of the fish which is defined as an on-demand stimulation of the sensory-motor responses and their quantitative analysis. This is challenging to achieve due to the complex, unpredictable and rapid movement patterns of the zebrafish larva. Conventional methods are inaccurate in evoking the response and monitoring it due to challenges associated with control over the stimuli and movement monitoring of the fish. In this paper, we present a miniaturized electro-fluidic device that utilizes electric and flow stimuli to evoke a longitudinal orientation response in 5-7 days post fertilization (dpf) zebrafish larvae which can be quantified accurately under a microscope. This method can be used for automated behavioral assays on zebrafish larvae in many applications such as the fundamental biological investigations, toxicology, and drug discovery.

KEYWORDS: Zebrafish, Microfluidics, Sensory-Motor Response, Electric Response, Flow Response, Behavioral Screening

INTRODUCTION

Zebrafish is the simplest vertebrate model organism used in behavioral screening[1], drug discovery[2], Central Nervous System (CNS) imaging and brain mapping[3]. It offers many advantages as a model for system biology investigations. For instance, zebrafish is endowed with genomic similarities to human which is appealing for biologists to investigate human diseases and neurological disorders[4,5]. The rapid growth rate of zebrafish provides the experimenters with a large number of fishes within a short period. Interestingly, it has been shown that only after 5 days post fertilization (dpf), the larva reaches an important developmental stage with a fully functional brain and ability to exhibit several types of behaviors[6–8]. Moreover, akin to the other biological models such as *Drosophila melanogaster* and soil-dwelling nematode *Caenorhabditis elegans,* the zebrafish's body at larval stage is optically transparent which enables the researchers to carry out a variety of *in-vivo* screening assays including CNS, brain and cardiovascular imaging to gain a better understanding of the sensory-motor activities[5,9]. Last but not least, zebrafish larva is relatively small which makes it an ideal model for high-throughput screening applications[5].

Precise stimulation of zebrafish larvae while monitoring their motor behavior output has been one of the key challenges in the previous studies. For instance, the on-demand longitudinal orientation of the zebrafish larvae is currently achieved with pipette loading[9,10] or magnetic particle-based manipulation[11]. Another example of zebrafish manipulation is the bimodal orientation of the zebrafish inside a plus maze environment which was recently demonstrated on adult fishes in a macro-scale platform[6]. These methods either lack a complete control over the orientation of the larvae[6] or require high levels of expertise for operation[7,10]. A precise, efficient, user-friendly, and on-demand method for zebrafish larvae manipulation and behavioral quantification is currently needed.

Microfluidic platforms could enhance our ability to precisely study model organisms such as zebrafish larvae by providing better control over microenvironments, automating manipulation and parallel processing, and enabling high throughput quantitative biological assays[12]. Some microfluidic devices have recently been developed to explore the response of zebrafish larvae to a variety of stimuli such as chemicals[13], oxygen deprivation[10], and optical stimuli[14]. For instance, a fish-trap technology was recently introduced for larval drug exposure with specially designed channels that allowed automatic positioning and handling of the zebrafish larvae. A computer-controlled photo-detection system was used to handle the larvae in the desired orientation, making the technique relatively complex for wide adoption by biologists[13]. Moreover, a majority of current microfluidic devices focus on zebrafish analysis while the larvae are fully[13] or partially[10,15] immobilized in a trap. However, in many of the standard behavioral assays, the larvae must be allowed to freely move in response to external cues such as electrical, optical and chemical stimuli.

In this paper, we introduce two simple, non-fatal, and accurate methods based on fluidic and electric stimulation in a microchannel to evoke zebrafish larvae's motor-response towards a desired direction in a microfluidic environment. In combination with optical microscopy, we were able to quantify zebrafish responses in a simple manner in our device.

Since the flow is already required to load and control the larvae inside the microchannel, we have benefited from the sensitivity of the fish to flow[16] to attain a directed longitudinal orientation response in 5-7 dpf larvae without adding any extra power consuming components to the device. Moreover, for scenarios where a stationary flow is encountered, we show the possibility of using electric current (EC) to stimulate zebrafish larvae's movement response in our microfluidic device which was inspired by our previously-reported observation of the *C. elegans* response to an electric signal in a microchannel[17]. These techniques to stimulate zebrafish larvae's sensory-motor responses and quantify them precisely in a controllable environment have a wide range of applications in fundamental biology studies, toxicology, and drug discovery.

MATERIALS AND METHODS

Zebrafish Larvae

The protocol of breeding and keeping the Tupfel long fin (TL) strain zebrafish larvae were followed in accordance with the guidelines of the Canadian Council for Animal Care (CCAC) after approval of the ACC protocol (GZ 2014-19 (R3)). The larvae were raised and kept at 28°C in an incubator.

Experimental Setup

The setup (Fig. 1a) consisted of a microfluidic device (Fig. 1b), syringe pumps (Legato 110, KdScientific, USA) to load the larvae and produce the flow stimulation, a power supplier (Model 2410 Sourcemeter, KEITHLEY) to generate the EC, and a microscope (BIM-500FLD, Bioimager Inc., Canada) equipped with a camera (GS3-U3-23S6M-C, Point Grey Research Inc., Canada) for observation and recording of behavioral responses of zebrafish larvae in the device. The EC and flow stimuli were manually tuned and controlled.

Microfluidic Chip Design and Fabrication

The microfluidic device (Fig. 1b) was made of a polydimethylsiloxane layer (PDMS, 10:1 base to agent ratio of Sylgard 184, Dow Corning, USA) patterned over a 3D-printed master mold and bonded to a flat glass slide. It consisted of a 45°-tilted inlet tube to smoothly load the larva, a side channel to help with convenient loading at inlet by carrying the larva to the main channel, two electrodes at side inlet and outlet reservoirs to apply the EC, and a U-shaped microchannel with a narrowing width from 1.6 mm to 0.9 mm that acted as a fluidic valve to keep the larva inside the device. The height of the main channel was 0.55mm. The electrical resistance of the channel was ~30MΩ.

Experimental Procedure and Data Analysis

After loading a larva into the main channel, it was given 60 seconds to recover from any possible loading stresses before exposure to the stimulus. For the flow experiments, the syringe pump connected to the titled inlet was used to apply a buffer flow rate of 1 mL/min in the channel. For electric experiments, the two electrodes were connected to the power supplier and inserted into the device reservoirs to apply an EC of 3 μA across the channel. The movement behavior of the larva in response to the flow or electric stimulus was recorded immediately after exposure under the microscope. Subsequently, the tested larva was ejected through the outlet, and the device was debubbled to perform the next experiment with a new fish and under a new stimulus setting.

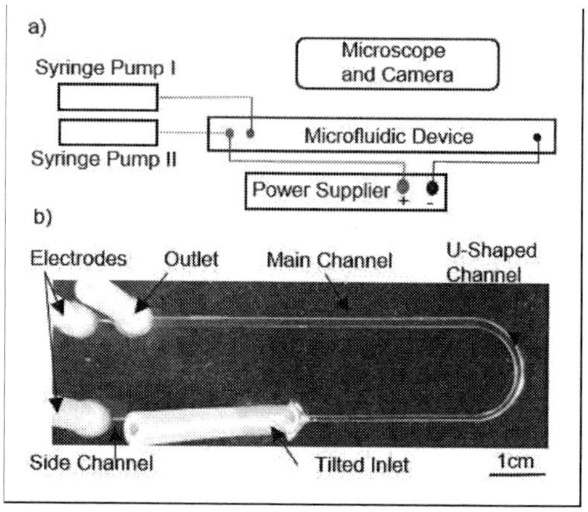

Figure 1. a) Schematic of the experimental set-up including the microfluidic device, two syringe pumps, power supplier, and a microscope equipped with a camera. b) The microfluidic device consisting of tilted inlet tube, side channels, U-shaped channel, main channel, and two electrodes. The flow in the side channel helped the tilted inlet flow to conveniently carry the zebrafish larva to the main channel, and U-shaped channel with different widths at its ends kept the larva inside the chip. The length, width, and height of the main channel were 63.3, 1.6, and 0.55 mm.

Post-Exposure Survival and Morphology Test

We conducted a viability test on the larvae after being loaded, exposed to stimuli, and ejected from our device. For this purpose, the survival and morphological changes of the tested larvae were studied[7]. Following each experiment, the larva was ejected through the tilted inlet tube using the withdrawal function of the syringe pump to prevent any possible damages to the fish. Three groups of larvae, i.e. control larvae that were not exposed to any stimuli but passed through the device and fishes exposed to fluidic and EC stimuli, were raised for four days and assessed on a daily basis. During this period, they were kept inside a 28 °C incubator and fed twice a day with paramecium while being monitored for viability and any morphological abnormalities which were mostly observed as an abnormal bending of the spine.

RESULTS AND DISCUSSIONS

Loading of Zebrafish Larva into the Microfluidic Chip

For each experiment, a single 5-7 dpf zebrafish larva was transferred to the tilted inlet tube of the device using a manual pipette. Then, using the syringe pumps connected to the tilted and side inlets, the larva was smoothly and conveniently transferred into the main channel. Fig. 2 shows sequential images of a single fish being loaded into the device at the inlet reservoir. The flow in the side channel prevented the larva from colliding with the walls of the channel which could potentially result in physiological damage to the larva. Once the larva reached the main channel, the syringe pumps were turned off to keep the larva inside the device. The animal was then left for 60 seconds without any exposure to flow or EC so that it could alleviate and adapt to the new environment.

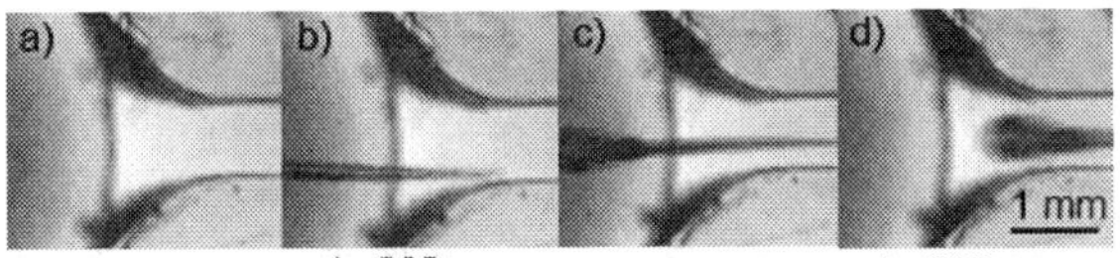

Figure 2. Loading of a zebrafish larva at the tilted inlet of the device using two flows, one at the tilted inlet and the other at the side channel from far left of the image to the right.

Response of Zebrafish Larvae to Flow Stimulus

After successful loading of zebrafish larvae into the main channel, we first investigated the response of fishes to a flow stimulus. For this, N=36 fishes were loaded into the channel and exposed to a flow rate of 1 mL/min which was in the direction of tail-to-head of the larva. The faster and slower rates were proved to be not as effective as this rate (data are not shown). Responses of animals were monitored and compared to a group of N=36 fishes which were loaded into the channel and not exposed to any flow stimulus (Control). The results are shown in Fig. 3. From the group of animals that were exposed to the flow stimulus, 68.4±9.4% responded to the flow and attempted to orient their bodies and swim against the flow direction within 15 seconds of exposure. Interestingly, only 19.4±10.1% of the control animals demonstrated an orientation response in the channel in the absence of the flow stimulus. The response of animals to flow was significantly different (p-value<0.01, two-tailed t-test) from the behavior of zebrafish larvae in the channel when no significant flow was imposed on them.

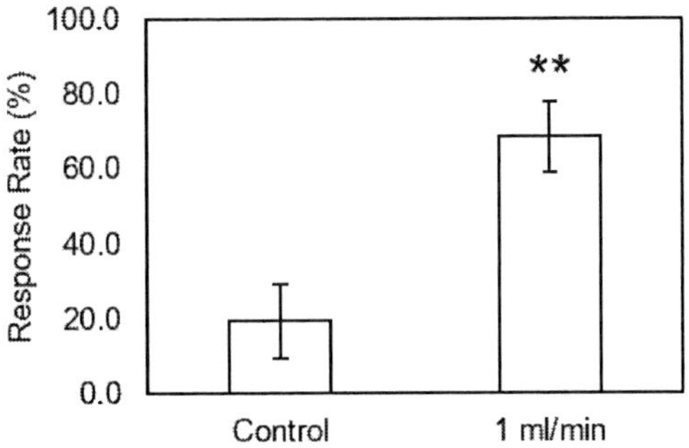

Figure 3. The response rate of 5-7 dpf zebrafish larvae (N=36) to a 1 mL/min flow stimulus compared with control animals (N=36) which were not exposed to any flow stimulus. Experiments were done in three trials of 12 animals each. Error bars: S.D. **: two-tailed t-test, p-value < 0.01

Response of Zebrafish Larvae to EC Stimulus

We also investigated the response of 5-7 dpf zebrafish larvae to EC applied across the channel after the larvae were successfully loaded into the device. A series of preliminary experiments showed that a minimum current of 3 μA was needed to evoke a response in the zebrafish larvae. The results of an experiment on a group of 30 larvae exposed to 3 μA current in comparison to a control group of animals that were not exposed to any EC are shown in Fig. 4.

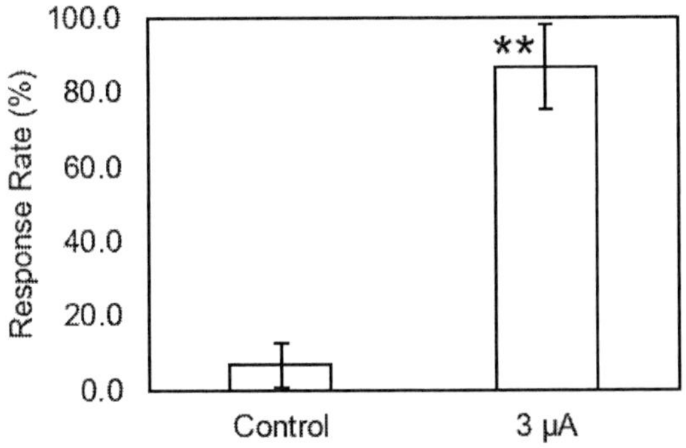

Figure 4. The response rate of 5-7 dpf zebrafish larvae stimulated by a 3 μA EC (N=30) in comparison to the control larvae (N=30) without any EC exposure. Experiments were done in three trials of 10 fishes each. Error bars: S.D. **: two-tailed t-test, p-value < 0.01

After applying the EC, 86.7±11.5% of the exposed larvae exhibited a response to the EC by showing a robust re-orientation of their bodies towards the anode electrode in the channel which was positioned initially at their tails. The control animals which were not exposed to the EC preferred to stay in their initial orientation along the channel (only 6.7±5.8% turned in Fig. 4). The response to EC was statistically different from the control group (p-value<0.01, two-tailed t-test) showing that the zebrafishes were able to sense the presence of EC in the channel and respond to it by rotating towards the anode electrode of the device.

Viability and Morphological Defects after Stimulation

We were interested to investigate the short and long term effects of flow and EC stimuli on morphology and viability of zebrafish larvae. For this, we conducted a series of experiments on groups of animals (N=15 each) that were exposed to these stimuli and compared the viability and morphological damage results with a control group of fishes. The results are shown in Fig. 5.

The control animals exhibited 80±20% survival rate while the flow- and EC-stimulated fishes could survive with the rate of 73.3±30.6% and 66.7±41.6% respectively. This demonstrates that these methods are not detrimental to the larvae since the experimented larvae lived with statistically analogous survival rates to the intact larvae (p-value>0.05, two-tail t-test). Their morphology was also assessed with 10±10%, 23.3±11.5%, and 13.3±5.8% damage rate, *i.e.* spine bending which has been visually identified by observers, to the control, flow-exposed and EC-exposed fishes, respectively. This proved that the proposed methods have no statistically significant negative developmental effects on the zebrafish larvae (p-value>0.05, two-tail t-test).

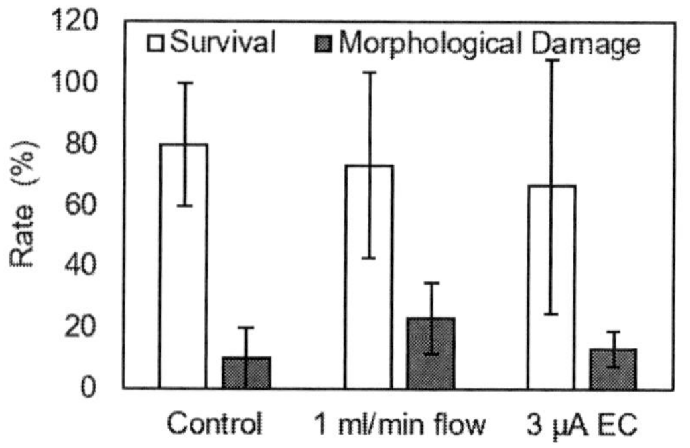

Figure 5. Survival and morphological damage of 5-7 dpf zebrafish larvae four days after exposure to flow and EC stimuli in comparison to a control group of fishes that were not exposed to any stimuli. N=15 animals tested for each condition. Experiments were conducted in three trials.

CONCLUSION

We report a microfluidic device for easy loading of 5-7 dpf zebrafish larvae into a microchannel followed by their exposure to fluidic and electric stimuli while monitoring their behavioral responses quantitatively. Our experiments demonstrated that zebrafish larvae were capable of sensing a flow rate of 1 ml/min or an EC of 3 μA along their body axes. The fishes showed a significant response to these stimuli by preferring to orient their bodies against the flow or towards the anode electrode in the microchannel. These on-demand and simple methods of stimulation did not affect the viability or morphology of the fishes significantly within 4-days post experiment. Accordingly, these methods can be used in any biological applications in which the zebrafish larvae's manipulation and behavioral investigation in a quantifiable manner are wanted. In addition to the behavioral studies, the biological pathways involved in sensing fluidic flow and EC can also be studied in zebrafish larvae using the presented microfluidic-based techniques.

REFERENCES

1. M. A. Wolman, R. A. Jain, L. Liss, and M. Granato, 2011 "Chemical modulation of memory formation in larval zebrafish.," *Proc. Natl. Acad. Sci. U. S. A.*, vol. 108, no. 37, pp. 15468–73.
2. P. McGrath and C. Q. Li, 2008 "Zebrafish: a predictive model for assessing drug-induced toxicity," *Drug Discov. Today*, vol. 13, no. 9–10, pp. 394–401.
3. M. Roberto, E. Pacheco, M. De Bem, A. Paula, G. Frazzon, R. Dutra, and C. Denise, 2006 "Exposure to Hg 2 + and Pb 2 + changes NTPDase and ecto-5 -nucleotidase activities in central nervous system of zebrafish (Danio rerio)," vol. 226, pp. 229–237.
4. J. Schweitzer, H. Löhr, A. Filippi, and W. Driever, 2012 "Dopaminergic and noradrenergic circuit development in zebrafish," *Dev. Neurobiol.*, vol. 72, no. 3, pp. 256–268.
5. F. Yang, C. Gao, P. Wang, G.-J. Zhang, and Z. Chen, 2016 "Fish-on-a-Chip: Microfluidics for Zebrafish Research," *Lab Chip*, vol. 16, pp. 1106-1125.
6. E. A. Osipova, V. V Pavlova, V. A. Nepomnyashchikh, and V. V Krylov, 2016 "Influence of magnetic field on zebrafish activity and orientation in a plus maze," vol. 122, pp. 80–86.
7. L. L. Bischel, B. R. Mader, J. M. Green, A. Huttenlocher, and D. J. Beebe, 2013 "Zebrafish Entrapment By Restriction Array (ZEBRA) device: a low-cost, agarose-free zebrafish mounting technique for automated imaging.," *Lab Chip*, vol. 13, no. 9, pp. 1732–6.
8. P. Blader and U. Strähle, 2000 "Zebrafish developmental genetics and central nervous system development.," *Hum. Mol. Genet.*, vol. 9, no. 6, pp. 945–951.
9. C. Pardo-Martin, T.-Y. Chang, B. K. Koo, C. L. Gilleland, S. C. Wasserman, and M. F. Yanik, 2010 "High-throughput in vivo vertebrate screening.," *Nat. Methods*, vol. 7, no. 8, pp. 634–6.
10. X. Lin, S. Wang, X. Yu, Z. Liu, F. Wang, W. T. Li, S. H. Cheng, Q. Dai, and P. Shi, 2015 "High-throughput mapping of brain-wide activity in awake and drug-responsive vertebrates," *Lab Chip*, vol. 15, no. 3, pp. 680–689.
11. M. Erickstad, L. A. Hale, S. H. Chalasani, and A. Groisman, 2015 "Lab on a Chip zebrafish larvae under acute hypoxia †," *Lab Chip*, vol. 15, pp. 857–866.
12. C. Y. Chen and C. M. Cheng, 2014 "Microfluidics expands the zebrafish potentials in pharmaceutically relevant screening," *Adv. Healthc. Mater.*, vol. 3, no. 6, pp. 940–945.
13. S. Padilla, D. L. Hunter, B. Padnos, S. Frady, and R. C. MacPhail, 2011 "Assessing locomotor activity in larval zebrafish: Influence of extrinsic and intrinsic variables," *Neurotoxicol. Teratol.*, vol. 33, no. 6, pp. 624–630.
14. B. D. Monesson-Olson, J. Browning-Kamins, R. Aziz-Bose, F. Kreines, and J. G. Trapani, 2014 "Optical stimulation of zebrafish hair cells expressing channelrhodopsin-2," *PLoS One*, vol. 9, no. 5, pp. 1–8.
15. R. Candelier, M. Sriti Murmu, S. Alejo Romano, A. Jouary, G. Debrégeas, and G. Sumbre, 2015 "A microfluidic device to study neuronal and motor responses to acute chemical stimuli in zebrafish," *Sci. Rep.*, vol. 5, p. 12196.
16. B. Y. S. Dijkgraaf, 1962 "Lateral-Line Organs," *Biol. Rev.* vol. 38, pp. 51–105.
17. P. Rezai, A. Siddiqui, P. R. Selvaganapathy, and B. P. Gupta, 2010 "Electrotaxis of Caenorhabditis elegans in a microfluidic environment.," *Lab Chip*, vol. 10, pp. 220–226.

Renewable Energy Systems

SUSTAINABLE USE OF BIOGAS FOR DISTRICT ENERGY SYSTEMS

Adreon Raymond Murphy, MASc Student
Mechanical & Industrial Engineering
Ryerson University
Toronto, Ontario, Canada
adreon.murphy@ryerson.ca

Dr. Alan S. Fung, PhD, PEng, FCSME
Associate Professor
Mechanical & Industrial Engineering
Ryerson University
Toronto, Ontario, Canada
alanfung@ryerson.ca

ABSTRACT

The energy available from the Ashbridges Bay Wastewater Treatment Plant in the form of digester gas was estimated, and the viability of building a direct pipe to three different district energy nodes was evaluated against inputting the gas into Enbridge's natural gas distribution network. An economic analysis showed that the Enbridge network scenario was the most viable, with an internal rate of return of 30.6% over six years. An alternative option is to create a direct pipe to the East Harbour future district energy node. The filtered digester gas is largely available in the summer, so the gas must be used for combined heat and power, due to the low heat demand.

Keywords – *biogas, district energy, wastewater treatment plant, renewable energy*

I. INTRODUCTION

Natural gas heating and electricity consumption in buildings are responsible for 24% of Ontario's equivalent carbon dioxide emissions [1]. Conventional natural gas fired boilers and electric generators can be run on a variety of fuels. Biogas is a sustainable alternative that can be used to reduce the amount of natural gas extracted from the earth, which therefore reduces resulting emissions from machinery and methane leakage during extraction. Connecting residential and commercial buildings to a district energy system can enable the use of biogas for heating and electricity production on a large scale.

A. *Project Goal*

This project proposes the use of anaerobic digester gas (biogas) as one of the many energy sources of a district energy node in Toronto. The project will evaluate the amount of digester gas energy available at the Ashbridges Bay Wastewater Treatment Plant (WWTP) and determine which district energy node is best suited for the utilization of this gas.

B. *Anaerobic Digestion*

Sewage contains approximately 10 times the energy required to treat it [2]. The largest portion of this energy is contained in digester gas. Digester gas production begins with raw sewage contacting water, which creates a process called hydrolysis, where the water breaks down the sewage. After the volume of the sewage is greatly reduced, processes called acetogenesis and acidogenesis create the molecules required for methanogenesis, producing methane gas as a by-product [3]. The sewage sludge goes through methonogenesis for approximately 15 days, in which the process is controlled to produce an even production of digester gas [3] [4].

Natural gas and digester gas have a very similar composition. Natural gas typically contains 95% methane, compared to approximately 64% methane and 30% CO_2 in digester gas. Methane is the heating fuel in both gases [5] [6].

C. *Wastewater Treatment Plants and Digester Gas*

In the USA there is approximately 1.1 GW available from all 544 WWTPs that have anaerobic digestion and have sewage flow rates higher than 5 million gallons per day [7]. This is because digester gas contains approximately 600 BTU/m^3 [7].

In 2004, 50% of the wastewater treatment plants in North America with over 200 million gallons (757,000 m^3) of sewage flow per day utilized their digester gas for plant processes [7]. However, there are opportunities for taking energy from these plants when digester gas production is greater than what the plant can consume with their conventional equipment.

The following is a calculation to determine how much energy is contained in digester gas compared to the amount of energy required to treat the sewage. According to the Eastern Research group, 4.4 million gallons per day of raw sewage can produce approximately 100 kW of electricity through anaerobic digestion, assuming 30% efficiency [7]. This is equivalent to 0.48 kWh/m^3 of total un-converted energy available from digester gas. On average WWTPs require 0.6 kWh to treat one m^3 of raw sewage from start to finish [8]. This means that digester gas contains enough energy to provide 80% of an average plant's needs, before considering equipment efficiencies.

A report by the Eastern Research Group, Inc., prepared for the U.S. Environmental Protection Agency, recommends that combined heat and power be used in WWTPs to enable the highest amount of consumption [7]. This directly translates to how district energy systems should use the gas.

II. ASHBRIDGES BAY WWTP

The Ashbridges Bay WWTP is the second largest WWTP in Canada and is only located 5km east of downtown Toronto. The plant receives an annual average of 818,000 m^3 of raw sewage per day and contains 20 anaerobic digesters on site.

These digesters produced an annual average of 64,560 m^3 of digester gas in 2015 [9].

A. Available Energy

The Ashbridges Bay WWTP uses their digester gas to fuel gas compressors and for plant heating, according to a plant superintendent [4]. All excess gas must be combusted to convert the methane to CO_2—a process known as flaring. This combustion process is carried out because methane has a global warming potential 28-36 times worse than that of CO_2 [10]. However, the amount of gas that is currently flared is equal to the amount of gas that can be extracted from the plant and used for district energy. Figure 1 compares the amount of gas flared each month to the amount of gas consumed each month. The total height of the bar is the amount of gas produced from the anaerobic digesters.

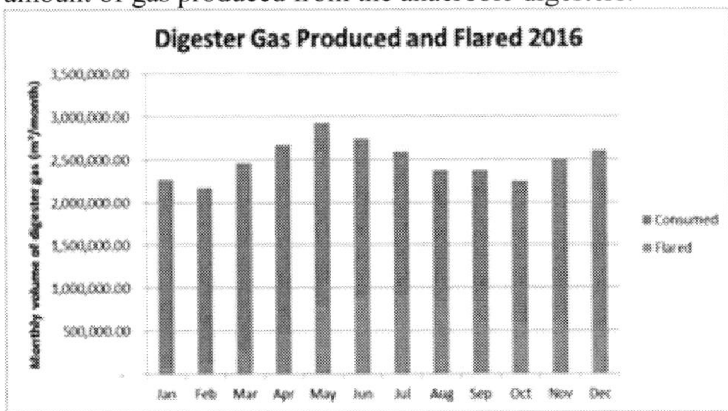

Figure 1: Digester gas produced and flared 2016.

The data presented in Figure 1 was provided from an Ashbridges Bay superintendent [4]. The amount consumed is the exact sum of the digester gas consumption of three gas compressors, on site. The total amount of gas produced by the three flaring stations is the sum of the consumption for compressors and the flaring stations [4].

If the digester gas that is normally flared was extracted, the annual energy available would be 149,000 GJ, assuming a net heating value (NHV, also known as lower heating value) of 569 BTU/ft^3 [11]. The NHV is calculated, assuming digester gas contains 62.5% of the energy due to the assumed 62.5% methane content. This available energy is equivalent to 4.7 million m^3 of natural gas, considering a natural gas NHV of 905 BTU/ft^3. The issue with this energy is that on average only 20% of the energy produced is available during the heating season (October to April) and even less (12%) is available from December to March. This is because of the plant heating requirements during the winter, and lack thereof in the summer; making the justification of the project difficult when only considering heating uses.

III. CONSUMPTION SITES

There are three possible locations for consumption of the Ashbridges Bay digester gas. First, a district energy node is proposed to be constructed at the bottom of the Don Valley Parkway at a location called the East Harbour, 2.5 km from the WWTP. This development will not be completed for approximately 10 years, but is planned to have 10-12 million ft^2 of building space.

The second consumption site is another proposed district energy node at the Redpath Sugar Refinery on Queens Quay, located 5 km from the WWTP. Over the next five to ten years, 15 million ft^2 of building space is proposed to be developed.

The third consumption site is the Enwave Energy Corporation Pearl Street Steam Plant, located 7 km from the WWTP. This is one of three plants that provide steam to 140 large buildings in downtown Toronto [12]. The plant consumes 66 million m^3 of natural gas annually and is generally used for peaking [13] [14].

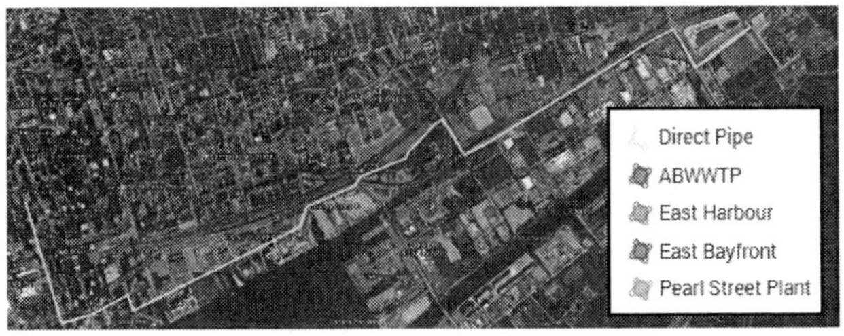

Figure 2: Ashbridges Bay WWTP's proximity to district energy nodes.

A. Direct Pipe Technical Process

There are two options for the transmission of carbon dioxide filtered digester gas. The first option is to build a dedicated pipe to any of the three consumption sites. This process involves six principle steps: extraction, filtration, compression, transmission, storage and, utilization. A molecular sieve will be needed for filtering the CO_2 molecules out of the biogas, accounting for approximately 30% gas' volume. The carbon dioxide can then be used to reduce the PH level in the WWTP's alkaline water, which will be a substitute for harmful mineral acids [15]. The compression process will require a large amount of electricity, because the gas will need to be compressed to fit into a small, cost effective polyethylene pipe.

1) Sizing

The molecular sieve will need to be sized for the peak flow rate. The highest monthly flow rate for the digester gas is 30,000 m^3/day in January. A variable production factor in the anaerobic digester was assumed to be 1.5, which results in an estimated peak flow rate of 1,000 CFM (1875 m^3/h).

The cost of installing a new pipe underground will increase significantly if the pipe is larger than six inches in inner diameter. This condition was used to determine the volume flow rate. The volume flow rate was calculated with the Spitzglass Formula, shown as Equation 1, where q is the natural gas flow capacity in ft^3/h, h is the pressure drop in inches Water Column, l is the length of the pipe in feet, d is the inside pipe diameter in inches and SG is the specific gravity [16].

The required volume flow rates for East Harbour, East Bayfront and Pearl respectively were 49 CFM for a 5" pipe 56 and 47 CFM for a 6" pipe, based on this formula. This condition was used to find the required increase in pressure to reduce the volume from 720 CFM (reduced from the original 1000 CFM due to filtering of CO_2) to the calculated value. The required pressure to achieve a flow rate of 53 CFM (approximately equal to the calculated values above) was calculated to be 75 PSI, with the ideal gas equation, PV=nRT on a per mole basis.

$$q = 3550 * k * \left(\frac{h}{l * SG}\right)^{1/2} \quad (1)$$

$$k = \left(\frac{d^5}{1+\frac{3.6}{d}+0.03*d}\right)^{1/2}$$

Now that the required pressure increase is determined, from 5 PSIG to 75 PSIG, the compressor can be sized. The compressor size was calculated with the help of a natural gas compression chart, from America Gas Compression Services [17]. The chart yielded a compressor size of 81 kW, with adjusted values to reflect the peak flow rate of the available gas at Ashbridges Bay.

The storage tank was sized to provide at least 24 hours of average, compressed gas storage. A tank size of 1000 m^3 can accommodate 27 hours of storage.

B. Enbridge Distribution Technical Process

The second transmission option is to use Enbridge's natural gas distribution network. This can be done by connecting to the nearest distribution pipe to the WWTP and paying a transportation fee. Once the fee is paid, the company that input the gas is permitted to use that same amount without a consumption charge [18]. The fee is based on the energy inputted to the network, not the distance from the point of input to the point of use, so this scenario will work equally for all three consumption sites. The digester gas will need to be filtered of CO_2, to meet Enbridge's standard of 36-40.2MJ/m^3 gross heating value, when injecting foreign gas [18]. A molecular sieve can be nearly 100% effective at removing CO_2, but approximately 2% of the methane will be lost to the filter [19].

IV. EMISSIONS SAVINGS

Emission savings for this project will be treated as follows. Since unused digester gas is combusted, only to prevent it from causing more pollution via methane, there are emissions savings if the gas is used to offset natural gas. The total emission savings potential is calculated by using net heating value of 910 BTU/ft^3 for methane and considering that only 62.5% of the gas is methane [11]. The available annual energy for an average of 19,250 m^3 per day of digester gas equates to 149,000 GJ. With a natural gas emission factor of 0.04903 tonnes CO_2/GJ, 7,300 tonnes of CO_2 can be offset [20]. An additional 3,500 tonnes can be saved annually if the filtered CO_2 is considered. This CO_2 will not be reemitted after filtration, but it would have been emitted from the flaring process that currently exists. The emissions savings from offsetting natural gas would reduce the Redpath Sugar Refinery's annual emissions by 12% and the Pearl Street Steam Plant's emissions by 7% [21] [13].

V. ECONOMIC MODEL

An economic analysis was conducted to determine the viability of bringing the direct pipe to the three consumption sites, as well as the scenario of using the Enbridge network.

A. Capital Costs

Table 1 shows the capital costs for the direct pipe scenarios.

Table 1: Capital cost inputs for economic model, assumed to be spent in 2018.

	Cost 2018 ($CAD)
Molecular Sieve (1875 m^3/h)	$2,350,000
Compressor (81 kW)	$47,900
Storage Tank (1000 m^3)	$1,000,000
Pipe Cost to East Harbour (2.5 km)	$1,000,000
Don River bridge pipe crossing	$100,000
Pipe Cost between East Harbour and East Bayfront (2.5 km)	$1,400,000
Pipe Cost from Redpath to Pearle Street (2 km)	$3,000,000
Compressor Annual Operation Cost (303,000 kWh)	$60,600
Compressor Annual Maintenance Cost	$5,000
Miscellaneous Annual Maintenance	$45,000

B. Annual Costs

The operation cost of the compressor was estimated by assuming the compressor operates at 35 kW. This is the load that the compressor must operate at to accommodate the annual average flowrate. The compressor was assumed to operate 8760 hours per year, resulting in an annual consumption of 303,000 kWh. The average unit cost of electricity is assumed to be $0.20/kWh, which results in an annual operating cost of $60,600.

C. Prices

The unit cost of natural gas was predicted up until 2026, with the breakdown of the unit cost shown in 2018 dollars.

Table 2: Breakdown of natural gas unit cost and predictions.

	Unit Cost $2018	Annual increase (6 years)
Commodity Cost	$4.66/GJ	5%
Transportation Cost	$1.70/GJ	-1.5%
Delivery Charge	$0.64/GJ	2%

The Enbridge special transportation charge was based on a generic gas agreement between companies that wants to input gas into their network [18]. This cost is lower than the ordinary transportation cost and was calculated based on the Niagara Indexed Price. This charge is used when calculating the cost of the Enbridge network scenario.

Ontario plans to start the price of carbon at $18/tonne on January 1, 2017. The price is assumed to increase by 5% annually until 2022 [22]. This price of carbon was considered as revenue in the economic model, considering the annual emissions savings of 7,300 tonnes.

$0.20/kWh was assumed to be the average unit electricity cost between 2018 and 2028.

D. Life Cycle Costs

The capital and annual costs for the three direct pipe scenarios are exactly the same, apart from the cost of the pipe. Figure 3 shows that the pipe cost drastically increases because of the increased distance and density.

The largest cost by far for the Enbridge network scenario is the molecular sieve, due to the short piping distance with no requirements for compressors and storage tanks.

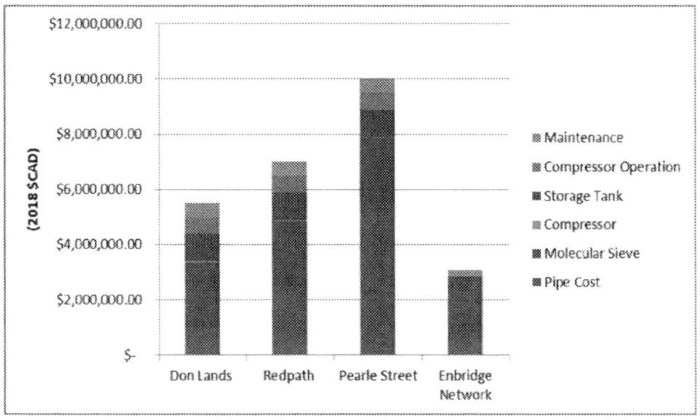

Figure 3: 10 year life cycle cost of four scenarios in 2018 $CAD.

E. Economic Analysis Results

All three direct pipe scenarios have equal access to savings from the cost of natural gas consumption, transportation and delivery, at an average of $1,155,000 over the first six years.

The Enbridge distribution network scenario has 21% less savings (revenue), due to the transportation charges. This scenario still has savings from the ordinary consumption and transportation costs, at an average of $915,000 over the first six years.

Cap and trade revenue had a significant impact on the project. There was an average of $172,000 in savings from Cap and Trade from 2018 to 2023.

The Enbridge network and East Harbour scenarios have a viable internal rate of return. The extra piping distance to the East Bayfront and especially through downtown Toronto to the Pearl Street Steam Plant made these scenarios too risky to justify.

Table 3: IRR of each scenario with an 8% cost of capital.

	East Harbour	East Bayfront	Pearl Street	Enbridge Network
IRR over 6 Years (8% WACC)	8.9%	-3.7%	-17.5%	30.6%

VII. CONCLUSIONS

For the feasible application of digester gas, the consumption site must be a combined heat and power plant. If the gas is only used for heating, there will not be nearly enough heat to justify it. Electricity is demanded all year round, so this free gas must be used to sell electricity to the grid, or used at the district energy site.

The Enbridge network scenario is the recommended option because yields the maximum IRR. The Enbridge distribution option requires significantly less piping material than the direct pipe scenario; this should be considered due to the added environmental benefits. The pipe from Ashbridges Bay to the local natural gas grid can also be installed in a timelier fashion than the other scenarios. Finally the gas input into the network can be used at any of the three locations and will ensure that the full capacity of the available gas is used.

VIII. REFERENCES

[1] Ontario Ministry of Environment, "Ontario's Five Year Action Plan 2016-2020," 8 June 2016. [Online]. Available: http://www.applications.ene.gov.on.ca/ccap/products/CCAP_ENGLISH.pdf. [Accessed 18 August 2016].

[2] Water Environment Research Foundation, "Energy Production and Efficiency Research – The Roadmap to Net-Zero Energy," August 2011. [Online]. [Accessed 18 August 2016].

[3] Clarke Energy, "Sewage Gas," [Online]. Available: https://www.clarke-energy.com/sewage-gas/. [Accessed 20 October 2016].

[4] P. Marto, Interviewee, *Superintendent Process & Operations, Ashbridges Bay Treatment Plant.* [Interview]. 12 August 2016.

[5] Engineering Toolbox, "Gaseous Fuels and their Chemical Composition," 2016. [Online]. Available: http://www.engineeringtoolbox.com/chemical-composition-gaseous-fuels-d_1142.html. [Accessed 1 November 2016].

[6] Union Gas, "Chemical Composition of Natural Gas," 2016. [Online]. Available: https://www.uniongas.com/about-us/about-natural-gas/Chemical-Composition-of-Natural-Gas. [Accessed August 2016].

[7] Eastern Research Group, Inc., "Opportunities for and Benefits of Combined Heat and Power at Wastewater Treatment Facilities," December 2006. [Online]. Available: https://www.americanbiogascouncil.org/pdf/EPA-Benefits%20of%20CHP%20at%20WWTPs.pdf. [Accessed 18 August 2016].

[8] P. L. McCarty, J. Bae and J. Kim, "Domestic Wastewater Treatment as a Net Energy Producer-Can This Be Acheived?," *Environmental Science and Technology,* no. 45, pp. 7100-7106, 2011.

[9] Toronto Water, "Ashbridges Bay Wastewater Treatment Plant 2015 Annual Report," 2015. [Online]. Available: http://www1.toronto.ca/City%20Of%20Toronto/Toronto%20Water/Files/pdf/W/wwtp-ashbridges-bay-annual-report-2015.pdf. [Accessed 18 August 2016].

[10] US Environmental Protection Agency, "Understanding Global Warming Potentials," 9 August 2016. [Online]. Available: https://www.epa.gov/ghgemissions/understanding-global-warming-potentials. [Accessed 1 November 2016].

[11] Engineering Toolbox, "Fuel Gases and Heating Values," 2016. [Online]. Available: http://www.engineeringtoolbox.com/heating-values-fuel-gases-d_823.html. [Accessed 6 October 2016].

[12] Power Mag, "Improving the Efficiency of Toronto's District Heating Plant," 1 August 2008. [Online]. Available: http://www.powermag.com/improving-the-efficiency-of-torontos-district-heating-plant/. [Accessed 6 October 2016].

[13] Government of Canada, "Enwave Pearl Street Steam Plant Facility and GHG Information," 2014. [Online]. Available: http://www.ec.gc.ca/ges-ghg/donnees-data/index.cfm?do=facility_info&lang=en&ghg_id=G10459&year=2012. [Accessed 6 October 2016].

[14] Enwave, "Enwave Technical Tour," 2011. [Online]. Available: http://www.districtenergy.org/assets/pdfs/2011Annual_Conf/Proceedings/Tour-ENWAVE-AW-IDEA-Presentation.pdf. [Accessed 6 October 2016].

[15] W.-T. Chang, M. Lee and W. Den, "Simultaneous Carbon Capture, Biomass Production, and Diary Wastewater Purification by Spirulina maxima Photobioreaction," *Industrial & Engineering Chemistry Research,* pp. 2046-2055, 2013.

[16] Engineering Toolbox, "Natural Gas - Pipe Sizing Calculator," [Online]. Available: http://www.engineeringtoolbox.com/natural-gas-pipe-calculator-d_1042.html. [Accessed 20 October 2016].

[17] American Gas Compression Services , "Quick Chart," [Online]. Available: http://www.agcsi.com/estimate/chart1A1.htm. [Accessed 15 November 2016].

[18] Enbridge, "Ontario Gas Purchase Agreement Generic," 2016.

[19] PermSelect, "Methane Purification, CO2 Removal," 2016. [Online]. Available: https://www.permselect.com/Platform_Technology/NG_CO2_Removal. [Accessed 6 October 2016].

[20] C. Thorn, Interviewee, *Energy Management Specialist, Golder Associates.* [Interview]. 12 September 2016.

[21] Government of Canada, "Redpath Sugar Refinery Facility and GHG Information," 2014. [Online]. Available: http://www.ec.gc.ca/ges-ghg/donnees-data/index.cfm?do=facility_info&lang=En&year=2011&ghg_id=G10537. [Accessed 18 August 2016].

[22] Government of Ontario, "How Cap and Trade Works," 30 November 2016. [Online]. Available: https://news.ontario.ca/ene/en/2016/11/how-cap-and-trade-works-1.html. [Accessed 1 December 2016].

Proceedings of the 26th CANCAM
Victoria, British Columbia, Canada, May 29 – June 1, 2017

THE EFFECT OF ICE ACCRETION ON WIND TURBINE AERODYNAMICS

Galal M. Ibrahim, Kevin Pope and Yuri S. Muzychka
Department Mechanical Engineering
Memorial University of Newfoundland
St. John's, NL, Canada
gmi588@mun.ca, kpope@mun.ca, yurim@mun.ca

ABSTRACT

Wind turbines are a promising technique to reduce diesel power generation in remote communities, however, the wind conditions in cold and harsh environments create several problems for the performance, longevity and safety of a wind turbine installation. This paper investigates ice accretion on wind turbine near the blade tip. New numerical simulations are conducted using FENSAP ICE to investigate different icing conditions and predict ice shapes. Flow angle of attack's effect on the ice formation and the effect of ice on aerodynamic performance are investigated.

KEYWORDS: Wind power, atmospheric icing, numerical predictions

I. INTRODUCTION

Wind farms located in cold climates can experience significant ice accretion diminishing its performance, safety and economics [1]. Atmospheric icing has two main types: glaze and rime ice. The type of ice is mainly based on the environmental and operational conditions during an icing event, such as temperature, pressure, flow speed over the object, liquid water content (LWC), flow angle of attack, median volume diameter of droplets (MVD), blade curvature, surface roughness, droplets impingement area, heat flux, icing time and collection efficiency [2]. Rapid changes in the atmospheric conditions during an icing event cause the prediction of ice type to be difficult [3].

Ice shape is significantly influenced by the operating conditions of the wind turbine. Ice shape / curvature has a major impact on power reduction rather than ice thickness [4]. In a study investigating icing on NREL Phase VI rotor using FENSAP-ICE solver [5] showed that power reduction percentage can be as high as 60% under certain conditions. Glaze ice accretion on the rotor blades can cause a greater loss in power than rime ice, as glaze ice accretion can correspond to greater ice shape with increased icing surface [5]. During high wind speeds and certain icing conditions, certain ice shapes can produce an unanticipated increase in aerodynamic characteristics [5-8].

A numerical investigation on the relation between atmospheric icing temperature and turbine size showed that ice accretion can be less severe for large wind turbines in terms of aerodynamic disturbance [9]. Other studies showed that the surface roughness caused by ice accretion can significantly degrade the aerodynamics and ice is mostly accreted at the leading edge of the blade [10, 11].

Ice mass distribution along the blade span is affected by the effect of centripetal forces, gravity, and flow angle over the blades. Centripetal force affects the water droplets impingement profile causing a greater ice formation at the blade tip [11]. In a study investigating ice accretion on five different sections of blade span showed that icing is less severe near the blade root section [7]. Other research indicated that rotor angular speed has a significant impact on the ice quantity and ice load increases with blade span [9, 12]. Combined changes in blade section length and relative flow velocity that correspond to blade span significantly affect ice accretion physics. In this paper different icing conditions are simulated to predict ice accretion shapes on a blade section at 80% of blade span, and then predicted ice shapes are used to predict the aerodynamic deterioration.

II. FORMULATION OF MULTIPHASE FLOW AND PHASE CHANGE

The air flow solution is computed by FENSAP solver using the 3D Navier-Stokes and energy equations [13]. The Navier-Stokes equation is written as:

$$\frac{\partial \rho_a \overrightarrow{V_a}}{\partial t} + \vec{\nabla}.(\rho_a \overrightarrow{V_a} \overrightarrow{V_a}) = \vec{\nabla}.\sigma_{ij} + \rho_a.g$$

where ρ_a is the air density, V_a is the velocity vector and σ_{ij} is the stress tensor. The conservation of energy equations can be written as:

$$\frac{\partial \rho_a E_a}{\partial t} + \vec{\nabla}.(\rho_a \overrightarrow{V_a} H_a) = \vec{\nabla}.[K_a(\vec{\nabla} T_a) + v_i \tau_{ij}] + \rho_a \vec{g}.\overrightarrow{V_a}$$

where E and H are the total internal energy and enthalpy, respectively. The air flow around blade section is mostly turbulent at high Reynolds numbers, thus one-equation of Spalart Allmaras turbulence model is used [14].

Water droplets trajectory, collection efficiency and impingement limits over the entire domain are obtained by the DROP 3D solver. Using the general Eulerian two-fluid model which consists of the Euler or Navier-Stokes equations augmented by the droplets continuity and momentum equations:

$$\frac{\partial(\alpha \overrightarrow{V_d})}{\partial t} + \vec{\nabla}(\alpha \overrightarrow{V_d} \otimes \overrightarrow{V_d}) = \frac{C_D R_{e_d}}{24K} \alpha(\overrightarrow{V_a} - \overrightarrow{V_d}) + \alpha(1 - \frac{\rho_a}{\rho_d}) \frac{1}{F_r^2}$$

where α and V_d are water volume fraction and droplet velocity, respectively. The droplet distribution is set as mono-disperse which indicates a calculation computed for a single diameter. The ICE 3D solver uses both mass conservation and energy conservation equations to predict the complex thermodynamics of ice accretion, including heat transfer generated by impinging super cooled water droplets, evaporation and ice accretion.

Computational grid discretization scheme

A blade section with a chord length of 1 m is fixed along 0.25 m blade span to represent a section near the tip. A NREL S809 profile is selected as it is widely used for wind turbines [15]. A fine 3D structured grid with 34,823 hexahedra cells is created with a minimum orthogonal quality of 0.835 (where values close to 0 correspond to low quality) and maximum orthogonal skew of 0.194 (where values close to 1 correspond to low quality). The grid's first boundary layer height is 1×10^{-6} m with a growth rate of 1.1 for 135 layers. As illustrated in Fig. 1a, a circular domain (O grid type) is created with a diameter of 20 m. The majority of ice typically accumulates at the leading edge due to flow stagnation. Therefore, as illustrated in Fig. 1b, smaller elements are created at the leading edge with equal spacing of 0.001 m to accurately evaluate both ice mass and shape. The cell size is gradually changing along the blade surface.

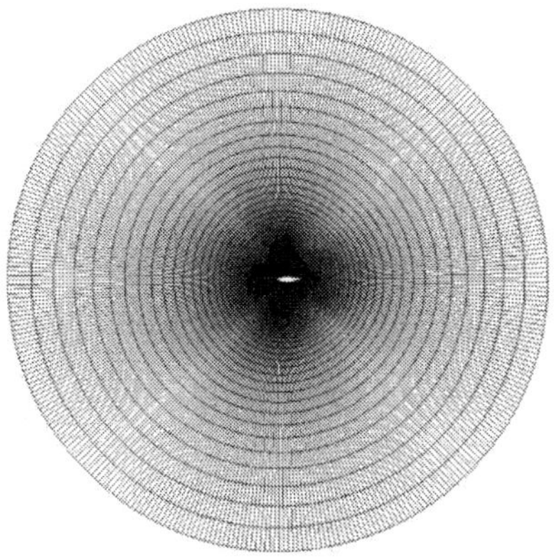

(a)

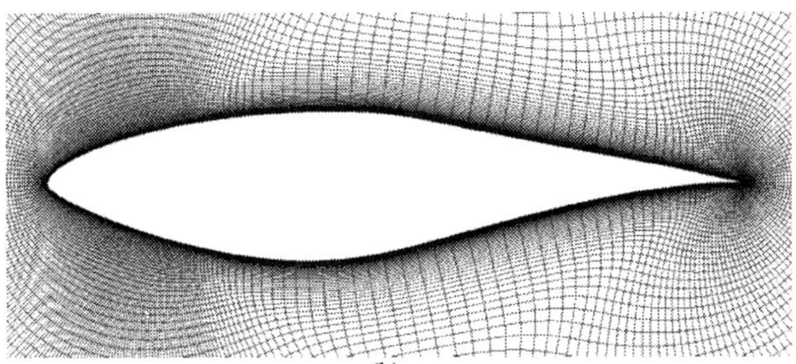

(b)

Figure (1) - Grid discretizations on a NREL S809 airfoil for the (a) entire flow domain and (b) close to the aerofoil surface.

III. Results and Discussion

A. The effect of angle of attack on ice shapes

In this section, ice accretion on the wind turbine blade is investigated at different flow angles of attack. As presented in Table (1), numerically predicted ice shapes on NREL S809 using multi-shot approach are obtained at specific icing conditions.

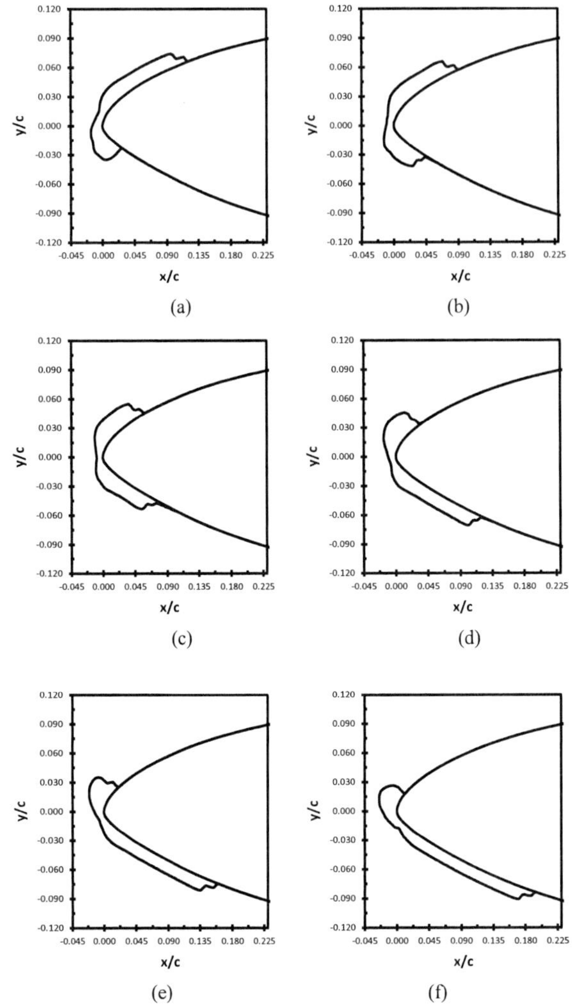

Figure (2) - Ice shapes at different flow angles of attack, (a) -4°, (b) -2°, (c) 0°, (d) 2° (e) 4° and (f) 6°.

As illustrated in Fig. (2), ice accretion is mainly found at the leading edge of a wind turbine blade. However, its location on the leading edge is depends on the flow angle during ice accumulation. Negative flow angles accumulate ice on the pressure side and positive angles generate ice on the suction side of the blade. This is likely due to the changing location in flow stagnation point with the flow's angle of attack. Furthermore, the droplet collection and impingement profiles are affected by the flow's angle of attack. At zero angle of attack, ice quantity on pressure and suction blade sides differs due to the asymmetry of airfoil used. Lower ice thickness can be found at the flow stagnation point. Ice propagation on blade sides can increase with the increase of flow attacking angle. However, for the same icing condition, total ice quantity is relatively consistent for different flow angles.

Table (1) - Variable conditions for Test 1

Parameter	Quantity
Wind speed (m/s)	75
Icing temperature (°C)	-5
MVD value (μm)	30
LWC value (g/m^3)	0.3
Angle of attack (degrees)	-4, -2, 0, 2, 4, 6
Icing time (minutes)	60

B. Aerodynamic performance of a wind turbine with iced blades

In this section, grids are created for each predicted ice shape. As presented in Table (2), the first ice shape is predicted at -5°C with 0.3 LWC and the second shape is predicted at -8°C with 1 LWC. The purpose of selecting the second icing condition (-8°C with a LWC of 1) is to obtain a large quantity of ice to study the aerodynamic response. Flow field solutions are computed using FENSAP solver at varied flow angle of attack ranges from -8° to 8°. For both simulations, the ice is accreted with a flow angle of attack of 2°, and then the angle of attack is altered without further ice accumulation. As illustrated in Fig. (3), the first icing condition (-5°C with a LWC of 0.3) is represented by a solid line and has a significantly lower quantity of ice compared to the second condition (-8°C with a LWC of 1), which is represented by the dashed line. The shapes are predicted for 60 minutes of ice accumulation using a multi-shot approach, at Reynolds numbers of 5.8 × 10^{-6} m.

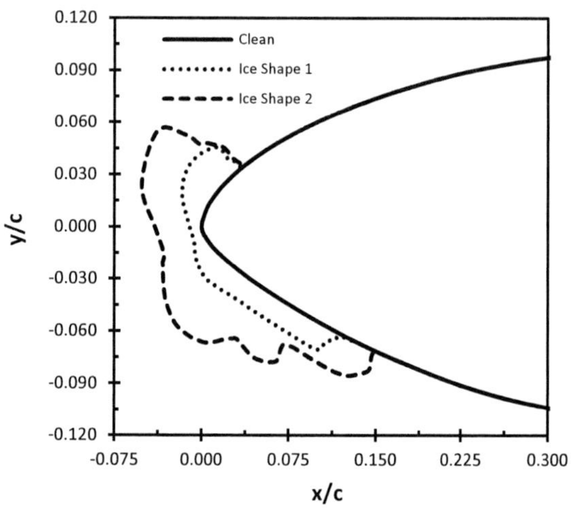

Figure (3) - Ice shapes caused by different icing conditions.

Lift and drag coefficients are computed for three cases: (i) clean, (i) Ice Shape 1, and (iii) Ice Shape 2. As illustrated in Fig. (4), a comparison between clean and iced blades is performed to study the aerodynamic effect of ice accretion. The second ice shape is responsible for higher reduction in lift

Table (2) - Variable conditions for Test 2

Parameter	Quantity
Wind speed (m/s)	75
Icing temperature (°C)	-5, -8
MVD value (μm)	30
LWC value (g/m^3)	0.3, 1
Angle of attack (degrees)	2
Icing time (minutes)	60

coefficient compared to the first ice shape. This is likely due to the higher flow separation and surface roughness caused by ice accumulation. For drag coefficient, the second ice shape has the highest drag values compared to the first ice shape. The predicted results agree well with the experimental results predicted by Broeren and Bragg [16].

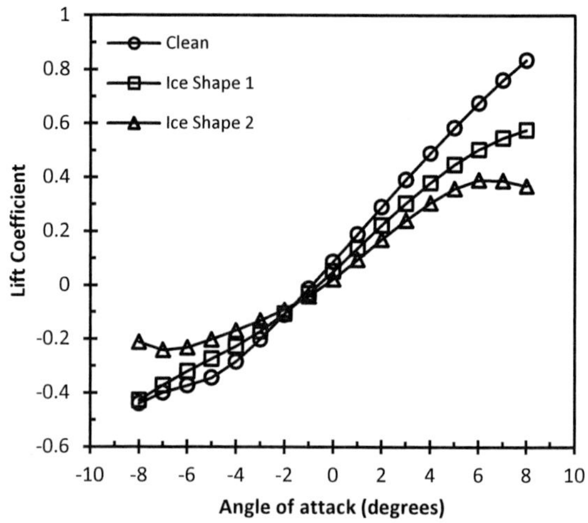

(a)

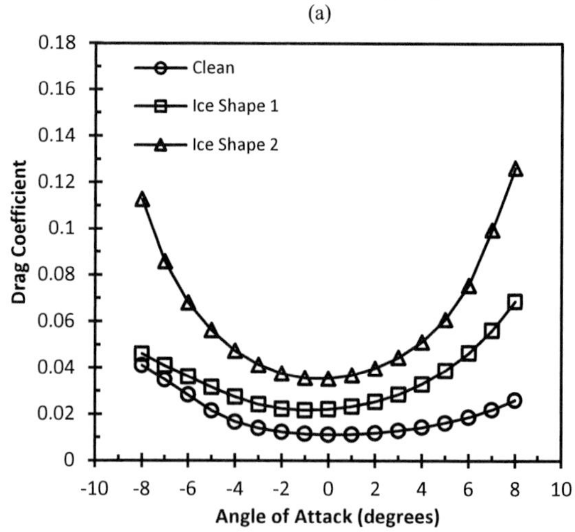

(b)

Figure (4) - Aerodynamic effect of ice accretion on (a) lift coefficient and (b) drag coefficient.

IV. CONCLUSIONS

In this paper, ice accretion on wind turbines was investigated numerically using FENSAP ICE. Ice shapes were predicted at different flow angles of attack. Light and heavy icing conditions were simulated, and then the iced airfoil's aerodynamic performance was investigated. Ice is mainly accreted at the blade's leading edge and its location whether on the top or the bottom is affected by the flow angle of attack. Minimum ice thickness can be found at the location of the flow stagnation point. For the same icing conditions, ice thickness does not change substantially with flow angle of attack, however, ice propagation on the blade surface increases with flow angle of attack. The results of this paper provide new insights and numerical predictions for ice accretion on wind turbines.

ACKNOWLEDGMENTS

The authors of this paper gratefully acknowledge the financial support of the Research and Development Corporation of Newfoundland.

REFERENCES

1. Virk, Muhammad S., Matthew C. Homola, and Per J. Nicklasson. "Atmospheric icing on large wind turbine blades." International Journal of Energy and Environment 3.1 (2012): 1-8.
2. Cattin, R. (2012). Icing of wind turbines. Vindforsk projects, a survey of the development and research needs Elforsk report, 12, 13.
3. Battisti, L. (2015). Wind turbines in cold climates: Icing impacts and mitigation systems. Springer
4. Fikke, S. M., Ronsten, G., Heimo, A., Kunz, S., Ostrozlik, M., Persson, P. E., ... & Laakso, T. (2006). COST 727: atmospheric icing on structures: measurements and data collection on icing: state of the art. Meteo Schweiz.
5. Homola, M. C., Wallenius, T., Makkonen, L., Nicklasson, P. J., & Sundsbø, P. A. (2010). Turbine size and temperature dependence of icing on wind turbine blades. Wind Engineering, 34(6), 615-627.
6. Fu, P., & Farzaneh, M. (2010). A CFD approach for modeling the rime-ice accretion process on a horizontal-axis wind turbine. Journal of Wind Engineering and Industrial Aerodynamics, 98(4), 181-188.
7. Bose, N. (1992). Icing on a small horizontal-axis wind turbine—Part 1: Glaze ice profiles. Journal of wind engineering and industrial aerodynamics, 45(1), 75-85.
8. Bose, N. (1992). Icing on a small horizontal-axis wind turbine—Part 2: Three dimensional ice and wet snow formations. Journal of Wind Engineering and Industrial Aerodynamics, 45(1), 87-96.
9. Virk, M. S., Homola, M. C., & Nicklasson, P. J. (2010). Effect of rime ice accretion on aerodynamic characteristics of wind turbine blade profiles. Wind Engineering, 34(2), 207-218.
10. Lamraoui, F., Fortin, G., Benoit, R., Perron, J., & Masson, C. (2014). Atmospheric icing impact on wind turbine production. Cold Regions Science and Technology, 100, 36-49.
11. Reid, T., Baruzzi, G., Ozcer, I., Switchenko, D., & Habashi, W. (2013, January). FENSAP-ICE simulation of icing on wind turbine blades, part 1: performance degradation. In 51st AIAA aerospace sciences meeting including the new horizons forum and aerospace exposition, Grapevine, Texas (pp. 7-10).
12. Homola, M. C., Virk, M. S., Wallenius, T., Nicklasson, P. J., & Sundsbø, P. A. (2010). Effect of atmospheric temperature and droplet size variation on ice accretion of wind turbine blades. Journal of wind engineering and industrial aerodynamics, 98(12), 724-729.
13. FENSAP ICE User Manual 2015.
14. Spalart, P. R., & Allmaras, S. R. (1992). A one equation turbulence model for aerodynamic flows. AIAA journal, 94.
15. Tangler, J. L., & Somers, D. M. (1995). NREL airfoil families for HAWTs (pp. 117-123). National Renewable Energy Laboratory.
16. Broeren, A. P., Bragg, M. B., & Addy, H. E. (2004). Effect of intercycle ice accretions on airfoil performance. Journal of Aircraft, 41(1), 165-174.

ROBUST LPV GAIN SCHEDULING FOR VARIABLE SPEED VARIABLE PITCH WIND TURBINE

Ali Farhangfar
Mechanical and Manufacturing Engineering Department
University of Calgary
Calgary, Alberta, Canada
Ali.farhangfar@ucalgary.ca

Jeffery Kurt Pieper
Mechanical and Manufacturing Engineering Department
University of Calgary
Calgary, Alberta, Canada
pieper@ucalgary.ca

Peyman Sindareh-Esfahani
Mechanical and Manufacturing Engineering Department
University of Calgary
Calgary, Alberta, Canada
peyman.sindarehesfah@ucalgary.ca

ABSTRACT

Presented is a state feedback gain scheduled controller for a Variable Speed Variable Pitch wind turbine. The wind turbine is represented by a non-linear time varying model that is modified to Linear Parameter Varying (LPV) form for control design purposes. The proposed LPV state feedback controller guarantees H_2 and H_∞ performance bounds and places closed-loop poles in desired locations via constraints on upper bounds of H_2 and H_∞ norms and radius of closed-loop poles location disk as described by Linear Matrix Inequalities (LMIs) formulas. The designed controller is implemented on a three blade horizontal axis 5 MW wind turbine and simulation results demonstrate the effectiveness of the proposed robust control strategy.

KEYWORDS: Wind turbine, linear parameter varying, H_2 , H_∞ , pole placement, linear matrix inequality

I. INTRODUCTION

In recent years, wind turbine technology has been widely developed as a clean and renewable source of electrical power. Control has an important role for improvement of the wind turbine performance and reduction of structural damage due to wind gusts and other external environmental factors. In this work we consider an advanced control strategy applied to a variable speed variable pitch wind turbine for maintaining rated power despite gusting and changing wind speeds [1].

Wind turbines have three general operational regions. Below the rated wind speed, the main control objective is to maximize the extracted power from the wind flow. This region is called Maximum Power Point Tracking (MPPT) during which generator torque controller is activated. Pitch is not usually used here as it would involve significant wear on the mechanical components and associated high maintenance costs. In high speed wind flow, the main control aim is to regulate the generated power. This region is known as Power Regulation (PR) region. The control objective during PR region can be achieved using generator torque and pitch blades controllers. The other region is a switching region between MPPT and PR regions. During this region, the objective is a smooth transition between MPPT and PR while the pitch angle controller is activated [2]. Fig. 1. shows the optimal power curve for different operation regions.

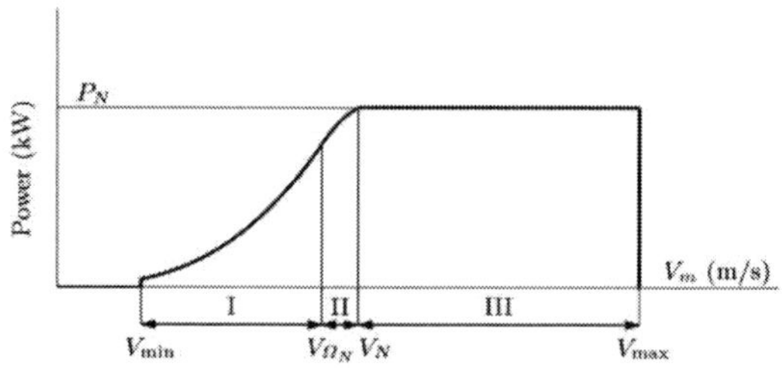

Figure 1: Operational regions of wind turbine in different wind speed [1]

Different control strategies have been applied on wind turbine systems such as neural network adaptive controllers [3], robust H_2 and robust H_∞ controllers [4]–[6]. One of the most common techniques applied to wind turbines is gain scheduling [6]-[8] which has an inherent capability of handling nonlinear systems. Typical gain scheduling involves linearizing the system at different operating points and designing controllers for each [9]. Then, the designed controllers are convexly interpolated to extend the control strategy for the whole operating range [6]. Recently, a new gain scheduling method was developed based on robust $H_2$and H_∞ and pole placement for LPV systems which can improve the performance through increasing robustness of the system [10].

This study considers implementation of robust mixed H_2 / H_∞ with pole placement in the LPV model of a wind turbine. The closed-loop system satisfies H_2 and H_∞ performance by applying constraints on norm upper bounds, and locates closed-loop poles in a circle of chosen radius and center. The constraints are inequalities and are organized in the form of LMIs. State feedback gains are obtained from solution of the LMIs in MATLAB LMI Toolbox. To show the performance and effectiveness of the proposed controller, it is implemented on a LPV model of a wind turbine.

The rest of this paper is organized as follow: Section 2 is modeling of wind turbine dynamics in LPV form. The controller strategy including constraints and desired conditions in LMIs formulas is presented in section 3. Simulation results of designed controller are shown in section 4, and conclusions are given in section 5.

II. LPV MODEL OF WIND TURBINE

This section introduces the LPV model of wind turbine. The general dynamic model of wind turbine is [1], [2]:

$$\begin{bmatrix} \dot{\theta}_s \\ \dot{\Omega}_r \\ \dot{\Omega}_g \end{bmatrix} = \begin{bmatrix} 0 & 1 & -1 \\ \frac{-K_s}{J_r} & \frac{-B_s}{J_r} & \frac{B_s}{J_r} \\ \frac{K_s}{J_g} & \frac{B_s}{J_g} & \frac{-B_s}{J_g} \end{bmatrix} \cdot \begin{bmatrix} \theta_s \\ \Omega_r \\ \Omega_g \end{bmatrix} + \begin{bmatrix} 0 & 0 \\ \frac{T_r}{J_r} & 0 \\ 0 & \frac{-T_g}{J_g} \end{bmatrix} \tag{1}$$

where θ_s ,Ω_r and Ω_g are torsion angle, rotor generator rotational speed respectively. K_s and B_s are stiffness and damping of the mechanical transmission, and J_r and J_g are inertias of rotor and generator.

Aerodynamic torque (T_r) and generator torque (T_g) are nonlinear equations of the wind turbine dynamic model. The generator torque is a function of generator speed and zero torque speed (Ω_z) that is considered as [1]:

$$T_g = B_g\left(\Omega_g - \Omega_z\right) \tag{2}$$

where B_g is a slope of the linear function of generator torque. Since the generator torque is considered as linear, only aerodynamic torque equation is nonlinear as given by [11]:

$$T_r = \frac{1}{2}\rho\pi R^3 \frac{C_p(\lambda,\beta)}{\lambda}(V)^2 \tag{3}$$

where λ is tip speed ratio with definition:

$$\lambda = \frac{\Omega R}{V} \tag{4}$$

In equations (3), (4) ρ is air density, R is the length of blades, and β, V are pitch angle and wind speed respectively, and C_p in Eq. (3) is the power coefficient.

In finding the LPV model we need to linearize the nonlinear equation as a function of its variables. Considering Eqs. (3), (4) wind speed, rotational speed and pitch angle are variable parameters of the wind turbine dynamic model which are called scheduling parameters: $q = \left[\bar{V}\ \bar{\Omega}\ \bar{\beta}\right]^T$

Then the aerodynamic torque equation can be rewritten as [1]:

$$T_r = -T_{r,\Omega}\left(\bar{\Omega},\bar{\beta},\bar{V}\right).\hat{\Omega}_r + T_{r,V}\left(\bar{\Omega},\bar{\beta},\bar{V}\right).\hat{V} + T_{r,\beta}\left(\bar{\Omega},\bar{\beta},\bar{V}\right).\hat{\beta} \tag{5}$$

$$T_{r,\Omega}\left(\bar{\Omega},\bar{\beta},\bar{V}\right) = -\frac{\partial T_r}{\partial \Omega_r},$$

$$T_{r,V}\left(\bar{\Omega},\bar{\beta},\bar{V}\right) = \frac{\partial T_r}{\partial V}, T_{r,\beta}\left(\bar{\Omega},\bar{\beta},\bar{V}\right) = \frac{\partial T_r}{\partial \beta}$$

In Eq. (5) T_r ,$\hat{\Omega}_r$,$\hat{V}$,$\hat{\beta}$ represent the steady state values of rotor torque, rotor speed, wind speed, and pitch angle.

Because in PR region pitch angle is activated as one of the control inputs, pitch actuator dynamic needs to be considered as [6]:

$$\dot{\beta} = -\frac{1}{\tau}\beta + \frac{1}{\tau}\beta_d \tag{6}$$

where β_d is the desired pitch angle, and τ is a time constant. In MPPT region pitch angle is constant and equal to zero, so scheduling parameters are: $q = \left[V\ \Omega\right]^T$

Because in PR region pitch angle is activated and rotor speed is constant and equal to its rated value $\left(\Omega = \Omega_N\right)$, the scheduling parameters are: $q = \begin{bmatrix} \beta \\ V \end{bmatrix}$

The LPV model of wind turbine is presented in the form [1]:

$$\begin{cases} \dot{x} = A(q)x + B_v(q)V + B(q)u \\ y = Cx + Du \\ \hat{T}_s = C_t x \end{cases} \tag{7}$$

where $\hat{T}_s$ is effective torque of shaft, and $A(q),B(q),B_v(q)$ are the parameter varying matrices.

Therefore by substituting Eq. (5) into Eq. (1) for PR region the LPV matrices are:

$$x = \begin{bmatrix} \theta_s \\ \Omega_r \\ \Omega_g \\ \beta \end{bmatrix}, u = \begin{bmatrix} \Omega_z \\ \beta_d \end{bmatrix}, B_v(q) = \begin{bmatrix} 0 \\ \frac{T_{r,v}(q)}{J_r} \\ 0 \\ 0 \end{bmatrix}, B(q) = \begin{bmatrix} 0 & 0 \\ 0 & 0 \\ 0 & \frac{B_g}{J_g} \\ \frac{1}{\tau} & 0 \end{bmatrix}$$

$$C = \begin{bmatrix} 0 & 0 & 1 & 0 \end{bmatrix}, D = 0,\ C_t = \begin{bmatrix} K_s & B_s & -B_s \end{bmatrix}$$

$$A(q) = \begin{bmatrix} 0 & 1 & -1 & 0 \\ -\frac{K_s}{J_r} & -\frac{B_s + T_{r,\Omega}(q)}{J_r} & \frac{B_s}{J_r} & \frac{T_{r,\beta}(q)}{J_r} \\ \frac{K_s}{J_g} & \frac{B_s}{J_g} & -\frac{B_s + B_g}{J_g} & 0 \\ 0 & 0 & 0 & -\frac{1}{\tau} \end{bmatrix} \tag{8}$$

In this operating region pitch angle is constant and equal to zero so the scheduling parameters for MPPT region are: $q = \left[V\ \Omega\right]^T$

III. MIXED H_2 / H_∞ WITH POLE PLACEMENT

This section describes a controller for LPV systems to satisfy norm bounds and place poles within a desired disk. Consider the following model as a general LPV system [10]:

$$\begin{cases} \dot{x} = A(q)x + B_w(q)w + B_u u \\ y_\infty = C_\infty(q)x + D_{\infty,w}(q)w + D_{\infty,u}u \\ y_2 = C_2(q)x + D_{2,u}u \end{cases} \tag{9}$$

LPV systems have 2^n vertexes according to number of their scheduling parameters, where n is number of scheduling parameters. For example if $q=[q_1 \quad q_2]$ while $q_{1,min} < q_1 < q_{1,max}$ and $q_{2,min} < q_2 < q_{2,max}$ the vertexes are:

$$\Gamma_1 = (q_{1,max}, q_{2,max}), \Gamma_2 = (q_{1,min}, q_{2,max}) \\ \Gamma_3 = (q_{1,min}, q_{2,min}), \Gamma_4 = (q_{1,max}, q_{2,min}) \tag{10}$$

And by using convex decomposition concept [6] the LPV matrices can be described in the form:

$$A(q) = \sum_{i=1}^{N} \eta_i(q) A(\Gamma_i) \tag{11}$$

In Eq. (11) N is number of vertexes, and $\eta_i(q)$ s are defined:

$$\eta_1(q) = \frac{(q_1 - q_{1,min})(q_2 - q_{2,min})}{(q_{1,max} - q_{1,min})(q_{2,max} - q_{2,min})}$$
$$\eta_2(q) = \frac{(q_{1,max} - q_1)(q_2 - q_{2,min})}{(q_{1,max} - q_{1,min})(q_{2,max} - q_{2,min})} \tag{12}$$
$$\eta_3(q) = \frac{(q_{1,max} - q_1)(q_{2,max} - q_2)}{(q_{1,max} - q_{1,min})(q_{2,max} - q_{2,min})}$$
$$\eta_4(q) = \frac{(q_1 - q_{1,min})(q_{2,max} - q_2)}{(q_{1,max} - q_{1,min})(q_{2,max} - q_{2,min})}$$

To guarantee H_∞ performance, the H_∞ norm of transform function needs to be less than its desired upper bound $(\Upsilon_\infty > 0)$ [12, 13]:

$$\|G_\infty(q)\|_\infty < \Upsilon_\infty \tag{13}$$

where $G_\infty(q)$ is transform function between disturbance (w) and output (y_∞).

The inequality to satisfy H_2 performance similar is structure to H_∞ performance is [12]:

$$\|G_2(q)\|_2 < \Upsilon_2 \tag{14}$$

The suggested controller is a state feedback $(u = Kx)$, for system (9) to guarantee conditions of Eqs. (13), (14) and put the closed-loop poles in the desired circle with radius r and centre $(-a,0)$. These constraints can be satisfied for the system in Eq. (9) if there exist symmetric matrices $X_{2,i}$, $X_{\infty,i}$, $X_{p,i}$, Y_i and matrices S_i with $i=1,\dots,N$ and a matrix H which satisfy the following LMIs [10]:

$$\begin{bmatrix} V_1 & V_2 \\ * & X_{p,i} - H - H^T \end{bmatrix} < 0, \quad \begin{bmatrix} Y_i & C_{2,i}H + D_{2,u}S_i \\ * & H + H^T - X_{2,i} \end{bmatrix} > 0$$

$$\begin{bmatrix} M_1 & M_2 & B_{w,i} & (C_{\infty,i}H + D_{\infty u}S_i)^T \\ * & -(H+H^T) & 0 & -(C_{\infty,i}H + D_{\infty u}S_i)^T \\ * & * & -I & D^T_{\infty w,i} \\ * & * & * & -\Upsilon_\infty^{\,2} I \end{bmatrix} < 0 \tag{15}$$

$$\begin{bmatrix} F_1 & F_2 & B_{w,i} \\ * & -(H+H^T) & 0 \\ * & * & -I \end{bmatrix} < 0$$

where:

$$\begin{aligned} V_1 &= (a^2 - r^2) X_{p,i} + a\left(A_i H + B_u S_i + (A_i H + B_u S_i)^T\right) \\ V_2 &= a(X_{p,i} - H^T) + A_i X_{\infty,i} + B_u S_i \\ M_1 &= A_i H + B_u S_i + (A_i H + B_u S_i)^T \\ M_2 &= -X_{\infty,i} + H^T - (A_i H + B_u S_i) \\ F_1 &= A_i H + B_u S_i + (A_i H + B_u S_i)^T \\ F_2 &= -X_{2,i} + H^T - (A_i H + B_u S_i) \end{aligned} \tag{16}$$

Then the feedback gains are [10]:

$$K_i = S_i H^T \text{ for } i = 1,\dots,N \tag{17}$$

In order to convert the feedback gains to a LPV feedback gain same as Eq. (11) we use

$$K(q) = \sum_{i=1}^{N} \eta_i(q) K_i \tag{18}$$

IV. SIMULATION RESULTS

The proposed controller is implemented on three blades horizontal type 5 MW variable speed variable pitch wind turbine operating in the PR region in SIMULINK MATLAB. The main parameters of this wind turbine are in Table 1.

Table 1: Wind Turbine Parameters.

Rated Power	5 MW
Radius of Blades	63 m
J_r	$38759228 kg.m^2$
J_g	$534.116 kg.m^2$
B_s	$6215000 Nms/rad$
K_s	$867637000 N/m$

Since there are two scheduling parameters in PR region the vertexes are same as Eq. (10). Calculations of $T_{r,\Omega}(\bar{\Omega},\bar{\beta},\bar{V})$, $T_{r,V}(\bar{\Omega},\bar{\beta},\bar{V})$, $T_{r,\beta}(\bar{\Omega},\bar{\beta},\bar{V})$ in Eq. (5) are (see Table 2.):

Table 2: $T_{r,\Omega}(\bar{\Omega},\bar{\beta},\bar{V})$, $T_{r,V}(\bar{\Omega},\bar{\beta},\bar{V})$, $T_{r,\beta}(\bar{\Omega},\bar{\beta},\bar{V})$ for Each Vertex.

Vertex	$T_{r,\Omega}(\bar{\Omega},\bar{\beta},\bar{V})$	$T_{r,V}(\bar{\Omega},\bar{\beta},\bar{V})$	$T_{r,\beta}(\bar{\Omega},\bar{\beta},\bar{V})$
(β_{max}, V_{max})	1.0186e+07	-3.1596e+04	-5.7188e+05
(β_{min}, V_{max})	-4.5030e+05	6.4838e+05	-3.4504e+05
(β_{min}, V_{min})	-2.4141e+06	1.0018e+06	-2.6652e+05
(β_{max}, V_{min})	1.5030e+07	-1.5715e+05	4.9419e+05

The feedback gains in Eq. (17) are attained by solving the inequalities in Eq. (15) with MATLAB LMI Toolbox. The wind speed for PR region which needs to higher than rated wind speed is generated by TurbSym software with mean value of 15 m/s (in Fig. 2.).

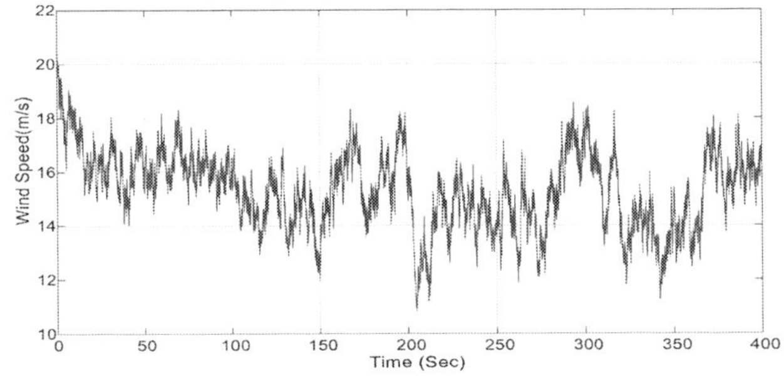

Figure 2: Wind speed with 15 m/s mean value

As mentioned in the Introduction main control goal in power regulation region is to keep generator power around its rated value (see Fig. 3.), for this this reason the pitch angle needs to activated as a control input (see Fig. 4.).

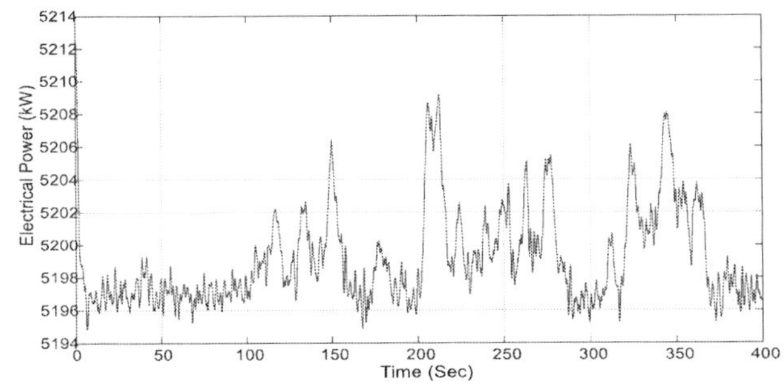

Figure 3: Generator power during PR region.

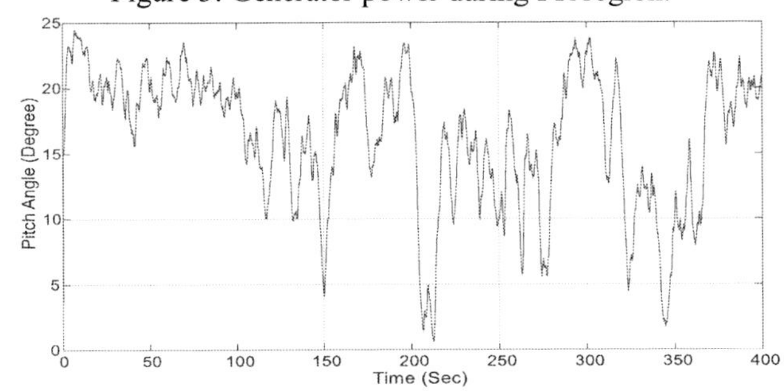

Figure 4: Pitch angle of blades during PR region.

V. CONCLUSION

This paper proposed a controller for variable speed variable pitch wind turbine. The suggested robust LPV controller guarantees the desired upper bounds of H_2 and H_∞ performances and locates the closed-loop poles of system in a desired circle. The control strategy applied on wind turbine in power regulation region. The simulation results show the high performance of closed-loop system during wide range of wind speed in power regulation region.

REFRENCES

1. Bianchi, F.D., Mantz, R.J. and De Battista, H., 2007. *The Wind and Wind Turbines,* Springer London.
2. Inthamoussou, F.A., Bianchi, F.D., De Battista, H. and Mantz, R.J., 2014. "LPV wind turbine control with anti-windup features covering the complete wind speed range". *IEEE Transactions on Energy Conversion,* 29(1), pp.259-266.
3. Johnson, K.E., Pao, L.Y., Balas, M.J. and Fingersh, L.J., 2006. "Control of variable-speed wind turbines: standard and adaptive techniques for maximizing energy capture". *IEEE Control Systems*, 26(3), pp.70-81.
4. Hassan, H.M., ElShafei, A.L., Farag, W.A. and Saad, M.S., 2012. "A robust LMI-based pitch controller for large wind turbines". *Renewable energy*, 44, pp.63-71.
5. Bakka, T. and Karimi, H.R., 2012, July. "Mixed H 2/H∞ control design for wind turbine system with pole placement constraints". In *Control Conference (CCC), 2012 31st Chinese* (pp. 4775-4780). IEEE.
6. Yao, X.J., Guo, C.C. and Li, Y., 2009, May. "LPV H-infinity controller design for variable-pitch variable-speed wind turbine". In *Power Electronics and Motion Control Conference, 2009. IPEMC'09. IEEE 6th International* (pp. 2222-2227). IEEE.
7. Ekelund, T., 1997. *Modeling and linear quadratic optimal control of wind turbines*. Chalmers University of Technology,.
8. Leithead, W.E., Leith, D.J., Hardan, F. and Markou, H., 1999. "Global gain-scheduling control for variable speed wind turbines". In *EWEC-CONFERENCE-* (pp. 853-856).
9. Bianchi, F.D., Mantz, R.J. and Christiansen, C.F., 2004. "Control of variable-speed wind turbines by LPV gain scheduling". *Wind Energy*, 7(1), pp.1-8.
10. Rotondo, D., Nejjari, F. and Puig, V., 2014. "Robust state-feedback control of uncertain LPV systems: An LMI-based approach". *Journal of the Franklin Institute*, 351(5), pp.2781-2803.
11. Jafarnejadsani, H., Pieper, J. and Ehlers, J., 2013. "Adaptive control of a variable-speed variable-pitch wind turbine using radial-basis function neural network". *IEEE Transactions on Control Systems Technology*, 21(6), pp.2264-2272.
12. Duan, G.R. and Yu, H.H., 2013. *LMIs in control systems: analysis, design and applications*. CRC press.
13. Ginter, V.J. and Pieper, J.K., 2011. "Robust gain scheduled control of a hydrokinetic turbine". *IEEE Transactions on Control Systems Technology*, 19(4), pp.805-817.

Proceedings of the 26th CANCAM
Victoria, British Columbia, Canada, May 29 – June 1, 2017

COMPARING PATHWAYS FOR DECARBONIZING A FOSSIL BASED POWER SYSTEMS THROUGH INTERTIE EXPANSIONS AND BIOENERGY

Victor Keller, Jeffrey English, Bryson Robertson, Peter Wild, Andrew Rowe.
Department of Mechanical Engineering
University of Victoria
Victoria, B.C., Canada
kellerv@uvic.ca

ABSTRACT

This study investigates intertie expansion and biomass as pathways for emission reductions in the power system of a fossil dominated jurisdiction. While previous research shows that both pathways may successfully lead to emission reductions, a cost comparison between the two alternatives has not yet been conducted. As a result, an assessment between cost comparison and potential emission reduction must be conducted to assess the potential of both pathways. In the present work, a long-term techno-economic model is used to evaluate the impacts of the two pathways on a fossil fuel dominated jurisdiction. While future research is needed, the results show that biomass may provide a lower cost option, although its impact is generally smaller than intertie expansion due to limitation on biomass resource availability.

KEYWORDS: Grid integration, Climate and energy policy, Techno-economic modelling, Coal retrofit, bioenergy

INTRODUCTION

To reduce carbon intensity and emissions in the power sector, the large scale deployment of variable renewable technologies such as wind and solar is frequently identified as a solution [1] [2]. However, these technologies face the limiting factor of natural intermittency leading to not entirely predictable outputs [3]. As a result, integration of these technologies may be technically limited or lead to system balancing issues [3] [4]. Interregional transmission or use of alternative renewable energy technologies such as bioenergy energy has been proposed as alternatives to either enable or substitute the use of wind and solar power. [5] [6]. In this study, the impact of two different approaches to reduce carbon intensity of a fossil fuel jurisdiction are analyzed. The first approach is by integration of the fossil based jurisdiction to a hydroelectricity dominated jurisdiction by transmission lines, while the second approach utilizes forest biomass in stranded coal generating units.

BACKGROUND: HYDROELECTRIC RESERVOIR

Hydroelectric reservoirs can be used to store excess renewable energy, increasing penetration of variable renewable power [2]. Reservoirs can provide flexibility to wind and solar power due to their fast ramping capabilities. At times when wind or solar output is high, hydro power may be reduced to balance supply and demand, effectively converting the reservoir into a battery. Moreover, when output from variable generators is decreased, hydro power may be ramped up to balance the system.

A number of studies have investigated the use of hydroelectric power with storage to enable higher penetration of variable renewable power [7] [8]. These studies find that by integrating storage hydro power with wind generating units reduces both system wide costs and emissions. The authors further conclude that this coupling of hydro storage with wind power also leads to fewer wind curtailment events leading to higher penetration of wind power [7] [8].

BACKGROUND: BIOMASS

A different approach to increase renewable penetration is to use dispatchable renewable power, such as bioenergy. A cost effective method for utilizing biomass is by re-purposing stranded or soon-to-be decommissioned coal units [9] [10]. As the capital investment cost has already been sunk, this technology offers a low cost option.

Although biomass may provide dispatchable renewable energy, not all sources of biomass are necessarily renewable and assumptions on carbon neutrality must be carefully considered [11] [12] [13]. It has been shown that dedicated harvest of forest wood for bioenergy leads to long time delays in carbon emission benefits [14] [15] [12]. The use of forest residue, however, has been shown to be an effective way of curbing emissions [12] [15] due to its shorter time delay in carbon sequestration parity [14].

CASE STUDY: BRITISH COLUMBIA - ALBERTA

Power generation in the province of Alberta is dominated by fossil generation. As of 2015, coal generation represented 51% of the energy generation in the province [16]. Natural gas amounted for 39 % of the generation while renewables only accounted for the remaining 10%. The province has recently set goals for phasing out all its coal generation while increasing its renewable energy share to 30% by 2030 [17]. This measure, while likely to greatly reduce emissions in the province, will lead to a number of coal units becoming stranded by 2030 [17].

In contrast, the province of British Columbia located to the west of Alberta, is dominated by hydroelectric generation. As of 2012, over 80% of the electricity generation in the province was provided by hydro power, while renewables accounted for 94% of the electricity generation [2] [18].

The current study examines two different pathways for Alberta to increase its renewable share. The first method to be analyzed is the potential gained benefit of increasing intertie capacity between Alberta and British Columbia, with a current capacity of 1200 MW. The intertie may provide the flexibility necessary to back up increasing wind generation in AB. The second method to be assessed is the conversion of stranded coal units to fire biomass. Biomass units are not constrained by variability, providing firm capacity and potentially reducing the need for backup capacity.

METHODS

The analysis is conducted with the Open Source Energy Modelling System (OSeMOSYS) [19]. OSeMOSYS is a long-term techno-economic optimization tool that minimizes net present cost of the system while meeting an exogenous demand [19]. More details on the formulation, inputs and outputs of OSeMOSYS can be found on [19] [20].

In the current work, two exogenous demands and resource options are defined for the two provinces, as seen in Fig. 1. Demand for AB is based on the Alberta Electric System Operator, while B.C.'s is based on B.C. hydro [21] [22]. Different technologies are available in the two provinces, due to resource availability. For example, the Alberta power system currently has 3.6 GW of cogeneration (COGEN) used in the oil sands for generation of heat, generating electricity as a by-product. Alternatively, British Columbia is home to large hydro resources, which are no present in Alberta on the same scale.

On top of meeting an energy demand, both systems are also subject to meeting a reserve margin constraint to ensure system reliability. Wind, solar and

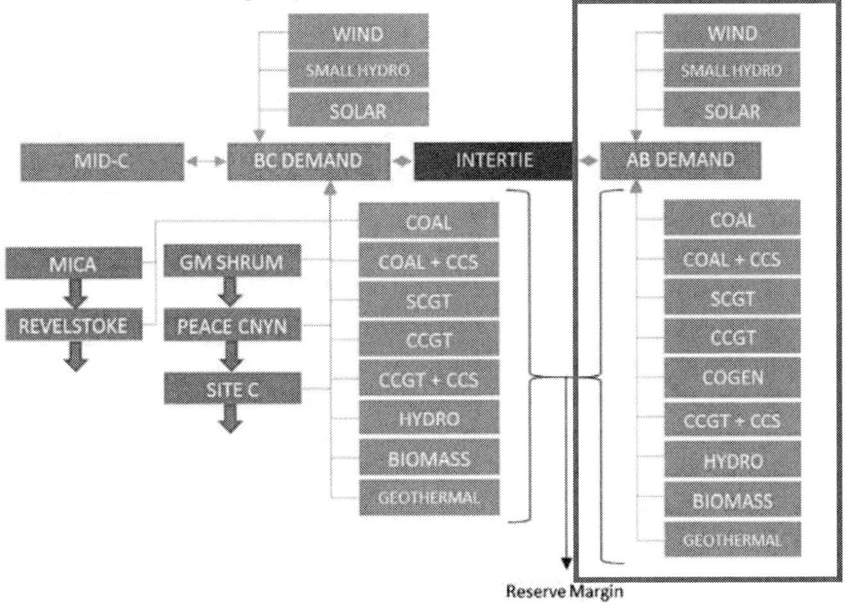

Figure 1. Diagram for modelled system. Green boxes represent demands, while grey boxes represent generating technologies and blue represent hydro reservoirs. British Columbia is allowed to purchase power from the United States (MID-C) with a maximum intertie capacity of 3.5 GW. The second pathway involving conversion of coal units to biomass analyzes the AB system in isolation, marked by a red square.

small hydro power may provide energy to meet the energy demand, but cannot provide capacity to meet the reserve margin constraint. As a result, extra capacity may have to be built to provide backup for intermittent renewables.

Modelled parameters are similar to those found in the work by Lyseng et al [20]. The electrical demand is divided into 36 annual time slices, representing peak, mid-peak and off-peak demand for each month. The model runs starting in the year 2010 until 2060. Assumptions on capital and operation and maintenance costs for generators are obtained from the U.S. Energy Information Administration [23]. Residual capacities along with expected decommissioning dates for existing generating portfolio in both provinces is also added to the model. All technologies are subject to either a maximum annual buildout and/or a maximum installed capacity cap to represent local resource limits and realistic capacity expansion. More details can be found at Lyseng et al. [20].

Biomass feedstock for coal retrofitted units in Alberta is modelled making use of forest residue available in both provinces. Data from NRCan was provided for estimation of resource availability and location [14] [6].

SCENARIOS

Six scenarios are modelled in the current work. Three for the connected British Columbia – Alberta system, and for the isolated AB system with the ability to convert coal units to biomass. The scenarios and its assumptions are listed in Table 1.

Table 1. Summary of modelled scenarios. Carbon caps in AB-B.C. intertie are modelled as a linear decrease from 2010 to the target year.

AB – B.C. Intertie		
Scenario	Intertie Cap [MW]	Imposed emission reduction
Reference H	1200	-
80 % by 2050	1200	80% below 2007 values by 2050
80 % by 2050 – intertie expansion	Endogenously modelled	80% below 2007 values by 2050
Coal to biomass		
Scenario	Renewable target by 2030	Biomass retrofit option
Reference B	-	No
Stranded	30	No
Convert	30	Yes

RESULTS

ALBERTA – B.C. INTERTIE

Shown in Figure 2 are the modelled results for the Alberta – British Columbia system. For brevity, only results for Alberta are shown. It is seen that when no standard on emission reductions or carbon taxes is imposed in the system, as in the *Reference H scenario (a),* the system continues its reliance on fossil based power. Federal regulations in Canada require improvement on intensity standards to less than 420 t CO_2/ GWh [24], new coal generation without carbon capture and sequestration is effectively prohibited past 2015. As a result, coal generation slowly transitions to an increased share of CCGT and Cogen as aging coal units retire. By the 2050's however, due to the high prices of natural gas and limits on cogeneration, hydro and wind expansion having already been reached, the system starts installing new coal-ccs capacity to meet the growing demand.

Seen in figure 2 (b) and (c) are the scenarios for an imposed emission reduction of 80% from 2007 values without and with the option for expanding intertie, respectively. In the third scenario Figure 2 (c), imported power from British Columbia displaces mostly coal – CCS generation. Imported power increases from 515 TWh (b) to 706 TWh (c) over the modelled period, to help meet the growing demand in the province. As a result, the intertie capacity is expanded from the current 1200 MW to 2500 MW, leading to an overall costs savings of 1%. As seen in Figure 2 (c), the intertie is utilized almost at maximum capacity for most of the modelling period, until the mid-2050s, at which point solar power starts being used in the province.

One similarity between scenarios (b) and (c) is the high utilization of the intertie. In both scenarios, utilization rates were above 80% for most of the modelling period. The results suggest that while intertie expansion may allow for further emission reduction, it is used mostly to import power and displace local generation, rather than being used for system balancing [2].

BIOMASS

Differently to the results in the Alberta-British Columbia model, all coal generation is phased out by 2030, following regulations established by the Alberta Climate Leadership Plan [1]. As a result, a total of 1.3 GW of coal capacity is available for retrofit to biomass by 2030. All three modelled scenarios, *reference B,*

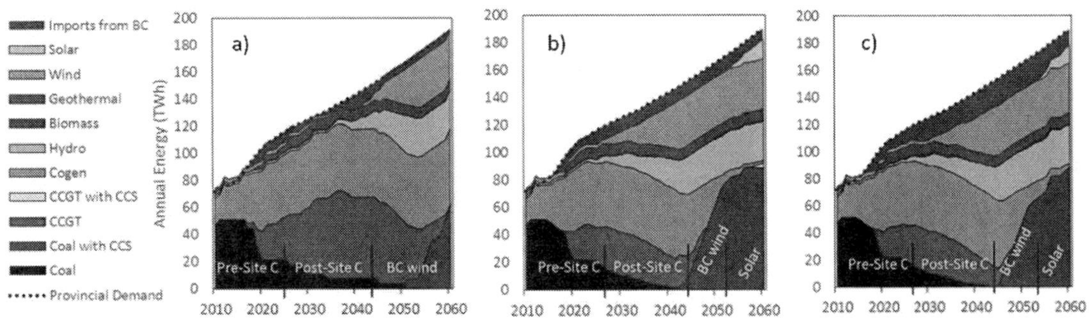

Figure 1. Modelled results for the Alberta – British Columbia intertie model. Model results only show modelled results for Alberta for the scenarios: a) reference scenario with intertie capacity of 1200 MW, b) 80% reduction of emissions from 2007 levels by 2050 with an intertie capacity limited to the current 1200 MW and c) 80% reduction by 2050 with expandable intertie capacity. Intertie capacity for c) is optimally defined by the model

retrofit and stranded, were modelled with a carbon tax of 6 $/tCO_2$, similar to current levels in the province [20]. Renewable energy credits (RECs) are offered to the system to meet reach a generation share of 30% from renewables by 2030.

As seen in Figure 3, in the reference scenario with no RECs, the system only reaches a renewable penetration of 10%, similar to levels at the start of the modelling period in 2010.

In the retrofit scenario, however, where the conversion to biomass is available, roughly 0.9 GW of coal capacity is converted to biomass, as seen in Figure 3 (b). This 900 MW of capacity accounts for roughly 7% of the electricity generation in the province by 2030, almost 25% of the mandated renewable share by the same date Figure 3 (a).

Also seen in Figure 3 (b), is the decrease in wind dependence with the adoption of biomass. The share of wind power drops from 24% in the *stranded* scenario to only 18% in the *retrofit* scenario. Due to the intermittency of wind, significantly more wind capacity is necessary to produce the 7% of the energy compared to wind. Furthermore, wind capacity required backup to meet reserve margin requirements, leading to a large buildout of OCGT. As a result, the overall capacity of the system is considerably smaller in the retrofit scenario than in the stranded scenario. This reduction in capacity investment would lead to a reduction in investment in new capacity from 13.4 to 10.8 $B from 2010 to 2030 or a 20% decrease in capital expenditure.

It was also found from the modelling results that the RECs necessary to reach the 30% renewable energy share by 2030 also dropped with the introduction of biomass. The REC dropped from 18.7 $/MWh in the *stranded* scenario to 16.5 $/MWh in the

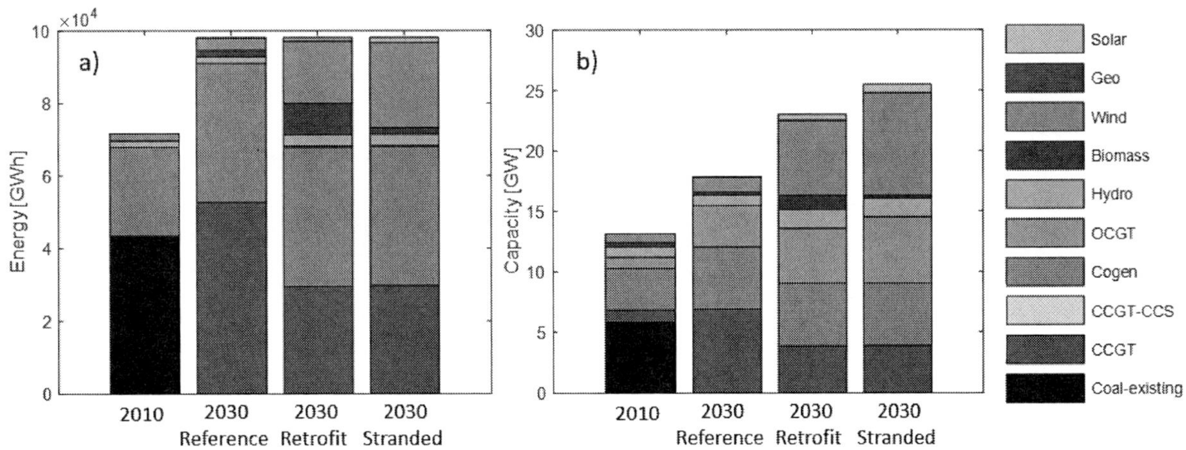

Figure 3. Modelled results for Alberta system for coal to biomass scenarios. a) stacked capacity bar plots for all three modelled scenarios at the year 2030 and starting levels at 2010 and b) stacked energy bar plots for three modelled scenarios at the year 2030 and starting levels at 2010. All coal generation is phased out by 2030, representing regulations of the Alberta Climate Leadership Plan [1].

convert scenario, leading to a cost savings of 0.6 $B in REC payments from 1020 to 2030 [6].

CONCLUSIONS

This study investigates the potential benefits in emissions and costs for the Alberta power system (fossil dominated system) to lower its carbon intensity by either connecting it to the British Columbia system (hydro dominated system) or by converting stranded coal capacity to biomass.

The results show that when considering linkage between the two systems, an intertie expansion was necessary to achieve the minimum cost. The intertie was expanded from 1.2 to 2.5 GW, or about 25% of installed capacity by 2030. However, it was found that with the intertie installed, utilization rates were always above 80%, showing that it may not be economical to only balance short term power imbalances and significant amount of energy trade is necessary to make it cost effective [2].

Use of forest biomass however, was found to be a cost effective option for displacing wind generation and reducing requirements for RECs and capacity investment. It was found that biomass could not only avoid a 2.6 $B in capacity investment, but would also reduce REC payments by a total of 0.6 $B from 2010 to 2030.

While the modelled results may suggest that adoption of biomass may be a lower cost option than intertie expansion, the results must be evaluated with caution. There is a limited amount of biomass available for electricity generation, and it is unlikely that this resource could provide over 10% of the energy needs in the province. However, for a small and relatively fast increment in renewable share in the province, biomass may offer a low cost option.

REFERENCES

References

[1] Government of Alberta, "Climate Leadership-report to the minister," 2015. [Online]. Available: http://www.alberta.ca/documents/climate-leadership-report-to-minister.pdf.

[2] J. English, T. Niet, T. Lyseng, K. Palmer-Wilson, V. Keller, I. Moazzen, L. Pitt, P. Wild and A. Rowe, "Impact of electrical intertie capacity on carbon policy effectiveness," *Energy Policy,* vol. 101, pp. 571-581, 2017.

[3] E. Varone and M. Ferrari, "Power to liquid and power to gas: An option for the German Energiewende," *Renewable and Sustainable Energy Reviews,* vol. 45, pp. 207-218, 2015.

[4] M. De Saint Jean, P. Baurens and C. Bouallou, "Economic assessment of a power-to-substitue-natural-gas process including high-temperature steam electrolysis," *International Journal of Hydrogen* , vol. 40, pp. 6487-6500, 2015.

[5] H. H., "Impact of hourly wind power variations on the system operation in the nordic countries.," *Wind energy,* pp. 197-218, 2005.

[6] V. Keller, B. Robertson, P. Wild and A. Rowe, "Using biomass at shuttered coal-fired plants could help provinces meet their climate targets for a lower cost," Policy Options, Montreal, 2017.

[7] M. K. Chang, J. Eichman, F. Mueller and S. Samuelsen, "Buffering intermittent renewable power with hydroelectric generation: a case study in California," *Applied Energy,* vol. 112, pp. 1-11, 2013.

[8] E. Ibanez, T. Magee, M. Clement, G. Birkman, M. Milligan and E. Zagona, "Enhancing hydropower modelling in variable generation integration studies," *Energy,* vol. 74, pp. 518-528, 2014.

[9] J. Cleary, D. Wolf and J. Caspersen, "Comparing the life cycle costs of using harvest residue as feedstock for small- and large-scale bioenergy systems (part II)," *Energy,* vol. 86, pp. 539-547, 2015.

[10] A. Cuellar and H. Herzog, "A path forward for low carbon power from biomass," *Energies,* vol. 8, pp. 1701-1715, 2015.

[11] M. Ter-Mikaelian, S. Colombo, D. Lovein, J. McKechnie, R. Reynolds, B. Titus, E. Larin, A.-M. Chapman, J. Chen and H. MacLean, "Carbon debt repayment of carbon sequestration parity? Lessons from a forest bioenergy case study in Ontario, Canada," *GCB Bioenergy,* vol. 7, pp. 704 - 716, 2015.

[12] J. McKechnie, S. Colombo, J. Chen, W. Mabee and H. MacLean, "Forest bioenergy or forest carbon Assessing trade-offs in greenhouse gas mitigation with wood-based fuels," *Environmental Science & Technology,* vol. 45, no. 2, pp. 789-795, 2011.

[13] CBC news, "Money doesn't grow on trees, but bioenergy might," 29 May 2009. [Online]. Available: http://www.cbc.ca/news/technology/money-

doesn-t-grow-on-trees-but-bioenergy-might-1.789017. [Accessed 23 02 2017].

[14] C. Smyth, G. Stinson, E. Neilson, T. Lemprieere, M. Hafer, G. Rampley and W. Kurz, "Quantifying the biophysical climate change mitigation potential of Canada's forest sector," *Biogeosciences,* vol. 11, pp. 3515-3529, 2014.

[15] M. Ter-Mikaelian, J. McKechnie, S. Colombo, J. Chen and H. MacLean, "The carbon neutrality assumption for forest bioenergy: A case study for northwestern Ontario," *The Forestry Chronicle ,* vol. 87, no. 5, pp. 644-652, 2011.

[16] Alberta Energy, "Electricity Statistics," 2017. [Online]. Available: http://www.energy.alberta.ca/electricity/682.asp . [Accessed 22 02 2017].

[17] Government of Alberta, "Pollution fomr coal-fired electricity generation will be phased out by 2030 under the Climate Leadership Plan," 2016. [Online]. Available: https://www.alberta.ca/climate-coal-electricity.aspx. [Accessed 16 02 2017].

[18] B. C. M. o. E. a. mines, "Electricity Supply and Generation," [Online]. Available: http://www.empr.gov.bc.ca/EPD/Electricity/supply/Pages/default.aspx.

[19] M. Howells, H. Rogner, N. Strachan, C. Heaps, H. Hunington, S. Kypreos, A. Hughes, S. Silveira, J. DeCarolis, M. Bazillian and A. Roehrl, "OSeMOSYS: The open source energy modelling system an introduction to its ethos, structure and development," *Energy Policy,* vol. 39, pp. 5850-5870, 2011.

[20] B. Lyseng, A. Rowe, P. Wild, J. English, T. Niet and L. Pitt, "Decarbonizing the Alberta power system with carbon pricing," *Energy strategy reviews,* vol. 10, pp. 40-52, 2016.

[21] AESO, "AESO 2016 Long-Term Outlook Update," [Online]. Available: https://www.aeso.ca/grid/forecasting/.

[22] B.C. Hydro, "B.C. Hydro Electric Load Forecast," 2013.

[23] EIA, "Updated Capital Cost Estimates for Utility Scale Electricity Generating Plants," http://www.eia.gov/forecasts/capitalcost/pdf/updated_capcost.pdf, Washington, DC, 2015.

[24] "Canadian Environmental Protection Act," 2015. [Online]. Available: http://laws-lois.justice.gc.ca/eng/regulations/SOR-2012-167/index.html.

MODELING OF DUAL STAGE PRESSURE RETARDED OSMOSIS FOR SCALED UP OPERATION

Roghayeh Soltani
Department of Mechanical Engineering
University of Victoria
Victoria, BC, Canada
ssoltani@uvic.ca

Henning Struchtrup
Department of Mechanical Engineering
University of Victoria
Victoria, BC, Canada
struchtr@uvic.ca

ABSTRACT

In order to decrease greenhouse gas emissions, the global efforts to utilize renewable sources of energy have been increased. Pressure retarded Osmosis (PRO) can be considered as a renewable source of energy released from the mixing of freshwater with saltwater using a semipermeable membrane. The estimated power density of PRO membrane to be economically feasible is said to be about 5 W/m^2. The current technology does not meet this criterion; hence the question is whether multi-staging might improve performance. A model for a scaled up dual stage PRO power plant is presented, where, in contrast with laboratory-scale dual stage PRO power plants, the introduced model considers the pressure and flow drop along the membrane as well as the change in salinity. A single stage system is compared for power output with dual stage one in the case of the same membrane area. A wide range of hydraulic pressures is applied in the draw side as well as draw and feed water flow rate are investigated. The results indicate that overall performance of the system could be improved in the case of dual stage PRO. It should be noted that, in order to improve the fidelity of such system in operation, the effects of reverse salt flux and concentration polarization were considered.

KEYWORDS: pressure retarded osmosis, dual stage system, scaled up, modeling, process characteristics.

Failure Mechanics and Design

FUNDAMENTAL SOLUTIONS AND ANALYSIS OF RECTANGULAR INTERFACE CRACK IN A 3D MAGNETOELECTROTHERMOELASTIC BI-MATERIAL

Huayang Dang
School of Mechanics and Engineering Science
Zhengzhou University
Zhengzhou, Henan, China
huayang@ualberta.ca

Minghao Zhao
School of Mechanics and Engineering Science
Zhengzhou University
Zhengzhou, Henan, China
memhzhao@zzu.edu.cn

Cuiying Fan
School of Mechanical Engineering
Zhengzhou University
Zhengzhou, Henan, China
fancy@zzu.edu.cn

Zengtao Chen
Department of Mechanical Engineering
University of Alberta
Edmonton, Alberta, Canada
zengtao.chen@ualberta.ca

ABSTRACT

This paper derives the fundamental solutions for uniformly distributed extended displacement discontinuities (EDDs) over a rectangular element in the isotropic plane of a 3D magnetoelectrothermoelastic bi-material, and investigates a rectangular interface crack subjected to remote thermal loadings. To eliminate the oscillatory singularity, the Delta function is approximated by the Gaussian distribution function. The extended stress intensity factors without oscillatory singularity, and energy release rate are all obtained in terms of the EDDs. The EDD boundary element method is proposed for the numerical simulation.

KEYWORDS: Magnetoelectrothermoelastic bi-material, Rectangular interface crack, Fundamental solutions, Extended displacement discontinuity, Boundary element method

INTRODUCTION

Due to their exceptional features such as flat frequency response and the ability to transform mechanical, electric, magnetic, and thermal energy from one form to another, magnetoelectothermoelastic materials have been widely used in smart structures. In engineering application, the interface cracks often exist in the material. Therefore, study on the fracture behavior of interface crack problems in magnetoelectrothermoelastic media is of great importance and is receiving increasing attention [1,2].

The displacement discontinuity method (DDM) was firstly proposed by Crouch [3] to study 2D crack problems in elastic media. Afterwards, this method was extended to interface crack problems [1,2,4]. Recently, Zhao et al. gave the fundamental solutions for unit-point extended displacement discontinuities (EDDs) and adopted the approximation of the Delta function with Gaussian distribution function to remove the oscillatory singularity [4,5]. The singularities of near-crack border fields and the extended stress intensity factors without oscillatory singularity were derived in terms of the EDDs, and the energy release rate can also be expressed by the intensity factors. In this paper, by integrating the fundamental solutions for unit point EDDs in Zhao et al. [1], we obtain the fundamental solutions for uniformly distributed EDDs over a constant rectangular element in isotropic plane of a 3D magnetoelectrothermoelastic bi-material and investigate a rectangular crack under combined loads.

FUNDAMENTAL SOLUTIONS OF UNIFORM EDDS OVER A RECTANGULAR ELEMENT

Consider a rectangular interface element of length $2a$ and width $2b$ in the oxy plane centered at the origin o, as shown in Fig. 1, where the rectangular side paralleling to the axis of the coordinate system, and the poling direction of the material is parallel to the z-axis.

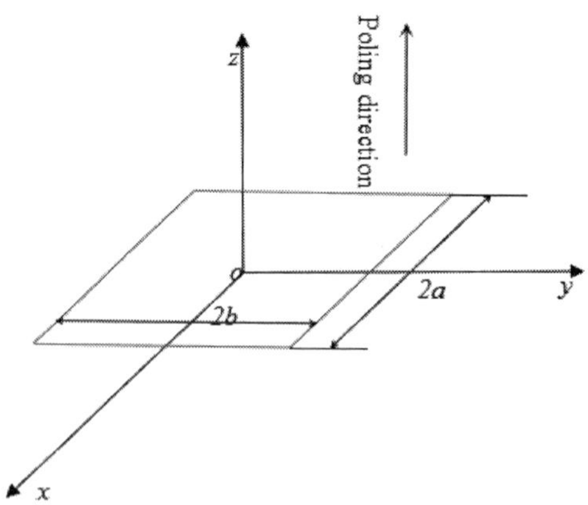

Figure 1: A rectangular element centered at the coordinate system.

Over the rectangular element, uniform EDDs are applied. Based on the definition of the Heaviside function, i.e.,

$$H(x)=\begin{cases}1, & x\geq 0,\\ 0, & x<0,\end{cases} \tag{1}$$

the EDDs within the constant rectangular element can be mathematically expressed as

$$\begin{aligned}
&\|u(x,y)\|=\|u^e\|S_\Omega,\\
&\|v(x,y)\|=\|v^e\|S_\Omega,\\
&\|w(x,y)\|=\|w^e\|S_\Omega,\\
&\|\varphi(x,y)\|=\|\varphi^e\|S_\Omega,\\
&\|\psi(x,y)\|=\|\psi^e\|S_\Omega,\\
&\|\theta(x,y)\|=\|\theta^e\|S_\Omega,
\end{aligned} \tag{2}$$

where

$$S_\Omega=[H(x+a)-H(x-a)][H(y+b)-H(y-b)], \tag{3}$$

As the Delta function is the derivative of Heaviside step function, the corresponding partial derivatives of S_Ω can be expressed as

$$\frac{\partial S_\Omega}{\partial x}=[\delta(x+a)-\delta(x-a)][H(y+b)-H(y-b)], \tag{4a}$$

$$\frac{\partial S_\Omega}{\partial y}=[H(x+a)-H(x-a)][\delta(y+b)-\delta(y-b)]. \tag{4b}$$

The fundamental solutions for the rectangular element can be derived by integrating the fundamental solutions of unit point EDDs in Zhao et al. [1]

$$\begin{aligned}\sigma_{zx}(x,y,0)=&G_{11}\|u^e\|+G_{12}\|v^e\|+G_{13}\|\theta^e\|\\&+G_{14}\|w^e\|+G_{15}\|\varphi^e\|+G_{16}\|\psi^e\|,\end{aligned} \tag{5a}$$

$$\begin{aligned}\sigma_{yz}(x,y,0)=&G_{12}\|u^e\|+G_{22}\|v^e\|+G_{23}\|\theta^e\|\\&+G_{24}\|w^e\|+G_{25}\|\varphi^e\|+G_{26}\|\psi^e\|,\end{aligned} \tag{5b}$$

$$\begin{aligned}\sigma_{z}(x,y,0)=&G_{31}\|u^e\|+G_{32}\|v^e\|+G_{33}\|w^e\|\\&+G_{34}\|\varphi^e\|+G_{35}\|\psi^e\|,\end{aligned} \tag{5c}$$

$$\begin{aligned}D_{z}(x,y,0)=&G_{41}\|u^e\|+G_{42}\|v^e\|+G_{43}\|w^e\|\\&+G_{44}\|\varphi^e\|+G_{45}\|\psi^e\|,\end{aligned} \tag{5d}$$

$$\begin{aligned}B_{z}(x,y,0)=&G_{51}\|u^e\|+G_{52}\|v^e\|+G_{53}\|w^e\|\\&+G_{54}\|\varphi^e\|+G_{55}\|\psi^e\|,\end{aligned} \tag{5e}$$

$$h_z(x,y,0)=G_6\|\theta^e\|, \tag{5f}$$

where the coefficients are

$$\begin{aligned}
&G_{11}=(L_{11}-L_{12})\int_{S_\Omega}\frac{3\cos^2\phi}{r^3}dS-(L_{11}-2L_{12})\int_{S_\Omega}\frac{1}{r^3}dS,\\
&G_{12}=(L_{11}-L_{12})\int_{S_\Omega}\frac{3\sin\phi\cos\phi}{r^3}dS,\\
&G_{13}=-L_{13}\int_{S_\Omega}\frac{\cos\phi}{r^2}dS,\quad G_{16}=-2\pi L_{16}\frac{\partial S_\Omega}{\partial x},\\
&G_{14}=-2\pi L_{14}\frac{\partial S_\Omega}{\partial x},\quad G_{15}=-2\pi L_{15}\frac{\partial S_\Omega}{\partial x},\\
&G_{22}=(L_{11}-L_{12})\int_{S_\Omega}\frac{3\sin^2\phi}{r^3}dS-(L_{11}-2L_{12})\int_{S_\Omega}\frac{1}{r^3}dS,\\
&G_{23}=-L_{13}\int_{S_\Omega}\frac{\sin\phi}{r^2}dS,\quad G_{24}=-2\pi L_{14}\frac{\partial S_\Omega}{\partial y},\\
&G_{25}=-2\pi L_{15}\frac{\partial S_\Omega}{\partial y},\quad G_{26}=-2\pi L_{16}\frac{\partial S_\Omega}{\partial y},\\
&G_{31}=-2\pi L_{1}\frac{\partial S_\Omega}{\partial x},\quad G_{32}=-2\pi L_{1}\frac{\partial S_\Omega}{\partial y},\\
&G_{33}=-L_{31}\int_{S_\Omega}\frac{1}{r^3}dS,\quad G_{34}=-L_{32}\int_{S_\Omega}\frac{1}{r^3}dS,\\
&G_{35}=-L_{33}\int_{S_\Omega}\frac{1}{r^3}dS,\quad G_{41}=-2\pi L_{2}\frac{\partial S_\Omega}{\partial x},\\
&G_{42}=-2\pi L_{2}\frac{\partial S_\Omega}{\partial y},\quad G_{43}=-L_{41}\int_{S_\Omega}\frac{1}{r^3}dS,\\
&G_{44}=-L_{42}\int_{S_\Omega}\frac{1}{r^3}dS,\quad G_{45}=-L_{43}\int_{S_\Omega}\frac{1}{r^3}dS,\\
&G_{51}=-2\pi L_{3}\frac{\partial S_\Omega}{\partial x},\quad G_{52}=-2\pi L_{3}\frac{\partial S_\Omega}{\partial y},\\
&G_{53}=-L_{51}\int_{S_\Omega}\frac{1}{r^3}dS,\quad G_{54}=-L_{52}\int_{S_\Omega}\frac{1}{r^3}dS,\\
&G_{55}=-L_{53}\int_{S_\Omega}\frac{1}{r^3}dS,\quad G_{6}=-L_{6}\int_{S_\Omega}\frac{1}{r^3}dS,
\end{aligned} \tag{6}$$

wherein the coefficients L_{ij} can be found in Zhao et al. [1], while the integral terms can be calculated according to Zhao et al. [4]. For brevity, the detailed derivation is omitted. As pointed out in the previous research[4,5], it is the Delta function that causes the solutions oscillatory. To remove the oscillatory singularity, the Gaussian distribution functions is adopted to approximate the Delta function[2,4,5]

$$\delta(x)=\frac{1}{\sqrt{2\pi}\kappa}\exp\left[-\frac{x^2}{2\kappa^2}\right], \tag{7}$$

where κ is the Gaussian parameter, and its selection is related to the geometry of the crack. According to the discussion about the parameter [4,5], a relative value of $\kappa / r_a = 0.0075$ is suitable with r_a referring to the length dimension of the crack.

As the Heaviside step function is the integral of the Delta function, the corresponding expression for the Heaviside step function is also determined

$$H(x) = \frac{1}{2}\left(1 + \mathrm{Erf}\left[\frac{x}{\sqrt{2\kappa}}\right]\right), \tag{8}$$

where Erf is the Error function. After singularity analysis of the extended stress fields near the crack front and the remove of the oscillatory singularity, the extended stress intensity factors can be obtained in terms of the EDDs with the classical singularity

$$K_{\mathrm{I}} = -\sqrt{2\pi}\pi \lim_{\rho\to 0}\left(L_{31}\|w\| + L_{32}\|\varphi\| + L_{33}\|\psi\|\right)/\sqrt{\rho}, \tag{9a}$$

$$K_{\mathrm{II}} = \sqrt{2\pi}\pi L_{11} \lim_{\rho\to 0}\|u\|/\sqrt{\rho}, \tag{9b}$$

$$K_{\mathrm{III}} = -\sqrt{2\pi}\pi L_{12} \lim_{\rho\to 0}\|v\|/\sqrt{\rho}, \tag{9c}$$

$$K_D = -\sqrt{2\pi}\pi \lim_{\rho\to 0}\left(L_{41}\|w\| + L_{42}\|\varphi\| + L_{43}\|\psi\|\right)/\sqrt{\rho}, \tag{9d}$$

$$K_B = -\sqrt{2\pi}\pi \lim_{\rho\to 0}\left(L_{51}\|w\| + L_{52}\|\varphi\| + L_{53}\|\psi\|\right)/\sqrt{\rho}, \tag{9e}$$

$$K_h = -\sqrt{2\pi}\pi L_6 \lim_{\rho\to 0}\|\theta\|/\sqrt{\rho}. \tag{9f}$$

The singularity of the stresses at the four corners of a rectangular crack is quite complicated, but its order of singularity is below 1/2 according to Zhao et al. [4]. As an important fracture parameter, the energy release rate is also obtained in terms of the extended stress intensity factors according Griffith's virtual work theory

$$G = A_1 K_{\mathrm{I}}^2 + A_2 K_D^2 + A_3 K_B^2 + A_4 K_I K_D + A_5 K_I K_B + A_6 K_D K_B + \frac{K_{\mathrm{II}}^2}{8\pi L_{11}} - \frac{K_{\mathrm{III}}^2}{8\pi L_{12}}, \tag{10}$$

where the involved material-related coefficients are

$$A_1 = \left(L_{42}L_{53} - L_{43}L_{52}\right)/L_{MDB}, \tag{11a}$$

$$A_2 = \left(L_{31}L_{53} - L_{33}L_{51}\right)/L_{MDB}, \tag{11b}$$

$$A_3 = \left(L_{31}L_{42} - L_{32}L_{41}\right)/L_{MDB}, \tag{11c}$$

$$A_4 = \left(L_{33}L_{52} - L_{32}L_{53} + L_{43}L_{51} - L_{41}L_{53}\right)/L_{MDB}, \tag{11d}$$

$$A_5 = \left(L_{32}L_{43} - L_{33}L_{42} + L_{41}L_{52} - L_{42}L_{51}\right)/L_{MDB}, \tag{11e}$$

$$A_6 = \left(L_{33}L_{41} - L_{31}L_{43} + L_{32}L_{51} - L_{31}L_{52}\right)/L_{MDB}, \tag{11f}$$

where

$$L_{MDB} = 8\pi\left[L_{31}\left(L_{43}L_{52} - L_{42}L_{53}\right) + L_{32}\left(L_{41}L_{53} - L_{43}L_{51}\right) + L_{33}\left(L_{42}L_{51} - L_{41}L_{52}\right)\right], \tag{12}$$

EDD BOUNDARY ELEMENT METHOD

Consider a rectangular interface crack S is divided into N rectangular elements of the same size, and the EDDs are uniformly distributed on each element. Using the fundamental solutions in Eq. (5), one can obtain the following algebraic equations

$$\begin{aligned}
&\sum_{e=1}^{N}\left(G_{11}\|u^e\| + G_{12}\|v^e\| + G_{13}\|\theta^e\| + G_{14}\|w^e\| + G_{15}\|\varphi^e\| + G_{16}\|\psi^e\|\right) = -p_x(x_i, y_i),\\
&\sum_{e=1}^{N}\left(G_{12}\|u^e\| + G_{22}\|v^e\| + G_{23}\|\theta^e\| + G_{24}\|w^e\| + G_{25}\|\varphi^e\| + G_{26}\|\psi^e\|\right) = -p_y(x_i, y_i),\\
&\sum_{e=1}^{N}\left(G_{31}\|u^e\| + G_{32}\|v^e\| + G_{33}\|w^e\| + G_{34}\|\varphi^e\| + G_{35}\|\psi^e\|\right) = -p_z(x_i, y_i),\\
&\sum_{e=1}^{N}\left(G_{41}\|u^e\| + G_{42}\|v^e\| + G_{43}\|w^e\| + G_{44}\|\varphi^e\| + G_{45}\|\psi^e\|\right) = -\omega(x_i, y_i),\\
&\sum_{e=1}^{N}\left(G_{51}\|u^e\| + G_{52}\|v^e\| + G_{53}\|w^e\| + G_{54}\|\varphi^e\| + G_{55}\|\psi^e\|\right) = -\gamma(x_i, y_i),\\
&\sum_{e=1}^{N}\left[G_6\|\theta^e\|\right] = -h_z(x_i, y_i),
\end{aligned} \tag{13}$$

It is observed from Eq. (13) that there are totally 6N equations, and corresponding 6N unknown discontinuities for the whole interface crack. When applied with a certain loads, one can obtain the EDDs for each rectangular element by solving these linear algebraic equations, then the induced extended stresses of any filed point in the bi-material space can be determined.

NUMERICAL SIMULATION OF AN RECTANGULAR INTERFACE CRACK

In the numerical simulation, the bi-materials made of different piezoelectric materials including BaTiO3, PZT-5H, and PZT-6B are chosen as the inclusion and piezomagnetic material CoFe2O4 as the matrix. The volume fraction of the inclusions is set as a fixed value of 0.5 for both the upper and lower spaces. The material constants of the involved materials can be found in Dang et al. [2], the rectangular interface crack is $2A \times 2B$.

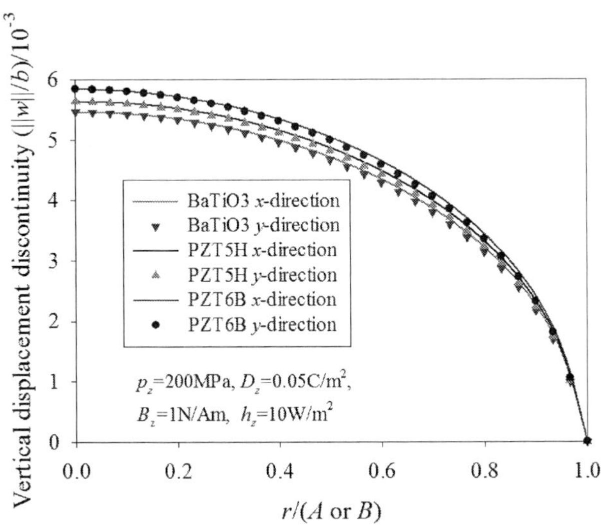

Figure 2. Distribution of the vertical displacement discontinuity across the crack faces along the x- and y-direction under fixed loadings for homogeneous and bi-materials.

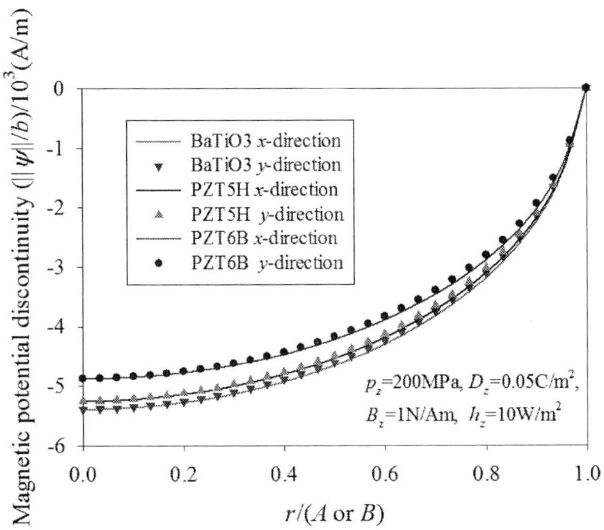

Figure 3. Distribution of the magnetic discontinuity across the crack faces along the x- and y- direction under fixed loadings for homogeneous and bi-materials.

Here two of the five EDDs across the rectangular crack faces along the the x- and y- direction respectively under fixed combined loads are picked as an example. Observing the Figures, one can conclude that the value of the bi-material is smaller than either of the homogeneous materials, and the value of the EDDs along the x- and y-direction is very close.

CONCLUSIONS

The closed-form fundamental solutions for the uniform EDDs over a rectangular element located in the isotropy plane of a magnetoelectrothermoelastic bi-material are derived. The oscillatory singularity of the stresses near the crack border is removed by approximating the Delta function with Gaussian distribution function. The extended SIFs and energy release rate are all obtained in terms of the EDDs. In the numerical simulation, an EDD boundary element method has been developed to study a rectangular interface crack in magnetoelectrothermoelastic bi-material.

ACKNOWLEDGMENTS

The work was supported by the National Natural Science Foundation of China (No. 11272290 and 11572289), the Specialized Research Fund for the Doctoral Program of Higher Education of China (No. 20124101110013), and the China Scholarship Council.

REFERENCES

1. Zhao, M.H., Dang, H.Y., Fan, C.Y., and Chen, Z.T. 2016, "Analysis of an interface crack of arbitrary shape in a three-dimensional transversely isotropic magnetoelectrothermoelastic bimaterial-part 1: Theoretical solution", *Journal of Thermal Stresses*, Accepted.
2. Dang, H.Y., Zhao, M.H., Fan, C.Y., and Chen, Z.T. 2016, "Analysis of an interface crack of arbitrary shape in a three-dimensional transversely isotropic magnetoelectrothermoelastic bimaterial-part 2: Numerical method", *Journal of Thermal Stresses*, Accepted.
3. Crouch, S.L. 1976, "Solution of plane elasticity problems by the displacement discontinuity method. I. Infinite body solution", *International Journal for Numerical methods in Engineering*, 10, 301-343.
4. Zhao, Y.F., Zhao, M.H., Pan, E., and Fan, C.Y. 2015. "Green's functions and extended displacement discontinuity method for interfacial cracks in three-dimensional transversely isotropic magneto-electro-elastic bi-materials", *International Journal of Solids and Structures*, 52, 56-71.
5. Zhang, A.B., and Wang, B.L. 2013. "An opportunistic analysis of interface crack based on the modified interface dislocation method", *International Journal of Solids and Structures*, 50, 15-20.

HIGH-SPEED INFRARED IMAGING FOR MATERIAL CHARACTERIZATION IN EXPERIMENTAL MECHANIC EXPERIMENTS

Marc-André Gagnon
Telops
Québec, Qc, Canada
marc-andre.gagnon@telops.com

Frédérick Marcotte
Telops
Québec, Qc, Canada
frederick.marcotte@telops.com

Marc-Antoine Langevin
Telops
Québec, Qc, Canada
marc-antoine.langevin@telops.com

ABSTRACT

Heat transfers are involved in many phenomena such as friction, tensile stress, shear stress and material rupture. Among the challenges encountered during the characterization of such thermal patterns is the need for both high spatial and temporal resolution. Infrared imaging provides information about surface temperature that can be attributed to the stress response of the material and breaking of chemical bounds. In order to illustrate this concept, high-speed infrared sequences were recorded during tensile and shear testing experiments carried out on steel and aluminum samples. Results from split-Hopkinson experiments carried out on a polymer material at high strain-rate are also presented. The results illustrate how high-speed infrared imaging in the midwave infrared (MWIR, 3 – 5 μm) spectral range can provide detailed information about the thermal properties of materials undergoing mechanical testing.

KEYWORDS: Infrared, Imaging, Hopkinson, tensile, heat.

INTRODUCTION

Characterization of mechanical properties such as Young's modulus, shear strain, viscosity and fracture toughness is very important in the development process of new materials. Researchers must typically carry out many different measurements like tensile displacement tests, compression tests and fatigue tests in order to determine these parameters. One of the most common way of characterizing materials consist in establishing a stress-strain curve. The stress-strain curve reflects the behaviour of the overall sample at the macroscopic level. It does not contain any information at the microscopic level on how the sample deforms or breaks locally during testing. For example, materials typically release heat as they undergo alterations because of elastic or plastic deformations (i.e., work). It is well known that thermal energy is released during the breaking of chemical bounds. Therefore, being able to monitor heat profiles across the sample during testing may provide complementary information about its mechanical properties. Depending on the extent of the applied constrains and the sample's properties, the material can switch from one regime to another (e.g., from elastic to plastic) very quickly. Therefore, measurement techniques with high temporal and/or spatial resolution are usually required for proper investigation. In this work, high-speed infrared imaging was carried out during tensile and shear stress tests on steel and aluminum respectively. In addition, results from a split-Hopkinson experiment carried out on a polymer material at high strain rate (>100 strain/s) are presented. This strain-rate regime is relevant for situations like impact and automotive crashes. The data collected in such experiments are used for validating finite element simulations for determining accurate material models. The results illustrate how infrared imaging can bring some additional insights for material characterization in experimental mechanics research.

EXPERIMENTAL

All samples were coated using a high-emissivity paint prior to testing. Consequently, the temperature values measured by infrared thermography are considered close to their actual surface thermodynamic temperature. A high-elongation extensometer from MTS was used for all experiments. A 12.5-mm gauge steel sample was pulled at 10 strain/s (125 mm/s). The aluminum sample was pulled under adiabatic shear conditions. The Telops FAST M2K is a cooled high-performance infrared camera featuring a 320×256-pixel indium antimonide (InSb) focal plane array (FPA) detector covering the 3 – 5.5 μm spectral range. A 50-mm Janos lens was used along with a ¼-inch extender ring for tensile and shear tests. For the tensile stress test carried out on the steel sample, a 128×256-pixel sub-portion of the FPA detector was used for imaging at 4350 frames per second. For the shear stress test carried out on the aluminum sample, a 192×192-pixel sub-portion of the FPA detector was used for imaging at 3350 frames per second. For the split-Hopkinson test carried out on a polymer (PVC-like) sample at about 100 strain/s. A 64×64-pixel sub-portion of the FPA detector was used for imaging at 12 480 frames per second.

RESULTS AND DISCUSSION

A tensile test was first carried out on a steel sample. Selected images recorded during the experiment, corresponding to different stages of a typical stress-strain curve are shown in Fig.1. In the first three frames (Fig.1a-c), the sample is still in the elastic deformation regime. The measured temperatures are slightly higher than room temperature (initial temperature of the sample prior to testing) and temperature increases are homogeneously distributed all across the sample. The infrared frames collected at a later stage (Fig. 1d-f) correspond to the necking stage, where localized deformations and important temperature increases occur. In the plastic deformation regime, the temperature rises locally and more rapidly than thermal exchanges (adiabatic conditions). Temperature rises on the order of +115 °C were measured, which is in good agreement with prior work on similar samples [1]. Finally, frames collected just before (g), during (h) and after (i) the fracture point are shown in Fig. 1. Beyond the fracture point, heat conduction through the sample and rapid cooling near the fracture area can be monitored (data not shown). The time labels in Fig. 1 illustrate how fast the sample switches from one regime to another and highlight the need for high temporal resolution in this kind of experimental testing.

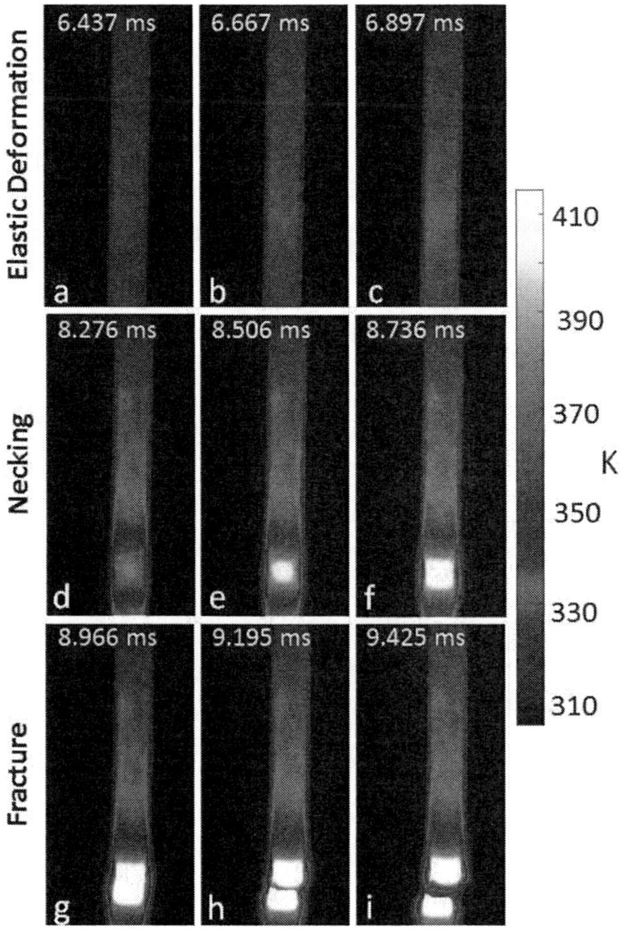

Figure 1 : Selected infrared images recorded during a tensile stress test carried out on a steel sample.

In order to demonstrate the versatility of high-speed infrared imaging, a shear test was carried out on an aluminum sample, as shown in Fig.2. In shear conditions, the constraint must be applied in a perpendicular plane. In order to perform such measurements with a high-elongation extensometer, a special cut-out was made in the sample, as shown in Fig.2. This special shape also strongly dictates where the fracture will likely happen.

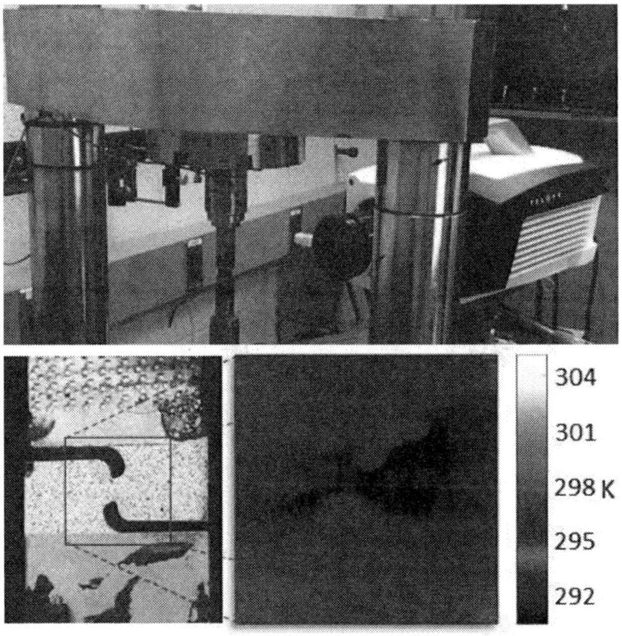

Figure 2 : A typical sample (before being painted) used for this test (bottom left) as well as a representative infrared image of the shear area (bottom right).

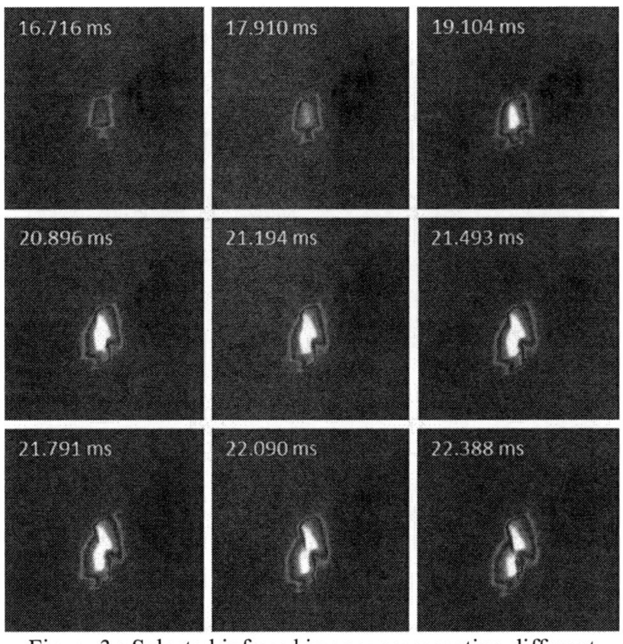

Figure 3 : Selected infrared images representing different stages of the shear experiment on aluminum

For this reason, this area was targeted for infrared imaging, as shown in Fig.3. Selected frames recorded during the shear stress experiment are presented in Fig.3A. In the early stage, the temperature rises rapidly within the area of the (eventual)

fracture. Once again, the experimental conditions ensure that adiabatic shear conditions prevail and that thermal equilibrium is not reached. Under such conditions, the sample mostly undergoes localized heating and softening. This favors stress release as the sample is being pulled. Therefore, moderate heat release occurs in the course of the fracture (approximately +30 °C). Successive frames recorded during the material's rupture are shown in Fig.3B. Once again, it can be seen that high-speed infrared imaging is needed in order to provide enough details to characterize the fracture's onset.

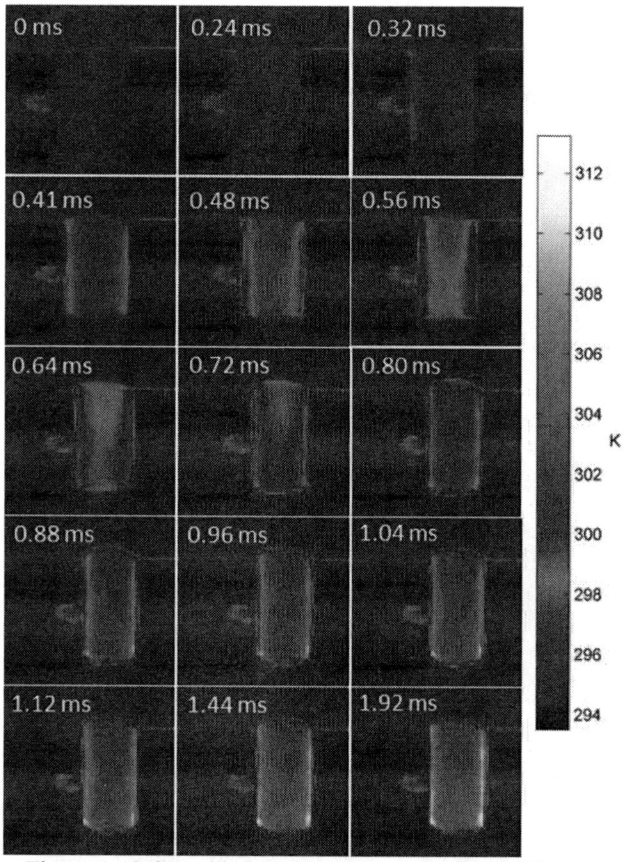

Figure 4 : Selected infrared images representing different stages of the split-Hopkinson experiment on a polymer sample.

Split-Hopkinson experiments were carried out at high strain-rate on PVC-like polymer material. Selected infrared images recorded during the experiment are shown in Fig.4. In the early stage after the incident bar strikes the material, the sample heats homogeneously, which is consistent with the elastic deformation regime. As the material switches to a plastic deformation regime, localized heating on both sides of the sample takes place. The sample keeps heating as the sample deforms. At some stage, the maximum stress point is reached and stress hardening occur [2]. No mechanical alteration of the sample occurs beyond this point. The sample axial length (L) relative to its initial length (L_i) was estimated from the image sequence and plotted against time, as shown in Fig.5. As expected, plastic deformation occurs as soon as the incident bar hits the sample. The sample axial length gradually decreases as a function of time until then stops.

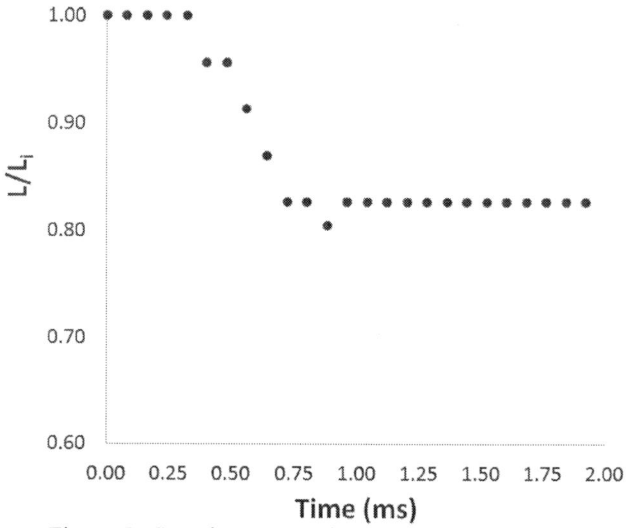

Figure 5 : Sample compression as a function of time.

Mechanical process and heat transfer take place on very different timescales. In order to visualize these very different phenomena, a 3D view of the net temperature increase at various stages of the experiment is shown in Fig.7. The thermal gradient gradually progresses from the sides toward the center of the sample. Significant temperature increase occurs after there is no more mechanical. Being able to track and measure these phenomena is important in order to calculate an accurate energy budget (thermal + work).

CONCLUSIONS

Heat release associated with tensile and shear testing can be successfully monitored using high-speed infrared imaging. Heat spots resulting from energy release in the course of the breaking of chemical bonds can be monitored with high-resolution infrared imaging. Infrared imaging was shown to be a useful tool for monitoring temperature profiles and an important asset for obtaining the most out of each experiment, especially in the case of sample-destructive testing experiments. Experiments at high strain-rate require very fast frame rates as the whole sequence of events takes place in less than a millisecond. Under these conditions, spatial and temporal resolution are both important assets for computing an energy budget and improve material models.

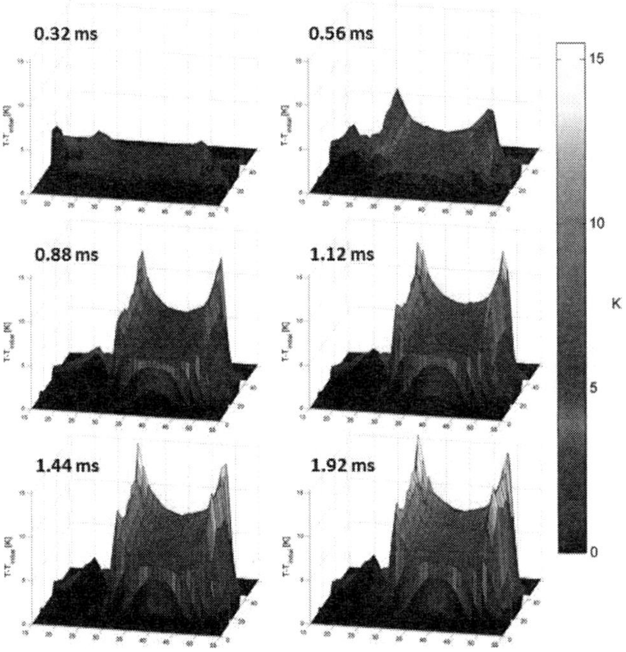

Figure 6 : 3D-views of the net temperature increase at different times.

ACKNOWLEDGMENTS

Telops would like to acknowledge the support of Clifford Butcher, Giovanni Montesano and Michael Worswick from the Mechanical Engineering laboratory of the University of Waterloo (Canada) as well as Arild Holm Clausen and Trond Auestad from the Structural Impact Laboratory (SIMLab) of the Norwegian University of Science and Technology in this work.

REFERENCES

1. Haneef, T., et al., 2014, "Study of the tensile behavior of AISI type 316 stainless steel using acoustic emission and infrared thermography techniques", *Journal of Material Research and Technology*, 4, pp. 241–253.

2. Ognedal, A.S., et al., 2014, "Void nucleation and growth in mineral-filled PVC – An experimental and numerical study", *International Journal of Solids and Structures*, 51, pp. 1494–1506.

DEVELOPMENT OF A MULTI-FAILURE SPECIMEN FOR VALIDATION AND CALIBRATION OF DAMAGE MODELS

Bruce W. Williams
CanmetMATERIALS
Natural Resources Canada
Hamilton, Ontario, Canada
bruce.williams@canada.ca

C. Hari. M. Simha
CanmetMATERIALS
Natural Resources Canada
Hamilton, Ontario, Canada
hari.simha@canada.ca

Mark Gesing
CanmetMATERIALS
Natural Resources Canada
Hamilton, Ontario, Canada
mark.gesing@canada.ca

Lucian Blaga
CanmetMATERIALS
Natural Resources Canada
Hamilton, Ontario, Canada
lucian.blaga@canada.ca

ABSTRACT

Damage models are increasingly used to describe the failure response of both thin and thick metallic materials over a wide range of loading conditions. The failure strain in the damage models has been shown to be dependent upon stress triaxiality and Lode angle. To calibrate and validate damage model coefficients, several mechanical tests that subject metal samples to a range of stress triaxialities and Lode angles are required. To reduce the number of mechanical tests required to calibrate/validate damage models, the current work presents a mechanical test specimen that undergoes multiple failure paths. Full-field displacement measurements are used to record strain fields in the specimen with multiple failure paths. On account of the arrangement and shapes of cut-outs in the sample, there were five failure locations with two ligaments undergoing uniaxial tension, two ligaments undergoing notched tension, and a single ligament subject to shear loading. Owing to the symmetry in the specimen, the results were the same for the two notched tension and two uniaxial tension ligaments. Specimens were fabricated from 2-mm thickness AA6061-T6 sheet and tested under quasi-state and intermediate rate loading. The stress triaxiality and Lode angle was predicted from numerical simulations. Future work is planned that will use strain data from the multi-failure specimen to aid in damage model development.

KEYWORDS: damage, failure, triaxiality, Lode angle

INTRODUCTION

Several damage models have been developed to describe failure of metal undergoing necking and fracture over a wide range of loading conditions. These models include the shear modified Gurson [1], the modified Mohr-Coulomb (MMC) [1], and Xue-Wierzbicki (XW) [2] fracture models. The MMC and XW models are damage-mechanics based models that use a scalar damage parameter to degrade strength upon damage initiation.

A commonality between these fracture models is that failure is dependent on stress triaxiality (mean stress divided by effective stress) and Lode angle (which can be thought of as a measure of the loading type). For uniaxial tension, the stress triaxiality, η, is equal to 1/3 with a Lode angle, θ_L=-30°. For pure shear, η=0 and θ_L =0° and for sharp notches higher positive triaxialites can be achieved and the Lode angle can be computed using stress analysis.

A number of mechanical tests are required to measure failure strains under a variety of triaxialities and Lode angles. Dunand and Mohr [2] describe a dual actuator system, sometimes referred to as the 'butterfly' rig, which uses vertical and horizontal actuators to generate various loading paths. Beese et al. [3] detail five fracture tests to generate failure data under uniaxial tension, plain strain, notched tension, equi-biaxial stretching, and 'butterfly' specimens as detailed in [2]. All of these test methods are intended to generate a single failure point under one loading condition. With the use of full-field displacement measurement using Digital Image Correlation (DIC) multiple paths can be measured in a single specimen.

Driemeier et al. [4] developed a bi-failure specimen that underwent notched tension and shear with the strain measured using DIC techniques. Tests were performed on an AA2024-T351 alloy with 9.72 mm thickness and it was concluded that the specimen is useful for the evaluation of failure criteria. The current work builds upon the geometry of Driemeier et al. [4] to include an additional failure path. It is not the intent of the present article to calibrate/validate damage models, but rather focus on the development of a specimen geometry that undergoes multiple failures under loading conditions relevant for the study of damage models.

TEST APPARATUS AND SPECIMEN GEOMETRY

Though the current work only presents results for AA6061-T6 sheet with 2-mm thickness, the test fixture was designed to be used for specimens with larger thickness. The fixture was designed to be used in a 1000 kN hydraulic press which has an hydraulic accumulator system capable of achieving punch displacement rates up to 80 mm/s. The press was designed to operate in compression but the test fixture was designed to subject the sample to tensile loading. It was also important to ensure that the specimen, hence grips, did not rotate, even if the ligaments did not break symmetrically. This was required to simplify corresponding numerical simulations and avoid modelling contact. Gripping was achieved by matching the curved segments of the sample with the corresponding curvature on the fixture. The text fixture is shown in the press in Fig. 1 and the components are labelled in Fig. 2. The middle of the multi-failure sample was visible to the DIC system, which was setup in a 2D configuration in Fig. 1 but could alternatively be used in a 3D configuration if significant through-thickness strains are expected.

Figure 1: Fixture for pulling tension samples in compression press

The top and bottom base-plates of the fixture were attached to the T-slots on the press. The top frame was bolted to the top plate and applied pressure on the bottom of the sample. The top frame strut prevented the top frame from spreading during the test. The bottom frame supported the loading pins which press on the top of the sample. The middle of the bottom frame was cut-away to give a line of sight for the DIC system. Two alignment blocks kept the front and back plates of the bottom frame parallel to each other. The bottom frame was not fixed to the base plate so that the assembly was not over-constrained. For the thin 2 mm samples, sample braces were bolted to the grip sections of the sample to prevent out-of-plane buckling during the test.

The multi-failure specimen geometry is detailed in Fig. 3 with the five failure ligaments labelled. The two uniaxial tensile ligaments (Ligaments 1 and 2) and the shear ligament (Ligament 3) were 2 x 2 x 4 mm. The diameter of the two notches (Ligament 4 and 5) were 8 mm, a value that could be reduced to produce higher triaxiality for a specimen with greater thickness. The cut-outs were machined.

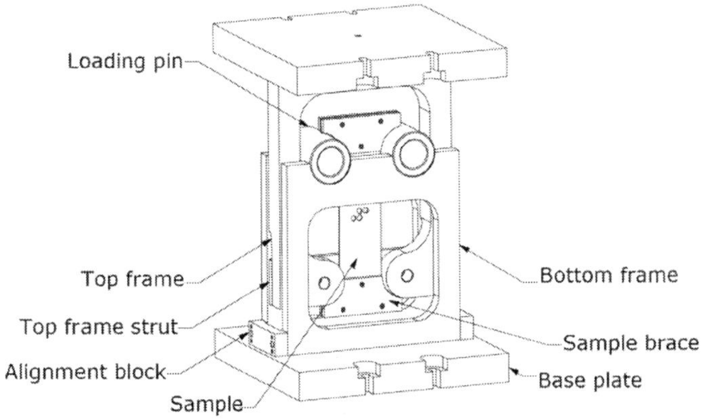

Figure 2: Schematic of fixture assembly

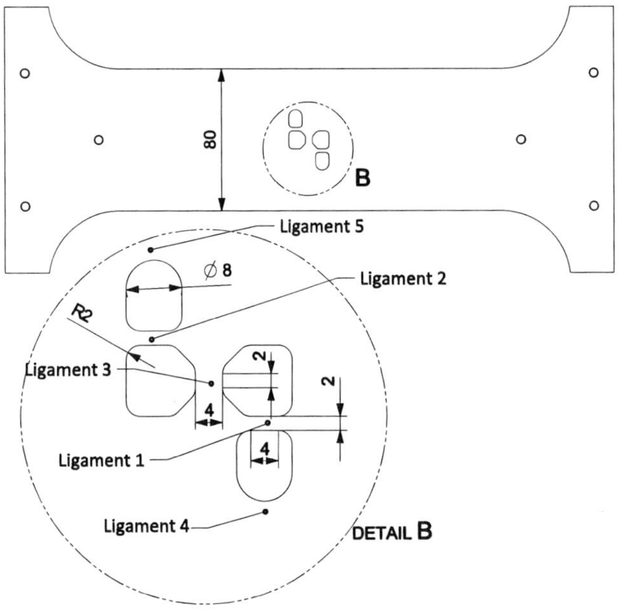

Figure 3: Multi-failure specimen geometry

EXPERIMENTAL RESULTS

Tests were performed at punch displacement rates of 0.016 and 40 mm/s. This corresponded approximately to strain-rates in the uniaxial tensile ligaments of 0.003 s^{-1} (quasi-static) and 3.2 s^{-1} (intermediate). The force versus (punch) displacement results are shown in Fig. 4. There was only a slight change in the peak force response indicating, that as expected for AA6061-T6, the hardening was strain-rate insensitive. The displacement at failure was greater for the fast test case, which indicates some strain-rate sensitivity in the fracture response. In both cases, the distance from the first major decrease in load to zero load is about 0.6 mm. For the fast case, oscillations after fracture were attributed to forces propagating in the load cell.

Ideally, a transition of failure from one ligament to the next would be seen in the force versus displacement curve. However, there was no major indication in Fig. 4 of a transition between failure of the uniaxial tension, notched tension, and shear ligaments. There was a small force discrepancy, as seen at Location 1 highlighted in the figure, for the slow test that could indicate a transition.

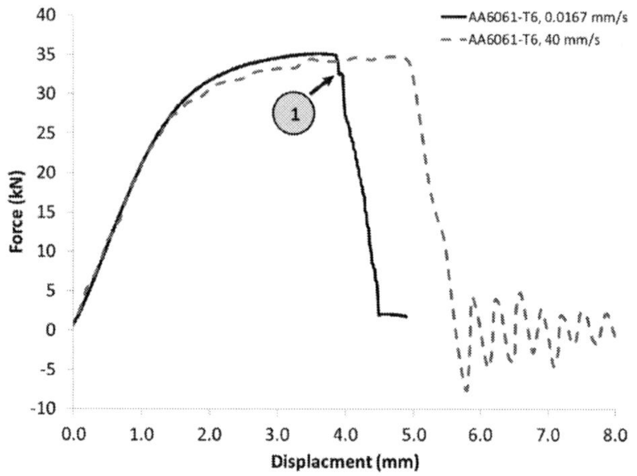

Figure 4: Force versus displacement response for AA6061-T6 multi-failure specimens

The progression of the failure for the fast case is seen in Fig. 5, in which the major strain is overlain with the (speckled) specimen image from DIC. The frame rate was set at 500 frames per second and failure is seen from 4 images over 23 stages (0.002 seconds/stage).

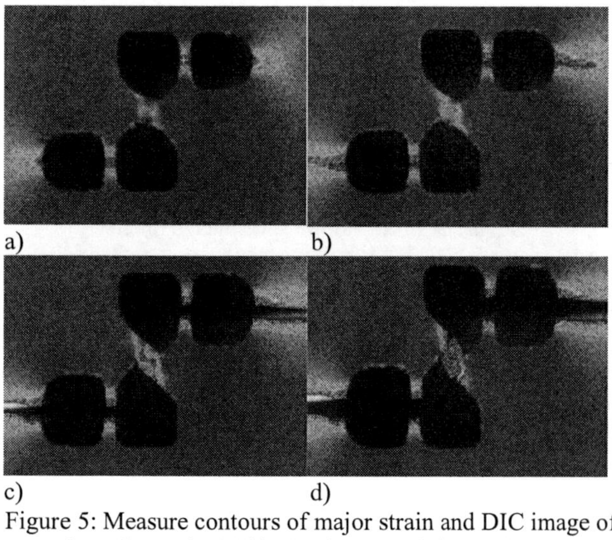

Figure 5: Measure contours of major strain and DIC image of sample at Stage a) 402 b) 405 c) 409 and d) 425 (maximum major strain of 0.45 given by the red zones)

Fig. 5a shows the state of strain just prior to failure of the tensile ligaments. Just 0.004 seconds after, failure has occurred in both the uniaxial tensile ligaments and at the notches, as seen in Fig. 5b. Fig. 5c indicates that for the shear ligament, the initiation of failure was not actually in shear but rather at a stress concentration, a typical issue with shear specimens. Complete failure of the shear ligament is seen in Fig. 5d. The entire progression of damage occurred in 0.046 seconds.

NUMERICAL SIMULATION

A numerical simulation of the multi-failure specimen was performed to determine the stress triaxiality and Lode angle for the failure ligaments in the specimen. The simulation was performed with von Mises yielding and the hardening law for AA6061-T6 detailed by Beese et al. [3] which is given by,

$$\bar{\sigma} = 438.0\left(0.00434 + \bar{\varepsilon}^{p}\right)^{0.07} \qquad (1)$$

where $\bar{\sigma}$ is the effective stress (from the rolling direction of the material) and $\bar{\varepsilon}^{p}$ is the effective plastic strain. The XW damage model was utilized with the parameters for AA2024-T351 detailed by Xue and Wierzbicki [1]. By using the damage parameters for AA2024-T351, accurate failure predictions of AA6061-T6 would not be expected from the simulation. However, the stress triaxialities and Lode angle would be precise in determination of the loading path which was the objective of the simulation effort.

There were 88,000 solid elements with 9 elements through the thickness in the model. The XW damage model detailed by Simha et al. [5] was implemented in the explicit solvers of DYNA3D and used in this study. The force versus displacement response predicted from simulation is shown in Fig. 6. Unlike the experimental profile, specific drops in the force response are predicted when the failure transitions from one ligament to another. Oscillations in the response were attributed to mass shedding upon element deletion in the time-scaled explicit simulation.

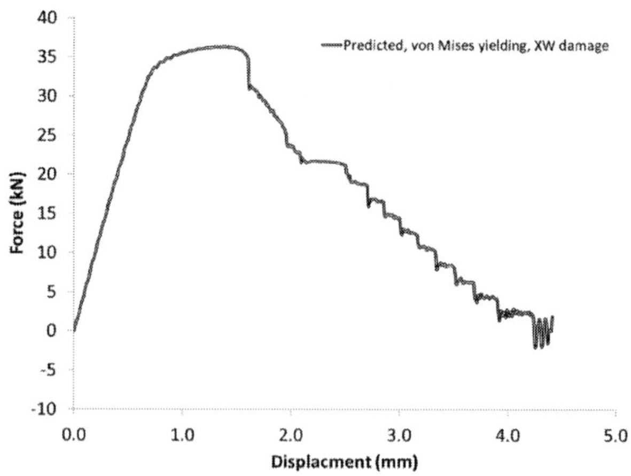

Figure 6: Predicted force versus displacement

Fig. 7 and Fig. 8 plot the effective strain versus stress triaxiality and Lode angle, respectively, for Ligaments 1, 3, and 5 (the results are equivalent for Ligaments 1 and 2; also for Ligaments 4 and 5). Both the uniaxial and notched tensile ligaments initially have $\eta \sim 1/3$, but then as the damage progresses the stress triaxiality changes going down to about η~0.20 just before failure of the tensile ligament and increasing to about η~0.74 at the notch. For the shear ligament, $\eta \sim 0$ with $\theta_L \sim 0°$. As per the definition of Lode angle in [1], both uniaxial and notch tensile ligaments have $\theta_L \sim -30°$ increasing slightly near failure. Though not achieved in the current loading configuration, an additional Lode angle of interest would be at 60° which could be achieved through equi-biaxial loading conditions.

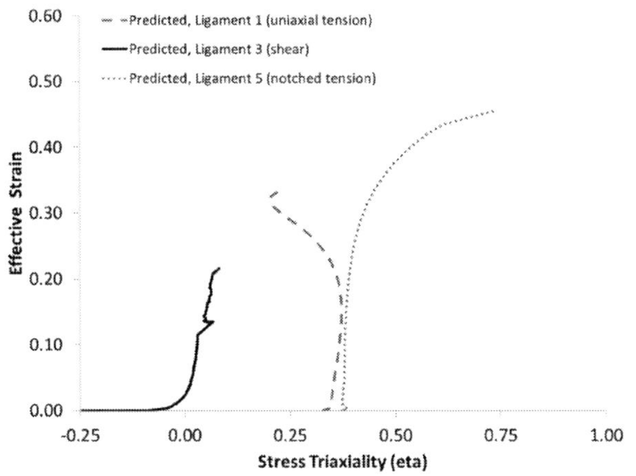

Figure 7: Predicted effective strain versus stress triaxiality

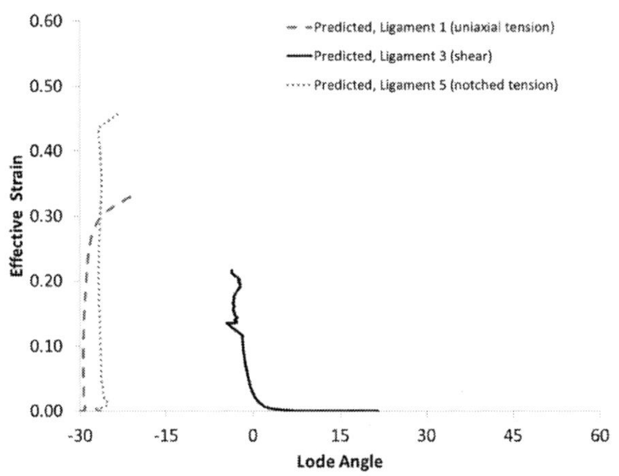

Figure 8: Predicted effective strain versus Lode angle

Fig. 9 shows the progress of damage predicted in the simulation. The undeformed mesh is shown in Fig. 9a and Fig. 9b shows failure starting to occur in the center of the tensile ligaments which have a slightly lower stress. The typical 'kidney' shaped stress distributed associated with a crack or notch can also be seen in the figure. Fig. 9c shows the stress state just after failure of the uniaxial and notched ligaments, while Fig. 9d shows the state after failure of the shear ligament.

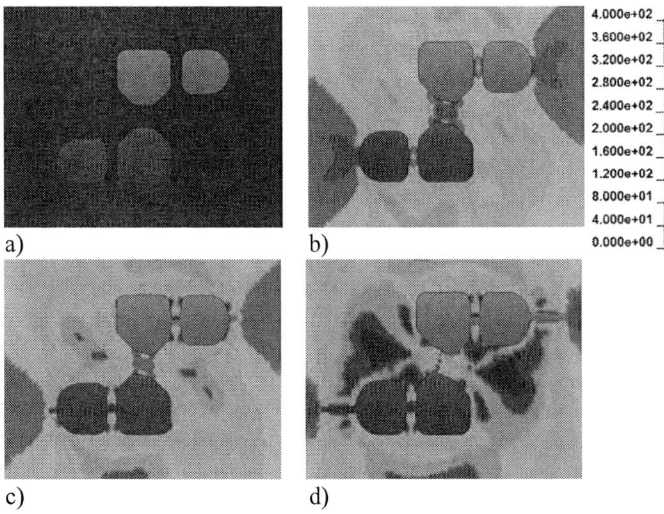

Figure 9: Predicted contours of effective stress a) Initial mesh b) just prior to failure of uniaxial tensile ligaments c) just after failure of the uniaxial and notched tension ligaments and d) failure of the shear ligament

CONCLUSIONS

A multi-failure specimen was designed that underwent three distinct paths of loading and failure: uniaxial tension, notched tension, and shear. The stress triaxiality ranged from zero to about 0.75 and the Lode angle ranged from zero to -30°. The results are planned to be used to aid in the calibration and validation of damage models that describe the full envelope of failure of metals.

ACKNOWLEDGMENTS

The authors gratefully acknowledge funding for this work provided through the Energy End Use programme managed by Dr. Mark Kozdras of CanmetMATERIALS.

REFERENCES

1. Xue, L. and Wierzbicki, T. 2008, "Ductile fracture initiation and propagation modeling using damage plasticity theory", *Eng Fract Mech*, Vol(75), pp. 3276-3293.
2. Dunand, M. and Mohr, D. 2011, "On the predictive capabilities of the shear modified Gurson and the modified Mohr–Coulomb fracture models over a wide range of stress triaxialities and Lode angles", *J Mech Phys Solids*, Vol(59), pp. 1374-1394.
3. Beese, A.M., Luo, M., Li, Y., Bai, Y., and Wierzbicki, T. 2010, "Partially coupled anisotropic fracture model for aluminum sheets", *Eng Fract Mech*, Vol(77), pp. 1128-1152.
4. Driemeier, L., Moura, R.T., Machado, I.F., Alves, M. 2015, "A bifailure specimen for accessing failure criteria performance", *Int J Plasticity*, Vol(71), pp. 62-86.
5. Simha, C.H.M, Xu, S., and Tyson, W.R. 2014, "Non-local phenomenological damage-mechanics-based modeling of the Drop-Weight Tear Test", *Eng Fract Mech*, Vol(118), pp. 66-82.

Proceedings of the 26th CANCAM
Victoria, British Columbia, Canada, May 29 - June 1, 2017

DEVELOPMENT AND FLIGHT OF AN IMAGING FABRY-PEROT SPECTROMETER FOR ATMOSPHERIC STUDIES

Jinjun Shan
Department of Earth and Space Science and Engineering
York University
Toronto, ON, Canada
Email: jjshan@yorku.ca

ABSTRACT

An imaging Fabry-Perot spectrometer (FPS) has been developed and validated through a stratospheric balloon mission in September 2016 in Kiruna Sweden. The purpose of this instrument is to obtain accurate, detailed, and collocated measurements of A-band oxygen at 20-40 km and to retrieve simultaneously surface pressure, aerosol information, and surface albedo from the measurements. The flight mission was successful. This paper summarizes the development of this instrument and some lessons learned from this mission.

Keywords: Fabry-Perot spectrometer (FPS), piezoelectric actuators, control system design, stratospheric balloon, atmospheric studies

INTRODUCTION

Piezoelectric actuators form the basis for many high-precision positioning stages and are finding use in many fields such as atomic force microscopy (AFM), space instrumentation. Piezoelectric actuators operate via the converse piezoelectric effect whereby an applied electric field generates strain in the material. As piezoelectric stack-actuators are monolithic in nature, they exhibit no stiction or friction effects, and thus have very precise positioning capabilities, typically only limited by the resolution and noise level of the system electronics. While the total displacement exhibited by the actuator is typically only on the order of micrometers, they have a large bandwidth, typically in the kHz to tens of kHz range, and are capable of generating forces in the kN range, making their application in micro/nano-positioning stages quite natural.

Recently, a 2-D imaging Fabry-Perot spectrometer (FPS) [1], shown in Fig 2, has been developed at the Spacecraft Dynamics Control and Navigation Laboratory (SDCNLab) of York University under CSA's FAST program. The key component of this FPS is a Fabry-Perot (or etalon), as shown in Fig. 1, which employs three piezoelectric actuators to adjust the gap spacing between two optical plates. The movement of the piezoelectric actuators results in a tuneable gap spacing, hence allowing the desired measurement band to be scanned incrementally as the actuators are moved.

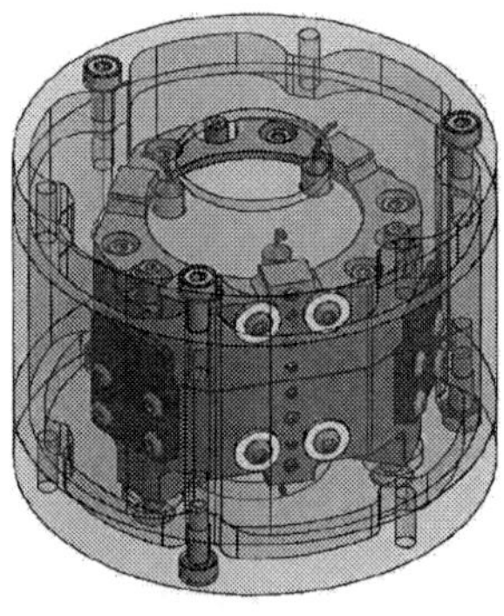

Figure 1: Fabry-Perot assembly (made by MPB Communications Inc.).

The FPS instrument is developed to obtain accurate, detailed, and collocated measurements of O_2 A-band at 20-40 km and to retrieve simultaneously surface pressure, aerosol information, and surface albedo from the measurements. The O_2 measurement provides information on the surface pressure, which is needed to develop accurate column measurements of gases like CO_2 and CH_4.

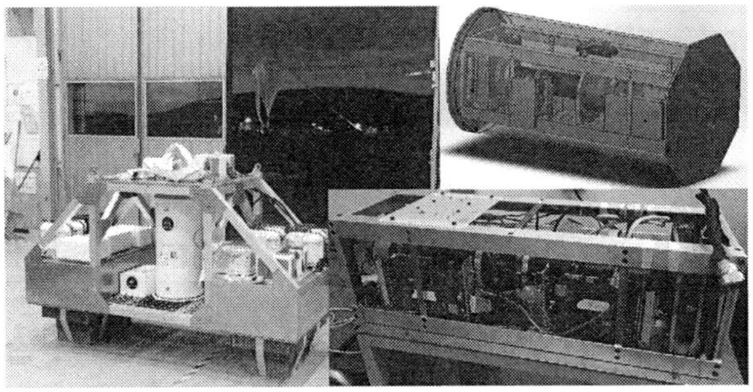

Figure 2: 2D imaging Fabry-Perot spectrometer for atmospheric research. Left: FPS instrument on balloon gondola before flight on September 3, 2016; Top right: FPS CAD drawing; Bottom right: Assembled FPS.

The FPS has flown on a stratospheric balloon launched from Esrange Space Center, Sweden on September 3, 2016. The flight was successful and we have obtained many meaningful observations. This paper summarizes the instrument development and lessons we learned through this project.

SYSTEM DEVELOPMENT

Due to the capacitive sensors on the F-P and capacitors using in piezo controllers, the operability of system in vacuum conditions is under question. Therefore we decided to build the entire system inside a sealed container, as shown in Fig. 2.

Control System Design

As a result of using three distributed actuators, synchronization of the motion of the actuators through both the transient response and in steady state is desired to maintain the parallelism between the two plates, which helps achieve a high finesse value. Another challenge of using piezoelectric actuators is its nonlinearities, especially hysteresis and creep [2], which can lead to a positioning error of as much as 10-15% in the operating range of a piezoelectric actuator [3]. A variety of models and also control strategies for the mitigation of these nonlinearities have been proposed [4, 5].

The theory of synchronization control is well developed and has been widely applied to many mechanical systems [6]. In [7] a simple yet powerful strategy for synchronization was presented and then augmented to provide adaptive compensation in [8]. Using the synchronization error presented in [7] an integral plus double integral synchronization controller was developed in [9] and simulations were conducted to prove the feasibility of the proposed control methodology with one experimental piezoelectric actuator and two mathematical models of the actuator. This controller was applied to the flight instrument to verify its effectiveness.

The linear dynamics of the parallel three-piezo system can be described as

$$M\ddot{q} + C\dot{q} + Kq = u(t) \tag{1}$$

where M is the mass matrix, C is the damping matrix, K is the stiffness matrix, and q is the generalized coordinate.

The integral plus double integral synchronization tracking control input u can be formulated as

$$u(t) = K_i \int_0^t e^*(\tau)d\tau \tag{2}$$

where the integral gain term $K_i = diag\{k_{i_1}, k_{i_2}, k_{i_3}\}$ and $e^*(t)$ is the synchronization error defined as [7]

$$e^*(t) = e(t) + BT^T T \int_0^t e(\tau)d\tau \tag{3}$$

where $e(t) = q_d - q$ is the position error, q_d is the desired position of each actuator, q is the actual position of each actuator, $B = diag\{B_1, B_2, B_3\}$ is a diagonal gain matrix that gives freedom in the choice of the double integral synchronization gain and T is the coupling matrix that is chosen such that $T^T T > 0$, which may be taken for example as

$$T = \begin{bmatrix} 1 & -1 & -1 \\ -1 & 1 & -1 \\ -1 & -1 & 1 \end{bmatrix} \tag{4}$$

To be precise, the exact problem that is under consideration in this paper, is when $q_d = r[\ 1\ \ 1\ \ \dots\ \ 1\ ]^T$, where r is the constant reference to be tracked, and the initial conditions of each of the systems are the same. The synchronization gain matrix K_s may also be defined as

$$K_s = K_i BT^T T \tag{5}$$

This control structure acts as a regular integral controller, hence the diagonal K_i term, with a cross-coupled double integral controller, created by the K_s term, where the tracking errors of each of the systems become coupled to each other. Finally $u(t)$ can be written as

$$u(t) = K_i \int_0^t e(\tau)d\tau + K_s \int_0^t \int_0^\tau e(\tau_1)d\tau_1 \tag{6}$$

which, with T taken as in Eq. (4), can be written in full as

$$u(t) = \begin{bmatrix} k_{i_1} & 0 & 0 \\ 0 & k_{i_2} & 0 \\ 0 & 0 & k_{i_3} \end{bmatrix} \int_0^t \begin{bmatrix} e_1(\tau) \\ e_2(\tau) \\ e_3(\tau) \end{bmatrix} d\tau +$$
$$\begin{bmatrix} 3k_{i_1}B_1 & -k_{i_1}B_1 & -k_{i_1}B_1 \\ -k_{i_2}B_2 & 3k_{i_2}B_2 & -k_{i_2}B_2 \\ -k_{i_3}B_3 & -k_{i_3}B_3 & 3k_{i_3}B_3 \end{bmatrix} \int_0^t \int_0^\tau \begin{bmatrix} e_1(\tau_1) \\ e_2(\tau_1) \\ e_3(\tau_1) \end{bmatrix} d\tau_1$$

For consideration of the stability of the controlled system, to start, consider only a single piezoelectric actuator, which can be described generally as the plant with the representation

$$\begin{bmatrix} \dot{x} \\ \dot{v} \end{bmatrix} = \begin{bmatrix} 0 & 1 \\ -\omega^2 & -2\zeta\omega \end{bmatrix} \begin{bmatrix} x \\ v \end{bmatrix} + \begin{bmatrix} 0 \\ g\omega^2 \end{bmatrix} u \tag{7}$$

$$y = \begin{bmatrix} 1 & 0 \end{bmatrix} \begin{bmatrix} x \\ v \end{bmatrix} \tag{8}$$

where x is the displacement of the actuator, v is the velocity, ω is the frequency, ζ is the damping coefficient, g is the gain, and the system represented by Eqs. (7–8) is assumed to be Hurwitz.

More details on control system design can be found in [10].

Thermal Control

The operational temperature of Fabry-Perot spectrometer should be at room temperature, i.e. 25-27°C since all the optics tests are done in laboratory. At 30 km the average temperature ranges from -20°C to -40°C. According to thermal simulations, the temperature inside the optical tube will range from -2°C to 16°C. Therefore, thermal control system is required. In addition, we also considered the following approaches:

- Localized heating and insulation around the spectrometer.
- From thermal simulations we found three 10" heater strips evenly spread around the spectrometer gave the best heating results.
- Using FR4 to isolate the spectrometer from rest of the system.
- Paint exterior surfaces white to prevent overheating.

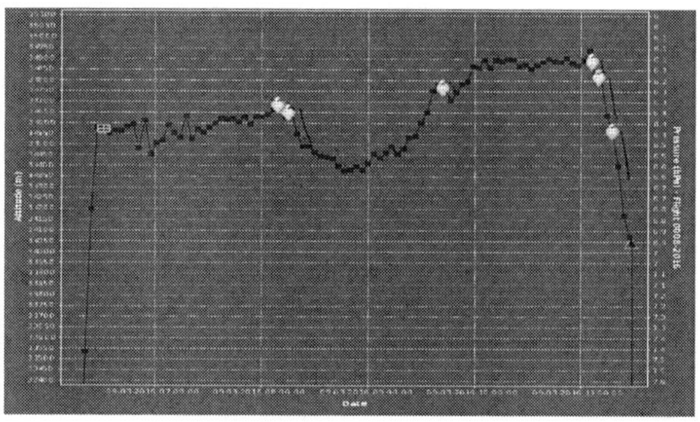

Figure 3: Profile of balloon flight.

Flight Results

The stratospheric balloon took off few minutes before 7am local time (1 am EDT) on Saturday September 3, 2016 at Esrange Space Center, Sweden. It landed around 2:15pm in Northern Finland and the instrument was successfully recovered before 8pm. The mission has been very successful. Figs. 3-4 shows the profile and trajectory of the balloon flight, respectively. During the 8-hour flight, key technologies of Fabry-Perot spectrometer have been demonstrated and validated. Fig. 5 shows the pressure of the tube, temperatures inside the tube and observation sequence. During the flight, we also tested two types of piezo controllers: PID controller and synchronization controller. The results are shown in Figs. 6-7. It can be seen that the synchronization controller did synchronize the motion among three piezoelectric actuators and makes the piezo response faster. However, due to some limitations, only the motion of piezo #1 and #2 were synchronized, but not piezo #3. The most excited moment was the successful observation of O_2 line, as shown in Fig. 8, during the balloon flight.

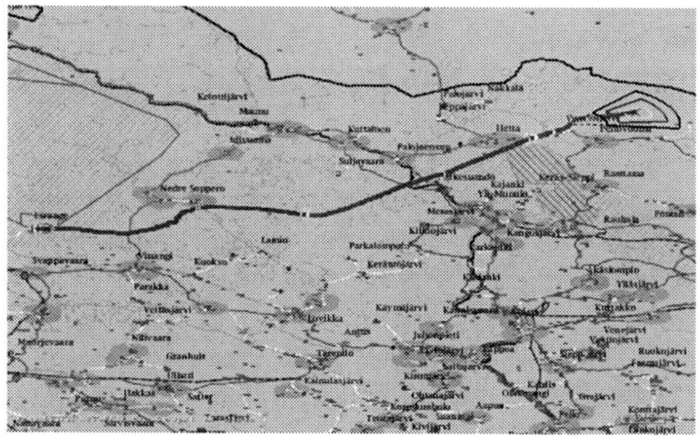

Figure 4: Trajectory of balloon flight.

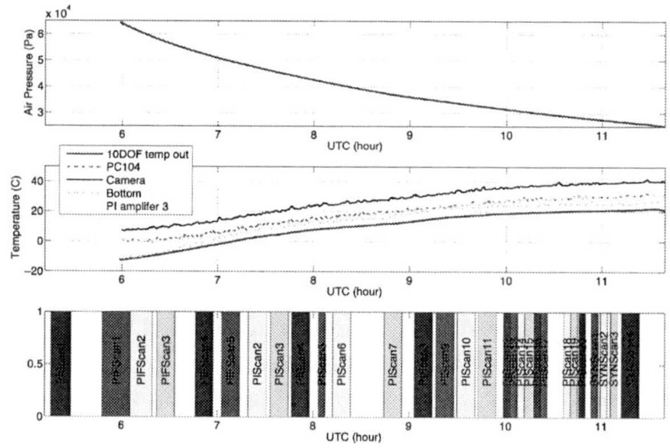

Figure 5: Results of balloon flight.

CONCLUSIONS

This paper presents the development of an imaging Fabry-Perot spectrometer (FPS) for measuring A-band oxygen from stratospheric balloon platform. This instrument is developed to obtain accurate, detailed, and collocated measurements of A-band oxygen and to retrieve simultaneously surface pressure, aerosol information, and surface albedo from the measurements. The balloon flight in Kiruna Sweden in September 2016 has been very successful and all scientific objectives have been achieved. We are working on the modifications of this instrument and expect to re-flight it soon.

ACKNOWLEDGMENTS

This work is supported by a grant through FAST (Flights for the Advancement of Science and Technology) program

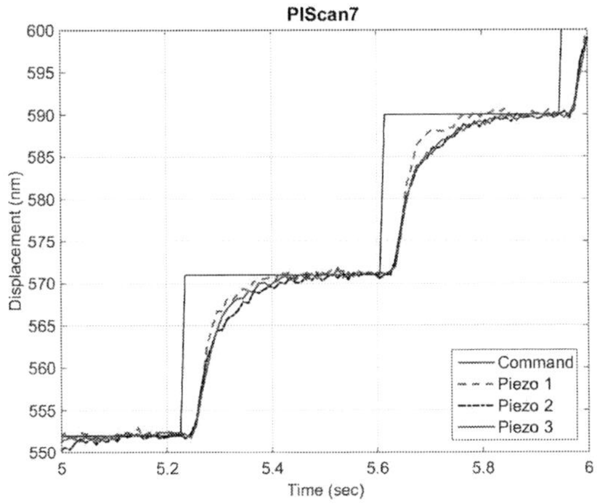

Figure 6: Response of three piezo actuators under PI controller.

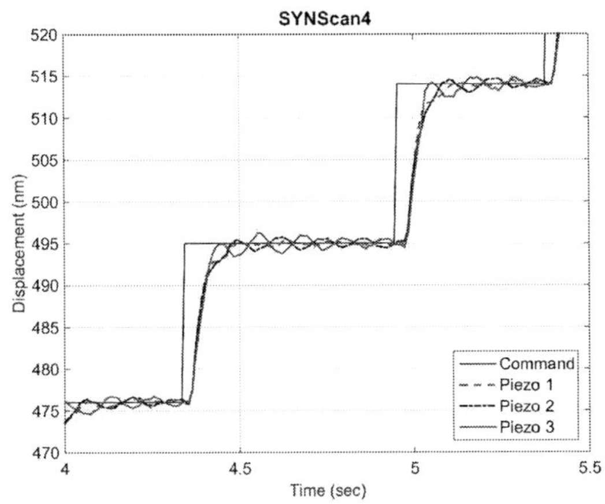

Figure 7: Response of three piezo actuators under synchronization controller.

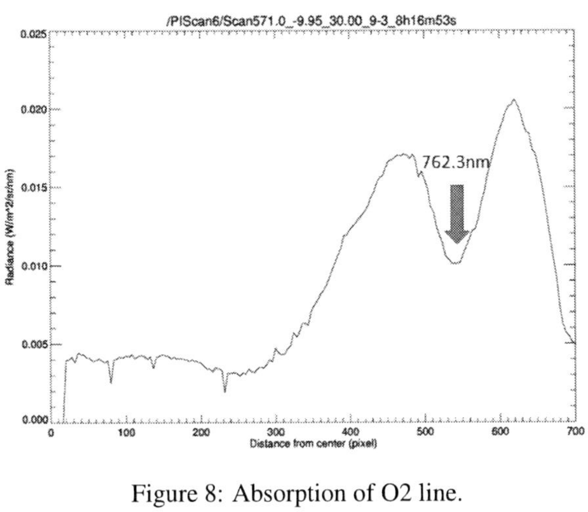

Figure 8: Absorption of O2 line.

under Canadian Space Agency (Project No.: 13STFAY007). The author would like to acknowledge MPB Communications Inc. and the following people for their contributions to this project: Gordon Shepherd, Chris Sioris, Yuan Ren, Brian Solheim, Ryan Orszulik, Mike Voutsogiannakis, Mohammed Kagalwala, Bosen Jiang, Young-Min Cho, Ganesh Ramkumar, Zishun Yu, Zhuo Li, Weihao Xu.

REFERENCES

[1] T. R. Hicks, N. K. Reay, and R. J. Scaddan, "A servo controlled fabry-perot interferomter using capacitance micrometers for error detection," *Journal of Physics E: Scientific Instruments*, vol. 7, pp. 27–30, 1974.

[2] M. Goldfarb and N. Celanovic, "A lumped parameter electromechanical model for describing the nonlinear behavior of piezoelectric actuators," *ASME. J. Dyn. Sys., Meas., Control*, vol. 119, no. 3, pp. 478–485, 1997.

[3] T. S. Low and W. Guo, "Modeling of a three-layer piezoelectric bimorph beam with hysteresis," *Journal of Microelectromechanical Systems*, vol. 4, no. 4, pp. 230–237, 1995.

[4] S. O. R. Moheimani, "Invited review article: Accurate and fast nanopositioning with piezoelectric tube scanners: Emerging trends and future challenges," *Review of Scientific Instruments*, vol. 79, no. 7, p. 071101, 2008.

[5] Y. Cao and X. Chen, "A survey of modeling and control issues for piezo-electric actuators," *ASME. J. Dyn. Sys., Meas., Control*, vol. 137, no. 1, pp. 014 001–014 001–13, 2014.

[6] H. Nijmeijer and A. Rodriguez-Angeles, *Synchronization of Mechanical Systems*. Singapore: World Scientific Publishing Co. Pte. Ltd., 2003.

[7] J. Shan, H.-T. Liu, and S. Nowotny, "Synchronised trajectory-tracking control of multiple 3-dof experimental helicopters," *Control Theory and Applications, IEE Proceedings* -, vol. 152, no. 6, pp. 683–692, 2005.

[8] H. Liu, J. Shan, and D. Sun, "Adaptive Synchronization Control of Multiple spacecraft formation flying," *Journal of Dynamic Systems, Measurement, and Control*, vol. 129, no. 3, pp. 337–342, 2007.

[9] R. Orszulik and J. Shan, "Integral plus double integral synchronization control for multiple piezoelectric actuators," in *European Control Conference (ECC), 2015*, July 2015, pp. 1082–1087.

[10] R. R. Orszulik and J. Shan, "An integral-based approach to the synchronization of multiple piezoelectric actuators," *Submitted to IEEE/ASME Transactions on Mechatronics*, 2016.

INTEGRATED ENGINEERING DESIGN APPROACH FOR UNDERGRADUATE TEACHING

Sharman Perera
Faculty of Energy Systems and Nuclear Science
University of Ontario Institute of Technology
Oshawa, Ontario, Canada
sharman.perera@uoit.ca

Talveen Labana
Faculty of Energy Systems and Nuclear Science
University of Ontario Institute of Technology
Oshawa, Ontario, Canada
talveen.labana@uoit.net

Edward Waller
Faculty of Energy Systems and Nuclear Science
University of Ontario Institute of Technology
Oshawa, Ontario, Canada
ed.waller@uoit.ca

Glenn Harvel
Faculty of Energy Systems and Nuclear Science
University of Ontario Institute of Technology
Oshawa, Ontario, Canada
glenn.harvel@uoit.ca

Robert Ulrich
Faculty of Energy Systems and Nuclear Science
University of Ontario Institute of Technology
Oshawa, Ontario, Canada
robert.ulrich@uoit.ca

Callan Brown
Faculty of Energy Systems and Nuclear Science
University of Ontario Institute of Technology
Oshawa, Ontario, Canada
callan.brown@uoit.ca

ABSTRACT

The objective of the Integrated Engineering Laboratory is to allow students to integrate the knowledge, tools, methods, and techniques learned from fundamental engineering courses such as Fluid Mechanics, Heat Transfer, Thermodynamics, Structures and Properties of Materials, and Engineering Graphics and Design towards the engineering design process. Students will generate 3D CAD models and manufacturing drawings of their selected designs using Siemens NX CAD software. These models are analyzed and optimized using multiphysics engineering tools such as Finite Element Analysis, Computational Fluid Dynamics and vibration analysis. The final designs are 3D printed and tested for reliability and durability using in-house testing facilities under the Faculty of Energy Systems and Nuclear Science teaching laboratories.

KEYWORDS: Integrated, Engineering, Laboratory, Design, Undergraduate, Teaching, CFD, FEM, 3D-printing, UOIT, FESNS

INTRODUCTION

The Integrated Engineering Laboratory (IEL) course is required for third year undergraduate students pursuing a Bachelor of Nuclear Engineering or Energy Systems Engineering degree within the Faculty of Energy Systems and Nuclear Science (FESNS) at the University of Ontario Institute of Technology (UOIT).

IEL invokes a project-based approach to practical experiences that cover multidisciplinary topics. The course material covers topics that include chemistry, fluid mechanics, materials, and structural analysis. Students are introduced to the engineering design process, developed through both numerical/simulation and experimental laboratories.

COURSE DELIVERY AND OUTCOMES

The course consists of a one-hour lecture delivered weekly in addition to a weekly three-hour laboratory session over the course of 12 weeks. Students who successfully complete this course will have gained extensive knowledge regarding the engineering design process, resulting in the preparation of a multidisciplinary technical report, and experience working within a team environment to complete an open-ended engineering design project. Students have to arrange meetings among team members to assign tasks, determine execution strategies, evaluate project progress throughout the semester, and take corrective measures when necessary.

LABORATORY TOPICS AND DELIVERABLES

The first five lab sessions are used to review and enhance analytical skills through numerical laboratory activities. In Session 1, a brief review of solid 3D modelling in the Siemens NX software environment is conducted. These skills are expanded upon through the introduction of numerical analytic methods in the following lab sessions that involve the introduction to finite element methods (FEM) in Session 2 and computational fluid dynamic (CFD) analysis in Session 3. Students are expected to apply these skills towards the development of their own miniature (170 mm long) wind turbine blades in Session 4. An example blade being analyzed in Siemens NX software is shown in Figure 1.

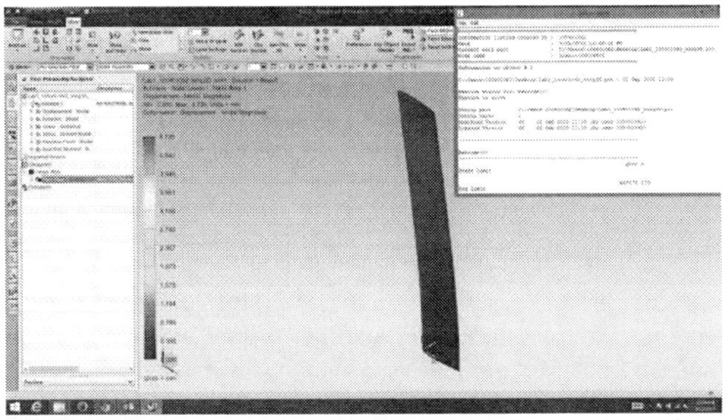

Figure 1: Screenshot of an FEM analysis being performed on a wind turbine blade.

The next four laboratory sessions are used to conduct practical experiments that are focused on material selection. In Session 5, students explore vibration and motion simulation and analysis. In Session 6, students model the wind chamber that will be used in the design competition at the end of the semester. The effects of stress and corrosion are examined in Session 7. Figure 2 illustrates the results of one such experiment.

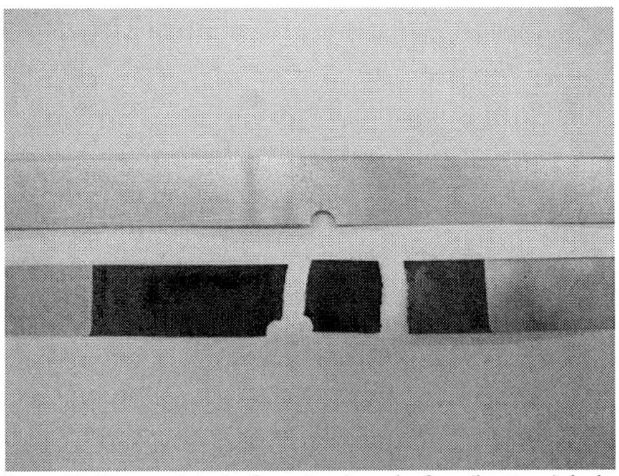

Figure 2: Brass strip before (top) and after (bottom) being subjected to stress-corrosion cracking conditions.

Students are given brass strips like the one pictured of varying thickness and deformity (a notch) along the length which expedites the corrosion process. These strips are then submerged in Mattson's solution, and a stress force is applied using a custom apparatus. Strain gauges are used to measure deformation. Students are then asked to analyze the results, and in the process, realize why material selection is an important consideration in the engineering design process. In Session 8, students study purification processes use an ion exchange column.

Sessions 9, 10 and 11 provide time for groups to work on their Engineering Design Project and miniature wind turbine blades. Students are able to seek assistance from the laboratory technicians, teaching assistants, and course instructor if necessary.

The purpose of the Engineering Design Project is to compile and apply the knowledge and skills gained in previous sessions, while following an engineering design process. In previous years, students have modeled components of a CANDU heat transport system and Rankine cycle.

Students present their Engineering Design Projects in the Session 12, then close out the semester with the testing of the miniature wind turbine blades in an on-site wind chamber. An example blade that was designed and printed by a previous student is shown in Figure 3.

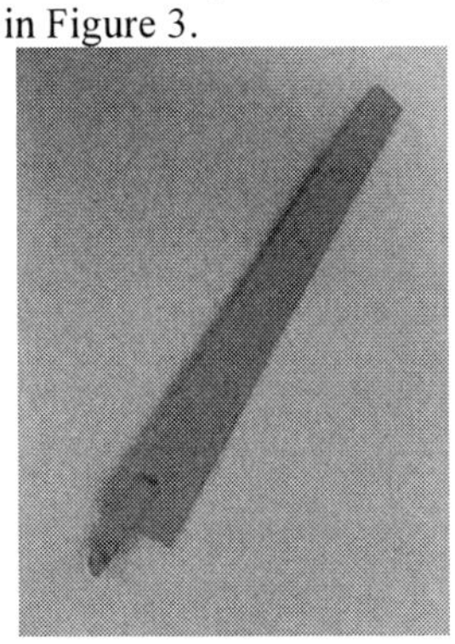

Figure 3: 3D-Printed wind turbine blade ready for testing

The blades are attached to prefabricated wind turbines and placed in the wind chamber, as shown in Figure 4.

Figure 4: A student places his turbine in the wind chamber for testing and evaluation.

Students are evaluated based on key performance criteria including stability, revolutions per minute, and electricity generated from varying wind conditions.

GRADUATE ATTRIBUTES

Having successfully completed this course, students in the program will have been introduced to working in a group setting to prepare a multidisciplinary engineering report as part of their Engineering Design Project.

The report introduces students to vital report-writing skills that involves the engineering design process that is widely used in the industry.

As a group, students generate sketches, perform design selection, analysis, determine economic viability, and make a final design decision. Within this process, students are introduced to generating project schedules, budgets, and execution strategies. Students are given the opportunity to perform individual tasks within a team while achieving a unified goal. This allows students to explore various team dynamics that may range from equal partnership among its members, to selecting a leader to guide the team towards its goal. In each case, all team members receive the same mark based on the quality of work output. Students are encouraged to discuss and settle creative differences within the group.

For students attending this course in 2015, this team dynamic was taken one step further by introducing a project that involved the entire lab section. Unlike previous years in which a team worked on its own project determined by its members, teams in that year were assigned a component that, when combined with other team's components, would come together as one functioning system. For the 2015 project, each team was randomly assigned a component of a simple Rankine cycle consisting of a pump, steam generator, turbine, condenser, and piping. This project introduced students to the idea of collaborating amongst other teams to achieve one goal. It was determined that this should be experienced by students because it is a common approach to completing an engineering project in the industry.

Collectively, these skills are sought after by several employers in the industry when looking to hire new engineering graduates.

LAB SAFETY

Maintaining a safe working environment is priority when conducting lab activities. Students are supervised by teaching assistants who are trained by lab technicians, and include an overview of lab safety practices at the start of the semester, as well as weekly briefings ahead of each lab session. Lab technicians, and the course instructor are also present during lab sessions to offer technical guidance and assistance.

Additionally, FESNS has prepared a document outlining the safety rules and procedures for the lab facilities, and must be signed by each student at the beginning of the semester. The document stipulates safety protocols and consequences of unsafe practices that may include remedies ranging from re-education to expulsion from the lab.

Workplace Hazardous Materials Information System (WHMIS) documentation is also provided in every lab for review for sessions that involve the use of chemicals. Teaching assistants are further instructed on how to handle chemical spills and contact between skin and chemicals.

The proper use of personal protective equipment (PPE) is required when conducting practical lab experiments. Students will not be permitted to attend the lab without having adequate PPE such as safety goggles, and proper attire and footwear. Other PPE such as gloves are provided in the lab.

EVALUATIONS

There are several areas where students are evaluated throughout the duration of a single lab session. Each session begins with a pre-lab quiz to ensure that students are aware of the core principles entailed in that lab. This generally includes multiple choice questions regarding theory, calculations, and safety. Each lab sessions also includes a list of deliverables that need to be submitted before the end of the session. This may include sketches, collected data, computer screenshots, and/or calculations. Pre-lab quizzes and in-lab deliverables are to be submitted on an individual basis. Lab reports are to be completed in groups, and submitted at the start of the following lab session.

Additional group evaluations are conducted in the form of the design project and turbine blade performance competition.

CONTINUOUS IMPROVEMENT

Additions and improvements are always being made to the lab to further enhance the learning experience. Apart from using the newest available version of the Siemens NX CAD software, hardware improvements are also being made regularly.

An example of this is the recently acquired 3D printer that can accommodate large print jobs, allowing for more flexibility for students in the completion of their design projects.

Another area that is currently being investigated for future implementation into the lab is the introduction of virtual reality. This would allow students to explore cutting edge technology into the engineering design process regarding model visualization and analysis.

CONCLUSIONS

The Integrated Engineering Laboratory at the University of Ontario Institute of Technology introduces third-year Nuclear Engineering and Energy Systems Engineering students to the concept of integrated engineering through the performance of laboratory activities that involve previously taught topics ranging from fluid dynamics to structural analysis.

This course is also provides students with guidance and experience in preparing multidisciplinary reports while enhancing teamwork skills and complementing academic knowledge attained in previous courses.

Vibration and Dynamics

NONLINEAR DYNAMICS OF A FREE-CLAMPED CYLINDER IN AXIAL FLOW

A.R. Abdelbaki
Mechanical Engineering Department
McGill University
Montreal, Quebec, Canada
ahmed.abdelbaki@mail.mcgill.ca

M.P. Païdoussis
Mechanical Engineering Department
McGill University
Montreal, Quebec, Canada
michael.paidoussis@mcgill.ca

A.K. Misra
Mechanical Engineering Department
McGill University
Montreal, Quebec, Canada
arun.misra@mcgill.ca

ABSTRACT

In this paper a nonlinear equation of motion is derived for the dynamics of a free-clamped cylinder in axial flow; i.e. with the flow directed from the free-end toward the clamped one. The cylinder is generally terminated by an ogival free-end. The forces due to the fluid are derived separately; an extension of Lighthill's slender-body analysis to third-order accuracy is applied to model the inviscid forces, while the viscous forces are formulated semi-empirically. The equation of motion is obtained via Hamilton's principle, with a separate derivation of the boundary conditions related to the free-end. The dynamics of the cylinder are explored by solving the discretized equations using AUTO and MATLAB as numerical tools. The stability of the system is investigated by means of bifurcation diagrams and phase-plane plots using a concrete example with parameters close to those in experiments from the literature. The theory qualitatively agrees with the experiments.

KEYWORDS: Nonlinear dynamics, free-clamped cylinder, axial flow, stability.

INTRODUCTION

The first study on the dynamics of cylinders in axial flow was undertaken by Hawthorne [1] in the early 1960s. He investigated the stability of the Dracone, a long flexible container towed behind a small craft, and used for the transportation of liquids lighter than seawater. This work was extended for cylinders subject to different boundary conditions, e.g. simply supported, cantilevered and so on, as described by Païdoussis [2]. The nonlinear dynamics of a cantilevered cylinder in axial flow was examined in [3, 4]. That system has certain physical and dynamical similarities to the problem at hand. Rinaldi and Païdoussis [5] investigated experimentally the stability of a free-clamped cylinder in confined axial flow. In addition, they derived a linear theoretical model which captures the essentials of the observed behaviour.

In the present study, the nonlinear dynamics of a free-clamped cylinder in axial flow is examined for the first time. A weakly nonlinear equation of motion is derived, which matches with the one obtained by Rinaldi and Païdoussis [5] in the linear limit. The results obtained using this model are presented and compared to the experimental data of [5].

DERIVATION OF THE EQUATION OF MOTION

The system under study consists of a flexible slender cylinder of diameter D, length L, flexural rigidity EI and mass per unit length m. The cylinder is located in a channel of diameter D_{ch} and is subjected to an axial flow velocity U directed from the free-end toward the clamped one.

The equation of motion is derived in this section making the following assumptions: (i) the motion is planar; (ii) the cylinder length-to-diameter ratio is high enough, so that Euler-Bernoulli beam theory can be applied; (iii) the cylinder centreline is inextensible; (iv) the strains of the cylinder are small, but the deflections can be large; and (v) the fluid is incompressible with constant mean flow velocity.

Two coordinate systems are used: the Lagrangian (X, Y, Z, T) and the Eulerian (x, y, z, t); the first one is associated with the undeformed state of the cylinder, while the second one is for the deformed state. The displacements are thus, $u = x - X$, $v = y - Y$, and $w = z - Z$, as shown in Fig. 1. For centreline motions in the (X-Y) plane, $Y = 0$ and $z = Z = w = 0$. Defining s as the curvilinear coordinate along the cylinder, X and s are related by $\partial X/\partial s = 1/(1+\tilde{\varepsilon})$, where $\tilde{\varepsilon}$ is the axial strain along the centerline. The cylinder centreline is assumed to be inextensible; hence $\tilde{\varepsilon} = 0$, $\partial X/\partial s = 1$ and $(\partial x/\partial X)^2 + (\partial y/\partial X)^2 = 1$.

The equation of motion is derived via Hamilton's principle, and it is correct to third order, $\mathcal{O}(\varepsilon^3)$, for $y = v \sim \mathcal{O}(\varepsilon)$. Hence the expression of the virtual work δW must be correct to $\mathcal{O}(\varepsilon^3)$, while the energy expressions to $\mathcal{O}(\varepsilon^4)$. In addition, the same nonlinear expressions for the kinetic and potential energies of the cylinder obtained in [3] can be used for the problem at hand.

Fluid-Related Forces

The forces due to the fluid are derived separately in a similar manner as in [2]. This approach, rather than the direct application of Navier-Stokes equations, simplifies the analysis considerably and generates acceptable results [2, 6].

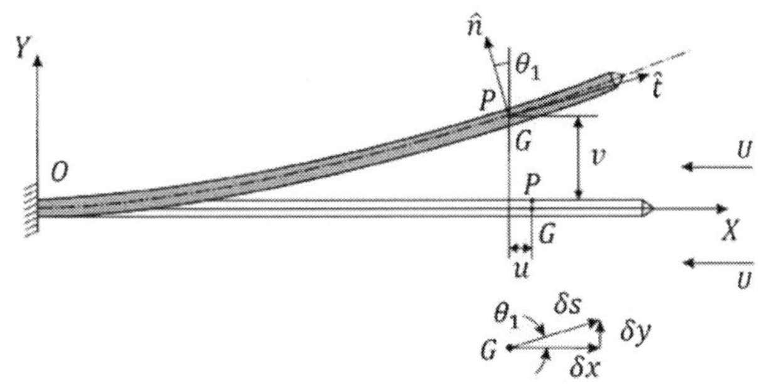

Figure 1: Geometrical sketch of the system.

The fluid-related forces acting on an element of the deformed cylinder are (i) the inviscid fluid dynamic force F_A; (ii) the normal and longitudinal viscous forces, F_N and F_L, respectively; and (iii) the hydrostatic forces F_{px} and F_{py} in the x and y directions, respectively, as shown in Fig. 2.

The inviscid fluid dynamic forces are derived in a similar manner as in [3] via slender-body potential flow theory, following Lighthill's formulation in [7]. The relative fluid-cylinder velocity V can be expressed as

$$V = \dot{y} - Uy' - \frac{1}{2}\dot{y}y'^2 + 2Uu'y' + \frac{1}{2}Uy'^3 - \dot{x}y' + O(\varepsilon^5), \quad (1)$$

where $(\ )' = \frac{\partial(\ )}{\partial s}$ and $(\dot{\ }) = \frac{\partial(\ )}{\partial t}$. The nonlinear expression of the inviscid hydrodynamic force becomes

$$F_A(X,T) = \left\{\frac{\partial}{\partial T} + \left[-U\left(1 - \frac{\partial u}{\partial X}\right) - \left(\frac{\partial u}{\partial T} - U\frac{\partial u}{\partial X}\right)\right]\frac{\partial}{\partial X}\right\} * \left[V - \left(\frac{\partial u}{\partial T}\frac{\partial v}{\partial X} - 2U\frac{\partial u}{\partial X}\frac{\partial v}{\partial X}\right) - \frac{1}{2}V\left(\frac{\partial v}{\partial X}\right)^2\right]M - \frac{1}{2}MV\frac{\partial v}{\partial X}\frac{\partial V}{\partial X} + O(\varepsilon^5), \quad (2)$$

where M is the virtual added mass, which is equal to ρA in the unconfined case and $\chi\rho A$ for the confined case, χ is the virtual mass coefficient, A is the cylinder cross-sectional area, and ρ is the fluid density.

The viscous forces are obtained on the basis of the semi-empirical expressions proposed by Taylor [8]. The following expressions are derived as in [3], but taking into account the new, reverse direction of U:

$$F_L = \frac{1}{2}\rho DU^2C_T\left[1 - \frac{1}{2}\left(y'^2 - 2\frac{y'\dot{y}}{U} + \frac{\dot{y}^2}{U^2}\right)\right] + O(\varepsilon^4), \quad (3)$$

$$F_N = \frac{1}{2}\rho DU^2\left[C_N\left(y' - \frac{\dot{y}}{U} - \frac{\dot{y}u'}{U} - u'y' + \frac{\dot{x}\dot{y}}{U^2} - \frac{1}{2}\left(y'^3 - \frac{\dot{y}^3}{U^3} - \frac{y'^2\dot{y}}{U} + \frac{y'\dot{y}^2}{U^2}\right)\right)\right] - k\dot{y} + O(\varepsilon^5), \quad (4)$$

where C_N and C_T are the friction coefficients; k is a viscous damping coefficient, the term $k\dot{y}$ represents a linearization of the quadratic viscous force at zero flow velocity, $\frac{1}{2}\rho DU^2C_{DP}\dot{y}|\dot{y}|$, as discussed in [6], and C_{DP} is the form drag coefficient.

The hydrostatic forces, which are the resultants of the steady-state pressure p acting on the cylinder, are derived by the procedure used in [6]. The pressure gradient can be expressed as in [3] by taking into account the reverse direction of the flow velocity, as follows:

$$A\left(\frac{\partial p}{\partial x}\right) = \frac{1}{2}\rho DU^2C_T\frac{D}{D_h} + \rho gA, \quad (5)$$

where D_h is the hydraulic diameter.

The nonlinear expressions of the hydrostatic forces are:

$$-F_{px} = -y'^2A\left(\frac{\partial p}{\partial x}\right) - y'y''Ap + O(\varepsilon^4), \quad (6)$$

$$F_{py} = (y' - u'y' - y'^3)\left(\frac{1}{2}\rho DU^2C_T\frac{D}{D_h} + \rho gA\right) + \left(y'' - u''y' - u'y'' - \frac{3}{2}y'^2y''\right)Ap + O(\varepsilon^5). \quad (7)$$

Equation of Motion

The virtual work can be expressed as

$$\int_{t_1}^{t_2}\delta W dt = \int_{t_1}^{t_2}\int_0^L\{[-F_{px} - F_L cos\theta_1 + (F_A - F_N)sin\theta_1]\delta x + [F_{py} - F_L sin\theta_1 - (F_A - F_N)cos\theta_1]\delta y\}\, dsdt, \quad (8)$$

where θ_1 is illustrated in Fig. 2, and can be expressed as in [3] as follows:

$$\theta_1 = y' - u'y' - \frac{1}{3}y'^3 + O(\varepsilon^5). \quad (9)$$

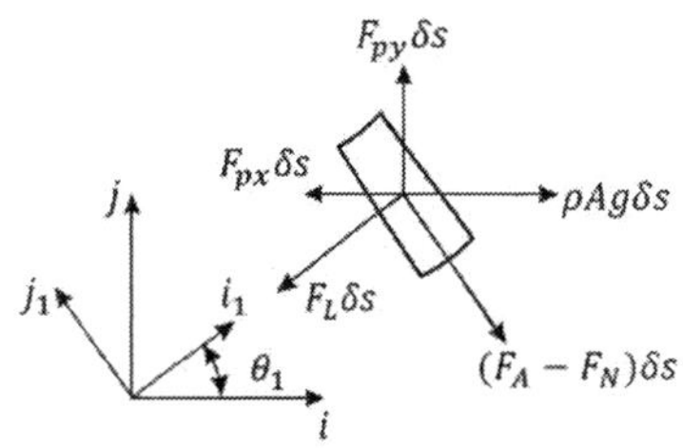

Figure 2: Fluid-related forces acting on an element of the cylinder δs.

By substituting Eqs. (2)-(4) and Eqs. (6)-(7) in Eq. (8), the following nonlinear equation of motion can be obtained via Hamilton's principle:

$$(m + M)\ddot{y} - 2MU\dot{y}'\left(1 - \frac{1}{4}y'^2\right) + MU^2y''(1 + 2y'^2) - \frac{3}{2}M\dot{y}y'(\dot{y}' - Uy'') - \frac{1}{2}\rho DU^2C_N\left(y' + \frac{1}{2}y'^3\right) + \frac{1}{2}\rho DU^2C_T(L - s)\left(y'' + \frac{3}{2}y'^2y''\right) - Ap(L)(y'' + y'^2y'') - \left(\frac{1}{2}\rho DU^2C_T\frac{D}{D_h} - mg + \rho gA\right)\left[y' + \frac{1}{2}y'^3 - (L - s)\left(y'' + \frac{3}{2}y'^2y''\right)\right] + EI(y'''' + 4y'y''y''' + y''^3 + y''''y'^2) + \frac{1}{2}\rho DC_N\dot{y}\int_0^s y'\dot{y}'\,ds + \frac{1}{2}\rho DU^2C_N\left(\frac{\dot{y}}{U} + \frac{1}{2}\frac{y'\dot{y}^2}{U^2} - \frac{1}{2}\frac{y'^2\dot{y}}{U} - \frac{\dot{y}^3}{2U^3}\right) + k\dot{y} - my''\int_s^L\int_0^s(\dot{y}'^2 + y'\ddot{y}')dsds + 2M(\dot{y}' -$$

$Uy'')\int_0^s y'\dot{y}'ds - My''\int_s^L(\ddot{y}y' - 2Uy'\dot{y}' + U^2y'y'')ds +$
$(m+M)y'\int_0^s(y'\ddot{y}' + \dot{y}'^2)ds - 3MUy'\int_0^s(y'\dot{y}'' +$
$y''\dot{y}')ds + y''\int_s^L\left\{Ap(L)y'y'' - \frac{1}{4}\rho DC_T\dot{y}^2\right\}ds -$
$$\frac{1}{2}\rho DU^2y''(C_T - C_N)\int_s^L\left(y'^2 - \frac{y'\dot{y}}{U}\right)ds = 0. \quad (10)$$

Boundary Conditions

The boundary conditions are derived to first order, i.e. correct to $O(\varepsilon)$, following the same procedure as in [3]. The virtual work by all the fluid dynamic forces acting on the free-end of the cylinder can be expressed as

$$\int_{t_1}^{t_2}\delta W dt = \delta\int_{t_1}^{t_2}\{-cos\theta_1\int_{L-l}^L F_{px}\,\delta u_L\delta s +$$
$$sin\theta_1\int_{L-l}^L F_{py}\,\delta u_L ds - \int_{L-l}^L F_L\,\delta u_L ds +$$
$$sin\theta_1\int_{L-l}^L F_{px}\,\delta u_N ds + cos\theta_1\int_{L-l}^L F_{py}\,\delta u_N ds -$$
$$f\int_{L-l}^L F_A\,\delta u_N ds + \int_{L-l}^L F_N\,\delta u_N ds\}\,dt, \quad (11)$$

where δu_L and δu_N are the virtual displacements in the longitudinal and transverse directions, as defined in [3], and l is the length of the end-piece.

Therefore, the boundary conditions can be expressed as follows:

$$-EIy''' + [fM(\ddot{y} - U\dot{y}') + m\ddot{y}]S_e + fMU(\dot{y} - Uy') +$$
$$(m - \rho A)gy'S_e - \frac{1}{2}\rho DU^2C_Thy'S_e + \frac{1}{2}\rho DUC_N(\dot{y} -$$
$$Uy')\bar{S}_e - \frac{1}{2}\rho D^2U^2C_by' = 0, \quad (12)$$

where f is a parameter; normally $0 \le f < 1$, $f \longrightarrow 1$ is taken for a well-streamlined end, while $f \longrightarrow 0$ for a blunt end; C_b is the base drag coefficient, $Ap(L) = \frac{1}{2}\rho D^2U^2C_b$, $S_e = \frac{1}{A}\int_{L-l}^L A(s)ds$, and $\bar{S}_e = \frac{1}{D}\int_{L-l}^L D(s)ds$. The reader is referred to [3, 5] for more details.

Methods of Analysis

The equation of motion and the boundary conditions can be expressed in dimensionless form, as follows, using the same dimensionless quantities presented in [3, 5], with $\xi = s/L$, $\eta = y/L$, $\tau = [EI/(m+\rho A)]^{1/2}t/L^2$, $u = (\rho A/EI)^{1/2}UL$, and $\kappa = kL^2/[EI(m+\rho A)]^{1/2}$:

$$F(\eta(\xi,\tau),u) + \delta(\xi - 1)\beta(\eta(\xi,\tau),u) = 0, \quad (13)$$

where the equation of motion is written as $F(\eta,u) = 0$ for simplicity, $\beta(\eta,u) = 0$ represents the boundary conditions, and $\delta(\xi - 1)$ is the Dirac delta function. Eq. (13) is discretized by Galerkin's technique with the cantilever beam eigenfunctions, $\phi_j(\xi)$, as comparison functions and with $q_j(\tau)$ as the corresponding generalized coordinate; thus,

$$\eta(\xi,\tau) = \sum_{j=1}^N \phi_j(\xi)q_j(\tau), \quad (14)$$

where N represents the number of modes of Galerkin scheme.

Substituting Eq. (14) into the final equation of motion, i.e. Eq. (13), multiplying by $\phi_i(\xi)$ and integrating from 0 to 1, leads to a set of ODEs.

NONLINEAR DYNAMICS

A system with the following parameters is considered: D = 0.0159 m, D_{ch} = 0.0508 m, EI = 0.401 N m^2, ρA = 2.43 × 10^{-4} kg m^{-1}, m = 0.213 kg m^{-1}, S_e = 0.00318 m, C_N = 0.0079, C_T = 0.0098, f = 0.8, and κ = 0.3. This corresponds to a physical system similar to the one considered in the theory and experiments of [5], where the surrounding fluid was air. The dissipation is also taken into account by replacing E by $[\bar{\alpha}(\partial/\partial t) + E]$ in the equation of motion. The ODEs obtained by Galerkin's technique were solved using AUTO [9] and MATLAB ODE solver (Mathworks, Inc.) as numerical tools. A bifurcation diagram is presented in Fig. 3, obtained using a six-mode Galerkin approximation with the first generalized coordinate q_1 being representative of the behaviour of the system. The model predicts a stable static solution at the original equilibrium state of the cylinder for u < 2.69; at u = 2.69, a branch point BP is predicted by AUTO which leads to divergence in the first mode via a supercritical pitchfork bifurcation; then a supercritical Hopf bifurcation point HB at $u \approx 3.1$ is predicted on each branch, which leads to flutter in the first mode, corresponding to stable periodic oscillations around the deformed, i.e. buckled, position, as shown in green in Fig. 3 (see also the phase-plane plot in Fig. 4(a)). At $u \approx 3.11$, the cylinder starts to oscillate around the origin (could not be followed by AUTO), as shown in the phase-plane plot in Fig. 4(b). The MATLAB ODE solver was used to follow the bifurcation diagram at $u \ge 3.11$ by plotting the maximum and minimum values of q_1 against u, after reaching the steady state, as shown in red in Fig. 3.

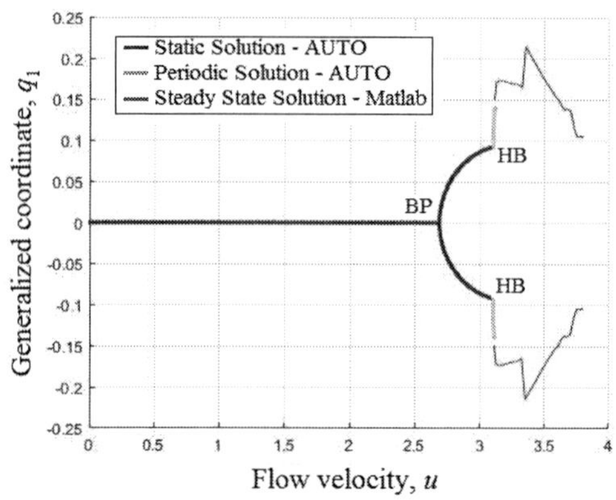

Figure 3: Bifurcation diagram.

The phase-plane plots at different flow velocities are presented in Fig. 4, which shows the transition between (a) the periodic oscillations around the buckled position and (b) the oscillations around the origin. By increasing the flow velocity to u = 3.3, new frequencies appear in the power spectrum (not shown in this paper) and the phase-plane plot becomes slightly irregular, as shown in Fig. 4(c). At higher flow velocities, 3.4 ≤ u ≤ 3.6, regular periodic motion resumes, as shown in Fig. 4(d)-(e). The phase-plane plot becomes more irregular with the possibility of having chaotic oscillations at u = 3.7, as shown in Fig. 4(f). At u > 3.8, the model fails to converge to any stable solution, perhaps because of the large deflection of the cylinder, exceeding the third-order accuracy assumption, $\mathcal{O}(\varepsilon^3)$, of this model.

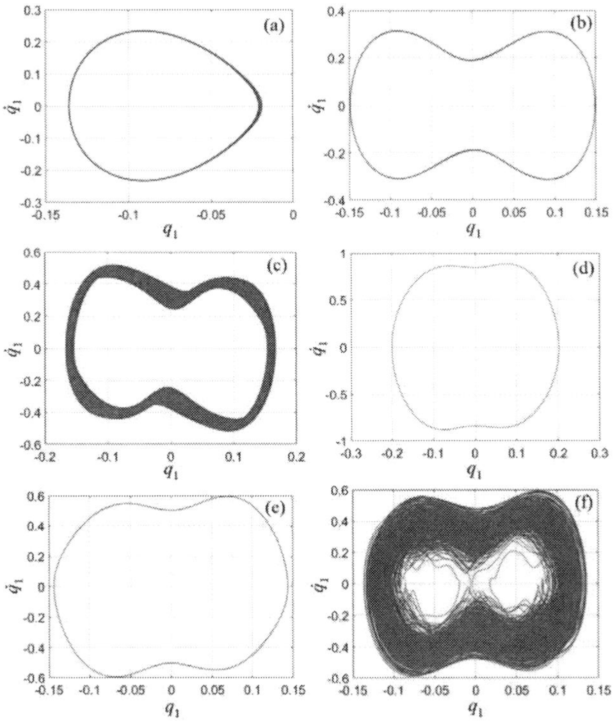

Figure 4: Phase-plane plots at: (a) $u = 3.1$, (b) $u = 3.11$, (c) $u = 3.3$, (d) $u = 3.4$, (e) $u = 3.6$, and (f) $u = 3.7$.

COMPARISON BETWEEN THEORY AND EXPERIMENTS

The values of the parameters used in the calculations were purposely chosen to allow comparison of the obtained results to the theoretical and experimental ones in [5]. The presented model is identical to the one derived in [5] in the linear limit, which means the weak damping of the first-mode is also found when applying linear analysis (not shown in this paper) at $0 < u < 2$. Thus, the experimentally observed oscillations at relatively low flow velocity, $U \approx 5$ m s^{-1}, can be justified as perturbation-induced vibrations. The nonlinear model predicts divergence as observed experimentally, but at higher flow velocity. It overestimates the critical flow velocity, $u_{cd} = 2.69$, which is almost the same as the linear model [5], while the experiments predict $u_{cd} = 1.7$. This may require further improvements in the presented model regarding the complexities of the flow at the free-end of the cantilever. There is no experimentally recorded data for the amplitude of the divergence observed; thus, the amplitude obtained by the nonlinear theory cannot be verified. Moreover, due to limitations in the maximum flow velocity available in [5], $u \approx 2$, the dynamical behaviour at higher flow velocities cannot be compared to the experiments.

CONCLUSIONS

This paper is the first nonlinear analytical study of a cantilevered cylinder subject to axial flow directed from the free-end to the clamped one. It provides fuller understanding of the dynamics of the system compared to linear theory; e.g. the amplitude and the frequency of oscillations can only be obtained by nonlinear theory. The nonlinear model predicts, for a cylinder in confined axial air-flow, fitted with a well-streamlined end-piece, divergence in the first mode at $u = 2.69$ and post-divergence flutter at $u \approx 3.1$. The model qualitatively agrees with experiments from the literature with the same system parameters.

ACKNOWLEDGMENTS

The financial support by SMRI, PRCI and NSERC is gratefully acknowledged. The first author is also grateful to McGill University for a MEDA scholarship.

REFERENCES

1. Hawthorne, W.R. 1961, "The early development of the Dracone flexible barge", *Proceedings of the Institution of Mechanical Engineers*, 175(1), pp. 52-83.
2. Païdoussis, M.P. 1966, "Dynamics of flexible slender cylinders in axial flow Part 1. Theory", *Journal of Fluid Mechanics*, 26(4), pp. 717-736.
3. Lopes, J.-L., Païdoussis, M.P., and Semler, C. 2002, "Linear and nonlinear dynamics of cantilevered cylinders in axial flow. Part 2: The equations of motion", *Journal of Fluids and Structures*, 16(6), pp. 715-737.
4. Semler, C., Lopes, J.L., Augu, N., Païdoussis, M.P. 2002, "Linear and nonlinear dynamics of cantilevered cylinders in axial flow. Part 3: Nonlinear dynamics", *Journal of Fluids and Structures*, 16(6), pp. 739-759.
5. Rinaldi, S. and Païdoussis, M.P. 2012, "Theory and experiments on the dynamics of a free-clamped cylinder in confined axial air-flow", *Journal of Fluids and Structures*, 28, pp. 167-179.
6. Païdoussis, M.P. 1973, "Dynamics of cylindrical structures subjected to axial flow", *Journal of Sound and Vibration*, 29(3), pp. 365-385.
7. Lighthill, M.J. 1960, "Note on the swimming of slender fish", *Journal of Fluid Mechanics*, 9(02), pp. 305-317.
8. Taylor, G. 1952, "Analysis of the swimming of long and narrow animals". *Proceedings of the Royal Society of London A: Mathematical, Physical and Engineering Sciences*, A214, pp. 158 - 183.
9. Doedel, E.J., Fairgrieve, T.F., Sandstede, B., Champneys, A.R., Kuznetsov, Y.A., and Wang, X. 2007, "AUTO-07P: Continuation and bifurcation software for ordinary differential equations".

Misalignment and Unbalance Faults Detection and Identification Using KNN Analysis

Ashkan Nejadpak, and Cai Xia Yang*

Mechanical Engineering Department, University of North Dakota
Grand Forks, ND
ashkan.nejadpak@und.edu; caixia.yang@engr.und.edu

Abstract—In this paper, K-Nearest Neighbor (KNN) analysis method is proposed for misalignment and unbalance faults detection and classification based on measured vibration data using four industrial accelerometers located at both bearing housings along radial (vertical and horizontal) directions while machinery is in operating. Measured signals were acquired by NI PCI-4498 and were processed and analyzed using MATLAB program. Three operation conditions, including normal, unbalance and misalignment were performed on a machine fault simulator. Features which distinguish different machine conditions were selected based on frequency spectrum distribution. The training dataset for KNN analysis was generated based on sixteen cases, and the machine health condition were grouped into three categories. Five measured data from unknown operation condition were used as testing data, results shown that proposed KNN analysis method achieve reliable classification of operation condition.

***Index Terms*— Fault detection and classification, k-nearest neighbor analysis, rotating machinery.**

I. Introduction

Rotating machinery are widely used in industry. Machine failures may result in costly downtime. Without effective fault diagnosis, it is impossible to predict the lead-time to failure. Therefore, conducting effective fault detection and identification are desirable and imperative. However, diagnosing faults in rotating machinery is often a labor-intensive and time-consuming practice. This makes conducting effective and efficient fault diagnosis a challenge for technicians and plant maintainers.

Misalignment and unbalance of rotating shafts are common cause of vibration in a rotating machinery [1-3]. Each fault generates unique signature in measured vibration data which corresponding to specific machine operating condition, therefore vibration analysis is widely used in machine condition monitoring.

By comparing the signals of a machine running in normal and faulty conditions, the detection of faults is possible. Fault detection is the common practice of identifying machine faults before they become a problem. Effective fault detection techniques can help industries reduce maintenance expense, increase equipment reliability and uptime, and reduce the usage of expensive test equipment [4-6]. Condition based predictive maintenance (PdM) provides reliable, cost-effective technique for monitoring and diagnosing machinery faults.

There have been several attempts to use supervised or unsupervised machine learning approaches for fault diagnoses of bearing with statistical and frequency based features [6-12]. Principal Component Analysis (APC) [6-8] and KNN method [9-12] have been used to fault classification. However, PCA makes assumption of the linearity of the data set, while KNN method is a nonlinear classifier and makes no assumption of the linearity of the data set. Using a simulation example generated from a nonlinear process model, He and Wang [9,10] demonstrated that nonlinearity make PCA less effective for fault detection, and KNN is capable for detecting all fault samples in their research.

Conducting controlled experiments help us to gain an in-depth understanding of different vibration signatures and to recognize the signatures of various machine faults. In this work, experimental investigation is conducted on a machinery fault simulator. KNN as a non-parametric supervised machine learning technique is performed for fault classification of misalignment and unbalance. Proposed KNN algorithm is based on Euclidean distance in same class, hence error due to improper selection of nearest neighbor K can be compensated.

KNN technique is used for identifying unbalance and misalignment faults from normal operation condition. The training dataset for KNN is generated through controlled predetermined amount of unbalanced mass and misalignment at different levels, as well as ideal situation. Several key features are selected based on results of spectrum analysis of measured vibration signals from ideal, unbalance and misaligned conditions. To improve classification accuracy, features selected from training dataset were weighted according to their influence levels on distinguishing fault types. The classification results shows that using proposed weighted KNN approach, improved effective and accurate of detecting and classifying for imbalanced and misaligned operation condition at early stage are achieved.

The remainder of this paper is organized as follows: fundamental background on fault detection and KNN method is presented in Section II, the experiment setup, and data collection process are provided in section III. Frequency analysis and KNN training dataset generation are presented in Section IV, the paper concludes with a summary recapping the main advantages of the proposed method and future work in Section V.

II. MISALIGNMENT, UNBALANCE AND KNN METHOD

Misalignment is a condition where the centerline of two shafts are not collinear. There are three basic types of misalignment: parallel offset, angular and combination of them. In frequency domain, misalignment show as peaks appear at 1x, 2x, 3x RPM due to the strain induced in the shaft. As severity of misalignment increase, more 1x RPM harmonics will show in the frequency spectra. In this research, two levels of parallel offset misalignment were artificial generated on machinery fault simulator.

Unbalance is the un-equal distribution of weight around the center of rotation. To study the unbalanced rotor condition, a screw with nut was mounted on different angles and locations of one rotor. This unbalanced mass generated unbalancing force induced an increased vibration, which shows as a large amplitude peak at 1x RPM. In the unbalanced condition, the disk is imposed to an increased force of:

$$F_U = m_U * e * \omega^2 \quad (1)$$

Where F_U denotes the unbalanced force, m_U denotes the unbalanced mass, e represents the distance from the unbalanced mass to the center of rotation, and ω is the running speed of the rotor. From Eq. (1), the amplitude of vibrations would increase with the operating speed of the motor, and the measured signal from the accelerometer and current transducer validate this statement. In frequency domain, unbalance will show a vibration frequency exactly at the rotational speed of the shaft with an amplitude in direct proportion to the centrifugal force.

KNN method classifies a new unknown sample by examining its distances to the *k* nearest neighboring training samples in the feature space [9-12]. This method has been used in statistical estimation and pattern recognition since 1970's, KNN stores all available cases as training data set, then classifies new cases based on a similarity measure by a majority vote of its neighbors, with the case being assigned to the class most common amongst its k nearest neighbors. Different distance functions are used for the similarity measure. In the paper, the Euclidean distance is used.

III. EXPERIMENT SETUP

In a condition monitoring system, acquire and process the data related to the condition of the machine is the most important stage. In this research, three operation conditions were studied in a series of experiments including normal condition, unbalanced load on one disk, and parallel misalignment of shaft with amount of 5 mils and 10 mils (1 mil is 0.001inches) were studied. Each of these experiments were conducted using the machinery fault simulator operating at 20Hz (1200 RPM) and 25Hz (1500 RPM). The setup consists of a three phase, 4 poles induction motor, an AC drive for the motor, an amplifier, an analog to digital converter, four accelerometers located at two bearing housing along vertical (Z axis) and radial (Y axis) directions. The vibration data is acquainted using National Instrument PCI-4498, analyzed by Sound and Vibration suit software and further graphed in MATLAB. Hanning window was selected to the acquired time-domain signal for Fast Fourier Transform (FFT). The image of experiment setup is shown in Figure 1.

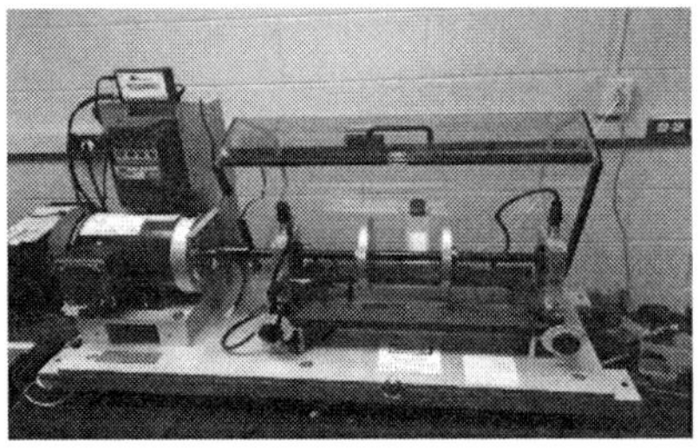

Fig. 1. Experiment setup

IV. FREQUENCY ANALYSIS AND KNN CLASSIFICATION

The goal of this study was to find evidence regarding vibration patterns associated with specific misalignment and unbalance faults. The severity and type of each fault condition can be assessed based on the amplitudes of the corresponding peaks as well as their respective locations on the frequency spectrum. Additionally, certain types of faults can be determined based on the location where data was recorded. Because some faults display a higher level of severity if the accelerometer was placed on various locations of the equipment. To determine the vibration spectra and signatures that occur due to different levels of unbalance and parallel misalignment, multiple controlled experiments were conducted on a machinery fault simulator.

a. Comparison of health and unbalance operation conditions

Two disks on the shaft are perfectly balanced, shafts are accurately aligned. The response of vibration under health/normal operating condition shows that the highest amplitude occurs at 1x RPM. Unbalanced rotors will have a large response amplitude peak at 1x RPM. Comparison results are shown in Figure 2.

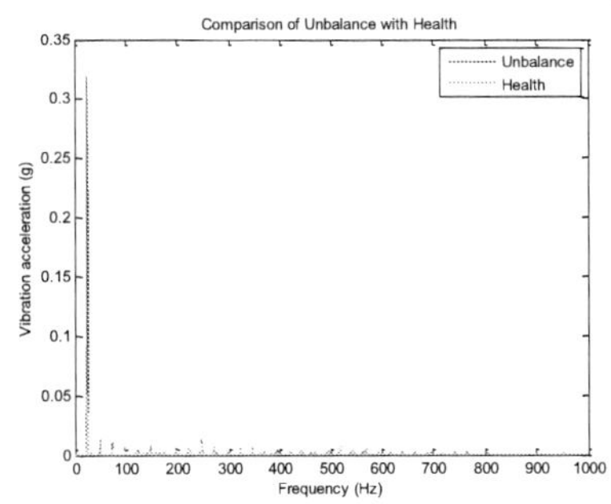

Fig. 2. Comparison of unbalance with health operation condition

Comparing unbalance condition with health operating condition in spectrum displacement, one can observe that the amplitudes of the acceleration at the exact operation speed (1X), three times of operation speed (3X) and seven times of operation speed (7X) have influence on classification on machine operation conditions. However, the most dominant difference happened at 1X, therefore, the amplitude value at 1X would be the relevant indicator to distinguish unbalance condition from health condition, the fault happened in system will be detected, and be identified as unbalance fault.

b. Comparison of health and misalignment condition

Experiment on parallel misalignments at amount of 5 mils and 10 mils were conducted, vibration signal measured by four accelerometers were recorded and analysis. Comparison results are shown in Figure 3.

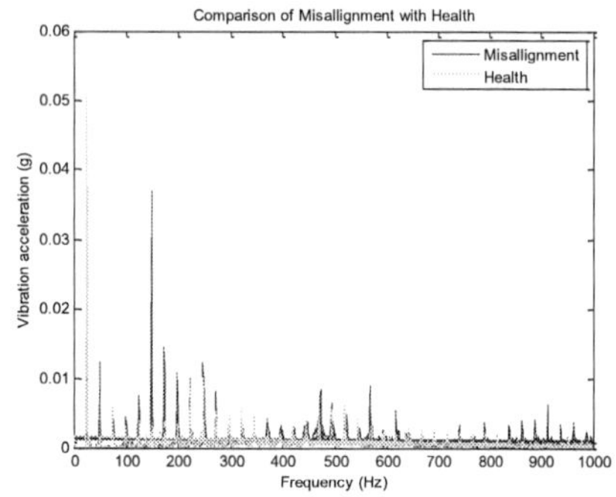

Fig. 3. Comparison of misalignment with health operation condition

Comparing misalignment condition with health condition in spectrum displacement, one can observe that the amplitudes of the acceleration at multi-times of operating speed (6X, 7X and 8X) have influence on classification on machine operation conditions for misalignment. However, the most dominant difference happened at 6X. Therefore, the amplitude value at 6X would be the relevant indicator to distinguish misalignment condition from health condition.

c. Comparison of three operation conditions

Comparison of the amplitude vs. frequency under three operation conditions are shown in Figure 4.

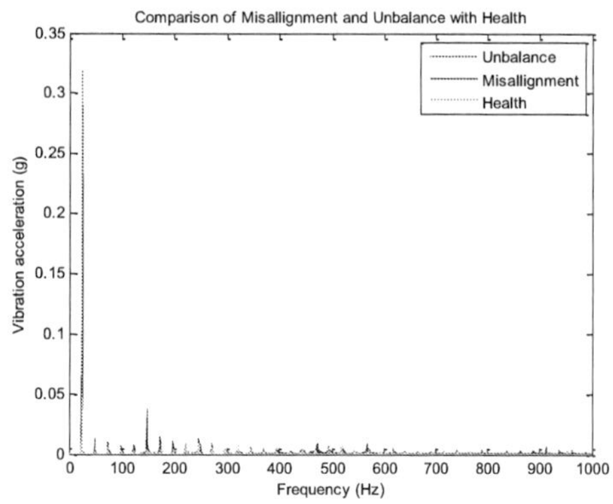

Fig. 4. Comparison of three operation conditions

d. Generation of training dataset

Features used for the classification or recognition task will play an important role in the choice of the classifier. For a specific set of features a classifier may perform better, but for another set, it might not provide the best solution.

The training dataset was generated from all features selected from each experiment. The amplitudes of acceleration at 1x RPM, 2x RPM, until 6x RPM are evaluated and imported to the training dataset. $k = 3$ is closed as the number of nearest neighbors, Euclidean distance of these three nearest neighbor to each sample from the training dataset were calculated. Given arbitrary testing sample, the proposed KNN algorithm provide fast and accurate classification based on majority vote rule. The classification results was shown in Table 1.

V. CONCLUSION AND FUTURE WORK

This paper presents a method for detecting and classifying unbalance and horizontal parallel misalignment faults for rotating machinery based on analysis of measured vibration signals using KNN technique from health condition. Operations under normal/health and fault conditions have been studied. It was shown that using the vibration profile, different level of faults can be detected and classified. The result of this evaluation will be used to classify the signature of fault on the machine vibration. Experimental results show potential improvement in classification with modified KNN and optimized feature selection. This can be used to acquire a complete machine vibration profile that may predict occurrence of damage in different parts of machine. This methodology can help to detect and identify the defective machineries at early stages.

Although KNN is not a new technique, to the best of our knowledge, there has not been reported in literature on detection and classification of unbalance and misalignment faults in rotary machine. Detection and identification of combined fault situations will be our future research direction. One limitation on misalignment detection is the impossibility to detect the severity of misalignment using vibration analysis, because the relation between the amount of misalignment and the level/amplitude of vibration has not been found.

ACKNOWLEDGMENT

The work reported in this paper was funded by ND EPSCoR New Faculty Start-up Award 43700-2725-UND0019805 and ND EPSCoR Advanced Undergraduate Research Award. The authors would like to thank Tristan Plante and Santosh Paudyal for conducted part of experiment.

REFERENCES

[1] SpectraQuest, Inc., Applied vibration analysis training manual & laboratory exercises, Richmond: SpectraQuest, Inc.

[2] S.S. Rao, Mechanical Vibrations, 6th Ed., Pearson – Prentice Hall, 2011

[3] Spectra Quest, InC., Machinery Fault Simulator Operation Manual.

[4] Q.H. Alsafasfeh, "Pattern Recognition for Fault Detection, Classification, and Localization in Electrical Power Systems," (2010). Dissertations. Paper 494.

[5] G. Verdier and A. Ferreira. Adaptive Mahalanobis distance and k-nearest neighbor rule for fault detection in semiconductor manufacturing. IEEE Trans. Semiconductor Manufacturing, (2010). Available:doi:10.1109/TSM.2010.2065531

[6] R. Isermann, "Fault deytection with Principal component analysis (PCA)" in Fault-Diagnosis System, Springer, 2005, 267-278.

[7] E. L. Russell, L. H. Chiang and R. D. Braatz, "Fault detection in industrial processes using canonical variate analysis and dynamic principal component analysis," Elsevier, 2000.

[8] T. Plante, L. Stanley, A. Nejadpak, and C. Yang, "Vibration Based Fault Detection using Principal Component Analysis," Proceedings of IEEE AUTOTESTCON 2016

[9] Q.P. He and J. Wang, "Fault detection using the k-nearest neighbor rule for semiconductor manufacturing processes," IEEE Trans. Semiconductor Manufacturing, vol.20, no.4, pp.345-354, Nov. 2007

[10] Q. He and J. Wang, "Fault Detection Using the k-Nearest Neighbor Rule for Semiconductor Manufacturing Processes," IEEE Transactions on Semiconductor Manufacturing (Volume: 20, Issue: 4, Nov. 2007) Page(s): 345 – 354, DOI: 10.1109/TSM.2007.907607

[11] J.M. Johnson and A. Yadav, "Fault detection and classification technique for HVDC transmission lines using KNN," International Conference on ICT for Sustainable Development ICT4SD, 2016

[12] D.H. Pandya, S.H. Upadhyay, S.P. Harsha, "Fault diagnosis of rolling element bearing with intrinsic mode function of acoustic emission data using APF-KNN," Expert Systems with Applications 40 (2013), 4137–4145.

Table 1 KNN for Fault Detection and Classification

Feature1	Feature2	Feature3	Feature4	Feature5	Feature6	Situation	Rank	Euclidean Distance	Category
0.0463	0.0041	0.0048	0.0028	0.0013	0.0022	Health	18	0.1893	Health
0.0461	0.0041	0.0045	0.0025	0.0009	0.0022	Health	12	0.1865	Health
0.0465	0.0041	0.0049	0.0021	0.0012	0.0021	Health	10	0.1858	Health
0.0465	0.0041	0.0051	0.0032	0.0015	0.0021	Health	5	0.1437	Health
0.0474	0.0042	0.0049	0.0023	0.001	0.0021	Health	10	0.1873	Health
0.0507	0.0041	0.0052	0.0025	0.0011	0.0022	Health	9	0.1859	Health
0.1517	0.0025	0.0047	0.0024	0.0015	0.0024	Unbalance	4	0.1192	Unbalance
0.0724	0.0034	0.0048	0.0022	0.0014	0.002	Unbalance	8	0.1878	Unbalance
0.2734	0.0047	0.0066	0.0035	0.0018	0.0026	Unbalance	7	0.1857	Unbalance
0.2048	0.001	0.0046	0.0023	0.0017	0.0022	Unbalance	3	0.0856	Unbalance
0.1129	0.0028	0.0039	0.0027	0.0013	0.0024	Unbalance	6	0.188	Unbalance
0.2423	0.0028	0.0056	0.0022	0.0021	0.0024	Unbalance	5	0.1857	Unbalance
0.0419	0.0065	0.0037	0.005	0.002	0.0067	Misalignment	2	0.0619	Misalignment
0.0432	0.0065	0.0035	0.0052	0.0012	0.0034	Misalignment	4	0.1886	Misalignment
0.0429	0.0061	0.0048	0.0053	0.0017	0.0045	Misalignment	3	0.1851	Misalignment
0.0411	0.0067	0.0037	0.0037	0.0029	0.0063	Misalignment	1	0.0422	Misalignment
0.0453	0.0033	0.0024	0.0041	0.0012	0.0033	Misalignment	2	0.1891	Misalignment
0.044	0.0052	0.0022	0.0034	0.002	0.0014	Misalignment	1	0.183	Misalignment

New measured vibration data available, need to classify the operation condition.

0.3401	0.0048	0.0059	0.0035	0.0029	0.0029	?

K	Category
1	Unbalance
2	Unbalance
3	Unbalance

Classification based on k = 3.

Operation condition is Unbalance

Proceedings of the 26th CANCAM
Victoria, British Columbia, Canada, May 29 - June 1, 2017

DYNAMIC POSITIONING OF AN AUV: A LYAPUNOV-BASED MODEL PREDICTIVE CONTROL APPROACH

Chao Shen
University of Victoria
Victoria, BC, Canada
shenchao@uvic.ca

Kunwu Zhang
University of Victoria
Victoria, BC, Canada
kunwu@uvic.ca

Yang Shi
University of Victoria
Victoria, BC, Canada
yshi@uvic.ca

Brad Buckham
University of Victoria
Victoria, BC, Canada
bbuckham@uvic.ca

ABSTRACT

This paper presents a Lyapunov-based model predictive control (LMPC) method for the dynamic positioning (DP) control of an autonomous underwater vehicle (AUV). As an optimization based control strategy, LMPC can make full use of the onboard thrust capability to obtain an enhanced DP control performance while guaranteeing the nominal stability of the closed-loop control system with an auxiliary Lyapunov-based nonlinear control law. The robustness of the DP control system is also improved by the receding horizon mechanism that is adopted in the LMPC strategy. Simulation studies based on the identified dynamic model of the Saab SeaEye Falcon open-frame ROV/AUV reveal the effectiveness of the proposed LMPC-based DP control.

Keywords: Underwater vehicle, model predictive control, dynamic positioning, Lyapunov-based control design, nonlinear control

INTRODUCTION

The autonomous underwater vehicle (AUV) has gained a growing interest in both industry and academia since it provides a safer and more efficient way of performing marine engineering projects. Dynamic positioning (DP), as one of the key functionalities to various AUV applications, has been extensively studied in the past several decades [1].

The most widely used DP controllers are of proportional-integral-derivative (PID) type for their easy implementation. Linear-quadratic-regulator (LQR) technique has also been applied to AUVs assuming that the linear approximation of the motion is sufficient for low-speed maneuvering applications such as DP. But obviously, DP controllers based on linear control technique only obtain local stability properties. Nonlinear control methods, such as Lyapunov-based backstepping and sliding mode control, therefore, are investigated for DP control of marine vehicles [2]. Recently, DP controller designs consider more practical issues. Observer-based controller design, control allocation and fault-tolerant capability have been discussed in the most recent work [1]. The system constraints is another type of practical factor which can be considered with the model predictive control (MPC) method [3]. But due to the implicit optimization procedure, the nominal stability is not easy to be proved within the standard MPC framework.

In this paper, we propose a novel LMPC-based DP control method for an AUV. Taking advantage of the existing Lyapunov-based DP control law, we impose a contraction constraint into the MPC formulation, which easily guarantees the closed-loop stability. The recursive feasibility is ensured by sufficiently small control gains of the existing control law, while the side-effect brought by the small control gains can be removed by the optimization procedure. Simulation results demonstrate much improved DP performance and robustness with the proposed LMPC method.

MODELING

The DP problem studied here considers the motion of a Saab SeaEye Falcon open-frame ROV/AUV in the local level plane. To facilitate the mathematical description of the AUV motion, two reference frames (shown in Figure 1) are introduced. The body-fixed reference frame (BRF) is affixed to the vehicle with the origin selected to be the center of gravity of the vehicle. Then the motion of the vehicle is considered as the motion of BRF with respect to the inertial reference frame (IRF) which reports the global information of the vehicle's position and orientation.

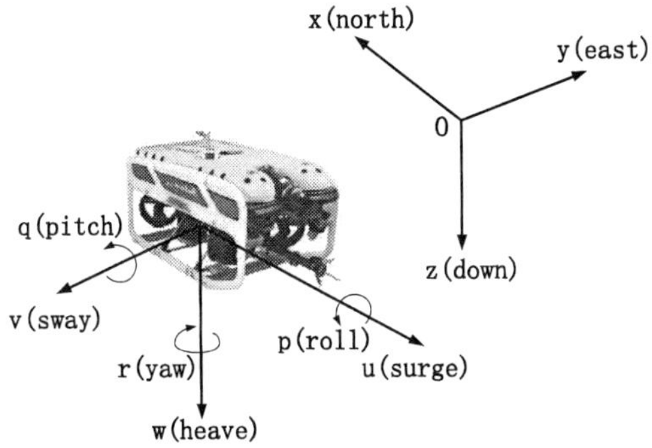

Figure 1: Reference frames, Falcon image courtesy of http://seaeye.com.

The 3 degree-of-freedom kinematic equations in the local level plane can be written in the following compact form [2]:

$$\dot{\boldsymbol{\eta}} = \mathbf{R}(\psi)\mathbf{v} \tag{1}$$

where $\boldsymbol{\eta} = [x, y, \psi]^{\mathrm{T}}$ is the position vector represented in the IRF, $\mathbf{v} = [u, v, r]^{\mathrm{T}}$ denotes the velocity vector represented in the BRF, and $\mathbf{R}(\psi)$ is the standard rotation matrix given by

$$\mathbf{R}(\psi) = \begin{bmatrix} \cos\psi & -\sin\psi & 0 \\ \sin\psi & \cos\psi & 0 \\ 0 & 0 & 1 \end{bmatrix} \tag{2}$$

The dynamic equations that analyze the cause of the motion are derived via Laws of Newton

$$\mathbf{M}\dot{\mathbf{v}} + \mathbf{C}(\mathbf{v})\mathbf{v} + \mathbf{D}(\mathbf{v})\mathbf{v} + \mathbf{g}(\boldsymbol{\eta}) = \boldsymbol{\tau} \tag{3}$$

where $\boldsymbol{\tau} = [F_u, F_v, F_r]^{\mathrm{T}}$ represents the thrust force. $\mathbf{M} = \mathbf{M}^{\mathrm{T}} > 0$ is the inertia matrix including added mass; $\mathbf{C}(\mathbf{v}) = -\mathbf{C}^{\mathrm{T}}(\mathbf{v})$ denotes the state dependent Coriolis and centripetal matrix which is skew-symmetric; $\mathbf{D}(\mathbf{v}) > 0$ is the damping matrix; and $\mathbf{g}(\boldsymbol{\eta})$ represents the restoring force.

Combining the kinematic equations and the dynamic equations, we can establish the control system model

$$\dot{\mathbf{x}} = \begin{bmatrix} \mathbf{R}(\psi)\mathbf{v} \\ \mathbf{M}^{-1}(\mathbf{u} - \mathbf{C}\mathbf{v} - \mathbf{D}\mathbf{v} - \mathbf{g}) \end{bmatrix} = \mathbf{f}(\mathbf{x}, \mathbf{u}) \tag{4}$$

where the state vector $\mathbf{x} = [x, y, \psi, u, v, r]^{\mathrm{T}}$ is consisted of the position and velocity of the vehicle, and the control vector $\mathbf{u} = [F_u, F_v, F_r]^{\mathrm{T}}$ is consisted of thrust force and moment.

LMPC-BASED DP CONTROL

The DP functionality refers to the capability of the vehicle to maintain its position and orientation by exclusively means of thrusters.

Consider the desired position and orientation denoted by $\boldsymbol{\eta}_d = [x_d, y_d, \psi_d]^{\mathrm{T}}$. The standard nonlinear MPC formulation can be easily established for the DP control

$$\begin{aligned} \min_{\hat{\mathbf{u}} \in S(\delta)} \quad & J = \textstyle\int_0^T \|\tilde{\mathbf{x}}(s)\|_Q^2 + \|\hat{\mathbf{u}}(s)\|_R^2 ds + \|\tilde{\mathbf{x}}(T)\|_P^2 \\ \text{s.t.} \quad & \dot{\hat{\mathbf{x}}}(s) = \mathbf{f}(\hat{\mathbf{x}}(s), \hat{\mathbf{u}}(s)) \\ & \hat{\mathbf{x}}(0) = \mathbf{x}(t_0) \\ & |\hat{\mathbf{u}}(s)| \leq \mathbf{u}_{\max} \end{aligned}$$

where $\hat{\mathbf{x}}(s)$ denotes the predicted state trajectory, and $\tilde{\mathbf{x}} = col(\tilde{\boldsymbol{\eta}}, \hat{\mathbf{v}})$ with $\tilde{\boldsymbol{\eta}} = \hat{\boldsymbol{\eta}} - \boldsymbol{\eta}_d$; $S(\delta)$ denotes the family of piecewise constant functions with sampling period δ. The optimal control signal $\hat{\mathbf{u}}^*(s)$ is determined by solving the above optimal control problem. $T = N\delta$ is the prediction horizon and Q, R, P are weighting matrices, positive definite.

However, it is well known that the optimality does not automatically ensures the closed-loop stability of the controlled system. Complex offline design procedure needs to be taken place [3]. Basically, the MPC design is based on local linearization in a neighborhood of equilibrium point. Therefore, only local stability property can be obtained. In view of that, many existing DP controllers are designed using the Lyapunov direct method and claim the global property in terms of closed-loop stability, we can take advantage of the existing DP controller to formulate the LMPC problem (P_0) that needs to be solved online:

$$\min_{\hat{\mathbf{u}} \in S(\delta)} J = \int_0^T \|\tilde{\mathbf{x}}(s)\|_Q^2 + \|\hat{\mathbf{u}}(s)\|_R^2 ds + \|\tilde{\mathbf{x}}(T)\|_P^2 \tag{5a}$$

$$\text{s.t.} \quad \dot{\hat{\mathbf{x}}}(s) = \mathbf{f}(\hat{\mathbf{x}}(s), \hat{\mathbf{u}}(s)) \tag{5b}$$

$$\hat{\mathbf{x}}(0) = \mathbf{x}(t_0) \tag{5c}$$

$$|\hat{\mathbf{u}}(s)| \leq \mathbf{u}_{\max} \tag{5d}$$

$$\tfrac{\partial V}{\partial \mathbf{x}} \mathbf{f}(\hat{\mathbf{x}}(0), \hat{\mathbf{u}}(0)) \leq \tfrac{\partial V}{\partial \mathbf{x}} \mathbf{f}(\hat{\mathbf{x}}(0), h(\hat{\mathbf{x}}(0))) \tag{5e}$$

where $h(\cdot)$ is the existing DP controller and $V(\cdot)$ is the corresponding Lyapunov function. Since the last contraction constraint (5e) is imposed, it can be shown that the LMPC inherits the stability properties of the state feedback control $h(\mathbf{x})$ [4]. Therefore, if $h(\mathbf{x})$ results in global stability of the closed-loop system, the LMPC control is also globally stable. Besides, due to the online optimization procedure, the LMPC controller naturally improves the DP control performance while respecting the system constraints.

The LMPC-based DP control will be implemented in a receding horizon fashion: (i) at the current sampling instant t_0, the optimal control problem (P_0) is solved given the system state $\mathbf{x}(t_0)$, and let $\hat{\mathbf{u}}^*(s)$ denote the optimal solution; (ii) the AUV uses $\hat{\mathbf{u}}^*(s)$ for only one sampling period: $\mathbf{u}(t) = \hat{\mathbf{u}}^*(s)$ for $s \in [0, \delta]$; (iii) at next sampling instant $t_0 + \delta$, new measurement of system state $\mathbf{x}(t_0 + \delta)$ is fed back, and (P_0) will be solved again using the new initial condition $\mathbf{x}(t_0 + \delta)$. Then repeat from (i).

The existing controller $h(\mathbf{x})$ under investigation is the universally used multivariable PID type, but any other Lyapunov-based nonlinear controller can, in principle, be used.

Consider the following nonlinear PD control:

$$h(\mathbf{x}) = \mathbf{g}(\boldsymbol{\eta}) - \mathbf{R}^{\mathrm{T}}(\psi)\boldsymbol{\tau}_{PD} \tag{6}$$

$$\boldsymbol{\tau}_{PD} = \mathbf{K}_p\tilde{\boldsymbol{\eta}} + \mathbf{K}_d\dot{\boldsymbol{\eta}} \tag{7}$$

where $\mathbf{K}_p$ and $\mathbf{K}_d$ are user specified control gains which are diagonal and positive definite matrices.

Then a Lyapunov function candidate is constructed in the following form

$$V = \frac{1}{2}\mathbf{v}^{\mathrm{T}}\mathbf{M}\mathbf{v} + \frac{1}{2}\tilde{\boldsymbol{\eta}}^{\mathrm{T}}\mathbf{K}_p\tilde{\boldsymbol{\eta}} \tag{8}$$

Calculate the time derivative of V along the trajectory of the closed-loop control system

$$\dot{V} = \mathbf{v}^{\mathrm{T}}\mathbf{M}\dot{\mathbf{v}} + \dot{\boldsymbol{\eta}}^{\mathrm{T}}\mathbf{K}_p\tilde{\boldsymbol{\eta}} \tag{9}$$

Substituting (1), (3), (6) and (7) into (11) gives

$$\dot{V} = -\mathbf{v}^{\mathrm{T}}[\mathbf{C}(\mathbf{v}) + \mathbf{D}(\mathbf{v}) + \mathbf{K}_d^*(\boldsymbol{\eta})]\mathbf{v} \tag{10}$$

where $\mathbf{K}_d^*(\boldsymbol{\eta}) = \mathbf{R}^{\mathrm{T}}(\psi)\mathbf{K}_d\mathbf{R}(\psi)$. Considering $\mathbf{v}^{\mathrm{T}}\mathbf{C}(\mathbf{v})\mathbf{v} = 0$ for all $\mathbf{v}$, we have

$$\dot{V} = -\mathbf{v}^{\mathrm{T}}[\mathbf{D}(\mathbf{v}) + \mathbf{K}_d^*(\boldsymbol{\eta})]\mathbf{v} \le 0 \tag{11}$$

since it can be easily shown $\mathbf{K}_d^*(\boldsymbol{\eta}) > 0$. Then by LaSalle's theorem [5], the closed-loop system with the nonlinear PD controller is globally asymptotically stable.

The detailed contraction constraint (5e) corresponding to the PD controller becomes

$$\begin{aligned}&\hat{\mathbf{v}}(0)^{\mathrm{T}}(\hat{\mathbf{u}}(0) - \mathbf{C}(\hat{\mathbf{v}}(0))\hat{\mathbf{v}}(0) - \mathbf{D}(\hat{\mathbf{v}}(0))\hat{\mathbf{v}}(0) - \mathbf{g}(\hat{\boldsymbol{\eta}}(0))\\ &+\mathbf{R}^{\mathrm{T}}(\hat{\psi}(0))\mathbf{K}_p\tilde{\boldsymbol{\eta}}(0)) \le -\hat{\mathbf{v}}(0)^{\mathrm{T}}[\mathbf{D}(\hat{\mathbf{v}}(0)) + \mathbf{K}_d^*(\hat{\boldsymbol{\eta}}(0))]\hat{\mathbf{v}}(0)\end{aligned} \tag{12}$$

For the recursive feasibility, the PD controller $h(\hat{\mathbf{x}})$ is always feasible for the LMPC problem (5) provided $|h(\hat{\mathbf{x}})| \le \mathbf{u}_{\max}$ can be satisfied. In (6), we should notice that the restoring force compensation $\mathbf{g}(\boldsymbol{\eta})$ is bounded and the gain of rotation matrix $|\mathbf{R}(\psi)|$ is also bounded. Therefore, if $|\boldsymbol{\tau}_{PD}|$ is small enough, $|h(\hat{\mathbf{x}})| \le \mathbf{u}_{\max}$ can be satisfied. From (7), it can be seen $\boldsymbol{\tau}_{PD}$ is determined by $\mathbf{K}_p\tilde{\boldsymbol{\eta}}$ and $\mathbf{K}_d\dot{\boldsymbol{\eta}}$. Obviously, $\tilde{\boldsymbol{\eta}}$ is bounded because $\tilde{\boldsymbol{\eta}}(0)$ is finite and the LMPC control is stable. And $\dot{\boldsymbol{\eta}}$ is also bounded because there exist damping $\mathbf{D}(\mathbf{v}) > 0$ and the thrust capability is finite. As we can choose arbitrarily small control gain matrices $\mathbf{K}_p$ and $\mathbf{K}_d$, $|h(\hat{\mathbf{x}})| \le \mathbf{u}_{\max}$ can be satisfied.

Note that although the asymptotical stability relies only on the positive definiteness of the control gain matrices $\mathbf{K}_p$ and $\mathbf{K}_d$, the control performance of the PD controller is determined by the magnitude of the control gains. A small control gain usually results in slow convergence, however, due to the optimization procedure in LMPC, the proposed controller can automatically make full use of the thrust capability to generate the best possible DP control performance with respect to the objective function (5a) even though we have selected a very small control gain for recursive feasibility consideration.

In addition, the receding horizon mechanism introduces feedback into the closed-loop system. Therefore, another merit of LMPC goes to the inherent robustness to uncertainties [6], which is attractive especially to marine control systems.

SIMULATION RESULTS

In this section, the proposed LMPC-based DP control is simulated on the dynamic model of Falcon which were identified through experimental data [7].

Without loss of generality, the desired position is chosen to be the origin of the IRF, i.e., $\boldsymbol{\eta}_d = [0, 0, 0]^{\mathrm{T}}$. The optimal control problem (5) is discretized and then solved by the sequential quadratic programming (SQP) method.

For the LMPC controller, we choose the parameters as follows: the sampling period $\delta = 0.1$ [sec], prediction horizon $T = 5\delta$, and the weighting matrices $Q = \mathrm{diag}(10^5, 10^5, 10^4, 10^3, 10^3, 10^3)$, $R = \mathrm{diag}(10^{-4}, 10^{-4}, 10^{-4}, 10^{-4})$ and $P = \mathrm{diag}(10^3, 10^3, 10^2, 10^1, 10^1, 10^1)$. The nonlinear PD control gain $\mathbf{K}_p = \mathbf{K}_d = \mathrm{diag}(1, 1, 1)$.

The AUV trajectories are plotted in Figure 2 and Figure 3, which shows that the LMPC-based DP control converges much faster than the PD control. This is because we have selected very small control gain matrices $\mathbf{K}_p$ and $\mathbf{K}_d$ for recursive feasibility consideration. The simulation results demonstrate the superior DP control performance brought by the online optimization.

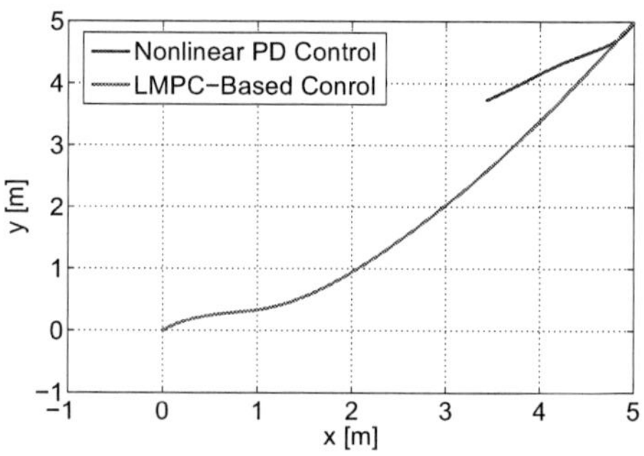

Figure 2: The AUV trajectory in xoy plane.

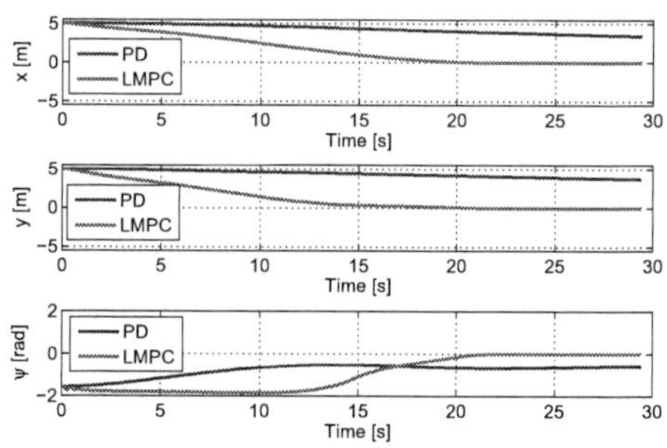

Figure 3: The state trajectories.

The thrust force and moment are shown in Figure 4. As we can see, the magnitude of each control input stays within the permitted range.

Also the robustness of the LMPC-based DP control has been investigated. In another simulation scenario, we assume that the system model is subject to 20% model error and there

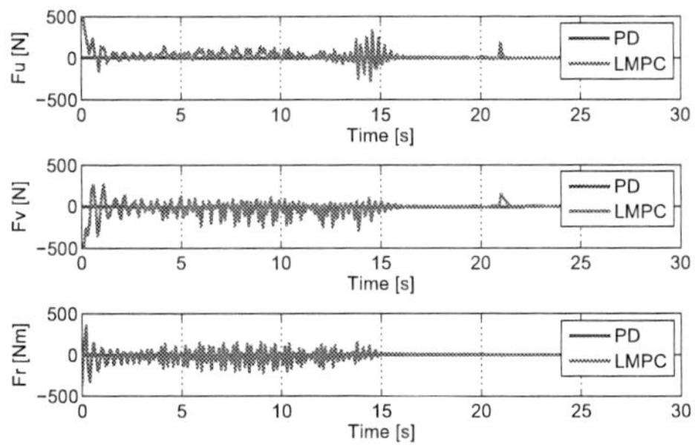

Figure 4: The control input signals.

is an irrotational ocean current with magnitude of $[4, 4, 0]^{\mathrm{T}}$ exerting on the vehicle all the time.

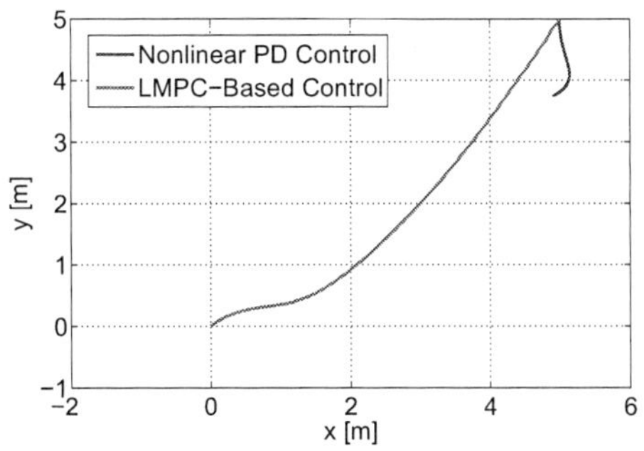

Figure 5: The AUV trajectory in xoy plane with disturbance.

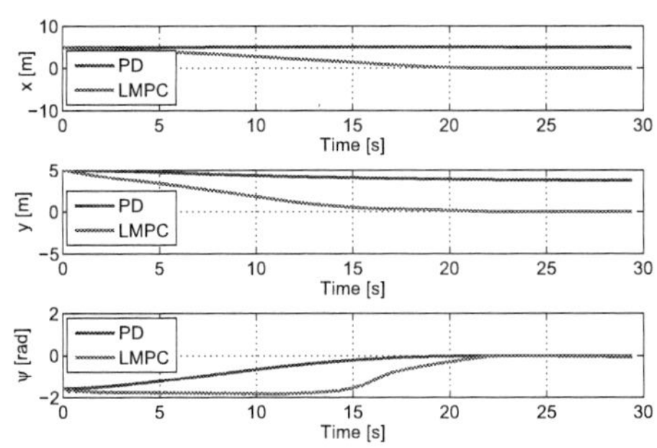

Figure 6: The state trajectories with disturbance.

From the simulation results shown in Figure 5 - Figure 7, we find that the LMPC-based control still gets converged to the desired position, while it is proved that the PD control only guarantees uniform ultimate boundedness (UUB) of the tracking error [1]. Therefore, the robustness of DP control is enhanced by the receding horizon mechanism as expected.

CONCLUSIONS

We have presented, in this paper, the LMPC method for the DP control of an AUV. The introduced online optimization improved the existing nonlinear PD controller significantly in terms of control performance and robustness. Meanwhile, the closed-loop stability and recursive feasibility of the LMPC can be guaranteed.

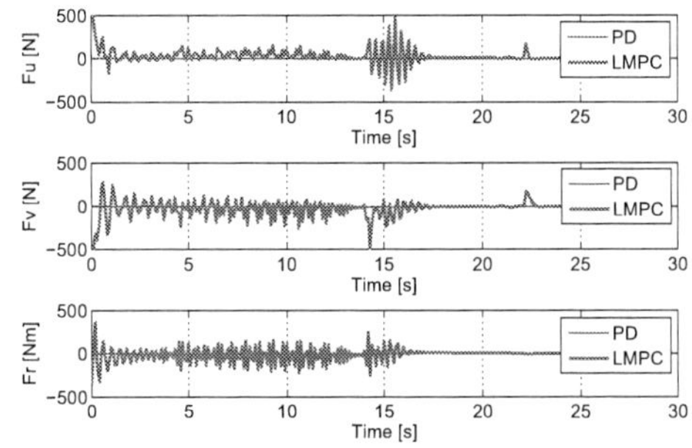

Figure 7: The control input signals with disturbance.

ACKNOWLEDGMENTS

This work was supported by the Natural Sciences and Engineering Research Council of Canada (NSERC) and the National Natural Science Foundation of China (61473116).

REFERENCES

[1] Sørensen, A. 2011, "A survey of dynamic positioning control systems," *Annual Reviews in Control*, Vol(35), pp. 123-136.

[2] Fossen, T. 2002, *Marine Control Systems: Guidance, Navigation and Control of Ships, Rigs and Underwater Vehicles*, Trondheim, Norway: Marine Cybernetics.

[3] Mayne, D., Rawlings, J., Rao, C., and Scokaert, P., 2000, "Constrained model predictive control: Stability and optimality," *Automatica*, Vol(36), pp. 789-814.

[4] Christofides, P., Liu, J., and Munoz de la Pena, D. 2011, *Networked and Distributed Predictive Control: Methods and Nonlinear Process Network Applications*, Springer-Verlag, London, England.

[5] Slotine, J., and Li, W. 1991, *Applied Nonlinear Control*, Prentice-Hall, NJ, United States.

[6] Pannocchia, G., Rawlings, J., and Wright, S. 2011, "Inherently robust suboptimal nonlinear MPC: Theory and application," *2011 50th IEEE Conference on Decision and Control and European Control Conference*, Orlando, FL, USA, pp. 3398-3403.

[7] Proctor, A. 2015, "Semi-automous guidance and control of a Saab SeaEye Falcon ROV," Ph.D. dissertation, University of Victoria.

Proceedings of the 26th CANCAM
Victoria, British Columbia, Canada, May 29 – June 1, 2017

A COROTATIONAL FORMULATION FOR GEOMETRICALLY NONLINEAR ANALYSIS OF 2D BEAMS AND FRAMES

Ahmed A. El-Erian
Department of Engineering Mathematics and Physics
Alexandria University
Alexandria 21544, Egypt
w2o@hotmail.com

Walied I. Hussien
Department of Engineering Mathematics and Physics
Alexandria University
Alexandria 21544, Egypt
walied_hussien@yahoo.com

Hesham A. Elkaranshawy
Department of Engineering Mathematics and Physics
Alexandria University
Alexandria 21544, Egypt
hesham_elk@alexu.edu.eg

ABSTRACT

A corotational finite element formulation for flexible 2-d beam element with geometrically nonlinear behavior is presented. The formulation separates the rigid body motion from the pure deformation which is always small relative to the corotational element frame. The axial displacement is evaluated using arc length integral formula. The stiffness matrices and the elastic force vector due to deformation are firstly computed in the local corotational coordinate system before transformed to the global system. The nonlinear equilibrium equation is developed using Hamilton's principle and is defined in the global coordinate system. A MATLAB code is developed for the numerical solution of the nonlinear equation. This code is developed using iterative method based on the conventional Newton-Raphson method without incremental loading. Two examples for flexible beams and frames with large displacements are presented to show the effectiveness and accuracy of the proposed method.

KEYWORDS: Corotational formulation, geometric nonlinearity, large displacement analysis, flexible 2d beam element.

1. INTRODUCTION

Over the past years the geometrically nonlinear finite element analysis of flexible beams and frames with large displacements draw the attention of many authors [1-18]. In literature, three approaches have basically been employed to describe the kinematics of a flexible beam element. They are the total Lagrangian, updated Lagrangian, and the corotational formulations. In the total Lagrangian formulation the system equations are defined in terms of a fixed global coordinate system. Consequently, these equations could be very complicated. This formulation is used by many authors like [1-3]. In the updated Lagrangian formulation the system equations are defined in terms of a coordinate system which is updated with the last accepted solution. However, this coordinate system is kept fixed during the solution steps [4-6]. The displacement from the current configuration to the last equilibrium configuration could be very large and there is an inconsistency domain problem. To avoid this problem, the corotational kinematical formulation had been developed. This formulation has been developed, improved and applied by many researchers, see for example [7- 17].

The corotational formulation is based on small deflection beam theory with the inclusion of the effect of axial force. It has two coordinate systems; the fixed global coordinate system and the element corotational local coordinate system. The corotational local frame rotates and translates with each element but does not deform with it. Therefore, this formulation easily separates the rigid body motion from the pure deformation which is always small relative to the corotational element frame.

In the present work, a corotational formulation is developed. The stiffness matrices and the internal elastic force vector due to deformation are firstly computed in the local corotational coordinate system. Subsequently, they are transformed to the fixed global coordinate system using standard procedure then assembled to the structure overall stiffness matrix and overall internal elastic force vector, respectively. Subsequently, all the terms of the nonlinear equilibrium equation of the structure are defined and solved in the fixed global coordinate system. The axial displacement is evaluated using the arc length integral formula. This integral is computed using Gauss integration scheme with five Gauss points. An iterative method based on the conventional Newton-Raphson method without incremental loading is used to obtain the numerical solution of the nonlinear equilibrium equation. A MATLAB code is developed for this purpose. The beam element kinematics and the displacement interpolation are defined in Section 2. The strain energy and the stiffness matrix are formulated in Section 3. Hamilton's principle is used to derive the nonlinear differential equation of the equilibrium of the system in Section 4. Section 5 contains the numerical algorithm. The effectiveness and the accuracy of the present method are demonstrated through working out two problems with large displacements in Section 6. The conclusions are presented in Section 7.

2. KINEMATICS OF A BEAM ELEMENT USING COROTATIONAL FINITE ELEMEN FORMULATION

2.1 COORDINATE SYSTEMS

Two coordinate systems used for this formulation. A global coordinate system associated with the fixed global frame (X, Z), and a local coordinate system associated with the corotational local frame (x_i, z_i). This corotational frame is updated, and attached to each beam element as shown in Fig. 1. It translates and rotates with the beam element but does not deform with it. After discretization of the structure into finite elements, the ith beam element can be defined with two end nodes (n=1, 2). Every node has three degrees of freedom and is defined in both coordinate systems. The displacement vector of the ith element in the fixed global coordinate system is

$$\mathbf{D}_i = [U_1 \quad W_1 \quad \theta_1 \quad U_2 \quad W_2 \quad \theta_2]^T \tag{1}$$

where U_n (n=1, 2) and W_n (n=1, 2) are the displacement components in X and Z directions, respectively, and θ_n (n=1, 2) are the counterclockwise rotations. The corresponding displacement vector in the element local coordinate system is

$$\mathbf{d}_i = [u_1 \quad w_1 \quad \overline{\theta_1} \quad u_2 \quad w_2 \quad \overline{\theta_2}]^T \tag{2}$$

where u_n (n=1, 2) and w_n (n=1, 2) are the displacement components in x_i and z_i directions, respectively, and $\overline{\theta_n}$(n=1, 2) are the counterclockwise rotations. The initial and current element chord lengths denoted as Lo and L are shown in Fig. 1. The initial and current locations of the corotational frame are specified by the two angles α_o, α that are measured counterclockwise from the fixed global X axis direction to the local xi axis. α_o and α are given by

$$\sin\alpha_o = (Z_2 - Z_1) / L_o \tag{3}$$

$$\sin\alpha = (Z_2 + W_2 - Z_1 - W_1) / L \tag{4}$$

The relation between $\theta_n, \overline{\theta_n}$ of the ith element is

$$\overline{\theta_n} = \theta_n - \beta \tag{5}$$

where β is the rigid body rotation and is given by

$$\beta = \alpha - \alpha_o. \tag{6}$$

The relation between the displacements vectors $\mathbf{D}_i$, $\mathbf{d}_i$ and the rigid body displacement of the corotational coordinate system $\mathbf{d}_i^r$ takes the form

$$\mathbf{D}_i = \mathbf{d}_i^r + \mathbf{T}\ \mathbf{d}_i. \tag{7}$$

where T is the ith beam element transformation matrix, expressed by

$$\mathbf{T} = \begin{bmatrix} \mathbf{A} & 0 \\ 0 & \mathbf{A} \end{bmatrix}, \ \mathbf{A} = \begin{bmatrix} \cos\alpha & -\sin\alpha & 0 \\ \sin\alpha & \cos\alpha & 0 \\ 0 & 0 & 1 \end{bmatrix}. \tag{8}$$

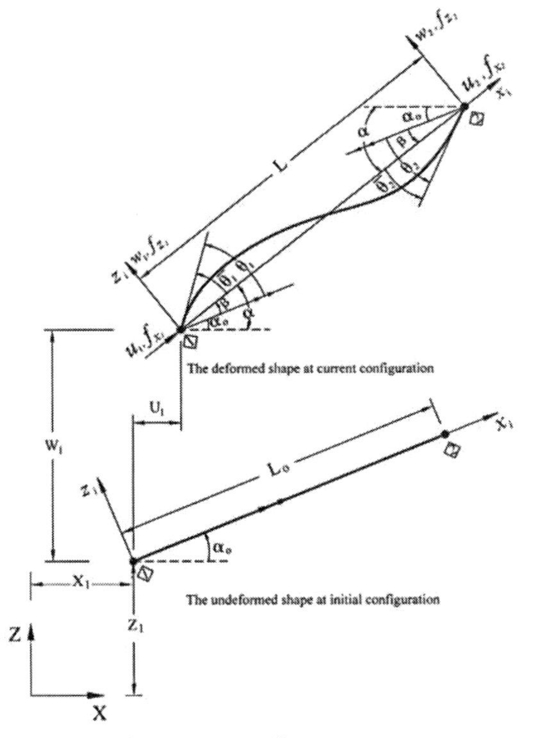

Figure 1: Coordinate systems.

2.2 DISPLACEMENT INTERPOLATION

The classical Hermitian shape functions are used to relate the element deformation vector $\boldsymbol{v}_i$ to the element nodal displacement vector $\mathbf{d}_i$ as

$$\boldsymbol{v}_i = \mathbf{N}\mathbf{d}_i. \tag{9}$$

The shape functions $\mathbf{N}$ components of ith beam element are given by

$$N_1 = 1 - \frac{1}{2}(1+\xi), \qquad N_2 = \frac{1}{4}(1-\xi)^2(2+\xi)$$

$$N_3 = \frac{L}{8}(1-\xi^2)(1-\xi), \ N_4 = \frac{1}{2}(1+\xi)$$

$$N_5 = \frac{1}{4}(1+\xi)^2(2-\xi), \quad N_6 = \frac{L}{8}(1+\xi^2)(1+\xi) \tag{10}$$

where ξ is given by

$$\xi = -1 + \frac{2x}{L}. \tag{11}$$

The transverse displacement w of the ith beam element takes the form

$$w(x) = N_2 w_1 + N_3\overline{\theta_1} + N_5 w_2 + N_6\overline{\theta_2}. \tag{12}$$

In the corotational formulation the frame (x_i, z_i) is updated each iteration. The x_i axis is always passing through the two end nodes of the element as shown in Fig.1 Therefore, the displacements components w_n (n=1, 2) are zeros. Thus Eq. (12) can be written as

$$w\ (x) = N_3\overline{\theta_1} + N_6\overline{\theta_2}. \tag{13}$$

One can get the approximate current arc length S_i

$$S_i = \frac{L}{2}\int_{-1}^{1}\sqrt{(\acute{w}^2+1)}\ d\xi \tag{14}$$

where $\acute{w}$ is the derivative of the function $w(x)$ with respect to x_i. This integral has evaluated using Gauss integration scheme with five Gauss points. The elongation of the ith beam element in x_i direction is

$$e_i = S_i - L_o. \tag{15}$$

The displacements w_n (n=1, 2) are zeros as mentioned before and the axial displacement of the first node is zero while that of second node is e_i. Therefore, $\mathbf{d}_i$ in Eq. (2) becomes

$$\mathbf{d}_i = [0 \quad 0 \quad \overline{\theta_1} \quad e_i \quad 0 \quad \overline{\theta_2}]^T. \tag{16}$$

3. THE STRAIN ENERGY AND STIFFNESS MATRIX

Considering isotropic elastic materials, the relation between the stress vector $\boldsymbol{\sigma}_i$ and the strain vector $\boldsymbol{\epsilon}_i$ for the ith beam element is given by

$$\boldsymbol{\sigma}_i = \mathbf{E}_i \, \boldsymbol{\epsilon}_i \quad (17)$$

where $\mathbf{E}_i$ is the symmetric matrix of the elastic coefficients. One can write the strain vector as

$$\boldsymbol{\epsilon}_i = \mathbf{D}_f \, \boldsymbol{v}_i \quad (18)$$

where $\mathbf{D}_f$ is the differential operator matrix. Substituting Eq. (9) into Eq. (18), the strain vector can be expressed by

$$\boldsymbol{\epsilon}_i = \mathbf{D}_f \, \mathbf{N}_i \, \mathbf{d}_i \, . \quad (19)$$

Combining Eqs. (17) and (19), the stress vector can be rewritten as

$$\boldsymbol{\sigma}_i = \mathbf{E}_i \, \mathbf{D}_f \, \mathbf{N}_i \, \boldsymbol{d}_i \, . \quad (20)$$

The strain energy for the ith beam element Π_i is given by

$$\Pi_i = \frac{1}{2}\int_{V_i} \boldsymbol{\sigma}_i^T \, \boldsymbol{\epsilon}_i \, dV_i \quad (21)$$

where V_i is the volume. Substituting Eqs. (19) and (20) into Eq. (21), the strain energy can be expressed by

$$\Pi_i = \frac{1}{2}\int_{V_i} (\mathbf{E}_i \, \mathbf{D}_f \, \mathbf{N}_i \, \mathbf{d}_i)^T \, \mathbf{D}_f \, \mathbf{N}_i \, \mathbf{d}_i \, dV_i \, . \quad (22)$$

The element local displacements vector $\mathbf{d}_i$ is independent of the volume V_i. Subsequently, Eq. (22) can be simplified to the following form:

$$\Pi_i = \frac{1}{2} \, \mathbf{d}_i^T \, \mathbf{k}_i \, \mathbf{d}_i \quad (23)$$

where $\mathbf{k}_i$ is the symmetric element stiffness matrix in the local coordinate system, defined as

$$\mathbf{k}_i = \int_{V_i} \mathbf{B}^T \mathbf{E}_i \, \mathbf{B} \quad (24)$$

and $\mathbf{B}$ is defined as

$$\mathbf{B} = \mathbf{D}_f \, \mathbf{N}_i . \quad (25)$$

Substituting Eq. (7) into Eq. (23), one can write Eq. (23) in the following form:

$$\Pi_i = \frac{1}{2} \, (\mathbf{T}^T(\mathbf{D}_i - \boldsymbol{d}_i^r))^T \, \mathbf{k}_i \, \mathbf{T}^T(\mathbf{D}_i - \boldsymbol{d}_i^r)$$
$$= \frac{1}{2}(\mathbf{D}_i - \boldsymbol{d}_i^r)^T \, \mathbf{K}_i \, (\mathbf{D}_i - \boldsymbol{d}_i^r) \quad (26)$$

where $\mathbf{K}_i$ is the symmetric element stiffness matrix in the fixed global coordinate system, defined as

$$\mathbf{K}_i = \mathbf{T} \, \mathbf{k}_i \, \mathbf{T}^T . \quad (27)$$

For the beam element used in this research work, the stiffness matrix $\mathbf{k}_i$ can be expressed as

$$\mathbf{k}_i = \mathbf{k}_1 + \mathbf{k}_2 \quad (28)$$

where $\mathbf{k}_1$ is the axial and bending stiffness matrix and $\mathbf{k}_2$ is the geometric stiffness matrix for the ith beam element.

4. THE NONLINEAR EQUILIBRIUM EQUATION

Hamilton's principle is used to derive the equilibium equation of the system. Mathematically, Hamilton's principle takes the form

$$\delta \int_{t_1}^{t_2} L_g \, dt = 0 \quad (29)$$

where L_g is the Lagrangian functional. L_g is defined as

$$L_g = K_e - P_o + W_i \quad (30)$$

where P_o is the potential energy, K_e is the kinetic energy and W_i is the work done by the external forces. In the structural applications, the potential energy P_o is the elastic strain energy Π_i given in Eq. (26). W_i is defined as

$$W_i = (\mathbf{D}_i - \boldsymbol{d}_i^r)^T \boldsymbol{F}_i^P \quad (31)$$

where $\boldsymbol{F}_i^P$ is the vector of the external applied forces on the ith beam element in the fixed global coordinate system. The variation and the integration operators are interchangeable. Therefore, one can write Eq. (29) in the following form

$$\int_{t_1}^{t_2} \delta \, (K_e - \Pi_i + W_i) \, dt = 0. \quad (32)$$

The strain energy term in Eq. (32) can be written as

$$\delta \, \Pi_i = \delta(\tfrac{1}{2}(\mathbf{D}_i - \boldsymbol{d}_i^r)^T \, \mathbf{K}_i \, (\mathbf{D}_i - \boldsymbol{d}_i^r))$$
$$= \delta(\mathbf{D}_i - \boldsymbol{d}_i^r)^T \, \mathbf{K}_i \, (\mathbf{D}_i - \boldsymbol{d}_i^r). \quad (33)$$

For static systems, the kinetic energy K_e in Eq. (32) vanished. The only way for the integration in Eq. (32) to be zero is the vanishing of the integrand itself. Therefore, Eq. (32) can be rewritten as

$$-\delta(\mathbf{D}_i - \boldsymbol{d}_i^r)^T \, \mathbf{K}_i \, (\mathbf{D}_i - \boldsymbol{d}_i^r) + \delta(\mathbf{D}_i - \boldsymbol{d}_i^r)^T \boldsymbol{F}_i^P = 0$$
$$\delta(\mathbf{D}_i - \boldsymbol{d}_i^r)^T \, (- \, \mathbf{K}_i \, (\mathbf{D}_i - \boldsymbol{d}_i^r) + \boldsymbol{F}_i^P) = 0. \quad (34)$$

Since the displacement variation is arbitrary, then

$$\mathbf{K}_i \, (\mathbf{D}_i - \boldsymbol{d}_i^r) - \boldsymbol{F}_i^P = 0 \quad (35)$$

According to Eqs. (7) and (27), one can write

$$\boldsymbol{F}_i^e = \mathbf{K}_i \, (\mathbf{D}_i - \boldsymbol{d}_i^r) = \mathbf{T} \, \mathbf{k}_i \, \mathbf{d}_i = \mathbf{T} \, \boldsymbol{f}_i^e \quad (36)$$

where $\boldsymbol{F}_i^e$ and $\boldsymbol{f}_i^e$ are the internal elastic force vector in the fixed global and the local coordinate systems, respectively. Substituting Eq. (36) in Eq. (35) gives

$$\boldsymbol{F}_i^e - \boldsymbol{F}_i^P = 0 \quad (37)$$

Eq. (37) is the geometric nonlinear equilibrium equation of the ith beam element. Assembling the element force vectors leads to the nonlinear equilibrium equation of the overall structure which takes the form

$$\boldsymbol{F}^e - \boldsymbol{F}^P = \mathbf{0}. \quad (38)$$

where $\boldsymbol{F}^P$ is the vector of the external applied forces of the entire structure and $\boldsymbol{F}^e$ is the internal elastic force vector of the entire structure. Both $\boldsymbol{F}^P$ and $\boldsymbol{F}^e$ are defined in the fixed global coordinate system.

5. NUMERICAL ALGORITHM

The conventional Newton-Raphson iterative method is employed to solve the nonlinear equilibrium equation of the overall structure, Eq. (38). Equation (38) can be written as

$$\boldsymbol{\varphi} = \boldsymbol{F}^e - \boldsymbol{F}^P = \mathbf{0} \quad (39)$$

where φ is the out of balance force. The iteration equilibrium convergence criterion is determined by

$$\|\boldsymbol{\varphi}\| \leq e_r \|\boldsymbol{\varphi}_f\| \quad (40)$$

where $\| \; \|$ is the Euclidean norm, $\boldsymbol{\varphi}_f$ is the reference out of balance force and e_r is the error tolerance. $\boldsymbol{\varphi}_f$ is considered to be the out of balance force obtained in the first iteration. The error tolerance e_r is chosen according to the problem. A MATLAB code is developed for the numerical solution. At the beginning, null displacement vectors are assumed. The iteration steps start with computing $\mathbf{k}_i$ and $\boldsymbol{f}_i^e$ for each element using Eqs. (28) and (36), respectively. Then, $\mathbf{K}_i$ and $\boldsymbol{F}_i^e$ are computed according to Eqs. (27) and (36), respectively. As a result of the assembly, $\boldsymbol{K}$ and $\boldsymbol{F}^e$ are obtained where $\boldsymbol{K}$ is the stiffness matrix of the entire structure. The out of balance force is obtained using Eq. (39). Iterations stop, if the convergence criterion satisfies the condition in Eq. (40). Otherwise, a displacement corrector $\mathbf{C}_R$ has to be calculated according to

$$\mathbf{C}_R = -\boldsymbol{K}^{-1}\,\boldsymbol{\varphi} \tag{41}$$

Before starting a new iteration, $\mathbf{C}_\mathrm{R}$ is added to the displacement vector of the entire structure, $\mathbf{D}$. Consequently, $\mathbf{D}_i$ can be extracted for each element. Using Eqs. (4)- (6) and (16), $\mathbf{d}_i$ can be determined. It has to be noticed that the stiffness matrices and elastic force vector are updated each iteration.

6. NUMERICAL EXAMPLES

6.1 CANTILEVER BEAM SUBJECTED TO A CONCENTRATED VERTICAL END LOAD

The first numerical example is a cantilever beam of length L and flexural rigidity EI, subjected to a vertical concentrated point load P at the free end as shown in Fig. 2.

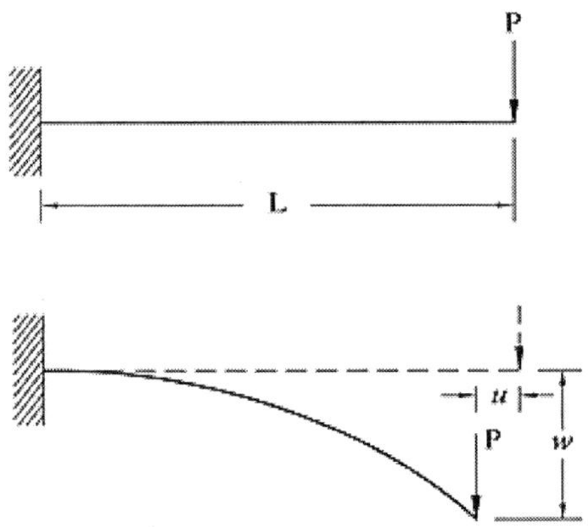

Figure 2: Cantilever beam with an end vertical point load.

In this example, five beam elements are used to obtain the solution and the error tolerance e_r is chosen to be 10^{-6}. The present results are compared with the results of the total Lagrangian formulation [2] and the elliptic integral numerical solutions [18] in Fig. 3. Where *u*/L, *w*/L are the normalized displacements and PL^2/EI is the normalized load. Considering the semi-analytical results of [18] the reference results, the comparison shows that the results of the proposed formulation is more accurate than the results of the formulation in [2].

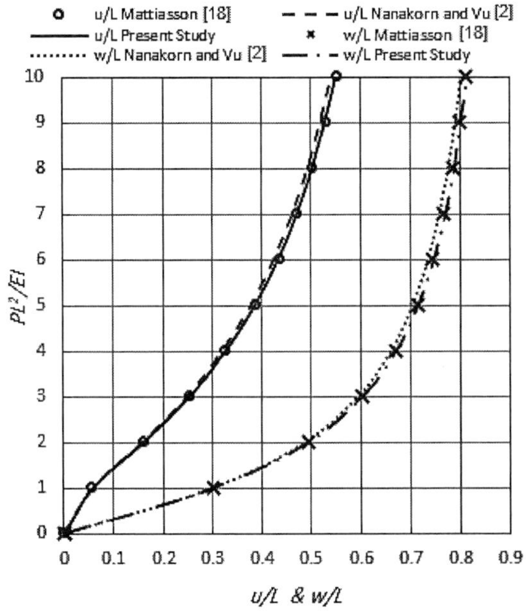

Figure 3: Dimensionless load-displacement curves of the cantilever beam subjected to end vertical point load.

6.2 PINNED-FIXED SQUARE DIAMOND FRAME

A pinned-fixed diamond frame of side length L and flexural rigidity EI loaded in tension as shown in Fig. 4 is considered in this example.

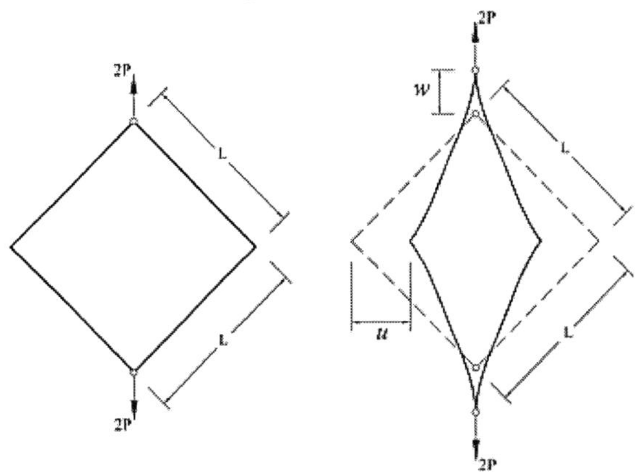

Figure 4: Pinned-fixed square diamond frame loaded in tension.

Two elements per member are used in the analysis and the error tolerance e_r is chosen to be 10^{-7}. The results of the present formulation are compared with the results of the total Lagrangian formulation [2] and the elliptic integral numerical solutions [18] in Fig. 5. *u*/L, *w*/L are the normalized displacements and PL2/EI is the normalized load. Good agreement can be noticed between the present results and the semi-analytical results in [18].

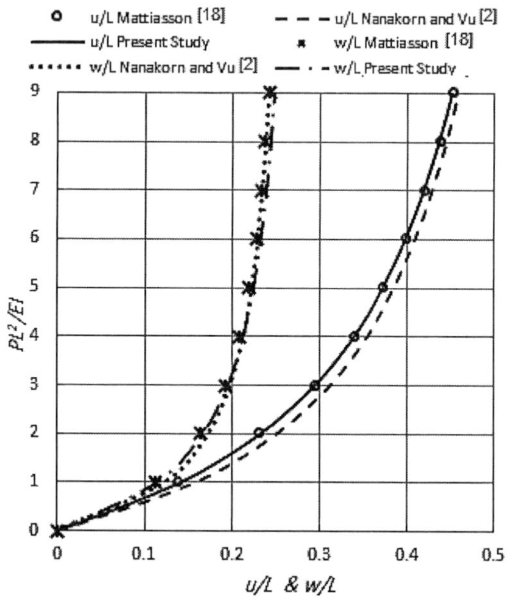

Figure 5: Dimensionless load-displacement curves of the pinned fixed diamond loaded in tension.

7. CONCLUSIONS

In this paper, a corotational finite element formulation for geometrically nonlinear analysis of planar beams and frames was developed. The corotational frame translates and rotates with the beam element but does not deform with it. Hence, the rigid body motion is separated from the pure deformation which is always small with respect to the corotational frame. Arc length integral formula is used to specify the axial displacement. Hamilton's principle is utilized to derive the nonlinear equilibrium equation. A MATLAB code was developed using iterative method based on the conventional Newton-Raphson method without incremental loading. Two numerical problems for beams and frames with large displacements were considered to show the effectiveness and the accuracy of the proposed formulation and the corresponding numerical algorithm. These problems were solved using small number of elements. Accurate results are obtained in comparison with the previously published results. With confidence, the proposed formulation is adequately accurate, simple and time saving.

REFERENCES

1. Beheshti, A. 2016, "Large deformation analysis of strain-gradient elastic beams", *Computers & Structures*, Vol(177), pp.162-175.
2. Nanakorn, P. and Vu, L. N. 2006, "A 2D field-consistent beam element for large displacement analysis using the total Lagrangian formulation", *Finite Elements in Analysis and Design*, Vol(42), pp.1240-1247.
3. Pai, P. F., Anderson, T. J. and Wheater, E. A. 2000, "Large-deformation tests and total-Lagrangian finite-element analyses of flexible beams", *International Journal of Solids and Structures*, Vol(37), pp.2951-2980.
4. Shi, G. and Voyiadjis, G. Z. 1991, "Geometrically nonlinear analysis of plates by assumed strain element with explicit tangent stiffness matrix", *Computers & Structures*, Vol(41), pp.757-763.
5. Yang, Y.-B. and Leu, L.-J. 1991, "Force recovery procedures in nonlinear analysis", *Computers & structures*, Vol(41), pp.1255-1261.
6. Yang, Y., Kuo, S. and Wu, Y. 2002, "Incrementally small-deformation theory for nonlinear analysis of structural frames", *Engineering structures*, Vol(24), pp.783-798.
7. Wempner, G. 1969, "Finite elements, finite rotations and small strains of flexible shells", *International Journal of Solids and Structures*, Vol(5), pp.117-153.
8. Belytschko, T. and Hsieh, B. 1973, "Non - linear transient finite element analysis with convected co - ordinates", *International Journal for Numerical Methods in Engineering*, Vol(7), pp.255-271.
9. Belytschko, T. and Glaum, L. W. 1979, "Applications of higher order corotational stretch theories to nonlinear finite element analysis", *Computers & Structures*, Vol(10), pp.175-182.
10. Oran, C. and Kassimali, A. 1976, "Large deformations of framed structures under static and dynamic loads", *Computers & Structures*, Vol(6), pp.539-547.
11. Battani, J. M. 2007, "A modified corotational framework for triangular shell element", *Computer Methods in Applied Mechanics and Engineering*, Vol(196), pp.1905-1914.
12. Crisfield, M. A. 1990, "A consistent co-rotational formulation for non-linear, three-dimensional, beam-elements", *Computer Methods in Applied Mechanics and Engineering*, Vol(81), pp.131-150.
13. Crisfield, M. A. and Moita, G. F. 1996, "A unified co-rotational framework for solids, shells and beams", *International Journal of Solids and Structures*, Vol(33), pp.2969-2992.
14. Hsiao, K. M., and Hou, F. Y. 1987, "Nonlinear finite element analysis of elastic frames", *Computers & Structures*, Vol(26), pp.693-701.
15. Hsiao, K. M. and Jang, J.-Y. 1991, "Dynamic analysis of planar flexible mechanisms by co-rotational formulation", *Computer Methods in Applied Mechanics and Engineering*, Vol(87), pp.1-14.
16. Elkaranshawy, H. A. and Dokainish, M. A. 1995, "Corotational finite element analysis of planar flexible multibody systems", *Computers & Structures*, Vol(54), pp.881-890.
17. Elkaranshawy, H. A. 1995, *Dynamic modelling of flexible multibody systems subjected to frictional impact,* Ph.D thesis, McMaster University, Hamilton, Canada.
18. Mattiasson, K. 1981, "Numerical results from large deflection beam and frame problems analysed by means of elliptic integrals", *International Journal for Numerical Methods in Engineering*, Vol(17), pp.145-153.

Proceedings of the 26th CANCAM
Victoria, BC, Canada, May 28 – June 1, 2017

POSTURE DEPENDENT CHATTER ANALYSIS IN ROBOTIC MILLING

Hamed Assadi
Mechanical Engineering Department
University Of Victoria
Victoria, BC, Canada
hamedassadi@uvic.ca

Keivan Ahmadi
Mechanical Engineering Department
University Of Victoria
Victoria, BC, Canada
kvahmadi@uvic.ca

Klaus B. Ørskov
Danish Advanced Manufacturing
Research Centre
Herning, Denmark
kbo@damrc.com

ABSTRACT

This paper reports the results of the experimental modal analysis that was conducted on a 6-DOF articulated milling robot. The shapes, frequencies, and damping of the dominant vibration modes of the robot are identified, and their variation in various postures are studied. To investigate the effect of the varying dynamics of the robot on the stability of vibrations during milling operations, the regenerative chatter stability diagrams corresponding with each of the identified modes are constructed using a frequency domain stability analysis method. The results of this research provide an insight into modeling the varying dynamics of machining robots, which is essential for predicting and avoiding chatter during robotic machining.

KEYWORDS: Robotic Machining, Modal Analysis, Chatter

INTRODUCTION

Compared to conventional machine tools, robots offer significant advantages such as lower price, multitasking, and larger workspace, but their application is limited by their higher structural flexibility causing static deflections, vibrations, and chatter (Pan et al.; Zhang et al.). In recent years, the advent of high performance machining methods has lowered the typical machining forces and therefore the application of robots to perform major machining operations such as milling and drilling has become a viable possibility (Iglesias et al.). Regenerative machining chatter is unstable vibrations that develop due to the feedback between subsequent cuts in machining(Tobias and Fishwick). Modeling and avoidance of regenerative chatter in machine tools is a well-established research (Munoa et al.), but the application of the available analysis methods to predict and suppress chatter in robotic machining requires accurate models of the structural dynamics of the robot as it moves from one posture to another (Law et al.).

The analytical methods of studying the varying dynamics of flexible manipulators require a knowledge of the flexibility and damping of the joints and the geometry of the assembly of the arms(Book, 1990; Dwivedy and Eberhard, 2006; Nadira and Mahil, 1986). Such information is not readily available for the users, therefore experimental methods that can be performed in industrial setup are advantageous (Behi and Tesar, 1991; Hearne, 2010). In this paper, we use experimental modal analysis to identify the shape, stiffness, and damping of the dominant vibration modes of a 6-DOF articulated robot with a milling end effector. The variation of the identified modes in various postures is then studied, and the chatter stability diagrams corresponding with each mode are constructed. The results of the presented study provide an understanding of the varying dynamics of the robot as it moves along the toolpath in milling operations. This understanding is essential to developing solutions to predict and avoid chatter during robotic machining processes.

MODAL ANALYSIS

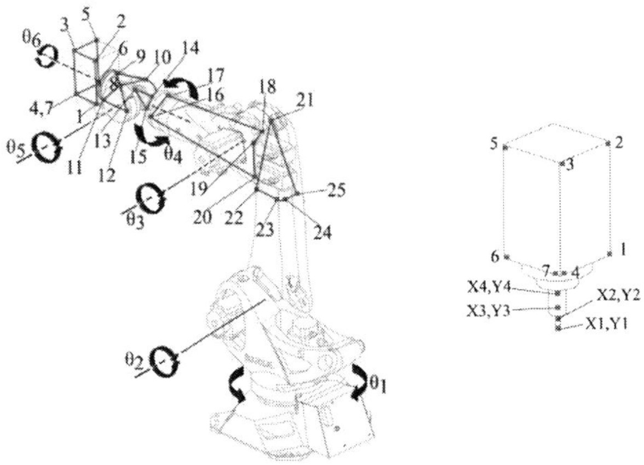

Figure 1. The KUKA R2500 robot and the measurement Degrees Of Freedom (DOFs)

Figure 1 shows the 6-DOF articulated robot (KUKA R2500) with a milling end effector. Experimental modal analysis is performed on the robot in the two postures shown in Figure 2.

(a) (b)

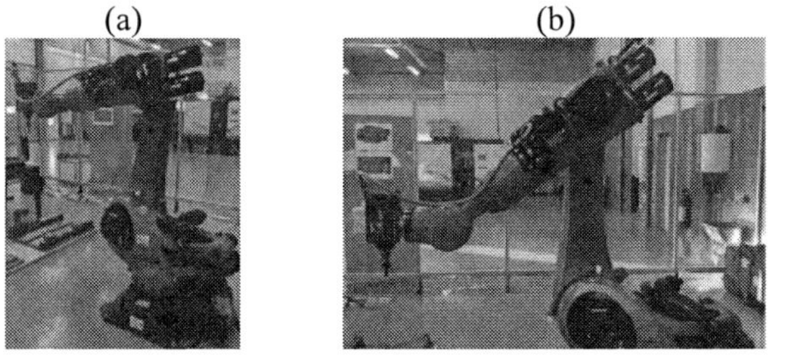

Figure 2: tested postures: (a) posture 1: $(\theta_1 = 0°, \theta_2 = 0°, \theta_3 = 0°, \theta_4 = 0°, \theta_5 = 0°, \theta_6 = 0°)$,(b) posture2: $(\theta_1 = 0°, \theta_2 = 0°, \theta_3 = -45°, \theta_4 = 0°, \theta_5 = 45°, \theta_6 = 0°)$

Direct and cross Frequency Response Functions (FRF) were measured at 33 DOFs; the measurement DOF are shown in

Figure 1. The DOFs X1-X4 and Y1-Y4 are located along the endmill and the holder and DOFs 1-25 are located on the arms of the robot and the end effector (spindle housing). The FRFs were measured by applying impact using an instrumented hammer at the 33 DOFs and measuring the resulting response at a fixed DOF using a piezoelectric accelerometer. In order to capture uncoupled modes in X and Y directions, two sets of experiments were performed. In the first set, the accelerometer was mounted at X1 and in the second set the accelerometer was mounted at Y1. Two types of instrumented hammer were used, Kistler 9722A500 and Kistler 9726A5000; the accelerometer was Kistler 8778A500. The measured force and acceleration signals were conditioned and transferred to a PC using a 24 bit combined DAQ-Signal Conditioning Unit (National Instrument 9234). The frequency domain analysis of the measured signals to obtain the corresponding FRFs was performed using Cutpro application. Figure 3 shows the measured direct FRFs at the tool tip in X and Y directions (i.e. at X1 and Y1) while the robot is in postures 1 and 2. As shown in this figure, by changing the robot's posture the amplitude and frequency of the low frequency peaks (~<500Hz) vary significantly, but higher frequency peaks remain relatively unchanged. Similar observation was made in all of the other measured FRFs.

The modal frequency, and damping of the dominant modes were extracted using SDOF curve fitting methods (Ewins, 1984) performed in Cutpro's Modal Analysis module. Table 1 shows the modal parameters of the first five dominant modes obtained from modal analysis. Also resulting from the performed modal analysis is the real shapes of the identified modes. The shapes of the modes at 11 Hz and 25.9 Hz are shown in Figure 4. As shown in this figure, both modes are associated with the movement of the 1st and 2nd joints of the robot. The first mode (11Hz) is associated with the movement of the links of the robot in XZ plane, but in the second mode (25.9 Hz) the robot links move in both XZ and XY planes. Since the first three modes are mostly associated with the deformations of the joints, they strongly vary with posture. Studying the shapes of the 4th and 5th modes shows that the robot links and joints are relatively rigid in those modes, and vibrations are mostly associated with the motion of the spindle shaft, tool holder, and tool assembly. In the 4th mode the end effector assembly vibrates in Y-direction, and in the 5th mode it vibrates in X-direction. Due to the symmetry of the spindle shaft, tool holder, and the endmill, the 4th and 5th modes are closely spaced in the FRFs. Robot posture does not affect the motion of the end effector assembly and thus the 4th and 5th modes do not vary with posture, unlike the first three modes that strongly depend on posture.

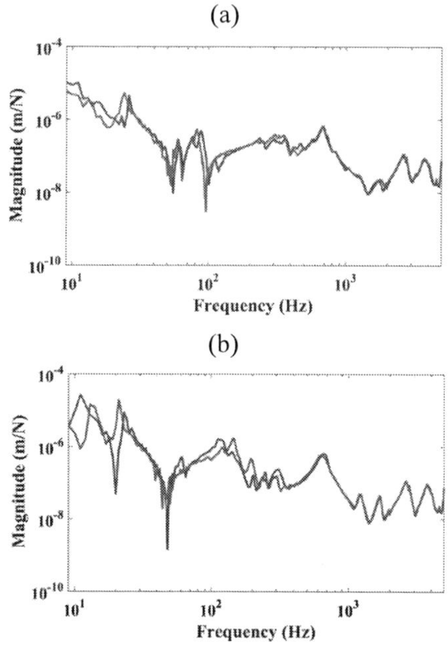

Figure 3: Measured FRFs at the tool tip: (a) FRF at X1, and (b) FRF at Y1

Table 1- Modal parameters of the first two dominant modes in. Response measured at Y1

mode	parameter	Posture 1	Posture 2
1st	Frequency (Hz)	11	13
	Damping ratio	0.5	1.1
	Residue at tool tip (x,y)	(0.0039,0.013)	(0,0.043)
2nd	Frequency (Hz)	25.9	23.5
	Damping ratio	1.7	2.5
	Residue at tool tip (x,y)	(0.031,0.047)	(0.079,0.074,)
3rd	Frequency (Hz)	112.5	145.5
	Damping ratio	6.0	3.8
	Residue at tool tip (x,y)	(0.028,0.277)	(0.041,0.0325)
4th	Frequency (Hz)	657	657
	Damping ratio	0.055	0.055
	Residue at tool tip (x,y)	(0,0.24)	(0,0.212)
5th	Frequency (Hz)	684	684
	Damping ratio	0.05	0.05
	Residue at tool tip (x,y)	(0.172,0)	(0.123,0)

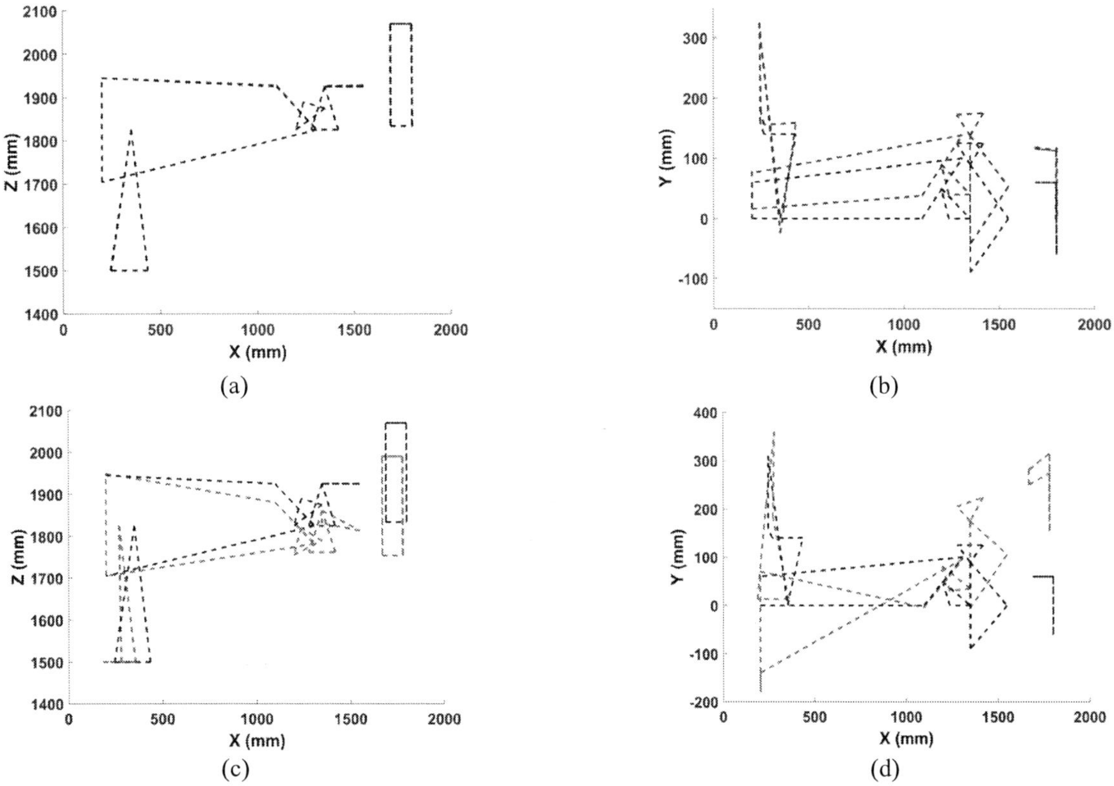

Figure 4: mode shapes in the first posture: (a) 1st dominant mode (11Hz) XZ view (b) 1st dominant mode XY view (c) 2nd dominant mode (25.9 Hz) XZ view (d) 2nd dominant mode XY view. Black dashed lines represent the un-deformed shape of the robot

MILLING STABILITY

Figure 5 shows the closed loop diagram that describes the dynamics of regenerative chatter in milling operations.

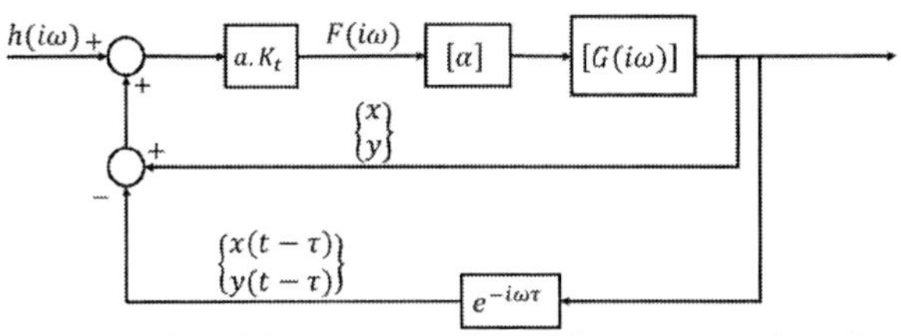

Figure 5: closed loop system describing regenerative chatter

At each spindle speed $(2\pi/NT)$, when the axial depth of cut (a) exceeds the border of stability $(a_{\lim})$, vibrations become unstable. Various time and frequency domain methods are available in the literature to obtain the stability border (Munoa et al., 2016), and usually the results of the stability analysis are presented in the form of stability lobe diagrams that determine the border of stable depth of cut at each spindle speed(Altintas, 2012). The zero-order frequency domain analysis of the milling stability presented in (Altintaş and Budak, 1995) is commonly used to establish the stability lobes in milling. In this method the stability borders are expressed using the following parametric equations:

$$a_{\lim} = -\frac{2\pi \operatorname{Re}(\Lambda)}{NK_t}\left(1+\kappa^2\right) \tag{1}$$

$$T = \frac{1}{\omega_c}(2k\pi + \varepsilon); \quad k = 1,2,... \\ \varepsilon = \pi - 2\tan^{-1}\kappa; \quad \kappa = \operatorname{Re}(\Lambda)/\operatorname{Im}(\Lambda) \tag{2}$$

where N is the number of the flutes of the endmill, T is the tooth passing period, K_t is the constant cutting force coefficient, and Λ are the eigenvalues of the oriented transfer function, $G_0(i\omega)$:

$$\det\left(I_{2\times2} + \Lambda G_0(i\omega_c)\right) = 0 \\ G_0(i\omega_c) = \begin{bmatrix} \alpha_{xx} & \alpha_{xy} \\ \alpha_{yx} & \alpha_{yy} \end{bmatrix} \begin{bmatrix} G_{xx} & G_{xy} \\ G_{yx} & G_{yy} \end{bmatrix} \tag{3}$$

The directional coefficients, α_{ij}, $i, j = x, y$, are constant coefficients depending on the tool-workpiece engagement geometry(Altintaş and Budak, 1995). In Equation (3)

, $G_{ij}, i, j = x, y,$ are the direct and cross FRFs at the tooltip in X and Y directions, and ω_c is chatter frequency. The details of constructing stability lobes is available in (Altintaş and Budak, 1995). In order to investigate the effect of the variation of the frequency and stiffness of various modes of vibration on chatter stability, the robot is modelled using a SDOF system in X and Y directions in each vibration mode, and the tooltip FRFs are expressed in terms of the modal parameters as follows:

$$G_{ij}\left(i\omega_c\right) = \frac{\varphi_i \varphi_j}{\omega_n^2 - \omega_c^2 + 2\zeta i \omega_n \omega_c}; \quad i, j = x, y \qquad (4)$$

Where $\omega_n, \varsigma,$ and φ_i are the modal frequency, damping and residue of each mode shown in Table 1. The tool tip FRFs corresponding with each mode identified in Table 1 were substituted in Eq. (1) and (2) and the obtained stability diagrams are shown in Figure 6. The milling operation is assumed to be full immersion milling, with four flutes, and the workpiece material is assumed to be Aluminum 7050. The stability diagram of the first mode (11Hz), as shown in Figure 6 (a), covers the low cutting speed region (<300 rev/min) due to its low natural frequency (Altintaş and Budak, 1995). Also, the damping of this mode varies significantly as the robot moves from posture 1 to 2. The variation of the modal damping causes a significant increase in the stability in the high spindle speed region. The low spindle speeds range that is shown in Figure 6 (a) is not used in common milling operations, and as seen in Figure 6 (a) the regenerative chatter stability limit increases constantly in higher spindle speeds. Although regenerative chatter in the first mode seems to be stable in commonly used spindle speeds and depth of cut values, mode coupling chatter may cause vibration instability in this mode cite (Gasparetto, 1998). Similar analysis applies to the second mode, except that in addition to damping, also the stiffness of this mode changes from posture 1 to 2, which results in the upward shift of the lobes, as shown in Figure 6(b). Due to the higher natural frequency of the third mode, its corresponding regenerative chatter stability diagrams shown in Figure 6(c) cover higher spindle speeds, where typical milling operations are conducted. Therefore, regenerative chatter caused by this mode has to be considered as a limiting parameter in selecting spindle speed and depth of cut values. The variation of the natural frequency and stiffness of the third mode results a substantial vertical and horizontal shift in stability lobes and thus the stable depth of cut and spindle speeds are different in various postures. Figure 6 (d) shows the stability diagrams of the 4th and 5th modes, in which the end effector vibrations dominate. Because these modes are independent of the posture of the robot, the stable spindle speed and depth of cut corresponding with this mode remain unchanged throughout the toolpath.

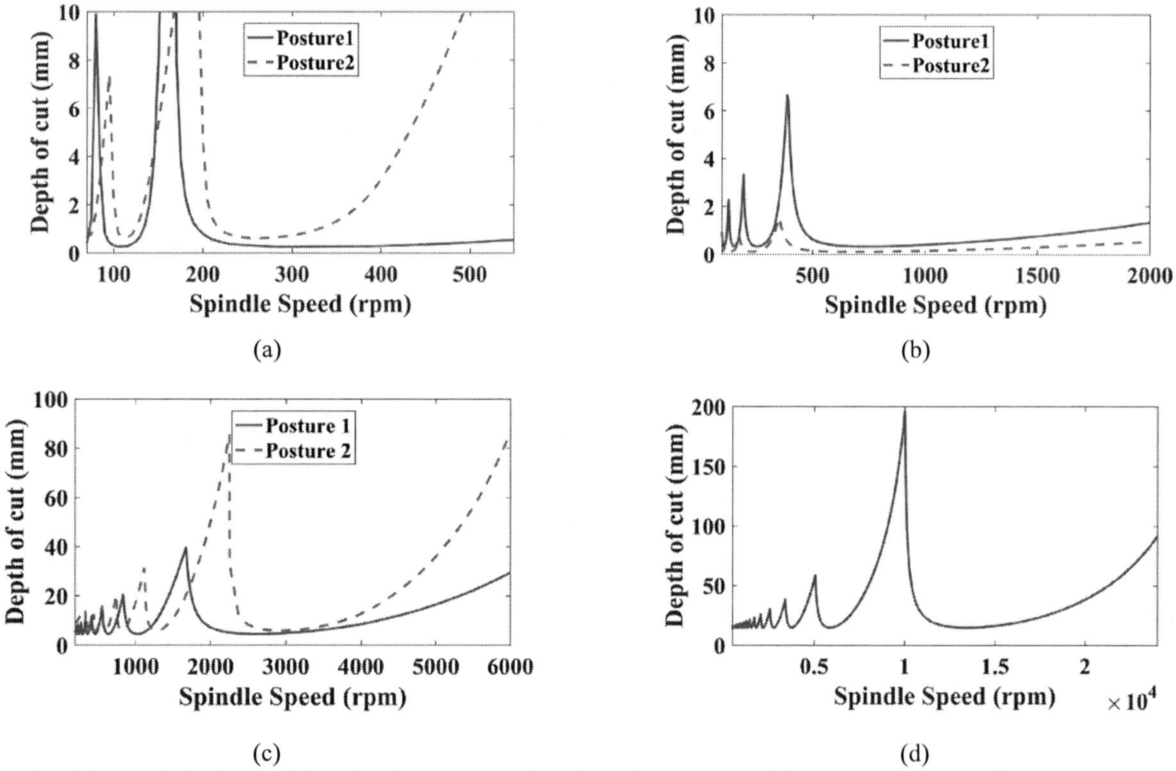

Figure 6: Milling stability lobes: (a)1st dominant mode (b) 2nd dominant mode (c) 3rd dominant mode (d) posture independent dominant mode (spindle and holder mode)

CONCLUSIONS

Experimental Modal Analysis was conducted on a 6-DOF articulated milling robot to identify the vibration modes that may cause chatter during milling. The variation of the structural dynamics of the robot in various postures and the effect of this variation on the resulting chatter stability diagrams was also studied. The results of this study shows that the vibration modes that are associated with the joints of the robot vary significantly with posture, but due to their low natural frequency, they are not likely to cause regenerative chatter in common machining operations. Low frequency robot modes however, due to their significantly higher flexibility, will cause excessive static deflections, forced vibrations, and mode coupling chatter, which needs to be considered in designing the toolpath and selecting the machining parameters.

ACKNOWLEDGMENTS

The authors thank the Danish Advanced Manufacturing Research Centre (DAMRC) for supporting this study by providing access to the milling robot.

REFERENCES

https://www.malinc.com/products/cutpro/.

Altintas, Y., 2012, Manufacturing automation: metal cutting mechanics, machine tool vibrations, and CNC design, Cambridge university press.

Altintaş, Y., and E. Budak, 1995, Analytical prediction of stability lobes in milling: CIRP Annals-Manufacturing Technology, v. 44, p. 357-362.

Behi, F., and D. Tesar, 1991, Parametric identification for industrial manipulators using experimental modal analysis: IEEE Transactions on Robotics and Automation, v. 7, p. 642-652.

Book, W. J., 1990, Modeling, design, and control of flexible manipulator arms: A tutorial review: Decision and Control, 1990., Proceedings of the 29th IEEE Conference on, p. 500-506.

Dwivedy, S. K., and P. Eberhard, 2006, Dynamic analysis of flexible manipulators, a literature review: Mechanism and machine theory, v. 41, p. 749-777.

Ewins, D. J., 1984, Modal testing: theory and practice, v. 15, Research studies press Letchworth.

Gasparetto, A., 1998, A system theory approach to mode coupling chatter in machining: Journal of dynamic systems, measurement, and control, v. 120, p. 545-547.

Hearne, J., 2010, Posture Dependent Vibration Resistance of Serial Robot Manipulators to Applied Oscillating Loads.

Iglesias, I., M. Sebastián, and J. Ares, 2015, Overview of the state of robotic machining: Current situation and future potential: Procedia Engineering, v. 132, p. 911-917.

Law, M., S. Ihlenfeldt, M. Wabner, Y. Altintas, and R. Neugebauer, 2013, Position-dependent dynamics and stability of serial-parallel kinematic machines: CIRP Annals-Manufacturing Technology, v. 62, p. 375-378.

Munoa, J., X. Beudaert, Z. Dombovari, Y. Altintas, E. Budak, C. Brecher, and G. Stepan, 2016, Chatter suppression techniques in metal cutting: CIRP Annals-Manufacturing Technology, v. 65, p. 785-808.

Nadira, R., and S. Mahil, 1986, A finite element/Lagrange approach to modeling lightweight flexible manipulators: Journal of dynamic systems, measurement, and control, v. 108, p. 199.

Pan, Z., H. Zhang, Z. Zhu, and J. Wang, 2006, Chatter analysis of robotic machining process: Journal of materials processing technology, v. 173, p. 301-309.

Tobias, S., and W. Fishwick, 1958, Theory of regenerative machine tool chatter: The engineer, v. 205, p. 199-203.

Zhang, H., J. Wang, G. Zhang, Z. Gan, Z. Pan, H. Cui, and Z. Zhu, 2005, Machining with flexible manipulator: toward improving robotic machining performance: Advanced Intelligent Mechatronics. Proceedings, 2005 IEEE/ASME International Conference on, p. 1127-1132.

Thermodynamics and Heat Transfer

Proceedings of the 26th CANCAM
Victoria, British Columbia, Canada, May 29 – June 1, 2017

PERFORMANCE EVALUATION OF WILDLAND FOREST FIRE CHEMICALS USING A CUSTOM BUILT THERMAL CANISTER

Razim Refai
Department of Mechanical Engineering
University of Alberta
Edmonton, Alberta, Canada
razim@ualberta.ca

Rex Hsieh
Wildfire Operations Research
FPInnovations
Hinton, Alberta, Canada
rex.hsieh@fpinnovations.ca

André G. McDonald
Department of Mechanical Engineering
University of Alberta
Edmonton, Alberta, Canada
andre2@ualberta.ca

ABSTRACT

A test methodology was developed to evaluate the performance of wildland forest fire chemicals. A thermal calorimeter was used to house treated vegetative fuel beds where a uniform heat load was applied to ignite the fuel. Heat release rates from the combustion of the vegetative fuel beds were estimated from the temperature distribution in the walls of the canister. The performance of the treated fuel beds at a fixed coverage level was compared. The results indicated that the test methodology was effective in evaluating the performance of the different fuel treatments. The two gel-treated fuel samples produced lower heat release rates during fifteen-minute burn durations, proving to be more effective than the water and foam-treated fuel samples. The test methodology can be considered as an alternative approach to evaluate fire chemicals.

KEYWORDS: Combustion, Fire, Heat Release Rate, Suppression Chemicals.

INTRODUCTION

The use of fire suppression chemicals to contain and extinguish wildland fires has become increasingly popular. These chemicals have been proven to be effective at retarding the rate of spread of fires as well as lowering fire intensities [1]. Fire chemicals can be used either in direct attack, where they are applied directly on burning fuels, or in indirect attack, where they are applied in advance of a fire front to create control lines [2].

There are several existing analytical methods that are used to evaluate the performance of wildland fire chemicals such as thermal gravimetric analysis, mass spectroscopy, and gas chromatography-Fourier transform infrared spectroscopy analysis, among others [3-5]. However, such methods use small test samples which can result in erroneous interpretation of the resulting data when scaled up to forest fuel flammability applications in real situations [6].

Several experimental flame spread tests have been developed to determine the effectiveness of fire chemicals [6]. These tests were found to be tedious and had low reproducibility. The cone calorimeter, a common apparatus that measures the heat release rate, has also been used to test fire chemicals [7-8]. However, the equipment, apart from its high cost, uses solid materials such as wood rather than foliar samples [6]. The wildfire industry uses the ASTM E1321-13 Standard Test Method for Determining Material Ignition and Flame Spread Properties [9]. The US Forest Services used this standard to develop the Lateral Ignition and Flame Spread (LIFT) test [10]. However, this LIFT test uses a plywood substrate for evaluation and, hence, is not indicative of field performance.

Anderson [11] developed a cost effective laboratory test methodology where the performance of fire chemicals is evaluated based on the heat release rate of treated foliar samples during combustion. Heat release rate is considered to be a more accurate property to evaluate the performance of various products in fire calorimetry [12]. The test methodology does not rely on time-to-ignition measurements, which are onerous and can cause health and safety issues due to the high heat fluxes generated during ignition [13]. This methodology was mathematically developed, but not experimentally validated. The purpose of this study is to fabricate the assembly and evaluate the performance of commercially available fire chemicals using the assembly.

MATHEMATICAL MODEL

A heat conduction model was developed by Anderson [11] to determine the temperature distribution in a custom-built calorimeter known as the "Thermal Canister." The model was based on a one-dimensional, finite length-scale problem developed on the assumption of uniform heating of a rectangular-shaped body. It was assumed that the heat transfer across the exposed walls of the canister was relatively uniform, and the high thermal diffusivity of aluminum facilitated the reduction of any thermal fluctuations. Anderson's model suggested the use of a Pitot tube to measure the pressure of the flue gas. However, Pitot tubes are ineffective in measuring low fluid velocities [14], and therefore, an alternate method of estimating heat loss was derived from Bernoulli's Principle. Equation (1) represents the modified mathematical model to estimate the heat release rate from the Thermal Canister:

$$q'_{Total}(t) = \rho C_p A_c V \Delta T_e - \sum_{i=1}^{10} \frac{A_s}{\varepsilon} \left\{ \left(\frac{k}{W} \Delta T_w + \sigma \varepsilon [T_2(W,t)^4 + T_1(0,t)^4 - 2T_\infty^{\ 4}] \right) \right\}_i , \quad (1)$$

where q' is heat release rate, t is time, ρ is the density of the flue gas, C_p is specific heat capacity of the flue gas, A_c is the area of the cross section of the exhaust pipe, V is the velocity of the flue gas, T_e is the temperature difference between that of the flue gas and the ambient air, A_s is the surface area of the canister walls, ε is emissivity of the canister wall, k is thermal conductivity of the canister wall, T_w is the differential canister wall temperature, T_2 is the temperature of the outer surface of the canister wall, T_1 is the temperature of the inner surface of the canister wall, and T_∞ is the ambient temperature. The properties of the flue gas were assumed to be that of air.

THERMAL CANISTER ASSEMBLY

The Thermal Canister was fabricated from 6061 aluminum plate. The dimensions of each vertical wall and the horizontal lid were 0.508 m long x 0.279 m wide x 0.01905 m thick and 0.295 m x 0.295 m x 0.020 m thick, respectively. An exhaust pipe of diameter 0.076 m and length 0.85 m was attached to the canister lid to enable transport of the flue gases from the Thermal Canister (Fig. 1). The dimensions of the exhaust pipe were modified from Anderson's original design [11] in order to minimize entrance effects and ensure fully-developed flow along most of the length of the pipe. The internal faces of the canister walls and lid were painted with a high-temperature black spray paint (Krylon 1618 BBQ and Stove Paint, The Sherwin-Williams Company, Cleveland, OH, USA) to enhance absorption of radiation energy. Holes of 1.98 mm diameter were drilled in the walls and lid of the Thermal Canister. The placement of the holes was similar to that of the Thermal Cube Heat Flux Sensor developed by Anderson *et al.* [15]. These holes served as housing for thermocouples, which were used to obtain transient temperature data at the front and back surfaces of the walls and lid. The heat source used in this experiment was an electric-powered radiant panel (Omegalux QH-121260, Omega Engineering Inc., Laval, Québec, Canada) that projected a uniform heat flux to the inside of the Thermal Canister. The radiant panel was programmed to reach and maintain a temperature of 500° C. Thirty (30) thermocouples were fitted in the walls and lid of the Thermal Canister and were connected to a multi-channel data acquisition unit (34970A Data Acquisition/Data Logger Switch Unit, Keysight Technologies Inc., Santa Rosa, California, USA). The thermocouple in the exhaust pipe was connected to a stand-alone data logger (DaqProTM 5300, Fourier Systems Inc., Mokena, IL, USA). An anemometer (2G-2948 EDRA Anemometer, Airflow Developments Ltd., Richmond Hill, Ontario, Canada) was placed above the exhaust pipe outlet to measure the velocity of the flue gas.

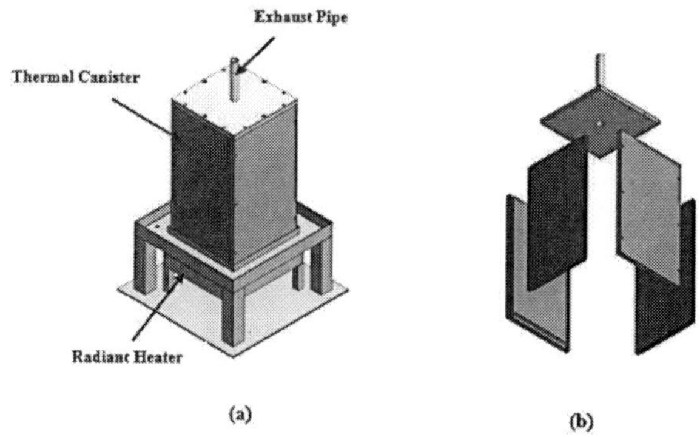

Figure 1: a) Thermal Canister assembled b) Exploded view of Thermal Canister

TEST METHODOLOGY

Red stemmed feather moss was dried to a moisture content of less than 5% in accordance with the ASTM D4442 oven dry standard method and used as the fuel in the experimental burns. Feather moss was chosen as the preferred fuel due to its ability to absorb and efficiently retain water and other water-based chemical products (foams, gels, and fire suppressants and retardants). A steel woven wire mesh (0.0075" wire diameter) of dimensions 25 cm x 25 cm was used as a surface to hold the fuel during the experiments. Forty-five (45) grams of feather moss was measured and placed on the steel mesh. The treatment of the fuel was applied prior to placement of the fuel in the Thermal Canister assembly.

The performance of five different types of treatments was evaluated using the Thermal Canister assembly. The names of the products used as treatments have been intentionally excluded for proprietary reasons. The different treatments were: zero coverage level (no water/fire chemicals applied), water, foam, and two gel (water enhancer) products, all at coverage level 0.8. The mixing ratios used have been presented in Table 1, and the ratios for the preparation of the foam and gel products were selected as per manufacturers' recommendations. Dry chemical concentrates were mixed by weight and wet chemical concentrates were mixed by volume. The treated fuels were placed in the Thermal Canister. The data loggers were programmed to record the temperatures and differential voltages at three-second intervals. The experiment was repeated three times for each type of fuel treatment under the same parametric conditions of mixing ratio, mass of fuel, and incident, heat flux. Data obtained from the experiments was extracted and Eq. (1) was used to calculate the heat release rate.

Chemical Product	Mixing Ratio
Foam Product	0.65%
Gel Product A	0.65%
Gel Product B	3%

Table 1: Fire chemical product mixing ratios

RESULTS AND DISCUSSION

Performance Evaluation of the Treatments

The performance of the different fuel treatments was evaluated by considering the calculated average heat release rates obtained from the three experimental burns for the different fuel treatments. Figure 2 shows five different average heat release rates, each corresponding to a different fuel treatment.

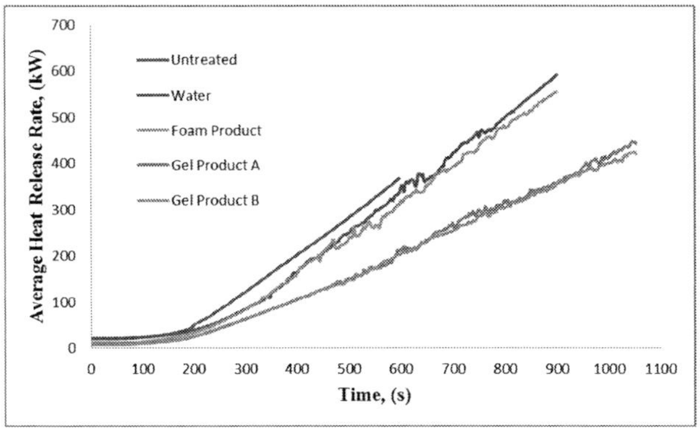

Figure 2: Comparison of average heat release rates of different fuel treatments

The experimental burns with untreated fuel lasted for a duration of 10 minutes. Water and foam-treated fuels burned for 15 minutes, and gel-treated fuels burned for 18 minutes. This difference was due to completion of the combustion process at different times for different treatments.

Figure 3 shows the difference in heat release rates between treated and untreated fuels for the time frame between 100 and 300 seconds after initial heating. At around 200 seconds, a sudden rise in average heat release rate is observed in the untreated fuel. This rise in the heat release rate can be seen within the black circle in Figure 3. This is indicative of the onset of the combustion process in the untreated fuel. In comparison, a sudden rise of average heat release rates is not observed for any of the treated fuels. This implies that the treatment of fuels results in the absence of a singular moment of combustion, wherein the treatment allowed for a more gradual combustion process when heat was applied. In addition, it can be seen in Fig. 2 that the average heat release rate for the untreated fuel at any given time between 200

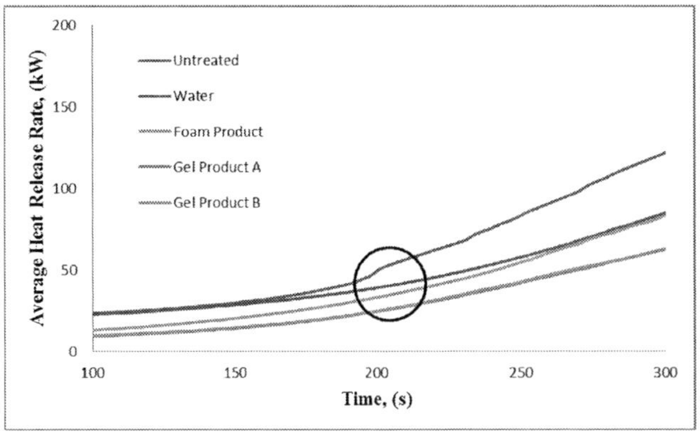

Figure 3: Onset of combustion process in the untreated fuel in comparison to the treated fuels

seconds and 600 seconds was higher than the average heat release rate for any of the treated fuels. This implies that the treatments were effective in reducing the heat release during the process of combustion, and were therefore effective.

Comparison of Treatments

Figure 2 presents the average heat release rates of the different fuel treatments: water, foam, gel product A, and gel product B. The average heat release rate of water at coverage level 0.8 was used as a control measure to compare the relative performance of the fire chemicals. The performance of the fire chemicals was evaluated in terms of percentage increase or decrease in average heat release rates in comparison to the average heat release rate of water-treated fuels. This percentage value was calculated at five different times, namely 300, 450, 600, 750, and 900 seconds. Table 2 shows the relative performance of the fire chemicals at the five different times.

From Table 2, it is observed that use of all three fire chemicals resulted in lower average heat release rates in comparison to the use of water, indicating that they were more effective than water in suppressing combustion. The foam product resulted in higher heat release rates in comparison to the two gel products. This suggests that the gel products were more effective at suppressing combustion of the fuel than the foam product.

Treatment	Foam Product	Gel Product A	Gel Product B
300 seconds	1.91%	26.58%	26.02%
450 seconds	-1.39%	37.45%	37.78%
600 seconds	8.92%	38.99%	42.03%
750 seconds	4.04%	34.60%	37.51%
900 seconds	6.24%	39.85%	40.27%

Table 2: Relative performance of fire chemicals in comparison to water-treated fuel beds

The gel product A and gel product B produced significantly lower heat release rates in comparison to water and foam product. This indicates that the gel product treatments were the most effective at suppressing the combustion of the fuel for the different fuel treatments that were tested. The performance of the gel product B was marginally better than the performance of gel product A.

CONCLUSIONS

The objective of this study was to develop a test methodology to evaluate the performance of wildland fire chemicals. A custom-built thermal calorimeter, known as the "Thermal Canister" was developed by Anderson [11] by using a one-dimensional heat conduction model based on uniform heating of a rectangular-shaped body. The purpose of the canister was to estimate the heat release rate produced during the combustion of vegetative fuels, which could be used as a means to evaluate the performance of the fire chemicals. Modifications to the mathematical model and experimental design were made in order to produce more accurate estimates of the heat release rate.

Feather moss was selected as the vegetative fuel to be used in the experiments due to its ability to absorb and retain water and water-based fire chemicals efficiently. Five different fuel treatments were selected to validate the performance of the experimental assembly as well as evaluate the relative performance of the treatments. The different fuel treatments selected were: untreated fuel (no water/chemicals), water, foam, gel product A, and gel product B, all at coverage level 0.8.

Experimental burns for each treatment type were carried out three times to evaluate the repeatability of the Thermal Canister. It was found that all fire chemicals produced lower heat release rates than water, implying that they were more effective at suppressing the combustion of the vegetative fuel. Data also showed that the two gel products performed better than the foam product. The gel products A and B produced significantly lower heat release rates than all other treatments, suggesting that the two gels were more efficient at suppressing the combustion of the vegetative fuel.

REFERENCES

1. Kaladokidis, K. 2000, "Effects of wildfire suppression chemicals on people and the environment – a review", *Global Nest: Int J*, Vol (2), pp. 129-137.
2. Swanson, D., and Helvig, T., "Effectiveness of direct and indirect attack on wildfire with air-delivered retardants", United States Depart of Agriculture (USDA) Forest Service, Wildland Fire Chemical Systems, https://www.fs.fed.us/rm/fire/wfcs/documents/Direct_Indirect_Attack_on_Wildfire.pdf
3. Liodakis, S., Bakirtzis, and D., Dimitrakopoulos, A.P. 2003, "Autoignition and thermogravimetric analysis of forest species treated with fire retardants", *Thermochim Acta,* Vol (399), pp. 31-42.
4. Lyons, J.W. 1987, *The Chemistry and Uses of Fire Retardants*, R.E. Krieger Publishing Company, Malabar, FL.
5. Wang, Q., Li, J., and Winandy, E. 2004, "Chemical mechanism of fire retardance of boric acid on wood", *Wood Sci Technol*, Vol (38), pp. 375-389.
6. Liodakis, S., Antonopoulos, I., and Kakardakis, T. 2010, "Evaluating the use of minerals as forest fire retardants," *Fire Safety J*, Vol (45), pp. 98-105.
7. Grexa, O., and Lubke, H. 2001, "Flammability parameters of wood tested on a cone calorimeter", *Polym Degrad Stabil*, Vol (74), pp. 427-432.
8. ISO 5660-01, 1993, Fire Tests: Rate of Heat Release for Building Products (Cone Calorimeter Method), International Standards Organization, Geneva, Switzerland.
9. ASTM E1321-12, 2013, Standard Test Method for Determining Material Ignition and Flame Spread Properties, ASTM International, West Conshohocken, PA, USA.
10. United States Department of Agriculture (USDA) Forest Service, Wildland Fire Chemical Systems, Standard test procedure, Section 2: Fire Tests, STP 2.1-2.2, https://www.fs.fed.us/rm/fire/wfcs/tests/index.htm
11. Anderson, S. 2015, "Quantification of performance of wildfire chemicals using custom-built heat flux sensors", University of Alberta, MSc. Thesis.
12. Janssens, M. L. 1995, "Methods and equations of fire calorimetry", M.M. Hirschler, R.E. Lyon (eds.), *Special Symposium on Fire Calorimetry*, Gaithersburg, MD, pp. 11-22.
13. NFPA 2113, 2015, Standard on Selection, Care, Use, and Maintenance of Flame-Resistant Garments for Protection of Industrial Personnel Against Short-Duration Thermal Exposures, National Fire Protection Association.
14. Klopfenstein, R. 1998, "Air velocity and flow measurement using a pitot tube", *ISA Trans*, Vol (37), pp. 257-263.
15. Anderson, S., McDonald, A. 2015, "Performance testing of wildland fire chemicals using a custom-built heat flux sensor", *J. Fire Sci*, Vol (33), pp. 473-492.

NUMERICAL SIMULATION OF FAST FILL OF COMPRESSED NATURAL GAS WITH PASSIVE HEAT REMOVAL

Guoyu Zhang, Joshua Brinkerhoff, Ri Li
School of Engineering
University of British Columbia, Okanagan campus
Kelowna, Canada
Email: Joshua.brinkerhoff@ubc.ca

Chris Forsberg, Todd Sloan
Agility Fuel Systems
Kelowna, Canada
Email: CForsberg@agilityfs.com

ABSTRACT

The fast fill process rapidly fuels natural gas vehicles by forcing compressed natural gas into the vehicle's fuel tank. The conventional fast fill process produces a significant amount of heat inside the fuel tank due to the recompression work applied to the natural gas. This heat lowers the gas density and thereby limits the mass of fuel that can be carried and reduces the vehicle's range. A passive heat removal method is proposed to increase the dispensed mass during fast refill of CNG vehicles. Two-dimensional axisymmetric computational fluid dynamics (CFD) simulations are presented of the unsteady, compressible, turbulent flow in a type III cylinder during fast fill with and without passive heat removal. Dynamic average temperature and mass curves as well as local temperature distribution in the cylinder are obtained at different time instances during the fill to illustrate the benefit of passive heat removal as a means of improving fill efficiency.

KEYWORDS: Compressed natural gas, fast filling, fill efficiency, passive heat removal, numerical study

INTRODUCTION

For the past few years, natural gas has played a prominent role in energy discussions around the world [1-2]. One reason is the emergence of abundant supplies of natural gas following the development of unconventional extraction methods. For example, the development of shale gas in USA has increased its production from 456 billion scf per year to 9198 billion scf per year from 2000 to 2012. Another driver is the low carbon content of natural gas relative to other fossil fuels, which has important implications for the transportation sector that is responsible for 20% of the global carbon dioxide emissions. Given the increasing energy demand and high greenhouse gas (GHG) emissions associated with conventional fossil fuels, Yeh [3] concluded that shifting the transportation sector to natural gas can significantly reduce GHG emissions.

Due to the low volumetric energy density of natural gas relative to liquid fuels, transportation systems typically store fuel either as a low-temperature liquefied natural gas (LNG) or a high-pressure compressed natural gas (CNG). Because of the large upfront investment, LNG requires large reserves of natural gas near the facilities to support an LNG project and to get acceptable investment returns. Therefore, its widespread use in commercial applications has been limited and CNG currently has higher market share [4]. Kountz [5,6] showed that because of the heat generation due to re-compression during fast fill process, underfilling of the cylinder is unavoidable. During the fill process, high pressure CNG passing through the dispenser throttling device and enters the cylinder with reduced temperature because of the Joule-Thomson effect. The filled natural gas is then gradually heated via recompression by the continuously incoming gas, stopping when the internal pressure reaches the allowed cylinder pressure. Due to the heat of recompression, the temperature at the end of fast fill is typically about 50°F above the ambient temperature [5]. As a result, the charged mass of CNG is less than the rated storage capacity of the cylinder, and as the filled gas cools to ambient temperature, its pressure will decrease below the rated pressure and vehicles will experience reduced driving range. Fill efficiency—defined as the ratio of the actual filled mass to the capacity at the rated pressure and ambient temperature—is used to evaluate CNG filling. Fast fill of CNG cylinders at 20°C ambient temperature typically achieves a fill efficiency of 80% [6].

Many studies have been conducted to investigate the thermal behaviour such as temperature rise and distribution during fast fill [5-11]. Kountz [5] and Farzaneh-Gord et al. [10] focused on mathematical theory and modelling of CNG fast fill. Shipley [8] experimentally measured the temperature rise of CNG with different ambient temperatures, and observed under-fill of test cylinders for all the fast fill tests. Farzaneh-Gord et al. [11] further studied the effect of initial conditions on the fill efficiency of CNG cylinders. Some researchers have investigated the transient thermal behaviour during gas filling by numerical methods. Dicken and Merida [12] used computational fluid dynamics (CFD) to simulate the filling process of hydrogen. Nahavandi et al. [13] also carried out CFD simulations of fast fill, comparing the temperature rise, pressure curves, and fill efficiency for natural gas and hydrogen.

Very few of the above studies have investigated potential solutions to improve the fill efficiency. Because heat generation during fast fill is inevitable, concepts that can actively or passively remove the generated heat during the fill process are required to achieve the needed improvements in fill efficiency. However, few studies have investigated any methods for active or passive cooling. Zhang et al [14] demonstrated the benefit of active heat removal through CFD of type-III cylinders containing cooling coils through which coolant circulates during the fill to transfer heat from the cylinder to the surroundings. The objective of the present study is to investigate the fast fill process of natural gas in Type-3 cylinders with passive heat removal in order to identify its effectiveness for improving fill efficiency. Here, active refers to a process where energy is expended during the fill to remove heat (i.e. coolant flow), while passive refers to a process where the energy expenditure occurs prior to the flow (i.e. a pre-chilled heat sink). Average gas temperature and mass filled at different instances during the refill have been presented to demonstrate the effectiveness of heat removal on increasing the fill efficiency.

NUMERICAL METHOD

Zhang .et al [14] built an analytical model of the CNG fast filling with heat removal to demonstrate its benefit. However, the analytical study suffers from the following assumptions: the cylinder contents are a single lumped mass, spatial variations in the cylinder conditions are neglected, and the convective heat transfer coefficient of the cooling coil was not calculated. To quantify the effect of these simplifications on the results of the analytical study and to validate the results against higher-order methods, a CFD model is developed of the fast fill process. ANSYS Fluent®, a commercial CFD package employing the finite-volume method to discretise the compressible form of the mass, momentum, and energy conservation equations. During the hydrogen filling process studied by Zhao et al. [15], the influences of gravity and buoyancy forces were found to be negligible if the mass flow rate exceeded 9 g/s. Since the mass flow rate in this case will be many times greater than 9 g/s, the flow inside the cylinder is modelled as an axisymmetric flow without buoyancy forces due to gravitational acceleration. The working fluid in the CFD model consists of pure methane with real gas properties determined using the Redlich-Kwong equation of state [16]. The k-ε turbulence model is applied with modified coefficients ($C_{1\varepsilon} = 1.52, C_{2\varepsilon} = 1.92, C_{\mu} = 0.09$) .

The computational domain is illustrated in Fig. 1. The gas enters the cylinder through a converging nozzle which has an entrance with diameter of 0.01 m and an exit with diameter of 0.005 m. This converging nozzle keeps the incoming gas flow in the subsonic regime. The geometry was divided into two domains, the gas domain and the solid (cylinder wall and nozzle) domain. Because the thermal resistance of aluminium is much smaller than that of carbon fiber, only the carbon fiber wrap (thermal conductivity k = 3.7 W/m·K, density ρ = 930 kg/m3, specific heat C_p = 1500 J/kg·m) is considered as cylinder wall for simplicity when building the geometry. The dimensions of the cylinder (length L = 893 mm, Outer diameter D = 396 mm, wall thickness t = 19 mm, etc.) are set to be the same as the model by Nahavandi [13].

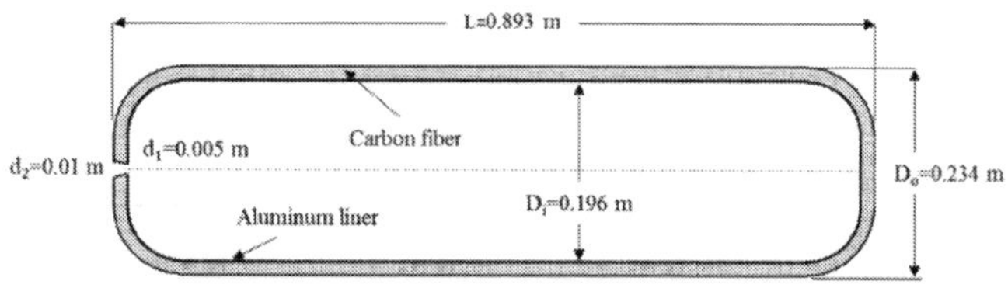

Fig. 1 The computational domain under investigation

Considering adding passive heat removal, three copper coils are placed in the back side of the cylinder since copper has high density, specific heat and easy to get. These coils are pre-chilled to a lower temperature than the surrounding gas prior to starting the filling process so they will keep bringing heat generated outside during the refill. It is assumed that the copper has the same temperature with the gas at the end of the filling process.

Applying conservation of energy and the equation of state to the system yields:

$$C_{p,NG} \frac{(1+\eta)PV_{tank}M}{1000RT_1z_1}(T_1 - T_2) = C_{p,Cu}\,\rho_{Cu}\,V_{Cu}\,(T_2 - T_i) \quad (1)$$

Where P is the final average gas temperature (20MPa), V_{tank} and V_{Cu} are the volume of the test cylinder (23.4L) and copper coils, T_1 and T_2 are the final average gas temperature without cooling (329K) and with cooling, T_i denotes the temperature of the pre-chilled copper coils, $C_{p,NG}$ and $C_{p,Cu}$ are the specific heat of natural gas and copper, M is the Molecular mass of natural gas, R is the universal gas constant, ρ_{Cu} denotes the density of copper, and η is the increase of the fill efficiency.

Conservation of mass is given by:

$$(1+\eta) \times \frac{PV_{tank}}{T_1z_1} = \frac{P(V_{tank}-V_{Cu})}{T_2z_2} \quad (2)$$

Where z_1 and z_2 are the average gas compressibility factor without and with cooling gained from NIST database.

By solving Eqs. (1) and (2), a correlation between V_{Cu}/V_{tank} and T_i can be obtained if η is given. Fig. 2 shows their relationship when η is fixed as 10%. It can be observed that in order to achieve a 10% increase on the fill efficiency, the temperature to which the copper has to be pre-chilled increases significantly with the volume percentage of the copper over the cylinder when this ratio is between 0% to 6%. The pre-chilling temperature almost remains constant if this ratio keeps increasing beyond this range. Therefore, pre-chilling the copper coils which accounts for 6% of the cylinder volume to 255K is selected to be the most effective and cost-optimal passive heat removal scenario.

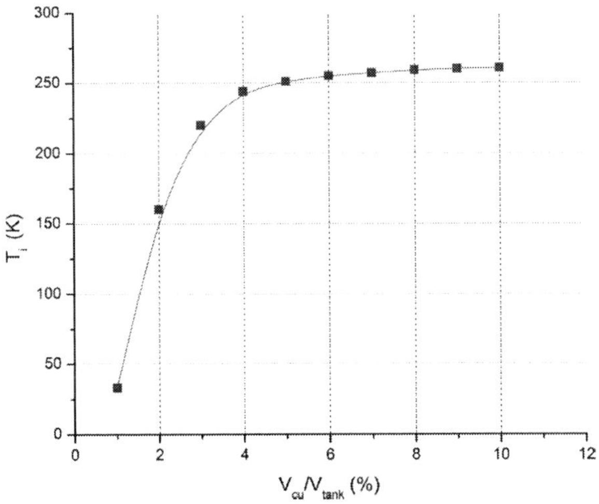

Fig. 2 The pre-chilling temperature vs. the volume percentage of the copper over the test cylinder

Unstructured grids have been mapped to the computational domain. Refined mesh has been utilized near the inlet and the copper coils where the biggest gas property changes are expected. The time-step size in this simulation varies depending on the maximum gas velocity, with $\Delta t = 10^{-7}$ being the smallest time step used.

RESULTS AND DISCUSSION

Figure 3, 4 compare the computed average gas temperature and total mass filled into the cylinder during fast refill with passive heat removal, without heat removal, and with active heat removal by Zhang .et al [14]. By imposing passive heat removal, the required filling time slightly increases as compared with the other cases. Figure 3 shows the dynamic change in average gas temperature during the fill. The results predict a significant temperature rise in the first 3 seconds in all cases. It could be realized that the different cooling scenarios have a significant effect on the temperature profile. For the case without cooling the mean temperature of natural gas increases monotonously throughout the fill as it gets compressed by the incoming gas and reaches a maximum at the end of the refill. On the contrary, the temperature profiles are divided into two parts when active cooling has been added. In the early part of those profiles (0 s < t < 5 s), the average gas temperature increases as the gas being compressed. In the later part (t > 5 s) the cooling effect becomes strong enough to overcome the compression and conversion of supply enthalpy energy to cylinder internal energy, so the gas mean temperature starts to decrease and reaches a balance in the end. For the filling process with passive cooling, the temperature increases at the beginning due to the heat generation converted from compression work. It turns to decrease when the cooling effect becomes dominant. As the copper keeps absorbing heat during the refill, its temperature gradually increases which results in lower and lower cooling rate. Therefore, the average gas temperature increases slowly and reaches 306 K in the end.

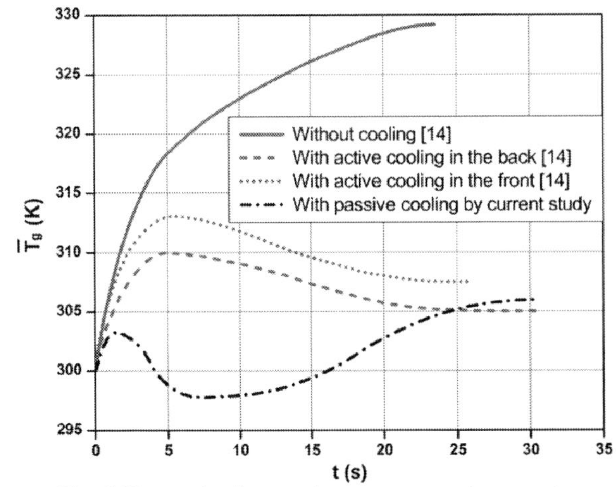

Fig. 3 Dynamic changes in gas average temperature

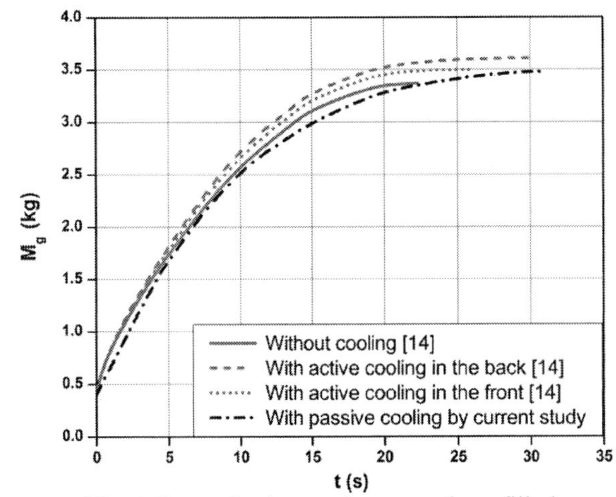

Fig. 4 Dynamic changes in mass of gas filled

Figure 4 shows the dynamic profile of the mass filled during the filling process. At the end, 4.5% more mass of gas can be filled into a cylinder with passive cooling because some heat has been removed, which results in a lower gas average temperature and higher gas density. This number disagrees with the proposed 10% fill efficiency increase when designing the copper volume and the pre-chilling temperature, which is because the copper still has lower average temperature than the gas at the end of the fill. As shown in Fig. 5, the temperature distribution of the copper is extremely non-uniform. The temperature of the inner part is much lower, which indicates that there's still further potential to achieve even stronger heat removal by using materials with higher thermal conductivity. 7% more mass of gas can be filled into a cylinder with active heat removal in the back, which is more beneficial for the fill efficiency. The result can provide some guidance on the future optimization of the cooling scenario design.

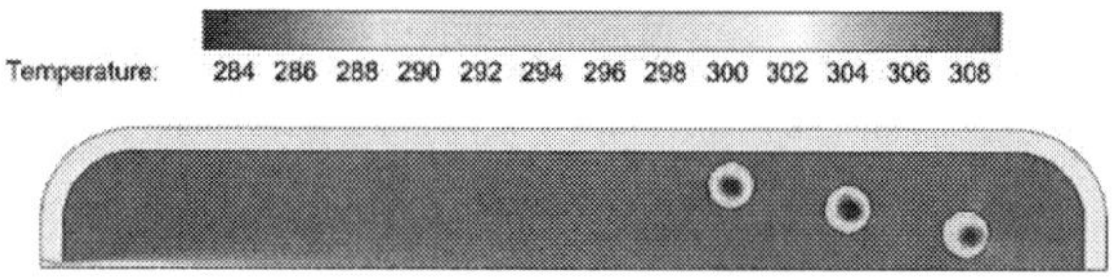

Fig. 5 Temperature distribution at the end of filling

CONCLUSIONS

A numerical study of fast fill of compressed natural gas was performed to identify the transient temperature and mass variation during the fill process with passive heat removal. The spatial variation of gas conditions in the cylinder were found through a CFD study of the fill process. The CFD study proves the trend that adding active heat removal brings a reduced gas average temperature increase so that more mass of natural gas can be filled throughout the filling process. It also indicates that this benefit strongly depends on the different cooling scenarios. Further potential has been found to achieve even lower gas temperature so that even more mass of natural gas can be filled.

ACKNOWLEDGMENTS

This work is supported by Agility Fuel Systems and Natural Sciences and Engineering Research Council of Canada(NSERC) (CRDPJ485290).

REFERENCES

1. Paltsev, S., Jacoby, H. D., Reilly, J. M., Ejaz, Q. J., Morris, J., O'Sullivan, Fin & Kragha, O. (2011). The future of US natural gas production, use, and trade. Energy Policy, 39(9), 5309-5321.

2. Moniz, Ernest J., et al. (2011). "The future of natural gas." Cambridge, MA: Massachusetts Institute of Technology.

3. Yeh, S. (2007). An empirical analysis on the adoption of alternative fuel vehicles: The case of natural gas vehicles. Energy Policy, 35(11), 5865-5875.

4. Economides, M. J., Sun, K., & Subero, G. (2006). Compressed natural gas (CNG): an alternative to liquefied natural gas (LNG). SPE Production & Operations, 21(02), 318-324.

5. Kountz, K. J. (1994). Modeling the fast fill process in natural gas vehicle storage cylinders (No. CONF-940301--42). Institute of Gas Technology, Chicago, IL (United States).

6. Kountz, K. J. (1996). Natural Gas Storage Conditions as a Function of Operating Environment. NASA, (19980008507).

7. Newhouse, Norman L., and William E. Liss (1999). Fast filling of NGV fuel containers. No. 1999-01-3739. SAE Technical Paper.

8. Shipley, E. (2002). Study of natural gas vehicles (NGV) during the fast fill process (Doctoral dissertation, West Virginia University).

9. Thomas, G., Goulding, J., & Munteam, C. (2002). Measurement, approval and verification of CNG dispensers. NWML KT11 Report.

10. Farzaneh-Gord, M. (2008). Compressed natural Gas-Single reservoir filling process. INTERNATIONAL GAS ENGINEERING AND MANAGEMENT, 48(6), 16-18.

11. Farzaneh-Gord, M., Eftekhari, H., Hashemi, S., Magrebi, M., & Dorafshan, M. (2007). The effect of initial conditions on filling process of CNG cylinders. In The Second International conference on Modeling, Simulation, And Applied optimization, Abu Dhabi, UAE, March (pp. 24-27).

12. Dicken, C. J. B., and W. Merida. (2007). "Modeling the transient temperature distribution within a hydrogen cylinder during refueling." Numerical Heat Transfer, Part A: Applications 53.7: 685-708.

13. Nahavandi, Navid NoorAliPour, and Mahmood Farzaneh-Gord. (2013). "Numerical simulation of filling process of natural gas onboard vehicle cylinder." Journal of the Brazilian Society of Mechanical Sciences and Engineering 35.3: 247-256.

14. Zhang, G.Y., Brinkerhoff, J.R., Li, R., Forsberg, C. (2016). Increasing fast refill of compressed natural gas by active heat removal. The 2nd Thermal and Fluids Engineering Conference and 4th International Workshop on Heat Transfer, Las Vegas, Nevada, USA, 2017 (Accepted)

15. Zhao, L., Liu, Y., Yang, J., Zhao, Y., Zheng, J., Bie, H., & Liu, X. (2010). Numerical simulation of temperature rise within hydrogen vehicle cylinder during refueling. International journal of hydrogen energy, 35(15), 8092-8100.

16. Li, Q., Zhou, J., Chang, Q., & Xing, W. (2012). Effects of geometry and inconstant mass flow rate on temperatures within a pressurized hydrogen cylinder during refueling. International Journal of Hydrogen Energy, 37(7), 6043-6052.

STABILITY AND THERMAL CONDUCTIVITY OF PARAFFIN WAX WITH MULTI-WALLED CARBON NANOTUBES

Vahit Saydam, Xili Duan
Department of Mechanical Engineering
Memorial University of Newfoundland
St. John's, NL, Canada
vs6205@mun.ca, xduan@mun.ca

Xiaoxu Fan
Energy Research Institute
Shandong Academy of Science
Jinan, Shandong, China
fanxiaoxu@sderi.cn

ABSTRACT

There have been growing interests in latent heat energy storage with nanoparticle enhanced phase change materials (NEPCMs). Various nanoparticles and base materials have been studied with different dispersion methods. However, stability and uncertainty regarding thermal conductivity enhancement are the major hindrances for engineering applications of NEPCMs. In this study, multi-walled carbon nanotubes were dispersed in paraffin wax using magnetic stirring and sonication. Stability of the prepared NEPCMs was investigated with a visualization method. Thermal conductivities were measured with 0.05wt%, 0.075wt% and 0.1wt% nanoparticle concentrations, after different sonication times, and at different temperatures. It was found that thermal conductivity of the NEPCMs did not show any clear trend of significant increase with the specified loadings of MWCNT. We also found that stability of the NEPCMs becomes progressively worse after a few thermal cycles, with significant aggregation of nanoparticles and separation from the base material.

KEYWORDS: Nanoparticle enhanced phase change materials, Multi-walled carbon nanotubes, paraffin wax, stability, thermal conductivity

INTRODUCTION

Growth of energy demand throughout the world requires efficient use of renewable energy sources. The intermittency of renewable energy sources such as wind and solar necessitates the usage of energy storage. Thermal energy storages that take advantage of the high latent heat of phase change materials (PCMs) have been used extensively [1]. Many PCMs have low thermal conductivities and several methods have been tried to improve that, for example by inserting metallic fins and matrix structure into PCM containers [2]. In recent years there is a growing interest of dispersing nanoparticles into PCMs for thermal conductivity enhancement [3].

One of the emerging nanoparticles for thermal conductivity enhancement is multi-walled carbon nanotubes (MWCNTs). Several studies have been done on the thermal properties of MWCNT dispersed paraffin wax. Wang et al. [4] studied paraffin wax with a melting temperature range of 52 to 54 °C dispersed with MWCNT of 0.2 to 2 wt% and reported thermal conductivity enhancement of 40% at 60 °C and 35% at 35 °C. Angayarkani and Philip [5] observed considerable thermal conductivity increase for hexadecane with MWCNT dispersed at 0.5 to 1 wt%. They reported that enhancement only happened up to a certain MWCNT concentration; higher concentration could lead to decrease in thermal conductivity due to agglomeration of nanoparticles. Another study done by Kumaresan et al. [6] showed that increasing volume concentration of MWCNT from 0.15% to 0.6% in paraffin not only improved thermal conductivity of the PCM but also mitigated the super cooling issue. They observed that samples maintained stability after three months as there was no visible agglomeration. However, many other studies showed different results. A research by Yu et al. [7] found insignificant improvement in thermal conductivity in liquid phase paraffin (with 1 to 4 wt% MWCNT) and highly clustered emulsions. Fan et al. [8] compared the effects of various nanoparticles on paraffin and found that increasing temperature resulted in lower thermal conductivity. They also found that long and short-walled CNTs did not provide notable enhancement of thermal properties. Wu et al. [9] also reached the same conclusion after dispersing MWCNT to paraffin.

Apparently there is still no agreement on the level of thermal conductivity enhancement of PCMs with MWCNT dispersions. The issue of NEPCM stability remains a major concern, with sometimes contradictory observations from different researchers. In particular, there is no study on the impact of dispersion methods on the thermal conductivity and stability of NEPCMs. This study is an attempt to address these issues. MWCNT dispersed paraffin samples prepared with mechanical dispersion methods, i.e., magnetic stirring then sonication, were investigated experimentally. The stability and thermal conductivity changes of the samples were analyzed.

PREPARATION OF NEPCMs

NEPCMs can be prepared with several methods, including mechanical and chemical dispersions. Mechanical dispersion methods include shear mixing and sonication. Magnetic stirring helps nanoparticles disperse at macro scale by spinning a stirring bar in the liquid medium. In sonication, cavitation is generated in the liquid in which micro bubbles form and collapse suddenly, leading to a good dispersion [10]. Significant amount of energy provided to the solution may decrease the aspect ratio of the CNTs and cause a detrimental effect on their conductive properties [11]. Therefore it is important to apply optimal sonication energy level and time in preparation of the NEPCMs. Another widely used dispersion method is to use surfactants, which is a chemical method. Surfactants help nanoparticles disperse better by altering the surface energy levels to make them less prone to agglomeration. However, addition of surfactant can also change the thermal properties of the nanoparticles [12]. In this study, magnetic stirring and sonication were used in the preparation of NEPCMs since the effects of mechanical dispersion methods are the main interest.

The MWCNTs and paraffin were purchased from Sigma Aldrich. MWCNTs have the following parameters: 6-9nm in diameter and 5 μm in length. The paraffin wax used in this study has a melting temperature range of 53-57 °C. In the preparation, a desired amount of paraffin and MWCNT was first weighed on electronic scales (Denver Instrument-Model P-214, Ohaus Corporation-Adventurer Pro AV8101C, respectively). Then the paraffin was melted in a beaker. The melted paraffin was placed on top of a hot plate stirrer (SP131320-33Q, Cimarec-Thermo Scientific). Temperature of the hot plate stirrer was kept above melting temperature at 70°C. Then sonication was applied to obtain a well dispersed NEPCM. The sonication probe (Sonifier Cell Disrupter, Branson) was set to average amplitude (50%). An environmental test chamber (Tenney Junior Test Chambers, Model T30RC) was used to provide the necessary stable temperatures for thermal conductivity measurement. At least four thermal conductivity measurements were taken for each condition.

THERMAL CONDUCTIVITY MEASUREMENTS

Thermal conductivity measurements were taken using a KD2 Thermal Conductivity Analyzer (Decagon Devices). The measurement technique is based on the transient hot wire principle. For the measurements of liquid samples, a KS-1 probe was used. Care was taken to ensure that there is no natural convection since that would cause errors [13]. For the solid phase, two probes could be used: the SH-1 sensor or the TR-1 sensor. Tight fitting of the probes to the medium is important to eliminate contact resistance due to air gaps. Thermal grease can be filled into the holes for more accurate results. Probes were calibrated with calibration samples provided by the manufacturer. This study found that the TR-1 sensor works better than the SH-1 for measuring thermal conductivity of solid MWCNT-paraffin samples so that probe was used for solid phase measurements. Measurement with this problem has an accuracy of 10%.

RESULTS AND DISCUSSIONS

The stability of the samples was observed visually in terms of settlement and agglomeration of nanoparticles within the paraffin wax. Three thermal cycles (melting-solidification) were carried out in the environmental chamber where samples were melted at 80°C and solidified to 25°C. Visual observation of each sample was made after fully melting.

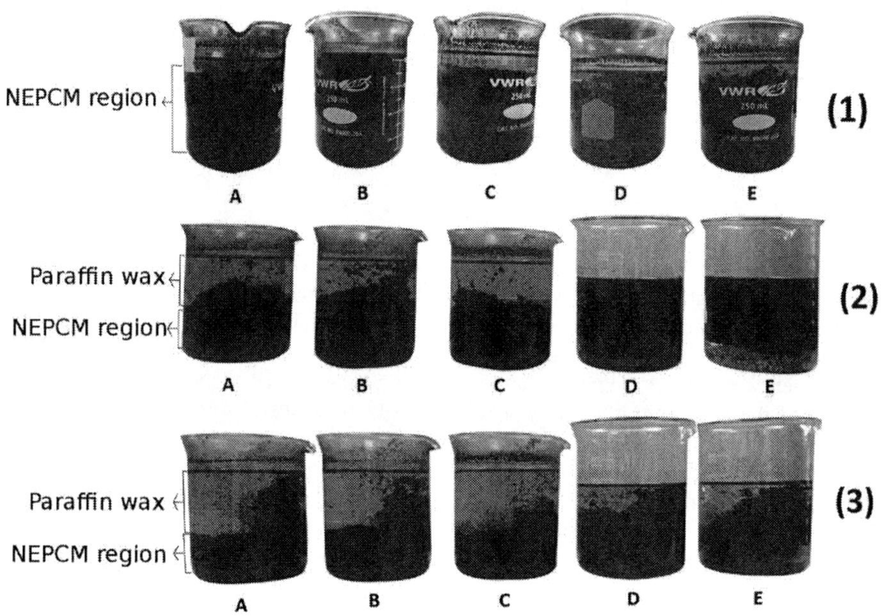

Figure 1: Observed stability of NEPCM after several melting-solidification thermal cycles, **(1)** after 1st thermal cycle; **(2)** after 2nd thermal cycle; **(3)** after 3rd thermal cycle; Different samples are: **A**: MWCNT 0.1 wt%, 100 min sonication, **B**: MWCNT 0.1 wt%, 40 min sonication, **C**: MWCNT 0.075 wt%, 70 min sonication, **D**: MWCNT 0.05 wt%, 100 min sonication, **E**: MWCNT 0.05 wt%, 40 min sonication.

Reasonably stable condition with insignificant settlement was attained for all the samples after the first thermal cycle as shown in Fig. 1 (1). However, considerable particle settlement took place with the second thermal cycle. After the third thermal cycle, significant particle settlement and agglomeration was observed. It can be clearly seen in Fig. 1 (2-3) that particles formed clusters on the lower portion of the container whereas the upper part is mainly composed of paraffin. It is evident after the third thermal cycle that up to a certain height, samples have dark color potentially indicating highly agglomerated regions. However, the region close to surface seems transparent showing that there is no particle network. This could be attributed to the fact that the density of the nanoparticles is higher than that of paraffin resulting in particles settled on the lower part.

It was noticed that different sonication times do not seem to have any effect on the improvement of stability at the investigated nanoparticle loadings. Significant agglomeration and settlement are unavoidable over these thermal cycles. It was also found that stability becomes worse dramatically for all the samples regardless of MWCNT content. It should be noted that temporary stable NEPCM was observed for sample D and E after the second thermal cycle. That was due to beaker replacement. However, sample D and E follow the same trend as sample A, B and C in terms of particle settlement as it is obvious after the third thermal cycle in Fig. 1 (3).

The accuracy and reliability of the KD2 Pro thermal conductivity analyzer were assessed before conducting the main measurements. For this purpose, thermal conductivity of distilled water and paraffin wax was measured at different temperatures and compared with the results from literature. These validation results are shown in Fig. 2. Thermal conductivity measurements were performed in the environmental chamber at different temperatures. It was found that results from present work for both distilled water and paraffin wax are in reasonable agreement with the results from other studies.

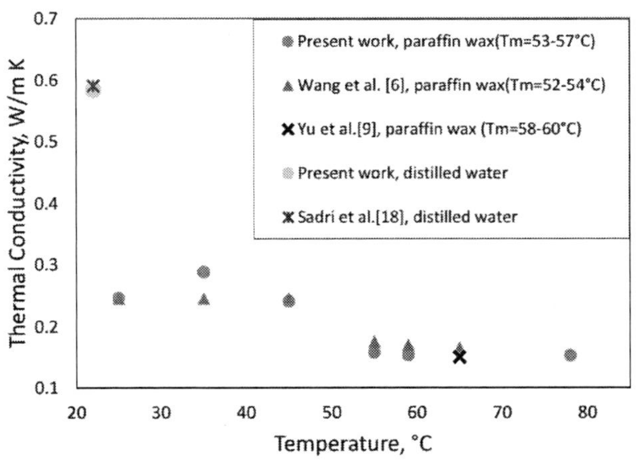

Figure 2: Comparison of thermal conductivity of distilled water and paraffin wax with literature (T_m, melting temperature range)

As can be seen from Fig. 3, measurements for MWCNT-paraffin wax nanocomposites revealed that thermal conductivity of the NEPCMs in liquid phase ranges between 0.15-0.16 W/m K indicating no improvement at the specified nanoparticle loadings, when compared to results in Fig.2. Similarly, results for solid phase NEPCMS did not show any notable enhancement. Minor thermal conductivity improvement was detected at 25°C for samples with longer sonication times. In Fig. 3, the thermal conductivity peaks at 35°C reaching up to 0.33 W/m K. This could be attributed to potential solid-solid phase transition [4, 14].

Overall this study found that dispersing MWCNT to paraffin wax at low loadings did not result in thermal conductivity enhancement when compared with pure paraffin wax results in Fig. 2. Besides, thermal conductivity enhancement does not present any dependency on sonication time. These outcomes could be associated with the inadequate particle network in the NEPCM, which disrupts thermally conductive paths at low loadings of nanoparticles. In addition, highly entangled nanotubes agglomerated in discontinuous clusters might have led to insignificant thermal conductivity improvement. We plan to investigate these mechanisms with detailed scanning electron microscope (SEM) measurements in the next step of this research.

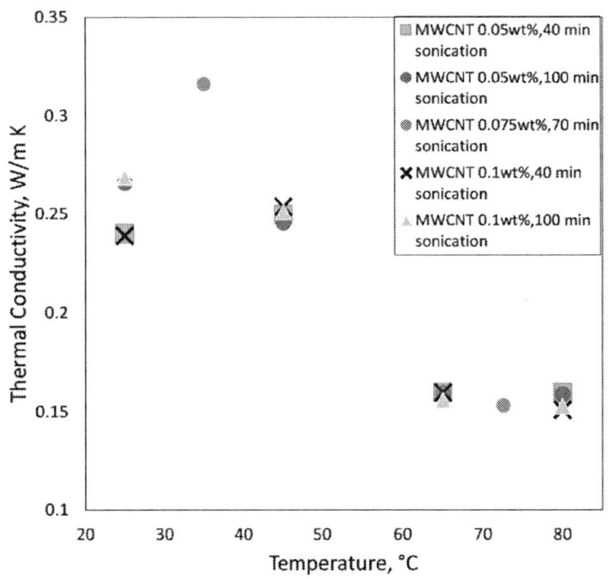

Figure 3: Thermal conductivity of MWCNT dispersed paraffin wax at various loadings and sonication time

It was previously reported that MWCNT has a high thermal conductivity up to 2000 W/m·K [15]. Therefore, MWCNTs were thought to be able to increase the thermal properties of the dispersed NEPCMs significantly. However, experimental results in this study, as shown in Fig. 3, have shown that the enhancement in thermal conductivity is not always as pronounced as it was foreseen.

CONCLUSIONS

MWCNTs dispersed paraffin wax NEPCMs were prepared using mechanical dispersion methods namely magnetic stirring and sonication. Thermal conductivity results showed that MWCNT content and sonication time do not have any notable impact on the enhancement in solid and liquid phases of the NEPCM. Stability observations demonstrated that samples become unstable with highly clustered settlement on the bottom after thermal cycles regardless of sonication time and MWCNT concentration.

This study makes contribution to show that MWCNTs might not be as promising as previously thought in enhancing thermal conductivity of PCMs at low nanoparticle loadings and mechanical dispersion methods seem to be not sufficient to achieve long term stability of MWCNT dispersed NEPCMs.

ACKNOWLEDGEMENTS

This study was financially supported by Natural Science and Engineering Council of Canada (NSERC), Research and Development Corporation (RDC), Shandong Natural Science Foundation (ZR2015YL001) and Shandong Academy of Sciences Basic Foundation (201506).

REFERENCES

1. Dincer I, Rosen M.A.,2011 *Thermal energy storage: systems and applications*, Wiley.
2. Kenisarin, M. and Mahkamov, K. 2007, "Solar energy storage using phase change materials", *Renew. Sustain. Energy Rev.*, Vol. (11), no. 9, pp. 1913–1965.
3. Khodadadi J. M., Fan L., and Babaei H. 2013, "Thermal conductivity enhancement of nanostructure-based colloidal suspensions utilized as phase change materials for thermal energy storage: A review," *Renew. Sustain. Energy Rev.*, Vol. (24), pp. 418–444.
4. Wang J., Xie, H., and Xin, Z. 2009, "Thermal properties of paraffin based composites containing multi-walled carbon nanotubes", *Thermochimica Acta,* Vol. (488), pp. 39–42.
5. Angayarkanni S.A. and Philip J. 2015, "Thermal conductivity measurements in phase change materials under freezing in presence of nanoinclusions", *Journal of Applied Physics*, Vol. (118), 094306.
6. Kumaresan V., Velraj R., and Das S. K. 2012, "The effect of carbon nanotubes in enhancing the thermal transport properties of PCM during solidification", *Heat Mass Transfer*, Vol. (48), pp. 1345–1355.
7. Z. Yu, Fang X., Fan L., Wang X., Xiao Y., Zeng Y., Xu X., Hu Y., and Cen K. 2012, "Increased thermal conductivity of liquid paraffin-based suspensions in the presence of carbon nano-additives of various sizes and shapes",*Carbon*, Vol. (3), pp. 1–9.
8. Fan L., Fang X., Wang X., Zeng Y., Xiao Y., Yu Z., Xu X., and Y. Hu 2013, "Effects of various carbon nanofillers on the thermal conductivity and energy storage properties of paraffin-based nanocomposite phase change materials", *Applied Energy*, Vol. (110), pp. 163–172.
9. S.Y. Wu, Tong X., Nie C.D., Peng D.Q., S. G. Gong, and Wang Z. Q. 2016, "The effects of various carbon nanofillers on the thermal properties of paraffin for energy storage applications", *Journal of Thermal Analysis and Calorimetry*, Vol. (124), pp. 181–188.
10. Digital Sonifier Models 250 & 450 User's Manual, Branson Ultrasonics Corportation.
11. S. M. Sabet., H. Mahfuz and J. Hashemi., M. Nezakat,J. A. Szpunar 2015 "Effects of sonication energy on the dispersion of carbon nanotubes in a vinyl ester matrix and associated thermo-mechanical properties", *Journal of Material Science*, Vol(50),pp. 4729–4740.
12. M. Kamalgharibi, F. Hormozi, S.A.H. Zamzamian, and M.M. Sarafraz 2015 "Experimental studies on the stability of CuO nanoparticles dispersed in different base fluids: influence of stirring, sonication and surface active agents" *Heat Mass Transfer*, Vol (50), pp. 55–62,
13. KD2 Pro Thermal Properties Analyzer, Operator's Manual, Decagon Devices
14. Tang Q, Sun J., Yu S., and G. Wang 2014, "Improving thermal conductivity and decreasing supercooling of paraffin phase change materials by carbon nanotubes", *RSC Advances*., Vol. (4), pp. 36584–36590.
15. Sadri R., Ahmadi G., Togun H., Dahari M., Kazi S.N., Sadeghinezhad E., Zubir N. 2014, "An experimental study on thermal conductivity and viscosity of nanofluids containing carbon nanotubes", *Nanoscale research letters*, Vol. (9),pp. 151.

LATTICE BOLTZMANN SIMULATION OF CONVECTION HEAT TRANSFER USING A THERMAL EXTERNAL BOUNDARY METHOD

Amin Parvan
Mechanical Engineering Department
Shahid Bahonar University of Kerman
Kerman, Iran
Amin.parvan@gmail.com

Mohammad Rahnama
Mechanical Engineering Department
Shahid Bahonar University of Kerman
Kerman, Iran
rahnama@uk.ac.ir

Ned Djilali
Mechanical Engineering Department
University of Victoria
Victoria, BC, Canada
ndjilali@uvic.ca

ABSTRACT

A novel thermal Lattice Boltzmann Method is introduced in which convection heat transfer from a solid particle surface is simulated using the Lattice Boltzmann Method. Fluid-solid interaction and its relevant heat transfer is simulated using an external boundary force method in which momentum and heat exchange at the interface is obtained by interpolation functions. This method is capable of simulating heat transfer from a particle in a fluid flow as well as particulate flows with many particles. It is also a computationally efficient method as all computations are performed on a regular uniform lattice. Results of local and averaged Nusselt number obtained for convection heat transfer from a stationary circular cylinder in a uniform flow as well as natural convection in an enclosure demonstrate the accuracy of the method compared to previous published works.

KEYWORDS: Lattice Boltzmann Method, External Boundary force, natural convection

INTRODUCTION

Analysis of heat transfer in a particulate flow is a challenging problem mainly due to the complex mechanism of momentum and heat exchange at the interface between solid and fluid. While various CFD methods have been used in simulating particulate flows, the Lattice Boltzmann Method (LBM) has attracted specific attention mainly due to its computational efficiency [1]. In spite of considerable progress in modelling the fluid flow aspect of particulate flows, simulations of heat transfer with thermal LBM have been limited..

Simulation of heat transfer using LBM was first performed successfully using a double distribution function model proposed by He, et. al. [2]. They proved that such a double distribution function leads to the correct derivation of the energy equation for an incompressible flow. Wang, et. al. [3] presented a LBM algorithm for fluid-solid conjugate heat transfer. They used a new generalized heat generation implementation which ensures temperature and heat flux continuity at the interface and tested their method in a micro-channel flow.

One of the main challenge in simulation of particulate flow is that solid-fluid interaction force plays an essential role in predicting correct momentum and heat transfer between solid and fluid. Goldstein et al. [4] proposed a method, called External Boundary Force (EBF), to obtain solid-fluid interaction forces. Their method which is similar to the Immersed Boundary Method (IBM) [5] uses an Eulerian grid for fluid and a Lagrangian one for the motion of particles which could be discretized with a boundary-fitted grid. Both standard IBM and EBF methods are second order accurate. The main difference between IBM and EBF is in the representation of motion of the solid body: the main idea of IBM is to use the physical boundary as a spring with high stiffness between internal solid nodes and calculate the acceleration or particle from the energy equation while the EBF method obtains acceleration from the dynamics of rigid body. To implement the no-slip condition at a particle surface, Wu and Aidun [6] added a body force term to the LB equation which is obtained from momentum difference between interpolated fluid velocity at solid boundary nodes and solid velocity at the same nodes, as a source term. They showed that EBF method is a reliable method for both deformable and non-deformable solid particle motion simulation in a fluid. They implemented the effect of fluid-solid interaction force by adding an additional term to the collision step in Boltzmann equation.

To the best of the authors' knowledge, the external boundary method has not been extended to heat transfer problems. The present paper presents the Thermal External Boundary Method (TEBM) and its application to forced and natural convection simulations.

EXTERNAL BOUNDARY FORCE

The External Boundary Force (EBF) is a method for simulating fluid-solid interaction force based on momentum exchange at the interface. Considering two sets of grid points, namely fluid Eulerian and solid Lagrangian ones, (cf. Fig. (1)) we note that solid nodes do not correspond to the fluid nodes in general, however, both fluid and solid velocities at boundary nodes need to be computed. Knowing these values, momentum exchange between fluid and solid can be computed. The fluid velocity at solid boundary nodes is obtained using an interpolation function which is explained in details below.

Figure 1 shows part of a computational domain consisting of a solid circular particle submerged in a fluid domain. A series of Lagrangian nodes located on the interface between solid and fluid are superimposed on the uniform Eulerian grids. The set of points are shown as $\mathbf{x}^b$ and $\mathbf{x}^f$ representing position vectors of solid nodes located on the boundary of particle and fluid nodes respectively.

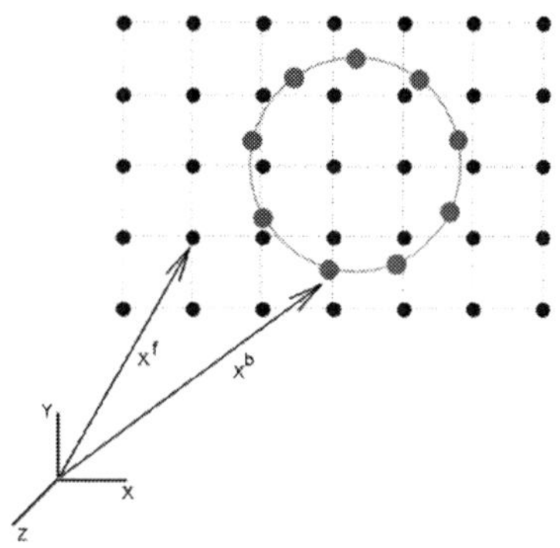

Figure 1: Solid and fluid nodes represented in external boundary force method.

In general, solid boundary nodes do not coincide with fluid nodes. Fluid velocity at solid boundary nodes, $\mathbf{U}_\alpha^\mathbf{f}$, is obtained through an interpolation scheme using discrete Delta function as:

$$\mathbf{U}_\alpha^\mathbf{f} = \sum_{k=1}^{n} \mathbf{U}_\mathbf{k}^\mathbf{f}\, \mathrm{D}(x_\alpha^b - x_k^f) \qquad (1)$$

where $\mathbf{U}_\mathbf{k}^\mathbf{f}$ is the velocity of fluid node "k" near the "α^{th}" solid boundary node. "n" determines the number of fluid nodes which are used in this interpolation. Two dimensional discrete Dirac delta function is expressed as:

$$\mathrm{D}(x) = \begin{cases} \frac{1}{64h^3}\left[1+\cos\left(\frac{\pi x}{2h}\right)\right]\left[1+\cos\left(\frac{\pi y}{2h}\right)\right] & if\ |x_\alpha^b - x_k^f| < 2h \\ 0 & otherwise \end{cases} \qquad (2)$$

Here 'h' is the unit lattice length. The accuracy of this function is of first order. The above relations are written for the present time. Knowing fluid velocity at boundary nodes in previous time step, it is possible to compute rate of momentum change and the subsequent fluid-solid interaction force. In EBF, Fluid velocity at a solid boundary node is considered to be equal to the velocity of solid boundary nodes at previous time step. It is worth mentioning that solid particle velocity is a known quantity at previous time step which starts from known initial conditions. It can be expressed as:

$$\mathbf{U}^\mathrm{f}(x^b, t-\Delta t^{lbm}) = \mathbf{U}^\mathrm{s}(x^b, t-\Delta t^{lbm}) \qquad (3)$$

here Δt^{lbm} and $\mathbf{U}^\mathrm{s}$ are Lattice Boltzmann time step (which is considered as 1) and particle velocity, respectively. The fluid-solid interaction force "$\mathbf{F}^{\mathrm{fsi}}$" acting on solid boundary nodes is obtained from rate of momentum exchange as:

$$\mathbf{F}^{\mathrm{fsi}}(x^b, t) = \rho_f\left[\mathbf{U}^\mathrm{f}(x^b, t) - \mathbf{U}^\mathrm{f}(x^b, t-\Delta t^{\mathrm{lbm}})\right]/\Delta t^{lbm} \quad (4)$$

where ρ_f is the density of fluid. The next step is to find interaction force at fluid nodes adjacent to solid boundary which is obtained through similar interpolation using Dirac delta function with a negative sign due to Newton's third law:

$$\mathbf{g}(x_\alpha^f, t) = -\sum_{k=1}^{m} \mathbf{F}^{\mathrm{fsi}}(x_k^b, \mathrm{t})\mathrm{D}(x_k^b - x_\alpha^f) \qquad (5)$$

"g" is introduced as External Boundary Force and will be used as source term in Lattice Boltzmann equation. Except fluid-solid interaction force, there might be other external forces such as gravitational, lubrication, buoyancy and magnetic forces. The total force acting on solid boundary nodes expressed as:

$$\mathbf{F}(x^b, t) = \mathbf{F}^{\mathrm{fsi}}(x^b, t) + \mathbf{F}^{\mathrm{ext}}(x^b, t) \qquad (6)$$

The force $\mathbf{F}(x^b, t)$ in equation 6 exerts on solid boundary nodes, while the total force and torque act on immersed particle boundary should be obtained as:

$$\mathbf{F}^\mathrm{T} = \sum_{\alpha=1}^{N} \mathbf{F}(x_\alpha^b, t) \qquad (7)$$

$$\mathbf{T}^\mathrm{T} = \sum_{\alpha=1}^{N} \mathbf{F}(\mathbf{x}_\alpha^\mathbf{b}, t)\,(x^{cg} - x_\alpha^b) \qquad (8)$$

here "N" is the number of nodes on the boundary. "x^{cg}" is the center of gravity of the particle. The Newtonian dynamics equations for a solid body are expressed as:

$$\mathbf{M}\frac{\mathrm{d}\mathbf{U}}{\mathrm{dt}} = \mathbf{F}^\mathrm{T} \qquad (9)$$

$$\mathbf{I}\frac{\mathrm{d}\boldsymbol{\Omega}}{\mathrm{dt}} + \boldsymbol{\Omega} \times (\mathbf{I}\,.\boldsymbol{\Omega}) = \mathbf{T}^\mathrm{T} \qquad (10)$$

where "**M**", "**I**", "**Ω**" and "**U**" are the mass, inertial tensor of the particle, angular and linear velocities of the solid body, respectively. Fourth-order Runge-Kutta method could be used to solve the equation (9) and (10) to obtain the linear and angular velocities of the particle.

THERMAL EXTERNAL BOUNDARY METHOD

Building on the discussion in the previous section, the thermal external boundary method is based on obtaining fluid temperature, $\mathbf{T}_\alpha^\mathbf{f}$, at solid boundary nodes through interpolation using Dirac delta function used in equation (2), i.e.:

$$\mathbf{T}_\alpha^\mathbf{f} = \sum_{k=1}^{n} \mathbf{T}_\mathbf{k}^\mathbf{f}\, \mathrm{D}(x_\alpha^b - x_k^f) \qquad (11)$$

Here, $\mathbf{T}_\mathbf{k}^\mathbf{f}$ is the fluid node temperature near the boundary node. Temperature of fluid nodes on solid boundary at previous time step are defined as:

$$\mathbf{T}^\mathrm{f}(x^b, t-\Delta t^{lbm}) = \mathbf{T}^\mathrm{s}(x^b, t-\Delta t^{lbm}) \qquad (12)$$

where $\mathbf{T}^\mathrm{s}$ is the temperature of solid boundary nodes which is known. The heat flux on the surface of solid particle is obtained from rate of change of temperature at a fluid node corresponds to a solid boundary point as:

$$\mathbf{Q}^{\mathrm{fsi}}(x^b, t) = \rho C_p\left[\mathbf{T}^\mathrm{f}(x^b, t) - \mathbf{T}^\mathrm{f}(x^b, t-\Delta t^{\mathrm{lbm}})\right]/\Delta t^{lbm} \quad (13)$$

This parameter is considered a heat source/sink term in temperature evolution equation as discussed later, but it has to be obtained for fluid nodes through interpolation. Therefore, heat source/sink for fluid neighboring-nodes is obtained as:

$$Q(x_\alpha^f, t) = -\sum_{k=1}^{m} \mathbf{Q}^{\mathrm{fsi}}(x_k^b, \mathrm{t})\mathrm{D}(x_k^b - x_\alpha^f) \qquad (14)$$

where $\boldsymbol{Q}(x^f_\alpha, t)$ which is defined as Thermal External Boundary source, is added to Thermal Lattice Boltzmann Equation as follows:

$$T_\alpha(\mathbf{x}^\mathbf{f} + \mathbf{c}_\alpha \Delta t, t + \Delta t) - T_\alpha(\mathbf{x}^\mathbf{f}, t) = -\frac{T_\alpha(\mathbf{x}^\mathbf{f},t) - T_\alpha^{eq}(\mathbf{x}^\mathbf{f},t)}{\tau_T} + \Delta t \mathbf{Q}_\alpha \quad (15)$$

The above equation is an evolution equation for the temperature field analogous to the Lattice Boltzmann equation which is expressed in the form:

$$f_\alpha(\mathbf{x}^\mathbf{f} + \mathbf{c}_\alpha \Delta t, t + \Delta t) - f_\alpha(\mathbf{x}^\mathbf{f}, t) = -\frac{f_\alpha(\mathbf{x}^\mathbf{f},t) - f_\alpha^{eq}(\mathbf{x}^\mathbf{f},t)}{\tau} + \Delta t F_\alpha \quad (16)$$

T_α is temperature distribution function in Eq. (15), T_α^{eq} is the equilibrium distribution function used for direction "α" and "τ_T" is thermal relaxation time. In order to solve the incompressible thermal flow, we substitute $\rho = \rho_0$ into the equilibrium distribution function as proposed by Peng [7]:

$$T_0^{eq} = -\frac{\rho_0 T}{3}\frac{\mathbf{u}.\mathbf{u}}{c^2} \quad (17a)$$

$$T_{1-4}^{eq} = \frac{\rho_0 T}{9}\left[\frac{3}{2} + \frac{3\mathbf{c}_\alpha.\mathbf{u}}{2c^2} + \frac{9(\mathbf{c}_\alpha.\mathbf{u})^2}{2c^4} - \frac{3\mathbf{u}.\mathbf{u}}{2c^2}\right] \quad (17b)$$

$$T_{5-8}^{eq} = \frac{\rho_0 T}{36}\left[3 + \frac{6\mathbf{c}_\alpha.\mathbf{u}}{c^2} + \frac{9(\mathbf{c}_\alpha.\mathbf{u})^2}{2c^4} - \frac{3\mathbf{u}.\mathbf{u}}{2c^2}\right] \quad (17c)$$

thermal diffusivity "χ" and macroscopic fluid temperature are calculated using:

$$\chi = 2(\tau_T - 0.5){c_s}^2 \Delta t \quad (18)$$

$$T(\mathbf{x}^\mathbf{f}, t) = \frac{1}{\rho_0}\sum_\alpha T_\alpha(\mathbf{x}^\mathbf{f}, t) \quad (19)$$

RESULTS

As a first validation test case, forced convective heat transfer from a heated circular cylinder placed in a uniform flow is studied. Figure 2 shows this flow geometry which consists of a cylinder with diameter of D and its computational domain of 30D×20D. The cylinder is placed at a distance of 8D downstream of inlet boundary. This test case has been studied by other researchers using different numerical methods [7].

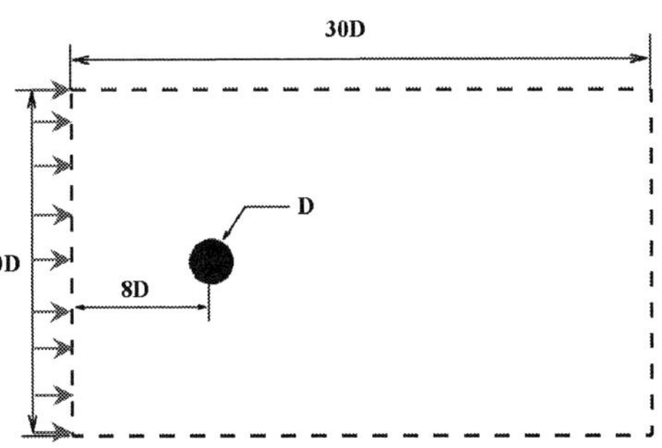

Figure 2. Computational domain considered for the flow geometry of a cylinder in a uniform flow.

Two important parameters in this problem are the Prandtl number, Pr, and Reynolds number, Re. The aim of this simulation is to obtain local and mean Nusselt numbers on the surface of the cylinder as a function of Reynolds number and Pradtl number. The local and mean Nusselt numbers are expressed as:

$$Nu = -\frac{\partial T}{\partial r}\Big|_{wall} \quad (32)$$

$$\overline{Nu} = \frac{1}{S}\int_0^S Nu\, ds \quad (33)$$

where "S" is the circumferential length of the cylinder surface. Local temperature gradient is computed by using second-order approximation as:

$$\frac{\partial T}{\partial n} \approx \frac{-3\theta_w + 4\theta(r+\delta x,t) - \theta(r+2\delta x,t)}{2\delta x} \quad (34)$$

Simulations were carried out for a regular rectangular domain of dimensions 30D×20D. Hydrodynamic boundary conditions used on top and bottom boundaries were free slip boundary conditions. Boundary conditions on inlet and outlet boundaries were applied based on known velocity $U = 1$ and zero velocity gradient, respectively. Thermal boundary conditions consist of adiabatic boundary condition for top and bottom walls, constant inlet temperature and zero temperature gradient for outlet boundary.

Simulations were done for three different Reynolds numbers of Re=10, 20, 40, all at a fixed Prandtl number of $Pr = 0.7$. The drag coefficient is defined as:

$$C_d = \frac{F_D}{\frac{1}{2}\rho U^2 D} \quad (35)$$

where F_D is the drag force computed from the fluid-solid interaction force. The drag coefficients obtained for Re=10, 20 and 30 are 1.863, 2.44 and 3.381 respectively which show an error of less than 2% compared to previous published data of Ref. [8]. Figure 3 shows temperature contours for uniform flow on a circular cylinder at Re=20. Higher temperature gradient is observed upstream of the cylinder compared to its downstream region which covers the recirculating flow region.

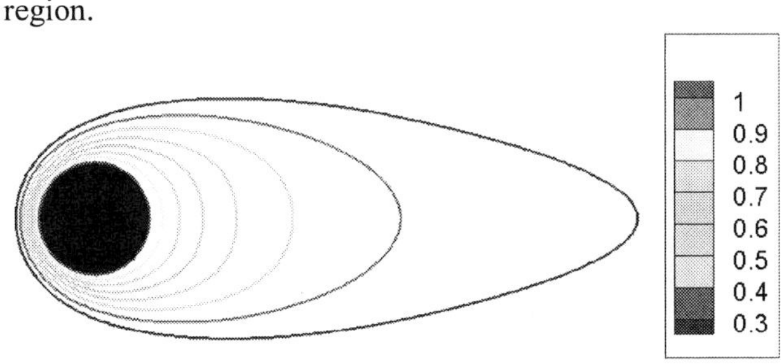

Figure 3. Temperature contours for flow over a circular cylinder at Re=20.

The second test case considers natural convection between a hot cylinder and a cold square enclosure. Here a cylinder with a diameter of 80 lattice units is placed at the center of a square domain consisting of 201×201 lattice nodes. The nob-dimensional temperature of the cylinder is prescribed as 1 while that of the square enclosure is 0. The computations are performed at a Rayleigh number of Ra=10^4. Figure 4 shows streamlines and temperature contours for this geometry.

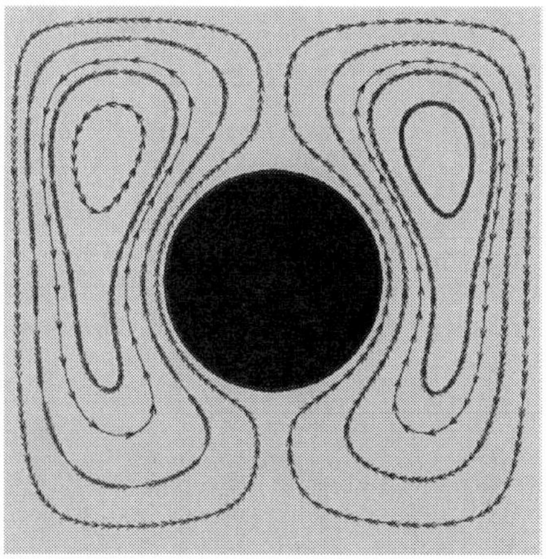

(a)

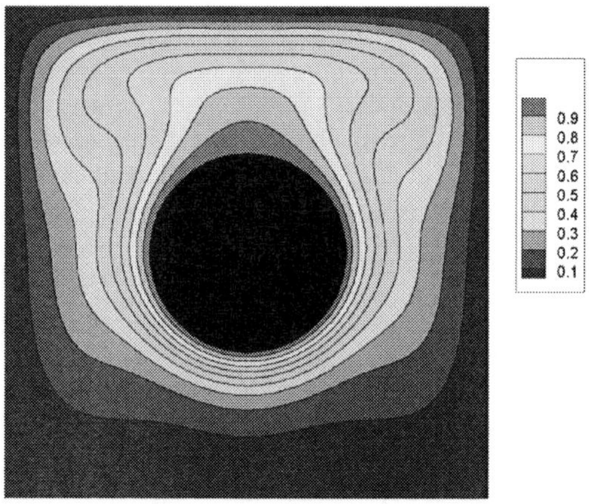

(b)

Figure 4. (a) Streamlines and (b) temperature contours for natural convection between a cylinder and a square enclosure.

Results of mean Nusselt number obtained for different cylinder diameter at three Rayleigh numbers of 10^4, 10^5 and 10^6 , are compared with previous published work in table 1.

Table 1. Mean Nusselt number vales for natural convection between cylinder and square enclosure.

Ra	R/L	Present	Seta[8]
10^4	0.1	2.296	2.0206
10^5	0.2	5.461	5.253
10^6	0.3	13.128	12.87

CONCLUSIONS

The Thermal External Boundary Method which is an extension of the External Boundary Force method in conjunction with the Lattice Boltzmann method, was shown to be capable of predicting temperature distribution and heat flux in natural and forced convection in simple geometries consisting of uniform flow over a circular cylinder and natural convection in a square enclosure from a heated cylinder placed at its center. The computed Nusselt numbers for these two geometries show good correspondence with previous published work. The TEBM formulation should provide a good and efficient computational platform for simulating convective heat transfer from a moving particle in single or multiple particle problems to be investigated in the future..

REFERENCES

1. Ladd, A.J.C., 2015, "Lattice Boltzmann methods for suspensions of solid particles", *Molecular Physics*, Vol (113), pp 2531–2537.
2. He, X, Chen, S., Doolen, G. D., 1998, "A novel thermal model for the Lattice Boltzmann method in incompressible limit", *J. Comput. Phys.* Vol(146), pp 282–300.
3. Wang, J., Wang, M. Li, Z., 2007, "A lattice Boltzmann algorithm for fluid–solid conjugate heat transfer", *Int. J. Thermal Sciences*, Vol (46), pp 228-234.
4. Goldstein, D., Handler, R., Sirovich, L., 1993, "Modeling a no-slip flow boundary with an external force field", *Journal of Computational Physics*, Vol (105), pp 354–366.
5. Feng, Z., Michaelides, E., 2004, "The immersed boundary-lattice Boltzmann method for solving fluid–particles interaction problems", *Journal of Computational Physics*, pp 195-602.
6. Wu, J. and Aidun, C. K., 2010, "Simulating 3D deformable particle suspensions using lattice Boltzmann method with discrete external boundary force", *Int. J. Numer. Meth. Fluids*, Vol (62), pp 765–783.
7. Peng, Y., Shu, C., Chew, Y. T., 2003, "Simplified thermal lattice Boltzmann model for incompressible thermal flows", *Phys. Rev. E.*, Vol (68), pp 026701-1.
8. Seta, T., 2013, "Implicit temperature-correction-based immersed-boundary thermal lattice Boltzmann method for the simulation of natural convection", *Physical Review E*, Vol (87), pp 063304-1.

Proceedings of the 26th CANCAM
Victoria, British Columbia, Canada, May 29 – June 1, 2017

MEASUREMENT OF ADIABATIC TEMPERATURE CHANGE IN A POROUS REGENERATOR USING FIBRE BRAGG GRATING

T.V. Christiaanse*, G. A. M. Burton, P. Govindappa, I. Niknia, R. Teyber, P.V. Trevizoli, A. Rowe
Mechanical Engineering
Victoria, B.C., Canada
theochri@uvic.ca

ABSTRACT

In Fibre Bragg Grating (FBG) is explored as an alternative to conventional temperature sensors in active magnetic regenerator devices. The complexity of varying magnetic fields, fluids and small space brings with it great difficulty to install temperature sensors. Fibre Bragg Grating allows for temperature sensing that is not impacted by magnetic fields and can be installed in small cavities. The fibre is installed in an AMR prototype device to measure the adiabatic temperature change of the active material as the magnetic field is varied. To validate the temperature sensing capabilities, a 2D COMSOL model is built to simulate expected temperature variation due to conduction of the assembly. The model matches well with measured values at low frequencies of 0.33Hz. At higher frequencies, the model over predicts the measured values. As a result, it is suggested that this method be used for low frequency acquisition (<0.33 Hz) but not for high.

KEYWORDS: Magnetic, refrigeration, heat pump, Fibre Bragg Grating, adiabatic temperature, AMR

INTRODUCTION

Magnetic refrigeration is a technology that could provide a more efficient heat pump as compared to vapor compression systems [1]. Magnetic refrigeration is based on the magnetocaloric effect (MCE). MCE is the reversible temperature change resulting from a change in applied magnetic field. For this work, Gadolinium (Gd) is used as magnetocaloric (MC) material. Gd is a soft magnetic material with no hysteresis so that the MCE is fully reversible [2]. A four-stage heat pump cycle can be built. First the field is applied increasing the temperature of the solid. Energy is removed to ambient temperatures. Then the magnetic field is removed lowering the temperature of the sample. Now, heat can be absorbed at lower temperatures. The MCE effect is on the order of several degrees per Tesla applied field which is not enough to build a useful temperature span. To solve this problem, a regenerative cycle is used and results in an active magnetic regenerator (AMR) constructed of Gd. The AMR is similar to a passive regenerator but has an active component due to the magnetic field induced temperature change. The AMR consists of a porous matrix through which, a heat transfer fluid can be oscillated. By timing the fluid and field waveforms, a temperature gradient can be created across the regenerator.

There are still several challenges to making magnetic refrigeration competitive with convention vapour compression devices [3]. One of the key issues is the understanding of the active magnetic regenerator thermodynamics where the operation is closely related to the temperature distribution. This regenerator is composed of porous magnetic material (in our case Gd) that reacts with the magnetic field source. The sensing of the temperature is made difficult due to several factors. First, the magnetic field interacts with thermocouples and the lead wires connected to the temperature sensors. Secondly, creating a magnetic field in a large space is costly, thus there is limited space to place for installing sensors. Fibre Bragg grating (FBG) is an interesting solution to these problems as it can be made small, more than one sensor can be used in the same fibre, and the magnetic field does not impact the temperature reading. The following paper describes the measurement of adiabatic temperature change inside an active magnetic regenerator prototype. No heat transfer fluid is used to simplify the validation. The FBG used is placed at the end of a fibre optic cable. The regenerator is constructed of Gd and the magnetic field is varied with a nested Halbach array [4].

PERMANENT MAGNET PROTOTYPE DEVICE

The FBG is installed in the PM1 prototype device [4] as shown schematically in Fig. 1. The fibre is cased in a small metal tube inserted into the regenerator housing. The tube is made of stainless steel and has a wall thickness of 0.004in. The regenerator housing is a G10 tube that connects the cold and hot heat exchangers. After the tube is installed in the G10 housing, the regenerator matrix material is added. The matrix material consists of 300 um gadolinium spheres. The length of the regenerator is approximately 55 mm and is situated at the center of the magnetic field source. The PM1 magnetic field is generated using a pair of nested Halbach magnets. The magnetic field is varied inside the bore of the magnets by rotating the inner Halbach array with respect to the outer Halbach array. The temperature of the gadolinium spheres changes with the varying magnetic field.

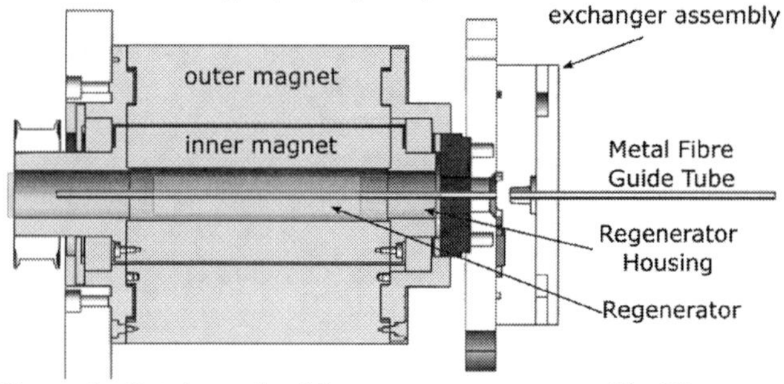

Figure 1: Cut through of the measurement assembly. The two segments in the center are the Halbach magnets. Two bearings suspend the inner magnet. The metal tube is inserted through the cold heat exchanger though the whole assembly. The hot heat exchanger is not displayed.

FIBRE BRAGG GRATING (FBG)

The FBG is constructed at the end of a fibre optic cable. To produce the light source and recording wavelength a SmartScan by Smart Fibres is used. The transition frequency is 125Hz and the acquisition frequency is 15Hz. The FBG reflects the light at a single wavelength. This reflected wavelength linearly varies with temperature and pressure applied at the fibre tip as expressed in Eq. 1 [5]:

$$\lambda_{\mathrm{FBG}} + \Delta\lambda = \lambda_{\mathrm{FBG}} + \alpha_T \Delta T + \alpha_p \Delta p \qquad (1)$$

where, λ_{FBG} is the reflected wavelength at room temperature and atmospheric pressure, $\Delta\lambda$ is the change in wavelength, ΔT is the change in temperature, Δp is the change in pressure and calibration constants, α_p and α_T are assumed fixed values in the region of interest. In our case the fibre is installed in a tube so as to prevent variations in pressure. The temperature calibration constant is found by submerging the fibre in a temperature controlled chiller bath and measuring the wavelength shift. This measurement can be found in the attachment 1.

MODELING OF HEAT TRANSFER

To determine the impacts of device configuration and imperfect thermal control on the measured temperature reading of the FBG, a 2D heat transfer model is built in COMSOL. A plane transverse to the regenerator axis is considered (see Fig. 2) so that only radial conduction is considered. The outer boundary condition is set to the ambient temperature, T_{amb}. The temperature at the core of the fibre, T_{FBG}, is taken to be the temperature sensed by the FGB. The driving force of the FBG temperature is the heat generated by the Gd spheres. The applied field is modeled using the measured field variation data of the Halbach magnets. The model determines the temperature variation in the fibre by solving the time dependent energy balance. The overall governing differential equation that is solved is given in Eq. 2.

$$d_z \rho C_p \frac{\partial T}{\partial t} + \nabla \cdot (d_z k \nabla T) = d_z Q_{gen} \qquad (2)$$

where, d_z is the partial depth of the regenerator, ρ is the density of the local element, C_p is the specific heat at the local element, T is the temperature at the local element, t is the time of the simulation, k is the conduction coefficient of the local element, and Q_{gen} is the heat generated by the MCE effect only in the regenerator domain, see Fig. 2. The conduction coefficient of the regenerator is a volumetric average of air and Gd. The adiabatic temperature change is extracted from the simulation by measuring the oscillating temperature, T_{FBG}. The steady state of the simulation is found when the cycle variation is less than 0.1K.

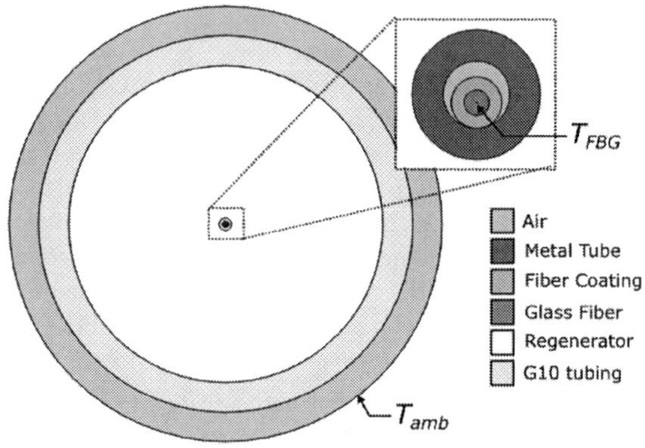

Figure 2: Graphical representation of the slice through of the regenerator.

The heat generated Q_{gen} $\left[\frac{W}{m^3}\right]$ by the spheres is calculated by using mean field theory (MFT) adiabatic temperature change to provide a function for the MCE; $\Delta T(T, H_i, H_f)$ and specific heat, $c_p(T, H_i)$ [6].

$$Q_{gen} = \frac{\rho c_p(T,H_i)\Delta T(T,H_i,H_f)}{\Delta t} \qquad (3)$$

where, ΔT is the adiabatic temperature change as function of the, T, initial temperature, H_i, initial field value, H_f, the final field value after, Δt, timestep. The specific heat is a function of temperature and initial field. The density for Gd is fixed at 7900 kg/m^3. The field data from the Halbach magnets is known along the center axis in the bore and as a function of the rotation of the inner magnet with respect to the outer magnet [4]. This data is corrected for demagnetizing field effects using Eq. 4 [7].

$$H = H_{apl} - N_d M(T, H_{apl}) \qquad (4)$$

The applied field is reduced by the magnetization, $M(T, H_{apl})$, of Gd and the demagnetization coefficient, N_d. This coefficient is a function of the shape, N_{sh}, and porosity, α, of the sample.

$$N_d = 1/3 + (1-\alpha)(N_{sh} - 1/3) \quad (5)$$

The shape demagnetizing factor is taken from Chen et al. who provides a table of values for various cylindrical shapes [8].

RESULTS

In Fig. 3 the results of the modeling and experiments are presented. The experiments are performed by moving the single ended fibre along the axis of the tube. Then the device is run at a set frequency. The temperature change is extracted using the calibration constant. Three experiments are performed: one at 1Hz and two at 0.33Hz, indicated in Fig. 3. After the second 0.33Hz experiment the fibre broke inside the tube. The experiment had to be disassembled to extract the fibre. The fibre was broken to close to the FBG so it could not be reattached to the fibre cable. The 0.33Hz simulated result is in good agreement with the measured results, but the model results of the 1Hz case to not match well.

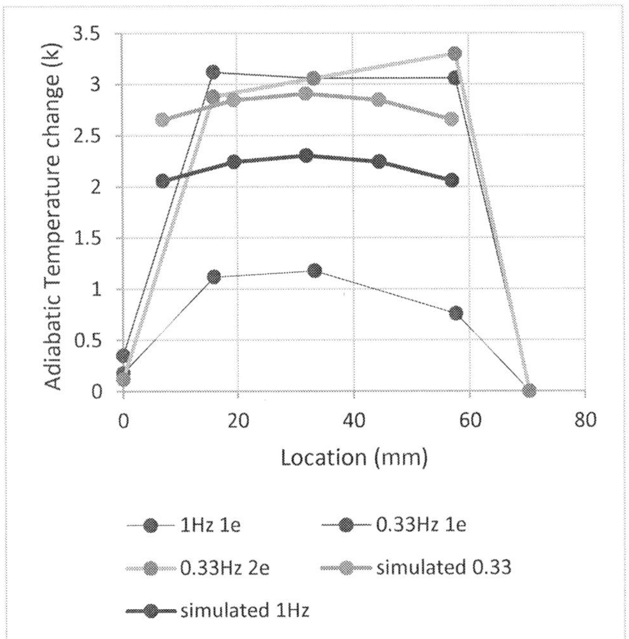

Figure 3: The results of the COMSOL simulations and the experimentally measured adiabatic temperature change.

CONCLUSIONS

Simulations and experimental results match well at 0.33Hz but the simulations overpredict the adiabatic temperature change at 1Hz. The simulations can be expanded to a 3D simulation to include the axial conduction. This could improve the accuracy of the simulation.

Thermal problem is not trivial for this measurement method and needs to be carefully considered; however, at low frequencies the fibre can find the correct temperature variation expected in the fibre.

The next phase is to use a fibre with multiple FBG's at different locations to measure the temperature along the axis of a working active regenerator.

ACKNOWLEDGMENTS

We would like to thank the Natural Sciences and Engineering Research Council of Canada and BASF New Business for their support of this research.

ATTACHMENT 1: CALIBRATION OF THE FBG

To use the Fibre Bragg Grating (FBG) it is calibrated. The fibre is inserted into a tube and submerged in a chiller. The chiller temperature is changed to several different temperature points. The peak wavelength is registered at the temperature settings. A K-type Omega thermocouple installed close to the FBG is used to measure the chiller bath temperature. Then the calibration constant is found by the slope in Fig. 4.

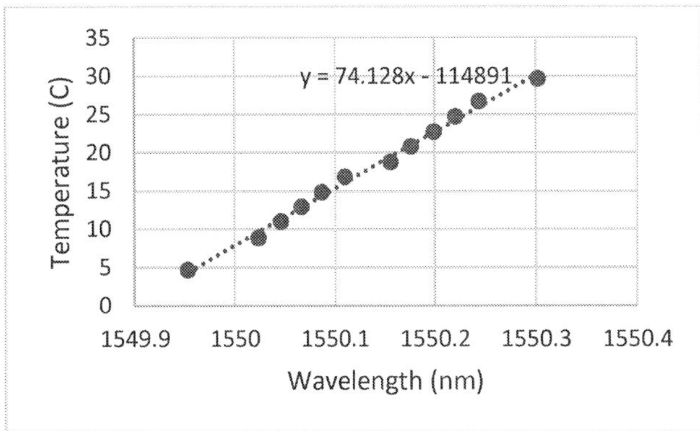

Figure 4: Calibration of the Fibre Bragg Grating.

REFERENCES

[1] I. Takeuchi and K. Sandeman, "Solid-state cooling with caloric materials," *Phys. Today*, vol. 68, no. 12, pp. 48–54, Dec. 2015.

[2] K. K. Nielsen, H. N. Bez, L. von Moos, R. Bjørk, D. Eriksen, and C. R. H. Bahl, "Direct measurements of the magnetic entropy change," *Rev. Sci. Instrum.*, vol. 86, no. 10, p. 103903, Oct. 2015.

[3] P. V. Trevizoli *et al.*, "Magnetic heat pumps: An overview of design principles and challenges," *Sci. Technol. Built Environ.*, vol. 22, no. 5, pp. 507–519, Jul. 2016.

[4] A. Tura and A. Rowe, "Permanent magnet magnetic refrigerator design and experimental characterization," *Int. J. Refrig.*, vol. 34, no. 3, pp. 628–639, May 2011.

[5] A. Othonos, "Fibre Bragg gratings," *Rev. Sci. Instrum.*, vol. 68, no. 12, pp. 4309–4341, Dec. 1997.

[6] J. M. D. Coey, *Magnetism and Magnetic Materials - Cambridge University Press*. 2010.

[7] R. Bjørk and C. R. H. Bahl, “Demagnetization factor for a powder of randomly packed spherical particles,” *Appl. Phys. Lett.*, vol. 103, no. 10, p. 102403, 2013.
[8] D.-X. Chen, J. A. Brug, and R. B. Goldfarb, “Demagnetizing factors for cylinders,” *IEEE Trans. Magn.*, vol. 27, no. 4, pp. 3601–3619, 1991.

Biomechanics and Biomaterials

Proceedings of the 26th CANCAM
Victoria, British Columbia, Canada, May 29 – June 1, 2017

A HYPER-VISCOELASTIC CONSTITUTIVE MODEL FOR NON-HUMAN PRIMATE SPINAL CORD WHITE MATTER

Shervin Jannesar
Mechatronic Systems Engineering
Simon Fraser University
Surrey, British Colombia, Canada
sjannesa@sfu.ca

Carolyn J. Sparrey
Mechatronic Systems Engineering
Simon Fraser University
Surrey, British Colombia, Canada
csparrey@sfu.ca

ABSTRACT

Computational models of spinal cord injury are a powerful tool for simulating the successful outcomes of animal trials before transferring to the clinical therapies. The accuracy of these models, however, is critically dependent on the pre-defined material constitutive models. Therefore, deriving the material properties from more human-like animals will help to improve the fidelity of the computational models. The objective of this study was to characterize the unconfined compressive behavior of non-human primate's spinal cord white matter. Samples were harvested from nine male adult non-human primates and were compresses up to 32.5% strain in three different strain rates. Tests were performed quickly after sacrifice to keep the post mortem time under one and a half hours. A quasi linear viscoelastic first order Ogden model with 3-term Prony series captured the experimental stress-relaxation data ($R^2 = 0.74 - 0.95$). This study is the first to characterize non-human primate spinal cord tissue.

KEYWORDS: Spinal cord injury, White matter, Unconfined compression, Soft tissue characterization

INTRODUCTION

As more successful animal spinal cord injury (SCI) therapeutic trials have emerged recently, the need for a reliable, accurate and cost effective platform for transferring these findings to clinical therapies have been critically increased. Computational models can provide a powerful tool for simulating the pre-clinical trials and examine their constructive outcomes for clinical trials [1,2]. However, computational models rely critically on pre-defined constitutive material models as input. Despite advances in characterizing spinal cord constituent material behavior using either rodents [3] or larger animals such as pig [4], the anthropological suitability of these models is unclear due to the substantial inhomogeneity between humans and these animals.

Non-human primates (NHPs) have a more similar central nervous system to humans than either rodent or pigs; therefore, it is presumable that mechanical material characteristics obtained from NHP tissues may more resemble human parenchyma.

Very few studies have characterized the spinal cord white matter using in-vitro compressive experiments [4], and for the most part, the white matter was characterized using tensile data [5]. Given that the majority of the SCIs are due to compressive loading on spinal cord [6], constitutive models characterized through compressive test deem to be more accurate and better resemble the white matter tissue response.

Of interest in this study was to determine the mechanical behavior of NHP spinal cord white matter undergoing unconfined compression; and present a hyper-viscoelastic constitutive material model for the tissue to be used in high accuracy computational models of SCI.

METHODS

Quickly following sacrifice (time post mortem less than 1:30 hours), spinal cords were harvested from cervical (C5-C6) section of male Macaca mulatta NHP (N=9). The dura mater was then removed from the segments with pia mater kept intact. 2.35±0.35 mm rostral-caudally cylindrical samples were cut from the lateral columns of the spinal cord white matters using a 3mm biopsy punch. Mechanical compression tests were performed using a Bose ElectroForce mechanical testing equipment (TA Instruments, New Castle, DE) and a 10N load cell. Compressive loads were applied in the rostral-caudal direction (parallel to the axonal fibers) in three different strain rates (namely 0.2, 2.0 and 20.0 [1/sec]) to a peak strain of 40% and then held for 60s. A total of 77 tests were performed.

A first order Ogden quasi-linear viscoelastic (QLV) strain energy function with 3-term Prony series was used to capture the time dependent material behavior of NHP white matter (Eq.1).

$$U = \int_0^t \left\{ \frac{\mu}{\alpha^2} \left(1 - \sum_{i=1}^{3} g_i \left(1 - e^{-(t-\tau)/\tau_i} \right) \right) \frac{d}{d\tau} \left(\lambda_1^\alpha + \lambda_2^\alpha + \lambda_3^\alpha - 3 \right) \right\} d\tau \qquad (1)$$

where μ, α, g_i's and τ_i's are material parameters, λ_i's are principal stretches and t is time.

To determine the QLV model parameters an optimization algorithm was used by linking MATLAB and ABAQUS. A finite element model of the compressive tests was simulated in ABAQUS containing the cylindrical sample and the equipment platens (Fig.1). The MATLAB code passed the material parameters to the ABAQUS model and simulations were run in the three different strain rates (0.2, 2.0 and 20.0 [1/sec]). The stress-relaxation responses from these simulations were fed back to the MATLAB code for determining a new set of parameters by simultaneous fitting to the experimental stress-relaxation responses in the three strain rates using fmincon.

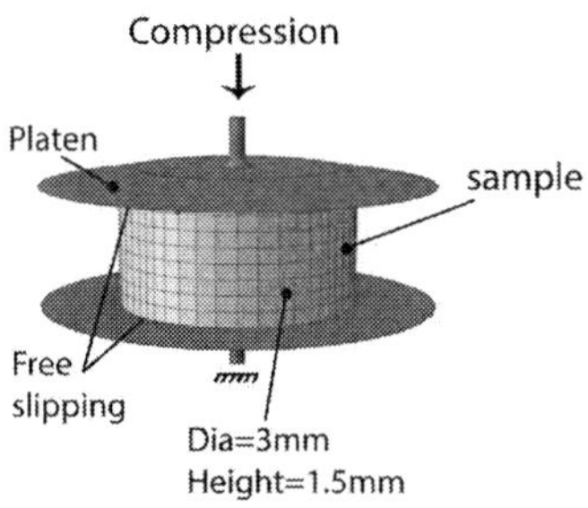

Figure 1: Representation of compressive test simulations in ABAQUS. Height and dimeter were obtained from averaging all the samples. Interaction between the sample and platens is considered frictionless.

RESULTS AND DISCUSSION

By increasing strain rate, mean stress response increased (Fig.2). Maximum stress (32.5% strain) at high strain rate ($20.0s^{-1}$) was significantly different from other strain rate groups ($p<0.05$).

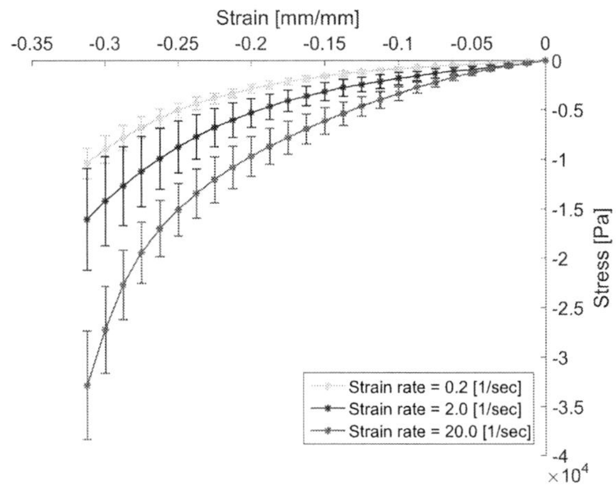

Figure 2: NHP white matter stress-strain response (mean and standard deviation) in three different strain rates.

Optimized material parameters resulted in the QLV model fitting all three strain rates well with R^2=0.87, R^2=0.74 and R^2=0.93 for 0.2, 2.0 and 20.0 [1/sec], respectively (Fig.3 and Table.1).

NHP spinal cord white matter was one order of magnitude stiffer than porcine spinal cord white matter [4], which may affect the SCI computational model outcomes significantly (Fig.4).

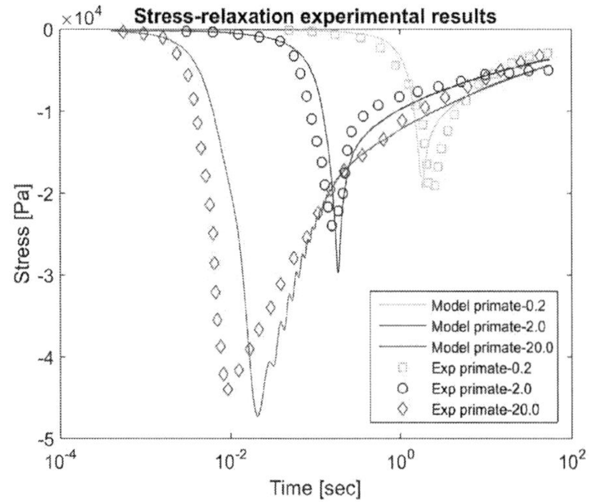

Figure 3: QLV fit to the experimental data.

Table 1: Optimized QLV constitutive model parameters obtained from the optimization process.

PARAMETER	*VALUE*
μ [Pa]	2969.3
α	4.41
g_1	05515
τ_1 [sec]	0.02
g_2	0.2809
τ_2 [sec]	0.2
g_3	0.0601
τ_3 [sec]	2.0

Furthermore, the viscoelastic model presented here was derived from high rate experimental data typical of traumatic injury and might improve the fidelity of the traumatic SCI finite element models.

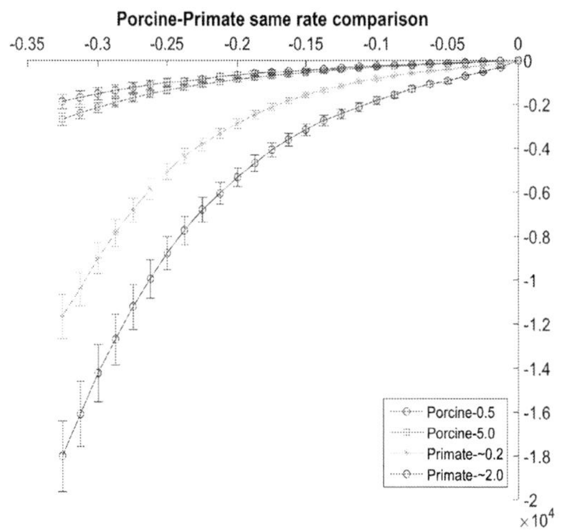

Figure 4: Comparison between primate tests in this study and results of porcine compression [4].

The significant difference between the results of primate and porcine study reveals the substantial effect of post mortem time in the stiffness of the spinal cord white matter material.

CONCLUSIONS

In conclusion, it was shown that the compressive response of NHP white matter is highly strain rate dependent. Furthermore, the compressive response of the NHP spinal cord white matter is one order of magnitude stiffer than the previously reported white matter properties. The QLV model has well represented the white matter material response in different strain rates and should improve the fidelity of the SCI computational models.

REFERENCES

[1] Salegio, E. A., Bresnahan, J. C., Sparrey, C. J., Camisa, W., Fischer, J., Leasure, J., Buckley, J., Nout-Lomas, Y. S., Rosenzweig, E. S., Moseanko, R., Strand, S., Hawbecker, S., Lemoy, M.-J., Haefeli, J., Ma, X., Nielson, J. L., Edgerton, V. R., Ferguson, A. R., Tuszynski, M. H., and Beattie, M. S., 2016, "A Unilateral Cervical Spinal Cord Contusion Injury Model in Non-Human Primates (Macaca mulatta).," J. Neurotrauma, **33**(5), pp. 439–59.

[2] Maikos, J. T., Qian, Z., Metaxas, D., and Shreiber, D. I., 2008, "Finite element analysis of spinal cord injury in the rat," J. Neurotrauma, **25**(7), pp. 795–816.

[3] Fiford, R. J., and Bilston, L. E., 2005, "The mechanical properties of rat spinal cord in vitro," J. Biomech., **38**(7), pp. 1509–1515.

[4] Sparrey, C. J., and Keaveny, T. M., 2011, "Compression behavior of porcine spinal cord white matter," J. Biomech., **44**(6), pp. 1078–1082.

[5] Ichihara, K., Taguchi, T., Sakuramoto, I., Kawano, S., and Kawai, S., 2003, "Mechanism of the spinal cord injury and the cervical spondylotic myelopathy: new approach based on the mechanical features of the spinal cord white and gray matter.," J. Neurosurg., **99**(3 Suppl).

[6] Metz, G. A., Curt, A., van de Meent, H., Klusman, I., Schwab, M. E., and Dietz, V., 2000, "Validation of the weight-drop contusion model in rats: a comparative study of human spinal cord injury.," J. Neurotrauma, **17**(1), pp. 1–17.

SIMULATION OF PARTICLE DEPOSITION IN EMPHYSEMATOUS HUMAN ACINI

Amitvikram Dutta, Joshua R. Brinkerhoff
School of Engineering
University of British Columbia
Kelowna, BC, Canada
joshua.brinkerhoff@ubc.ca

André Phillion
Materials Science & Engineering
McMaster University
Hamilton, ON, Canada

ABSTRACT

To contribute to the understanding of early stage emphysema, computational fluid dynamics (CFD) simulations are performed on a section of a mathematically derived model of the human acinus. Emphysematous conditions are simulated through changes in the computational domain, boundary conditions, and motion of the spatial grid mimicking respiration. The transport of 1 and 3 micron particles is computed over multiple breath cycles with two orientations of the gravitational acceleration vector. The changes in deposition patterns of particles with increasing levels of emphysema are studied to identify the effect of individual pathological features of the disease on particle deposition. The results suggest that emphysematous destruction increases particle deposition in the acinus. The location of the increased deposition within the acinus depends both on the particle size and the relative orientation of the airway to the direction of gravitational acceleration.

Keywords: CFD, Emphysema, Particle Deposition, Acinus

INTRODUCTION

Emphysema is a disease that leads to the permanent enlargement of air spaces within the parenchyma of the lung. Along with chronic bronchitis, emphysema forms one of the two components of chronic obstructive pulmonary disease, a serious condition that is responsible for severe limitation of expiratory airflow. Due to limitations in existing *in vivo* methods of studying this condition, little is known how emphysema affects airflow within the acinus especially in the early stages of the disease. Computational fluid dynamics (CFD) simulations thus offer an alternative approach to this problem. This study simulates progressive emphysema in a single section of the human acinus. Physiologically realistic wall motions and flow rates have been incorporated as well as the expected decline in both of these parameters with the advancement of the disease. This study aims to determine the effects of the disease on particle deposition in the acinus, an important question considering that particle deposition is hypothesized as a primary factor in the pathogenesis of emphysema. Oakes et. al. [1] performed similar simulations on particle depositions based on data from healthy and emphysematous acinar rat models. while Hofemeir and Sznitman [2] have performed particle deposition simulations on healthy models of the human acinus. This study focuses on how the progression of emphysema affects particle deposition in a single section of the human acinus, thereby concentrating on how particle deposition is affected during at the very early stages of emphysema in the terminal airways of the human lung.

METHODOLOGY

To accurately capture the flow through the airways in a CFD study, physiologically realistic boundary conditions and wall deformations must be used. The closely packed alveolar openings in the acinar region are modeled using a geometry composed of 14-hedrons [3, 4]. The collection of 14-hedron alveolar spaces that are ventilated by a single airway are referred to as a duct; the computational domain of the present study consists of three ducts across two generations of the pulmonary acinus, one occurring proximal to the dichotomous branching and associated with the fourth generation (Duct 1),

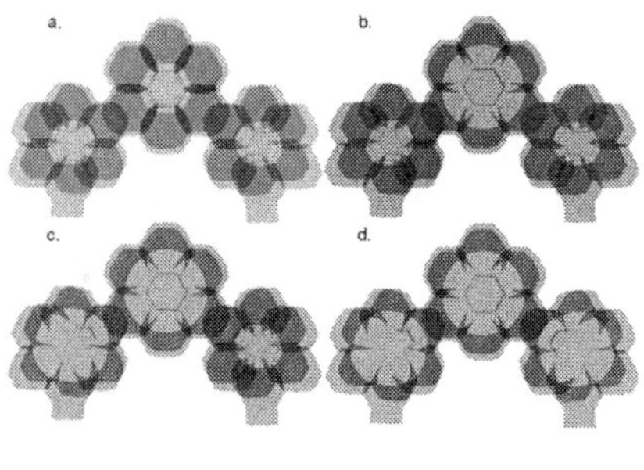

Figure 1: Progressive destruction of alveolar septa in the computational domain. (a) Healthy case (b) Case I (c) Case II (d) Case III.

and two occurring distal to the dichotomous branching and associated with the fifth generation (Ducts 2 and 3). The destruction of tissue walls (i.e. septa) between the alveolar spaces due to emphysematous progression is modelled by progressively removing septa in the three ducts of the computational geometry, as shown in Fig. 1; the progressive nature of emphysema is incorporated by simulating separate cases with increasing levels of septal destruction.

Boundary conditions for the computational domain consist of impermeable, zero-velocity walls for all surfaces except the inflow and outflow boundaries. The boundaries defined at the entrances and exits to the ducts, labelled in Fig. 1, are defined as either inlet or outlet boundaries depending on the time of the simulation within the breath cycle. The boundary information defined at the inlet and outlet boundaries must reflect the actual physiology of the pulmonary acinus. Because the *in vivo* measurements at this depth of the lung are impossible, the volume/flow rate relationships measured at the mouth are used to infer flow rates in the pulmonary acinus. A typical volume/flow rate relationship at the mouth is given in Fig. 3. Figure 3 shows that the inspiratory flow rate dependence on volume can be approximated as a sinusoid (OAC). Normal exhalation by contrast is a passive process; as the intercostal muscles relax, the natural elasticity of the lung parenchymal tissues recoil the fibres attached to the airways, compressing them and driving the air out. During the elastic recoil, the flow rate in Fig. 3 rises linearly with decreasing volume (OB), until it reaches a maximum expiratory flow rate (B). Thereafter it falls linearly to zero (BC), signalling the end of expiration.

The inspiratory and expiratory volume/flow rate relationships depicted in Fig. 3 can be approximated as follows. The flow rate during inspiration is

$$Q_{in} = Q_{in,\, max} \sin(\omega_p V) \quad \text{for} \quad 0 \leq V < V_{max} \qquad (1)$$

and during expiration is

$$Q_{exp} = -\beta V + \kappa \quad \text{for} \quad V_{max,\, exp} \leq V < V_{max}, \qquad (2)$$

$$Q_{exp} = \gamma V \quad \text{for} \quad 0 \leq V < V_{max,\, exp} \qquad (3)$$

where $Q_{in,\, max}$ is the maximum inspiratory flow rate, V_{max} is the maximum volume excursion of the lung/alveolus, ω_p is the breathing frequency (reciprocal of the breath period),

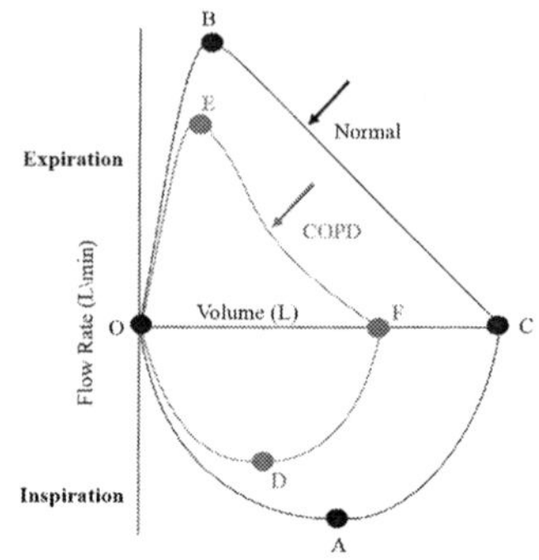

Figure 2: Variation of respiratory flow rate versus lung volume for healthy and COPD affected lungs.

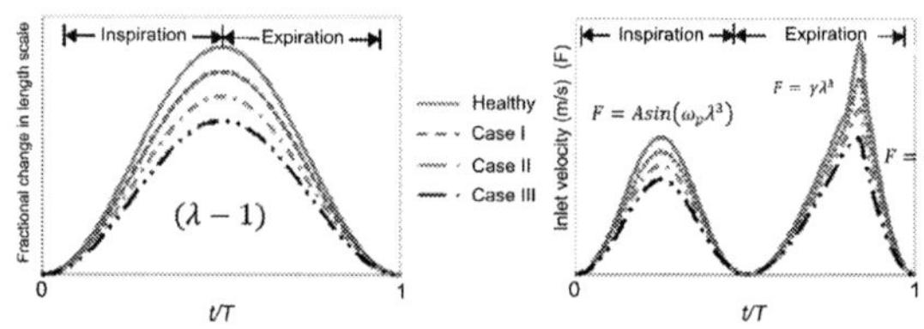

Figure 3: Displacement of the airway wall from rest during inspiration and expiration for the healthy and diseased cases and temporal variation of flow rate into the pulmonary acinus during inspiration and expiration for the healthy and diseased cases

$V_{max,\, exp}$ is the volume corresponding to maximum expiratory flow rate, and β, γ, and κ are constants. A typical flow rate/volume relationship for a patient with COPD is also illustrated in Fig. 3. Although, the effects of COPD on the lung volume/air-flow rate relationship is not exclusively due to emphysema (small airways disease may also be present), due to the lack of available *in vivo* data from patients suffering exclusively from early stage emphysema, the boundary conditions imposed on the simulation have been obtained from the interpolating the values between healthy and COPD observations in Fig. 3. Emphysematous degradation of the tissue matrix supporting the airways leads to the loss of elastic recoil of the lung parenchyma. As a result, airways do not inflate as effectively and the maximum inspiratory flow rate (D) and volume excursion at the end of inspiration is reduced (F). Due to the decreased elastic recoil, the expiratory flow rate is also reduced (E). This reduction in flow rate due to COPD can be modelled by altering the parameters β , κ, γ, V_{max} and $Q_{in,max}$ in Eqns. 1-3. The parameters for the healthy and the most diseased cases were obtained from [5], with the intermediate values being obtained by interpolation.

The particle dynamics in the computational domain are governed by a balance of convective and gravitational forces. These equations are integrated over time in order to determine the trajectories of inhaled particles. The equation for the particle velocity may be restated in a simplified form as

$$\frac{du_p}{dt} = \frac{(u_p - u_f)}{\tau} + a \qquad (4)$$

where τ represents the drag force coefficients and a represents all other external forces, which in this case consists of the gravitational force only but in general may include other forces related to Brownian motion if the particle is very small. The position of a particle is given by

$$\frac{dx_p}{dt} = u_p. \tag{5}$$

Equations 5 and 4 are therefore coupled and need to be integrated simultaneously in order to determine the particle trajectories. Particles are introduced into the domain from the inlet from $\tau = 0$ to $\tau = 0.5$, with τ defined as the instantaneous time normalised over one breath-period, $\tau = t/T$. Since the orientation of the gravity vector has been shown to have considerable influence on the particle trajectories and hence deposition, simulations were conducted for two cases where the gravity vector was oriented normal and tangential to the entrance of the solution domain respectively.

ANSYS Fluent (v. 16) is used to solve the mass and momentum conservation equations, a transport equation for oxygen concentration, and one-way (uncoupled) transport of solid particles. The solutions were initialized with a zero velocity field and at atmospheric pressure mimicking conditions immediately preceding the start of inspiration. The initial mass fraction of oxygen inside the alveolar ducts is 0.15, which corresponds to the oxygen-poor condition that exists at the completion of expiration. Each breath period was 3 seconds long corresponding to a breathing frequency of 20 breaths per minute. The computations were performed in parallel using a message-passing interface (MPI) on 6 virtual cores of an Intel Xeon CPU with a clock-speed of 2.50 GHz such that the average simulation time for one complete breath period was approximately 4.5 hours.

RESULTS

In order to characterize the deposition of particles in healthy and emphysematous geometries, the deposition fraction (f) quantity is defined as follows

$$f = \frac{N_D}{N} \tag{6}$$

where N_D is the number of particles that deposit and N is the total number of released particles. A particle is assumed to have been deposited when its path intersects the wall of the computational domain. The simulations are conducted for two particle sizes with diameters of 1 and 3 μm and for two separate orientations of the gravitational vector. In the normal orientation, the gravity vector is co-axial with the inflow boundary of Duct 1 (normal), while in the tangential orientation, it is directed tangentially. The particle transport simulations are conducted for the four cases (Healthy, Case I, Case II and Case III). The results are grouped into four categories according to particle size and orientation of the gravity vector: 1 μm Normal, 3 μm Normal, 1 μm Tangential and 3 μm Tangential. The simulations are conducted for five successive breath cycles and the particle trajectories and the deposition fractions were monitored at the end of each cycle.

Plots showing the variation in steady-state deposition fraction in the three ducts, for all of the simulated cases are shown in Fig. 4. The convective and gravity forces on the particles reinforce each other when they are aligned as in the normal cases. In the tangential cases the gravity vector is perpendicular to the bulk incoming flow, thus directing part of it towards the acinar walls. This effect is magnified in the emphysematous cases where the convective flow is reduced. From Fig 4(a), it is apparent that for the 1 μm particles with a normal gravity vector the steady-state deposition fraction in Duct I increases with increasing levels of emphysema. While the normal case shows a f value of only 0.15, this increases in Case III to 0.32. The deposition fraction values also increase uniformly in Ducts II and III for the healthy, Case I, and Case III simulations. In Case II it is observed that Duct II has a higher steady state deposition fraction (0.1) than Duct III (0.05). Since in Case II, Duct II has septal degradation while Duct III does not, it suggests that septal degradation preferentially enhances particulate deposition between two ducts of the same generation. This pattern is observed in all simulated cases.

Fig 4(b) shows steady state deposition fraction values for 3 μm particles with the gravity vector oriented in the normal direction. While the steady-state f value in Duct I increase with advancing emphysema like the 1 μm particle cases, the deposition is much higher; the Healthy case has a steady state deposition fraction value of 0.3 while the corresponding value for Case III is 0.6. This may be attributed to the greater effect of gravity on the larger and heavier particles. Also, a greater proportion of particles deposit in Duct I than Ducts II and III for the diseased cases than for the corresponding simulations with 1 μm particles. In fact, the deposition in the proximal duct is so greatly enhanced, that the more distal Ducts II and III have lower deposition fraction values than the ones corresponding to the Healthy case under identical simulation conditions. Moreover, the increase in deposition fraction for Duct I is not uniform as in the 1 μm case; Case I and Case II have very similar steady state values of f for Duct I (0.51 and 0.52 respectively). This is due to the larger effect of gravity on the 3 μm particle in Case II; while the deposition is increased due to emphysematous degradation, the normal direction of the gravity vector (collinear to the incoming flow) directs particles away from the walls and deeper into the acinus, thus lowering the deposition fraction in Duct I so that is comparable to Case I. Case III does show an increase in deposition fraction for Duct I as compared to Cases I and II, attributed to a further lowering of the flow rate, which prevents particles from being convected into the more distal generations and instead deposit at the branching point between Ducts I, II and III.

Fig. 4(c) shows the steady state f values for 1 μm particles with the gravity vector oriented in the tangential direction. On comparing with the simulations with an identical particle size but a normal gravity vector, it is seen that the deposition fraction values in Duct I are consistently greater for the present

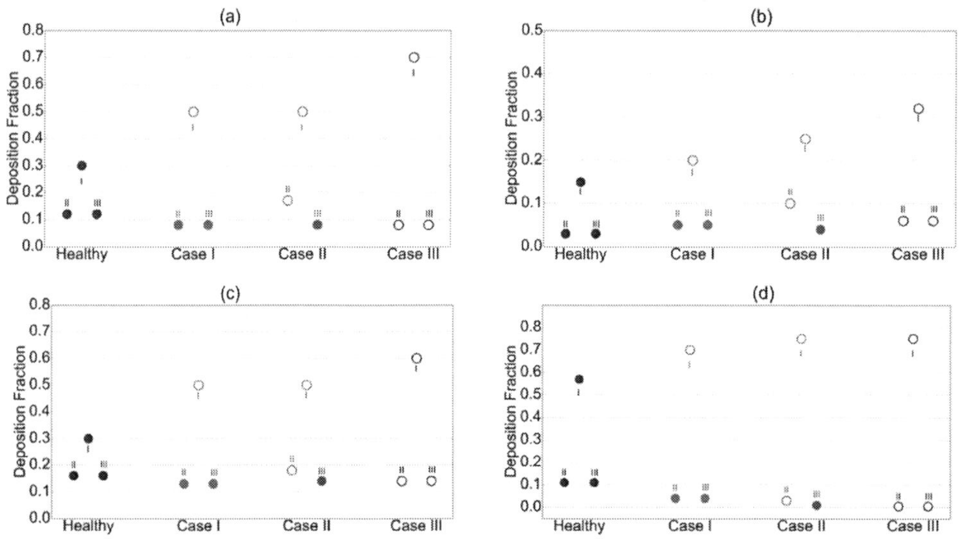

Figure 4: Plot of final deposition fraction after five breath cycles. The labels near each symbol indicate the duct. Filled symbols indicate healthy acini; empty symbols indicate emphysematous acini.

cases. This is explained by the fact that the tangential gravity vector directs the particles towards the walls of the acinus instead of along the bulk flow, thereby enhancing the effects of emphysematous destruction. This also leads to a decrease in deposition fraction in the more distal ducts as compared to the healthy case. The pattern of increased deposition in Duct II as compared to Duct III in Case II is also observed here.

Fig. 4(d) shows the steady state f values for 3 μm particles with the gravity vector oriented in the tangential direction. These cases show the highest deposition fraction of all the cases due to the tangential gravity and heavy particles, which drives particles towards the walls. There's a marked increase in deposition faction values between the Healthy case and Case III (0.5 and 0.75 respectively). However there is very little increase in deposition fraction values for Cases I, II and III. This is because the gravitational effects completely overwhelm the convective effects and since the geometrical conditions in Duct I are the same for Case I, II and III, they have identical deposition fraction values. The pattern of increased deposition in Duct II as compared to Duct III in Case II is also observed here.

CONCLUSIONS

This study show that the deposition fraction increases in emphysematous ducts due to both a decrease in airway wall motion and emphysematous destruction. The increase in deposition fraction in ducts distal to the first emphysematous ducts depend on the particle size and orientation of the gravitational vector. Smaller particles and gravitational vectors aligned with the the flow tend to penetrate and deposit deeper into the acinus, while larger particles and tangential gravitational vectors promote the deposition of particles in the proximal ducts of the acinus. For two ducts in the same generation, particles preferentially deposit in the one with emphysematous destruction as compared to one without. Thus, while emphysema notably enhances the deposition in the first duct, deposition patterns in the subsequent ducts depend on particle size, gravitational vector orientation and the presence of emphysematous destruction in the distal airways.

REFERENCES

[1] Oakes, J. et. al. 2015, "Aerosols in healthy and emphysematous in silico pulmonary acinar rat models," *Journal of Biomechanics*, Vol(49), pp. 2213-2220.

[2] Hofemeir P., Sznitman J. 2015, "Revisiting pulmonary acinar particle transport: convection, sedimentation, diffusion, and their interplay," *Journal of Applied Physiology*, pp. 1375-1385.

[3] Sznitman, J. et. al. 2009, "Respiratory flow phenomena and gravitational deposition in a three-dimensional space-filling model of the pulmonary acinar tree,"*Journal of Biomechanical Engineering*

[4] Kumar, H. et. al. 2009, "The effects of geometry on airflow in the acinar region of the human lung," *Journal of Biomechanics*, Vol(42), pp. 1635-1642.

[5] West, J.B. 2005, "Pulmonary Pathophysiology : The Essentials," *Lippincott Williams and Wilkins*

NUMERICAL INVESTIGATION OF ASCENDING THORACIC AORTIC ANEURYSM USING FLUID-STRUCTURE INTERACTION METHOD

Han Hung Yeh
Department of Mechanical Engineering
Biomedical Engineering Program
University of British Columbia
Vancouver, BC, Canada
han.hyeh@alumni.ubc.ca

Simon W Rabkin
Department of Medicine (Cardiology)
University of British Columbia
Vancouver, BC, Canada
rabkin@mail.ubc.ca

Dana Grecov
Department of Mechanical Engineering
Biomedical Engineering Program
University of British Columbia
Vancouver, BC, Canada
dgrecov@mail.ubc.ca

ABSTRACT

This study presents the simulations of ascending thoracic aortic aneurysm (TAA) with patient-specific parameters. The focus was to investigate the effect of hypertensive blood pressures on TAA using Newtonian and non-Newtonian Carreau models with a fully coupled fluid-structure interaction method. TAA was considered to behave as an anisotropic material because of its composition specifically collagen fibers. Our model reproduced comparable prediction in cyclic strain found in the literature. The influences from the Carreau model in predicting the motion and the stress in aneurysm wall were minimum; however, variations were found when comparing the wall shear stress generated by blood flow between the two rheological models. In addition, the distribution of Von Mises stress indicated that the elevated stress under hypertension could be geometrically dependent. This geometrical influence in TAA could be a risk factor predicting further aortic expansion.

KEYWORDS: Ascending thoracic aortic aneurysm, fluid-structure interaction, hypertension, wall shear stress, non-Newtonian

INTRODUCTION

There are several well-known risk factors including age, gender, smoking, and genetics that affect the probability of aortic aneurysm formations and expansion; however, the precise causes of aneurysm formations whether physically or biochemically are still unclear [1]. Currently, the values of blood pressure and the sizes of aneurysm diameter are critical parameters used for managing patients and recommending surgical options [2]. Additional biomechanical parameters such as the curvature of ascending thoracic aorta, the stiffness of the blood vessel, as well as the peak magnitude of aortic wall stress and wall shear stress (WSS) could provide further assistant in disease predictions [3-5]. Since, in clinical practice, geometrical characteristic of the diseased aorta would be the most accessible information and given that the value of other parameters would be difficult to obtain directly *in vivo*, an alternative method would be of value to estimate the aortic wall stress, blood flow velocity, and WSS.
Numerical methods are widely adopted for analyzing the biomechanics of aneurysm formations and interactions since the patient-specific model can easily be generated based on computed tomography (CT) imaging or magnetic resonance imaging (MRI). More specifically, fluid-structure interaction (FSI) method can provide analysis of both blood flow and arterial wall deformation based on how blood pressure affects the general motion of an aortic aneurysm and vice versa. However, it is necessary to provide an accurate numerical model for predicting the behavior of blood rheology and aneurysm structure.
The objective of current study was to develop an accurate fully coupled FSI model that accounted for the aortic wall anisotropy and the physiological parameters of a specific patient. Evaluations of the biomechanics of ascending thoracic aortic aneurysm will focus on the differences between normotensive and hypertensive scenario and the effect of Newtonian and non-Newtonian Carreau model.

METHOD

Idealized 3D finite element models of the human ascending aorta using measured patient-specific data were created with COMSOL Multiphysics® (V5.2, Stockholm, Sweden). Fully coupled fluid-structure interaction (FSI) method was inhered and upgraded from our previous FSI study [6]. Two clinical cases were investigated with the unique geometrical characteristic (Fig.1). The area of focus was based on the available geometrical data of specific patient cases. The inlet of the current computational domain was started distally from aortic sinus while the outlet was terminated proximally to brachiocephalic artery. Both patient cases were associated with dilated aorta where case 1 had a similar inlet (45 mm) and outlet diameter (44 mm) and case 2 had a smaller inlet (42 mm) than the outlet diameter (53 mm). The curvature of the ascending thoracic aorta was assumed to be identical between the two cases and the radius of curvature was approximated to be 55 mm [7]. Both the geometrical parameters as well as the peak velocity magnitudes were measured via echocardiogram. Peak velocity measured for case 1 was 1.6 m/s and for case 2 was 4.36 m/s. These measured parameters were then applied to construct the geometrical domain and the associated boundaries. Based on this, the inlet boundary was prescribed with fully developed velocity profile and the outlet boundary was prescribed with an averaged normotensive (120/80 mmHg) and hypertensive (160/90 mmHg) blood pressure.

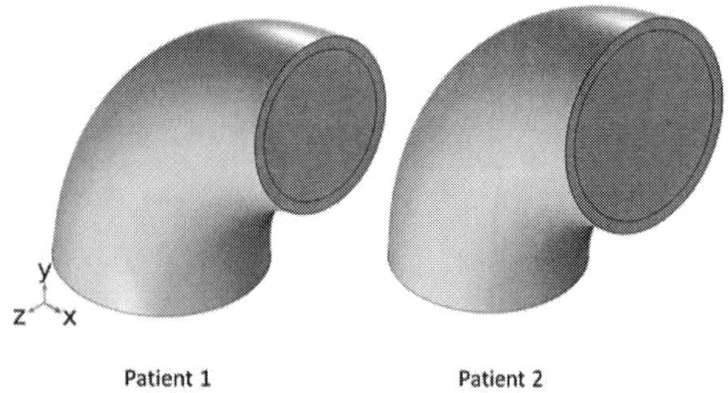

Figure 1: Computational Domain for Two Patient Cases

The rheology of blood was modelled using both Newtonian and the non-Newtonian Carreau model (Eq.1) with constant blood density of 1060 kg m^{-3}. Blood was assumed to be within the laminar region while the ascending thoracic aorta was approximated as single layer anisotropic hyperelastic nearly incompressible material. The effect of gravity as assumed to be negligible due to the small vertical distance from inlet to the outlet. The constitutive model for the ascending thoracic aorta was constructed based on studies conducted by Holzapfel *et. al* [8], and Weisbecker *et. al* [9].

$$\eta(\dot{\gamma}) = \eta_\infty + (\eta_0 - \eta_\infty)[1 + (\lambda\dot{\gamma})^2]^{\frac{(n-1)}{2}} \qquad (1)$$

Where the viscosity at zero shear rate, η_0, and the viscosity at infinite shear rate, η_∞, is 0.056 Pa-s and 0.0035 Pa-s, respectively. The relaxation time, λ, is 3.313s with a fluid index, n, of 0.3568. These constants used in Carreau model are widely accepted for the modelling of normal blood at body temperature.

The strain energy density function $\bar{\Psi}$, shown in Eq. 2, was used for evaluating the strain energy stored to account for arterial wall deformation. Eq. 2 is calculated based on the summation of the isotropic and the anisotropic component.

$$\bar{\Psi} = \bar{\Psi}_{iso}(\bar{I}_1) + \bar{\Psi}_{aniso}(\bar{I}_4 + \bar{I}_6) \qquad (2)$$

The consideration of material anisotropy is required because of the collagen fibers network within the adventitia layer of the aorta. Currently, it is assumed that only two sets of symmetrically orientated fiber crossing at 51° are governing the anisotropic response and Eq. 2 can be rewritten into Eq. 3 [9]. The material parameters and the thickness of the aortic wall were kept as constants such that the effect of blood pressure and the patient-specific geometry could be evaluated.

$$\bar{\Psi} = \frac{C}{2}(\bar{I}_1 - 3) + \frac{k_1}{2k_2}\sum_{i=4,6}\{e^{[k_2(\bar{I}_i-1)^2]} - 1\} \qquad (3)$$

Where $\bar{I}_i$ are the invariants and C and k_i are material parameters (C is 0.017 MPa, k_1 is 0.56 MPa, and k_2 is 16.21) [9].

To properly constrain the ascending thoracic aorta, the arterial wall at the inlet position was applied with a fixed boundary condition relative to the rest of the domain. The arterial wall at the outlet position was only constrained its displacement normal to the cross-sectional plane while maintained the freedom of sliding. With these prescribed boundary conditions, current model would be constrained minimally to allow maximum flexibility under pulsatile pressure.
Also, to account for numerical instability and errors due to prescribed initial conditions, a total of five cardiac cycles (0.8 second each or 75 beats per minute) were simulated per scenario. The solution difference was calculated between each cardiac cycle and it was found that the solution converged after fourth cardiac cycle. Therefore, the fifth cycle, from 3.2s to 4.0s, was taken for final analysis. Mesh convergence study was also performed with a solution variation within ±1%.

RESULTS AND DISSCUSION

The current analysis focused on the hemodynamics of blood flow within ascending thoracic aneurysm as well as the structural behavior of the aorta. Our investigation started with the fluid dynamics of blood flow. The most important finding related to the differences in wall shear stress between Newtonian and non-Newtonian models. Differences were also found in hypertension compared to the normotensive conditions. Interestingly there were only minimum differences in blood flow distribution and wall stress distribution if the Carreau model was employed.

Table 1: Average Percent Difference in Maximum WSS between Normotension and Hypertension

% Difference	Case 1			Case 2		
	YZ	XZ	XY	YZ	XZ	XY
Newtonian	13.4	6.1	4.2	14.2	12.6	5.7
Carreau	9.7	5.8	3.1	8.4	13.7	5.7

The difference, in percentage, of the maximum WSS between normotension and hypertension are summarized for both the Newtonian and the Carreau models in Table 1. A significant difference can be seen for all three components of WSS using both rheological models. However, the use of Carreau model predicted a smaller difference in maximum WSS for almost all components except the XZ component in case 2. A full analysis of maximum WSS for all six components are shown for the fifth cardiac cycle simulated using Carreau model (Fig. 2). Clearly, the magnitudes of WSS for case 2 were much higher than case 1. This would be due to a higher blood inlet velocity to start with for case 2. However, the magnitude of WSS for both YZ and XZ components in case 1 were relatively smaller than the YX component while the XZ and YX components in case 2 were relatively higher, this may suggest that in addition to the upstream blood flow velocity, additional factors, such as the geometrical difference between the two cases, could be involved for the elevated maximum WSS. This information would be potentially useful for investigating blood cell damages under shear flow and/or thrombosis formation.

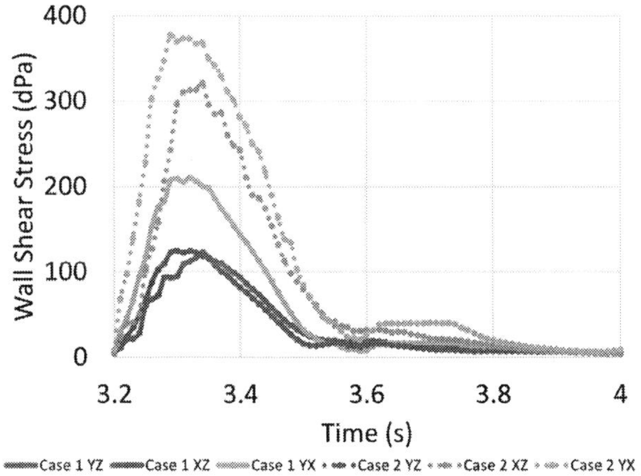

Figure 2: Components of Maximum Wall Shear Stress under Hypertensive Scenario for Case 1 and Case 2 with Carreau Model

The introduction of non-Newtonian Carreau model was insignificantly affecting the structural responses of the ascending thoracic aortic aneurysm. It was found that the average differences in aortic expansion under cardiac cycle were less than 1% for both blood pressure scenarios between two rheological models; therefore, only the Carreau model would be discussed for the analysis of aortic biomechanics.
To evaluate the expansion of the aorta, the circumferential cyclic strain, $\varepsilon_{\theta\theta}$ as shown in Eq. 4, was used and compared with the study conducted by Morrison *et. al* [10].

$$\varepsilon_{\theta\theta} = \frac{1}{2}\left(\frac{C_s^{\,2}}{C_d^{\,2}} - 1\right) = \frac{1}{2}\left(\frac{D_s^{\,2}}{D_d^{\,2}} - 1\right) \tag{4}$$

Where C_s and C_d are the circumferences during systolic phase and diastolic phase and D_s and D_d are the corresponding diameters

Approximately, the values of cyclic strain for case 1 were 2.6% and 3.9% for normotension and hypertension, respectively. The values of cyclic strain for case 2 were 2% and 3.2% for normotension and hypertension, respectively. These values were calculated at the outlet of current models and were in agreement with the average strain value of 3.4±1.5% by Morrison *et. al* at the same location of interest [10]. Of course, when the blood pressure elevated, both cases indicated a significant increase in cyclic strain of approximately 52% for case 1 and 60% for case 2. If the calculations of the increase of cyclic strain were conducted at the middle of the TAA between the inlet and outlet, the increase in $\varepsilon_{\theta\theta}$ due to elevated pressure dropped to 46% for case 1 while case 2 remained the same increase at 60%. This further supported the idea that the geometrical differences would affect the biomechanics of ascending thoracic aneurysm.

Similarly, the magnitude of wall stress was processed from current models for the analysis of stress distribution along a thoracic aneurysm. Specifically, the Von Mises stress along the inner surface of the aneurysm models was plotted in Fig. 3. As indicated in the figures, the wall stress under the influence of hypertension had a stress concentration toward to the aortic arch along the inferior side of an aneurysm. A closer inspection, the elevated wall stress distributed more widely for case 2 (Fig. 3b) under the hypertensive scenario. Interestingly, these areas under stress concentration would also migrate toward to the upstream if the aneurysms were under hypertension.

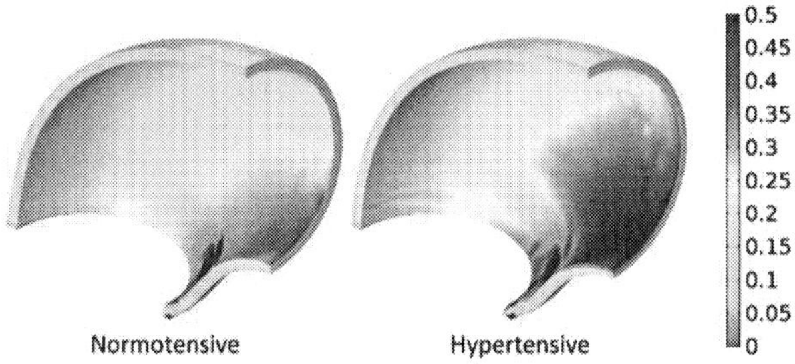

(a) Case 1

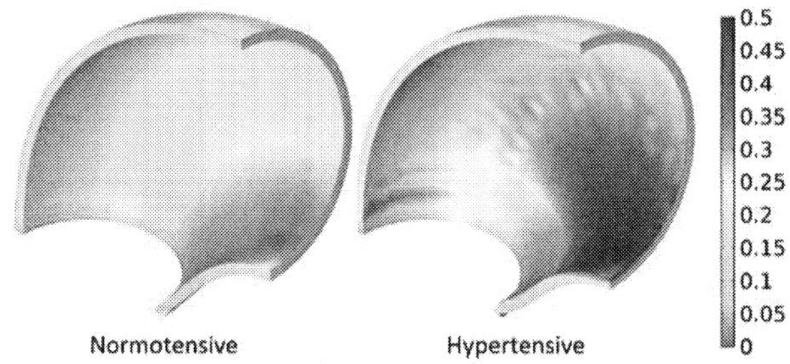

(b) Case 2

Figure 3: Von Mises Stress (unit in MPa) Contour Between Normotensive and Hypertensive Scenario for (a) Case 1 and (b) Case 2 using Carreau model

To quantify the level of increased wall stress under hypertension, the total surface area of the interior surface of an aneurysm was calculated. Shown in Fig. 4, a threshold of 0.25 MPa was applied to segment the surface area, which was calculated as a percentage, depending on the stress level. While both cases shared a similar level of stress distribution under normotension, the increase in elevated wall stress was greater for case 2 under hypertension. The high-stress area was increased by 60% for case 2 whereas only 20% of the interior region increased above the threshold for case 1.

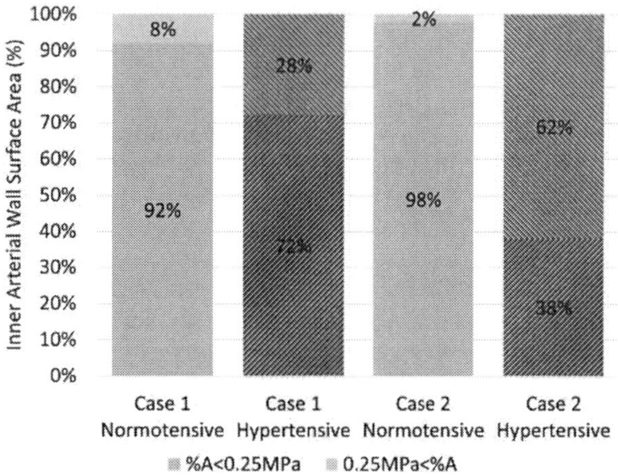

Figure 4: Total Surface Area (unit in %) Along the Interior Surface of Ascending Thoracic Aneurysm under Normotensive and Hypertensive Scenario for both Cases using Carreau Model

Based on the results analyzed, there were geometrical influences in the biomechanics of ascending thoracic aortic aneurysm. According to the stress distribution analysis in Fig. 4, case 2 should deserve additional attention in blood pressure management to avoid suddenly elevated blood pressure, which lead to larger increase in wall stress distribution to the aorta.

CONCLUSIONS

In summary, a fully coupled FSI method was used with the anisotropic behavior of ascending thoracic aorta accounted for. The investigation was extended with the addition of non-Newtonian Carreau model that was compared with the Newtonian model under both normotensive and hypertensive conditions.

There were definite differences in wall shear stress between Newtonian and non-Newtonian models in hypertension compared to the normotensive conditions. The current model predicted that with an increase of blood pressure from normotensive scenario to hypertensive scenario, the aortic diameter, and the cyclic strain would increase and importantly, the geometrical characteristic would also have a significant influence regarding the change in overall biomechanics. Therefore current modelling technique suggests that the geometrical configurations of the ascending thoracic aorta influences aortic expansion, wall stress distribution, and WSS. These insights will be further investigated for the development of additional risk criterion for patient evaluation and management.

ACKNOWLEDGMENTS

The financial support of the Natural Sciences and Engineering Research Council (NSERC) of Canada is gratefully acknowledged.

REFERENCES

1. Rabkin S.W., 2015, "Accentuating and Opposing Factors Leading to Development of Thoracic Aortic Aneurysms Not Due to Genetic or Inherited Conditions", *Front Cardiovasc Med*, 2:21. doi: 10.3389/fcvm.2015.00021

2. Hiratzka, L. F., Bakris, G. L., Beckman, J. A., Bersin, R. M., Carr, V. F., Casey, D. E., ... & Kouchoukos, N. T., 2010, "2010 ACCF/AHA/AATS/ACR/ASA/SCA/SCAI/SIR/STS/SVM guidelines for the diagnosis and management of patients with thoracic aortic disease", *Journal of the American College of Cardiology*, *55*(14), e27-e129.

3. Alberta H.B., Secor J.L., Smits T.C., Farber M.A., Jordan WD., Matsumura JS., 2013, "Differences in aortic arch radius of curvature, neck size, and taper in patients with traumatic and aortic disease", *J Surg Res*, 184:613–618. doi: 10.1016/j.jss.2013.05.098

4. Roccabianca S., Figueroa CA., Tellides G., Humphrey JD., 2014, "Quantification of regional differences in aortic stiffness in the aging human", *J Mech Behav Biomed Mater*, 29:618–634. doi: 10.1016/j.jmbbm.2013.01.026

5. Bürk J., Blanke P., Stankovic Z., Barker A., Russe M., Geiger J., Frydrychowicz A., Langer M., Markl M., 2012, "Evaluation of 3D blood flow patterns and wall shear stress in the normal and dilated thoracic aorta using flow-sensitive 4D CMR", *J Cardiovasc Magn Reson*, 14:84. doi: 10.1186/1532-429x-14-84

6. Yeh HH., Grecov D., Karri S., 2014, "Computational modelling of bileaflet mechanical valves using fluid-structure interaction approach", *J Med Biol Eng*, 34:482–486. doi: 10.5405/jmbe.1699

7. Vasava P, Jalali P, Dabagh M, Kolari PJ. 2012, "Finite element modelling of pulsatile blood flow in idealized model of human aortic arch: Study of hypotension and hypertension", *Comput Math Methods Med* 2012:1–14. doi: 10.1155/2012/861837

8. Holzapfel G.A., Gasser T.C., Ogden R.W., 2000, "A new constitutive framework for arterial wall mechanics and a comparative study of material models", *J Elast*, 61:1–48. doi: 10.1023/a:1010835316564

9. Weisbecker H., Pierce D.M., Regitnig P., Holzapfel G.A. 2012, "Layer-specific damage experiments and modeling of human thoracic and abdominal aortas with non-atherosclerotic intimal thickening", *J Mech Behav Biomed Mater* 12:93–106. doi: 10.1016/j.jmbbm.2012.03.012

10. Morrison T.M., Choi G., Zarins C.K., Taylor C.A., 2009, "Circumferential and longitudinal cyclic strain of the human thoracic aorta: Age-related changes", *J Vasc Surg*, 49:1029–1036. doi: 10.1016/j.jvs.2008.11.056

DRUG RELEASING SCAFFOLDS FOR NEURONAL DIFFERENTIATION OF HUMAN INDUCED PLURIPOTENT STEM CELLS (HIPSCS)

Bahram Mirani
Department of Mechanical Engineering
University of Victoria
Victoria, British Columbia, Canada
bmirani@uvic.ca

Tara Styan
Department of Mechanical Engineering
University of Victoria
Victoria, British Columbia, Canada
tstyan@uvic.ca

Nima Khadem Mohtaram
Department of Mechanical Engineering
University of Victoria
Victoria, British Columbia, Canada
nkhadem@gmail.com

Sarah Wong
Department of Mechanical Engineering
University of Victoria
Victoria, British Columbia, Canada
smwong@uvic.ca

Erik Pagan
Department of Mechanical Engineering
University of Victoria
Victoria, British Columbia, Canada
erikpm@uvic.ca

Armin Bayati
Department of Mechanical Engineering
University of Victoria
Victoria, British Columbia, Canada
arminbay@uvic.ca

Daniel Pedde
Department of Mechanical Engineering
University of Victoria
Victoria, British Columbia, Canada
dpedde@uvic.ca

Stephanie Willerth
Department of Mechanical Engineering
University of Victoria
Victoria, British Columbia, Canada
willerth@uvic.ca

Mohsen Akbari
Department of Mechanical Engineering
University of Victoria
Victoria, British Columbia, Canada
makbari@uvic.ca

ABSTRACT

Cellular response to the physical and biochemical cues present in their environment has been well established in previous works.[1] Here we investigate the impact of topography and morphogenetic signaling by using 3-dimentional (3D) bioprinted hydrogel scaffolds and incorporation of retinoic acid (RA) on neuronal differentiation of human induced pluripotent stem cell (hiPSC) aggregates. Hydrogel scaffolds were fabricated using a bioink composed of alginic acid sodium salt and gelatin methacryloyl (GelMA). RA was encapsulated in polycaprolactone (PCL) microspheres which were loaded into the bioink before scaffold fabrication process. The synergistic effect of topography and biochemical signaling by RA was studied by conducting immunohistochemistry assays. Upregulated neuronal marker expression was achieved with the incorporation of RA and fiber-based directional topography of scaffolds.

KEYWORDS: 3D bioprinting, hiPSC, peripheral nerve injury, retinoic acid, bioink

INTRODUCTION

Injuries and ailments affecting the spinal cord and brain are among the most debilitating known to mankind. For example, damage to the spinal cord can lead to permanent and complete paralysis. These traumatic injuries lead to neuronal cell death and atrophy. After damage occurs in the central nervous system, glial scar forms around the remaining neurons to protect them from any further damage.[2] This scarring leads to a decrease in regeneration of neurons.

To address the need for the treatment of nervous injuries, tissue engineering techniques that combine advanced biofabrication approaches with biomaterials and stem cells of the nerve tissue using have emerged. Among different biofabrication strategies, 3D bioprinting holds a great promise as it enables fabrication tissue-like constructs with exquisite control over their microstructure and physicochemical properties. Challenges encountered when treating ailments of the central nervous system have led researchers to look toward stem cells to regenerate lost neurons. Stem cells can differentiate into multiple cell types. Factors that influence this differentiation include growth factors, signals from their surrounding cells, and the topography in which they are growing.

In this study, 3D bioprinting was employed to generate hydrogel scaffolds on top of which hiPSC aggregates were seeded. This type of stem cell is formed by reverting adult cells back into stem cells using the Yamanaka factors (Figure 1). Cellular response in terms of neuronal differentiation to the fiber-based topography of scaffolds along with the impact of encapsulated morphogenic molecule, RA was studied. RA was encapsulated in PCL microspheres using a single emulsion method.[3] Bioink comprised of alginate and GelMA was laden with drug loaded microspheres to induce neuronal differentiation of hiPSCs.

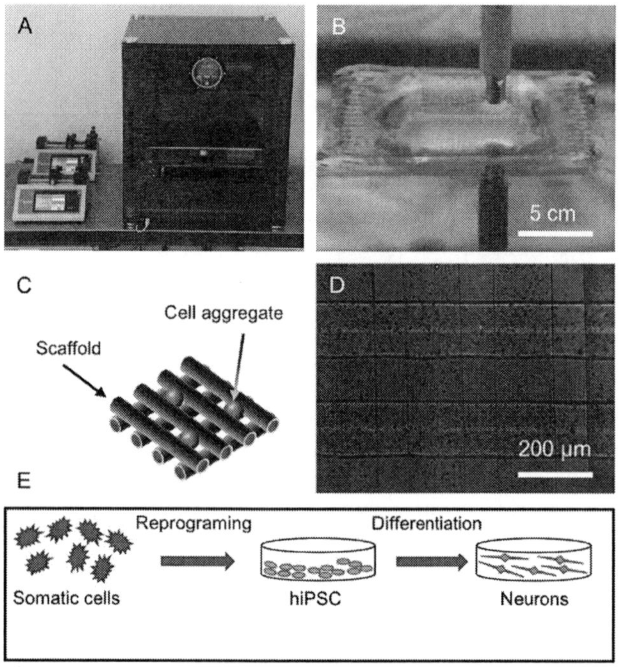

Figure 1. (A) 3D bioprinting system, consisting of a commercial 3D printer and two syringe pumps to provide the bioink. (B) 3D bioprinted scaffold, fabricated by the deposition of bioink on the bed of the 3D printer. (C) Schematic of a scaffold and cell aggregates on top of it. (D) Microscopic image of a two-layer scaffold, showing the distance between fibers. (E) Reverting differentiated cells into stem cells and then back into neurons.

EXPERIMENTAL METHODS

PCL microspheres loaded with RA were fabricated by a previously developed single emulsion technique.[3] In this method, 500 mg of PCL with molecular weight (M_n) of 45,000 Da dissolved in 3 ml of dichloromethane (DCM) and after adding 3 ml of RA solution in ethanol (0.5 mg/ml), this oil phase was emulsified with 3 ml of 2% w/v polyvinyl alcohol (PVA) aqueous solution by vortex mixer. The resulting emulsion was added to 100 ml of 0.3% w/v PVA solution and was stirred at 34°C for 4 hours to let the organic solvent evaporate. This process led to the hardening of microspheres. Particles were subsequently washed by centrifugation and replacing deionized water 4 times and freeze-dried for 24 hours.

Encapsulation efficiency of these RA loaded PCL particles was determined by dissolving them in DCM and precipitating PCL out by adding ethanol and pelleting by centrifugation. The content of RA in the remaining solution was determined by measuring its absorbance at 354 nm using an Infinite M200Pro plate reader and subtracting the absorbance of solution derived from the same process with blank PCL particles. Encapsulation efficiency (EE) was calculated according to Eq(1) by dividing the RA content in the microspheres to the amount of RA used during their fabrication process.

$$EE = \frac{RA\ content\ in\ microsphere}{RA\ used\ to\ fabricate\ microspheres} \times 100 \qquad (1)$$

For determining the release kinetics of RA from PCL particles, samples consisting of 5 mg of microspheres were suspended in 1 ml of PBS and incubated at 37°C on a microtube rotator. Their supernatants were collected periodically and replaced by fresh PBS. The diffusion rate of RA to the medium was obtained similarly by plate reading the supernatants at 354 nm and accumulating the released amounts over time.

Bioink in this project was composed of 5% w/v GelMA, 0.5% w/v of a photoinitiator (2-hydroxy-4'-(2-hydroxyethoxy-2-methylpropiophenone), and 2% w/v sodium alginate. PCL microspheres with two different concentration – 0.2 mg/ml as low and 2 mg/ml as high concentration condition – was loaded in the bioink prior to 3D bioprinting process. Scaffolds without microspheres were also included as negative control conditions regarding the impact of RA on cell behavior. Additionally, the effect of fiber-based morphology in printed scaffolds was studied by including gel sheets made of the same material composition and comparing the cellular response.

As demonstrated in Figure 2, 3D bioprinting was performed by incorporating a microfluidics-based extruder composed of coaxial core-sheath needles to a commercial 3D printer. Core and sheath flows, which were supplied by two syringe pumps, were bioink and calcium chloride, respectively. In this process, bioink was ionically crosslinked at the tip of the extruder and deposited on the printer's bed. Printed scaffolds were then post-photo-crosslinked by ultraviolet (UV) irradiation to promote the integrity of the final construct. In this previously developed method[4], alginate contributes to the generation of a semi-integrated structure during 3D bioprinting and GelMA as a modified form of gelatin, a denatured product of collagen, contributes to the final integration of different layers in the printed scaffold and provide a biologically active environment for cell adhesion due to its arginine-glycine-aspartic acid sequences.[5]

Undifferentiated hiPSCs (1-DL-01 line obtained from WiCell Research Institute) were cultured in defined conditions on 6 well plates coated with Vitronectin XF with daily media changes (TeSR-E8 media). Cells were passaged using the ReLeSR reagent to maintain high levels of pluripotency. To induce neural differentiation, these cells were passaged onto AggreWellTM800 plates and cultured in the presence of STEMdiff™ Neural Induction Medium (NIM) for 9 days. After 9 days, aggregates or embryoid bodies (EBs) were seeded onto the scaffolds. The engineered tissues were then cultured for 11 days in NIM media before fixation and staining. To evaluate the differentiation, the cultures were stained for the neuronal marker βIII-tubulin and counterstained with DAPI to label cell nuclei.

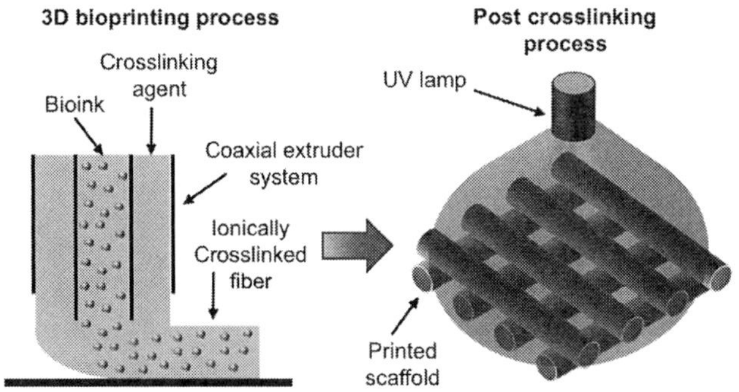

Figure 2. 3D bioprinting process and post crosslinking by UV irradiation.

RESULTS

Fabricated PCL microspheres were characterized in terms of morphology, encapsulation efficiency of RA, and drug release kinetics. Scanning electron microscopy (SEM) and post processing of SEM images by ImageJ were used to determine the size and surface morphology of the particles, as shown in Figure 3A,B. Analyzing the size distribution of these microspheres determined their diameter as 5.19 ± 2.69 μm.

Encapsulation efficiency was 58.1% $\pm$4.1% based on Eq(1) and this value was considered for determining the actual content of RA in the microspheres for the release study. Release kinetics of RA from PCL microspheres were characterized over 20 days (Figure 3D). This release trend showed an initial burst release of RA at the beginning of the process and slowed down as it proceeded to the end of the period, reaching to 21.5% $\pm$ 1.7% of the total RA content in the microspheres.

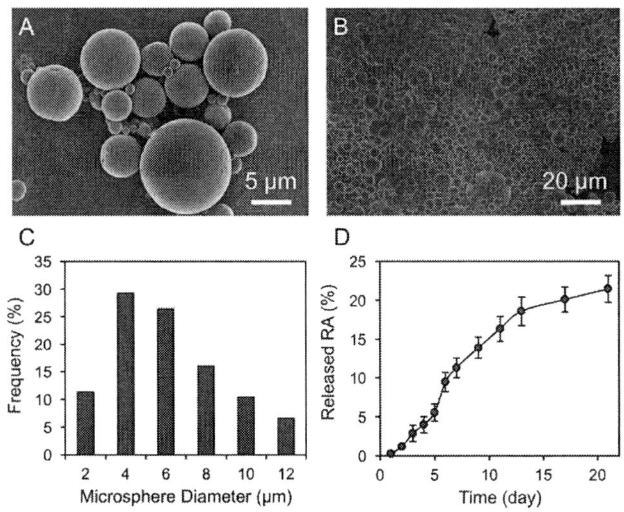

Figure 3. (A,B) SEM image of RA loaded PCL microspheres, showing their surface morphology and size. (C) Size distribution of microspheres obtained by the analysis of SEM images in ImageJ. (D) Release kinetics of RA from PCL microsphere in a 20-day period.

3D bioprinted scaffolds were characterized in terms of hydrogel fiber diameter and the spacing between adjacent fibers to provide a directional topography in the aggregates size order. For this purpose, fibers with about 200 μm of diameter were printed with 100 μm of spacing. This architecture was generated by adjusting the bioink flow rate and printing speed to produce fiber with desired diameter and applying required printing pattern to regulate the distance between fibers.

Cell aggregates responded differently to 3D bioprinted scaffolds compared to gel sheets. As indicated in Figure 4A, after 7 days of incubation, fiber-based topography and directionality in printed scaffolds contributed to the alignment of cell aggregates with the direction of fibers. In contrast, they maintained a spherical shape on gel sheets. For this part of the study, RA was not used in both 3D bioprinted and sheet-shaped scaffolds to determine the impact of topography independent from the biochemical effect of RA.

Immunocytochemistry (ICC) was performed on the cells after 11 days of incubation to identify their viability after being cultured on the scaffolds. LIVE/DEAD stain from ThermoFisher Scientific was used to identify live and dead cells. It can be seen in Figure 4B that a large majority of the cells were alive after 11 days of culture.

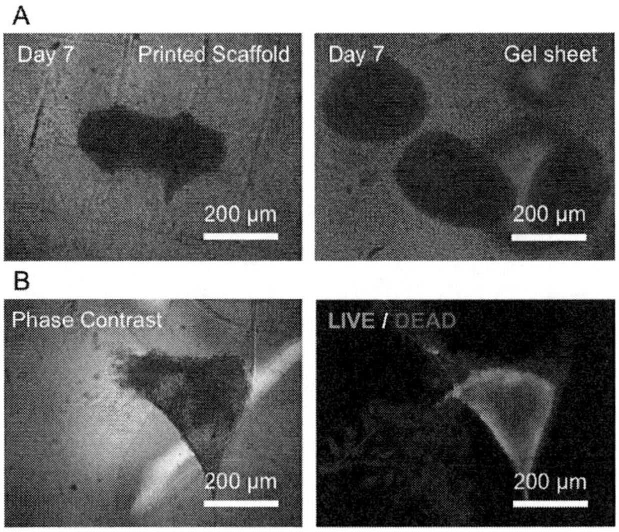

Figure 4. (A) embriod body on 3D bioprinted scaffold tends to extend along with hydrogel fibers. As opposed to 3D bioprinted scaffolds, cell aggregates maintain a spherical shape on hydrogel sheets with the same materials composition. Both hydrogel structure did not have RA loaded PCL microspheres. (B) Phase contrast and ICC images of cells on scaffold with no content of microspheres.

Cellular differentiation was identified by Tuj1 staining. TUJ1 detection is an early neuronal marker which indicates differentiation into neuronal cell types. Furthermore, DAPI marker was used to distinguish cell nuclei. Staining with TUJ1 and DAPI allows a comparison of the number of cells and the amount of neuronal differentiation.

As demonstrated in Figure 5, all of the EBs were positive for TUJ1 however, the scaffolds with low microsphere concentration appeared to have a lower ratio of TUJ1 to DAPI. In contrast, scaffolds with high microsphere loading induced more TUJ1 positive cells in comparison to DAPI positive cells, demonstrating the induction of neuronal differentiation due to the incorporation of RA. Furthermore, EBs on scaffolds with high concentration of shown a directionality in Tuj1, implying the presence of directionality in neurite outgrowth. This phenomenon can result from the integration of physical (topographical) and biochemical cues. The similar synergistic effect that triggers signaling pathways toward governing cell behaviors, especially cellular differentiation, was identified in previous studies for human mesenchymal stem cells.[6]

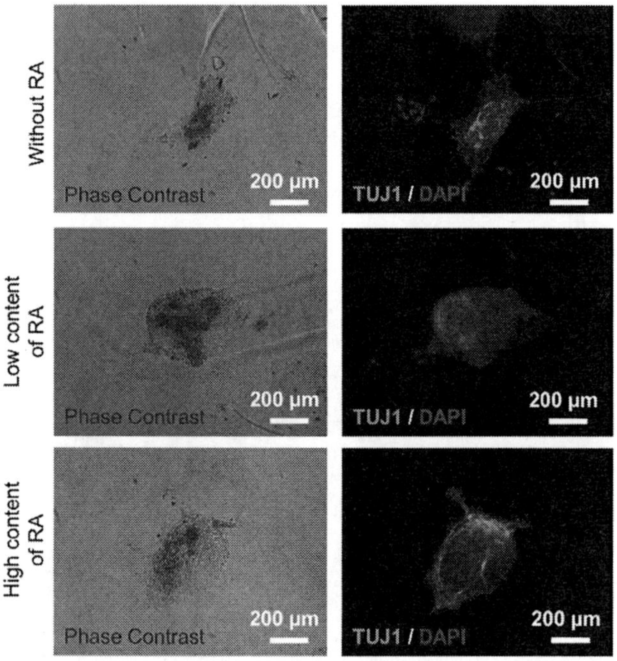

Figure 5. Phase contrast and ICC images of cell aggregates on top of scaffolds with three different content of RA: 0, 0.2 mg/ml PCL/hydrogel as low content, and 2 mg/ml of PCL/hydrogel as high content of RA.

CONCLUSION

In this study, the impact of biochemical and topographical induction toward neuronal differentiation of hiPSCs was investigated by using 3D bioprinted hydrogel scaffolds along with hydrogel sheets, loaded with RA. Cellular response to the scaffolds was studied for a 11-day period and immunohistochemistry assays showed an enhanced neuronal differentiation and enhanced neurite outgrowth in the presence of RA. Topography also contributed to the alignment of cell aggregates with the direction of fibers on 3D bioprinted scaffolds; however, cell aggregates remained spherical on gel sheets. As a future direction, studying the effect of using proteins such as laminin which is also present in the native nervous tissue in recommended to promote cell adhesion to hydrogel scaffolds.

ACKNOWLEDGMENTS

M.A. would like to thank the Natural Sciences and Engineering Research Council of Canada (NSERC) and Canadian Foundation for Innovation for supporting this work. S.M.W. receives support from the Canada Research Chairs program, the Stem Cell Network, NSERC, and the International Collaboration on Repair Discoveries.

REFERENCES

1. Murphy, W. L., McDevitt, T. C. & Engler, A. J. 2014, Materials as stem cell regulators. *Nat. Mater.* **13,** 547–557.

2. Fawcett, J. W. & Asher, R. 1999, The glial scar and central nervous system repair. *Brain Res. Bull.* **49,** 377–391.

3. Jeong, Y.-I. *et al.* 2003, Preparation of poly(DL-lactide-co-glycolide) microspheres encapsulating all-trans retinoic acid. *Int. J. Pharm.* **259,** 79–91.

4. Colosi, C. *et al.* 2016, Microfluidic Bioprinting of Heterogeneous 3D Tissue Constructs Using Low-Viscosity Bioink. *Adv. Mater.* **28,** 677–684.

5. Yue, K. *et al.* 2015, Synthesis, properties, and biomedical applications of gelatin methacryloyl (GelMA) hydrogels. *Biomaterials* **73,** 254–271.

6. Yim, E. K. F., Pang, S. W. & Leong, K. W. 2007, Synthetic nanostructures inducing differentiation of human mesenchymal stem cells into neuronal lineage. *Exp. Cell Res.* **313,** 1820–9.

IRREVERSIBLE STRAINS AND EVOLUTION OF FIBRE PATTERN IN HYDRATED, FIBRE-REINFORCED SOFT TISSUES

Salvatore Di Stefano
Dept Mathematical Sciences
"G.L. Lagrange"
Politecnico di Torino
Torino, Italy
salvatore.distefano@polito.it

Kotaybah Hashlamoun,
Salvatore Federico
Dept Mechanical and
Manufacturing Engineering
The University of Calgary
Calgary, Alberta, Canada
kwhashla@ucalgary.ca,
salvatore.federico@ucalgary.ca

Alfio Grillo
Dept Mathematical Sciences
"G.L. Lagrange"
Politecnico di Torino
Torino, Italy
alfio.grillo@polito.it

ABSTRACT

Within the framework of finite-strain poroplasticity, we present a model of structural evolution in hydrated, fibre-reinforced, soft tissues. Our model accounts for both production of irreversible strains and deformation-driven reorientation of collagen fibres. The irreversible strains are assumed to be concomitant with the formation and propagation of microcracks in the tissue as a consequence of ageing, injuries, and interaction with the tissue's mechanical environment.

Keywords: Fibre-reinforced tissues, Poroplasticity, Tissue remodelling, Damage, Ageing

INTRODUCTION

The scope of this contribution is to present a poroplastic model of the structural reorganisation of hydrated, soft, fibre-reinforced biological tissues, in which the collagen fibres are oriented statistically. Although the tissue which we will be referring to in the sequel is articular cartilage, the theory we report on may also be extended to other types of tissues, with some slight adaptations.

Our point of departure is the work by Garcia et al. [1], in which the authors develop an elastoplastic constitutive framework for addressing the onset and evolution of damage in bone tissue. The mechanical framework established in [1] relies on three descriptors. These are the standard deformation, and two "internal variables": a scalar damage parameter $D \in [0, 1[$ [1], which represents the density of microcracks appearing as a consequence of severe loading conditions or injuries, and the tensor of plastic strains, which keeps track of the irreversible strains accompanying the formation of the microcracks. In [1], bone is regarded as a monophasic medium whose elastoplastic damage behaviour is encompassed by a nonsmooth, convex free energy density, expressed as the sum of a quadratic function of the elastic strains and a quadratic function of the plastic strains.

The model we propose here is inspired by that in [1], but with two substantial differences. Firstly, we formulate our model within the framework of biphasic media (indeed, in our idealisation, a hydrated soft tissue is a porous medium saturated with a fluid, which can be macroscopically regareded as a biphasic mixture). Secondly, we consider a nonlinear elastoplastic constitutive framework, in which both the strain energy density and the permeability of the medium are defined on the basis of the multiplicative decomposition of the deformation gradient tensor. At the moment, however, we consider a simplified setting in which only plastic distortions, without damage, are taken into account. Moreover, since our target tissue is articular cartilage, which is transversely anisotropic, we study the onset and evolution of plastic distortions in the case of anisotropic material. This constitutes a generalisation of the results obtained in [2] for isotropic media.

MATHEMATICAL MODEL

The standard, Darcy-based model of hydrated soft tissues relies on the balance laws (cf. e.g. [2, 3])

$$\dot{J} = \mathrm{Div}\,[\boldsymbol{K}\mathrm{Grad}\,p]\,, \tag{1a}$$

$$\mathrm{Div}\left(-Jpg^{-1}\boldsymbol{F}^{-\mathrm{T}} + \boldsymbol{P}_{\mathrm{sc}}\right) = \boldsymbol{0}, \tag{1b}$$

which describe the material forms of the balance laws of mass and linear momentum, respectively. In (1a) and (1b), $J = \det \boldsymbol{F} > 0$ is the volume ratio, i.e., the determinant of the deformation gradient tensor $\boldsymbol{F}$ of the solid phase motion, p is referred to as *pore pressure*, $\boldsymbol{K} := J\boldsymbol{F}^{-1}\boldsymbol{k}\boldsymbol{F}^{-\mathrm{T}}$ is the material permeability tensor, i.e., the backward Piola transform of the spatial permeability $\boldsymbol{k}$, $\boldsymbol{g}$ is the spatial metric tensor, and $\boldsymbol{P}_{\mathrm{sc}}$ is said to be the constitutive part of the solid phase first Piola-Kirchhoff stress tensor. We recall that (1a) and (1b) are defined in the reference configuration $\mathcal{B}$ of the solid phase, and are obtained under the hypotheses that the mass densities of the interstitial fluid, matrix, and fibres are constant, that the viscous part of the fluid stress tensor is negligible with respect to the hydrostatic part, and that *all* body forces can be disregarded.

To account for the production of irreversible strains, we invoke the theory of finite strain elastoplasticity. To this end, we enforce the multiplicative decomposition of the deformation gradient tensor into an elastic and a plastic part, i.e., $\boldsymbol{F} = \boldsymbol{F}_{\mathrm{e}}\boldsymbol{F}_{\mathrm{p}}$, where $\boldsymbol{F}_{\mathrm{e}}$ and $\boldsymbol{F}_{\mathrm{p}}$ describe the elastic and plastic distortions, respectively (see [4] for a review on the subject). Within this setting, $\boldsymbol{C}_{\mathrm{p}} = \boldsymbol{F}_{\mathrm{p}}^{\mathrm{T}}.\boldsymbol{F}_{\mathrm{p}}$ is the Cauchy-Green tensor of plastic distortions, and $\boldsymbol{E}_{\mathrm{p}} = \frac{1}{2}(\boldsymbol{C}_{\mathrm{p}} - \boldsymbol{G})$ is the Green-Lagrange strain tensor generated by $\boldsymbol{F}_{\mathrm{p}}$. Note that $\boldsymbol{G}$ represents the metric tensor associated with the reference configuration of the tissue. Once $\boldsymbol{F}_{\mathrm{p}}$ is known, the irreversible strains discussed in the Introduction can be identified with $\boldsymbol{E}_{\mathrm{p}}$.

The multiplicative decomposition of the deformation gradient tensor evokes the concept of "natural state" of the solid material, i.e., a collection of stress-free body elements (see [4]). We call $\mathcal{C}_{\nu}$ such collection and, compatibly with the time scales associated with the deformation processes undergone by the tissue, we assume that each material point X enjoys hyperelastic mechanical behaviour starting from its relaxed state. Hence, by adapting the hyperelastic potential used in previous works [3] to the present theoretical setting, we write

$$\hat{W}_{\nu}(\boldsymbol{C}_{\mathrm{e}}) = \Phi_{\mathrm{s}\nu}\hat{U}(J_{\mathrm{e}}) + \Phi_{0\mathrm{s}\nu}\hat{W}_0(\boldsymbol{C}_{\mathrm{e}}) + \Phi_{1\mathrm{s}\nu}\hat{W}_{\mathrm{en}}(\boldsymbol{C}_{\mathrm{e}}). \tag{2}$$

In (2), $\Phi_{0\mathrm{s}\nu} = J_{\mathrm{e}}\phi_{0\mathrm{s}}$, $\Phi_{1\mathrm{s}\nu} = J_{\mathrm{e}}\phi_{1\mathrm{s}}$, and $\Phi_{\mathrm{s}\nu} = \Phi_{0\mathrm{s}\nu} + \Phi_{1\mathrm{s}\nu}$ are, respectively, the volumetric fractions of matrix, fibres, and solid phase as a whole, evaluated in the natural state (here, $\phi_{0\mathrm{s}}$ and $\phi_{1\mathrm{s}}$ are the volumetric fractions of matrix and fibres in the spatial configuration of the tissue), $J_{\mathrm{e}} = \det \boldsymbol{F}_{\mathrm{e}} > 0$ is the volume ratio assoaciated with the elastic distortions, and $\boldsymbol{C}_{\mathrm{e}} = \boldsymbol{F}_{\mathrm{e}}^{\mathrm{T}}.\boldsymbol{F}_{\mathrm{e}} = \boldsymbol{F}_{\mathrm{p}}^{-\mathrm{T}}\boldsymbol{C}\boldsymbol{F}_{\mathrm{p}}^{-1}$ is referred to as the elastic part of the Cauchy-Green deformation tensor. The energetic contributions to $\hat{W}_{\nu}(\boldsymbol{C}_{\mathrm{e}})$ are given by

$$\hat{U}(J_{\mathrm{e}}) = \alpha_0 \mathcal{H}(J_{\mathrm{cr}} - J_{\mathrm{e}})\frac{[J_{\mathrm{e}} - J_{\mathrm{cr}}]^{2q}}{[J_{\mathrm{e}} - \Phi_{\mathrm{s}\nu}]^{r}}, \tag{3a}$$

$$\hat{W}_0(\boldsymbol{C}_{\mathrm{e}}) = \alpha_0 \frac{\exp\left(\alpha_1[I_1 - 3] + \alpha_2[I_2 - 3]\right)}{[I_3]^{\alpha_3}}, \tag{3b}$$

$$\hat{W}_{\mathrm{en}}(\boldsymbol{C}_{\mathrm{e}}) = \hat{W}_{1\mathrm{i}}(\boldsymbol{C}_{\mathrm{e}}) + \langle\!\langle \hat{W}_{1\mathrm{a}}(\boldsymbol{C}_{\mathrm{e}}, \boldsymbol{A}_{\nu})\rangle\!\rangle, \tag{3c}$$

$$\hat{W}_{1\mathrm{a}}(\boldsymbol{C}_{\mathrm{e}}, \boldsymbol{A}_{\nu}) = \mathcal{H}(I_4 - 1)\tfrac{1}{2}c[I_4 - 1]^2. \tag{3d}$$

In (3a), $\hat{U}(J_{\mathrm{e}})$ is a penalty term introduced to prevent J_{e} from violating the unilateral constraint $J_{\mathrm{e}} \geq \Phi_{\mathrm{s}\nu}$, α_0 [MPa] is a material parameter defining the energy scale, $J_{\mathrm{cr}} \in]\Phi_{\mathrm{s}\nu}, 1]$ is a critical threshold of J_{e} that establishes when the Heaviside function $\mathcal{H}(J_{\mathrm{cr}} - J_{\mathrm{e}})$ is active (we set $\mathcal{H}(J_{\mathrm{cr}} - J_{\mathrm{e}}) = 1$ for $J_{\mathrm{cr}} > J_{\mathrm{e}}$, and $\mathcal{H}(J_{\mathrm{cr}} - J_{\mathrm{e}}) = 0$ for $J_{\mathrm{cr}} \leq J_{\mathrm{e}}$), $q \geq 2$ is an integer number, and $r \in]0, 1]$ is a material parameter expressing the rate with which $\hat{U}(J_{\mathrm{e}})$ diverges positively in the limit $J_{\mathrm{e}} \to \Phi_{\mathrm{s}\nu}$. In (3b), $\hat{W}_0(\boldsymbol{C}_{\mathrm{e}})$ is the isotropic hyperelastic potential of the matrix, α_1, α_2, and α_3 are material parameters chosen in such a way that $\alpha_3 = \alpha_1 + 2\alpha_2$ [5], and I_1, I_2, and I_3 represent the first three principal invariants of $\boldsymbol{C}_{\mathrm{e}}$, i.e., $I_1 = \mathrm{tr}(\boldsymbol{C}_{\mathrm{e}})$, $I_2 = \frac{1}{2}\{[\mathrm{tr}(\boldsymbol{C}_{\mathrm{e}})]^2 - \mathrm{tr}(\boldsymbol{C}_{\mathrm{e}}^2)\}$, and $I_3 = \det \boldsymbol{C}_{\mathrm{e}}$. In (3c), the energy density $\hat{W}_{\mathrm{en}}(\boldsymbol{C}_{\mathrm{e}})$ is referred to as *ensemble energy potential* [6], and it accounts for the elasticity of the fibres. The term $\hat{W}_{1\mathrm{i}}(\boldsymbol{C}_{\mathrm{e}})$ is the isotropic contribution of the fibres, whereas the summand $\langle\!\langle \hat{W}_{1\mathrm{a}}(\boldsymbol{C}_{\mathrm{e}}, \boldsymbol{A}_{\nu})\rangle\!\rangle$ is the directional average of the anisotropic energy density $\hat{W}_{1\mathrm{a}}(\boldsymbol{C}_{\mathrm{e}}, \boldsymbol{A}_{\nu})$ defined in (3d), where $\boldsymbol{A}_{\nu} = \boldsymbol{M}_{\nu} \otimes \boldsymbol{M}_{\nu}$ is the *structure tensor*, and $\boldsymbol{M}_{\nu}$ is a unit vector defining the direction of fibre orientation at a given material point in the natural state. In (3d), c [MPa] is an elastic coefficient, and $I_4 = \boldsymbol{C}_{\mathrm{e}} : \boldsymbol{A}_{\nu} = \boldsymbol{F}_{\mathrm{p}}^{-\mathrm{T}}\boldsymbol{C}\boldsymbol{F}_{\mathrm{p}}^{-1} : \boldsymbol{A}_{\nu}$ denotes the fourth invariant of the elastic part of the overall deformation.

Equations (2) and (3a)–(3d) allow to determine $\boldsymbol{P}_{\mathrm{sc}}$ as

$$\boldsymbol{P}_{\mathrm{sc}} = \boldsymbol{F}\left[J_{\mathrm{p}}\boldsymbol{F}_{\mathrm{p}}^{-1}\left(2\frac{\partial \hat{W}_{\nu}}{\partial \boldsymbol{C}_{\mathrm{e}}}(\boldsymbol{C}_{\mathrm{e}})\right)\boldsymbol{F}_{\mathrm{p}}^{-\mathrm{T}}\right], \tag{4}$$

where $J_{\mathrm{p}} = \det \boldsymbol{F}_{\mathrm{p}} > 0$ is the volume ratio assoaicted with the plastic distortions. Along with $\hat{W}_{\nu}(\boldsymbol{C}_{\mathrm{e}})$, we provide a constitutive expression for the material permeability tensor $\boldsymbol{K}$. By adapting the results reported in [3] to our purposes, we obtain

$$\boldsymbol{K} = J\,k_0\boldsymbol{C}^{-1} + J^{-1}k_0\boldsymbol{F}_{\mathrm{p}}^{-1}\boldsymbol{Z}\boldsymbol{F}_{\mathrm{p}}^{-\mathrm{T}}, \tag{5a}$$

$$k_0 = k_{0\nu}\left[\frac{J - \Phi_{\mathrm{s}\nu}J_{\mathrm{p}}}{J_{\mathrm{p}} - \Phi_{\mathrm{s}\nu}J_{\mathrm{p}}}\right]^{\kappa_0}\exp\left(\frac{m_0}{2}\frac{J^2 - J_{\mathrm{p}}^2}{J_{\mathrm{p}}^2}\right), \tag{5b}$$

$$\boldsymbol{Z} = \left\langle\!\!\left\langle \frac{\boldsymbol{A}_{\nu}}{I_4} \right\rangle\!\!\right\rangle, \tag{5c}$$

$$k_{0\nu} = k_{0\nu}^{(0)}\left[\frac{e_{\nu}}{e_{\nu}^{(0)}}\right]^{\kappa_0}\exp\left(\frac{m_0}{2}\left[\left(\frac{1 + e_{\nu}}{1 + e_{\nu}^{(0)}}\right)^2 - 1\right]\right), \tag{5d}$$

where $k_{0\nu}^{(0)}$ is a referential (constant) value of permeability, $e_{\nu} = (1 - \Phi_{\mathrm{s}\nu})/\Phi_{\mathrm{s}\nu}$ is the void ratio. The numerical values of the model parameters used for the shown simulations are $\alpha_0 = 0.125$ MPa, $\alpha_1 = 0.7778$, $\alpha_2 = 0.111$, $\alpha_3 = 1$, $c = 0.075$ MPa, $q = 2$, $r = 0.5$, $\kappa_0 = 0.0848$, $m_0 = 4.638$, $k_{0\nu}^{(0)} = 0.03\ \mathrm{mm^4N^{-1}s^{-1}}$, $e_\nu^{(0)} = 4.0$. Moreover, the volumetric fractions of matrix, fibres, and solid phase as a whole in the natural state are given by

$$\Phi_{0s\nu} = \Phi_{0s\nu}(\xi) = -0.062\xi^2 + 0.038\xi + 0.046, \quad (6a)$$

$$\Phi_{1s\nu} = \Phi_{1s\nu}(\xi) = +0.062\xi^2 - 0.138\xi + 0.204, \quad (6b)$$

$$\Phi_s = \Phi_s(\xi) = -0.100\xi + 0.250. \quad (6c)$$

We remark that the directional averages of $\hat{W}_{1a}(\boldsymbol{C}_e, \boldsymbol{A}_\nu)$ and $\boldsymbol{A}_\nu / I_4$ are computed as

$$\langle\!\langle \hat{\mathfrak{F}}(\boldsymbol{C}_e, \boldsymbol{A}_\nu) \rangle\!\rangle = \int_{\mathbb{S}^2\mathcal{B}} \hat{\mathfrak{F}}(\boldsymbol{C}_e, \boldsymbol{A}_\nu) \Psi(\boldsymbol{M}_\nu), \quad (7)$$

with $\hat{\mathfrak{F}}(\boldsymbol{C}_e, \boldsymbol{A}_\nu)$ being $\hat{W}_{1a}(\boldsymbol{C}_e, \boldsymbol{A}_\nu)$ or $\boldsymbol{A}_\nu/I_4$, and $\Psi(\boldsymbol{M}_\nu)$ being a suitably chosen probability density of finding a fibre aligned along $\boldsymbol{M}_\nu$ at a given material point.

To close the mathematical model, it is necessary to determine an evolution law for $\boldsymbol{F}_p$. For our purposes, we restrict our investigation to the case of isochoric plastic distortions, thereby meaning $J_p = \det \boldsymbol{F}_p = 1$. Under this assumption, the dissipation inequality can be cast in the form [2]

$$\mathcal{D} = \boldsymbol{K} : \mathrm{Grad}\, p \otimes \mathrm{Grad}\, p + \boldsymbol{C}_e \boldsymbol{S}_\nu : \boldsymbol{L}_p \geq 0, \quad (8)$$

where $\boldsymbol{S}_\nu$ is given by $\boldsymbol{S}_\nu = 2(\partial \hat{W}_\nu / \partial \boldsymbol{C}_e)(\boldsymbol{C}_e)$ and $\boldsymbol{L}_p = \dot{\boldsymbol{F}}_p \boldsymbol{F}_p^{-1}$ is the rate of plastic distortions. Note that $\boldsymbol{L}_p$ is deviatoric because the constraint $J_p = 1$ applies. By invoking the relation $\boldsymbol{S}_\nu = J_p^{-1} \boldsymbol{F}_p \boldsymbol{S}_{sc} \boldsymbol{F}_p^{\mathrm{T}}$, with $\boldsymbol{S}_{sc} = \boldsymbol{F}\boldsymbol{P}_{sc}$, and following the line of thought suggested in [7], we perform a change of variables, which consists of setting $\boldsymbol{\Pi} = \boldsymbol{F}_p^{-1}$, and rephrasing (8) in terms of the rate of distortion $\boldsymbol{\Lambda} = \dot{\boldsymbol{\Pi}}\boldsymbol{\Pi}^{-1} = -\boldsymbol{\Pi}\boldsymbol{L}_p\boldsymbol{\Pi}^{-1}$. Hence, the dissipation inequality becomes

$$\mathcal{D} = \boldsymbol{K} : \mathrm{Grad}\, p \otimes \mathrm{Grad}\, p - \mathrm{Dev}^*(\boldsymbol{C}\boldsymbol{S}_{sc}\boldsymbol{C}) : \mathrm{sym}(\boldsymbol{\Lambda}\boldsymbol{C}^{-1}) \geq 0, \quad (9)$$

where we introduced the symmetric measure of stress $\boldsymbol{C}\boldsymbol{S}_{sc}\boldsymbol{C}$ [7], which is obtained by making $\boldsymbol{S}_{sc}$ fully covariant through the deformed metric $\boldsymbol{C}$, and we defined the *pulled-back deviatoric operator* [8] as

$$\begin{aligned}\mathrm{Dev}^*(\boldsymbol{C}\boldsymbol{S}_{sc}\boldsymbol{C}) &= \boldsymbol{C}\boldsymbol{S}_{sc}\boldsymbol{C} - \tfrac{1}{3}\mathrm{Tr}^*(\boldsymbol{C}\boldsymbol{S}_{sc}\boldsymbol{C})\boldsymbol{C} \\ &:= \boldsymbol{C}\boldsymbol{S}_{sc}\boldsymbol{C} - \tfrac{1}{3}\mathrm{Tr}[\boldsymbol{C}^{-1}(\boldsymbol{C}\boldsymbol{S}_{sc}\boldsymbol{C})]\boldsymbol{C}. \quad (10)\end{aligned}$$

We choose for $\boldsymbol{\Pi}$ an evolution law of the type [9, 10]

$$\mathrm{sym}(\boldsymbol{\Lambda}\boldsymbol{C}^{-1}) = -2\zeta_p\, \mathbb{I}^{\sharp *} : \mathrm{Dev}^*(\boldsymbol{C}\boldsymbol{S}_{sc}\boldsymbol{C}), \quad (11)$$

with $\mathbb{I}^{\sharp *} = \frac{1}{2}\boldsymbol{C}^{-1}\overline{\underline{\otimes}}\boldsymbol{C}^{-1}$ [8], and ζ_p being defined by [10]

$$\zeta_p = \lambda \left[\frac{\|\mathrm{dev}\, \boldsymbol{\sigma}_{sc}\| - \sqrt{2/3}\sigma_y}{\|\mathrm{dev}\, \boldsymbol{\sigma}_{sc}\|} \right]_+, \quad (12)$$

where $\boldsymbol{\sigma}_{sc} = J^{-1}\boldsymbol{F}\boldsymbol{S}_{sc}\boldsymbol{F}^{\mathrm{T}}$ is the constitutive part of the solid phase Cauchy stress tensor, $\lambda = 0.5\ \mathrm{MPa\,s^{-1}}$ is a material parameter, $\sigma_y > 0$ is referred to as *yield stress*, and the operator $[\cdot]_+$ extracts the positive part of the quantity to which it is applied. Note that, in the presented model, σ_y is assumed to be constant and is set equal to $\sigma_y = 0.002$ MPa.

RESULTS AND DISCUSSION

In summary, the unknowns of the model are identified with the motion of the solid phase, χ, pressure, p, and the tensor $\boldsymbol{\Pi}$. These are determined by the set of coupled, and highly non-linear equations given by (1a), (1b), and (11), which we solve for the case of an unconfined compression test in displacement control. This experiment is performed *in silico* by enforcing the boundary conditions

$$\text{on } \Gamma_U, \quad \begin{cases} \chi^3 = \mathfrak{g}, \\ (-\boldsymbol{K}\mathrm{Grad}\, p).\boldsymbol{N} = 0, \end{cases} \quad (13a)$$

$$\text{on } \Gamma_B, \quad \begin{cases} (-Jp\boldsymbol{g}^{-1}\boldsymbol{F}^{-\mathrm{T}} + \boldsymbol{P}_{sc}).\boldsymbol{N} = 0, \\ p = 0, \end{cases} \quad (13b)$$

$$\text{on } \Gamma_L, \quad \begin{cases} \chi(X,t) - \chi(X,0) = \boldsymbol{0}, \\ (-\boldsymbol{K}\mathrm{Grad}\, p).\boldsymbol{N} = 0, \end{cases} \quad (13c)$$

where $\boldsymbol{N}$ is the unit vector normal to the surface of the sample, Γ_U, Γ_B, and Γ_L represent, respectively, the upper, lateral, and lower boundary of the sample, and $\mathfrak{g}$ is the loading ramp

$$\mathfrak{g}(t) = \begin{cases} L - \frac{t}{T_{\mathrm{ramp}}} u_T, & \text{for } t \in [0, T_{\mathrm{ramp}}], \\ L - u_T, & \text{for } t \in\,]T_{\mathrm{ramp}}, T_{\mathrm{end}}], \end{cases} \quad (14)$$

with $u_T = 0.15$ mm, $T_{\mathrm{ramp}} = 30$ s, and $T_{\mathrm{end}} = 80$ s.

Some selected results of the performed numerical simulations are shown in Figs. 1–3. More specifically, in Figs. 1 and 2 we report, respectively, the time trend of the magnitude of Darcy velocity, $\|\boldsymbol{q}\|$, and pore pressure, p, at a given point on the sample for the case in which no plastic distortions occur, and for the case in which plastic distortions are triggered off. For both cases, we notice a rather pronounced influence of plasticity on the tissue's response. Indeed, as visible in Fig. 1, in the purely elastic case (i.e., when the tissue deforms in the absence of irreversible strains, which alter its internal structure), the velocity of the fluid is markedly higher than that characterising the elastoplatstic case, and it relaxes more slowly towards the stationary state. An analogous result holds true for the pressure (see Fig. 2), which is lower when plastic distortions arise, and tends towards zero faster than it does in the elastic case. We remark that, before plastic distortions commence, both the curves of Darcy's velocity and pressure overlap. However, the onset of plastic distortions produces a deviation in the behaviour of the tissue and, it leads to the formation of a kink in the curve of the magnitude of Darcy velocity (cf. Fig. 1). Finally, we report in Fig. (3) the evolution in time of the norm of the plastic strain tensor, i.e., $\|\boldsymbol{E}_p\| = \sqrt{\mathrm{tr}(\boldsymbol{E}_p.\boldsymbol{E}_p)}$, with $\boldsymbol{E}_p = \frac{1}{2}(\boldsymbol{\Pi}^{-\mathrm{T}}.\boldsymbol{\Pi}^{-1} - \boldsymbol{G})$, and we notice that this quantity is an increasing function of time, which is zero as long as plastic distortions are not activated, and then grows monotonically towards an asymptotic value, determined by the magnitude of the imposed target boundary displacement u_T.

In a future study, we would like to refine the model, and study the combined effect of damage and plasticity, as done in [1] for the case of bone.

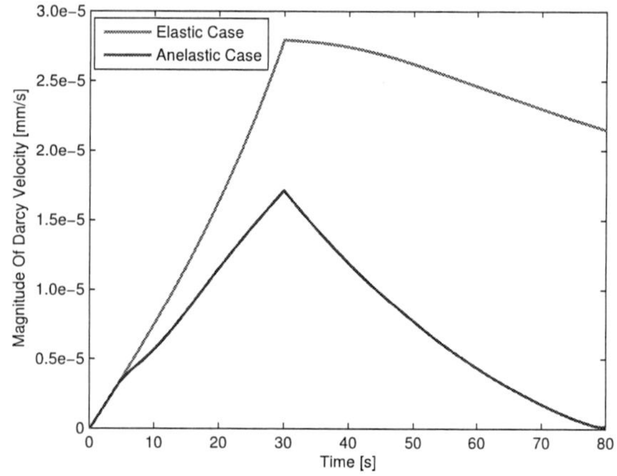

Figure 1: Time trend of the magnitude of Darcy's velocity at the point of Cartesian coordinates $(0, 0, 0.25)$ for both the elastic and the elastoplastic case.

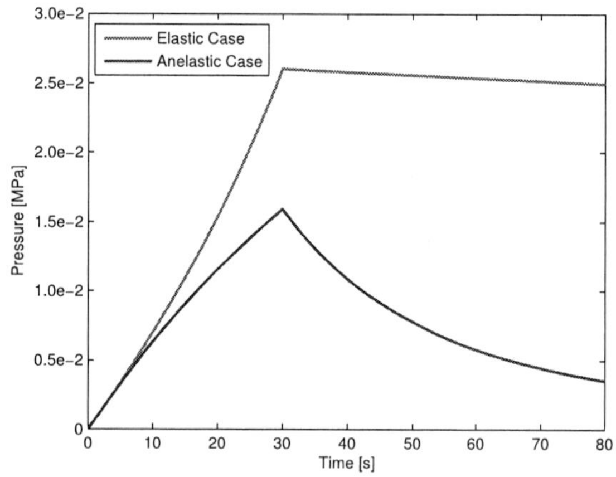

Figure 2: Time trend of the pressure p, evaluated at the point of Cartesian coordinates $(0, 0, 0.25)$ for both the elastic and the elastoplastic case.

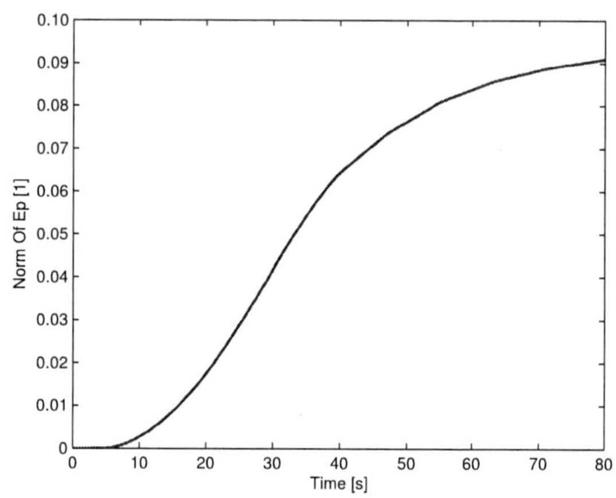

Figure 3: Time trend of the norm of the plastic strain $\|\boldsymbol{E}_{\mathrm{p}}\|$, evaluated at the point of Cartesian coordinates $(0, 0, 0.25)$.

ACKNOWLEDGMENTS

This work has been supported in part by the *Politecnico di Torino* (Italy) [SDS, AG, KH], in part by the *Fondazione Cassa di Risparmio di Torino* (Italy), through *La Ricerca dei Talenti* ("HR Excellence in Research") programme [SDS, AG, and KH] and in part by the Natural Sciences and Engineering Research Council of Canada, through the NSERC Discovery Programme [SF and KH].

REFERENCES

[1] Garcia, D., Zysset, P.K., Charlebois, M., and Curnier, A. 2009, "A three-dimensional elastic plastic damage constitutive law for bone tissue," *Biomech Model Mechanobiol*, Vol(8), pp. 149-165.

[2] Grillo, A., Prohl, R., Wittum, G. 2016 "A poroplastic model of structural reorganisation in porous media of biomechanical interest," *Continuum Mechanics and Thermodynamics*, Vol(28), pp.579-601.

[3] Tomic, A., Grillo, A., Federico, S. 2014 "Poroelastic materials reinforced by statistically oriented fibres—numerical implementation and application to articular cartilage," *IMA Journal of Applied Mathematics*, Vol(79), pp.1027-1059.

[4] Mićunović, M.V. 2009. *Thermomechanics of Viscoplasticity — Fundamentals and Applications*, Springer, Heidelberg.

[5] Holmes, M.H., Mow, V.C. 1990. "The nonlinear characteristics of soft gels and hydrated connective tissues in ultrafiltration," *J. Biomech.*, Vol(23/11), pp.1145-1156.

[6] Federico, S., Herzog, W. 2008. "Towards an analytical model of soft tissues," *J. Biomech.*, Vol(41), pp.3309-3313.

[7] Maugin, G.A., Epstein, M. 1998. "Geometrical material structure of elastoplasticity," *International Journal of Plasticity*, Vol(14/1-3), pp.109-115.

[8] Federico, S. "Covariant formulation of the tensor algebra of non-linear elasticity," *International Journal of Non-Linear Mechanics*, Vol(47/2), pp.273-284.

[9] Grillo, A., Federico, S., Wittum, S. 2012. "Growth, mass transfer, and remodeling in fiber-reinforced, multi-constituent materials," *International Journal of Non-Linear Mechanics*, Vol(47/2), pp.388-401.

[10] Giverso, C., Preziosi, L. 2012. "Modelling the compression and reorganization of cell aggregates," *Math. Med. Biol.*, Vol(29), pp.181-204.

Nanostructures and MEMS

A MULTI-AXIS PIEZORESISTIVE TACTILE SENSOR ARRAY

Shichao Yue
Mechanical Engineering Department
University of Alberta
Edmonton, AB, Canada
shichao2@ualberta.ca

Walied Moussa
Mechanical Engineering Department
University of Alberta
Edmonton, AB, Canada
walied.moussa@ualberta.ca

ABSTRACT

Touchscreens or tactile panels on electronic devices have been prevalent for modern daily life. However, the tactile force information has been the missing part for most of these devices. To improve user experience and diversify touchscreen functions, this work reports a multi-axis piezoresistive tactile sensor array, which senses three-axis tactile force and location information of a touch event generated by a finger or stylus and reduces the amount of tactile cells within the sensor array. The sensor array utilizes 2×2 miniaturized MEMS piezoresistive force sensors, achieving the sensitivity of tactile forces in both normal and shear directions over an area of 60mm×60mm with spatial resolution of 1 mm. The prototype of the sensor array demonstrated promising potentials in various tactile applications.

KEYWORDS: tactile, sensor array, piezoresistive, force sensor, MEMS

INTRODUCTION

A conventional touchscreen is a tactile panel sensing location information, which has been widely applied and prevalent for daily life in portable digital devices like cell phones, tablets and other Human-Machine-Interfaces (HMIs), owing to its intuitive interactions between users and smart systems. However, the force information is the missing part of a touch event generated by a finger or stylus on machines in this interaction. Researchers have been working on implementation of adding force feedback onto conventional tactile panels[1], or implementing touch screen panels with digitized tactile cells[2]. These studies are mostly limited with tactile force in normal direction, and the amount of tactile cells would dramatically grow in order to cover a large area, which increases the complexity for signal acquisition and processing. This work initiated in[3] and built on modifications and optimizations of a multi-axis sensor developed for biomedical implants reported in [4]. The tactile cell in current work was optimized to a more compact design with increased sensitivity comparing to the previous, and fabricated by Silicon-on-Insulator (SOI) wafers for better film conformity and higher performance. The sensor array prototype demonstrated the ability of sensing tactile force in both normal and shear directions, with potential spatial resolution of 1mm across a transparent area of 60mm×60mm.

SENSOR ARRAY DESIGN AND FABRICATION

The deformation of a simply supported plate subjected to a lateral force can be written as following equation [5]:

$$w = \frac{4Pb^3}{\pi^4 aD}\sum_{m=1}^{\infty} S_m \sin\frac{m\pi\xi}{a} \qquad (1)$$

Where $S_m = \sum_{n=1}^{\infty}\frac{\sin\frac{n\pi\eta}{b}\sin\frac{n\pi y}{b}}{\left(\frac{m^2b^2}{a^2}+n^2\right)^2}$, a and b denote the width and length of the plate, ξ and η stand for the location of the lateral force.

Inspired by this fundamental, a 2×2 piezoresistive sensor array packaged underneath a glass film is able to sense the film deformation caused by a tactile force as shown in Figure 1 c), which would enrich tactile information of the HMI for better user experience and provide more application-program-interface (API) choices. Each piezoresistive tactile cell has four sensing elements on the membrane of the device. The distance between two individual sensors, which is defined as the *pitch*, is assigned to be 60 mm for this prototyping. Both the bottom substrate and cover film are chosen to be glass in order to demonstrate the transparency ability of the sensor array, and the top cover film can be replaced by display panel in future for practical applications.

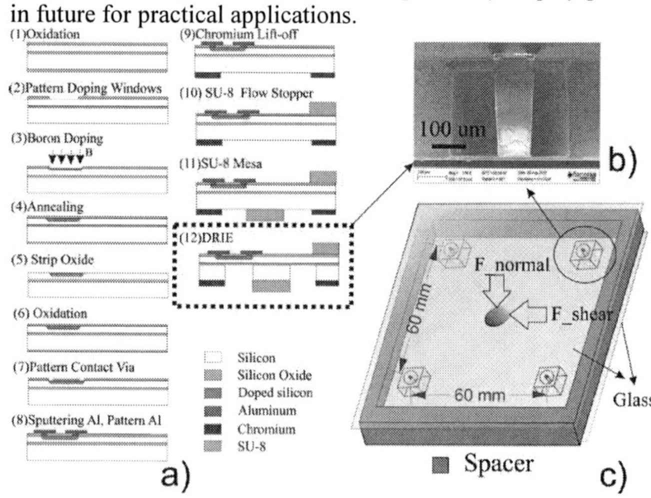

Figure 1 a) Fabrication process flow; b) SEM image of the device; c) Structure of packaged sensor array

As shown in figure 1 a), the tactile cell, which is the individual device, has been fabricated using SOI wafers following the process flow. The process flow started with an SOI wafer, with a layer of 1um silicon oxide between device layer and handle layer. A wet thermal oxidation was done to create a layer of silicon oxide, which was patterned for doping windows. The wafer was put into a furnace with Boron source and kept at 950°C for 45 minutes and annealed at 1000 C for 3 hours without the Boron source to achieve the resistance level at 50 ohm/sq. After doping, the mask layer of silicon oxide was stripped off by buffered oxide etching (BOE) and another layer of silicon oxide was created using the same wet oxidation method as mentioned above. This new layer of silicon oxide was patterned to form contact vias for following metallization procedure. A layer of 500nm aluminum was sputtered on top of the isolation oxide and patterned to be the electronic connections and pads of the device. At the back side of the wafer, a procedure named Chromium lift-off was done to create a Chromium mask layer, leaving backside etching region exposed. And on top of the wafer, flow stoppers were formed by doing lithography with SU-8. Similarly, the mesa structure was built by developing SU-8 at the central area where without the Chromium mask. The membrane of the device was formed by deep reactive ion etching by using isolation oxide layer of the SOI wafer as etching stop. The cross sectional view of the device has been shown in figure 1 b).

SENSOR ARRAY TEST

A test scheme has been developed to evaluate the function of the tactile sensor array as depicted in figure 2. The packaged sensor array was mounted on top of a microbalance, which has the resolution of 1mN. The normal force was applied onto the surface of the sensor array through a tungsten probe on an extended silicon arm. Underneath of the transparent sensor array, a piece of white paper printed with a 10mm grid was placed flatly. The force was applied over the cross point of two lines of the printed grid. And the output voltages were collected at each cross point of the printed grid over the working area of the sensor array, with corresponding force read from the microbalance. Similarly, a polymer block was mounted on top of the cover film at locations where the shear force to be applied. And a uniaxial force gauge was fixed on top of a multi-axis stage to apply shear force on the blocks. To overcome the momentum generated due to the setup gap between the contact point of the block and extended arm, a preload bolt was placed on top of the block, and the bolt weight was equal to the applied normal force in previous test. Since the response of the normal force has been tested and the shear force response can be calculated by subtracting the value form the normal preload if assuming the response would be superposition of both normal and shear scenarios. In order to maintain the simplicity of the test, only one sensor in the 2X2 sensor array has been evaluated in this paper.

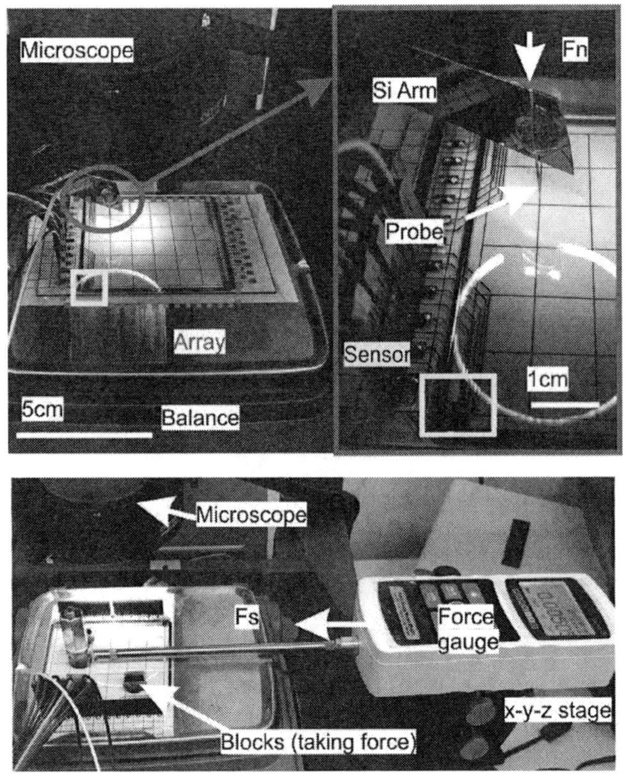

Figure 2 Test setup on the normal force **Fn**(top two) and shear force **Fs**(bottom)

RESULTS AND DISCUSSION

Numerical simulations have been done by applying a 300mN normal force at different locations across the 60mmX60mm area of the sensor array, powered by 3 volts in ANSYS. As shown in figure 3, the outputs from one element of one tactile cell increase as the location of the tactile force applied closer to this sensor and decrease vice versa. Notably, the sensitivity of the sensor drops to a low level at further corner on the sensor array. But by overlapping all the response from four sensors at four corners, this issue can be solved in future characterization procedures, which means the sensibility of the sensor array is able to cover every piece of the claimed working area due to response redundancy.

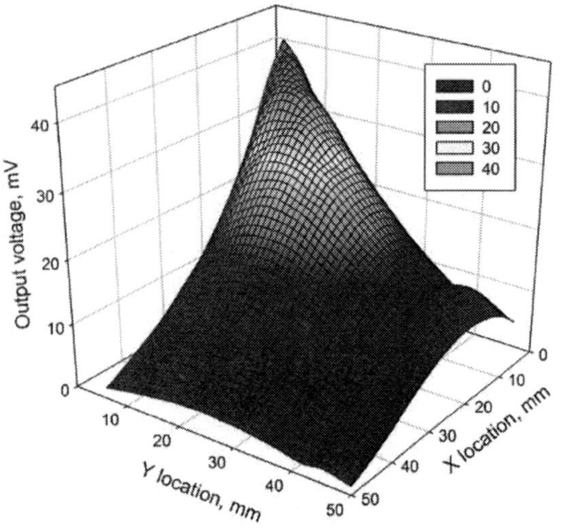

Figure 3 Simulated response of one element in one sensor to a normal force

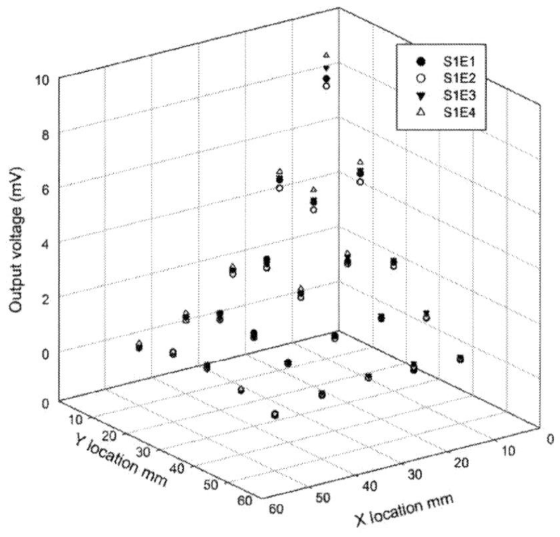

Figure 4 Test results of four elements in one sensor to a normal force

Based on the test scheme described in previous section, in figure 4, the test results of the sensor array showed similar response pattern as of the simulation. This test was performed by applying a 100 mN normal force and 1 volt power supply as a conservative manner. The results showed comparable sensitivity, comparing with the simulation done under 300mN with 3 volts power supply. Form the test data, the tactile sensor array demonstrated a peak sensitivity of 0.072mV/V•mN without any amplification in the peripheral circuits. With the combination of three other sensors in the sensor array, the 60X60 mm area which was the functional size of the prototype could be achieved.

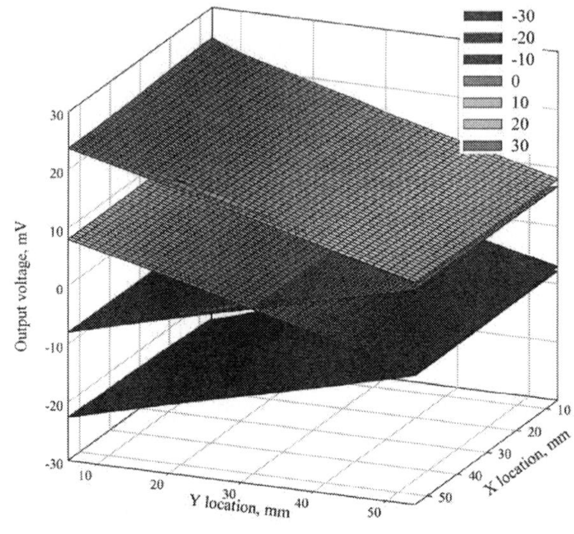

Figure 5 Simulated responses of four elements from one sensor in the array to a shear force

The evaluation of the response of the sensor array in terms of shear force has also been done by both numerical simulations and experimental tests. Comparing figure 5 and figure 6 which are the simulation and test results to a shear force at three different locations (*a:(20mm, 50mm), b:(20mm,30mm) and c:(50mm, 10mm)*) on the sensor array, similar response patterns are achieved, however, with about 40% sensitivity reduction in actual test from figure 1 c), due to packaging variations and elasticity difference of the spacer material. Results shown in this work demonstrated promising potential of the sensor array for multi-axis tactile applications.

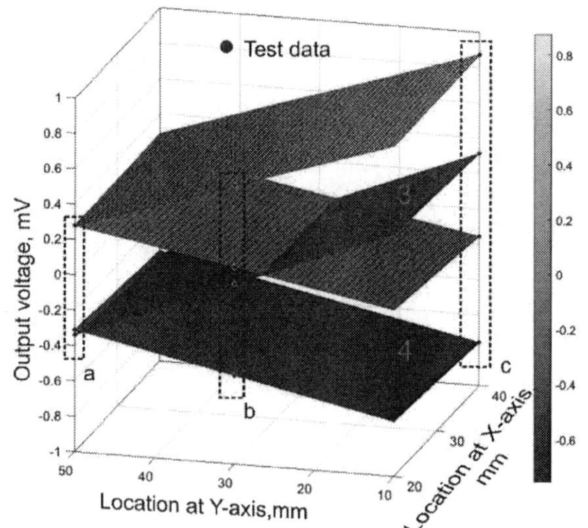

Figure 6 Test responses from four elements of one sensor in the array and fitting surfaces to a shear force

CONCLUSIONS

To improve user experience and diversify functions of touchscreens for portable digital devices, a piezoresistive tactile sensor array has been designed, fabricated, packaged and tested, demonstrating great potential for various applications. This sensor array utilized MEMS devices to accomplish multi-axis force feedback with four tactile cells, which reduced the amount of tactile cells comparing with conventional sensor array configurations, resulting simplified conditioning circuits and signal processing workload. For future work, more research will be done on larger sensor array configuration and complete characterizations for four sensors in the array and dynamic response with stylus.

ACKNOWLEDGMENTS

Authors would like to acknowledge CMC microsystems and nanoFAB for fabrication supports.

REFERENCES

[1] Y. Nakai, S. Kudo, R. Okazaki, H. Kajimoto, and H. Kuribayashi, "Detection of tangential force for a touch panel using shear deformation of the gel," presented at the CHI '14 Extended Abstracts on Human Factors in Computing Systems, Toronto, Ontario, Canada, 2014.

[2] H.-K. Kim, S. Lee, and K.-S. Yun, "Capacitive tactile sensor array for touch screen application," *Sensors and Actuators A: Physical,* vol. 165, pp. 2-7, 1// 2011.

[3] S. Yue, W. Moussa, and L. R. Williston, "Feasibility on a piezoresistive tactile normal force sensor array," *Applied Physics A,* vol. 116, pp. 409-414, 2014/08/01 2014.

[4] B. David, Y. Shichao, L. Edmond, and A. M. Walied, "Design and calibration of a six-axis MEMS sensor array for use in scoliosis correction surgery," *Journal of Micromechanics and Microengineering,* vol. 24, p. 085008, 2014.

[5] S. Timoshenko, "Theory of plates and shells," S. Woinowsky-Krieger, Ed., 2d ed. ed. New York :: McGraw-Hill, 1959.

Proceedings of the 26th CANCAM
Victoria, British Columbia, Canada, May 29 – June 1, 2017

EXPERIMENTAL INVESTIGATION ON A T-BEAM MICRORESONATOR EXHIBITING HALF SUBHARMONIC RESONANCE

Atabak Sarrafan, Oldooz Pooyanfar, Behraad Bahreyni, Farid Golnaraghi
Mechatronic Systems Engineering
Simon Fraser University
Surrey, British Columbia, Canada
asarrafa@sfu.ca, opooyanf@sfu.ca, behraad@ieee.org, mfgolnar@sfu.ca

ABSTRACT

We present design and experimental investigation on an electrostatically actuated MEMS T-shaped resonator exhibiting half subharmonic resonance. The aim of this work is to explore the possibility of utilization of a T-beam structure exhibiting one-half subharmonic resonance in design of resonant sensors. The microresonator is fabricated in SOIMUMPs process from MEMSCAP. The device is excited by an AC harmonic force with a frequency tuned near twice the first natural frequency (subharmonic excitation of order one-half). The influences of the AC excitation amplitude and the sweeping type are investigated experimentally. In case of subharmonic excitation, it is shown that, at 1/2 subharmonic resonance, all frequency response curves show softening stiffness behavior and a jump phenomenon. The results show the location of sudden jumps and regions of hysteretic behavior. Sharp amplitude jumps in the T-beam design due to half subharmonic resonance have a great potential in application of MEMS devices, especially for highly sensitive sensors.

KEYWORDS: MEMS, Microresonators, Nonlinear vibrations, Subharmonic Resonance, T-beam

INTRODUCTION

Popularity of microresonators is growing due to their potential and existing applications such as inertial and chemical sensors and radio frequency (RF) resonators. Microresonators are typically designed to oscillate at or near their primary resonance frequency through a capacitive, piezoelectric, or thermal actuation mechanism. Capacitive transduction is preferred over other methods in many microelectromechanical systems (MEMS) because of its simplicity and efficiency [1]. Often, high actuation voltages are needed for capacitive actuation/sensing to amplify the response of resonators for strong actuation forces and output signals. However, these high voltages applied in capacitive mechanism can result in unwanted nonlinear behaviors such as hysteresis, jumps, and dynamic instabilities [2].

Interplay between nonlinearities arising from multiple sources such as large mechanical displacements and capacitive actuation mechanism govern the overall dynamic behavior of resonators. The nonlinearities arising from the geometry or structure often stiffen the MEMS structure causing a shift in the natural frequency to higher values, while nonlinearities associated with the electrostatic forces soften the structure, in a way that the natural frequency tilts toward lower values. These interactions are inevitable, particularly in case of large displacements and AC harmonic load applied to the device. Therefore, it is crucial for designers to know how to deal with various sources of nonlinearities in a resonator operation, and potentially exploit them to enhance the resonator performance.

In many cases, researchers utilized these nonlinearities in order to enhance the performance of the resonators in a variety of MEMS applications. Jin and Wang indicated that driving a microbeam of a resonant micro sensor at superharmonic excitation of order one-half increases the signal-to-crosstalk ratio as compared to driving it at primary resonance [3]. Erbe et al investigated the use of a two-frequency AC excitation voltage to realize a nano bandpass RF filter [4]. Shaw et al suggested the use of principal parametric resonance to realize a MEMS bandpass RF filter on an improved stopband and a sharp response roll-off [5]. Souayeh and Kacem investigated nonlinear internal resonance between cantilever arrays using a multi-modal Galerkin discretization coupled with a perturbation technique for multi-mass sensing applications [6]. Vyas et al theoretically demonstrated the application of a pedal-type MEMS based filter using the internal resonance of its flexural mode and torsional modes [7]. Younis and Alsaleem explored the utilization of subharmonic resonance on a clamped-clamped and cantilever beams [8]. They proposed to employ this nonlinear phenomenon to design novel mass sensors and switches of improved characteristics.

The T-shaped structure has been under increasing focus in recent years, especially as filters, actuators, mass sensors and gyroscopes [7], [9]–[11]. The schematic of a T-shaped resonator is shown in Fig.1. The T-beam structure comprises of two beams: a clamped-clamped (CC) beam whose ends are anchored to fixed supports and a clamped-free (CF) beam attached orthogonally to the middle of the CC beam. This structure possesses a nonlinear behavior due to presence of quadratic and cubic nonlinearities, arising from its geometry. The earlier work mainly focused on a macro-scale T-beam structure actuated by piezoelectric patches to investigate 2:1 internal resonance [11]. To the best of our knowledge, the experimental studies on T-beam design have been on macro models, and the investigation has remained understudied at small scales. This work is motivated by earlier investigations on subharmonic resonance and T-shaped beam.

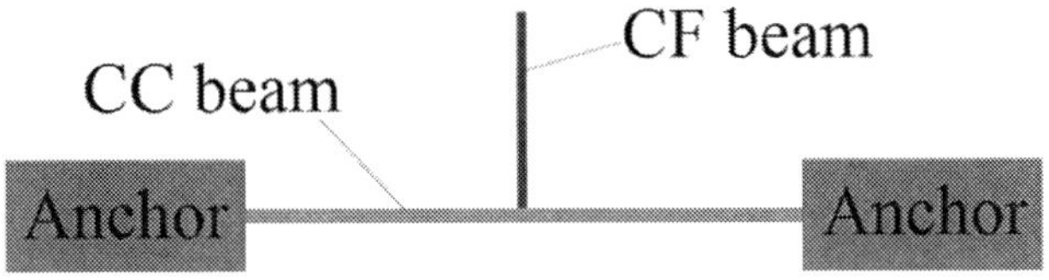

Figure 1: Schematic of the T-shaped resonator

A suitable micro device was fabricated and manufactured to study the nonlinear dynamics of a MEMS T-beam resonator excited around twice its first-mode natural frequency (subharmonic resonance of order one-half). The objectives of this paper are to present an insight into nonlinear subharmonic resonance of the electrostatically actuated T-shaped resonator, and to provide experimental results, reflecting this phenomenon.

HALF SUBHARMONIC RESONANCE

Subharmonic resonances are secondary resonances as a result of nonlinear spring forces, which generate large responses at a fraction of the excitation frequency [12]. For instance, when the excitation frequency is $N\Omega$, the system responds at Ω, where Ω is the natural frequency of the system. This means that this system has subharmonic resonances at the order $1/N$, where N is an integer greater than zero. Quadratic nonlinearity triggers subharmonic resonance of order 1/2. Therefore, the system in the presence of nonlinear electrostatic forces, which are quadratic in nature, can exhibit pure softening behavior at large deflection due to 1/2 subharmonic resonance [13]. The activation of subharmonic resonance requires the AC amplitude and quality factor to exceed specific threshold levels.

The electrostatic force induced between two parallel plates is expressed as

$$F=\frac{V^2}{2}\frac{\partial}{\partial x}\left(\frac{\varepsilon_0 A}{g-x}\right)=\frac{\varepsilon_0 A}{2g^2}V^2\left(1+2x+3x^2+4x^3\right) \quad (1)$$

where ε_0 is the permittivity of free space; A is overlapping area between the plates; g is spacing gap; V is applied voltage and x is displacement. This equation represents the Taylor's expansion of the electrostatic force equation. It consists of a variety of nonlinear spring forces including quadratic and cubic forces. In this equation, in case of 1/2 subharmonic resonance, quadratic term causes the jump, hysteresis and softening behavior. Similarly, the cubic spring force triggers 1/3 subharmonic resonance only if the voltage value V and quality factor, i.e., environment pressure, exceeds a threshold value. In subharmonic resonance, therefore, the excitation level and damping make significant contributions.

FABRICATION PROCESS

The T-shaped resonator was fabricated in SOIMUMPs process from MEMSCAP [14]. Figure 3 presents a cross-sectional view of this process. The SOIMUMPs process starts with a silicon-on-insulator (SOI) wafer which consists of a 25 μm top device layer, a 1 μm oxide layer, and a 400 μm substrate. The device layer is patterned in a deep reactive ion etching (DRIE) step to define the mechanical structure. The minimum feature size on device layer (i.e., gap width) is 2 μm in this process. Next, the wafer is selectively backside-etched through the substrate to realize trenches. The trenches are used not only to release the movable structures from device layer, but also to suppress the substrate parasitics. There are two metal layers available in the process: the "pad metal" used for fine electrical interconnections and the "blanket metal" used for large metal lines such as substrate contacts. The micrograph of the fabricated device is shown in Fig. 4. Electrostatic force is applied through the drive electrode (DE) to drive the resonator into resonance along Y-axis. Vibrations along X-axis are detected through the sense electrodes 1 and 2 (SE 1 and SE 2).

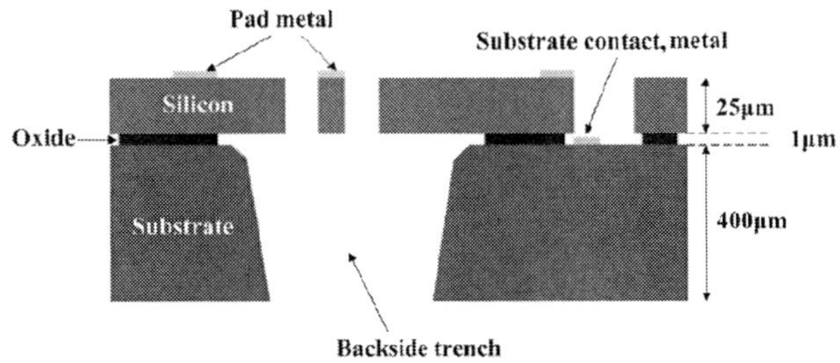

Figure 3: Cross sectional view of different layers in SOIMUMPs process [14]

The geometry of the device are as follows: the length of the CC beam is 452 μm; the width of the CC beam is 14 μm; the length of the CF beam is 139 μm; the width of the CF beam is 5 μm; the length of drive electrode (Ld) facing the CC beam is 388 μm; the length of the sense electrodes (Ls) facing the CF beam is 126 μm; the gap width separating the CC and CF beams from the electrodes is 2 μm. The natural frequencies were measured with the DC voltage of 40 V applied to the device structure to be 318.5 kHz for the first mode and 659.3 kHz for the second mode. The measured Q factors of the first and second modes of vibration using network analyzer are $Q_1 = 1540$ and $Q_2 = 458$.

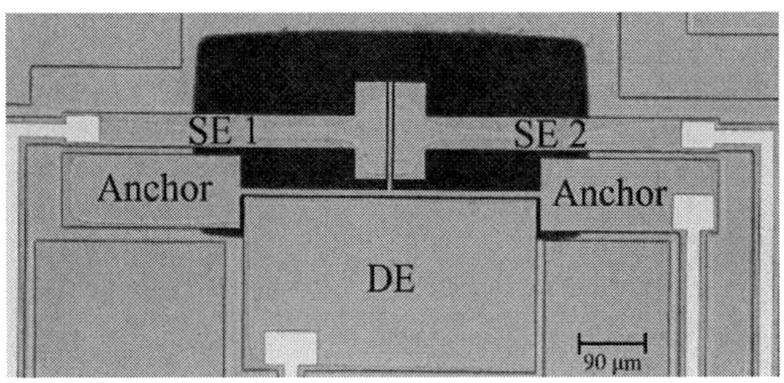

Figure 4: Micrograph of the T-beam resonator

EXPERIMNETAL SETUP

The experimental setup used for measurements is illustrated in Figure 5. The setup consists of a signal generator, used to apply the required excitation signal V_{AC}; a spectrum analyzer to study the sensed signal; a DC power source to provide DC bias voltage; a current pre-amplifier. The device is tested within a vacuum chamber at about 8 mTorr with electrical feedthroughs. Testing under vacuum reduces damping. The DC polarization voltage V_{DC} is directly applied to the device. The AC signal V_{AC} from the signal generator is connected to the drive electrode. The signal generator is used to perform forward- and backward frequency sweeps. As the structure vibrates, an output current (i_s) is produced through the DC-biased time-varying capacitance formed between the CF beam and the sense electrodes. This signal is then amplified by the current amplifier and analyzed on the spectrum analyzer. The MAX-HOLD function of the spectrum analyzer is activated to achieve the nonlinear frequency response curves during forward and backward sweeps.

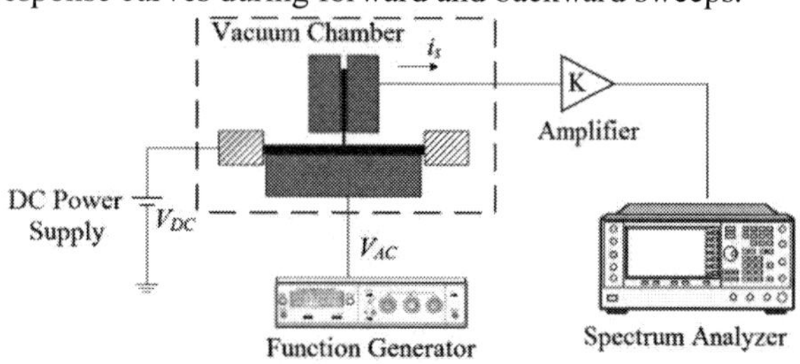

Figure 5: Experimental setup

EXPERIMENTAL RESULTS

The fabricated device was used to study the nonlinear dynamics of the structure at subharmonic resonance of order one-half. A series of forward and backward frequency sweeps are conducted to reveal various nonlinear dynamic behaviors under large excitation forces. Figure 6 shows experimental results for sweep up and down resonant frequency curves for a T-beam system driven with the DC bias amplitude of 40 V, and various AC voltage signals. Frequency sweep tests are performed near the frequencies close to twice the first natural frequency, in the frequency interval of 626 and 638 kHz, with sweep period of 60 seconds.

Figure 6a shows the sweep up resonant curves under the excitation setting of $V_{DC} = 40$ V and $V_{AC} = 2.5$, 3.5 and 4 V. Due to subharmonic resonance the CF beam oscillates at a frequency that is half the frequency of the excitation signal. This data confirms the presence of one-half subharmonic resonance, which occurs at twice the resonant frequency of the first mode. For a one-half subharmonic resonance, the fundamental frequency in the system is half the excitation frequency. However, the subharmonic resonance requires the AC amplitude to exceed a specific threshold level. Below this threshold level subharmonic resonance is not activated. It follows from Figure 6a that exciting micro T-beam near subharmonic resonance of order one-half of its first natural frequency yields frequency response curve with sharp jump and softening behavior. This is a desirable characteristic in the design of MEMS mass sensors [8]. Figure 6b shows a softening behavior with a sharp transition in the resonator amplitude in backward sweep frequency response curve. The loading conditions are similar to the upward frequency sweep setting. The bending-to-lower-frequency exhibits softening effect. As V_{AC} is further increased in Fig. 6, the curve increasingly bends to lower frequencies, eventually giving rise to the discontinuity (jump) due to frequency hysteresis. In Figure 7, the measured responses for forward sweep (FS) and backward sweep (BS) experiments are shown. It is seen that the response for higher V_{AC} becomes increasingly nonlinear. Starting from lower frequencies in forward frequency sweep, a sudden upward jump occurs, and starting from higher frequencies in backward sweep, a sudden downward jump takes place. Sharp jump in response amplitudes can be beneficial in the MEMS devices, especially highly sensitive sensors.

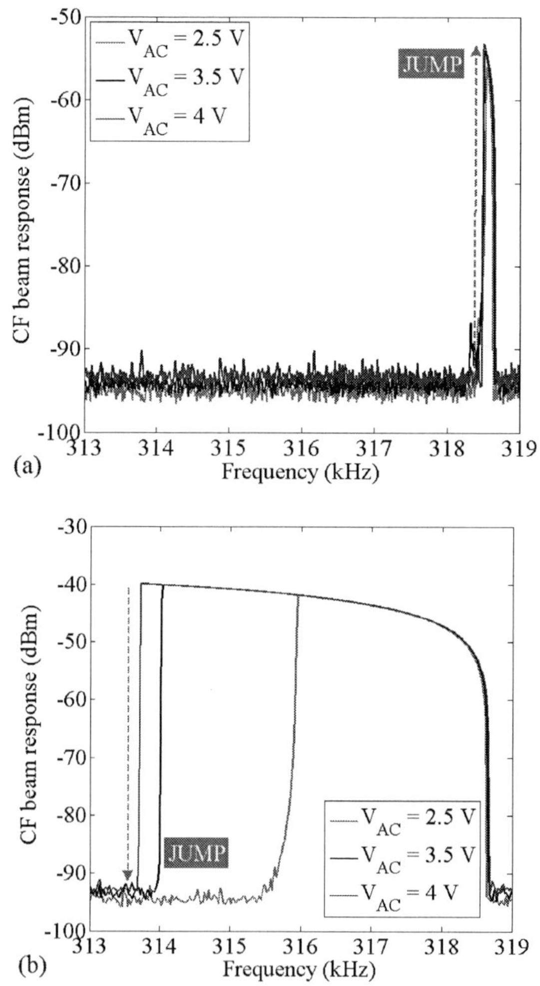

Figure 6: Experimentally generated nonlinear frequency response curves in case of (a) forward and (b) backward sweep for $V_{DC} = 40$ V and different values of V_{AC}

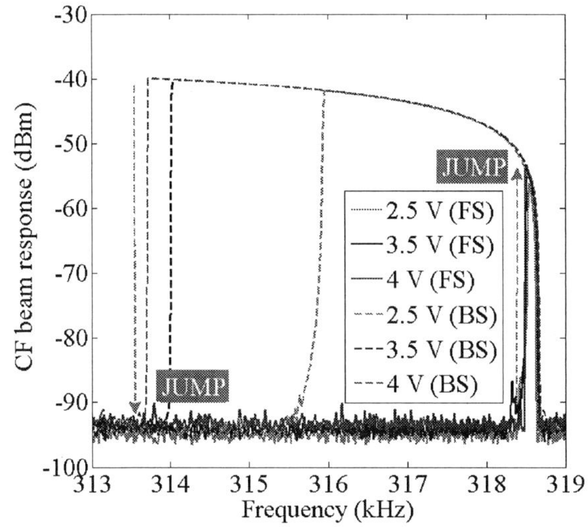

Figure 7: Experimentally generated forward sweep (FS) and backward sweep (BS) response for V_{DC} = 40 V and different values of V_{AC}

CONCLUSIONS

In this paper, an experimental investigation has been conducted to explore one-half subharmonic resonance in an electrostatically actuated T-beam microresonator. This concept relies on actuating a MEMS T-shaped structure by a DC load superimposed to an AC harmonic load of frequency near twice first natural frequency. In 1/2 subharmonic resonance, frequency-response curves for forward and backward sweeps have been demonstrated experimentally. Interesting nonlinear phenomena, jump, hysteresis, and softening behavior, have been detected. Sharp jumps in response amplitude of the T-beam can be utilized to design highly sensitive sensors including mass sensors.

ACKNOWLEDGMENTS

The authors would like to acknowledge Natural Sciences and Engineering Research Council of Canada (NSERC) for providing financial support and CMC Microsystems for the provision of products and services that facilitated this research. We would also like to thank The Integrated Multi-Transducer Systems Laboratory (IMuTS Lab) for providing essential equipment for testing the device.

REFERENCES

[1] A. H. Nayfeh and M. I. Younis, "Dynamics of MEMS resonators under superharmonic and subharmonic excitations," *J. Micromechanics Microengineering*, vol. 15, no. 10, p. 1840, 2005.

[2] A. H. Nayfeh, M. I. Younis, and E. M. Abdel-Rahman, "Dynamic pull-in phenomenon in MEMS resonators," *Nonlinear Dyn.*, vol. 48, no. 1–2, pp. 153–163, Apr. 2007.

[3] Z. Jin and Y. Wang, "Electrostatic resonator with second superharmonic resonance," *Sens. Actuators Phys.*, vol. 64, no. 3, pp. 273–279, Jan. 1998.

[4] "Mechanical mixing in nonlinear nanomechanical resonators," *Appl. Phys. Lett.*, vol. 77, no. 19, pp. 3102–3104, Nov. 2000.

[5] S. W. Shaw, K. L. Turner, J. F. Rhoads, and R. Baskaran, "Parametrically Excited MEMS-Based Filters," in *IUTAM Symposium on Chaotic Dynamics and Control of Systems and Processes in Mechanics*, Springer, Dordrecht, 2005, pp. 137–146.

[6] S. Souayeh and N. Kacem, "Internal resonances in nonlinear nanocantilever arrays under electrostatic actuation," in *1st Euro-Mediterranean Conference on Structural Dynamics and Vibroacoustics (MEDYNA 2013)*, France, 2013, pp. 1–4.

[7] A. Vyas, D. Peroulis, and A. K. Bajaj, "Dynamics of a nonlinear microresonator based on resonantly interacting flexural-torsional modes," *Nonlinear Dyn.*, vol. 54, no. 1–2, pp. 31–52, Oct. 2008.

[8] M. I. Younis and F. Alsaleem, "Exploration of New Concepts for Mass Detection in Electrostatically-Actuated Structures Based on Nonlinear Phenomena," *J. Comput. Nonlinear Dyn.*, vol. 4, no. 2, pp. 021010-021010-15, Mar. 2009.

[9] A. Vyas, D. Peroulis, and A. K. Bajaj, "A Microresonator Design Based on Nonlinear 1 : 2 Internal Resonance in Flexural Structural Modes," *J. Microelectromechanical Syst.*, vol. 18, no. 3, pp. 744–762, Jun. 2009.

[10] A. Vyas, A. K. Bajaj, A. Raman, and D. Peroulis, "Nonlinear micromechanical filters based on internal resonance phenomenon," in *Digest of Papers. 2006 Topical Meeting on Silicon Monolithic Integrated Circuits in RF Systems*, 2006, p. 4 pp.-.

[11] A. Marzouk, "Utilization of Internal Resonance in Gyroscope Design," Thesis, Applied Sciences: School of Mechatronic Systems Engineering, 2014.

[12] M. Ozdogan and S. Towfighian, "Nonlinear Dynamic Behavior of a Bi-Axial Torsional MEMS Mirror with Sidewall Electrodes," *Micromachines*, vol. 7, no. 3, p. 42, 2016.

[13] F. M. Al Saleem and M. I. Younis, "Theoretical and Experimental Investigation of Dynamic Instabilities in Electrostatic MEMS," in *Proceedings of the XIth International Congress and Exposition*, Orlando, Florida, USA, 2008.

[14] A. Cowen, G. Hames, D. Monk, S. Wilcenski, and B. Hardy, *SOIMUMPs design handbook*. MEMSCAP Inc, 2011.

Proceedings of the 25th CANCAM
London, Ontario, Canada, May 31 - June 4, 2015

INSIGHTS INTO LAYER DEPENDENT ADHESION AND FRICTION ON ATOMICALLY THIN FILMS INTERPRETED THROUGH SIMULATIONS SOLVED USING FINITE ELEMENTS

Philip Egberts
Department of Mechanical and Manufacturing Engineering
University of Calgary
Calgary, Alberta, Canada
philip.egberts@ucalgary.ca

Peng Gong
Department of Mechanical and Manufacturing Engineering
University of Calgary
Calgary, Alberta, Canada
peng.gong@ucalgary.ca

ABSTRACT

Adhesive contacts simulating an the apex of an atomic force microscopy tip interacting with few layer graphene sheets were developed using simulations solved using the finite element method (FEM). In these simulations, a small increase in the pull-off force was observed as the number of graphene layers between the tip and the supporting silicon substrate was increased. The slight increase in pull-off force is within experimental error of pull-off forces on graphene covered substrates, which typically show a layer independence in the pull-off force. Investigation into the stored elastic energy and other parameters indicate that the greater van der Waals interaction between the tip and the graphene layers, compared to the substrate, results in the increased adhesive interaction as the number of graphene layers increases. Additionally, FEM simulations were also performed on suspended graphene to isolate the influence of the substrate on the simulated pull-off forces. In these suspended graphene simulations, a similar trend in the pull-off force was observed compared with the supported graphene sheets, except that a much smaller increase was observed when the graphene thickness was increased from two to four layers of graphene. Examination of the simulation snapshots shows that local delamination of the graphene sheets during pull-off occurs when there are fewer layers of graphene present, but that the cohesive stresses between the graphene layers in multilayer graphene effectively arrest delamination from occurring. Implications of the change in surface energy, resulting from increasing the number of graphene layers, are also observed in experimental measurements, and will be discussed in context of the observations made in the FEM simulations.

Keywords: graphene, finite element method (FEM), adhesion, van der Waals, atomic force microscopy (AFM)

A macro scale design criterion for gecko-inspired dry adhesives

Mattia Bacca [a,*], Jamie A. Booth [b], Kimberly L. Turner [b], and Robert M. McMeeking [b,c,d,e]

[a]*Mechanical Engineering Department, University of British Columbia, Vancouver BC V6T1Z4, Canada*
[b]*Mechanical Engineering Department, University of California, Santa Barbara CA 93106, USA*
[c]*Materials Department, University of California, Santa Barbara CA 93106, USA*
[d]*School of Engineering, University of Aberdeen, King's College, Aberdeen AB24 3UE, UK*
[e]*INM – Leibniz Institute for New Materials, Campus D2 2, 66123 Saarbrücken, Germany*

Abstract
Bio-inspired dry adhesives rely on the utilization of short-range intermolecular forces harnessed by intimate contact at the tips of small peripheral structures called fibrils. The quality of the adhesive bond in these systems is related to the load distribution at the interface and it should be observed at multiple length scales. The goal for an optimal adhesive strength is to obtain equal load sharing (ELS) across the whole interfacial area. The physical and geometrical properties of each single fibril and the mutual interactions among them are key features to determine the global adhesive strength of the system. While there have been extensive investigations on obtaining ELS within the tip of a single fibril, very limited attention has been drawn to obtain ELS among fibrils. Bacca *et al.* in 2016 developed a simple mathematical model to calculate the ideal distribution of fibril properties to obtain ELS at the length scale an array of fibrils. They also considered the presence of an interfacial misalignment and its impact on the final result. With this paper we propose an alternative derivation of the optimal stiffness distribution among the array.

Introduction The possibility of creating robust adhesion via Van der Waals molecular interaction forces has long been noted from the observation of the locomotion mechanisms developed by various species of gecko, spider and some insects [1-3]. These animals have evolved a particular microstructure at the interface of their feet, which led to the development of a multitude of micrometric pillars called fibrils [4,5]. The utilization of universal intermolecular forces renders the influence of surface chemistry low, for this reason multi-scale mechanics has proven to be a valuable tool in revealing how nature has harnessed weak and short-ranged molecular forces to produce robust adhesion. Scaling arguments based on the Johnson-Kendall-Roberts (JKR) theory of contact mechanics suggest an inherent increase in adhesive strength (force per unit projected contact area) can be achieved through increasingly refined subdivision of terminal structures [6]. This argument is based on the observation of an increasing sensitivity to tip geometry and presence of interfacial flaws when the length scale of the adhesive surface is significantly larger than the molecular scale. This flaw sensitivity converts into a deviation from the ideal condition of equal load sharing (ELS) for which optimal adhesive strength is achieved. Practical limitations on the fibril size, which can hardly be manufactured to be smaller than a few microns, have risen the necessity of formulating improved design criteria to overcome flaw sensitivity, and therefore to approach ELS, at the single fibril length scale [7,8]. Limited attention has been drawn to the achievement of ELS at the length scale of an array of fibrils, [9,10]. We propose here an alternative derivation for the optimal distribution of fibril stiffness across the array to achieve ELS proposed by Bacca *et al.* [10] and define it as a macro scale design principle for fibrillar dry adhesives.

* Corresponding author.
E-mail address: mattia.bacca@engineering.ucsb.edu.

Mathematical Model The model is composed by two objects: (*i*) the adhesive, which includes the fibrils and the backing layer (BL), and (*ii*) the rigid support (RS), at which the fibrils' tips are adhering. The adhesive is made of homogeneous linear elastic material, and the BL thickness is considered to be ideally infinite, so that it can be modeled as an elastic half space. A sketch of the model is reported in Figure 1. We describe the interfacial misalignment between the adhesive and the RS with the angular coefficients A_x and A_y, directed with respect to the axes x and y parallel to the surface of the BL. The displacement at the tip of each fibril then becomes

$$u_i = \bar{u} - A_x x_i - A_y y_i \qquad (1)$$

where $\bar{u}$ is the prescribed separation between RS and BL, and x_i and y_i define the position of the i-th fibril with respect to the center of the array. The displacement created at the tip of the i-th fibril by a unit force applied at the tip of another (j-th) is given by [11],

$$c_{ij} = \frac{\left(1-\nu^2\right)}{\pi r_{ij} E}, \qquad i \neq j \qquad (2a)$$

$$c_{ij} = \frac{h_i}{\pi a^2 E} + \frac{16\left(1-\nu^2\right)}{3\pi^2 a E}, \qquad i = j \qquad (2b)$$

where E and ν are the Yong modulus and the Poisson modulus of the adhesive, respectively, r_{ij} is the mutual distance between the i-th and the j-th fibril, a is the radius of the circular fibril tip and h_i is the i-th fibril length. Eq. (2a) relies on the approximation of considering the load applied to the j-th fibril as point force, while Eq. (2b) consider a uniform pressure over the circular area defined by the fibril tip. The total displacement measured at the tip of the i-th fibril is then

$$u_i = c_{ij} f_j \qquad (3)$$

where f_j is the force applied at the j-th fibril. Picturing the critical condition at which total separation occurs, for a critical displacement u_c, we replace Eq. (1) into Eq. (3) obtaining

$$c_{ij} f_j^c = u_c - A_x x_i - A_y y_i \qquad (4)$$

where f_j^c indicates the forces applied to the fibrils at total separation. The sum over j is done among the fibrils that are still attached at the moment the separation reaches the value u_c. ELS indicates that we have simultaneous detachment of all the fibrils of the array at the separation displacement u_c, and also $f_j^c = f_{\max}$ for each j-th fibril. The latter condition imposed to Eq. (4) can only be obtained for a specific set of coefficients c_{ij}, which corresponds, via Eq. (2), to a specific distribution of fibril length h_i. The fibril length h_i can include both the geometrical and the physical properties of the i-th fibril by incorporating the ratio of the Young modulus of the BL to the Young modulus of the fibril. The distribution of h_i satisfying ELS is obtained as

$$h_i = \langle h \rangle + a\left(1-\nu^2\right)\left\{\langle \psi \rangle - \psi_i + \frac{\pi a E}{\left(1-\nu^2\right) f_{\max}}\left[\left(\langle x \rangle - x_i\right) A_x + \left(\langle y \rangle - y_i\right) A_y\right]\right\} \quad (4)$$

with

$$\psi_i = \sum_{j=1}^{N} \frac{a}{r_{ij}} \quad (5)$$

and $\langle \xi \rangle = \frac{1}{N}\sum_{j=1}^{N} \xi_i$, where ξ is one of the functions h, ψ, x, y.

The contour plot of the result reported in Eq. (4) and (5) is reported in Figure 2 where the optimal distribution of h giving ELS is shown for the case of a square array of fibrils and $A_x = A_y = 0$. The figure evidences stiffer fibrils (shorter h) at the center of the square and softer fibrils (longer h) at the perimeter.

Conclusions

Robust design of dry adhesive systems must be done at multiple length scales and the suboptimal behavior of them, associated with deviation from ELS at the fibril's length scale, can be overcome by proper distribution of fibril stiffness within the array. This strategy proves the possibility of exploiting BL interaction to counteract the effect of interfacial misalignment, responsible for tremendous reduction of the resulting adhesive strength. This feature is extensively discussed in Bacca *et al.* [10] and for briefness is here neglected.

References

1. S.N. Gorb and M. Varenberg, (2007) "Mushroom-shaped geometry of contact elements in biological adhesive systems". *Journal of Adhesion Science and Technology*. **21**(12-13) pp. 1175-1183.
2. S. Gorb, "Attachment devices of insect cuticle". 2001, New York, NY: Springer Science & Business Media.
3. K. Autumn, "Properties, Principles, and Parameters of the Gecko Adhesive System", in "Biological Adhesives", A. Smith and J. Callow, Editors. 2006, Springer Berlin Heidelberg. pp. 225-256.
4. K. Autumn and A.M. Peattie, (2002) "Mechanisms of adhesion in geckos". *Integrative and Comparative Biology*. **42**(6) pp. 1081-1090.
5. K. Autumn, M. Sitti, Y.A. Liang, A.M. Peattie, W.R. Hansen, S. Sponberg, T.W. Kenny, R. Fearing, J.N. Israelachvili, and R.J. Full, (2002) "Evidence for van der Waals adhesion in gecko setae". *Proceedings of the National Academy of Sciences*. **99**(19) pp. 12252-12256.
6. E. Arzt, S. Gorb, and R. Spolenak, (2003) "From micro to nano contacts in biological attachment devices". *Proceedings of the National Academy of Sciences*. **100**(19) pp. 10603-10606.
7. A. del Campo, C. Greiner, and E. Arzt, (2007) "Contact Shape Controls Adhesion of Bioinspired Fibrillar Surfaces". *Langmuir*. **23**(20) pp. 10235-10243.
8. A.V. Spuskanyuk, R.M. McMeeking, V.S. Deshpande, and E. Arzt, (2008) "The effect of shape on the adhesion of fibrillar surfaces". *Acta Biomaterialia*. **4**(6) pp. 1669-1676.

9. H. Yao and H. Gao, (2006) "Mechanics of robust and releasable adhesion in biology: Bottom–up designed hierarchical structures of gecko". *Journal of the Mechanics and Physics of Solids*. **54**(6) pp. 1120-1146.
10. M. Bacca, J.A. Booth, K.L. Turner, R.M. McMeeking (2016), …Cambridge; New York: Cambridge University Press.
11. K.L. Johnson, "Contact Mechanics". 1985, Cambridge; New York: Cambridge University Press.

Figures

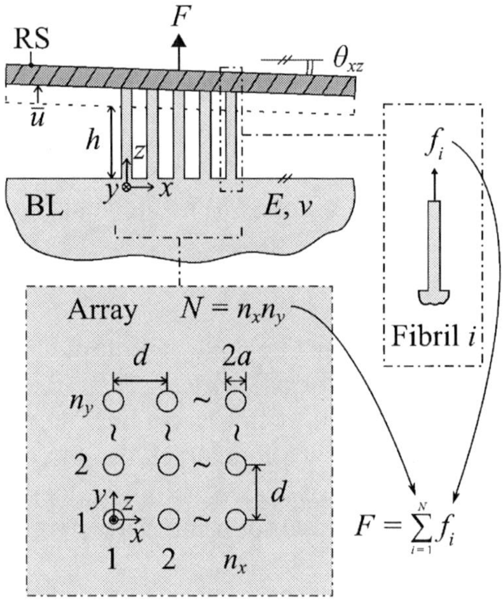

Figure 1. Schematic of the adhesive system. Fibrils are arranged in a square array, with the center to center distance in each direction being *d*. The rectangular array has n_x and n_y fibrils along the *x*- and *y*-axes, respectively, with the total number of fibrils being *N*. A pulling force *F* is applied to generate a separation displacement $\bar{u}$ between the rigid support (RS) and the backing layer (BL), at a fixed angle of misalignment θ_{xz} for which $A_x = \tan(\theta_{xz})$.

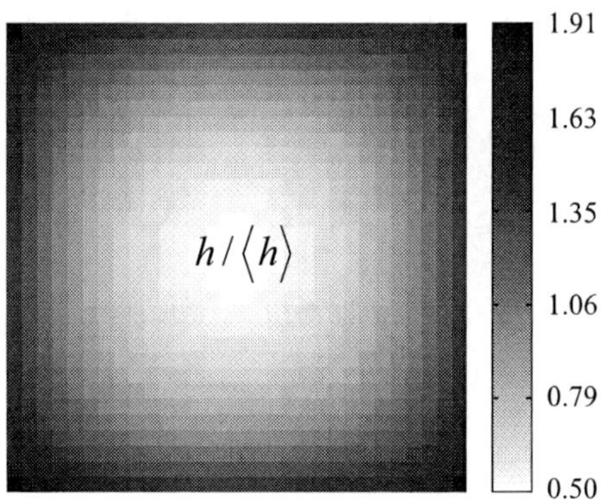

Figure 2. Fibril length distribution for obtaining ELS. Each small square represents a fibril and its color is associated with its length, through the scale bar on the right hand. The value reported is normalized with respect to the mean length $\langle h \rangle$.

NUMERICAL SIMULATION OF A MICROFIBER FABRICATION SYSTEM

Bahram Mirani
Department of Mechanical Engineering
University of Victoria
Victoria, British Columbia, Canada
bmirani@uvic.ca

Gillian Nixon
Department of Mechanical Engineering
University of Victoria
Victoria, British Columbia, Canada
gnixon@uvic.ca

Mohsen Akbari
Department of Mechanical Engineering
University of Victoria
Victoria, British Columbia, Canada
makbari@uvic.ca

ABSTRACT

This study demonstrates the numerical simulation of a continuous microfiber fabrication process using a microfluidic device with on-the-fly polymerization of a photo-crosslinkable polymer by ultraviolet (UV) irradiation. Gelatin methacryloyl (GelMA), photo-crosslinkable gelatin, is used as the polymer to generate fibers. Polymer concentration in the microfluidic system, prior to crosslinking, is determined by using a multiphysics model implemented in the commercially available COMSOL 5.1 to include different physics such as laminar flow and diffusion phenomenon. Additionally, we considered the non-Newtonian viscoelastic properties of the polymer in the simulations. The diameter and pore size of the fabricated fibers are characterized as a function of polymer concentration. Moreover, our results suggest that a considerable difference in the final results was achieved by modeling the biopolymer as Newtonian and non-Newtonian fluids.

KEYWORDS: fiber fabrication, microfluidics, numerical simulation, gelatin methacrylate

INTRODUCTION

Polymeric fibers enable the generation of 3-dimentional (3D) structure with different methods such as textile technology and microextrusion 3D bioprinting.[1,2] These structures can be employed as scaffolds in tissue engineering applications. Hydrogel fibers among others have the advantage of high water content similar to native tissue and enhanced biocompatibility. Structures made of hydrogel fibers can mimic the microenvironment of natural extracellular matrix to promote desired cellular function.

Different approaches such as wetspinning, electrospinning, meltspinning, and microfluidic method have been developed to fabricate polymeric fibers. Microfluidic method facilitates on demand polymerization, generation of multicomponent fibers, and better control on fiber properties, namely porosity. Numerical simulation of the fiber fabrication via microfluidic systems have been studied in the previous works[3]; however, an assumption of Newtonian fluid was considered for polymers to simplify the simulation. On the other hand, the dependence of dynamic viscosity on shear rate in polymers and therefore their non-Newtonian behavior was investigated in the literature.[4] Accordingly, there is a need for developing a mathematical model for polymeric fiber fabrication system with including the polymer non-Newtonian behavior.

In addition to the biochemical functionality of the fabricated fibers, Their porosity plays an important biological role as it affects the diffusion of oxygen and nutrients throughout the scaffold[5]. Additionally, interconnected porous channels facilitate cell migration and attachment [6]. Cell response to its environment can also be controlled by altering the mechanical properties of engineered tissue via the impact of porosity to the stiffness of final construct [7].

The system considered in this study continuously fabricates microfibers using a microfluidic device and a UV light with an adjustable position. As demonstrated in figure 1, microfluidic device creates a coaxial flow in which diluted GelMA solution in phosphate-buffered saline (PBS) is the core, and pure PBS is the sheathing flow. With regards to the geometry of the system, a 2-dimensional axisymmetric model is introduced to simulate the 3-dimensional fiber fabrication process and numerical simulation is employed to determine the concentration of GelMA in the final photo-crosslinked hydrogel. Knowledge of the concentration can contribute to control over the porosity and diameter of the fiber, through movement of the position of UV exposure. Additionally, characterizing the fabricated fiber with a multiphysics model alleviates the need for characterization through lengthy experimental trial and error processes.

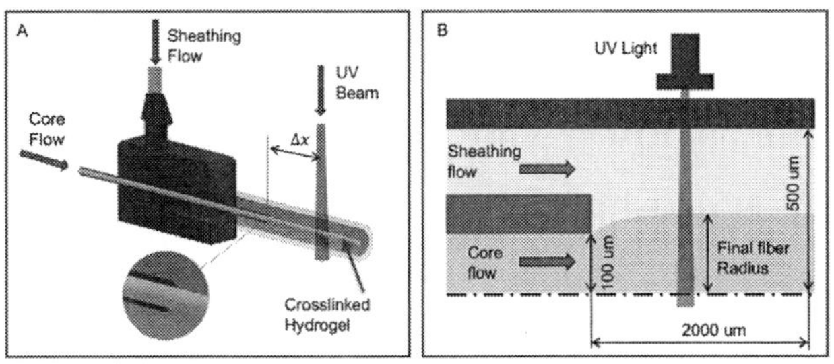

Figure 1. (A) 3-dimensional schematic diagram of the continuous fiber fabrication system. Δx indicates the position of UV exposure. (B) the system can be modeled with a 2-dimensional axisymmetric geometry.

COMPUTATIONAL ANALYSIS

The mentioned system in isothermal conditions can be described by simultaneously solving continuity (1), momentum (2) and mass transport (3) equations. The transport equation describes the diffusion of polymer from higher concentration to lower concentration regions (from core flow to sheathing flow). With the assumptions of steady state condition, incompressible fluid, and laminar flow in horizontal direction, these equations can be written as:

$$\boldsymbol{\nabla}.\boldsymbol{V} = \boldsymbol{0} \quad (1)$$

$$\boldsymbol{\rho}(\boldsymbol{V}.\boldsymbol{\nabla})\boldsymbol{V} = \boldsymbol{\nabla}.\{-\boldsymbol{PI} + \boldsymbol{\mu}(\boldsymbol{\nabla V} + (\boldsymbol{\nabla V})^{\boldsymbol{T}})\} + \boldsymbol{F} \quad (2)$$

$$\boldsymbol{\nabla}.(\boldsymbol{D\nabla C}) - \boldsymbol{V}.\boldsymbol{\nabla C} = \boldsymbol{0} \quad (3)$$

where V is velocity field, P is pressure, ρ is density, μ is dynamic viscosity, F is body force, and C and D are the concentration and diffusion coefficient of GelMA in PBS respectively. As the horizontal direction is assumed for laminar flow, gravity (body force) is neglected in the governing equations. Additionally, owing to the fact that PBS is a water-based salt solution with very low concentrations of sodium hydrogen phosphate and sodium chloride, its fluid dynamic properties are assumed to be the same as water.

Most biomaterials that are used in tissue engineering applications such as GelMA are non-Newtonian due to their molecular structure and therefore their viscosity depends on the shear rate. Accordingly, based on previously reported studies, GelMA is a shear thinning polymer, showing lower viscosity in higher shear rates (Figure 2A, B) [4]. Regarding the momentum equation, viscosity μ depends on the strain rate tensor $(\nabla V + (\nabla V)^T)$ and, therefore, its variation in the flow should be considered. Radial diffusion of the polymer results in variable concentration and viscosity of the polymer. Here we assumed a linear relation between concentration of GelMA and its dynamic viscosity (Figure 2C).

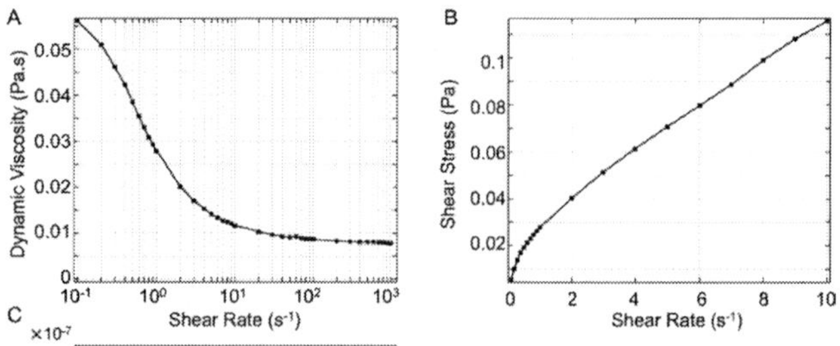

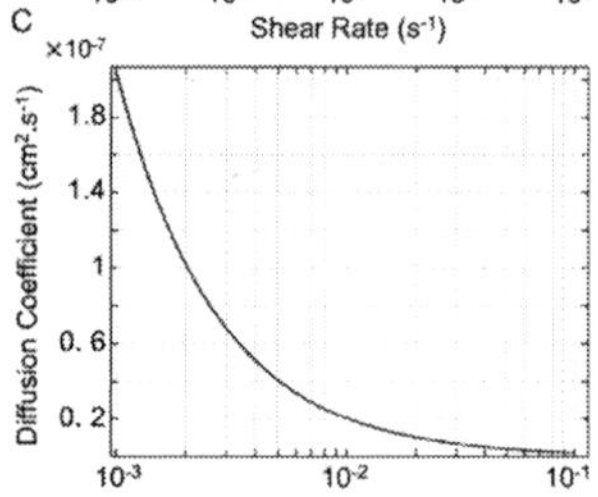

Figure 2. (A) Dynamic viscosity of GelMA 4.5% w/v in PBS. Reprinted with permission[4]. Copyright 2015, John Wiley and Sons. (B) shear stress in GelMA solution with respect to shear rate. (C) Diffusion coefficient of GelMA 4.5% w/v in PBS.

Furthermore, the diffusion coefficient of GelMA in PBS should be defined in a way that describes the diffusive behavior of polymer in sheathing flow. For this purpose, Stokes-Einstein[8] relation for diffusion coefficient is considered.

$$D = \frac{A}{\mu} \quad (4)$$

In this relation, diffusion coefficient is inversely related to the viscosity of the fluid domain by the constant A, which depends on the absolute temperature and hydrodynamic radius of polymer chains. This relation is obtained by calculating constant A, using experimental data for gelatin [9] at one viscosity point (Figure 2C).

By coupling laminar flow and transport equations, viscosity, diffusion coefficient, and concentration are determined in each iteration of numerical solution until convergence. The solution procedure is defined in a way that in each iteration, continuity and momentum equations are solved for velocity and pressure, and then the velocity field is imported to the transport equation to solve for concentration.

This system is modeled for simultaneous simulation of the laminar flow and diffusion processes. As indicated in Figure 3A, the computational domain is a 2-dimensioanl axisymmetric geometry from 200 μm before and 2000 μm after the point that core and sheathing flows meet (origin). Boundary conditions are defined as two laminar flows in the inlets with defined velocities and GelMA concentration fractions of one and zero for core and sheathing flow respectively. Concentration fraction, C_f, is a dimensionless variable and defined as the ratio of the concentration of GelMA at any point to the concentration of GelMA in the core flow inlet. Atmospheric pressure is considered for outlet boundary condition.

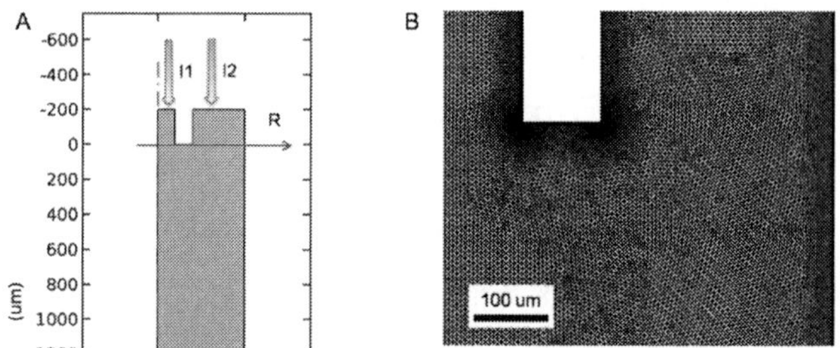

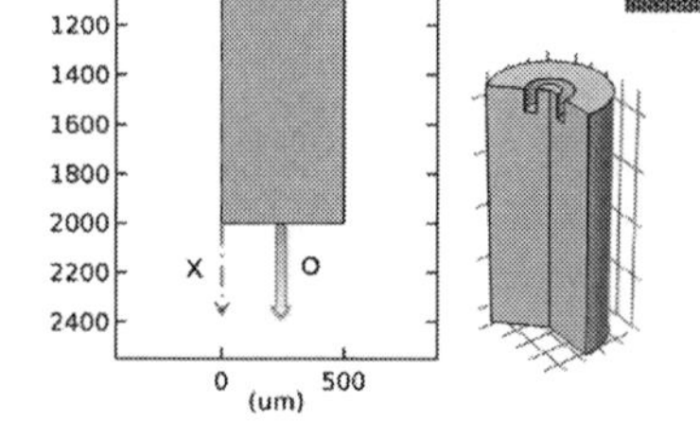

Figure 3: (A) Computational domain, modeled in COMSOL Multiphysics with an axisymmetric geometry. I1 and I2 are the inlets of the system at core and sheathing flow respectively. Concentration fraction is set to 1 for core and 0 for sheathing flow. O is the outlet of the system with defined atmospheric pressure. (B) Mesh layout in the computational domain with triangular elements.

COMSOL Multiphysics uses finite element analysis (FEA) to solve the governing differential equations numerically. In solving momentum and transport equations, it can be mathematically proven that an element Peclet number of more than 1 leads to numerical instability. Peclet number is the ratio of the contribution of convection and diffusion to mass transport, and is defined according to equation (5). The Peclet number can be adapted for both momentum and transport equations.

$$Pe = \frac{\|V\|h}{2D} \tag{5}$$

In this equation, h is the mesh element size and D is a coefficient of diffusion. To avoid numerical instability, which impedes the convergence of numerical solution, mesh element size can be forced to keep the Peclet number equal or less than one.

$$h \leq \frac{2D}{\|V\|} \tag{6}$$

This goal can be achieved by either choosing a very small mesh size or employing a stabilization method. Diffusion of polymers such as GelMA is typically slow, necessitating a very small mesh size which is computationally expensive. An alternative way to model the diffusion is by defining an elemental artificial diffusion coefficient to decrease the Peclet number of the elements. This process can be implemented in COMSOL using "Consistent Streamline Stabilization," which applies the essential artificial diffusion coefficient in the direction of streamlines to reduce computational effort. The amount of additional diffusion coefficient in the element depends on element size and velocity field, in which case, finer mesh layout would reduce the additional diffusion coefficient. Here we need a balance between accuracy and computational effort, which has been acquired by running the simulation with different mesh sizes and comparing the results.

Finally, the discretization of the domain is carried out using 780k triangular elements with a maximum size of 3.35 µm and refining the mesh in regions near the boundaries (Figure 3B). Regarding stability, reducing the mesh size did not affect the results, implying the fact that stabilization does not apply any significant error to the calculation (data not shown). The discretization of the governing equation was also implemented using first order element for velocity, first order element for pressure, and linear discretization for concentration.

RESULTS AND DISCUSSION

The aim of solving governing equations in this work was to find the concentration of GelMA in different cross sections away from the point that core and sheathing flow meet. Different cross sections imply different positions of UV exposure. Here, concentration fraction C_f was used to demonstrate the concentration distribution inside the domain at different cross sections (Figure 4A). Numerical simulation was also conducted with constant (average) viscosity (and, therefore, constant diffusion coefficient) in order to identify the effect of non-Newtonian behavior and variable diffusion coefficient of the polymer on the results.

The results obtained when GelMA was considered Newtonian was significantly different than when the polymer was considered to be non-Newtonian (Figure 4A & B). This difference was more pronounced at the center of the fiber and further downstream of the flow. As indicated in Figure 5, this discrepancy is due to the variation of shear rate from its minimum at the center of the core flow to its maximum at outer radii and therefore the significant variation of viscosity, diffusion coefficient and finally the diffusive flux which determines the concentration distribution in fluid domain.

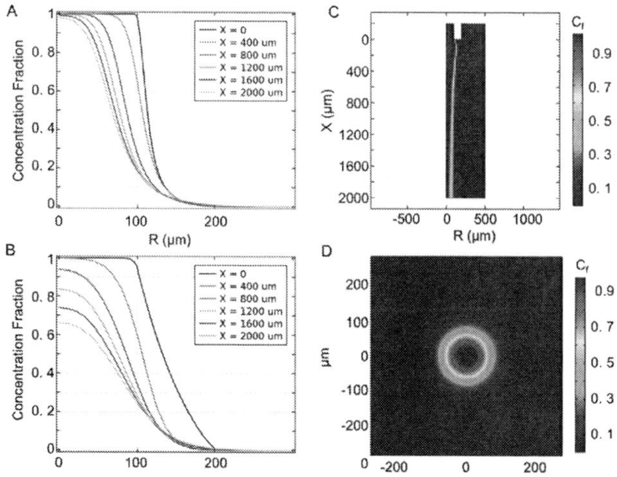

Figure 4. Data derived from numerical simulation for GelMA 4.5% w/v and $V_{average} = 0.1\, m/s$ for the core and sheath flows. (A) Concentration fraction C_f distribution in different cross sections defined by the distance from the point where core and sheathing flow meet. (B) Concentration distribution in different cross sections by assuming constant viscosity and diffusion coefficient for GelMA. (C) Concentration fraction C_f in the axisymmetric domain. (D) Cross-sectional Concentration fraction C_f at the outlet ($200\mu 0$ m away from the origin.

In order to determine the final fiber diameter, concentration isolines are needed to determine the points with equal concentration value. A common way to determine fiber diameter is relating a concentration value to the region where the boundary of the fiber is developed in the UV crosslinking process. This value can be obtained by conducting an experiment in one position of UV exposure to calibrate the results derived from the numerical simulation. In other words, one of the concentration isolines will demonstrate the fiber boundary (Figure 6A).

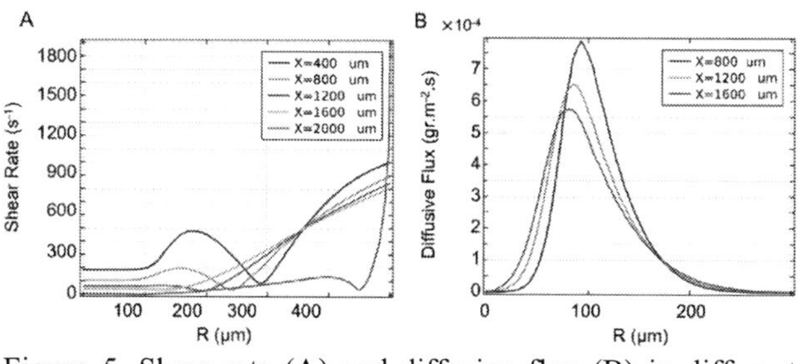

Figure 5. Shear rate (A) and diffusive flux (B) in different cross-sections away from the origin.

Porosity, or average pore size, is one of the important parameters affecting the biological properties of the fiber. Pore size in the fiber is inversely related to the concentration of the polymer. Importing the simulated concentration to an experimental function (pore size as a function of concentration) which was developed in a previous study [10], allows identification of the average pore size in different radial positions in the fiber cross section. Here, by assuming different position for UV exposure, X, different pore size condition in the fiber was achieved (Figure 6B). It can be observed that increased distance X of the UV exposure from the origin results in decreased pore size gradient. Additionally, since the regions with higher pore size gradient refer to the points with higher concentration gradient, porosity at the center of the fiber is independent of the position of UV exposure.

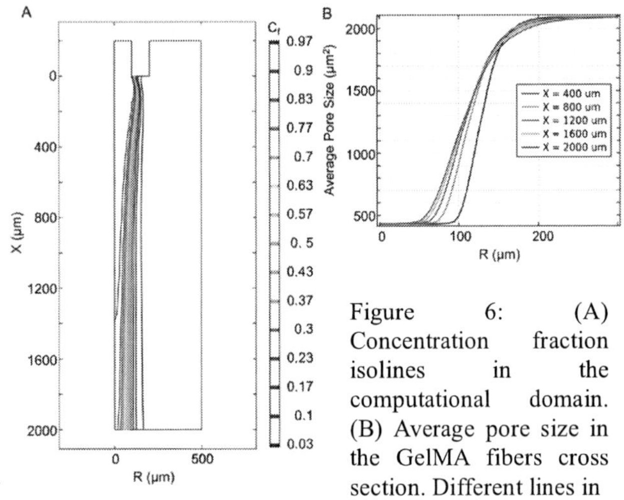

Figure 6: (A) Concentration fraction isolines in the computational domain. (B) Average pore size in the GelMA fibers cross section. Different lines in the plot show different position for UV exposure.

CONCLUSIONS

In this report, numerical simulation of a continuous microfiber fabrication system consisting of a microfluidics device and UV light to photo-crosslinkable GelMA polymer was carried out. The analysis determined the concentration of GelMA in the fiber cross-section with regards to the position of UV exposure. The numerical results can be interpreted as diameter and average pore size in the final fabricated fiber. This simulation was carried out in the commercially available Multiphysics tool, COMSOL 5.1, to consider laminar flow and diffusive transport phenomenon. Developing this multiphysics model for mentioned system will ease the arduous experimental trial and error process needed for characterizing the fabricated fiber. As a further step to improve the characterization of fiber fabrication, including the effect of variation of sheathing flow rate with fixed core flow can contribute to control on the diameter and average pore size in the fiber independently.

ACKNOWLEDGMENTS

M.A. would like to thank the Natural Sciences and Engineering Research Council of Canada (NSERC) and Canadian Foundation for Innovation for supporting this work.

REFERENCES

1. Tamayol, A. *et al.* Fiber-based tissue engineering: Progress, challenges, and opportunities. *Biotechnol. Adv.* **31,** 669–687 (2013).
2. Pedde, R. D. *et al.* Emerging Biofabrication Strategies for Engineering Complex Tissue Constructs. *Adv. Mater.* **In press,** (2017).
3. Ju, J. Il, Oh, H. J., Kim, S. R. & Lee, S. H. CFD-based modeling of the hydrodynamic production process of polymeric microfibers and an experimental evaluation. *J. Micromechanics Microengineering* **15,** 601–607 (2005).
4. Colosi, C. *et al.* Microfluidic Bioprinting of Heterogeneous 3D Tissue Constructs Using Low-Viscosity Bioink. *Adv. Mater.* **28,** 677–684 (2016).
5. Yang, S., Leong, K.-F., Du, Z. & Chua, C.-K. The Design of Scaffolds for Use in Tissue Engineering. Part I. Traditional Factors. *Tissue Eng.* **7,** 679–689 (2001).
6. Hollister, S. J. 2005, Porous scaffold design for tissue engineering. *Nat. Mater.* **4,** 518–524.
7. Murphy, W. L., McDevitt, T. C. & Engler, A. J. 2014, Materials as stem cell regulators. *Nat. Mater.* **13,** 547–557.
8. Einstein, A. 1905, Über die von der molekularkinetischen Theorie der Wärme geforderte Bewegung von in ruhenden Flüssigkeiten suspendierten Teilchen. *Ann. Phys.* **322,** 549–560.
9. Masuelli, M. A. & Sansone, M. G. 2012, Hydrodynamic Properties of Gelatin – Studies from Intrinsic Viscosity Measurements. *InTech* (2012).
10. Benton, J. A., DeForest, C. A., Vivekanandan, V. & Anseth, K. S. 2009, Photocrosslinking of gelatin macromers to synthesize porous hydrogels that promote valvular interstitial cell function. *Tissue Eng. Part A* **15,** 3221–30.

Advanced Manufacturing 2

DEVELOPMENT OF A MULTI-AXIS CNC GRINDING WHEEL GROOVING DEVICE

Cameron Forbrigger, Andrew Warkentin, Robert Bauer
Department of Mechanical Engineering
Dalhousie University
Halifax, NS, Canada
cameron.forbrigger@dal.ca, andrew.warkentin@dal.ca, robert.bauer@dal.ca

ABSTRACT

As interest in grooved and textured grinding wheels increases, the technology used to form grooves on the surface of grinding wheels must be improved to broaden the applications of grooved wheels. These improvements will open new areas of potential research. The current authors present a summary of work performed as a part of improving a servo-actuated, single-point diamond method of grooving grinding wheels to extend its application to profiled grinding wheels. A brief analysis of the effect of the profile geometry on the kinematic requirements of the design is described. An experimental investigation into the magnitudes of forces experienced by a single-point diamond tool during a grooving process is also presented. The results of this work were then used to design and build a prototype "Enhanced Dressing and Grooving Robot" (*EDGR*).
Keywords: profiled grinding wheel, grooved grinding wheel, textured grinding wheel

INTRODUCTION

Grinding wheels with macroscopic textures on their cutting surface have been shown to have benefits over non-textured grinding wheels. In a recent survey paper on the topic of textured grinding wheels, Li and Axinte define textured grinding wheels as "...[grinding wheels] that have both specially–designed active and passive grinding areas on their geometrically active surfaces." [1] Macroscopic texturing of grinding wheels is often referred to as grooving or slotting, where parts of the grinding wheel's active surface are removed by machining or by assembling multiple abrasive segments along a structural wheel rim. An example of a grooved and ungrooved wheel are shown in Figure 1.

Figure 1: Photos of the surfaces of ungrooved (left) and grooved (right) grinding wheels.

Grooved grinding wheels have been shown to reduce the specific energy of machining, process forces, and process temperatures compared to conventional (un-grooved) grinding wheels. Most authors have also concluded that grooved wheels result in an increased surface roughness compared to conventional wheels. Textured grinding wheels, including grooved wheels, have been shown to be useful for imparting a texture or pattern to the workpiece. [2]

Multiple methods have been developed for creating grooves on the surface of grinding wheels, each with their own strengths and weaknesses. Various authors have produced grooves using milling machines, laser machining, crushing roll dressing tools, single-point diamond dressing tools, and by arranging abrasive segments around a structural wheel rim. [2] The current authors present a design improvement to the grooving method developed by Mohamed et al. [3] This method uses a single-point diamond dressing tool, actuated along a linear stage, to scribe a groove onto the surface of a grinding wheel rotating at constant angular velocity.

The primary advantages to the single-point, servo-actuated dressing tool method of grooving grinding wheels are that it can create a variety of groove geometries by simply re-programming the device and that it can regenerate existing grooves on a grinding wheel. In most other grooving methods, if the existing grooves wear to the point where they are no longer useful, the grinding wheel must be dressed and then re-grooved. Therefore the single-point servo method reduces the grooving time and reduces the amount of grinding wheel material consumed by the grooving process. The disadvantages of the single-point method are that, in its current state, it cannot produce groove geometries with discontinuities or produce grooves on profiled wheels. In fact, most grinding wheel grooving methods are incapable of producing grooves on profiled grinding wheels. [2]

Profiled grinding wheels are used for creating high-precision, low-surface roughness profiles or channels in a workpiece. Some examples of applications for profiled grinding wheels include cutting the races for ball bearings, forming involute gear teeth [4] or creating micro-riblets on the surface of compressor blades in gas turbines [5].

The current authors propose that the capabilities of the single-point dressing tool grooving method can be significantly improved by adding a second, vertical axis to create a computer-numerical-control (CNC) grooving device with a planar workspace. For a single-point diamond grooving device, the tool must be actuated to make contact with the surface of the grinding wheel at controllable positions and speeds. The design requirements for sizing the actuators in a single-point grooving device have not been covered in the literature. Therefore, a description of the methods used to determine the design requirements for an Enhanced Dressing and Grooving Robot (*EDGR*) is presented in this paper.

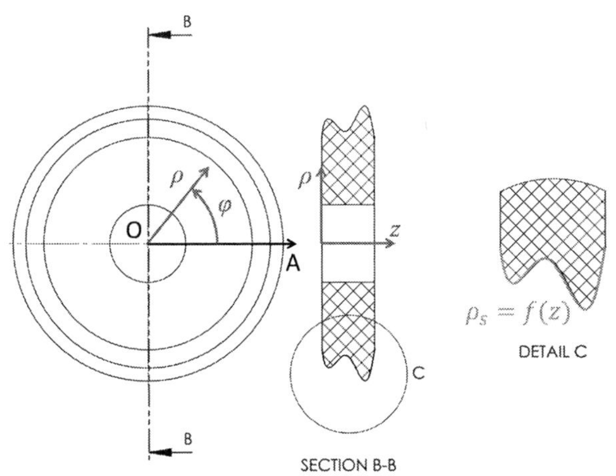

Figure 2: A diagram of a profiled grinding wheel showing the cylindrical coordinates used to define the profile geometry.

METHODS

The kinematic requirements for a profiled wheel grooving device are a function of the desired geometry of the groove, the geometry of the wheel profile and the speed at which the wheel is rotating. It was assumed that the grooving device would be producing helical grooves on a profiled wheel.

The active surface of a non-grooved grinding wheel can be described in cylindrical coordinates as having a radius ρ_s and a slope ρ_s' that are a function of the distance along the direction of the wheel axis (z–axis) as shown in Figure 2.

$$\rho_s = f(z) \tag{1}$$

$$\rho_s' = f'(z) \tag{2}$$

For helical grooves on a profiled wheel, the speed of the grooving tool in the z–axis $\dot{z}_t$ is constant and is determined by

$$\dot{z}_t = L\omega_s \tag{3}$$

where ω_s is the constant angular velocity of the grinding wheel in revolutions per minute, and L is the constant lead of the groove in m/revolution. The maximum speed of the grooving tool in the ρ–axis $\dot{\rho}_{t,max}$ can be determined by

$$\dot{\rho}_{t,max} = \dot{z}_t|\rho_s'|_{max} \tag{4}$$

where $|\rho_s'|_{max}$ is the maximum absolute slope of the grinding wheel profile.

The following constraints were chosen for the design of *EDGR*:

- $\omega_s = 155$ rev/min
- $L = 0.600$ m/rev when $|\rho_s'|_{max} = 0$ m/m
- $L = 0.005$ m/rev when $|\rho_s'|_{max} = \tan(70°) \approx 2.75$

which results in $\dot{z}_{t,max} = 1.55$ m/s and $\dot{\rho}_{t,max} = 0.0355$ m/s.

The dynamic requirements for the grooving device had to be determined experimentally. The experimental setup is described in Figure 3.

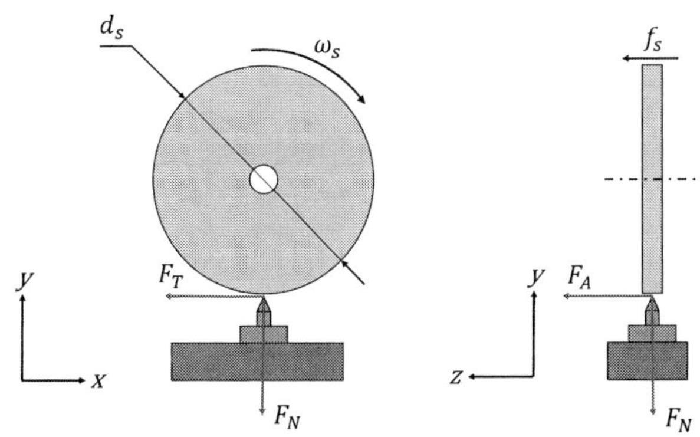

Figure 3: A diagram of the experimental setup to determine the forces on a single-point diamond dressing tool during grooving.

A single-point diamond dressing tool was attached to a Kistler 9257B Multicomponent Dynamometer connected to a Kistler 5019B Charge Amplifier. The data were acquired with a National Instruments PCI–MIO–16XE–10 at a sample rate of 250 Hz. The grooving experiments were performed on an aluminum oxide grinding wheel (Radiac Abrasives WR–A–60–J–5–V). The single-point dressing tool used was a conical tool with a 60° cone angle and a 0.508 mm nose radius.

The angular velocity of the wheel was held constant for all trials at 155 rev/min and a dressing tool feed of $f_s = 787$ mm/min, thereby producing grooves with $L = 5.08$ mm/rev. Lubricant was not used during the grooving tests in order to determine the forces present during a "worst-case scenario" grooving operation. The geometry of the grooves for these experiments was chosen based on the geometry of grooves used by Mohamed et al. [6]. The grooves were removed by dressing the grinding wheel with lubricant with several passes, each at a dressing depth of 0.0127 mm, a dressing feed rate of 25.4 mm/min, and a wheel surface speed of 22.35 m/s (4400 ft/min). Grooves were produced in the grinding wheel with a single pass at three different depths: 0.1, 0.2 and 0.5 mm. The maximum depth of 0.5 mm was chosen because it is the maximum groove depth produced by a single pass of a single-point diamond grooving device in literature. [2] This is a very aggressive depth of cut when compared to normal dressing applications where the dressing depth would be equal to or less than 0.025 mm.

RESULTS AND DISCUSSION

Figure 4 shows an example of the raw tangential force data collected for a single trial at a depth of cut of 0.2 mm. It can be seen from Figure 4 that there is a significant amount of noise present in the signal that is much larger than the average force being applied to the dressing tool.

The force data for each trial and for each direction (**N**ormal, **T**angential, **A**xial) were averaged over the region of interest

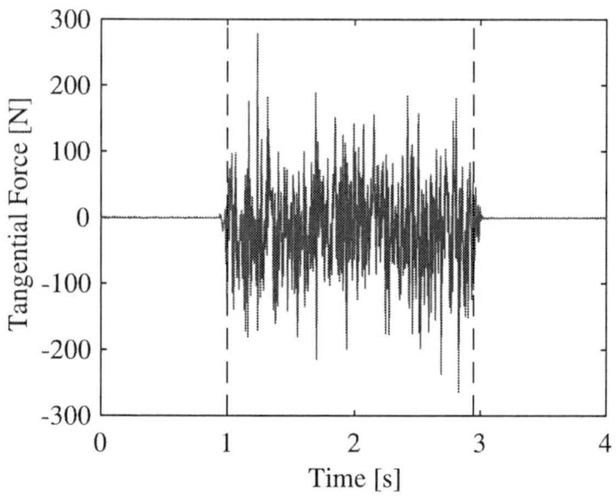

Figure 4: Example of raw tangential force data collected for a depth of cut of 0.2 mm.

(shown as dashed lines in Figure 4) and the results are shown in Figure 5 plotted against the cross-sectional area of the generated grooves. The Root-Mean-Square (RMS) amplitude of the signals after subtracting the average values were also calculated and are shown in Figure 6.

A Discrete Fourier Transform was used to approximate the power spectral density (PSD) of the measured force signals. The estimated PSD showed approximately evenly-distributed signal powers across all measured frequencies. It is proposed,

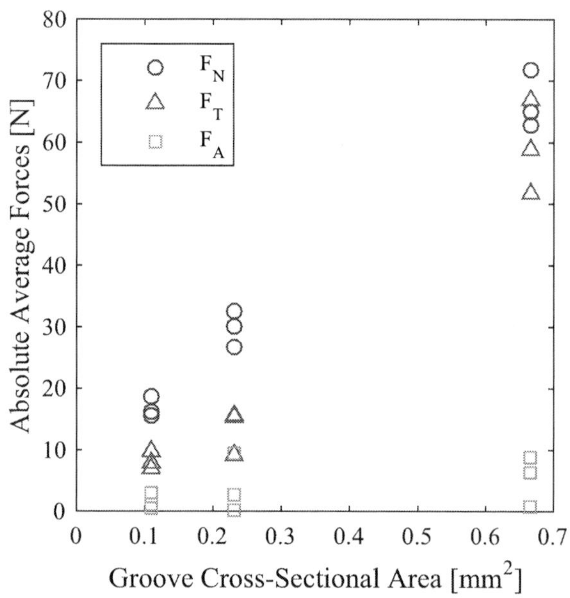

Figure 5: Absolute average forces during grooving.

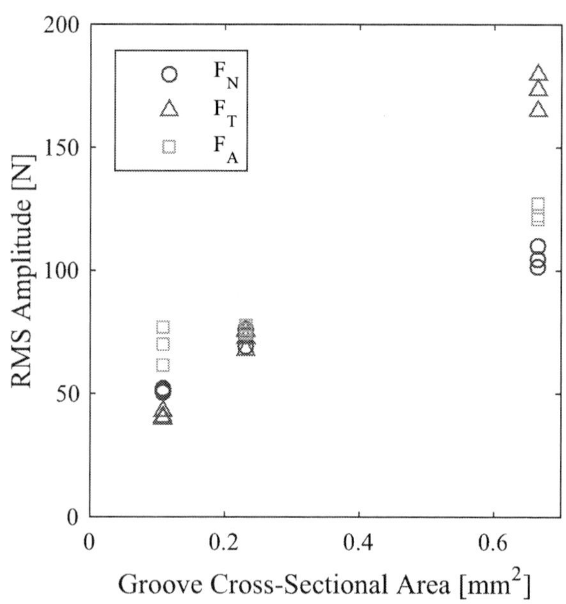

Figure 6: RMS amplitude of forces during grooving.

therefore, that the noise on the force signals for each axis could be modelled as band-limited Gaussian white noise with a pass-band between 0.5 Hz and 120 Hz.

There are too few data points to draw any significant conclusions about the relationship between average forces on a dressing tool during grooving and the cross–sectional area of the groove being produced. However, it is an interesting result that the average normal and tangential components of the force on the tool increase with increasing groove cross-sectional area (deeper grooves) whereas the average axial component does not increase significantly. It should also be noted that for all components of the force on the tool, the RMS amplitude of the forces increased with increasing groove cross-sectional area. Furthermore, all RMS amplitudes are similar in their magnitudes, unlike the average forces where the axial force had a significantly smaller magnitude than the other directions.

The results of this experiment show that there is a significant advantage to designing a grooving device that can take multiple cuts at shallow depths to produce deeper grooves. This would allow the grooving device to use smaller actuators and lighter components, thereby reducing the cost and power requirements of the device.

CONCLUSIONS AND RECOMMENDATIONS

When designing a device for creating grooves in the surface of profiled grinding wheels it is important to consider the following conclusions.

Assuming a single-point diamond grooving device that performs a helical grooving operation on a grinding wheel rotating at constant speed, the maximum radial and axial feeds of the tool depend only on the speed of the wheel, the desired lead of the helical groove, and the maximum slope of the grinding wheel profile.

The forces experienced by a single-point diamond dressing tool during grooving of an aluminum oxide wheel without lubricant were determined experimentally. It was shown that the force experienced by the tool increased with increasing groove cross-sectional area. The average magnitudes of the forces experienced by the tool reached maximum values of 72 N, 67 N and 9 N in the normal, tangential and axial directions, respectively. The forces on the tool during grooving are highly dynamic, resembling band-limited Gaussian white noise. The RMS amplitudes of the noise on the average forces during grinding reached maximum values of 110 N, 179 N and 127 N in the normal, tangential and axial directions, respectively. These results are significant because they indicate that any single-point diamond grooving device needs to be capable of maintaining accuracy while under significant dynamic loads. Furthermore, the results indicate that it is advantageous for a grooving device to be capable of creating grooves by running several shallow grooving passes instead of taking a single deep pass.

The conclusions of this work were used by the present authors to design and construct a multi-axis single-point diamond grooving device (*EDGR*), continuing the work from

Figure 7: *EDGR* inside Blohm Planomat grinding machine without protective cover.

[3]. A photo of the finished design is shown in Figure 7.

ACKNOWLEDGMENTS

The authors would like to thank the Natural Sciences and Engineering Research Council of Canada (NSERC) and the Canadian Foundation for Innovation (CFI) for their financial support of this research.

REFERENCES

[1] Li, H. N. and Axinte, D., 2016, "Textured grinding wheels: A review," *International Journal of Machine Tools and Manufacture*, 109, pp. 8–35.

[2] Forbrigger, C., Bauer, R., and Warkentin, A., 2016, "A review of state-of-the-art vitrified bond grinding wheel grooving processes," *The International Journal of Advanced Manufacturing Technology*, pp. 1–10.

[3] Mohamed, A.-M., Bauer, R., and Warkentin, A., 2014, "A novel method for grooving and re-grooving aluminum oxide grinding wheels," *The International Journal of Advanced Manufacturing Technology*, 73(5), pp. 715–725.

[4] You, H.-Y., Ye, P.-Q., song Wang, J., and Deng, X.-Y., 2003, "Design and application of {CBN} shape grinding wheel for gears," *International Journal of Machine Tools and Manufacture*, 43(12), pp. 1269 – 1277.

[5] Denkena, B., de Leon, L., and Wang, B., 2009, "Grinding of microstructured functional surfaces: a novel strategy for dressing of microprofiles," *Production Engineering*, 3(1), pp. 41–48.

[6] Mohamed, A.-M. O., Bauer, R., and Warkentin, A., 2013, "Application of shallow circumferential grooved wheels to creep-feed grinding," *Journal of Materials Processing Technology*, 213(5), pp. 700–706.

ELECTROFORMING OF NON-CONDUCTIVE 3D PRINTED PARTS

Panteha Fallah
Department of Mechanical & Industrial Engineering
Concordia University
Montreal, Quebec, Canada
p_fallah@encs.concordia.ca

Lucas A. Hof, Rolf Wüthrich
Department of Mechanical & Industrial Engineering
Concordia University
Montreal, Quebec, Canada
l_hof@encs.concordia.ca
rolf.wuthrich@concordia.ca

ABSTRACT

Additive Manufacturing Technologies (AMT) are extensively used to build complex 3D parts by accumulating thin individual layers. By using AMT more complex shapes can be achieved as compared to subtractive processes like conventional machining methods (e.g. milling, turning).
Electroplating of non-conductive 3D-printed parts like ABS (Acrylonitrile Butadiene Styrene) is a method to obtain low-weight parts with high precision.
In the presented study, copper electroplating followed by electro-polishing (EP) was performed on 3D printed ABS parts with different geometries which can have a variety of applications in fields such as the electronic industry, automotive industry, and computer body parts.
The effect of polishing settings is investigated, and it is shown that EP eliminates surface asperities effectively and canbrighten the part resulting in a significant decreased surface roughness.

KEYWORDS: Additive Manufacturing, Electroplating, non-conductive 3D parts, low weight parts, Electro-polishing

1. INTRODUCTION

Additive Manufacturing Technologies (AMT) / 3D printing technologies allows the making of products with virtually any shape by converting a digital 3D model into thin slices (typically 25μm up to 250 μm depending on which AMT is used) and building the 3D parts by accumulating these thin layers. Whereas polymer parts are mostly used for prototypes and esthetical purposes, metal parts are required for fully functional parts in high precision industries such as electronic, automotive and computer body part industries. However, metal parts can usually be 3D printed with lower precision and less fine structures then polymer parts.
Electroforming on non-conductive high precision 3D printed polymer parts is a promising approach for the manufacturing of high precision complex metal parts. This approach enables higher control over 3D structure width, compared to direct metal 3D printing methods, due to the initial size of the used metallic micro powders (25 – 75 μm).

Electroforming is a basic manufacturing process in which metal is deposited on a substrate (model) by electrodeposition. When specific requirements are needed, such as tight tolerances, complex and precise geometries, electroforming is a good candidate to use as economical manufacture process [1]. Electroforming is a particular form of electroplating. The principle of electroforming is that metal ions, dissolved in an electrolytic solution, are reduced and deposited at a cathode by charge transfer reactions. In electroplating the deposited metallic layer becomes part of the object whereas in electroforming the deposited metal (patterned surface) will be removed from the substrate. This enables manufacturing of very thin (several tens of microns up to hundreds of microns) and complex metallic parts depending on the 3D model used as substrate.
The characteristics of electroforming, compared to traditional metal forming processes, play a significant role to obtain a variety of advantages which determine its applications [1]. Its principal advantages are [1]:

- high dimensional conformity
- precise reproduction of surface details
- production of thin-walled components
- possibility to deposit composite materials
- Extensive range of size (essentially only be limited by the electrical power of the plating equipment)

By electroforming, components can be produced which would be either difficult or even impossible to manufacture by conventional methods.

Adding thin metallic layers by electroplating on polymer 3D printed parts have already proven to increase the mechanical properties as stiffness and tensile strength [2].
Complex hollow metal 3D geometries have been accomplished by Williams et al. [3] upon using AMT and electroforming. However, issues of rupture and crack formation were found to occur when burning the model material (polymer 3D part) to obtain the final metal part. ABS seems a good candidate as polymer for the 3D model as it is a well-studied substrate material for electroplating [3, 4]. Nevertheless, uniform electroforming (constant thickness) on complex shapes is still difficult to obtain. The proposed study aims to contribute on overcoming this challenge.

In the proposed manufacturing method (see Fig. 1), 3D polymer parts are printed by AMT in a first step, then an initial conductive layer is applied in a second step (electroless plating or applying a conductive paint). Subsequently, metal is deposited onto the part by electroplating until sufficient thickness is obtained. In a last step, the polymer model is removed by e.g. dissolving in a solvent. To obtain a low surface roughness, electro-polishing was conducted as final post-processing step.

2. PRINCIPLES AND METHODOLOGY

Electroforming of copper (Cu) was studied firstly on a metal substrate to achieve smooth, uniform layers. Secondly, the best settings were used on 3D printed ABS parts (Cu painted). In a last step, EP experiments were carried out on the electroplated metal plate sample to optimize surface roughness reduction. These EP settings were transferred to the electroformed sample (see Fig. 4).

Electrodeposition of metals or alloys is defined as the reduction of metal ions from electrolytes combined with deposition of them onto the working electrode. In its simplest form, the reaction in an aqueous medium at the cathode follows the equation: $M^{n+} + ne^- \rightarrow M$, with a similar anode reaction. The anode material can be either a sacrificial anode or an inert anode. For the sacrificial anode, the anode reaction is: $M \rightarrow M^{n+} + ne^-$

The principle of electroplating is shown in Fig. 1. A computer controlled potentiostat operated in galvanostic mode controls the electron transfer reactions responsible for the deposition process.

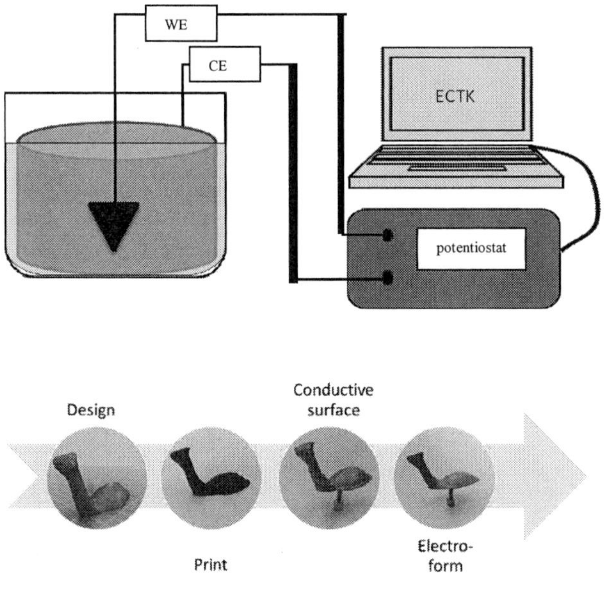

Figure 1: *Above* Experimental Electroplating Setup. *Below* Process flow from 3D printing a non-conductive ABS part up to electroplating (Cu) the copper painted part.

In the presented study, a stainless steel (SS) substrate was copper electroplated with a solution of 0.5 M $CuSO_4$, 1 M H_2SO_4, 1.5 mM HCl, 500 ppm polyethylene glycol (PEG) and 2mM 3-mercapto-1-propane sulfonic acid (MPSA) following the methodology described in section 2 [5]. Furthermore, the thickness difference of deposited layer was investigated for various deposition times.

Electropolishing was performed afterward in a solution of 0.5M H_3PO_4 and 0.5M H_2O for various times. The applied polishing potentials were deducted from the current limiting region of the *E-I* curves (see Fig.6). Polishing results were analyzed qualitatively by microscopy and confocal laser scanning microscope (CLSM) to show the feasibility of this approach on a metal part [6]. The influence of the polishing time on the surface roughness was investigated quantitatively by CLSM on a SS copper plated part.

The same condition was applied on an ABS part produced by an Ultimaker 2+ 3D printer. 3D model parts were painted by copper conductive paint to enable electroplating.

3. EXPERIMENTAL RESULTS

Thickness differences of the deposited layer were identified by both measurement (*Mitutoya* micrometer) and calculation. The effect of deposition time on thickness growing can be obviously seen in Fig. 2. Increasing in deposition time leads to linear increasing of the layer thickness. Layer thickness can be controlled starting from a few microns up to several millimeters with high precision (~ few microns).

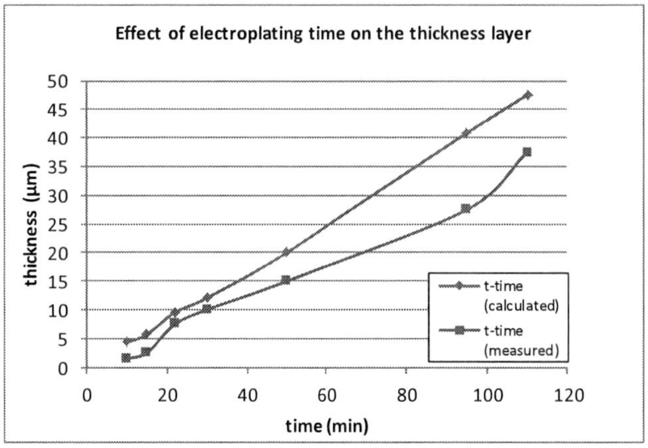

Figure 2: Effect of electroplating time on thickness of the deposited layer

The deposition thickness that can be achieved was calculated by the following relation (Eq. 1):

$$t = \Delta m / d\,W\,L \quad (1)$$

Where t is thickness of the layer, Δm is the measured mass difference at a specific time, d is the density of the deposited metal, W and L are the width and length of the part, respectively.

Effect on the surface roughness of the investigated different times for electropolishing is summarized in Fig. 3.

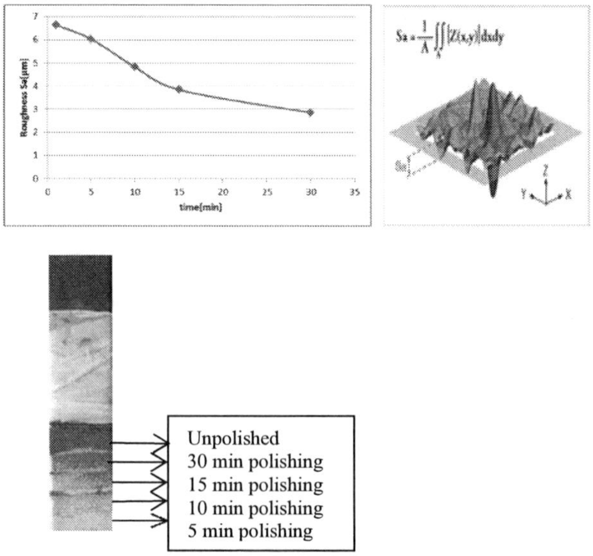

Figure 3: Effect of time on electropolishing process on the arithmetic mean surface roughness Sa of a copper plated sample with an initial roughness of Sa = 5.22µm

Fig. 4 presents CLSM images of electropolishing on SS copper plated part. A difference in surface roughness can be observed, showing smooth surfaces after electropolishing process with longer duration.

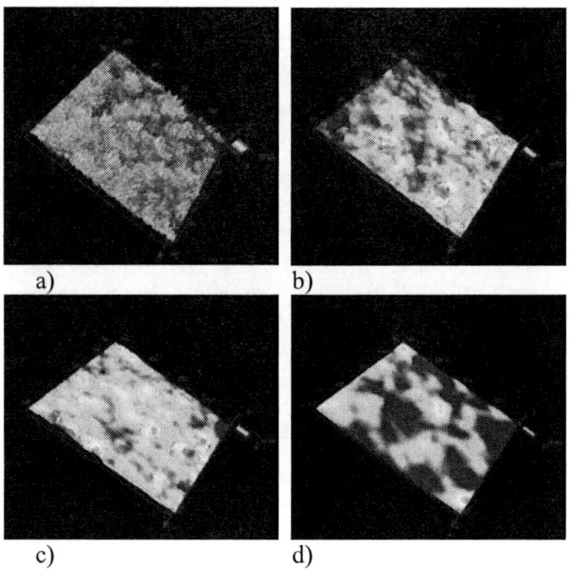

Figure 4: Effect of EP time on surface roughness 5min b) 10min c) 15min d) 30min polishing

To demonstrate the feasibility of the approach for manufacturing complex parts, AM structures were copper electroplated and electro-polished (see Fig. 4) with the following parameters, respectively: current density= 5 mA/Cm^2, time= two days and voltage=1.04 V, time = 30 min.

The rough surface can be clearly observed in Fig. 5. The beads are greatly reduced and a smooth and shiny surface is achieved after electro-polishing treatment.

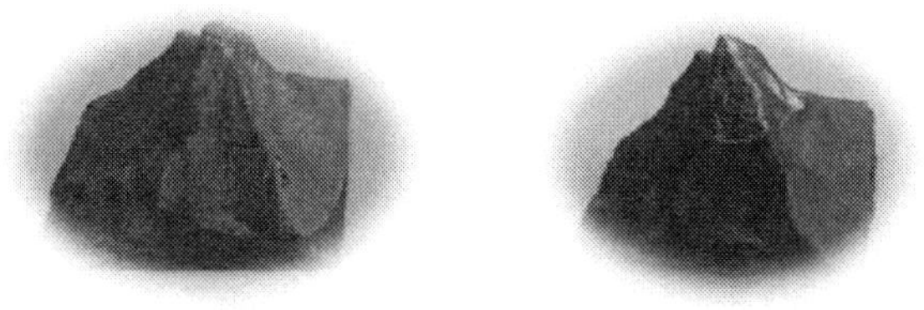

a) Unpolished b) Polished (30 min)

c) View from inside

Figure 5: Effect of EP on ABS 3D printed parts.

a) unpolished electroformed part, b) polished electroformed part, c) electroformed (hollow) part

Fig. 6 demonstrates the current (I) vs. potential (E) curve obtained for the Cu^{2+} solution at a sweep rate of 0.001V/s.

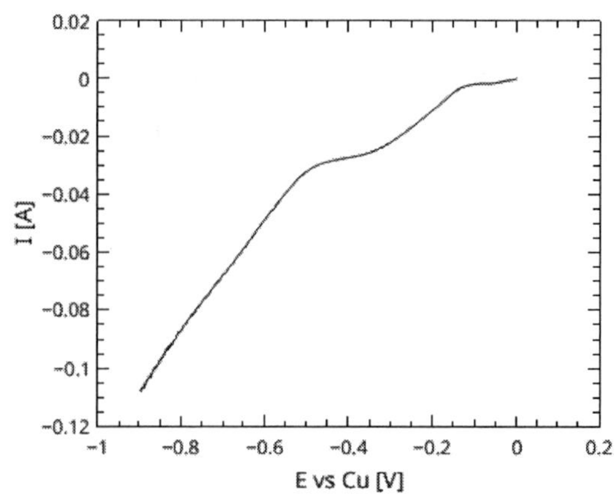

Figure 6: Linear voltammetric curve obtained from solution containing 0.5 M $CuSO_4$, 1 M H_2SO_4, 1.5 mM HCl, 500 ppm polyethylene glycol (PEG) and 2mM 3-mercapto-1-propane sulfonic acid (MPSA)

CONCLUSIONS

Electroforming on non-conductive high precision 3D printed parts is a promising approach for the manufacturing of high precision complex metal parts. This approach enables higher control over 3D structure width, compared to direct metal 3D printing methods, due to the initial size of the used metallic micropowders (25 – 75 µm).
Subsequent electro-polishing of the electroformed parts in a phosphoric acid solution is an effective way to eliminate surface asperities on AM electroplated parts. A significant decrease in surface roughness could be easily obtained after EP.

ACKNOWLEDGEMENTS

This work was supported by the Natural Sciences and Engineering Research Council of Canada (NSERC). L.A. Hof would like to thank the Ministère de l'Education, de l'Enseignement supérieur et de la Recherche (MEESR) for the bourse d'excellence pour étudiants étrangers (V1). Furthermore, the authors would like to acknowledge Mr. G. Desharnais from AxisPrototypes Inc. for the fruitful discussions and guidance on additive manufacturing technologies.

REFERENCES

1. J. A. McGeough et al. Electroforming process and application to Micro/Macro manufacturing. CIRP Annals - Manufacturing Technology 50(2), pp. 499-514. 2001. DOI: http://dx.doi.org/10.1016/S0007-8506(07)62990-4.
2. N. Saleh, N. Hopkinson, R. J. M. Hague, and S. Wise, "Effects of electroplating on the mechanical properties of stereolithography and laser sintered parts," Rapid Prototyp. J., vol. 10, no. 5, pp. 305–315, 2004.
3. D. L. M. and C. B. Williams, "Creating Complex Hollow Metal Geometries Using Additive Manufacturing and Electroforming," FFF Work., pp. 108–120, 2012.
4. S. Olivera, H. B. Muralidhara, K. Venkatesh, K. Gopalakrishna, and C. S. Vivek, "Plating on acrylonitrile–butadiene–styrene (ABS) plastic: a review," J. Mater. Sci., vol. 51, no. 8, pp. 3657–3674, 2016.
5. T. O. Drews, J. C. Ganley and R. C. Alkire. Evolution of surface roughness during copper electrodeposition in the presence of additives comparison of experiments and monte carlo simulations. J. Electrochem. Soc. 150(5), pp. C325-C334. 2003.
6. B. Du and I. I. Suni. Mechanistic studies of cu electropolishing in phosphoric acid electrolytes. J. Electrochem. Soc. 151(6), pp. C375-C378. 2004.

Proceedings of the 27th CANCAM
Victoria, BC, Canada, May 29 – June 1, 2017

ON USING NANO-CUTTING FLUID WHEN MACHING AUSTEMPERED DUCTILE IRON (ADI)

A. A. Eltaggaz
Advanced Manufacturing Laboratory (AML)
School of Engineering
University of Guelph
Guelph, ON, Canada
aeltagga@uoguelph.ca

H. A. Hegab
Machining Research Laboratory (MRL)
Faculty of Engineering and Applied Science
University of Ontario Institute of Technology
Oshawa, ON, Canada
Hussien.Hegab@uoit.ca

I. M. Deiab
Advanced Manufacturing Laboratory (AML)
School of Engineering
University of Guelph
Guelph, ON, Canada
Ideiab@uoguelph.ca

H. A. Kishawy
Machining Research Laboratory (MRL)
Faculty of Engineering and Applied Science
University of Ontario Institute of Technology
Oshawa, ON, Canada
Hossam.Kishawy@uoit.ca

ABSTRACT

The use of Austempered Ductile Iron (ADI) is rapidly increasing in many engineering applications namely aerospace, automation and biomedical due to its unique and promising properties such as high strength to weight ratio, high wear and corrosion resistance, high yield stress, and high toughness. However, other properties such as low thermal conductivity and high chemical reactivity negatively affect ADI machinability and accelerate cutting tool failure. Additionally, other issues associated with cutting ADI are the high cutting temperature, high pressure and dynamic loads, and tendency of chip to adhere to cutting tool face. Minimum Quantity Lubrication (MQL) serves as the best alternative to flood cooling from an environmental perspective as it minimizes the amount of cutting fluid. To improve the cooling and lubricating efficiency of MQL, Aluminum Oxide (Al_2O_3) gamma nanoparticles are used in this work and its effect on the tool wear behavior during cutting of ADI is investigated.

KEYWORDS: Austempered Ductile Iron (ADI), Minimum Quantity Lubrication (MQL), Machinability, Nano-cutting fluid

INTRODUCTION

Although many commercial materials are available, some of them are not easy to machine by conventional cutting operations. Over the last several decades cutting tools and cutting processes have been greatly improved; therefore, the machining of many materials has become much easier. However, other materials such as titanium and Austempered Ductile Iron (ADI) are still considered difficult to machine. Thus, these metals are identified as difficult-to-cut materials. Recently, these difficult-to-cut materials have attracted research attention due to some remarkable properties such as excellent corrosion resistance and high strength-to-weight ratio. Hence, such materials are used widely in many engineering applications, including automotive and aerospace designs. For instance, ADI is currently used in vehicle parts and agricultural machinery manufacturing, railroad technology and mining [1].

Recently, the interest by the manufacturing industry in Austempered ductile iron (ADI) has rapidly increased, particularly in fabricating vehicle parts such as suspensions and crankshafts, which must meet exact specifications in terms of fatigue strength [2]. In fact, ADI is a novel type of difficult-to-cut material which has outstanding mechanical properties namely high strength-to-weight ratio, wear resistance, ductility, fatigue, and toughness compared to other ductile irons and steel [3].

These outstanding properties can be attributed to the unique ausferrite microstructure of ADI, which is made as a result of heat treatment operation [4]. However, since the first use of ADI in engineering applications, many challenges have been experienced in attempting to machine this material. The hardness of ADI is relatively high which poses a challenge for any machining process. Furthermore, as a result of transformation induced plasticity (TRIP) during machining of ADI, a lot of stresses are loaded to ADI microstructure due to the formation of martensite, which leads to poor machinability [5].

Machining difficult-to-cut materials such as ADI can be improved by addressing the principal issues such as tool life and machining cycle time associated with the cutting process. On the other hand, Minimum Quantity Lubricant (MQL) has become widely used in machining of difficult-to-cut materials. The MQL is a suitable alternative coolant because it has the potential to close the gap between flood coolant and dry machining. Therefore, several experiments revealed that

MQL using vegetable oil improved the tool life and surface quality during machining titanium [6-8].

Although the MQL technique provided promising machining outcomes, enhancing its heat removal capacity is still required. A previous work [9] has focused on controlling the friction behavior in the grinding process as the contact between the abrasive grains and work-piece is highly affected the machining quality characteristics such as (e.g. the wheel wear and grinding forces). Different volume concentrations of Al2O3 and CuO nanoparticles have been added to water as a base fluid. The grinding process has been performed on Ti-6Al-4V titanium alloy using MQL system and the previous nano-cutting fluid. The wheel morphology, surface integrity of ground surface, grinding forces, the coefficient of friction, and chip formation have been investigated. The nano-cutting fluid and MQL have showed promising results in reducing the coefficient of friction, and tangential forces. Moreover, the short C-type chip formation has been observed; hence, it could indicate the cooling effect of the proposed lubrication system [9].

Additional work [10] has evaluated the enhancement in MQL technique performance by dispersing MoS2 nanoparticles into a base fluid (i.e. soybean oil) during grinding of 45 steel. They compared the performance of nano-fluid with other cooling strategies including flood and MQL. It has been observed that MQL-nanofluid shows a significant enhancement in removable heat capacity compared to other cooling strategies, thereby improves the cutting process characteristics. In addition, it has claimed that the optimal mass concentration of MoS2 nanoparticles into the base cutting fluid is 6 wt. %, as it is an important factor in order to avoid the agglomeration of nanoparticles.

The proper dispersion of nanoparticles into the base cutting fluid shows a promising advantage in cutting processes as it helps to reduce the induced friction between the workpiece and cutting tool [13]. Thus, nano-cutting fluids can contribute in facing the heat dissipation challenge during cutting processes as it offers a highly observed thermal conductivity value in comparison with the base lubricants due to the remarkable properties of the added Al_2O_3 nanoparticles (e.g. wear resistance, and excellent thermal conductivity) as have been discussed in a previous study [14]. Additionally, it has been shown in another previous study that nano-cutting fluids have superior cooling properties due to its good heat extraction capabilities [15]. The previous investigations discussed the main reasons behind the promising results which have been offered using MQL-nanofluid technique in the current study. The focus of this work is mainly to study the effects of cooling strategies on the machinability of ADI and investigate the influence of dispersed Al_2O_3-nanofluid on the MQL performance. Two main cooling techniques have been used in this work; MQL and MQL with dispersed gamma aluminum oxide (Al_2O_3) nanoparticles into the conventional fluid (vegetable oil). Investigations are carried out to study and analyse the tool wear behaviour under employed Al_2O_3-nano-fluid.

EXPERIMENTATION & METHODOLOGY

In order to investigate the influence of dispersed Al_2O_3-nanofluid on the cutting quality performance, experiments are carried out by turning ADI. The tests were performed on a CNC lathe machine using standard coated carbide turning inserts (CNMG120416MR) with 1.6 mm nose radius. Cutting tests were performed using MQL strategy, at different levels of cutting speed and feed rate. The nano-additives have been dispersed into the base cutting fluid with volume fraction of 4%. The depth of cut for each cutting pass is 0.5 mm with cutting length of 50 mm. Regarding the MQL system, the air-oil mixture was supplied by the stand-alone booster system (Eco-Lubric) with a nominal oil flow rate of 40 ml/min and air pressure of 0.5 MPa. In addition, ECOLUBRIC E200 is the vegetable oil used (base cutting fluid). The used oil, rapeseed, is environmentally harmless, suitable for industrial used, and has a biodegradability of 90% in 28 days [11]. The position and the angle of the MQL nozzle were adjusted by experimental observation to avoid being blocked by chips. The tool flank wear is measured using a tool maker's microscope. 0.4 mm tool wear was selected as the tool life criterion [12].

Al_2O_3 nanopowder having 20 nm average diameter, 95% purity, and 138 m^2/g specific surface area are used for nano-cutting fluid preparations. Dispersion of Al_2O_3 into the vegetable oil is an important aspect, which affects the resultant nano-cutting fluid thermal conductivity and viscosity. The dispersion of Al_2O_3 into the base cutting fluid is performed using an ultrasonic machine (AQUASONIC-50HT) for 3 hours at 60 °C. On the other hand, dispersion of nano-additives into the base fluid is considered as a difficult challenge due to the strong van der Waals interactions which results in nano-additives agglomeration, clogging, and sedimentation. Thus, using a physical or chemical treatment such as surfactants is recommending in order to achieve powerful forces on the clustered nano-additives to ensure sufficient dispersion and accordingly achieve observed enhancements in the resultant thermal conductivity and viscosity [16-18]. In this work, Sodium Dodecyl Sulfate (SDS) has been used in preparing the nano-fluid as a surfactant (i.e. 0.2 gm). The purpose of using a surfactant is enhancing the nanoadditives performance to be more hydrophilic, and increasing the nanoadditives surface charges; hence, the repulsive forces between the nano-additives could be increased as mentioned in previous studies [19, 20]. In current study, three design parameters have been studied (i.e. cutting speed, cutting feed and added nano-additives percentage). Table 1 shows the design parameters, their corresponding levels and tests performed.

RESULTS & DISCUSSIONS

A maximum flank tool wear of 0.5 mm is used as the tool life criteria. Fig 1 shows the tool wear progress versus the total cutting length. It can be clearly seen that Al_2O_3-nanofluid provides the better results compared to classical MQL. Test 7 shows the optimal flank tool wear (the highest tool life period). The cutting conditions at test 7 are cutting speed of 120 m/min and feed rate of 0.2 mm/rev using MQL-nanofluid technique. Test 7 shows an improvement percentage in tool wear reduction of 32.5% in comparison with test 1 (without nano additives) which has been performed at same cutting conditions. The worst flank tool wear was obtained at tests 3 (without nano additives) while this test has been performed at cutting speed of 240 m/min and feed rate of 0.2 mm/rev. The MQL-nanofluid technique shows an improvement percentage of 12.3% compared to MQL at the same these cutting conditions.

Table 1: The studied parameter levels and the plan of experiments

Test #	Lubrication technique	Feed rate (mm/rev)	Cutting speed (m/min)
1	MQL	0.2	120
2	MQL	0.2	180
3	MQL	0.2	240
4	MQL	0.3	120
5	MQL	0.3	180
6	MQL	0.3	240
7	MQL-nanofluid	0.2	120
8	MQL-nanofluid	0.2	180
9	MQL-nanofluid	0.2	240
10	MQL-nanofluid	0.3	120
11	MQL-nanofluid	0.3	180
12	MQL-nanofluid	0.3	240

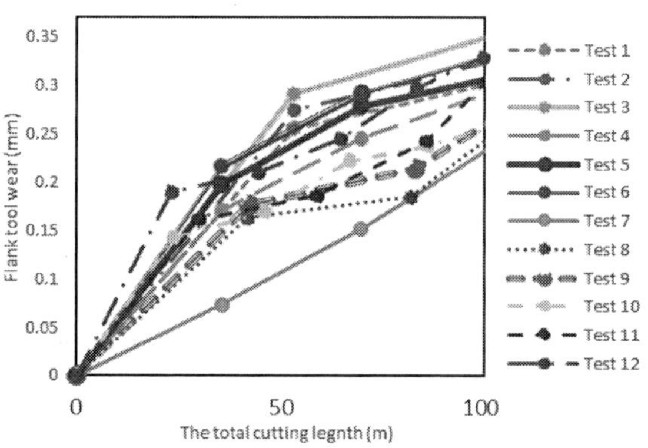

Figure 1: Tool wear results

It is also an important aspect to understand the MQL nano-fluid mechanism/ the technique schematic is shown everywhere at Fig.2. It has been demonstrated from a previous study [21] that the resultant nano-fluid is atomized in the MQL system through the applied compressed air, forming a fine mist surrounded with a base cutting fluid film (see Fig.2) as have been discussed in a previous study. This mist has the ability to penetrate into the chip-tool interfere zone and forming a tribo-film which helps in absorbing the generated cutting heat as well as reducing the induced friction. The mentioned characteristics enormously improve the cooling and lubrication properties that retain the cutting tool hardness for a longer time as have been similarly confirmed in a previous study [22]. Thus, the MQL-nanofluid offers promising results in terms of tool life measurement in comparison with tests were done using classical MQL without any added nano-additives.

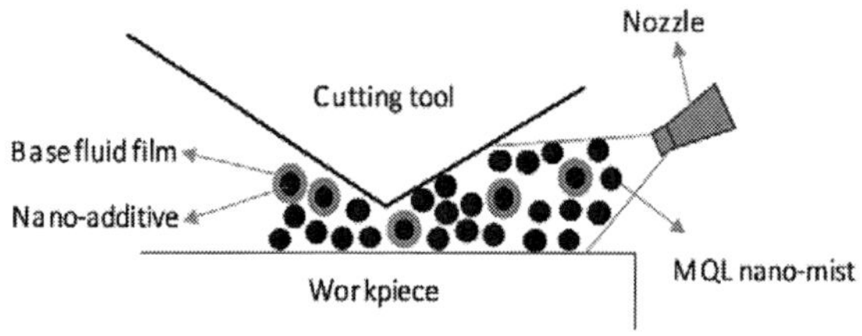

Figure 2: The MQL-nanofluid mechanism schematic

In addition, the tool wear modes were investigated for both best and worst tests as shown in Fig.3. It can be seen for test 3 (classical MQL) that chipping is the dominant tool wear mode, and it takes place usually due to the overload of mechanical tensile stresses which are generated because of the hard inclusions presence inclusions (e.g. 2.5 % of Si). Meanwhile, the MQL-nanofluid technique shows less abrasive wear (see Fig.3-left side) as the proposed nanofluid helps in reducing the rubbing between the cutting tool and workpiece because of the tribo-film formation, which helps in reducing the induced friction and the generated tensile stresses accordingly. Thus, the friction and wear behaviour can be improved using MQL-nanofluid technique.

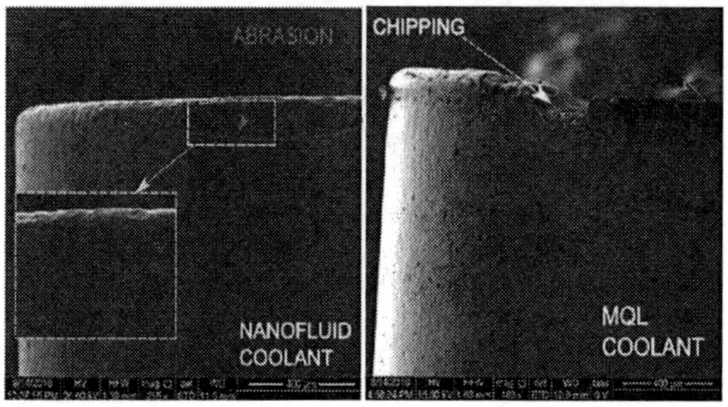

Figure 2: Scanning Electron Microscope (SEM) images for both test 3 (right side) and test 7 (left side).

CONCLUSIONS

In this study, MQL-nanofluid technique offered promising results in improving the tool wear behavior during cutting of ADI in comparison with tests performed without adding Al_2O_3 nanoparticles. The combination of MQL-nanofluid at cutting speed of 120 m/min and feed rate of 0.2 mm/rev showed the optimal tool life, while test 3 which has been performed at cutting speed of 240 m/min and feed rate of 0.2 mm/rev using classical MQL provides the worst tool wear value. Due to the improved convection, wettability, and conduction characteristics of the proposed nanofluid, the tool wear behavior has been significantly enhanced because of the tribo film formation along the chip-tool interface zone. Finally, more investigations need to be performed in order to understand the effects of the added nano-additives percentage and size on the coefficient friction and the induced tool wear.

ACKNOWLEDGMENTSThe authors acknowledge the support of the Natural Sciences and Engineering Research Council of Canada (NSERC) and Ontario centers of Excellence (OCE).

REFERENCES

1. Polishetty, A., Machinability and microstructural studies on phase transformations in Austempered Ductile Iron. 2011, PhD thesis, Auckland University of Technology: Auckland.
2. Klocke, F. and C. Klopper, Machining of ADI. WZL Laboratory for Machine Tools and Production Engineering: Aachen, Germany.
3. Meena, A. and M. El Mansori, Study of dry minimum quantity lubrication drilling of novel austempered ductile iron (ADI) for automotive applications. Elsevier, Wear, 2011. 271(9-10): p. 2412-2416.
4. Brandenberg, K., Successfully Machining Austempered Ductile Iron (ADI). 2001, Applied Process Inc.: Michgen, USA.
5. Nofal, A., Advanced in the Metallurgy and Applications of ADI. Journal of Metallurgical Engineering (ME), 2013. 2(1).
6. Pervaiz, S., et al., An experimental investigation on effect of minimum quantity cooling lubrication (MQCL) in machining titanium alloy (Ti6Al4V). International Adv. Technology 2016. 87: p. 1371-1386.
7. Sadeghi, M.H., et al., Minimal quantity lubrication-MQL in grinding of Ti–6Al–4V titanium alloy. International Journal Adv. Manufacture Technology, 2008. 44: p. 487-500.
8. Pervaiz, S., Investigation Cooling and Lubrication Strategies for Sustainable Machining of Titanium Alloys: Impact on Machinability and Environmental Performance. 2014, KTH Royal Institute of Technology: Stockholm.
9. Setti, D., et al., Performance evaluation of Ti–6Al–4V grinding using chip formation and coefficient of friction under the influence of nanofluids. International Journal of Machine Tools and Manufacture, 2015. 88: p. 237-248.
10. Zhang, Y., et al., Experimental evaluation of MoS 2 nanoparticles in jet MQL grinding with different types of vegetable oil as base oil. Journal of Cleaner Production, 2015. 87: p. 930-940.
11. Deiab, I., S.W. Raza, and S. Pervaiz, Analysis of lubrication strategies for sustainable machining during turning of titanium Ti-6Al-4V alloy. Procedia CIRP, 2014. 17: p. 766-771.
12. Maity, T.K., Development and Study of the Mechanical Properties of Silica Glass Based Composite. 2014.
13. Alberts, M., K. Kalaitzidou, and S. Melkote, An investigation of graphite nanoplatelets as lubricant in grinding. International Journal of Machine Tools and Manufacture, 2009. 49(12): p. 966-970.
14. Hsieh, S.-S., H.-H. Liu, and Y.-F. Yeh, Nanofluids spray heat transfer enhancement. International Journal of Heat and Mass Transfer, 2016. 94: p. 104-118.
15. Srikant, R., et al., Applicability of cutting fluids with nanoparticle inclusion as coolants in machining. Proceedings of the Institution of Mechanical Engineers, Part J: Journal of Engineering Tribology, 2009. 223(2): p. 221-225.
16. Hwang, Y., et al., Thermal conductivity and lubrication characteristics of nanofluids. Current Applied Physics, 2006. 6: p. e67-e71.
17. Li, X., et al., Thermal conductivity enhancement dependent pH and chemical surfactant for Cu-H 2 O nanofluids. Thermochimica Acta, 2008. 469(1): p. 98-103.
18. Zhu, H., et al., Preparation and thermal conductivity of suspensions of graphite nanoparticles. Carbon, 2007. 45(1): p. 226-228.
19. Hwang, Y., et al., Stability and thermal conductivity characteristics of nanofluids. Thermochimica Acta, 2007. 455(1): p. 70-74.
20. Huang, J., et al. Influence of pH on the stability characteristics of nanofluids. in Photonics and Optoelectronics, 2009. SOPO 2009. Symposium on. 2009. IEEE.
21. Mao, C., et al., Investigation on the effect of nanofluid parameters on MQL grinding. Materials and Manufacturing Processes, 2013. 28(4): p. 436-442.
22. Khandekar, S., et al., Nano-cutting fluid for enhancement of metal cutting performance. Materials and Manufacturing Processes, 2012. 27(9): p. 963-967.

Proceedings of the 26th CANCAM
Victoria, BC, Canada, May 29 – June 1, 2017

OPTIMIZATION OF METAL CUTTING THIN-BLADE GEOMETRY FOR SLOTTED LINER MANUFACTURING

Yazdan Kordestany
Mechanical Engineering Department
University of Alberta
Edmonton, Alberta, Canada
yazdan@ualberta.ca

Yongsheng Ma
Mechanical Engineering Department
University of Alberta
Edmonton, Alberta, Canada
yongsheng.ma@ualberta.ca

ABSTRACT

The demanding slotting process for the tubular liners in Steam Assisted Gravity Drainage (SAGD) oil extraction production was studied in this paper with the main objective of optimizing the unique tool geometry. Typical cutting tool and workpiece properties were used as the case for this study. Maximum applied force applied to the overall circular blade was obtained from numerical simulation. The results were used as the boundary conditions for static analysis to conduct geometry optimization to reduce blade displacement. After comparing different geometries that resulted from optimization, the outer diameter, relief angle, tooth height, and number of teeth, bore diameter, rake angle, gullet radius, and tooth pitch were modified based on the criterion of having lower displacement than the initial blade.

KEYWORDS: Circular saw blade, thin-blade slotting, slotted liners, finite element analysis, optimization, ABAQUS, NX

1. INTRODUCTION

Slotted liners for sand control in SAGD typically include a pipe casing on which rectangular apertures (slots) are made by methods such as cutting with circular saw blades. These slots will provide the required open area to steam injection and oil production while preventing sand particles to enter the pipe. It has been proven that slotted liners are one of the most popular methods of choice for sand control in SAGD due to their mechanical integrity for long horizontal well completions [1]. Slot geometry plays important role in preventing entry of sands.

Circular saw blades can be used to manufacture slotted liners in numeric control (NC) milling machines. There are several industries that apply multi spindle NC machines to produce slotted liners. Having multi-spindle slotting machine reduces production cost and time in such a way that it allows simultaneous cutting of slots as many as the number of spindles. Oscillation of the blades could be seen as the major concern in the cutting process since it affects surface roughness, tool life, and vibration. When the blade rotation reaches a critical value or is close to its natural frequency, it will no more be stable and actions are required to avoid considerable vibration leading to shorter tool life. Although rotational speeds of saw blades are higher at lower thicknesses, this affects the lateral stiffness [2]. High vibrations can be expected for a thin blade especially working at the greater speeds.

The blade experiences deformation and displacement due to applied cutting force during slotting process, especially when thin circular blades are used. These displacements result in cutting vibration, reduce the surface quality, and lead to inaccurate slot geometry. There have been a number of studies to eliminate the vibration, such as Wang et al. [3] where the effect of rotational speed on blade mechanical properties. Andelic et al. [4] studied the influence of rotational speed on blade natural frequency with and without slots. Reiter et al. [5] applied experimental methods to study the effect of blade geometry and kinematic parameter on circular saw noise in an idling condition and concluded that blades having slots had a better noise control.

The first step in this study is calculation of applied force in the slotting process. Having determined the maximum number of teeth that are in touch with the workpiece during cutting and the corresponding depth of cut, the applied force can be obtained from finite element simulations. Direct experimental measurements to study cutting processes are usually costly and time consuming. The mechanics of machining cannot be also fully explained by using only analytical approaches. Therefore, numerical approaches in metal cutting processes have been interesting to researchers using various available explicit and implicit software packages such as ABAQUS, ANSYS, AdvantEdge and etc. For instance, Iwata et al. [5] adapted a rigid-plastic finite element model with an Eulerian formulation to simulate steady-state orthogonal metal cutting. Komvopoulos and Erpenbeck [6] applied constitutive material behavior models and interfacial friction mechanisms representing real application scenarios for metal cutting simulation of AISI 4340 steel with ceramic-coated tools to investigate the effect of cutting parameters. The effect of tool edge geometry was also studied by Lei Wan et al. [7] in orthogonal metal cutting by adoption of arbitrary Lagrangian-Eulerian (ALE) approach for simulation in ABAQUS/Explicit and considering P20 as the workpiece material.

Finite element methods along with the obtained force are then applied to simulate the static force distribution for a circular saw blade. The determined blade deformation and a

CAD/CAE software are used for the optimization process. Finally, it is possible to compare the resulting deformations for both optimized and original blade geometries.

2. METHODOLOGY

2.1. Force calculation

Figure 1 shows the manufacturing process for creating a slot on the pipe. The circular blade rotates at the set revolutions per minute (rpm) to the arbor at the distance H and moves downwards by having feedrate (v_f) to cut through the workpiece. It should be noted that the feedrate varies during the cutting process as follows: (1) starting with v_{f1}, reaching the outer diameter of the pipe with v_{f2}, (2) starting cutting through the inner diameter of the pipe with v_{f3}, and completing the process by having v_{f4} at last position. In the slotting process, the mechanics on each tooth on the circular saw blade follows orthogonal cutting theory since the cutting edge of each tooth is parallel to the surface of the workpiece and perpendicular to the cutting orientation [8].

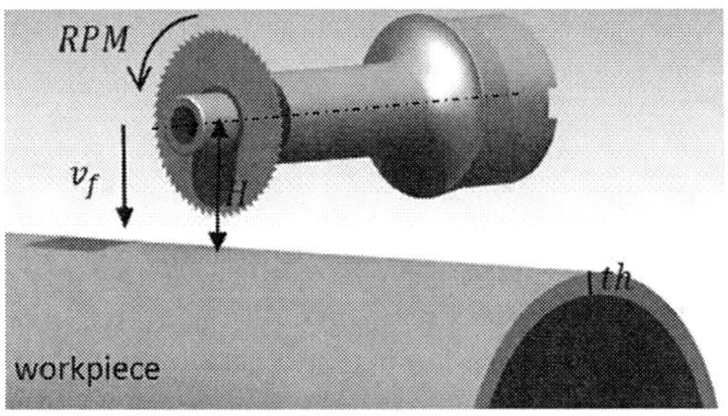

Figure 1: Slotting process using a circular saw blade

The number of teeth involved in the slotting process at a transient time point is defined as the effective teeth which is a function of time. When the blade starts to cut the pipe, the number of effective teeth on the blade which remove the material during cutting increases until the blade reaches inner diameter of the pipe (time t_2). The number of teeth that are in touch with the workpiece decreases by moving the tool to its last position. Therefore, the greatest number of teeth that are in contact with the pipe and its corresponding time step during slotting process needs to be calculated.

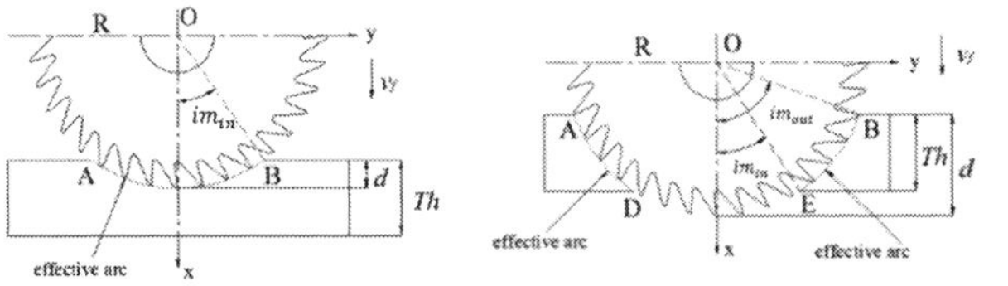

(a) $t_1<=t<=t_2$ (b) $t_2<=t<=t_3$

Figure 2 : Effective teeth before and after time t_2

Figure 2 shows the effective arcs to represent the number of effective teeth before and after time t_2, where im and Th are the tooth immersion angle and pipe thickness, respectively. The highest cutting force is applied to the tool at this time.

Because of time varying immersion angle, the uncut chip thickness in slotting process is similar to that in a milling process and varies periodically [9]. It is based on feed-per-tooth but changes due to the tool rotation. The instantaneous uncut chip thickness, dt_0 in milling can be calculated as:

$$dt_0 = v_f \, \theta_{ju}. \qquad (1)$$

where, v_f is the feed rate (mm/rev-tooth) and θ_{ju} is the instantaneous angle of immersion (rad). The uncut chip thickness is determined for the teeth that are in touch with the workpiece and then used as a boundary condition for force calculation in numerical simulation.

The orthogonal cutting process described above is simulated by using a 3D model in ABAQUS/Explicit with Lagrangian formulation. The cutting parameters, tool and workpiece geometries used are given in Table 1.

Table 1: Input parameters for the numerical simulation

Blade diameter (in)	3.07	Pipe thickness (in)	0.362
Blade rake angle (degree)	2	Pipe diameter (in)	7
Blade relief angle (degree)	30	Slot length (in)	1.8
Teeth edge radius (mm)	0.01	Friction coefficient	0.5
Blade teeth number	56	Feedrate (in/min)	2
Cutting speed (in/sec)	43	RPM	255

Johnson-Cook plasticity and damage model with element deletion technique is applied to this study as the separation criteria in finite element simulation of cutting [10], [11] and [12]. It considers the separated effects of strain hardening, strain-rate (viscosity) and thermal softening. It is a relatively simple model and defines the general response of material deformation [13]. The material properties for the workpiece AISI 4140 can be found from [11]. The tool material is selected as HSS M2 based on typical industry standards. The Coulomb friction law with a constant friction coefficient of 0.5 is chosen here.

2.2. natural frequency of blade

Natural frequency of the considered circular saw blade is obtained by using analytical model and finite element software. The analytical equations used for natural frequency are shown in Eq. (2) and Eq. (3) which are the results from classical vibration theory of circular blade as shown in [14].

$$\omega = \frac{h}{D^2}\sqrt{\frac{E\,\alpha}{12\rho(1-\nu)}} \qquad (2)$$

$$f_{s(n)} = \frac{\omega}{2\pi} \quad (3)$$

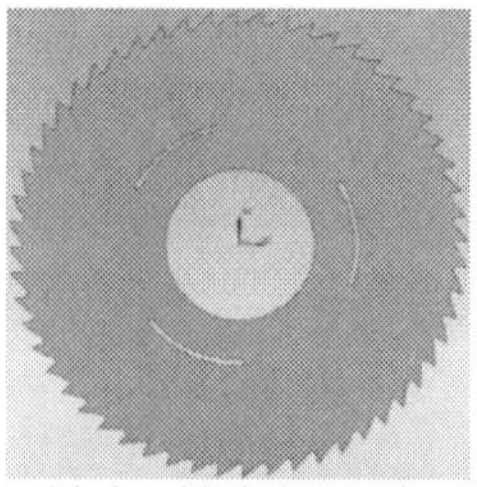
(a) Blade with three annular slots

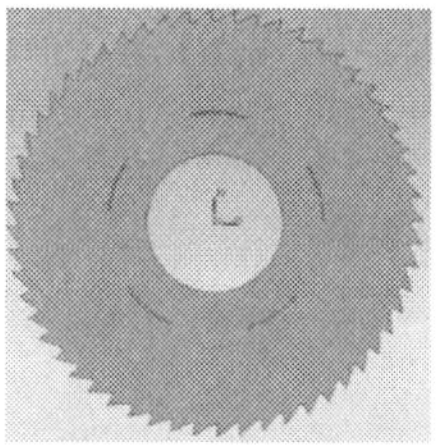
(b) Blade with five annular slots

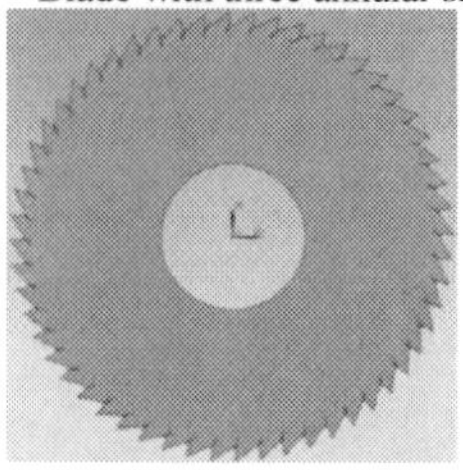
(c) Blade with radial slots

(d) Blade with annular and radial slots

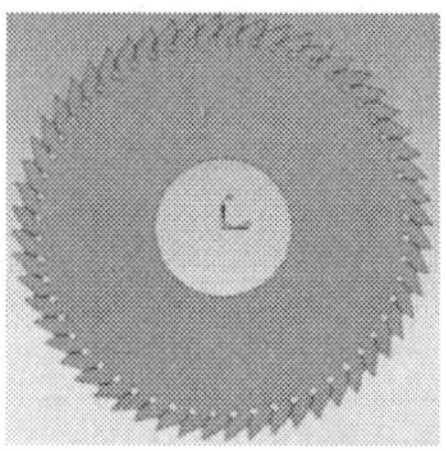
(e) Blade with larger radial slots

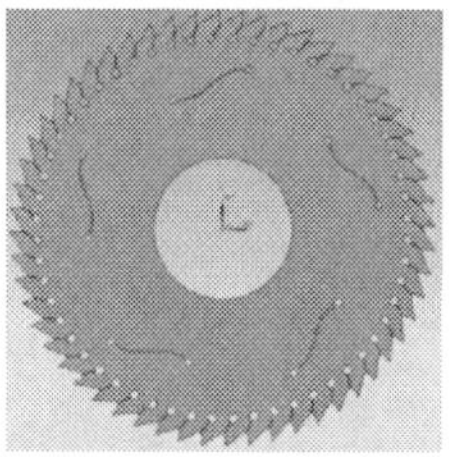
(f) Blade with annular and radial slots

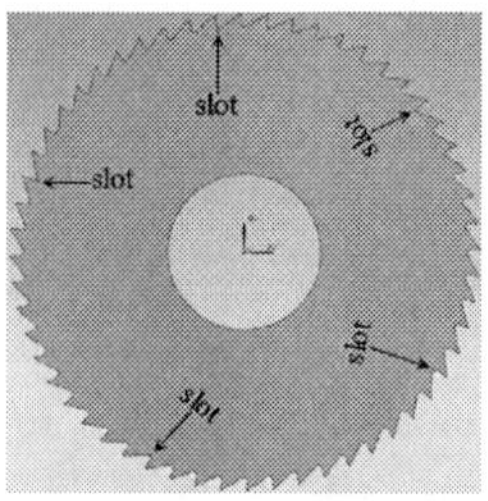

(g) Blade with five radial slots

Figure 3: Different slots on the circular saw blade

After creating the CAD model and assigning material to the part model in NX, TETRAL 10 element type has been chosen to apply the mesh using NX SOL 103. The boundary condition is a fixed boundary condition for which the surface of inner diameter of the blade or bore diameter has been selected.

Circular blades that have radial or annular slots as additional features can also be used. The main purposes of using these blades to make slots that are symmetrical about the blade center are to reduce vibration and noise as well as increasing thermal dispersion efficiency especially when operating at high rotational speeds [15]. The different slots geometries that are made on the circular saw blade for this analysis are shown in Figure 3. Table 2 shows the natural frequencies that resulted from the analysis of the different cases where annular and/or radial slots were made on the blade. It should be noted that increased natural frequency is desired so that vibration and noise during the cutting process are reduced.

Of the considered cases, the natural frequency for case e had better desirability with respect to its potential to reduce undesired tool vibration. Based on this criterion, the original design (that has no added features) and case e which consisted of radial slots have been chosen to further conduct optimization.

Table 2 : Natural frequencies for blade having slots (Hz)

Case	First mode	Second mode	Third mode	Fourth mode
No slots	705.99	831.658	1291.19	2073.48
Case a	600.8	782.6	1223.4	2037.9
Case b	658.3	797.4	1270.1	2057.8
Case c	707.2	824.3	1246.1	2875.3
Case d	668.3	787.7	1136.1	1731.3
Case e	718.9	824.4	1182.2	2482.2
Case f	682.1	791.5	1141.4	1707.6
Case g	706	831.3	1289.8	2070.1
Analytical	700.4	835.7	1328.5	2277.2

2.3. optimization of blade Geometry

The circular blade is selected to conduct static analysis. NX Nastran with same mesh type that is used for analysis of natural frequency is used to perform the static analysis having the blade fixed at its inner surface and applying the resulting force obtained from numerical simulations.

The aim of this section is to investigate the possibility of changing the blade geometry to reduce the blade teeth displacement due to the applied force. This can be achieved by using software packages such as NX. A local sensor should first be created to measure the tooth deformation. It is preferred to place the sensor where the maximum displacement is experienced. This sensor can be parametrized and considered as an objective function for a minimization problem. Parameters for all features used to create the CAD model of the blade have to be created. These parameters in the current study are the outer diameter of blade, bore diameter, tooth height, tooth pitch, gullet radius, rake angle, and relief angle. To design all teeth for the blade, pattern feature is the most common tool which uses one tooth to create all teeth along the outer diameter of the blade. Therefore, the formula for the number of teeth in the pattern feature should be used to automatically update the teeth number when other dimensions are changed.

There are several constraints that should be included in the problem. The blade has to have enough material to keep its stiffness; it can reach the plunge and cut the workpiece (for 0.362 in of pipe thickness and 0.3 in plunge). The upper limit for the outer diameter must be specified such that blades are separated by enough distance when they are mounted on the arbor. The gullet area has to have enough space to maintain the chips during cutting.

The upper limit for the relief angle is not defined because the rake angle is small and it can only be increased in the optimization. It is therefore important that as the rake angle is increased, the relief angle does not increase with it so as to maintain the material angle on the blade. The optimization process can thus be represented by Eq. (4).

$$objective: \min(deformation)$$

$$cstr: \begin{cases} outer\ diameter - bore\ diameter > 1.5in \\ 2in < outer\ diameter < 4in \\ 0.6in < bore\ diameter < 2in \\ 2deg < rake\ angle < 40deg \\ 3deg < relief\ angle < 30deg \\ 0.06in < tooth\ height < 0.15in \\ 0.14in < tooth\ pitch < 0.2in \\ 0.01in < gullet\ radius < 0.05in \end{cases} \quad (4)$$

3. RESULTS

Figure 4 shows the deflection enlarged by a factor of 1 for the blades with and without radial slots obtained from the static analysis as result of applying force to 12 teeth in both x and z directions.

(a) Blade deformation without slots (b) Blade deformation with slots

Figure 4 : Deflection of the blades in static analysis

Table 3 shows the maximum displacement the blade undergoes and that stress applied to it due to the highest cutting force during slotting process. Circular blades having radial slots have greater natural frequencies whereas blades without any radial slots experience less deformation and stress during cutting. This is very important as the cutting force has a direct impact on the tool life. The reason for blade deformation when having radial slots is the opening area between the teeth which, upon an applied force, makes is the blade material easier to break.

Table 3 : Deformation for blades with and without slots

	Maximum displacement (mm)			
Blade	*magnitude*	*x*	*y*	*z*
Without slots	0.037	0.036	0.0002	0.011
With slots	0.107	0.104	0.00063	0.022

Updated design geometries after optimization process for both blade types are given in Table 4. It is observed that increase in rake angle reduce the blade deformation which improves the tool life. It is also important to decrease the relief angle as rake angle rises to maintain blade stability during cutting.

Table 4 : Geometry dimensions for modified blades

Parameter	**Blade without radial slots**	**Blade with radial slots**
Outer diameter (in)	2.93	3.616
Bore diameter (in)	1.57	1.486
Rake angle (degree)	5	4
Relief angle (degree)	28	30
Gullet radius (in)	0.027	0.018
Tooth pitch (in)	0.184	0.167
Tooth height (in)	0.073	0.077
Number of teeth	50	67
Radial length (in)	-	0.075
Radial width (in)	-	0.001
Radial radius (in)	-	0.016

The maximum number of effective teeth has to be calculated for the new designs which were determined as 14. It is required to conduct another static analysis for the new blade geometries including the forces applied to the mentioned 14 teeth. Figure 5 shows the deformation of both modified blades with and without slots from static analysis. The highest displacements experienced by blades' teeth are given in Table 5.

(a) Blade without slot (b) Blade with slots

Figure 5 : Deflection of the blades in static analysis

Table 5 : Displacement results for modified blades

	Maximum displacement (mm)			
Blade	magnitude	x	y	z
Modified without slots	0.0264	0.0264	0.0002	0.006
Modified without slots	0.057	0.0052	0.000464	0.03

The displacement of both blades as a result of static analysis is reduced after geometry optimization. However, the modified blade without radial slots has lower deformation and tool life than modified blade with radial slots.

4. CONCLUSION

The natural frequency of circular blades having different shapes was obtained applying both analytical and numerical methods. The natural frequency resulting from simulation and mathematical model were in good agreement especially for the first mode.

Optimization was conducted to identify the new blade geometry for improved performance. By making use of the results from the force analysis different cases for blade geometry with and without added features were studied to determine the optimal blade geometry and dimensions. Table 4 and Figure 5 give the specifications for the optimal blade geometry that is identified. It was concluded that, if changing the blade geometry for better tool life is considered, the specifications given in Table 4 shall be followed.

The same procedure for optimization of geometry shall be conducted by including the arbor dimensions in the analysis. Since the blade is mounted on an arbor, the geometry of arbor should be changed along with the bore diameter of the blade. Therefore, the cost related to the material change of both blades and arbors may also be considered in future works.

5. ACKNOWLEDGEMENT

The authors remain grateful for the financial support from RGL Reservoir Management Inc. Special thanks is hereby extended to Mr. Michael Leitch (P.Eng) for his invaluable contributions during the entire course of the research.

REFERENCES

1. D. Matanovic, M. Cikes, and B. Moslavac, *Sand Control in Well Construction and Operation*. Springer, 2012.
2. A. Skoblar, N. Andjelic, and R. Zigulic, "DETERMINATION OF CRITICAL ROTATIONAL SPEED OF CIRCULAR SAWS FROM NATURAL FREQUENCIES OF ANNULAR PLATE WITH ANALOGOUS DIMENSIONS," *Int. J. Qual. Res.*, vol. 10, no. 1, pp. 177–192, 2016.
3. X. L. Wang and Q. Zhang, "The Mechanical Property Analysis of Circular Saw Blades under Different Rotational Speeds," *Adv. Mater. Res.*, vol. 145, pp. 365–370, 2010.
4. N. Anđelić, A. Pavlović, and S. Braut, "Variation of Natural Frequencies by Circular Saw Blade Rotation Variation of Natural Frequencies by Circular Saw Blade Rotation," no. April 2016, 2015.
5. W. F. Reiter and R. F. Keltie, "On the nature of idling noise of circular saw blades," *J. Sound Vib.*, vol. 44, no. 4, pp. 531–543, 1976.
6. K. Komvopoulos1 and S. A. Erpenbeck, "Finite Element Simulation of Orthogonal Metal Cutting," *J. Eng. Ind.*, vol. 113, no. August, pp. 253–267, 1991.
7. L. Wan, D. Wang, and Y. Gao, "Investigations on the Effects of Different Tool Edge Geometries in the Finite Element Simulation of Machining," *J. Mech. Eng.*, vol. 61, no. 3, pp. 157–166, 2015.
8. N. Balihodzic, "A Numerical Investigation of Orthogonal Machining," University of New Brunswick, 2002.
9. Y. Altintas, *Manufacturing automation: metal cutting mechanics, machine tool vibrations, and CNC design*, 2nd ed. Cambridge: Cambridge University Press, 2012.
10. I. Escamilla, O. Zapata, B. Gonzalez, N. Gámez, and M. Guerrero, "3D Finite Element Simulation of the Milling Process of a TI 6AL 4V Alloy," *2010 SIMULA Cust. Conf.*, pp. 1–10, 2010.
11. M. Agmell, A. Ahadi, and J. E. Ståhl, "The link between plasticity parameters and process parameters in orthogonal cutting," *Procedia CIRP*, vol. 8, pp. 224–229, 2013.
12. N. Sawarkar and G. Boob, "Finite Element based Simulation of Orthogonal Cutting Process to Determine Residual Stress Induced," *Int. J. Comput. Appl.*, pp. 33–38, 2014.
13. Y. Zhang, J. C. Outeiro, and T. Mabrouki, "On the selection of Johnson-Cook constitutive model parameters for Ti-6Al-4V using three types of numerical models of orthogonal cutting," *Procedia CIRP*, vol. 31, pp. 112–117, 2015.
14. J. Svoreň, Ľ. Javorek, A. Droba, and D. Paulíny, "Comparison of Natural Frequencies Values of Circular Saw Blade Determined by Different Methods," *Drv. Ind.*, vol. 66, no. 2, pp. 123–128, 2015.
15. W. H. Gau, K. N. Chen, and Y. L. Hwang, "Model updating and structural optimization of circular saw blades with internal slots," *Adv. Mech. Eng.*, vol. 2014, 2014.

Proceedings of the 26th CANCAM
Victoria, British Columbia, Canada, May 28 - June 1, 2017

FINITE STRIP MODELLING OF THE VARYING DYNAMICS OF SHELL-LIKE STRUCTURES DURING MACHINING PROCESSES

Josiah Stefani
Department of Mechanical Engineering
University of Victoria
Victoria, BC, Canada
josiahstefani@gmail.com

Keivan Ahmadi
Department of Mechanical Engineering
University of Victoria
Victoria, BC, Canada
kvahmadi@uvic.ca

ABSTRACT

The efficiency of the finite strip method (FSM) in modelling the varying dynamics of shell-like structures during machining operations is investigated. The workpiece is modelled as a shallow, doubly curved, helicoidal, cantilever shell, and the natural modes are computed using both FSM and the finite element method (FEM). In the FSM solution, the membrane and bending displacement fields of the shell in the spanwise direction are approximated by a set of basis function that satisfy clamped-free boundary conditions. The displacement fields in the chordwise direction are approximated using polynomial functions. The efficiency of the presented FSM is investigated by comparing the computed natural vibration modes against the ones that are obtained using FEM.

Keywords: Shell dynamics, finite strip method, self-excited vibrations

INTRODUCTION

Predicting the optimum machining parameters that lead to a high material removal rate without causing excessive and unstable vibrations (i.e. chatter) requires accurate models of the workpiece structure [1]. The finite element method (FEM) is commonly used for structural modelling of the workpiece [2, 3, 4], but as the material is removed from the stock, the mass and stiffness matrices in the FEM need to be updated, which requires an inhibitive computational load [5]. Therefore, computationally efficient approaches are to be developed to model the varying dynamics of the workpiece.

FEM has been used in the literature to study the varying dynamics of the workpiece during machining operations [2, 3, 4]. Budak et al. [6] used a structural dynamics modification technique to reduce the computational load that is needed to update finite element models of the workpiece. Kersting and Bierman [7] used mesh-free and modal models to approximate the dynamics of the workpiece during milling operations. Meshreki et al. [8] used a 2D Multi Span Plate (MSP) model to compute the natural modes of pocket structures that are commonly used in aerostructure. Ahmadi [9] modelled the walls of pocket structures using Kirchhoff plate theory [10] and used FSM to compute the natural modes of the structures at different stages of machining. In FSM, the plate structure is discretized only in the transverse direction and the displacement field in the longitudinal direction is approximated using a set of trigonometric functions that satisfy the boundary conditions at the two ends. The number of DOF that is required in FSM is substantially smaller than FEM and thus the computational load is significantly lower.

In this paper, the application of FSM is extended to shell-like structures to address the varying dynamics of blades that are machined during the manufacturing of gas turbine parts such as fans, impellers, blisks, etc. The blade is assumed to be a shallow, doubly curved, helicoidal structure, and the coupling between membrane and bending deflections is considered in modelling. The structure is divided into helicoidal strips in the chordwise direction, and the displacement field in the spanwise direction is approximated using a set of basis functions that satisfy the given boundary conditions. Trigonometric and spline basis functions are used in the spanwise direction, and their performances are compared against one another. The efficiency of FSM in computing the varying natural modes of the blade during machining is compared against FEM.

SHELL MODEL

Turbomachinary blades were modelled by shallow, doubly curved, helicoidal shells in a paper by Walker [11], and the same model is used in this study. The blade geometry (see Fig. 1) is described in a helicoidal orthogonal coordinate system (ξ^*, η^*, ζ^*), where the chordwise axis η^* rotates around the spanwise axis ξ^* at the twist angle of $\theta(\xi^*)$. The geometry of the shell is fully defined by expressing the ζ^* coordinate of any point on the middle surface of the shell in terms of its ξ^* and η^* coordinates as $\zeta^*(\xi^*, \eta^*)$, and by determining the thickness of the blade at that point as $t(\xi^*, \eta^*)$. Transverse shear deformations are neglected for thin shells and the deflection of any arbitrary point on the middle surface is described by its membrane deformations in ξ^* and η^* directions, and its bending deformation in ζ^* direction. Deformations in ξ^*, η^*, and ζ^* directions are denoted by u, v, and w, respectively.

Membrane strains ($\varepsilon_{\xi\xi}$, $\varepsilon_{\eta\eta}$, and $\varepsilon_{\xi\eta}$) and bending strains ($\kappa_{\xi\xi}$, $\kappa_{\eta\eta}$, and $\kappa_{\xi\eta}$) are related to deformations u, v, and w through the following equation:

$$\varepsilon = \begin{Bmatrix} \varepsilon_{\xi\xi} \\ \varepsilon_{\eta\eta} \\ \varepsilon_{\xi\eta} \\ \kappa_{\xi\xi} \\ \kappa_{\eta\eta} \\ \kappa_{\xi\eta} \end{Bmatrix} = \begin{Bmatrix} \frac{1}{A}\frac{\partial u}{\partial \xi} + \frac{1}{A}\frac{\partial A}{\partial \eta} v + w k_{\xi\xi} \\ \frac{\partial v}{\partial \eta} + w k_{\eta\eta} \\ \frac{1}{A}\frac{\partial v}{\partial \xi} + \frac{\partial u}{\partial \eta} - \frac{1}{A}\frac{\partial A}{\partial \eta} u + 2 w k_{\xi\eta} \\ \frac{1}{A}\frac{\partial \psi}{\partial \xi} + \frac{1}{A}\frac{\partial A}{\partial \eta}\varphi + \varpi k_{\xi\eta} \\ \frac{\partial \varphi}{\partial \eta} - \varpi k_{\xi\eta} \\ \frac{1}{A}\frac{\partial \varphi}{\partial \xi} + \frac{\partial \psi}{\partial \eta} - \frac{1}{A}\frac{\partial A}{\partial \eta}\psi \end{Bmatrix} \tag{1}$$

The rotational deformations of the surface normal about the ξ and η axes (ψ and φ, respectively) and the rotational deformation about this normal ϖ are defined in terms of the translational deformations:

$$\psi = -\frac{1}{A}\frac{\partial w}{\partial \xi} + v k_{\xi\eta}, \quad \varphi = -\frac{\partial w}{\partial \eta} + u k_{\xi\eta}$$
$$\varpi = \frac{1}{2}\left(\frac{1}{A}\frac{\partial v}{\partial \xi} - \frac{1}{A}\frac{\partial A}{\partial \eta} u - \frac{\partial u}{\partial \eta}\right) \tag{2}$$

In Eqs. 1 and 2, A is a Lamé parameter, and $k_{\xi\eta}$, $k_{\eta\eta}$, and $k_{\xi\xi}$ are the curvature components of the middle surface of the shell. These parameters are defined in terms of $\zeta^*(\xi^*, \eta^*)$, $\theta^*(\xi^*)$, and their derivatives [11, 12]. Note that the thin-shell assumption made in this analysis neglects the transverse shear strain components $\varepsilon_{\xi\zeta}$, $\varepsilon_{\eta\zeta}$, and $\varepsilon_{\zeta\zeta}$. The strain energy (U) and kinetic energy (T) resulting from the membrane and bending strains are expressed using the following integrals [11]:

$$U = \frac{E}{2(1-\nu^2)} \int_0^a \int_{b_1(\xi)}^{b_2(\xi)} t(\xi,\eta)\, I_1 A(\xi,\eta)\, d\eta d\xi$$
$$I_1 = \varepsilon_{\xi\xi}^2 + \varepsilon_{\eta\eta}^2 + 2\nu\varepsilon_{\xi\xi}\varepsilon_{\eta\eta} + \frac{1}{2}(1-\nu)\varepsilon_{\xi\eta}^2 + \frac{t^2(\xi,\eta)}{12}\left[\kappa_{\xi\xi}^2 + \kappa_{\eta\eta}^2 + 2\nu\kappa_{\xi\xi}\kappa_{\eta\eta} + \frac{1}{2}(1-\nu)\kappa_{\xi\eta}^2\right] \tag{3}$$

$$T = \frac{1}{2}\rho\omega^2 \int_0^a \int_{b_1(\xi)}^{b_2(\xi)} t(\xi,\eta)\, I_2 A(\xi,\eta)\, d\eta d\xi$$
$$I_2 = \left(u^2 + v^2 + w^2\right) + \frac{t^2(\xi,\eta)}{12}\left(\psi^2 + \varphi^2\right) \tag{4}$$

where ω is the circular frequency of vibration, a is the spanwise length of the blade, and $b_1(\xi^*)$ and $b_2(\xi^*)$ are chordwise integration limits at ξ^*. The Young's modulus, Poisson's ratio, and density are denoted by E, ν, and ρ.

FINITE STRIP METHOD (FSM)

In FSM, the middle surface of the blade is discretized in the chordwise direction using narrow strips, as shown in Fig. 1. The edges of these strips are treated as nodal curves, and the displacement field along the nodal curves is approximated by a linear combination of a set of basis functions. The variation of the displacement field between two adjacent nodal curves is approximated by simple polynomials:

$$\begin{Bmatrix} u \\ v \\ w \end{Bmatrix} = \sum_{p=p_i}^{p_f} [\mathbf{N}_\eta][\Xi_p]\{\mathbf{d}_p\} \tag{5}$$

The local helicoidal coordinates of the strip (ξ and η) are related to the coordinates of the strip origin (η_0 shown in Fig. 1) and the global coordinates (ξ^* and η^*) as follows:

$$\xi^* = \xi, \quad \eta^* = \eta + \eta_0 \tag{6}$$

$[\mathbf{N}_\eta]$ contains the the cubic shape functions that are used in the η direction:

$$[\mathbf{N}_\eta] = \begin{bmatrix} N_1 & N_2 & 0 & 0 & 0 & 0 & N_3 & N_4 & 0 & 0 & 0 & 0 \\ 0 & 0 & N_1 & N_2 & 0 & 0 & 0 & 0 & N_3 & N_4 & 0 & 0 \\ 0 & 0 & 0 & 0 & N_1 & N_2 & 0 & 0 & 0 & 0 & N_3 & N_4 \end{bmatrix} \tag{7}$$

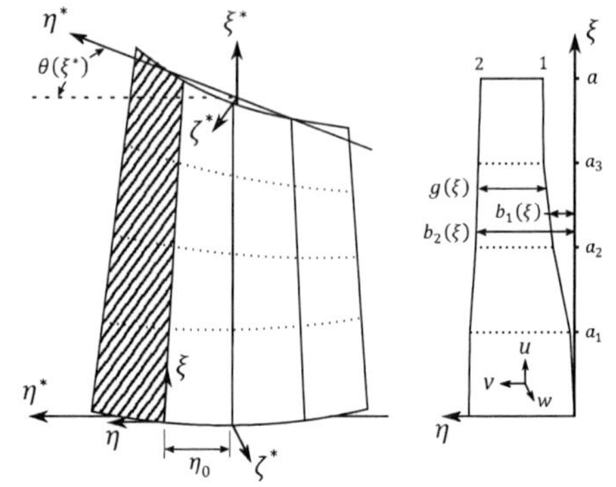

Figure 1: Shell discretization using helicoidal strip elements

$$N_1 = \frac{2\eta^3}{g^3} - \frac{3\eta^2}{g^2} + 1, \quad N_2 = \eta\left(\frac{\eta^2}{g^2} - \frac{2\eta}{g} + 1\right)$$
$$N_3 = \frac{3\eta^2}{g^2} - \frac{2\eta^3}{g^3}, \quad N_4 = \eta\left(\frac{\eta^2}{g^2} - \frac{\eta}{g}\right) \tag{8}$$

and $[\Xi_p]$ is a diagonal matrix containing the pth member of the set of basis functions approximating u, v,and w:

$$[\Xi_p] = diag(\Xi_{p,u}, \Xi_{p,v}, \Xi_{p,w}, \Xi_{p,u}, \Xi_{p,v}, \Xi_{p,w}) \tag{9}$$

The degrees of freedom corresponding to the pth segment of the the strip are denoted by $\{\mathbf{d}_p\}$:

$$\{\mathbf{d}_p\}^T = \left(u_{1p}, \frac{\partial u_{1p}}{\partial \eta}, v_{1p}, \frac{\partial v_{1p}}{\partial \eta}, w_{1p}, \frac{\partial w_{1p}}{\partial \eta}, u_{2p}, \frac{\partial u_{2p}}{\partial \eta}, v_{2p}, \frac{\partial v_{2p}}{\partial \eta}, w_{2p}, \frac{\partial w_{2p}}{\partial \eta}\right) \tag{10}$$

The conventional choice for these functions is a series of continuously differentiable trigonometric functions that satisfy the boundary conditions at the two ends of the strip, but more complex geometries and boundary conditions can be modelled by using more localized functions such as splines.

Spline basis functions

The spline function chosen for this analysis is the cubic B_3-spline of equal section length h, which requires each strip to be divisible into m equal sections. Cubic splines are used to approximate all three deflections and thus all of the components in Eq. 9 are equal:

$$\Xi_{p,u} = \Xi_{p,v} = \Xi_{p,w} = \Xi_p = \frac{1}{6h^3}\begin{cases} (\xi - \xi_{p-2})^3, & \xi_{p-2} \leqslant \xi \leqslant \xi_{p-1} \\ h^3 + 3h^2(\xi - \xi_{p-1}) + 3h(\xi - \xi_{p-1})^2 - 3(\xi - \xi_{p-1})^3, & \xi_{p-1} \leqslant \xi \leqslant \xi_p \\ 4h^3 - 6h(\xi - \xi_p)^2 + 3(\xi - \xi_p)^3, & \xi_p \leqslant \xi \leqslant \xi_{p+1} \\ h^3 - 3h^2(\xi - \xi_{p+1}) + 3h(\xi - \xi_{p+1})^2 - (\xi - \xi_{p+1})^3, & \xi_{p+1} \leqslant \xi \leqslant \xi_{p+2} \\ 0, & \text{otherwise} \end{cases} \tag{11}$$

The boundary conditions at the two ends of each strip are applied by altering or removing local splines near these boundaries. This requires the addition of auxiliary spline segments at the boundaries and thus the p index in Eq. 5 starts from $p_i = -1$ and ends at $p_f = m+1$. The modifications necessary for the relevant boundary conditions can be found in [13]. Note that because the derivative of the longitudinal displacement at a clamped edge is not zero, the u displacement at a clamped edge is best represented by simply supported boundary conditions.

Trigonometric basis functions

Trigonometric basis functions in the spanwise direction are not localized, but make up a continuously differentiable series of m trigonometric functions that satisfy the boundary conditions. Clamped-free boundary conditions are used for v and w displacements, while the u displacements at the boundaries are better described by simply supported-free conditions. As a result, the form of the basis functions used to approximate w and v are different than the ones used to approximate u:

$$\Xi_{p,u} = \frac{p-\frac{1}{2}}{p}\sin\left(\frac{\pi\xi}{a}\left(p-\frac{1}{2}\right)\right) \tag{12}$$

$$\Xi_{p,v} = \Xi_{p,w} = 1-\cos\left(\frac{\pi\xi}{a}\left(p-\frac{1}{2}\right)\right) \tag{13}$$

Mass and stiffness matrices

The mass and stiffness matrices of the strips are obtained by substituting the displacement approximations from Eq. 5 into the energy equations in Eqs. 3 and 4, and rearranging into a matrix form. The procedure of deriving the strip mass and stiffness matrices and then assembling them to obtain the mass and stiffness matrices of the blade are similar to general FEM, and the details can be found in [13]

CASE STUDIES

Three case studies are conducted to compare the efficiency of FSM against FEM. The case studies include a basic cylindrical fan blade, a complex Rolls Royce compressor blade, and a modified version of the first case involving thickness discontinuity and progressive material removal.

I. Uniform cylindrical blade

The first case to be undertaken is a steel ($E = 30 \times 10^6$ psi, $\rho = 0.284$ lb/in^3, $\nu = 0.3$) cantilever cylindrical blade with uniform thickness and a square planform. The radius of the cylinder is $R_{\eta\eta} = 24.0$ inches, its thickness is 0.12 inches, and its overhang length and width are both 12 inches. The first five natural frequencies of the blade were computed using a 4x4 FEM grid as presented in [11], using FSM with 4 strips and 4 spline segments, and using FSM with 4 strips and a set of 4 sinusoidal basis functions. These models had 205, 155, and 120 degrees of freedom (DOF), respectively. The results are summarized in Table 1.

Table 1: Natural frequencies of the cylindrical blade [Hz]

Mode	FEM	Splines FSM	Sinusoidal FSM
1	86.6	86.2	86.4
2	138.9	138.7	139.2
3	250.1	249.4	251.1
4	351.6	344.8	348.3
5	398.2	389.6	392.8

The natural frequencies obtained from both sinusoidal and spline FSM are fairly close to those obtained from FEM, while sinusoidal FSM is substantially more efficient than the other two methods because it requires fewer DOF. Fig. 2 shows the nodal lines of the mode shapes that are obtained using FEM and the presented sinusoidal and spline FSM.

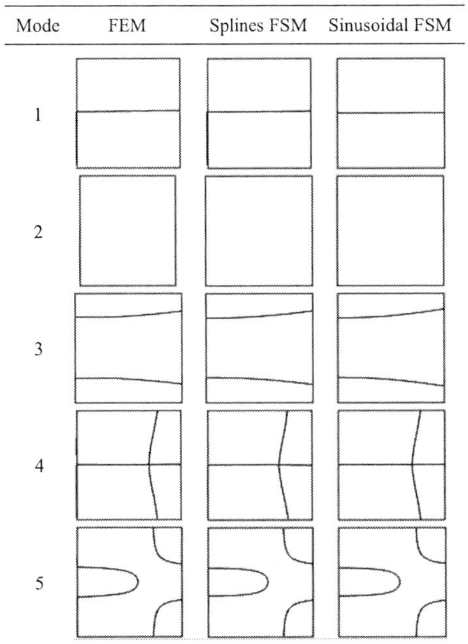

Figure 2: mode shapes of the uniform cylindrical shell

II. Fan blade

The second case to be studied is a more complex compressor blade with variable ζ^* offset from the baseplane, variable thickness, tapering width, and a non-zero rate of twist. The ζ^* offset is a global function of the helicoidal base plane coordinates, expressed in inches as

$$\zeta^*(\xi,\eta) = -\eta^{*2}\left(0.122 - 0.0172\xi^* + 0.00088\xi^{*2}\right) \tag{14}$$

The thickness is assumed to vary linearly between 9 known points arranged in a 3×3 grid, and the width of the blade varies linearly from 3.16 inches at $\xi^* = 7$ in to 2.70 inches at the blade's tip ($\xi^* = 10.66$ in). The Young's modulus, density, and Poisson's ratio of the blade are $E = 10.5 \times 10^6$ psi, $\rho = 0.0985$ lb/in^3, and $\nu = 0.25$ respectively.

FEM was applied to this blade using a 2x2 grid with 63 DOF, while 2-strip FSM was applied with 2 and 4 sinusoidal functions, resulting in 36 and 72 DOF respectively. The natural frequencies resulting from various solutions are shown in Table 2. The natural frequencies of a similar blade were measured experimentally by Walker et al. in [11], and the resulting frequencies are also shown in Table 2. As shown in this table, using only two sinusoidal functions to approximate the nodal lines is clearly too restrictive to achieve accurate results, while increasing up to 4 yields values much closer to the values obtained from FEM and the experimental measurements.

Table 2: Natural frequencies of the fan blade [Hz]

Mode	2x2 FEM	FSM (m=2)	FSM (m=4)	Exp.
1	135.1	292.3	145.9	128
2	568.6	819.6	499.6	426
3	827.3	1105.3	755.6	714
4	1198.5	1961.7	1173.0	1108
5	1680.0	2371.7	1418.4	1655

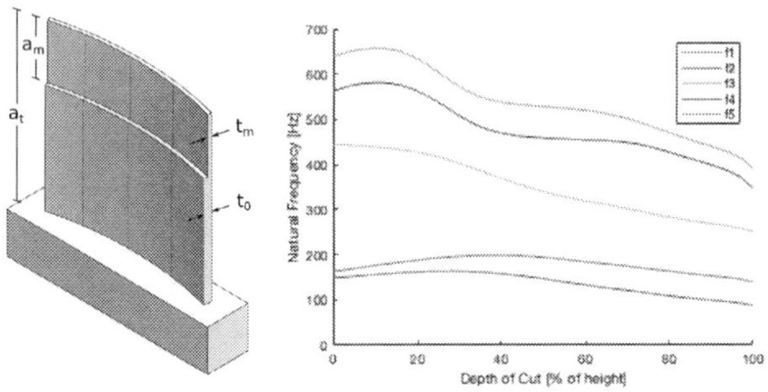

Figure 3: Dimensions of step blade and variation of natural frequencies during material removal

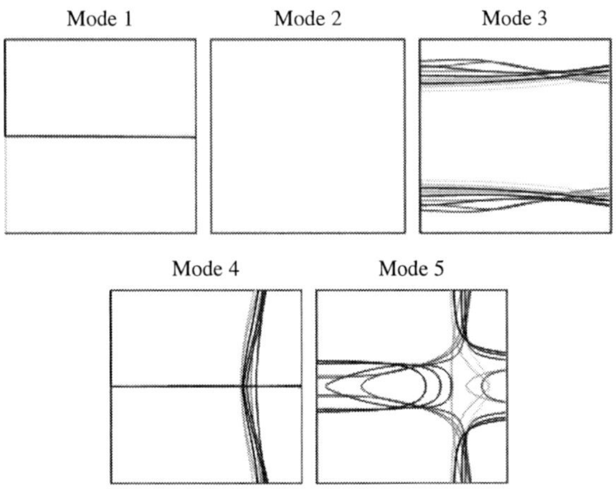

Figure 4: Variation of the modes shapes of the cylindrical shell as its thickness reduces. Initial mode shape is shown in light grey and darker lines are used as more material is removed

III. Step shell

In the third case study, the variations of the natural modes of the cylindrical blade from Case I are investigated as material is removed during the finishing machining pass. This finishing pass is typically done in a series of sweeping horizontal cuts, thus the workpiece can be approximated as a shell with a thickness discontinuity at the horizontal edge between the roughed and finished sections of the blade, as shown in Fig. 3. The uniform cylindrical blade from Case I is revisited in this section, but the blade is divided into two sections in the spanwise direction. The section adjacent to the clamped edge is the thicker, unfinished section of thickness t_0, while the section adjacent the free edge is of finished thickness t_m. For this current analysis, it is assumed that the workpiece begins as double the intended thickness ($t_0 = 0.24$ in and $t_1 = 0.12$ in), and the material is removed around the perimeter in small incremental steps beginning from the top. The resulting changes in the first five natural frequencies are shown in Fig. 3, while Fig. 4 shows the progression of the mode shapes during the machining process. The initial patterns are drawn in light grey, with the subsequent shapes getting darker as more material is removed.

CONCLUSIONS

The Finite Strip Method (FSM) was shown to be an efficient and versatile method for modelling the dynamics of thin, shallow, shell-like structures such as blades. The accuracy of FSM was verified using experimental data and compared with FEM to show its greater computational efficiency. FSM was also easily modified to accommodate discontinuous thickness at an arbitrary location in the longitudinal direction. By being able to change the location of this discontinuous thickness without re-meshing, FSM was used to model the changing natural frequencies of a basic cylindrical fan blade during the machining process. The ability to model these natural frequencies is useful for manufacturing, as it allows the manufacturer to identify and avoid cutting settings that may cause unstable vibrations (i.e. chatter). The analysis could also be expanded to predict dynamic deflections and chatter stability, which can be used in further process optimization methods.

REFERENCES

[1] Y. Altintas. *Manufacturing Automation: Metal Cutting Mechanics, Machine Tool Vibrations, and CNC Design, 2nd Edition*. Cambridge University Press, 2012.

[2] L. Arnaud, O. Gonzalo, S. Seguy, H. Jauregi, and G. Peigné. Simulation of low rigidity part machining applied to thin-walled structures. *The International Journal of Advanced Manufacturing Technology*, 54(5-8):479–488, 2011.

[3] M. A. Elbestawi and R. Sagherian. Dynamic modeling for the prediction of surface errors in the milling of thin-walled sections. *Journal of materials processing technology*, 25(2):215–228, 1991.

[4] V. Thévenot, L. Arnaud, G. Dessein, and G. Cazenave-Larroche. Integration of dynamic behaviour variations in the stability lobes method: 3d lobes construction and application to thin-walled structure milling. *The International Journal of Advanced Manufacturing Technology*, 27(7-8):638–644, 2006.

[5] S. Ratchev, S. Nikov, and I. Moualek. Material removal simulation of peripheral milling of thin wall low-rigidity structures using FEA. *Advances in Engineering Software*, 35(8):481–491, 2004.

[6] E. Budak, L. T. Tunç, S. Alan, and H. N. Özgüven. Prediction of workpiece dynamics and its effects on chatter stability in milling. *CIRP Annals-Manufacturing Technology*, 61(1):339–342, 2012.

[7] P. Kersting and D. Biermann. Modeling techniques for simulating workpiece deflections in nc milling. *CIRP Journal of Manufacturing Science and Technology*, 7(1):48–54, 2014.

[8] M. Meshreki, H. Attia, and J. Kövecses. Development of a new model for the varying dynamics of flexible pocket-structures during machining. *Journal of Manufacturing Science and Engineering*, 133(4):041002, 2011.

[9] K. Ahmadi. Finite strip modeling of the varying dynamics of thin-walled pocket structures during machining. *The International Journal of Advanced Manufacturing Technology*, pages 1–9, 2016.

[10] S. S. Rao. *Vibration of continuous systems*. John Wiley and Sons, 2007.

[11] K. P. Walker. Vibrations of cambered helicoidal fan blades. *Journal of Sound and Vibration*, 59(1):35–57, 1978.

[12] A. L. Gol'Denveizer. *Theory of Elastic Thin Shells: Solid and Structural Mechanics*, volume 2. Elsevier, 2014.

[13] S. C. Fan and Y. K. Cheung. Analysis of shallow shells by spline finite strip method. *Engineering Structures*, 5(4):255–263, 1983.

Experimental Methods

AXIAL STRAIN VARIATIONS THROUGH CARTILAGE DEPTH

Amin Komeili
Faculty of Kinesiology
The University of Calgary
Calgary, Alberta, Canada
amin.komeili@ucalgary.ca

Ziad Abusara
Faculty of Kinesiology
The University of Calgary
Calgary, Alberta, Canada
zabusara@ucalgary.ca

Salvatore Federico
Department of Mechanical and Manufacturing Engineering
The University of Calgary
Calgary, Alberta, Canada
salvatore.federico@ucalgary.ca

Walter Herzog
Faculty of Kinesiology
The University of Calgary
Calgary, Alberta, Canada
wherzog@ucalgary.ca

ABSTRACT

Quantifying transient and steady state deformations of the extracellular matrix in cartilage is an important step toward understanding the mechanical properties of the environment of chondrocytes. A novel combination of a custom-built loading device, specimen preparation, and microscopy technique enabled us to measure local tissue deformations in the surface, middle, and deep zones of pig cartilage explants subjected to physiological levels of compressive strains. A nominal 15% surface-to-surface transient and equilibrium strain produced up to 40% strains in the superficial zone while strains in the middle and deep zones were smaller than the average nominal strain. These results confirm that the compressive stiffness of the superficial zone is significantly smaller than that of the middle and deep zones. The strain distribution across the cartilage thickness was more uniform in the transient than the steady state phase because the matrix swells due to fluid pressurization.

KEYWORDS: Cartilage, Material property, Depth dependent strains, Compression test

INTRODUCTION

Articular cartilage is the connective tissue covering the ends of articulating bones in diarthrodial joints. The main function of cartilage is to provide frictionless contact and transfer of loads between the articulating surfaces. The non-homogeneous and anisotropic material properties, along with the porosity of the cartilage matrix, give cartilage a unique and complex structure to withstand forces. Under normal loading conditions, weight-bearing joints experience forces as high as ten times body weight (Torzilli et al. 1999). The preservation of cartilage is highly dependent on maintaining its homeostasis, a task that is allocated to chondrocytes as the only living cells in the cartilage. Chondrocyte synthesis is regulated by a variety of factors that are affected by the chemical and mechanical environment of the cells. Measuring the stresses and strains in the extracellular matrix may help in understanding the mechanical properties of chondrocytes and their environment.

The deformation behavior of the extracellular matrix has been studied globally, i.e., the deformation of the matrix at different cartilage depth has been expressed by the surface-to-surface strain of the cartilage (Mow, Holmes, and Lai 1984; Holmes and Mow 1990; Herberhold et al. 1999; Athanasiou et al. 1991; Froimson et al. 1997; Suh and Spilker 1994).

These experiments have been a valuable source for theoretical simulations and provided insight into the overall function of articular cartilage. However, there exist considerable differences in the local properties and local behavior of cartilage across its depth. Chondrocyte deformations have been measured in many different preparations, including: isolated chondrocytes (Nguyen et al. 2010), chondrocytes embedded in agarose gel constructs (Freeman et al. 1994; Knight et al. 1998; Lee et al. 2000), chondrocytes in tissue explants (Guilak, et al. 1995) and in situ (Madden et al. 2013). Using second harmonic generation microscopy, we can image cartilage with a high temporal and spatial resolution. A high temporal frequency of imaging minimizes artifacts in real-time, 3D imaging of chondrocytes under dynamic conditions. However, imaging of real-time chondrocyte deformations under dynamic loading conditions has not been made, although there is good evidence that the dynamic deformations of cells are much more important in regulating cartilage homeostasis than the corresponding steady-state chondrocyte deformations that are typically shown in the literature.

Furthermore, most studies on chondrocytes and matrix deformations have focused on the superficial zone mechanics. The logic of considering superficial zone could be its crucial role in protecting cartilage from contact wear and shear deformations. However, the superficial zone constitutes only 5 to 10 percent of the total cartilage depth in large animals including humans. Because of the highly non-homogeneous mechanical and chemical properties of cartilage, the mechano-biology of the superficial zone does not represent well the local mechanical and biological properties in other zones of the cartilage and does not represent well the entire cartilage properties. Determining the mechanical behavior of the deep zones in cartilage remains a challenge. The objective of this study was to measure the depth-dependent axial compressive strains of cartilage subjected to dynamic loads.

METHODS

Cylindrical core plugs with a diameter of 6 mm (N=6) were extracted from the femoral condyle of pig knee joints using a dental drill. To expose the entire depth of the cartilage to the microscope objective, the samples were cut along a plane passing through the axis of the explant sample, which resulted in two semi-cylindrical samples. One of the semi-cylindrical sample was used for mechanical testing, and the other was preserved for histology. The samples used for mechanical testing were fixed in a custom designed compression testing device (Fig1). The compression device was actuated with a piezo-electric motor, together with a flat impermeable compression plate, to apply displacements. A 0.17mm thick cover slip was used to cover the flat surface of the sample, in order to minimize lateral bulging of the sample. Fluorescent microbeads (FluoSpheres Polystyrene Microspheres, 1.0 µm, orange fluorescent (540/560), Thermo Fisher Scientific, USA) were placed at the edge of the compression plate as viewed under a dissection microscope (Zeiss SteREO Discovery V8, Carl Zeiss Inc., Germany). The cartilage was then secured in the loading device beneath a multi-photon microscope (Olympus FVMPE-RS, Olympus Corp., Japan) equipped with a 25x/1.05 NA water-immersion objective (XLPLN25XWMP2, Olympus Corp., Japan). Two filters (FF01 400/40 and FF01 520/60, Semrock Inc., USA) permitted the collection of SHG and TPF signals upon excitation by two multi-photon lasers tuned to 800 nm and 940 nm respectively. Prior to compression, selected areas of the cartilage were burned with a laser for tracking during compression and the calculation of local strains across the cartilage depth (Fig.1). A 15% linear ramp compression was applied over 50 seconds, while the flat surface of the sample was imaged using confocal microscopy. Imaging started with the onset of loading and continued for 10 minutes after reaching a nominal tissue strain of 15%. Strains at two time points were selected for presentation here: (i) a transient time, immediately after the 15% strain was reached, and (ii) a steady-state time, when 15% strain had been maintained for 10 minutes. Transient and steady-state strains were calculated from the laser burn markers. Axial strains and depth were normalized with respect to the surface-to-surface strain and the cartilage depth, respectively. The surface and tidemark of the samples were represented with $\xi=0$ and $\xi=1$, respectively. Normalized strains (ε) were plotted against normalized depth (ξ) at steady and transient times.

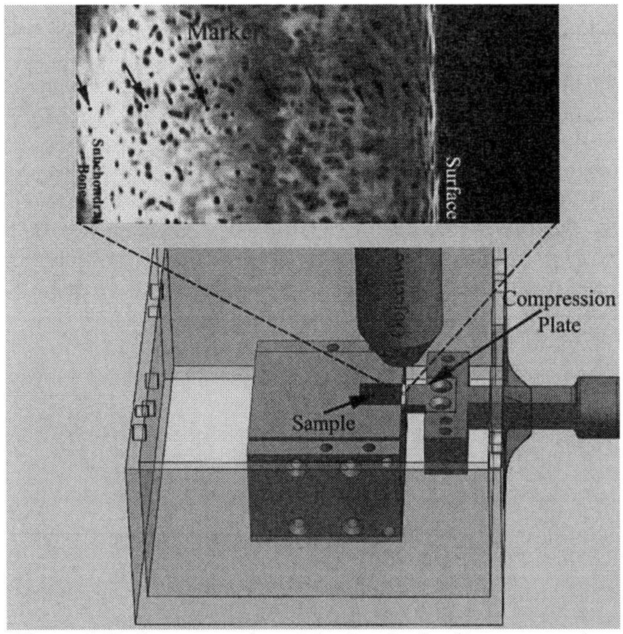

Figure 1-Bottom: experimental setup for in situ compression testing. Top: burned markers along the depth of the cartilage.

RESULTS

Figure 2 shows the normalized strain variations through the normalized depth of the cartilage for one of the samples. The axial strains in the superficial region with $\xi<0.1$ were greater than the surface-to-surface strains for the transient and steady-state conditions in all samples. On average, compressive strains in the superficial zone were 2.8 times greater than the surface-to-surface strains. The normalized axial strains decreased continuously along the depth of the cartilage, so that for $\xi>0.2$ normalized axial strains were smaller than the total normalized strains ($\varepsilon=1.0$) in all samples. The steady state strains in the superficial and middle zones with $\xi<0.2$ were greater than the corresponding strains for the transient states. However, no significant differences were observed for strains in the deep cartilage regions ($\xi>0.2$) for the transient and the steady state conditions.

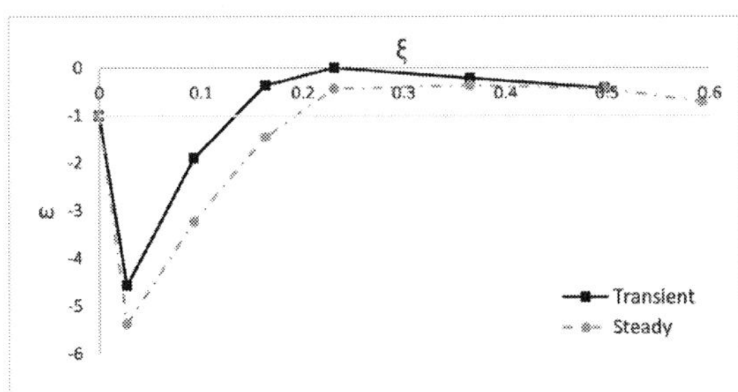

Figure 2- Variation of matrix normalized strains through the depth of cartilage for the transient and steady state conditions.

DISCUSSION

The findings of this study are in general agreement with the results of previous investigations in which greater tissue strains were measured in the superficial zone compared to other regions at steady state (Guilak et al. 1995; Madden et al. 2013; Gore et al. Minns 1983; Lyyra et al. 1999; Schinagl et al. 1996; O'Connor et al. 1988). Guilak et al. (1995) measured normalized tissue strains of 1.2±0.3 in the superficial zone and 1.0±0.2 in the middle and deep zones of the patellofemoral groove of skeletally mature mongrel dogs at steady state. Madden et al. (2013) reported 1.6±0.15 normalized strains in the superficial zone of patellae of New Zealand white rabbits at steady-state condition. Schinagl et al. (1996)) reported normalized strain variations from 2.43±0.11 in the superficial zone to 0.4±0.06 in the deep zone of cartilage samples. Strains in our study were less homogeneous across the cartilage depth than reported by others. Normalized steady-state strains measured in our study varied from 2.8±1.2 in the superficial zone to 0.7±0.3 in the deep zone. Despite the agreement of these studies on the depth-dependent distribution of strains, there are several reasons for the discrepancy of our results and other published works. First, the experimented samples were harvested from different animals with different cartilage structure and composition, and thus, likely different mechanical properties. Moreover, the experimental technique in our study was different than those used by Guilak etl al. (1995), Madden et al. (2013), and Schinagl et al. (1996). They calculated strains through the change in edge-to-edge distance of nearby cells, while we measured the change in distance between tissue material points. Because of the difference in mechanical properties between chondrocytes and the extracellular matrix, deformation of the chondrocytes may have affected the strain measurements made by by Guilak etl al. (1995), Madden et al. (2013), and Schinagl et al. (1996). Furthermore, chondrocytes are not perfectly bonded to the extracellular matrix and sliding of chondrocytes relative to the matrix would affect the tissue strain measurements.

The increase in compressive strains in the superficial zone from the transient to the steady-state conditions could be caused by a reduction in the fluid pressure. In the transient state, the low permeability of the superficial zone cartilage may cause a high matrix pressurization which may partly counteract the compressive deformation of the matrix swell. After 10 minutes of stress relaxation, the viscoelastic behaviour of matrix is known to be reduced (Abusara et al. 2011), and the overall mechanical stiffness of the material is primarily defined by the elastic properties of the matrix. Due to the lower proteoglycan content in the superficial zone compared with other regions in cartilage, compressive strains in the superficial zone increased in the relaxation phase. The reduction in compressive strain for transient conditions may have important functional implications as articular cartilage in a joint is mostly loaded dynamically. Therefore, under dynamic loading conditions, a more uniform strain distribution through the cartilage matrix protects cells and proteoglycan bonds from excessive local deformations.

CONCLUSIONS

Local axial strains of articular cartilage in transient and steady state conditions were measured using a nominal 15% linear ramp compression. Superficial zone cartilage deformed three times more than other regions in pig femoral condyle cartilage. This result implies that the compressive modulus of the surface zone was approximately one third of that of the middle and deep zones. The excessive deformations in the superficial zone could be harmful to chondrocyte and tissue integrity, and may result in degeneration of the cartilage and lead to the onset of osteoarthritis. The low permeability of the cartilage in the superficial zone increases the internal fluid pressure, which helps decrease compressive tissue strains for transient loading conditions. The strain in deep cartilage regions ($\xi>0.2$) was smaller than the surface-to-surface strain, which is a result of the low permeability and the high concentration of proteoglycans in the matrix.

ACKNOWLEDGMENTS

Financial support of this project was provided by the University of Calgary Eyes High Postdoctoral Scholarship, The Canadian Institutes of Health Research through a Foundation Scheme Grant, The Canada Research Chair Program, and the Killam Memorial Chair for Interdisciplinary Research at the University of Calgary.

REFERENCES

Abusara, Z., R. Seerattan, a. Leumann, R. Thompson, and W. Herzog. 2011. "A Novel Method for Determining Articular Cartilage Chondrocyte Mechanics in Vivo." *Journal of Biomechanics* 44 (5). Elsevier: 930–34. doi:10.1016/j.jbiomech.2010.11.031.

Athanasiou, K. A., M. P. Rosenwasser, J. A. Buckwalter, T. I. Malinin, and V. C. Mow. 1991. "Interspecies Comparisons of in Situ Intrinsic Mechanical Properties of Distal Femoral Cartilage." *Journal of Orthopaedic Research* 9 (3): 330–40. doi:10.1002/jor.1100090304.

Freeman, P. M., R. N. Natarajan, J. H. Kimura, and T. P. Andriacchi. 1994. "Chondrocyte Cells Respond Mechanically to Compressive Loads." *Journal of Orthopaedic Research* 12 (3): 311–20. doi:10.1002/jor.1100120303.

Froimson, M I, A Ratcliffe, T R Gardner, and V C Mow. 1997. "Differences in Patellofemoral Joint Cartilage Material Properties and Their Significance to the Etiology of Cartilage Surface Fibrillation." *Osteoarthritis and Cartilage* 5 (6): 377–86. http://www.ncbi.nlm.nih.gov/pubmed/9536286.

Gore, D M, G R Higginson, and R J Minns. 1983. "Compliance of Articular Cartilage and Its Variation through the Thickness." *Physics in Medicine and*

Biology 28 (3): 233–47. http://iopscience.iop.org/article/10.1088/0031-9155/28/3/004/meta;jsessionid=D991E8BF1B1B6141E7DECBBE6706D71A.ip-10-40-2-119.

Guilak, Farshid, Anthony Ratcliffe, and Van C. Mow. 1995. "Chondrocyte Deformation and Local Tissue Strain in Articular Cartilage: A Confocal Microscopy Study." *Journal of Orthopaedic Research* 13 (3): 410–21. doi:10.1002/jor.1100130315.

Herberhold, C, S Faber, T Stammberger, M Steinlechner, R Putz, K H Englmeier, M Reiser, and F Eckstein. 1999. "In Situ Measurement of Articular Cartilage Deformation in Intact Femoropatellar Joints under Static Loading." *Journal of Biomechanics* 32 (12): 1287–95. http://www.ncbi.nlm.nih.gov/pubmed/10569707.

Holmes, M H, and V C Mow. 1990. "The Nonlinear Characteristics of Soft Gels and Hydrated Connective Tissues in Ultrafiltration." *Journal of Biomechanics* 23 (11): 1145–56. http://www.ncbi.nlm.nih.gov/pubmed/2277049.

Knight, M. M., S. a. Ghori, D. a. Lee, and D. L. Bader. 1998. "Measurement of the Deformation of Isolated Chondrocytes in Agarose Subjected to Cyclic Compression." *Medical Engineering and Physics* 20 (9): 684–88. doi:10.1016/S1350-4533(98)00080-0.

Lee, D A, M M Knight, J F Bolton, B D Idowu, M V Kayser, and D L Bader. 2000. "Chondrocyte Deformation within Compressed Agarose Constructs at the Cellular and Sub-Cellular Levels." *Journal of Biomechanics* 33 (1): 81–95. http://www.ncbi.nlm.nih.gov/pubmed/10609521.

Lyyra, T, J P A Arokoskik, N Oksala, A Vihko, M Hyttinen, and J S Jurvelin{ and I Kiviranta†. 1999. "Experimental Validation of Arthroscopic Cartilage Stiffness Measurement Using Enzymatically Degraded Cartilage Samples." *Physics in Medicine and Biology* 44 (2): 525–35. doi:PII: S0031-9155(99)96397-2.

Madden, Ryan, Sang-Kuy Han, and Walter Herzog. 2013. "Chondrocyte Deformation under Extreme Tissue Strain in Two Regions of the Rabbit Knee Joint." *Journal of Biomechanics* 46 (3): 554–60. doi:10.1016/j.jbiomech.2012.09.021.

Mow, V C, M H Holmes, and W M Lai. 1984. "Fluid Transport and Mechanical Properties of Articular Cartilage: A Review." *Journal of Biomechanics* 17 (5): 377–94. http://www.ncbi.nlm.nih.gov/pubmed/6376512.

Nguyen, Bac V, Qi Guang Wang, Nicola J Kuiper, Alicia J El Haj, Colin R Thomas, and Zhibing Zhang. 2010. "Biomechanical Properties of Single Chondrocytes and Chondrons Determined by Micromanipulation and Finite-Element Modelling." *Journal of the Royal Society, Interface* 7 (53). The Royal Society: 1723–33. doi:10.1098/rsif.2010.0207.

O'Connor, P, C R Orford, and D L Gardner. 1988. "Differential Response to Compressive Loads of Zones of Canine Hyaline Articular Cartilage: Micromechanical, Light and Electron Microscopic Studies." *Annals of the Rheumatic Diseases* 47 (5): 414–20. doi:10.1136/ard.47.5.414.

Schinagl, Robert M, Michael K Ting, Jeffrey H Price, and Robert L Sah. 1996. "Video Microscopy to Quantitate the Inhomogeneous Equilibrium Strain Within Articular Cartilage During Confined Compression." *Annals of Biomedical Engineering* 24: 500–512.

Suh, J K, and R L Spilker. 1994. "Indentation Analysis of Biphasic Articular Cartilage: Nonlinear Phenomena under Finite Deformation." *Journal of Biomechanical Engineering* 116 (1): 1–9. http://www.ncbi.nlm.nih.gov/pubmed/8189703.

Torzilli, P. A., R. Grigiene, J. Borrelli, and D. L. Helfet. 1999. "Effect of Impact Load on Articular Cartilage: Cell Metabolism and Viability, and Matrix Water Content." *Journal of Biomechanical Engineering* 121 (5). American Society of Mechanical Engineers: 433. doi:10.1115/1.2835070.

Proceedings of the 26th CANCAM
Victoria, BC, Canada, May 29 – June 1, 2017

Experimental Investigation of Droplet Oscillation and Impact on Hydrophilic and Hydrophobic Surfaces with Varying Wettability

Yuntao Pan, Kewei Shi, Xili Duan, Greg Naterer
Faculty of Engineering and Applied Science
Memorial University of Newfoundland
St. John's, NL, Canada
yp8367@mun.ca

ABSTRACT

A liquid repellent surface can potentially be used to reduce ice accretion on surface and improve asset integrity and safety in harsh environment. Icing on a hydrophobic surface is affected by the oscillation of water droplets impacting on the surface. This paper uses a statistical design of experiments method to investigate the effects of surface temperature (A), impact speed (B), droplet temperature (C), and surface wettability (D) on the oscillation time of a droplet. The Design Expert software was used to analyze 38 data points. It was found that the impact speed and surface wettability are significant factors. The interaction effect between surface temperature and droplet temperature, and that between surface temperature and surface wettability are also significant. The oscillation time increased significantly on a hydrophobic surface. Visualization with a high-speed camera showed that the liquid droplet started to freeze at the moment when the oscillation fully stopped.

KEYWORDS: hydrophobic, anti-icing, wetting, freezing, design of experiment

INTRODUCTION

In arctic and harsh environments, icing accumulation on power transmission lines, wind turbines, and offshore structures causes significant hazards. Extensive research has been directed to develop hydrophobic metallic surfaces with water repellent and anti-icing properties for applications in these harsh environments. Two main approaches are investigated to make a hydrophobic surface, i.e., applying coatings with low surface energy and machining micro or nano-structures to reduce liquid-solid contact area. Recently our research group has developed hydrophobic steel surfaces with micron-scale and sub-micron structures from laser fabrication [1]. The surfaces achieved static contact angle higher than 120°. Surface roughness plays an important role in making hydrophobic and super-hydrophobic surfaces, and many designs have been investigated [2]. Vaibhav et al. [3] reported a droplet impact dynamics model for spreading and retraction dynamics of a droplet hitting a superhydrophobic surface. Another group did extensive experiments about icing dynamics and delay time on different hydrophobic surfaces [4]. They predicted ice formation on superhydrophobic surfaces based on droplet impact dynamics, heat transfer, and ice nucleation [5].

The oscillation time of a droplet is the total time for the droplet contacting a surface until its spreading and retraction fully stops. This research examines the relationship between oscillation time and icing delay time. For a better hydrophobic surface, the elastic potential energy of the droplet is higher, so the droplet may bounce off the surface after retraction and a longer oscillation time is expected. The objective of this paper is to find the key factors and their effects on droplet oscillation time while impacting a cooled hydrophobic surface. Moreover, it presents a baseline test for future research on the effects of salinity and other factors on droplet dynamics and icing.

In this paper, the design of experiments (DOE) technique and the Design Expert software [7] (Version 8) are used to complete experiment design and data analysis. A two-level four-factor face-centered composite design (2^4) from the response surface methodology was used to exam the effects of four factors: surface temperature (A), impact speed (B), droplet temperature (C), and surface wettability (D) on the response of the droplet oscillation time.

EXPERIMENTAL APPARATUS AND PROCEDURE

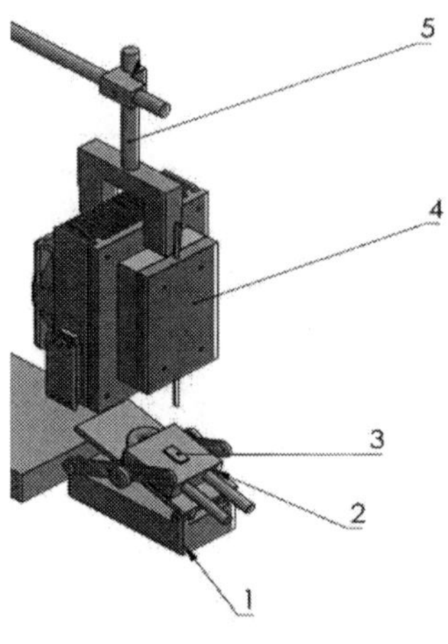

Figure 1: Liquid droplet impact and icing apparatus

Figure 1 shows a schematic of the experimental setup. It consists of a stand with a tilt base (1) of adjustable inclination angles, and only horizontal surface is tested for all the experiments in this paper. A cold plate (2) that is mounted on the tilt base and controls the temperature of sample materials (3) surfaces, and a thermal electric cooler (4) where a small copper tube is used for generating liquid droplets with controllable temperatures. To generate small water droplets, a needle of 0.25 mm ID was used at the bottom end of the drop generator tubing. The liquid comes from a syringe pump which is not shown on this diagram, and the cold plate is connected to a thermal bath that provides control of surface temperature of the material sample. High speed imaging of droplet dynamics and icing phenomena was conducted with a high-speed camera (Phantom V611). The measurement was taken carefully by reliable devices each run to ensure the accuracy of each factor and response. The sample material used in this study is a17-4 PH stainless steel. The hydrophobic surface was laser machined on the stainless steel sample with microscale pillars (detailed in Ref. [1]) resulting in a static contact angle of 120°. A non-machined hydrophilic surface was also tested (with a contact angle of 67°) for comparison purpose.

DESIGN OF EXPRIMENT

There are many methodologies of experimentation. The traditional one-factor-at-a-time (OFAT) method has very limited ability when there are many factors for the response parameter. Using an OFAT method in this case requires large number of experiments and it's hard to find interactions between the factors [8]. The statistical design of experiments (DOE) methods provide better efficiency and capabilities. The analysis of variance (ANOVA) method was firstly used in experimental design by Fisher in 1918 [9]. The response surface methodology (RSM) was first investigated in 1951 and is now widely used by experimentalists and designers [10-11].

This study uses a face-centered composite design (FCCD) method, one type of the RSM, to design the experiments and analyze the droplet oscillation time. A two-level four-factor FCCD with five center points was chosen and the factors with levels, shown in Table 1, were entered into Design Expert to generate the run sheet.

Table 1: Summary of design factors and levels

Factor	Name	Low Actual	High Actual	Low Coded	High Coded
A	Surface Temperature	-7 °C	-1 °C	-1	1
B	Impact Speed	0.63 m/s	1 m/s	-1	1
C	Droplet Temperature	5 °C	15 °C	-1	1
D	Surface Wettability	67°	120°	NS	SS

The speed of the droplet when it first touches the metal surface is called impact speed (Vs). Neglecting the air drag force, the impact speed can be calculated from the distance (H_d) between the droplet generator needle tip and the metal surface. The impact speed is calculated as 0.63 m/s and 1 m/s respectively based on Eq. 1 shown below.

$$Vs = \sqrt{2gH_d} \tag{1}$$

RESULTS AND DISCUSSION

An ANOVA table was generated by Design Expert after completing a total of 38 runs of experiments. The results in Table 2 shows the following significant factors: the impact speed, the surface wettability, the interaction effects between surface temperature and droplet temperature, and the interaction effect between surface temperature and surface wettability, since their p-values are all less than 0.05.

Table 2: ANOVA results

Source	Sum of Squares	df	Mean Square	F Value	P-value
Model	34.72	10	3.47	14.97	<0.0001
A-Surface Temp.	0.031	1	0.031	0.13	0.7191
B-Impact speed	4.06	1	4.06	17.49	0.0003
C-Droplet Temp.	0.019	1	0.019	0.082	0.7768
D-Surface Wett.	26.68	1	26.68	115.01	<0.0001
AB	0.00769	1	0.00769	0.033	0.857
AC	1.53	1	1.53	6.61	0.0160
AD	1.28	1	1.28	5.51	0.0264
BC	0.49	1	0.49	2.11	0.1579
BD	0.054	1	0.054	0.23	0.6320
CD	0.58	1	0.58	2.49	0.1259
Residual	6.26	27	0.23		
Lack of Fit	4.50	19	0.24	1.07	0.4880
Pure Error	1.77	8	0.22		
Cor Total	40.99	37			

A droplet experiences impact, spreading and retraction on the surface. Fig. 2 shows a series of images of a droplet with interesting variations of droplet shapes in its oscillation.

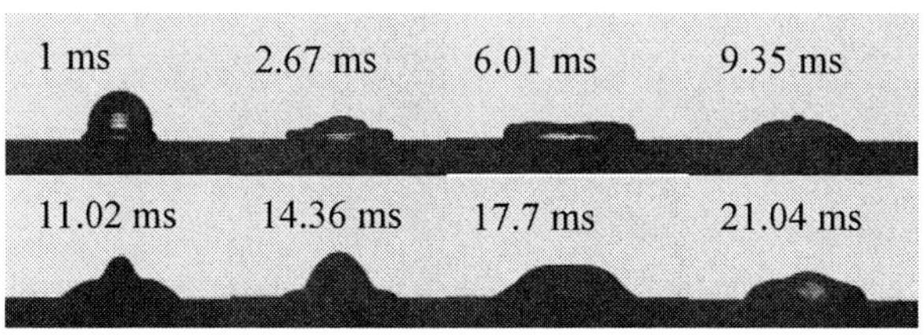

Figure 2: droplet (with impact speed = 0.77 m/s) oscillation on surface of the hydrophilic sample. Surface and droplet temperatures are at room temperature of 18.5 °C.

DOE methods were used to analyze the effects of operational factors. Figure 3 indicates that the oscillation time increases as the droplet temperature increases on a hydrophobic surface. The oscillation time doesn't change much with temperature on the hydrophilic surface. Figure 4 shows that higher surface temperature leads to increase of oscillation time for a hydrophobic surface, and it doesn't affect the oscillation time on hydrophilic surfaces.

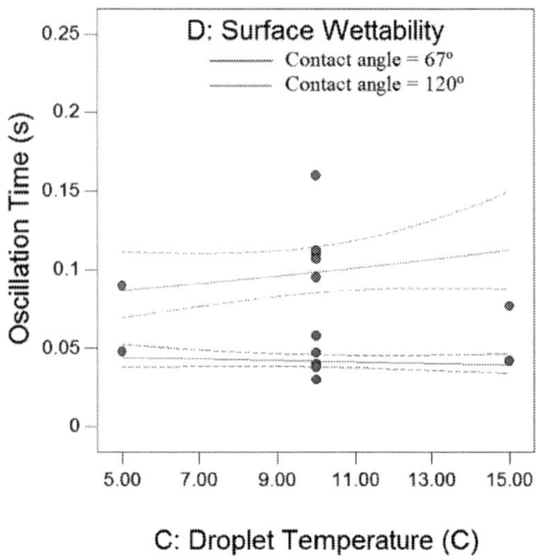

Figure 3: The interaction plot between droplet temperature and surface wettability. The graph is plotted under surface temperature = -10 °C and impact speed = 0.81 m/s.

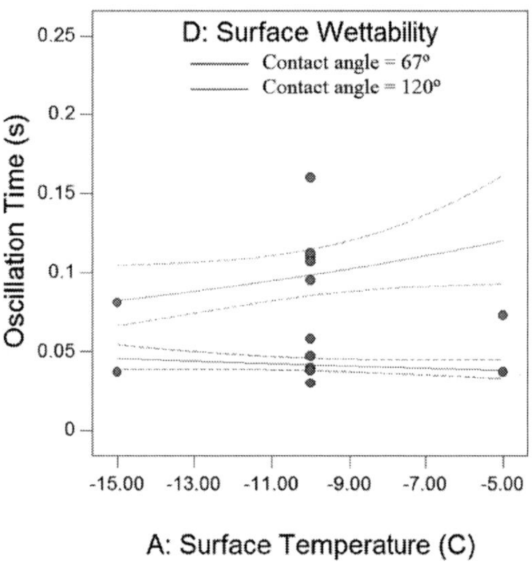

Figure 4: The interaction plot between surface temperature and surface wettability. The graph is plotted under droplet temperature = 10 °C and impact speed = 0.81 m/s.

Figure 5 shows that the oscillation time is longer on a hydrophobic surface than a hydrophilic surface under the same impact speed and temperature. As the impact speed decreases, the oscillation time increases for both hydrophobic and hydrophilic surfaces. This occurs because the higher impact speed leads to a larger spreading area, and results in a faster icing speed. This result is more reasonable after considering the icing with the droplet spreading together. Figure 6 shows images of droplets with different impact speed but at approximately the same time of 1ms after impact. It is clear that a higher impact speed leads to less oscillation time. The reason is that the larger spreading area results from a higher impact speed that creates greater contact area with a solid base, and the freezing time is significantly decreased.

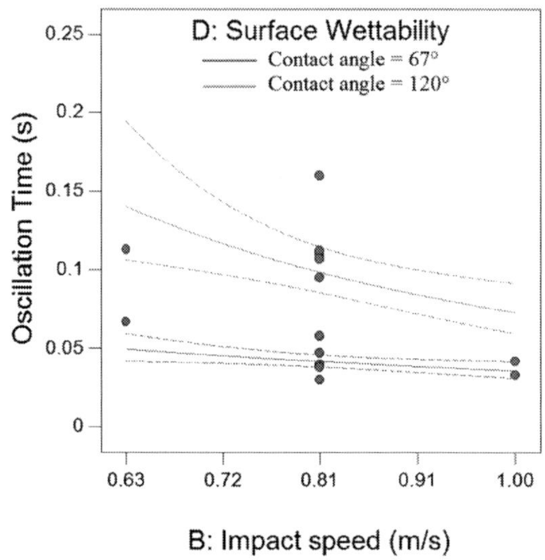

Figure 5: The interaction plot between impact speed and surface wettability. The graph is plotted under surface temperature = -10 °C and droplet temperature = 10 °C.

a) Vs = 0.44 m/s OT = 0.5916 s   b) Vs = 0.77 m/s OT = 0.3240 s   c) Vs = 1 m/s OT = 0.2366 s

Figure 6: droplet impact with the 17-4 PH stainless steel sample at different impact speeds. Surface and droplet temperatures are at a room temperature of 18.5 °C.

Initial nominal diameters of the droplets are in the range of 2 mm to 5 mm. To analyze droplet geometry variations on the surfaces, the lateral and vertical dimensions of a droplet are defined as 2a and b, as shown in Fig. 7. Droplet dynamics can be studied by analyzing these geometrical parameters, as shown in Fig. 8. In the first few mili-seconds (<10 ms), the droplet experienced a fast spreading process, indicated by increasing lateral dimension (a) and decreasing vertical dimension (b). Then there is a retraction process indicated by the shrinking of the lateral dimension and increasing vertical dimension. This oscillation repeats for several cycles but with rapidly dampening amplitude until the droplet becomes steady and freezes in low temperature cases.

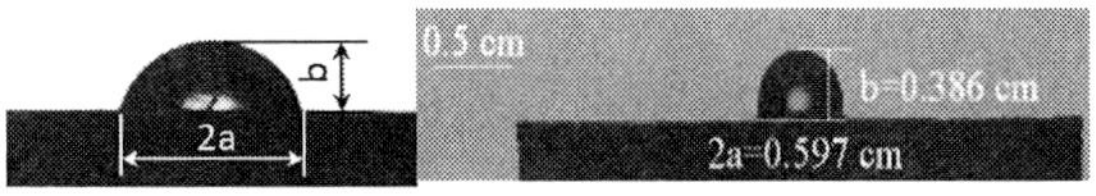

Figure 7: Droplet geometry and imaging scale

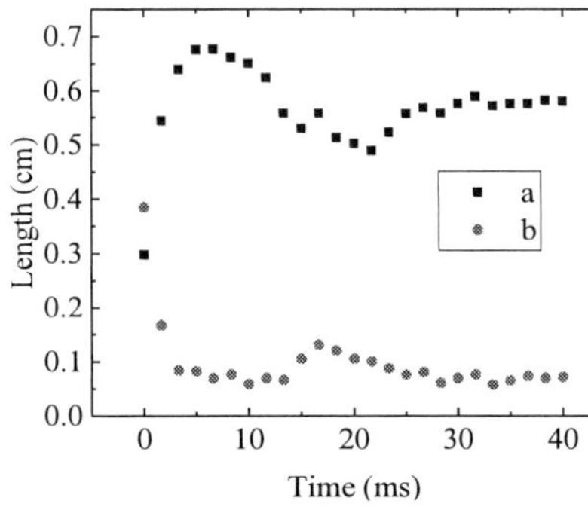

Figure 8: Droplet oscillations on the hydrophilic stainless steel surface, with droplet initial temperature 18.5 °C, impact speed 1 m/s, steel surface temperature -7 °C.

CONCLUSIONS

This study used statistical design of experiments for determining the effect of surface temperature (A), impact speed (B), droplet temperature (C) and surface wettability (D) on the droplet oscillation time. A two-level four-factor face-centered composite design was used to investigate the effect of all four factors. Based on the DOE analysis, the surface wettability (D) is the most important factor on the oscillation time. The droplet oscillates longer on a hydrophobic surface than hydrophilic surface in a cold environment. The sample surface temperature and the droplet temperature have a positive effect on the oscillation time for hydrophobic surfaces, but not for a hydrophilic surface. Moreover, the impact speed has a negative impact on the oscillation time because a larger spreading area under greater height induces a faster solidification process. A longer oscillation time increases icing delay time during a freezing process, and bouncing off from the surface is expected for even more hydrophobic surfaces.

ACKNOWLEDGMENTS

Financial support for this research by Petroleum Research Newfoundland & Labrador (PRNL) and the Natural Sciences and Engineering Research Council of Canada (NSERC) are gratefully appreciated.

REFERENCES

1. Cui, C., Collier, B., Duan, X., Poduska, K., "hydrophobic Steel Surfaces with Micron-scale and Sub-micron Structures from laser Fabrication", 2nd Thermal and Fluid Engineering Conference, April 2-5, 2017, Las Vegas, NV, USA.
2. Won Gyu Bae, Ki Young Song, Yudi Rahmawan, Chong Nam Chu, Dookon Kim, Do Kwan Chung, and Kahp Y. Suh, 2012. "One-Step Process for Superhydrophobic Metallic Surfaces by Wire Electrical Discharge Machining", ACS Applied Materials & Interfaces, Vol(4), pp. 3685-3691.
3. Bahadur, V., Mishchenko, L., Hatton, B., Taylor, J. A., Aizenberg, J., Krupenkin, T. 2011, "Predictive Model for Ice Formation on Superhydrophobic Surfaces", Langmuir Vol(27), pp. 14143-14150.
4. Mishchenko, L., Hatton, B., Bahadur, V., Taylor, A., Krupenkin, T., and Aizenbery, J., 2010. "Design of Ice-free Nanostructured Surfaces Based on Repulsion of Impacting Water Droplets", ACS Nano 4(12), pp. 7699-7707.
5. Bahadur, V., Mishchenko, L., Hatton, B., Taylor, A. J., Aizenberg, J., and Krupenkin, T., 2011. "Predictive Model for Ice Formation on Superhydrophobic Surfaces" Langmuir 27, pp. 14143-14150.
6. Piotr Tourkine, Marie Le Merrer, and David Quéré. 2009, "Delayed Freezing on Water Repellent Materials", Langmuir, 25(13), pp. 7214-7216.
7. Design Expert version 8 [computer software]. Stat-Ease, Inc. 2016. Retrieve fromhttp://www.statease.com/software.html
8. Czitrom, V. 1999, "One-Factor-at-a-Time versus Designed Experiments", The American Statistician, 53(2), pp. 126-131.
9. Montgomery, D.G. 2012, Design and Analysis of Experiments, 8th Edition, Wiley.
10. Myers, R. H., Montgomery, D. C., Cook, C. M. A. 2016, Response Surface Methodology: Process and Product Optimization Using Designed Experiments, 4th Edition, Wiley.
11. Lye, L. M. 2002, "Design of Experiments in Civil Engineering: Are we still in the 1920's?", Annual Conference of the Canadian Society for Civil Engineering, Montréal, Canada.

The Correlation between pulp properties and bar forces in LC refining using a piezoelectric sensor

R. Harirforoush[1], **J. Olson**[2], **P. Wild**[1]

[1] University of Victoria, Department of Mechanical Engineering, P.O. Box 1700 STN CSC, V8W 2Y2, Victoria, BC, Canada

[2] University of British Columbia, Department of Mechanical Engineering, Point Grey Campus, 5000-2332 Main Mall, V6T 1Z4, Vancouver, BC, Canada

Abstract:

In Mechanical pulping, refiner control variables affect the pulp, and ultimately the paper properties. These variables also influence the forces that a pulp experiences during its time in the refiner. The objective of this study is to understand the relation between pulp properties and bar forces by using a custom-designed piezoelectric force sensor that measures shear and normal forces applied to pulp fibers by the refiner bars. Trials are performed in an AIKAWA/Advanced Fiber Technologies Inc. 16'' single-disc low consistency refiner. The trials are run using mechanical softwood SPF pulp with 378 ml CSF at 2.5% and 3.5% consistency at rotational speeds of 800 rpm, 1000 rpm, and 1200 rpm. Pulp properties including curl, kink, fine percentage, and length-weighted fiber length are measured and correlations to bar forces are studied. For each operating condition, mean peak forces, and Weibull scale parameter of coefficient of friction are calculated and pulp properties are plotted as a function of these parameters. The results show that length-weighted fiber length correlates to bar forces. Moreover, frequency analysis of the sensor data shows that the magnitude of the dominant frequency correlates to length-weighted fiber length. These correlations appear as distinct transitions which corresponds to the degradation of mechanical properties of pulp.

Introduction

Low consistency (LC) refining of mechanical pulp has been shown to be more energy efficient than conventional high consistency refining [1–4]. Moreover, LC refining is effective for removal of shives, external fibrillation of fibers, removal of latency and improving fiber networking and bonding capabilities [5,6]. To maximize the efficiency of LC refining, understanding the relation between refiner operating conditions, bar forces (*i.e.* the force applied to pulp fibers by the refiner bars), resulting pulp properties, and paper quality is crucial.

The relationship between refiner control variables, fiber properties, and paper properties has been the focus of many studies. Luukkonen et al. found that the plate gap is a key link between the refiner process variables, and resulting changes in pulp properties [7]. In a separate study, Luukkonen et al. [8] proposed a methodology to correlate operating conditions to pulp quality by relating the refiner operating conditions (*i.e.* power, flow rate, rotational speed and plate gap) to fiber quality (*i.e.* fiber length and freeness) and then relating fiber quality to pulp handsheet quality (*i.e.* tensile, tear, and bulk).

In a recent study, the relation between operational variables and pulp properties in a AIKAWA 16'' pilot-scale LC refiner is investigated by Elahimehr et al. [9]. Trials were run at three rotational speeds, with four refiner plate patterns, and at 3% consistency. Good correlations were found to identify key operating variables effecting pulp and paper quality development.

Harirforoush et al. [10] found that, as the plate gap is reduced, the length-weighted fiber length remains relatively constant while net refiner power and the mean peak normal and shear forces increase up to threshold values which correspond to critical gaps.

A piezoelectric sensor has been developed to measure bar forces. This sensor, referred to here as the refiner force sensor (RFS), has a sensing probe which replaces a short length of a refiner bar enabling measurement of forces normal to and tangential to the refiner bars. The RFS has been used in a number of studies in both low consistency [11,12] and high consistency refining [13,14]. This sensor has been used to measure shear and normal forces during individual bar crossing events but has not been applied to investigate the relation between pulp properties and bar forces.

The objective of this study is to investigate the correlation between pulp properties and bar forces measured with the RFS. Trials are performed in an AIKAWA/Advanced Fiber Technologies Inc. 16'' single-disc low consistency refiner using mechanical softwood SPF pulp with 378 ml CSF. Data is collected over a wide range of rotational speeds and consistencies. Distributions of the peak normal and shear forces, and peak coefficient of friction (i.e. peak shear force divide by peak normal force) are determined for each operating condition. Length-weighted fiber length shows strong correlations to mean peak forces as well as Weibull scale parameter of coefficient of friction. Moreover, frequency analysis of the sensor data shows that the magnitude of the dominant frequency correlates to length-weighted fiber length.

Materials and Methods

Trials were conducted using the AIKAWA 16'' single-disc refiner at the Pulp and Paper Center at the University of British Columbia, Canada. The refiner is equipped with a power meter, plate actuation and a variable speed drive. The plate used in these trials is described in [10].

A RFS-type sensor was custom designed and fabricated based on the design previously used in refiner trials [15]. A photograph of the RFS used in this work is shown in [10]. The sensor is installed in a custom recess that is machined into the back of the stator plate. The sensor wires pass from the back side of the sensor and through a custom gland located in the door of the refiner. The sensor recess is located so that the tip of the sensor replaces a short length of the refiner bar (i.e. 5 mm). The sensor measures force normal to the plate surface (*i.e.* parallel to the axis of the refiner bar) and the shear force perpendicular to the major axis of the refiner bar and normal to the bar edge. The position of the sensor is explained in [10].

Calibration and environmental tests were performed as described in [10,16]. Modal analyses were performed to ensure that the first natural frequency of the sensor is well above the bar-passing frequency of the refiner so that phase and magnitude distortion due to sensor resonance is avoided.

A custom charge amplifier converts the high impedance charge output of the sensors to a low impedance voltage signal which is read by a National Instruments (Austin, TX, USA) PXI 1042 Q high-speed board. National Instrument BNC-2110, Shielded Connector Blocks are used to connect the high-speed DAQ board to the charge amplifier. A custom LabVIEW™ interface is used to control data acquisition. A sampling rate of 150 kHz is used which is more than twenty times the maximum bar-passing frequency that occurs during the trials.

Mechanical SPF (spruce, pine, and fir) softwood pulp, produced at Quesnel River Pulp Mill, with 378 ml CSF at 2.5% and 3.5% consistency was used in all trials. The operating conditions for the trials are tabulated in [17]. The details of recording sensor signals and collecting pulp samples are described in [17].

The *length-weighted fiber length* (L_w), curl index, kink Index, and fine percentage were measured for all pulp samples using a Fiber Quality Analyzer (HiRes FQA, Optest Equipment Inc.). These parameters are defined in [18,19].

An algorithm was created in MATLAB to identify *bar-passing events* in the force data. A bar-passing event is defined to occur when the passage of a rotor bar over the sensor results in a maximum or peak value in the force data that exceeds a predefined threshold value. The algorithm and its verification are described in [10].

For each bar-passing event, peak force is defined as the difference between the force at the base of the local valley that precedes the peak and the force at the peak [12]. The peak coefficient of friction is calculated by dividing the peak shear force by the peak normal force.

The Weibull distribution function is widely used because of its ability to characterize a wide variety of distributions and it has been used in previous HC and LC refining studies [11,13]. The Weibull distribution function is explained in [20].

The effects of Weibull parameters on the Weibull distribution are discussed by Lee [21]. The two-parameter Weibull distribution function is used in this study. In this paper, the Weibull scale and shape parameters of the peak normal, shear force, and peak coefficient of friction distributions are calculated for each operating condition. Each of these parameters are then correlated to pulp properties.

Results

Figures 1a-1c show the relationship between pulp properties and mean peak normal force for three rotational speeds (*i.e.* 1200 rpm, 1000 rpm, and 800 rpm) at 3.5% consistency and 1200 rpm at 2.5% consistency. A distinct trend in the relation between L_w and mean peak normal force was previously shown by authors in [10,17]. As it is depicted in Fig 1a-1c, fine percentage, curl index and kink index do not correlate to mean peak forces.

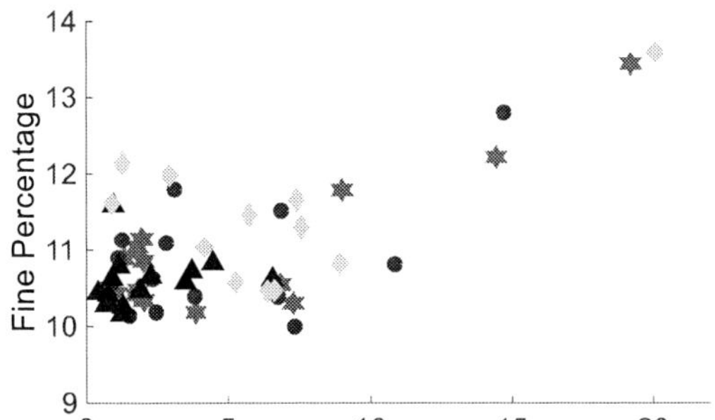

(a)

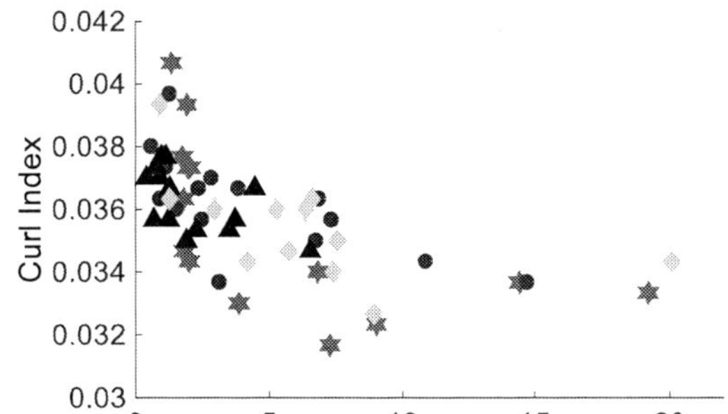

(b)

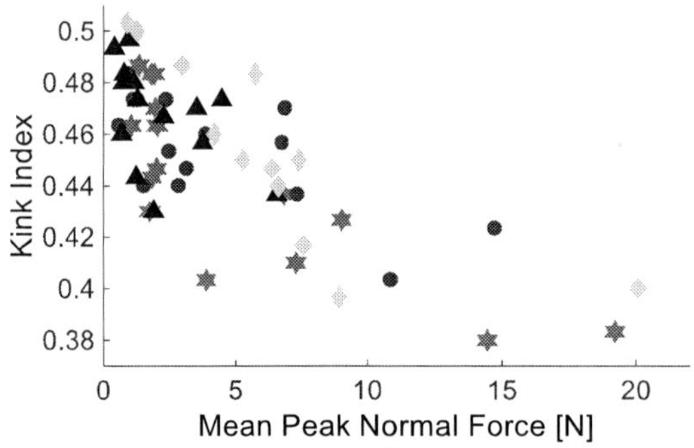

(c)

Fig 1. (a) Fine percentage, (b) curl index, and (c) kink index versus mean peak normal force for three rotational speeds (i.e. 1200 rpm, 1000 rpm, and 800 rpm) at 3.5% consistency. Also data is shown for 1200 rpm at 2.5% consistency.

The relation between pulp properties and the Weibull scale parameter of the distribution of peak coefficient of friction for three rotational speeds (i.e. 1200 rpm, 1000 rpm, and 800 rpm) at 3.5% consistency and 1200 rpm at 2.5% consistency is shown in Fig 2a-2c. As it is shown in Fig 2a-2c, there is not a distinct transition in the plots of Fig 2a-2c.

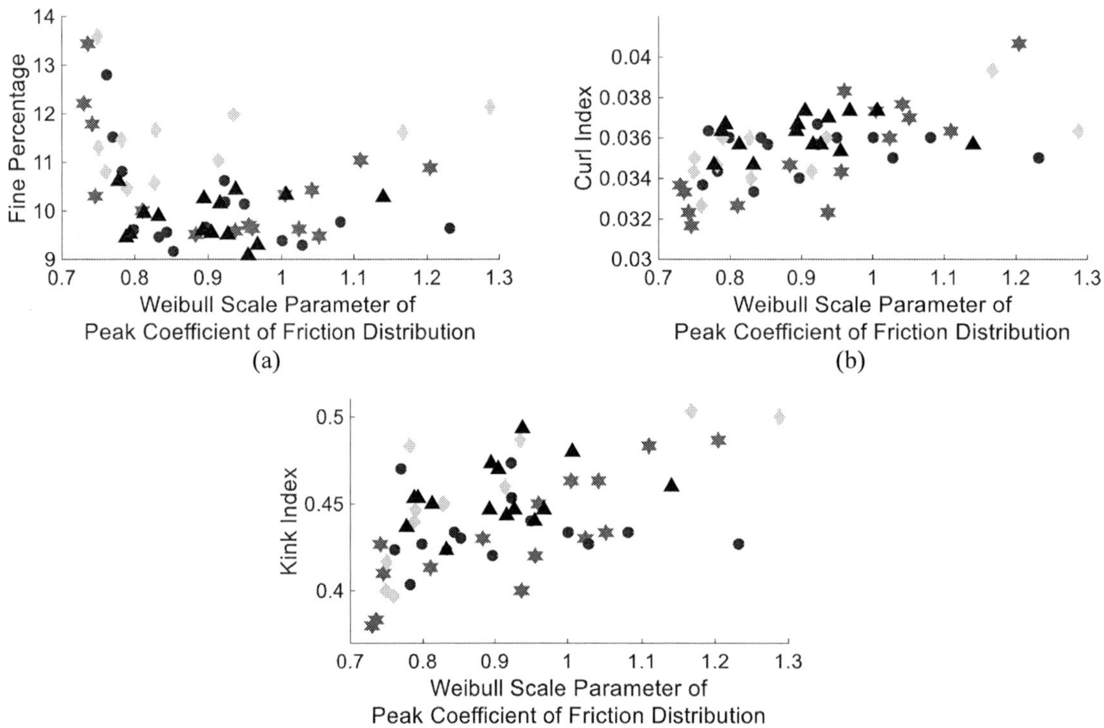

(c)

Fig 2. (a) Fine percentage, (b) curl index, and (c) kink index versus Weibull scale parameter of peak coefficient of friction distribution for three rotational speeds (i.e. 1200 rpm, 1000 rpm, and 800 rpm) at 3.5% consistency. Also data is shown for 1200 rpm at 2.5% consistency.

Figures 3a-3d show the magnitude of the dominant frequency versus pulp properties for three rotational speeds (i.e. 1200 rpm, 1000 rpm, and 800 rpm) at 3.5% consistency and 1200 rpm at 2.5% consistency. Dominant frequency corresponds to the passage of equivalent edges in adjacent three-bar clusters described in [10]. This frequency appears in the spectra for all of the operating conditions. As shown in Fig 3a, L_w remains relatively unchanged as magnitude increases up to a threshold value. The L_w value decreases in an approximately linear manner beyond this threshold. No correlation is found in Fig 3b-3d.

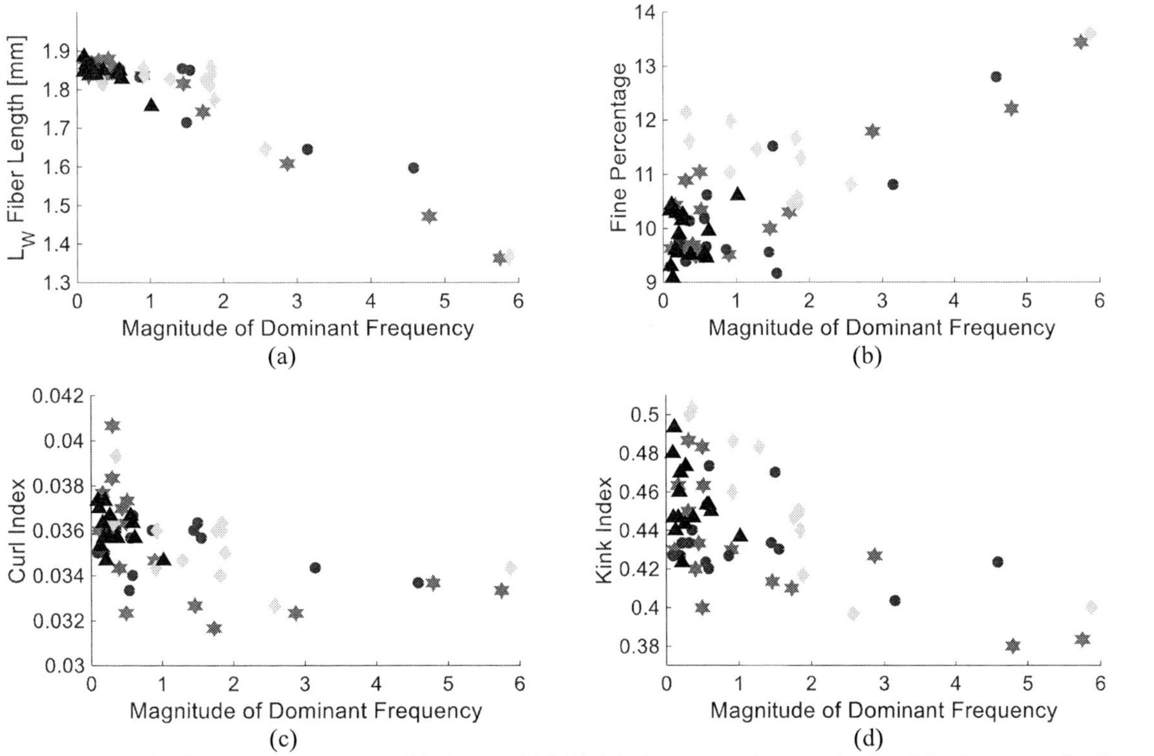

Fig 3. (a) L_w, (b) fine percentage, (c) curl index, and (d) kink index versus the magnitude of the frequency for three rotational speeds (i.e. 1200 rpm, 1000 rpm, and 800 rpm) at 3.5% consistency. Also data is shown for 1200 rpm at 2.5% consistency.

Conclusion

This study investigates the relation between pulp properties and bar forces using a custom designed piezoelectric sensor measuring normal and shear forces in an AIKAWA/Advanced Fiber Technologies Inc. 16'' single-disc LC refiner. Trials were run using mechanical SPF softwood pulp with 378 ml CSF at 2.5% and 3.5% consistency for three rotational speeds (1200, 1000, and 800 rpm) and a wide range of plate gaps. The pulp was sampled at regular intervals and curl, kink, fine percentage, and length-weighted fiber length were determined for each sample. For each operating condition, mean peak forces, and Weibull scale parameter of coefficient of friction are calculated and pulp properties are plotted as a function of these parameters. The results show that length-weighted fiber length correlates to bar forces. Moreover, the magnitude of the dominant frequency correlates to length-weighted fiber lengths. These correlations appear as distinct transitions which corresponds to the degradation of mechanical properties of pulp. This result suggests that the sensor has potential application to be used for in-process detection of the onset of fiber cutting.

Acknowledgment

This work is funded by a Collaborative Research and Development grant provided by National Sciences and Engineering Research Council of Canada (NSERC) and the following partners, who we thank for their ongoing support: AB Enzymes, Alberta Newsprint Company, Andritz, BC Hydro, Canfor, Catalyst Paper, FPInnovations, Holmen Paper, Meadow Lake Pulp (Paper Excellence), Millar Western, NORPAC, West Fraser, Westcan Engineering, and Winstone Pulp International.

A special thanks to Meaghan Miller, Emilia Jahangir, Elizabeth Trudel, and George Soong for their assistance during the preparation and execution of the refining trials at the Pulp and Paper Centre at the University of British Columbia.

References

[1] Muenster, H.; Ferritsius, O.; Lecourt, M.; Petit-Conil K. Energy savings in TMP by high temperature LC/MC refining. Int. Mech. Pulping Conf., Oslo, Norway, 2005, p. 219–23.

[2] Zha, Q.; Lanouette, R.; Law, K.N.; Bousquet, J.P.; Bussières, S. Refining of long fibre fractions after fractionation. Prepr. 94th Ann. Mtg., PAPTAC, Montreal, Canada, 2008, p. B481–7.

[3] Musselman, R.; Letarte, D. ; Simard, R. Third stage low consistency refining of TMP for energy savings and quality enhancement. 4th Int. Refin. Conf., Fuiggi, Italy, Fuiggi, Italy, 1997, p. 139–48.

[4] Sandberg, C.; Sundström, L.; Nilsson, L. Potential of low consistency refining of TMP – mill evaluation. Int. Mech. pulping Conf., Sundsvall, Sweden, 2009, p. 186–9.

[5] Kappel J. Mechanical pulps: From wood to bleached pulp. Tappi press, Atlant, 1999.

[6] Smook G.A. Handbook for pulp and paper technologists. 2nd Edition. Vanouver, BC, Angus Wilde Publications, 1992.

[7] Luukkonen, A.; Olson, J.A.; Martinez, D.M. Low consistency refining of mechanical pulp, effect of gap, speed and power. J Pulp Pap Sci 2010;36:28–34.

[8] Luukkonen, A.; Olson, J.A.; Martinez, D.M. Low Consistency Refining of Mechanical Pulp: A Methodology to relate operating conditions to paper properties. J Pulp Pap Sci 2010;36:107–11.

[9] Elahimehr, A.; Olson, J.A.; Martinez, D.M. Understanding LC refining:The effect of plate pattern and refiner operation. Nord Pulp Pap Res J 2013;28.

[10] Harirforoush, R.; Wild, P.; Olson, J. The relation between net power, gap, and forces on bars in low consistency refining. Nord Pulp Pap Res J 2016;31:71–8.

[11] Prairie, B.; Wild, P.; Byrnes, P.; Olender, D.; et al. Forces During Bar-Passing Events in Low- Consistency Refining: Distributions and Relationships to Specific Edge Load. Journal of Pulp and Paper Science, 2008, 34(1), 1-8.

[12] Prairie, B.; Wild, P.; Byrnes, P.; Olender, D.; Francis, D. W.; Quellet, D. Forces during bar-passing events in low consistency refining: Effects of refiner tram. Pulp Pap Canada 2007;108:34–7.

[13] Olender, D.; Wild, P.; Byrnes, P; Ouellet, D; Sabourin, M. Forces on bars in HC mill-scale refiners:Trends in primary and rejects stage refiners. J Pulp Pap Sci 2007;33(3), 163-171.

[14] Olender, D.; Wild, P.; Byrnes, R.; Ouelle,t D.; Sabourin, M. Forces on bars in high-consistency mill-scale refiners: Effect of Consistency. J Pulp Pap Sci 2007;33:218–23.

[15] Olender, D.; Wild, P.; Byrnes P. A piezoelectric force sensor for mill-scale chip refiners. J Process Mech Eng 2008;222:115–22.

[16] Prairie, B. Measurement of forces in a low consistency refiner. M.Sc. Thesis, University of Victoria, Canada, 2005.

[17] Harirforoush, R.; Wild, P.; Olson, J; Measurement of forces on refiner bars for in-process detection of fiber cutting in low consistency refining. International Mechanical Pulping Conference, Florida, USA, 2016.

[18] Graca Carvalho, M.; Ferriera, P.J.; Martins, A.A.; Figueiredo MM. A comparative study of two auto-mated techniques for measuring fiber length. Tappi J 1997;80:137–42.

[19] Page, D.H.; Seth, R.S.; Jordan, B.D.; Barbe M.C. Curl, crimps, kinks and microcompressions in pulp fibres: Their origin, measurement and significance. In Papermaking Raw Materials,Trans. VIIIth Fund. Res. Symp. Oxford, 1985, (V. Punton, ed.),pp 183–227.

[20] D. N. Prabhakar Murthy; Min Xie R.J. Weibull Models.Wiley Series in Probability and Statistics, 2003.

[21] Lee S. Estimation and Comparison of Weibull Parameters for Reliability Assessments of Douglas-fir wood. Int. J Basic Appl Sci 2014;14:18–21.

Generating non-spherical alginate microparticles by electrohydrodynamic spraying

Morteza Jeyhani
Department of Mechanical and Industrial Engineering
Ryerson University
Toronto, ON, Canada
mjeyhani@ryerson.ca

Sze Yi Mak
Department of Mechanical Engineering
University of Hong Kong
Hong Kong
h10symak@connect.hku.hk

Stephen Sammut
Department of Electrical and Computer Engineering
Ryerson University
Toronto, ON, Canada
ssammut@ryerson.ca

Ho Cheung Shum
Department of Mechanical Engineering
University of Hong Kong
Hong Kong
ashum@hku.hk

Dae Kun Hwang
Department of Chemical Engineering
Ryerson University
Toronto, Canada
dkwhwang@ryerson.ca

Scott S. H. Tsai
Department of Mechanical and Industrial Engineering
Ryerson University
Toronto, ON, Canada
scott.tsai@ryerson.ca

ABSTRACT

We describe a simple fabrication method, based on electrohydrodynamic spraying, to generate non-spherical biodegradable alginate particles. Specifically, we investigate the dependence of the resulting microparticle morphology on several experimental parameters.

KEYWORDS: Microparticles, non-spherical, electrospray, alginate, drug delivery

INTRODUCTION

Biodegradable and biocompatible microparticles are of substantial interest for biomedical drug delivery applications due to their unique features including small size, drug encapsulation ability, bioavailability, and reduced toxicity and adverse effects in comparison to the microparticles are not biocompatible and biodegradable.[1] Biodegradable polymers such as alginate are among the main materials used in applications involving the encapsulation of drugs, proteins and cells in particles.[2]

The shape and morphology of the drug-carrying particles are important. Morphology impacts the outcome of drug delivery in different ways, including the kinetics of drug release and transport, and the circulation of the particles inside body.[1,3,4] Despite the importance of morphology and shape on the efficacy of alginate-based drug delivery particles, there are few manufacturing methods developed to controllably tune the shape of alginate microparticles. Here, we describe a new method to control the shape of non-spherical alginate microparticles using an electrospray.

EXPERIMENTAL AND RESULTS

A. Electrospray technique

Electrohydrodynamic atomization, also known as electrospraying, is a method of liquid atomization. This technique uses a high voltage electric field to deform the interface of a liquid droplet. The electric field overwhelms the surface tension of the droplet and causes the primary droplet to break apart into smaller droplets. Electrosprayed droplets fall into a calcium chloride bath underneath the spray, and chemically react to solidify into microparticles.

B. Materials

We use alginic acid sodium salt solution (2 wt%) as the emulsion phase and calcium chloride dihydrate solution for the continuous phase with concentrations of 2.5, 5, 7.5 and 10 wt%.

C. Effective parameters

For the purpose of controlled generation of non-spherical particles, we tune several electrospray experimental parameters. Namely, we adjust the concentration of calcium chloride in the bath, the distance from the electrospray ring to the free surface of the bath, and the liquid depth of the bath.

We observe that by changing the abovementioned parameters we can control the morphology of the non-spherical microparticles. We also develop a phase diagram of the final particle shape based on the calcium chloride concentration and the distance from the electrospray ring to the free surface of the bath.

CONCLUSIONS

Fabrication of biocompatible and biodegradable microparticles have been significantly important for biomedical studies in drug delivery applications. Various studies show that morphology of particles is considerably relevant to the efficiency of drug delivery application. We demonstrate a new method to generate non-spherical alginate microparticles with a controllable manner to change the shape of particles.

ACKNOWLEDGMENTS

S. S. H. Tsai and D. K. Hwang both acknowledge funding support from the Canadian Natural Sciences and Engineering Research Council (NSERC), and association with the Institute for Biomedical Engineering, Science, and Technology (iBEST)—a partnership between Ryerson University and St. Michael's Hospital. H.C. Shum and S.Y. Mak both acknowledge funding support from the Research Grants Council of Hong Kong through the General Research Fund (Nos. HKU 719813E and 17304514).

REFERENCES

1. Almería, B., Deng, W., Fahmy, T. M. & Gomez, A. 2010, "Controlling the morphology of electrospray-generated PLGA microparticles for drug delivery", *J. Colloid Interface Sci.,* Vol(343), pp. 125–133.

2. Lambert, G., Fattal, E. & Couvreur, P. 2001, "Nanoparticluate system for the delivery of antisense oligonucleotides", *Adv. Drug Deliv. Rev.* Vol(47), pp. 99–112.

3. Champion, J. A., Katare, Y. K. & Mitragotri, S. 2007, "Particle shape: a new design parameter for micro- and nanoscale drug delivery carriers", *J. Control. Release* Vol(121), 3–9.

4. Simone, E. A., Dziubla, T. D. & Muzykantov, V. R. 2008, "Polymeric carriers: role of geometry in drug delivery", *Expert Opin. Drug Deliv.* Vol(5), 1283–1300.

CRACK DETECTION USING PAVEMENT DEPTH IMAGE BASED ON BLOCK SEGMENTATION

Lei Wang, Ze Chen, Mingming Zhou, Chenglong Xiao, Mengmeng Yang, Hongda Ma, Tongqi Chen, Li He*
Department of Electrical and Electronic Engineering
Hubei University of Technology
Wuhan, Hubei Province, China
***heli.edu@hotmail.com**

ABSTRACT

The aim is to investigate a crack detection algorithm for extract crack from pavement depth, which includes outlier removal, image enhancement and crack segmentation. The zero-value points were eliminated from the depth image, and edge-preserving smoothing was performed on the images using bilateral filters. Based on block segmentation, Sobel operator multi-direction edge detection was introduced to improve the crack-segmentation algorithm. The results showed that the proposed algorithm can achieve higher of recognition and recall for pavement depth images than the traditional bidirectional scanning algorithm.

KEYWORDS: Crack detection, depth image, salient feature, block segmentation

INTRODUCTION

Pavement cracks are one of the most important indices to evaluate pavement condition. It is very important for pavement maintenance and management to find an automatic and accurate method for crack detection and identification [1-5]. Because of the environmental noise, traditional 2D crack-detection technology is difficult to meet the needs of pavement maintenance and management [6].

Crack-detection technology uses both depth images and gray images for crack recognition. Depth-image data have the obvious advantages of being free from light, oil pollution and other influencing factors, and the recognition result is more accurate [7]. However, the huge amount of data obtained by depth images impacts the processing speed, and the rich texture information in the depth images increases the probability of misidentification. Therefore, various new algorithms have been proposed to identify the cracks.

It is taken into consideration that crack edges in the images are signals with high frequency and high amplitude, while the random textures in the pavement surface are signals with high frequency but low amplitude [8]. The cracks display relatively larger depth values than the other areas in pavement depth images. Combing with global contrast and enhancement of surrounding spaces, an enhanced algorithm for crack significance suitable for depth images was designed in the paper.

1 METHOD

1.1 OUTLIER REMOVAL

Analysis of the actual pavement depth image data shows that the outliers include not only zero-value points, but also many discrete points close to zero. As the experiment used structured light to obtain data, and each line of data was taken from a laser line, the relevance of the adjacent column during data collection was not obvious. Therefore, we removed the zero-value point of each row of data. A median filter is used to replace the coordinate value of a zero-value point with the median of the z-coordinate values of the normal pavement points around it.

1.2 IMAGE ENHANCEMENT

The road surface image data often contain complex information. The edge of the crack in the depth image exhibits signals with high frequency and high amplitude, while texture displays signals with high frequency but lower amplitude. Therefore, we used frequency domain feature calculation to approximate the background of the image.. A Gaussian filter can reduce the high-frequency part to a certain extent. The role of the Gaussian kernel function $G(x,y,\rho)$ is to blur the image edge but retain the overall picture trend, thus approximating the image background. Denote the original image by $I_0(x,y)$ and the approximate image obtained from the original image through Gaussian filtering by $I_g(x,y)$. Then the image after enhancement processing is defined as

$$H(x,y) = I_0(x,y) - I_g(x,y) \tag{1}$$

Where

$$I_g(x,y) = I_0(x,y) * G(x,y,\rho) \tag{2}$$

And

$$G(x,y,\rho) = \frac{1}{2\pi\sigma^2} e^{-(x^2+y^2)/(2\sigma^2)} \tag{3}$$

The enhanced image preserves the details of the cracks and textures on the same horizontal plane, but the rough texture possibly contained in the image may affect the segmentation effect.

1.3 SEGMENTATION OF THE CRACK IMAGE

Since cracks are linear and long, while textures are relatively small and intensive, we propose to first determine the existence of crack information through segmentation, so as to differentiate cracks from large textured areas.

We divided the image into n rectangular windows. Analyzing the image features after correction revealed that the amplitude dispersion of the window containing cracks was higher than that of the window without cracks. According to this feature, the variance of the amplitude of all the pixels of the window was selected to reflect the window dispersion, and the whole image was divided into n windows, with the variance of each window recorded as $\delta_i (i = 1,2,3 \cdots n)$. Analyzing the δ_n histogram, the maximum interclass variance threshold δ_0 was taken as a threshold for determining the window dispersion. The windows with values below δ_0 were classified as plain-texture windows, which were marked as normal areas, and those with values greater than δ_0 were marked as areas to be divided. This method can extract the windows with cracks and large textures, thus reducing the complexity of the algorithm and shortening the computing time. Let the preliminary decision matrix be $T(x, y)$. Then

$$T(x,y) = \begin{cases} 1 & \delta_i > \delta_0 \\ 0 & \delta_i \le \delta_0 \end{cases} . \tag{4}$$

2 RESULTS

We selected two representative images to experiment. Image 1 is pavement with a very smooth texture. The biphasic scanning algorithm was used to compare with the proposed method in the paper. The results showed that our proposed segmentation algorithm enhanced images and extracted the crack edges very well from images with large textures.

In order to facilitate the evaluation and comparison, this algorithm evaluates the performance through the accuracy rate and the recall rate. Let C represent the number of cracks identified correctly, Y the number of cracks mistakenly identified, and F the number of undetected cracks. Then the recognition accuracy rate P is the proportion of calculation results that are consistent with the real results, and it is calculated with the formula.

$$P = \frac{C}{C+Y} \tag{1}$$

Recall that the rate R represents the ratio of the calculated positive cases to all positive samples. It is calculated with the equation

$$R = \frac{C}{C+F} \tag{2}$$

The rate of accuracy and recall of Bi-Directional Scanning method are 80.9% and 74.3% respectively, while the proposed algorithm in the paper reach to 84.8% and 79.5% respectively.

CONCLUSIONS

This paper proposed an automatic crack recognition algorithm for high-precision depth images of pavement with different texture types, which removes noise such as outliers in images, enhances images based on salient features, and improves the segmentation algorithm to achieve the regional segmentation of cracks. Tests showed that the proposed algorithm can achieve higher accuracy and recall rates, and they show a general advantage over the bidirectional scanning algorithm. However, there are many problems worthy of further study.

ACKNOWLEDGMENT

Financial support from China Scholarship Council under Grant No. 201608420056 and Major Science and Technology Innovation Project of the Ministry of Science and Technology of Hubei Province under Grant No. 2013AEA001 is gratefully acknowledged. Valuable research support from Department of Mechanical Engineering at University of Victoria is gratefully acknowledged. We are indebted to the editor and referees of the Conference for helpful comments.

REFERENCES

1. Gao, J., Ren, M., Tang, Z. Automatic road crack detection and identification [J]. Computer Engineer, 2002, 29(2): 149-150.
2. Xu, W., Tang, Z., Lv, J. Pavement crack detection based on image saliency [J]. Journal of Image and Graphics, 2013, 18(1): 69-77.
3. Yan, M., Bo, S., He, Y. A method of image detection and analysis for pavement crack based on morphology [J]. Journal of Engineering Graphics, 2008, 29(2): 142-147.
4. Lou Jing, Zang Aiqin, "Pavement Crack Distress Detection Based on Image Analysis," in 2010 International Conference on Machine Vision and Human-machine Interface (2010).
5. Subirats, P., J. Dumoulin, V. Legeay and D. Barba. Automation of Pavement Surface Crack Detection Using the Continuous Wavelet Transform. In Image Processing, 2006 IEEE International Conference on, 3037-3040, 2006.
6. Guan Haiyan et al., "Automated Road Information Extraction from Mobile Laser Scanning Data," Intelligent Transportation Systems, IEEE Transactions on 16, no. 1 (2015).
7. Peng, B., Wang, K.C.P., Chen, C., Jiang, Y. Automatic recognition algorithm for crack seeds based on 1 mm resolution 3D pavement images [J]. China Journal of Highway and Transport, 2014, 27(12): 23-32.)
8. Feng, C., Dai, S. The depth map filter algorithm based on salient features in telepresence [J]. Journal of Harbin Engineering University, 2014, 11:1364-1368.

Mechanical Optimization

EVALUATION OF LATTICE STRUCTURE OPTIMIZATION SOFTWARE USING 3D PRINTING

Mohamed Abdelhamid
Department of Mechanical Engineering
York University
Toronto, Ontario, Canada
mahamid@yorku.ca

Aleksander Czekanski
Department of Mechanical Engineering
York University
Toronto, Ontario, Canada
alex.czekanski@lassonde.yorku.ca

ABSTRACT
Recent advances in additive manufacturing technologies have simplified the implementation of lattice structures for different applications, such as mechanical load-carrying components and energy-absorption, among others. Hence, lattice structure optimization (LSO) is gaining more attention recently. LSO in the commercial software, Altair's HyperWorks, is evaluated in comparison to a continuum-based analytical model of the octet-truss lattice. Comparison samples are verified using finite element analysis and further validated using stereolithography-based 3D printing and mechanical testing.

KEYWORDS: Lattice structure optimization, octet-truss, macroscopic stiffness tensor, 3D printing, mechanical testing.

INTRODUCTION
Metamaterials refer to materials consisting of a unit cell tessellated in three directions with dimensions scaling down to the nano and micro length scales [1,2]. Applications include lightweight structures, energy absorption, photonics and acoustics, among others [2]. A subclass of metamaterials is mechanical metamaterials, which refer to materials with certain mechanical properties defined by their structure/geometry rather than their composition [1]. The deformation mode of the unit cell's individual members defines its type. The two main types are (i) stretch-dominated, where individual members deform by tension/compression and (ii) bend-dominated where they deform by bending [1,3]. Maxwell's criterion for static determinacy is generally used to categorize the type of the unit cell. For stretch-dominated, it is given by Eq. (1) for 2D structures and Eq. (2) for 3D structures as follows:

$$b - 2j + 3 \geq 0 \tag{1}$$

$$b - 3j + 6 \geq 0 \tag{2}$$

where b is the number of members and j is the number of nodes within the unit cell [4,5]. For a 3D structure to be rigid (i.e. not to collapse when loaded), it must have a connectivity of $Z = 6$ at the nodes. A connectivity of $Z = 12$ is required for stretch-dominated structures, and a connectivity of $6 \leq Z < 12$ for bend-dominated ones [4,6].

The macroscopic mechanical performance of architected materials depend on the unit cell configuration, relative density and the constituent material properties [6]. Studies have shown that for a certain unit cell configuration and constituent material, there are coupling relations between the mechanical properties (stiffness and strength) and the relative density. Stretch-dominated structures are preferable to bend-dominated ones since they have better coupling relations (i.e. specific stiffness and strength do not degrade substantially with decreasing density) [1,2,6,7]. Hence, the octet-truss, a stretch-dominated lattice, is the unit cell of choice in this work.

MODELLING OF THE OCTET-TRUSS LATTICE
In order to describe the macroscopic behavior of the octet-truss lattice, continuum-based models have been developed. In previous work [8,9], the macroscopic stiffness tensor of the octet-truss lattice was obtained through the development of a continuum-based model. A tetrahedron substructure was selected as the least substructure representing the whole octet-truss unit cell as shown in Fig. 1. Next, its individual members were transformed from their local coordinates to the global coordinates (x, y, z), and finally homogenization was applied to relate the local stresses and strains to the global stresses and strains. The stiffness tensor includes a number of geometric parameters as shown in Eq. (3) [8]:

$$C_{ijkl} = \frac{E_s\ A_c}{L^2} * \tag{3}$$

$$\begin{bmatrix} \frac{m+2\sqrt{2}\,m^4}{2\sqrt{2}\,n} & \frac{1}{2\sqrt{2}\,p} & m^2.n & 0 & 0 & 0 \\ \frac{1}{2\sqrt{2}\,p} & \frac{m+2\sqrt{2}\,m^4}{2\sqrt{2}\,n} & m^2.n & 0 & 0 & 0 \\ m^2.n & m^2.n & 2n^3 & 0 & 0 & 0 \\ 0 & 0 & 0 & m^2.n & 0 & 0 \\ 0 & 0 & 0 & 0 & m^2.n & 0 \\ 0 & 0 & 0 & 0 & 0 & \frac{1}{2\sqrt{2}\,p} \end{bmatrix}$$

where E_s, A_c and L are the constituent material elastic modulus, the individual member cross-sectional area and the tetrahedron base side length respectively [8]. m, n and p are $cos(\theta)$, $sin(\theta)$ and $tan(\theta)$ respectively with θ as the lattice angle defined as the angle between the members and horizontal midplane as shown in Fig.1. It's worth noting that since the octet-truss is a stretch-dominated type, only the tensile/compressive stiffness of the individual members contributes to the macroscopic stiffness, while the bending resistance of the members and the nodes is negligible [6,8,10]. Hence, modelling the octet-truss lattice as 1D truss elements is an accurate approximation [5].

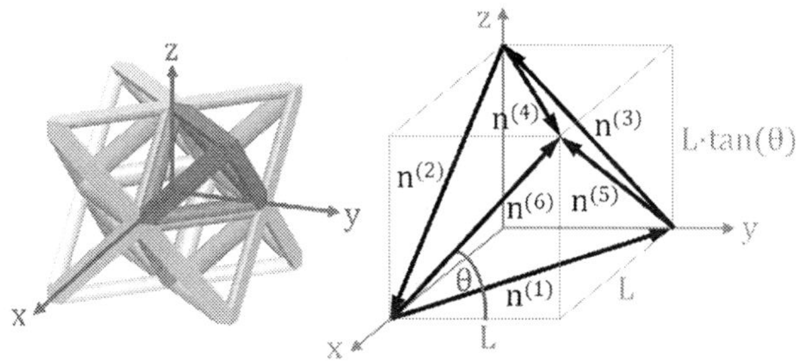

Figure 1: Octet-truss unit cell (left) and the tetrahedron substructure with its six members (right).

LATTICE STRUCTURE OPTIMIZATION IN ALTAIR HYPERWORKS

Lattice structure optimization (LSO) has been introduced recently in the software package HyperWorks by Altair. It enables the creation of blended solid and lattice structures [11]. It is worth noting that HyperWorks uses 1D tapered beam elements (CBEAM) to represent the individual members of the lattice structure. LSO is inherently similar to topology optimization, albeit the design domain can now include intermediate density elements. LSO is carried out in two main phases: (i) Regular topology optimization is performed first with reduced penalty options in order to retain more intermediate density elements. At the end of phase I, these elements are replaced with lattice structures such that the volume fraction of the lattice structure corresponds to the intermediate element density, (ii) Size optimization is then performed to define the diameters of each element so that the homogenized lattice properties correspond to the initial intermediate element stiffness. Some anisotropy is integrated in the lattice structure during phase II in order to increase efficiency. An important remark about LSO in HyperWorks is that each initial mesh element is replaced with a single unit cell, hence the size, orientation and lattice angle are dependent upon the initial model mesh [11].

In the next two sections, the 3D printing and mechanical testing equipment are discussed in order to pinpoint the limitations and considerations to be accounted for when defining the problem employed to compare HyperWorks LSO to the continuum-based analytical model.

3D PRINTING & MECHANICAL TESTING

Form 2.0, a stereolithography-based 3D printer by Formlabs Inc., was utilized to produce sample lattice structures using a proprietary material called Tough with a reported elastic modulus of 1.7 GPa for the green part without additional treatments. Form 2.0 requires the addition of support material to the manufactured part, the distribution and size of which can be controlled within their software [12]. The minimum available layer thickness of 50 microns was used to obtain the most geometrically-accurate results. Sample octet-truss structures were manufactured using this 3D printer (Fig. 2) to check the suitable range of element diameters and support material settings. In order to check the validity of this 3D printer as a manufacturing platform for the problem specimens, we needed to ensure the material properties were consistent in all directions. Hence, ASTM D638-14 Type IV tension samples were manufactured in three orthogonal directions [13]. These samples were subjected to a destructive tensile test on MTS C43 Universal Testing System. Testing results showed inconsistency in the elastic modulus values for different directions rendering Form 2.0 unreliable for such purpose, albeit the tensile strength showed more consistency. However, given the high quality and the low cost of the printed samples, Form 2.0 can be used for demonstration purposes.

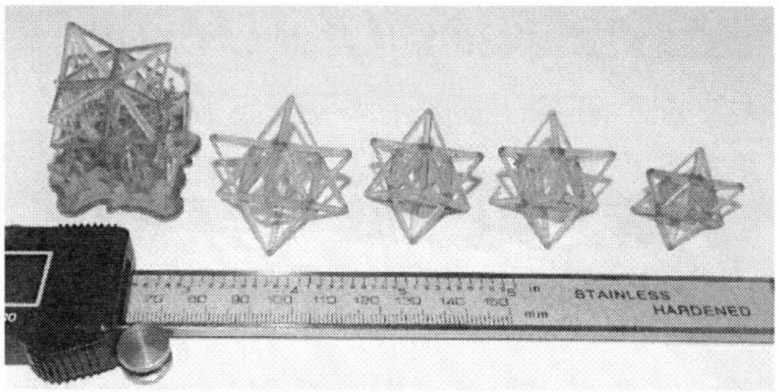

Figure 2: Octet-truss samples using Form 2.0, unit cell width 15-25mm, elements diameter 2mm, lattice angle 45-50 deg. Support material is still attached to the leftmost lattice.

Another 3D printer was utilized, Objet260 Connex3 by Stratasys. This 3D printer implements a process called PolyJet 3D Printing, where the printer head jets drops of curable photopolymer onto a build tray, which are subsequently cured using UV radiation. A proprietary material called Digital ABS was used, with a reported elastic modulus of 2.6-3.0 GPa [14]. Objet260 requires the use of support material, which could be removed later using water jetting. Similarly to Form 2.0, sample octet-truss structures were manufactured to investigate the appropriate settings that would produce the highest quality as shown in Fig. 3. ASTM D638-14 Type IV samples were manufactured in three orthogonal directions. Tensile testing results showed consistent results, with a mean elastic modulus of 2.3 GPa and a standard deviation of 0.08 GPa which is within the ranges suggested in ASTM D638-14 [13]. A laser extensometer was utilized to achieve high accuracy with reflective tapes attached to the top and bottom cylindrical platens as shown in the compression testing setup in Fig. 4. Given that this is a compression test and the polymer test material has a much lower elastic modulus than the machine building materials, the crosshead should give accurate displacement readings as well. The 3D printing and mechanical testing considerations for the problem definition can be summarized as follows: (i) The suitable element diameter should be 1.5-3 mm in order to obtain the finest 3D printing quality without unnecessary costs and (ii) The sample compressed area should not exceed a square of 20 x 20 mm.

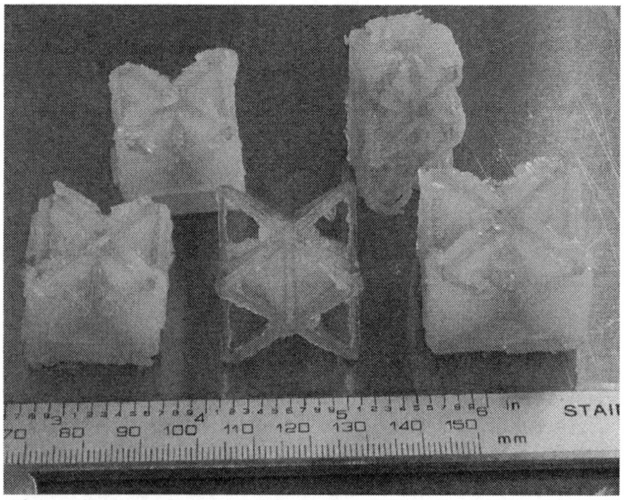

Figure 3: Octet-truss samples using Objet260, unit cell width 15-25mm, elements diameter 2mm, lattice angle 40-45 deg. Support material is partially removed.

Figure 4: Compression test setup on MTS C43 showing a laser extensometer and a video camera. A close-up of the cylindrical compression platens with the sample in place is shown in the top righ corner.

PROBLEM DEFINITION

For HyperWorks, the input geometry is a square rod with dimensions of *71.5 x 20 x 20 mm*. These dimensions are selected to enable meshing with three hexahedral elements at a lattice angle of *50* degrees. LSO is performed on this sample to generate lattice structure that would achieve the lowest volume fraction for a certain displacement ΔL under a certain compressive force F. The elastic modulus of the resultant lattice structure can be calculated per Eq. (4):

$$E_{lattice} = \frac{F * 71.5}{20 * 20 * \Delta L} \tag{4}$$

The resultant lattice structure is shown in Fig. 5. It's worth noting that HyperWorks applied fixed boundary conditions (BC's) for the uppermost and lowermost nodes of the structure while leaving the middle nodes free to move in all directions. To simulate these BC's in the compression experiment, two plates of 2 mm thickness were added at the top and bottom of the structure as shown in Fig. 5.

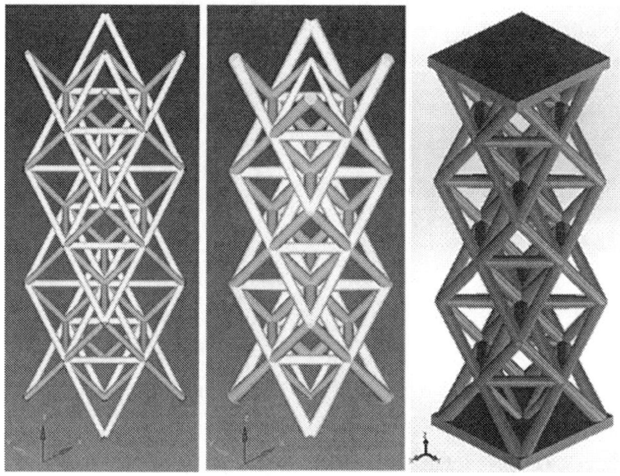

Figure 5: Lattice structure after phase I (left), elements' radii range of 0.56-1.35 mm after phase II (center), and CAD model for 3D printing (right).

The stiffness tensor from the continuum-based model was used to generate a similar lattice structure for comparison. The same input geometry and material properties were used to obtain the elements' diameter for the same elastic modulus from LSO. A multiplier had to be introduced into the stiffness tensor equation to account for the difference in defining the BC's in the analytical model and HyperWorks. Similarly, two plates were added at the top and bottom of the structure to simulate the same BC's in the compression test.

COMPRESSION TESTING RESULTS

The volumes of the two structures (analytical model vs HyperWorks) were calculated through SolidWorks, and they were found to be almost the same. Given the same input mesh geometry, HyperWorks and the analytical model can perform at the same level of efficiency. The two structures were manufactured using the Objet260 3D printer as shown in Fig. 6. Cyclic compression testing of the two structures resulted in an elastic modulus of 57.6 MPa for HyperWorks and 63.1 MPa for the analytical model, with a percentage error of 8.7%. A possible explanation for this error could be the size/scaling effects in 3D printed polymers, in other words elements of different diameters have different mechanical properties. To investigate this phenomenon, a 2-mm diameter tension sample was manufactured using the Objet260 and mechanically tested. The mechanical behavior was significantly different from the standard Type IV sample as shown in Fig. 7. The 2-mm diameter sample showed more ductility and strain hardening, and lower elastic modulus and tensile strength than the standard Type IV sample. Further extensive experimentation is required to fully characterize this phenomenon.

CONCLUSIONS

LSO in HyperWorks is performed in two phases, topology

optimization then size optimization. Comparison between LSO in HyperWorks with a continuum-based analytical model showed that they can perform on the same level of optimization given the same input geometry. Decoupling the lattice unit cell size and orientation from those of the initial mesh size could potentially increase the optimization efficiency and the user convenience of HyperWorks' LSO. Objet260 3D printed samples demonstrated size/scaling effects, different mechanical behavior for different sample sizes. However, the analytical model vs HyperWorks comparison still holds as it was verified using finite element analysis.

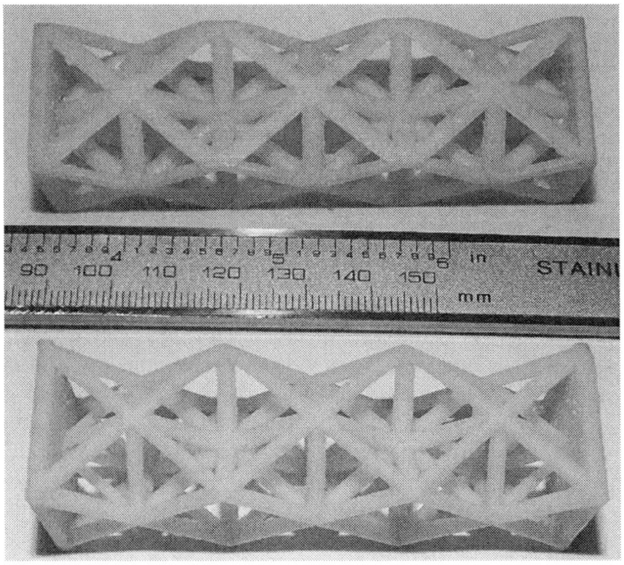

Figure 6: Lattice structures generated using analytical model (top) vs HyperWorks' LSO (bottom).

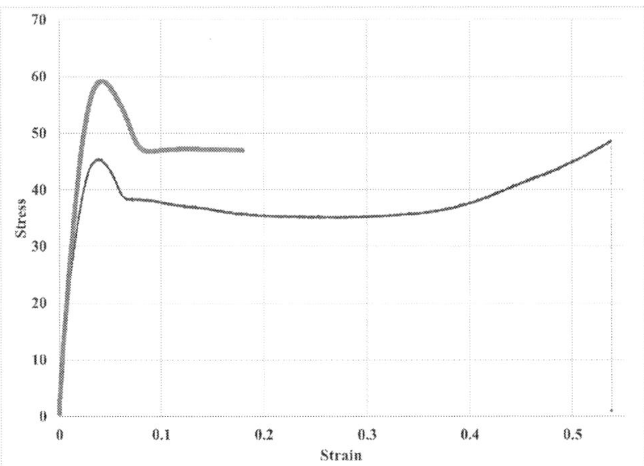

Figure 7: Mechanical behavior of standard Type IV specimen (orange) vs 2 mm diameter sample (blue) under tension.

ACKNOWLEDGMENTS

Part of this work was funded by an NSERC-Engage grant in collaboration with Altair Engineering, Inc. The authors thank Mr. Salman Chaudhry (York University) for his help with the mechanical testing and Mr. Greg Delbridge (Altair Canada) for his technical support with HyperWorks.

REFERENCES

1. Zheng, X., Lee, H., Weisgraber, TH., Shusteff, M., DeOtte, J., Duoss, EB., Kuntz, JD., Biener, MM., Ge, Q., Jackson, J a., Kucheyev, SO., Fang, NX., and Spadaccini, CM. 2014, "Ultralight, ultrastiff mechanical metamaterials.", *Science*, Vol(344), pp. 1373-1377.
2. Torrents, A., Schaedler, TA., Jacobsen, AJ., Carter, WB., and Valdevit, L. 2012, "Characterization of nickel-based microlattice materials with structural hierarchy from the nanometer to the millimeter scale", *Acta Mater*, Vol(60), pp. 3511-3523.
3. Bauer, J., Hengsbach, S., Tesari, I., Schwaiger, R., and Kraft, O. 2014, "High-strength cellular ceramic composites with 3D microarchitecture.", *Proc Natl Acad Sci U S A*, Vol(111), pp. 2453-2458.
4. Fleck, NA., Deshpande, VS., and Ashby, MF. 2010, "Micro-architectured materials: past, present and future", *Proc R Soc A Math Phys Eng Sci*, Vol(466), pp. 2495-2516.
5. Deshpande, VS., Fleck, NA., and Ashby, MF. 2001, "Effective properties of the octet-truss lattice material", *J Mech Phys Solids*, Vol(49), pp. 1747-1769.
6. Montemayor, LC., and Greer, JR. 2015, "Mechanical Response of Hollow Metallic Nanolattices: Combining Structural and Material Size Effects", *J Appl Mech*, Vol(82), pp. 1-10.
7. Schaedler, TA., Jacobsen, AJ., Torrents, A., Sorensen, AE., Lian, J., Greer, JR., Valdevit, L., and Carter, WB. 2011, "Ultralight Metallic Microlattices", *Science (80-)*, Vol(334), pp. 962-965.
8. Abdelhamid, M., and Czekanski, A. 2016, "On the Effective Properties of 3D Metamaterials", *ASME, Volume 9: Mechanics of Solids, Structures and Fluids; NDE, Diagnosis, and Prognosis*, ASME, Phoenix, USA.pp. V009T12A062.
9. Abdelhamid, M., and Czekanski, A. 2016, "Effects of the Lattice Angle and the Loading Direction on the Effective Properties of the Octet-Truss Lattice Structure", Manuscript submitted for publication.
10. Wan Hu, J., and Park, T. 2013, "Continuum Models for the Plastic Deformation of Octet-Truss Lattice Materials Under Multiaxial Loading", *J Eng Mater Technol*, Vol(135), pp. 21004.
11. Altair. 2016, *HyperWorks 14.0 Desktop User Guides*, http://www.altairhyperworks.ca/.
12. Formlabs. 2017, *Online Documentation*, https://formlabs.com/.
13. ASTM. 2014, "D638 Standard Test Method for Tensile Properties of Plastics", *Annu B ASTM Stand*,pp. 1-15.
14. Stratasys. 2017, *Online Documentation*, http://www.stratasys.com/.

Proceedings of the 26th CANCAM
Victoria, BC, Canada, May 29 – June 1, 2017

A NEW META-MODELING GLOBAL OPTIMISATION ALGORITHM FOR SOLVING COMPLEX PROBLEMS WITH IMPROVED SEARCH EFFICIENCY

Abdulbaset Elhadi Saad
Mechanical Engineering
University of Victoria
Victoria, BC, CANADA
asaad@uvic.ca

Khalifa Ezaedi
Mechanical Engineering
University of Victoria
Victoria, BC, CANADA
kezaedi@uvic.ca

Zuomin Dong
Mechanical Engineering
University of Victoria
Victoria, BC, CANADA
zdong@uvic.ca

Salem Jliedi
Computer Science
Institute of Health
Tripoli , LIBYA
salem_4460@yahoo.com

ABSTRACT

Today computer modeling, virtual prototype experiments, numerical simulation and analysis become important tools in engineering design. However, many of these computer experiments and analyses are associated with black-box problems that also require intensive computation. In this work, a new Kriging-Differential Evolution (K-DE) meta-modeling based search algorithm is introduced for solving global optimization problems with limited computational budget and expensive black-box objective functions. The backbone of the new approach is Kriging meta-modeling model coupled with a differential evolution scheme to assist the DE algorithm reaching the global solution faster with less computation. During the search, Latin hypercube sampling (LHS) is employed to construct the Kriging model, which is used for identifying the promising region. Based upon this strategy, better sample points are obtained to supplement the sample population that is used to guide the DE search. The search process continues until the global optimum is found. The introduced K-DE algorithm was tested using various optimization benchmark problems. The test results revealed noticeable advantages of the new algorithm in dealing with computationally intensive black-box problems. In comparison with original DE algorithm, K-DE showed improved search efficiency and capability.

KEYWORDS: Global Optimization, Metamodel, DE-Kriging, DE, Expensive Black Box Problems

INTRODUCTION

Most of today's challenging design optimization problems with complex nonlinear objective function is complex and time consuming in computation. Some global optimization methods can identify the global optimum for many problems efficiently. For instance, Genetic Algorithm (GA), Simulated Annealing (SA), Particle Swarm Optimization (PSO) and Differential Evolution (DE) are very powerful methods for solving many tough global optimization problems [1-6]. Among them, GA is a search global algorithm that utilizes by natural evolution. The algorithm starts with an initial set of points/population and gradually improves based on the evaluation of the fitness function that measures the "quality" of the points on how good the candidate is at improving the problem towards the global optimum. A new set of points is created from previous population through the processes of crossover, mutation, and selection that are similar to a biology system where two parents in one generation mix their genes to create their children. There are many advantages of GA over conventional optimization methods. One of the most well-known is its ability to deal with complex problems [7]. SA algorithm received its name from the annealing process of metal, identifying the global optimum through a controlled search process [8], similar to the annealing process for modifying the properties of heated metal using well controlled slow cooling rate from high temperature. SA search in the design space is controlled by two parameters, temperature and cooling rate, and the method has been successfully applied in many real life application problems [9]. PSO is another global optimization method inspired by swarm behavior in nature, such as schools of fish and flocks of birds, and is one of the most widely used optimization methods due to its flexibility and simplicity of implementation [9]. DE is a vector–based random search algorithm, considered as modification of GA, due to its use of crossover and mutation, and is used for solving a wide range of real world optimization problems. DE was considered as one of the most powerful random search optimization algorithms today [9]. Many of the existing global optimization search algorithms require a significant number of evaluations of the objective/constraint functions through numerical experiments or simulations to reach the global optimum, and each of these evaluations on computationally expensive functions might take minutes or hours to complete, making accurate global search beyond reach. One way to address this problem is to use easy to calculate approximation models, known as meta-models or surrogate models, to replace the expensive functions in the global optimization. Typical design optimization problems of this type include multiple objectives, multiple disciplines, and complex simulations of the designed products, such as the wing of aircraft or automotive body. For instance, Ford Motor Company needs about 36–160 hours to run a car crash simulation [10], global design optimization using the simulation result as objective is unaffordable in reality. Nowadays, computationally intensive models are very common in engineering design optimization and many real-world problems involve conflicting objectives or constraints functions that are computationally expensive to be evaluated. A meta-modeling model mimics the original model (expensive function) using data point fitted surrogate models to guide the search with much reduced number of simulations, to efficiently and quickly locate the global optimum. A number of these approaches have been introduced in recent years, and the most popular and representative one is Kriging which is a flexible statistical modeling method, particularly useful for high-dimensional problems involving interactions and curvature. The method can identify global optimum using a smaller number of sample points, which has some similarities with the Gaussian basis function. The objective of this work is to develop a new method using metamodeling integrated with DE conventional algorithms to solve intensive black-box problems. The new algorithm Kriging-DE considerably outperformed the original DE algorithm for low budget optimization problems.

DIFFERENTIAL EVOLUTION ALGORITHM

Differential Evolution (DE) algorithm was developed by Rainer Storn and Kenneth Price [10]. It is a branch of evolutionary global optimization methods developed for optimization problems over continuous design space [10]. DE uses mutation, crossover as the search mechanism, and selection to direct the search toward the promising regions in the design space. The main difference between GA and DE is that GA in based on the crossover mechanism to exchange useful information among solutions to trace better solutions, while DE scheme uses mutation as the primary search mechanism [11]. DE is one of the best evolutionary algorithms for solving highly nonlinear optimization problems and has been used in a number of engineering applications. On the other hand, the performance of DE relays heavily on the selected mutation and crossover mechanisms and their associated parameters. In this work, a metamodel based Kriging-DE algorithm was proposed to overcome trial and error issue and to solve computationally expensive optimization problems with higher efficiency.

METAMODELING MODEL

Computational cost and mathematical formulation are two main concerns in engineering optimization applications [12]. Metamodel is the "model of model", also called surrogate model or approximation model. It has been widely used in the recent years to solve optimization problems with complex computationally expensive objective/constrains functions. Metamodeling techniques are mostly used for replacing the actual analysis/simulation models, such as FEA and/or CFD models, with an explicit approximation model to reduce computationally intensive function evaluations in the optimization search process. Most engineering analyses and simulations cannot be represented as explicit mathematical form, referred as black-box functions. Different metamodels, such as radial basis function (RBF), Kriging, multivariate adaptive regression splines (MARS), and polynomial regression (PR) models, have been used in metamodel based global optimizations, although their search efficiency is case dependent. In this work, metamodeling will be used as a surrogate of the original objective function to guide DE algorithm to find the global optimum with a limited number of sample points due to computation cost.

KRIGINH MODEL

Kriging is an interpolation method widely used for estimating the objective function of black-box problems. The Kriging interpreter and its estimated mean square error (MSE) function have the following forms:

$$\hat{f}(\boldsymbol{x}) = \hat{\mu} + r^T(\boldsymbol{x})\boldsymbol{R}^{-1}(\boldsymbol{y}^{-1}\hat{\mu}) \tag{1}$$

$$\hat{s}^2(\boldsymbol{x}) = \hat{\sigma}^2\left[1 - \boldsymbol{r}^T(\boldsymbol{x})\boldsymbol{R}^{-1}\,\boldsymbol{r}(\boldsymbol{x}) + \frac{(1-\mathbf{1}^T\,\boldsymbol{R}^{-1}\boldsymbol{r}(\boldsymbol{x}))^2}{\mathbf{1}^T\boldsymbol{R}^{-1}\mathbf{1}}\right] \tag{2}$$

where, $\hat{f}(\boldsymbol{x})$ is the approximation of the true function $f(\boldsymbol{x})$, $\boldsymbol{R}$ is a correlation matrix with $M{\times}M$ elements, $\hat{s}^2(\boldsymbol{x})$ reflects the uncertainty of the predictor $\hat{f}(\boldsymbol{x})$, and $\boldsymbol{r}(\boldsymbol{x})$ is a M-dimensional vector.

THE PROPOSED KRGING-DE ALGORITHM

The proposed K-DE approach uses a Kriging-based SM, DE algorithm, and alternating sampling over the design space. Kriging SM and DE are used to accelerate the global optimization search. The local optima obtained from the SM are stored in a database as "current population for DE to start as initial population". In each iteration, several better sample points are selected from these "potential sample points by Kriging-SM" to refine the SM to allow it to have a better approximation in the close neighborhood of the global optimum. The detailed search algorithm is given as:

Step 1: Generate an initial number of sampling points in the whole design space using LHS
Step 2: Calculate and evaluate the original objective function
Step 3: Construct the SM using Kriging to replace the original objective function by using all sampling points in the sample set
Step 4: Calculate and evaluate the constructed model built using Kriging.
Step 5: Use the best objective function points from the cheap model (Kriging) as initial population
Step 6: Apply the DE optimization algorithm to find the optimum using the initial population from the SM in the reduced search space, and add the obtained optimum to the sample set
Step 7: Repeat the previous steps until the global optimum is identified or the stopping criteria is met.

TEST RESULTS AND DISCUSSION

The main objective of this newly introduced method was to increase the search efficiency of the DE algorithm for global optimization problems with tight computation time/effort constraint. The superior performance of the novel Kriging-DE approach is compared with the original DE algorithm using several benchmark optimization functions with different numbers of variables. For this purpose, commonly used benchmark optimization functions are used as test examples to reveal the pros and cons of the new global optimization method for solving expensive black-box problems.

Eight global optimization functions with 2 to 5 design variables were used as test examples, including five two-dimensional functions (Banana, Peaks, GP, SC, Beale), one three-dimensional function (Sum square), one four-dimensional function (Sargan), and one five dimensional function (Sphere) [13, 14].

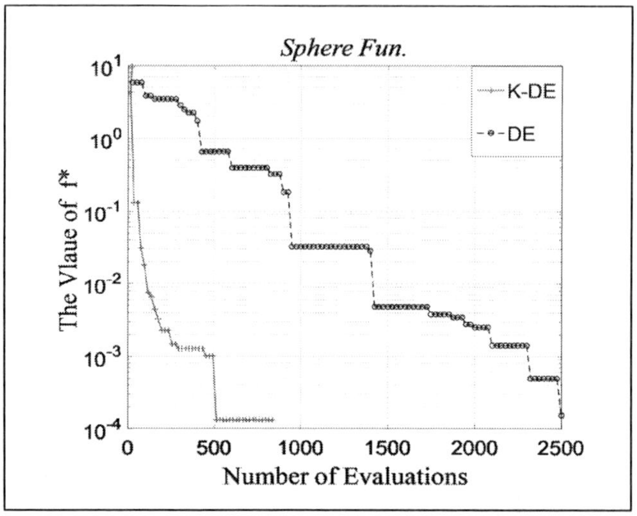

Figure 2: Convergence Plot of f_8 (Sphere)

Summary of test results and the performance comparison

Fun	Design Space	Analytic minimum	Mothed						NFE Saved %
			K-DE			DE			
			f^*	NFE	CPU	f^*	NFE	CPU	
Banana	[-2,2]	0	3.502E-5	819	1.3438	9.425E-5	1500	0.0937	45%
GP	[-2,2]	3	3	843	1.250	30001	1500	0.0468	44%
SC	[-2,2]	−1.0320	-1.0316	784	1.2969	-1.0309	1500	0.1250	48%
Peaks	[-3,4]	−6.5511	-6.5511	75	0.8906	-6.5511	1000	0.0625	92%
Beale	[-4.5,4.5]	0	2.126E-6	816	0.8125	1.542E-6	1500	0.0609	46%
Sum. S	[-10,10]	0	4.092E-4	658	0.7187	1.098E-4	2000	0.1250	67%
Sargan	[-4,4]	0	3.327E-4	467	0.6406	5.462E-4	1500	0.0937	69%
Sphere	[-5.12,5.12]	0	1.320E-4	834	1.2813	1.545E-4	2500	0.1562	67%

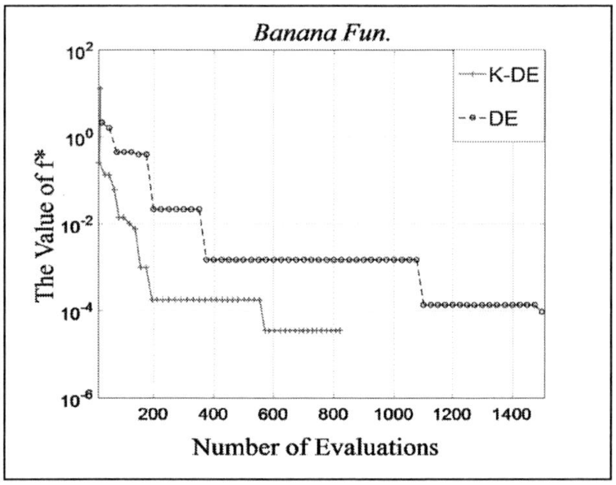

Figure 1: Convergence Plot of f_1 (Banana)

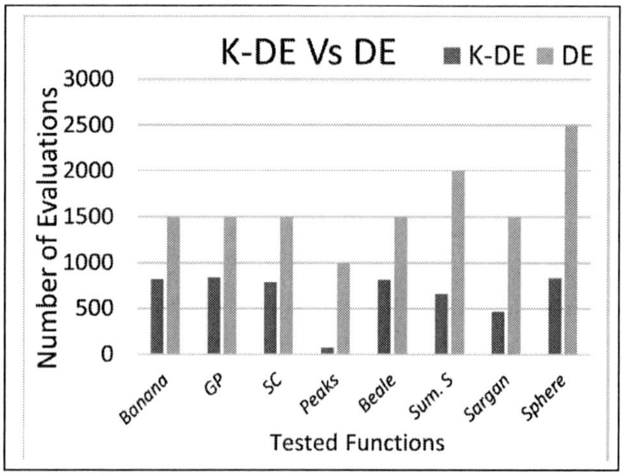

Figure 3: NFE DE versus K-DE for tested functions

A series of modifications has been made, leading to the superior search efficiency of Kriging-DE with considerably less number of function evaluations (NFE). A notable improvement has been achieved for each of the eight tested functions. The history of convergence in both methods can clarify the efficiency of the proposed modification algorithm. Table 1 summarizes the test results, and Figures 1 and 2 show the convergence trend of benchmark test problems. The same graphs can be plotted for the other six benchmark functions. As it can be seen from Table 1 that in order to reach the same accuracy, the NFE required by the Kriging-DE is significantly smaller than NFE for the original DE algorithm. Figure 3 illustrates the required NFE in DE versus Kriging-DE for all test functions. The introduced method needed the least number of function evaluations and appeared to be most robust, presenting to be the most promising low budget global optimization problems

CONCLUSIONS

In this work, a new surrogate based optimization algorithm, called Kriging-DE (K-DE), has been introduced. As an extension of the original DE algorithm, the K-DE algorithm were tested and compared to DE method to show its pros and cons. Both original DE and K-DE metamodel based global optimization algorithms were tested to explore the efficiency and capability of the new method using a variety of benchmark test problems. DE algorithm yields high accuracy results with high computational cost. The meta-modeling based algorithm Kriging-DE presents comparable search accuracy and superior computation efficiency, which make it an excellent approach for solving computationally intensive, computer simulation based global design optimization problems with computation time/expense constraint.

ACKNOWLEDGMENTS

Supports from the Libyan Ministry of Higher Education, University of Victoria and authors' families are gracefully acknowledged.

REFERENCES

1. Goldberg, D., 1989, "Genetic Algorithms in Search, Optimization and Machine Learning", Addison-Wesley, Reading, MA.
2. Deep, K., Bansal, J. C., 2009, "Mean particle swarm optimization for function optimization", Int. J. Compute. Intel. Studies, 1, 72-92.
3. Geem, Z.W., Kim, J. H., Loganathan, G. V., 2001 "A new heuristic optimization algorithm: Harmony search Simulation", 76, 60-68.
4. Holland, J. H., 1975, "Adaptation in Natural and Artificial Systems", University of Michigan Press, Ann Arbor.
5. Kennedy, J. and Eberhart, R.1995 "Particle swarm optimization", Proc. IEEE Int. Conf. Neural Networks. Perth, Australia, 1942-1945.
6. Storn, R. and Price, K., 1997, "Differential Evolution - A Simple and Efficient Heuristic for Global Optimization over Continuous Spaces", Journal of Global Optimization, 11, pp. 341–359.
7. Yang, X.-S., 2008, "Nature-inspired Metaheuristic algorithms", Luniver Press.
8. Kirkpatrick, S., Gelatt, C. D., and Vecchi, M. P., 1983, "Optimization by simulated annealing", Science, 220, 671-680 (1983).
9. Yang, X.-S., 2012."Nature-inspired Optimization Algorithms", Elsevier Insights.
10. S. Das, P.N. Suganthan, 2011, "Differential evolution: a survey of the state-of-the-art", IEEE Trans. Evol. Comput. 15, 4–31.
11. K.V. Price, R.M. Storn, J.A. Lampinen, 2005, " Differential Evolution: A Practical Approach to Global Optimization", Springer, Berlin.
12. Wang, L., Shan, S., and Wang, G., 2004. "Mode-Pursuing Sampling Method for Global Optimization on Expensive BlackBox Functions." Engineering Optimization, 36(4): 419-438.
13. Younis, A., and Dong, Z., 2010,"Metamodeling Search Using Space Exploration and Unimodal Region Elimination for Design Optimization," Engineering Optimization, 42(6), pp. 517-533.
14. Gu J, Li GY, Dong Z., 2012, "Hybrid and adaptive meta-model-based global optimization". Eng Optim 44(1):87–104.
15. Long T, Wu D, Guo X, Wang GG, Liu L., 2015, "Efficient adaptive response surface method using intelligent space exploration strategy", Struct. Multidiscip Optim 51(6):1335–1362

DESIGN OPTIMIZATION OF AUTOFRETTAGED PRESSURE VESSEL

Dr. Abu Rayhan Md. Ali*
Department of Mechanical Engineering
Bangladesh University of Engineering & Technology
Dhaka, Bangladesh
Email: armali@me.buet.ac.bd

Aminur Rahman, Nidul Ch. Ghosh , Tanvir-E-Alam
Department of Mechanical Engineering
Bangladesh University of Engineering & Technology
Dhaka, Bangladesh

ABSTRACT

Finite element analysis is a computational technique used to obtain the effect of autofrettage process in thick-walled pressure vessels. Equivalent von-Mises stress is used as yield criterion to evaluate the optimum autofrettage pressure and the optimum radius of elastic-plastic junction. It has been remarked that the optimum autofrettage pressure increases along with the working pressure. It is also observed that to reduce the flow stress within the wall of the cylinder, the autofrettage pressure must be greater than the working pressure. For two different working pressures, the effect of the ratio of outer to inner radius (b/a=K) value on the optimum autofrettage pressure is also noticed. The Optimum autofrettage pressure solely depends on K value rather than on the inner or outer radius. Furthermore, percentage reduction of von-Mises stresses is compared for different working pressures and different K values.

KEYWORDS: Working pressure, optimum radius, von-Mises stress, thick-walled cylinder, elastic-plastic junction

1. INTRODUCTION

Pressure vessels are used to store and transmit liquids, vapors, and gases under pressure in general. For these cases, fluid will exert pressure equally in all direction on the walls and ends of the pressure vessels. Because of the internal pressure, stresses are developed on certain sections of the cylinder (pressure vessel) walls. Different techniques are applied to increase the capacity of pressure vessel [1]. To contain a high pressure would typically require a very thick - walled tube due to the concentration of tensile hoop stresses at the inner diameter (ID). The magnitude of pressure is also limited by the material yield stress, which must not be exceeded in normal use. Autofrettage is a well- known elastic-plastic technique to increase the pressure capacity of thick-walled cylinders. In this technique, the cylinder is subjected to internal pressure to cause plastic expansion of some or the entire tube wall. The pressure is then released and residual compressive hoop stresses are created in the near-bore region while residual tensile hoop stresses are created in the outer-bore region. The resulting residual stress leads to a decrease in the value of maximum von-Mises stress in the next loading stage. A key problem in the analysis of autofrettage process is to determine the optimum autofrettage pressure and corresponding radius of elasto-plastic boundary where the maximum equivalent von-Mises stress in the cylinder becomes minimal. To compute optimum autofrettage pressure ANSYS software is employed for numerical simulation.

2 . ANALYTICAL APPROACH

Bi-Linear elasto-plastic behavior has been considered in this work.

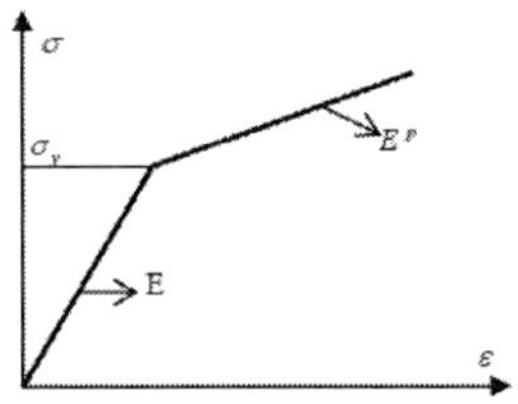

Figure 1: Bi-linear stress strain curve

The model, shown in figure 1 is described as follows:

$$\sigma = \sigma_y + \varepsilon E^p \quad (1)$$

In which σ is the effective stress, σ_y is the initial yield stress, E^p is the slope of the strain hardening segment of the stress strain curve, and ε is the effective strain.

2.1 RESIDUAL STRESS PATTERN

A simple case study: Let's consider an Aluminum cylinder where internal radius a= 0.01 m, and external radius, b = 0.02m. Material properties are summarized in table 1.

Table 1: Material properties

Mat.	σ_y(MPa)	E(GPa)	E^p(GPa)	ν
Al	90	72	1.75	0.33

This cylinder is subjected to an internal pressure (known as autofrettage pressure) so that its wall becomes plastic up to r/inner radius =1.56 and the pressure is then released. Ghomi&Majzoobi [2] proposed set of equations for determining radial and hoop stresses at different location along the cylinder wall in autofrettaged cylinder. By using the equations the resulting residual stress pattern is shown in figure 2:

*All correspondence should be made to Professor Dr. Abu Rayhan Md. Ali
Email: ***armali@me.buet.ac.bd***

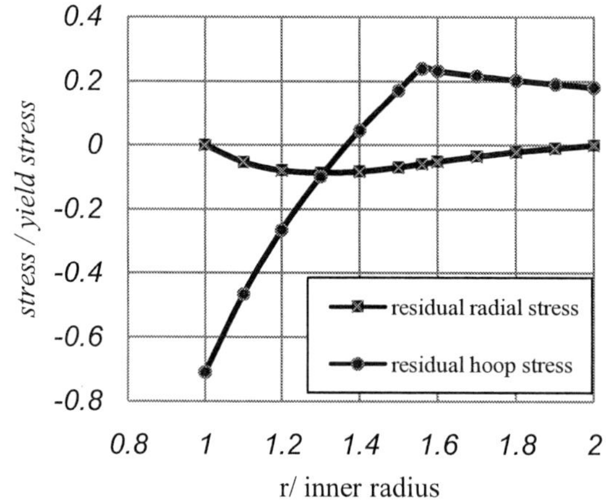

Figure 2: Residual Stress Distribution

From figure 2, it is observed that residual compressive hoop stress occurs in near-bore region, while residual tensile hoop stress occurs at outer portion. The resulting residual compressive hoop stress leads to a decrease in the maximum value of the von-Mises stress in the next loading stage.

2.2. COMPARISON OF STRESSES WITH AND WITHOUT AUTOFRETTAGE

By using Lame's equation for thick wall cylinder, the stress pattern is obtained for non autofrettaged cylinder. If the same cylinder undergoes autofrettage process then the overall stress pattern will change, which is shown in Figure 3. Here the working pressure is 46 MPa.

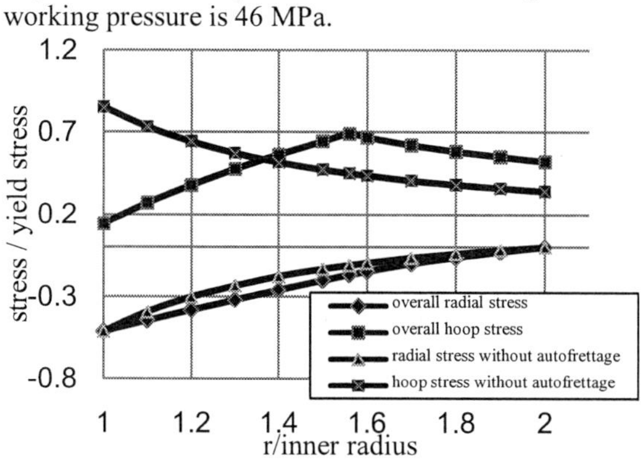

Figure 3: Comparison of Stresses with and without autofrettage.

From figure 3, following points are observed: 1. Because of residual compressive hoop stress at inner bore, the resultant hoop stress becomes significantly lower in the autofrettaged cylinder than the original hoop stress developed at the same cylinder. 2. Radial stress doesn't vary significantly after autofrettage process. 3. The cylinder which undergoes autofrettage process has maximum stress occurring at the point of elasto-plastic junction rather than at the inner bore.

2.3. OPTIMUM ELASTIC PLASTIC RADIUS

For different radius of elasto plastic junction developed, von-Mises stresses are calculated using Ghomi&Majzoobi's [2] proposed set of equations. Then the results are shown in Figure 4.

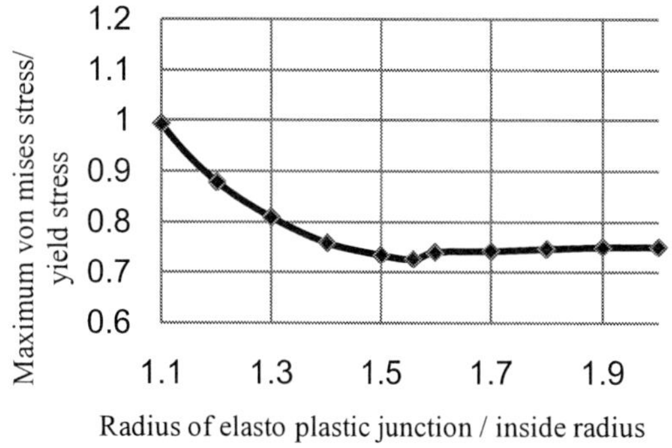

Figure 4: Maximum von-Mises Stress at Different Radius of Elasto Plastic Junction

As the figure suggests, Maximum von-Mises stress starts to decrease as the radius of elastic plastic junction increases. After attaining a certain value of elastic plastic junction, maximum von-Mises stress started to increase. The point at which maximum von-Mises stress is minimal is the optimum radius of elasto-plastic junction.

3. NUMERICAL RESULTS

Commercially available software ANSYS 10.0 has been employed for finite element modeling of the Autofrettaged vessel. The element Quad 4 Node PLANE 42 with the capacity of elastic and plastic material modeling has been used for the modeling [3].

Single cylinder with the dimensions; a=0.1 m, b=0.2 m and an elastic plastic material's model with σ_y= 800 MPa; Modulus of elasticity E = 207 GPa; Slope of the strain hardening segment E^p = 4.5 GPa; ν = 0. 29; were used for numerical modeling. The two pressures limits Py1 and Py2 can be computed as follows [4]: Here Py1 is the pressure at which yielding commences at inner surface and Py2 is the pressure at which plasticity has spread throughout the cylinder.

$$P_{y1} = \sigma_y (1-1/k^2)/\sqrt{3} = 347 \text{ MPa}$$
$$P_{y2} = \sigma_y \ln(k) = 555 \text{ MPa}$$

If the autofrettage pressure is lower, then 347MPa then there will be no autofrettage effect. If the pressure is higher than 555MPa then there will be converse effect. That means the pressure capacity of the cylinder will decrease instead of increasing.

In this paper, effects of following factors are considered in autofrettage process. The considered factors are 1. Working pressure; 2. Value of K (b/a).

3.1 WORKING PRESSURE

The cylinders were subjected to autofrettage pressures ranging from 350 MPa to 650 MPa. After removing the autofrettage pressure (AP), the cylinders were subjected to the working pressures of 100, 200 and 300 MPa. From the numerical simulation with ANSYS software, the curve of the von-Mises stress distribution was obtained for each autofrettage and working pressure (WP). From the curve, the value and the position of maximum von-Mises stress (MVS) were extracted. This stress and its position were then plotted against autofrettage pressure for each working pressure. The results are shown in figure 5.

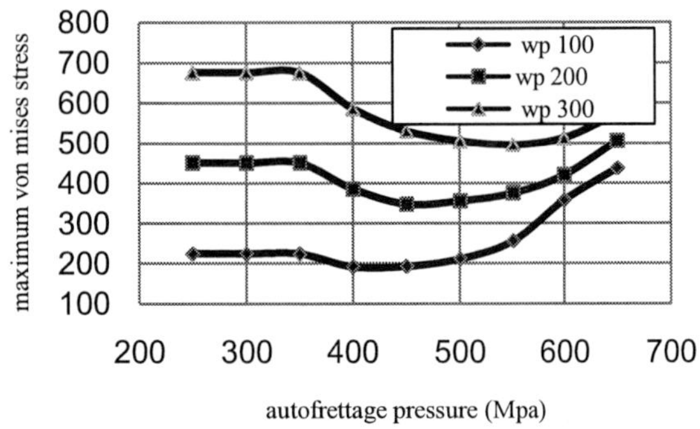

Figure 5.Variation of MVS vs.Autofrettage Pressure at different Working Pressure.

It is observed that for each working pressure, the MVS remains constant up to an autofrettage pressure which is nearly equal to P_{y1}. The curve then begins to decline to a certain point thereafter begins to rise or remains constant. It can be seen that for all working pressures, the rising portion of the curves end at a point which is nearly equal to P_{y2}.

From the numerical results, it can be concluded that: (i) the MVS depends on the working pressure and for any WP, the best A_P lies between P_{y1} and P_{y2}; (ii) for autofrettage pressures lower than P_{y1} the MVS remains unchanged.

In the preceding case the working pressure is lower than P_{y1} but if the working pressure itself is higher than P_{y1} then the effect of autofrettage does not start to occur until exceeding the working pressure. In the figure 6 two working pressure of 400 and 450 MPa are considered.

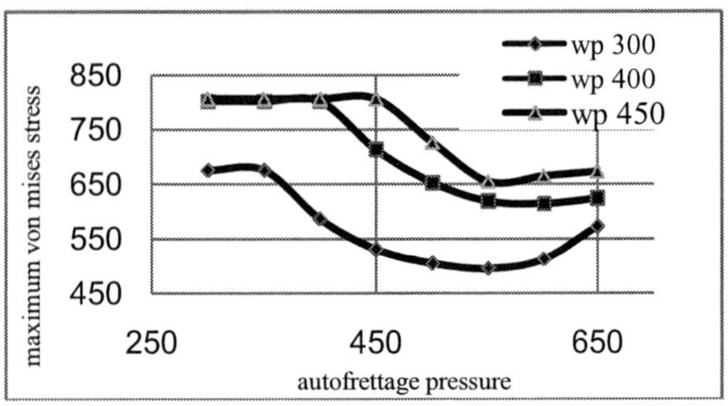

Figure 6: variation of MVS versus autofrettage pressure at different working pressure

In case of 400 and 450 MPa the developed von-Mises stress starts to decrease only after exceeding working pressure. If the autofrettage pressure is lower than the working pressure then the flow stress remains unchanged thus there is no autofrettage effect.

For a constant value of K percent reduction of MVS is calculated for different working pressures.

Table 2: Effect of working pressure at maximum von-Mises stress

WP (MPa)	MVS Without autofrettage (MPa)	MVS With autofrettage (MPa)	%Reduction inMVS
100	225	193	14.22
200	450	348	22.67
300	676	496	26.62
400	840	615	26.78

From table 2, it is observed that percent reduction of MVS is higher at higher working pressure. This means the autofrettage effect is more beneficial at higher working pressure.

2.2 VALUE OF K (b/a)

For a constant inner radius a=0.1m, cylinder with different K values (K= 2, 2.5, 3.0) are subjected to autofrettage process. Here the working pressure remains constant at N_w=0.25. Where N_w= Working pressure P_w/ yield stress σ_y.

For K=2.0: P_{y1} = 347 MPa, P_{y2} = 555 MPa.
For K=2.5: P_{y1} = 388 MPa, P_{y2} = 733 MPa.
For K=3.0: P_{y1} = 411 MPa, P_{y2} = 879 MPa.

From numerical simulation, the value and the position of MVS were extracted. This MVS is then plotted against the autofrettage pressure for each K value. The results are shown in Figure 7.

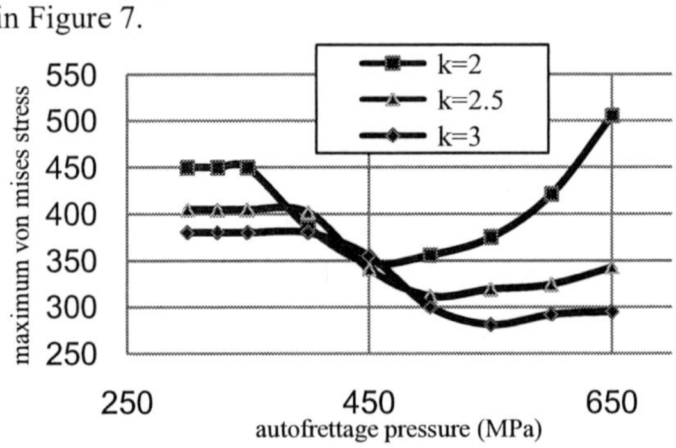

Figure 7: Effect of K (b/a) On Optimum Autofrettage Pressure (constant inner radius = 0.1m)

As the figure suggests, for different values of K, the optimum autofrettage pressure obtain a higher value for thicker cylinder.

For a constant working pressure (N_w=0.25), percent reduction of MVS is calculated for different K values.

Table 3: Effect of K (b/a) on MVS

K(b/a)	MVS Without autofrettage (MPa)	MVS With autofrettage (MPa)	%Reduction in MVS
2.0	450	348	22.67
2.5	405	312	22.97
3.0	380	281	26.05

From table 3 it is observed that, percent reduction of MVS is higher for higher value of K. That means the autofrettage effect is more beneficial with the increase of the thickness of the cylinder wall.

In the preceding case the inner radius of the cylinder was kept constant. Now by assuming the outer radius constant (b = 0.3m), cylinders with three K values of 2.0, 2.5 and 3.0 were considered. From the numerical simulations, the curve of von-Mises stress distribution was obtained for each K value. The developed von-Mises stresses are then plotted against autofrettage pressure in Figure 8.

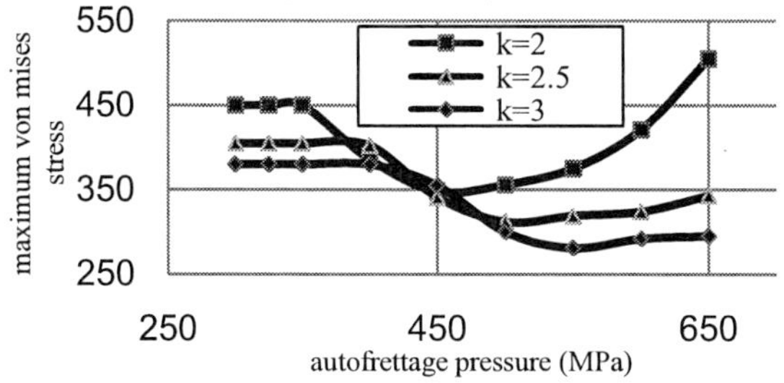

Figure 8: Effect of K (b/a) on Optimum Autofrettage Pressure (constant outer radius = 0.3m)

From figure 8, it is noticed that for same K value, developed MVS & the optimum autofrettage pressure is same though the value of inner and outer radius have been changed. Thus, the optimum autofrettage pressure depends on K value only. If the inner and outer radius are changed keeping the K value constant, then there will be no change in the optimum autofrettage pressure.

Now for a constant working pressure of N_w=0.375, cylinder with different K values (K= 2, 2.5 and 3.0) are subjected to autofrettage process. From numerical simulation the value and position of MVS were extracted and then plotted against the autofrettage pressure. The results are shown in Figure 9.

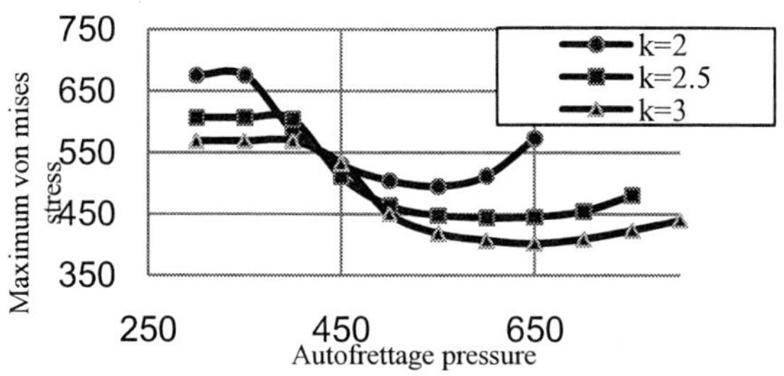

Figure 9: Effect of K (b/a) on Optimum Autofrettage Pressure (working pressure 300 MPa)

From figure 9, it is also noticed that for same K value, developed MVS & the optimum autofrettage pressure is same though the value of inner and outer radius have been changed. Thus, the optimum autofrettage pressure depends on K value only.

CONCLUSION

From the analytical and numerical results the following conclusions can be drawn:

1. In autofrettaged cylinder, maximum stress does not occur at inner bore instead of that, it occurs at the radius of elastic plastic junction. As the autofrettage pressure increases the point of maximum stress move towards the outer bore.
2. Optimum autofrettage pressure is not a constant value rather it depends on the working pressure. The optimum autofrettage pressure increases along with the increase of working pressure.
3. For same working pressure, increasing the ratio of outer to inner radius (K) leads to an increase in the optimum autofrettage pressure.
4. Optimum autofrettage pressure depends on K value rather than on the internal or external radius.

ACKNOWLEDGEMENT

The authors deeply acknowledge the financial and technical support of the Department of Mechanical Engineering, Bangladesh University of Engineering and Technology, Dhaka, Bangladesh.

REFERENCES

1. Ali, Abu Rayhan Md., Rahman, Aminur., Amin, Md. Tanjin., Optimum Design of Compound Pressure Vessel, Global Journal of Researches in Engg. Mech. and Mechanics Engg. (USA), Volume 13, Issue 11, Version 1.0, Year 2013.
2. GH Majzoobi, GH Farrahi, AH Mahmoudi, A finite element simulation and an expt. study of autofrettage for strain hardened thick-walled cylinders, J. Mater. Sci. Eng. A., vol. 359 pp. 326-31, 2003.
3. Ghosh, Nidul Ch., Tanvir-E-Alam, Theo and numerical opt of autofrettage in strain hardened thick wall cylinders, B.Sc. Engg thesis, ME. BUET, Dhaka, Bangladesh, 2009.
4. GH. Majzoobi, A. Ghomi, Opt of comp. pre cylinder, , Journal of Achievements in Materials and Manu Engg., In press, 2006.

NOMENCLATURE

Symbol	Meaning	Unit	Symbol	Meaning	Unit
a	Internal radius	(m)	σ_{rr}	Residual radial stress	(MPa)
b	External radius	(m	σ_{rf}	Overall radial stress	(MPa)
$\sigma_{\theta r}$	Residual hoop stress	(MPa)	P_w	Working Pressure	(MPa)
$\sigma_{\theta f}$	Overall hoop stress	(MPa)	N_w	Working pressure,P_w/yield stress, σ_y	Nill

A REVIEW OF PHEV CONTROL, ENERGY MANAGEMENT AND THEIR REAL-TIME OPTIMIZATION

Haijia Zhu
Department of Mechanical Engineering
University of Victoria
Victoria, BC, Canada
hjzhu@uvic.ca

Zuomin Dong
Department of Mechanical Engineering
University of Victoria
Victoria, BC, Canada
zdong@uvic.ca

ABSTRACT

Using the UVic EcoCAR2 prototype Plug-in Hybrid Electric Vehicle (PHEV) as an example, a quick introduction and comparison of various HEV/PHEV powertrain architectures and the need of real-time optimal control are presented. The method for modeling PHEV powertrain system and power control/energy management development are discussed. The advantages and limitations of various HEV/PHEV real-time energy management techniques are outlined to provide hints for HEV/PHEV optimal energy management research.

KEYWORDS: PHEV, Multi-regime PHEV, ECMS, DP, PMP

INTRODUCTION

Hybrid electric vehicles (HEVs) showed rapid developments in recent years due to their improved energy efficiency and reduced emission through the optimized control of coordinating mechanical and electric drive components. Plug-in hybrid electric vehicles (PHEVs) presented room for additional improvement with larger on-board electric energy storage system (ESS) and capability of grid charging, and also demanded more advanced energy management strategies.

The optimal power control and energy management of HEV and PHEV seek the best trade-off among vehicle energy efficiency, performance, drivability, and utility. In this work, we examine the representative optimal control methods of HEVs introduced over the past 20 years, and the more recent optimal management techniques for PHEVs.

Different HEVs are classified into three major groups based upon their powertrain architectures: series, parallel and power-split (or series/parallel), as shown in Figures 1-3.

Series hybrid powertrain allows the internal combustion engine (ICE) to operate at its more efficient, ideal operating conditions, relies on the larger electric machine to propel the vehicle as an electric vehicle (EV), and uses the on board ESS to store surplus energy or fill in the energy gap during vehicle speed and load variations. The design often experiences mechanical-electrical-(chemical-electrical)-mechanical energy conversion losses. PHEV's charge depletion (CD) mode operates as an EV while the used energy from ESS is likely from grid charge, possibly from renewable energy sources.

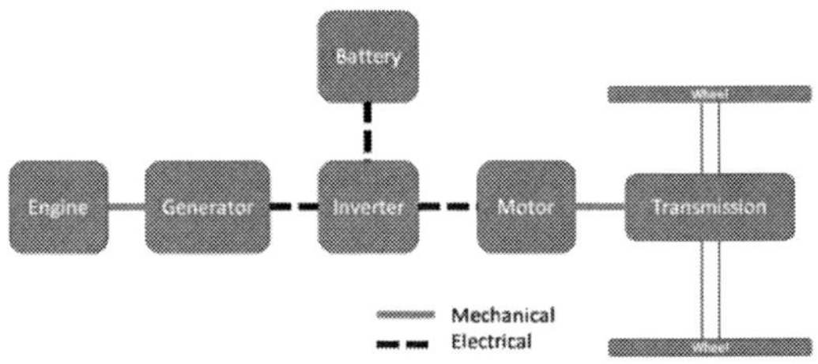

Figure 1: Series architecture

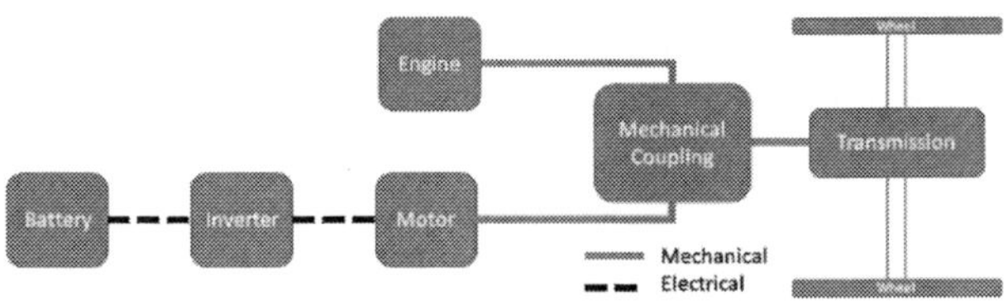

Figure 2: Parallel architecture

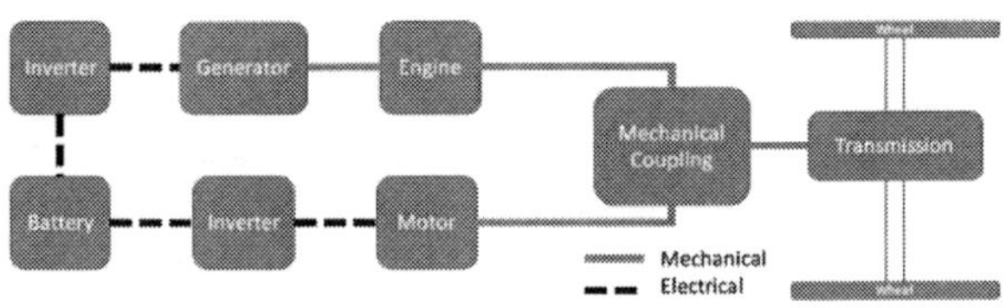

Figure 3: Series-parallel architecture

The more popular parallel hybrid powertrain allows the mechanical and electric drives to propel the vehicle jointly. The ICE and motors are coupled to the final drive through direct connection or some special mechanism. The wasteful energy conversion is avoided, while operation condition of the ICE is tied to the vehicle speed and load with less chance to operate optimally.

The series-parallel powertrain architecture, also known as power split, is one of the most successful architecture in the midsize hybrid vehicle market, this architecture combines the advantage of both parallel and series architectures to operate in series or parallel mode to best meet vehicle operation conditions. The market leading HEV Toyota Prius and our UVic EcoCAR2 are in this category. The advantage of power split architecture lies on its more degree of freedom for vehicle operation control to allow more advanced control methods for achieving better energy efficiency, performance, and drivability. However, the system and its control are much more complex and warrant for in-depth review and analysis. In this paper, the powertrain system model of our UVic EcoCAR2 is introduced using the US ANL's powertrain system modeling tool AUTONOMIE [1], [2] to illustrate the powertrain architecture and components. A literature review on power control and energy management strategies is then presented.

UVIC ECOCAR2 POWERTRAIN SYSTEM MODEL

The vehicle was designed as a series/parallel PHEV with multi-regime of operations. It can operate as a rear wheel drive (RWD) EV using the a Motor/Generator (M/G) as rear traction motor (RTM), or as a front wheel drive (FWD) and RWD parallel hybrid vehicle using the ICE and the super belted alternator starter (BAS) motor. It also possesses all-wheel drive (AWD) capability. Its powertrain system and major components are illustrated in Figure 4 and Table 1. The key powertrain components include an ICE, two electric machines, transmission, and ESS. The introduced model contains the *following* main parts, *as shown in Figure 5.*

Driver Model

A forward vehicle simulation model receives the speed demand and deduce the required torque to meet the desired vehicle speed, the information flow from the pedals to engine and motors to gearbox to final drive to tire. To determine the torque demand that send to the vehicle controller, a driver model, essentially a PI controller, is needed. The driver model controls vehicle speed according to the given driving cycle, then deduces a pedal position signal and sends it to the supervisory controller.

Supervisory Controller

The supervisory controller commands several sub-controllers that are based different control strategies and algorithms [3].

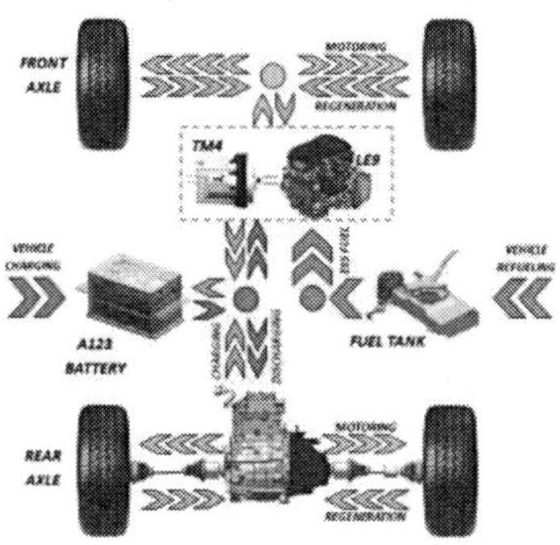

Figure 4: UVic EcoCAR2 Powertrain

Table 1: Powertrain Components

Retrofitted Vehicle	2013 Chevrolet Malibu
GM LE9 2.4L EcoTEC Engine	130kW @5800rpm, 2300Nm @5000rpm
GM 6T40 Transmission	6-Speed AT
TM4 Drive(BAS)	80kW peak 37kW continuous
Magna E-Drive	90kW peak 45kW continuous
A123 System ESS	152kW peak 51kW continuous 14.5kWh usable capacity

Powertrain Model

The engine and M/G are modeled using dyno test data embedded in look up tables and efficiency maps for fact execution in optimal control. Specifically, the engine is modelled using a fuel consumption map in terms of engine speed and torque.

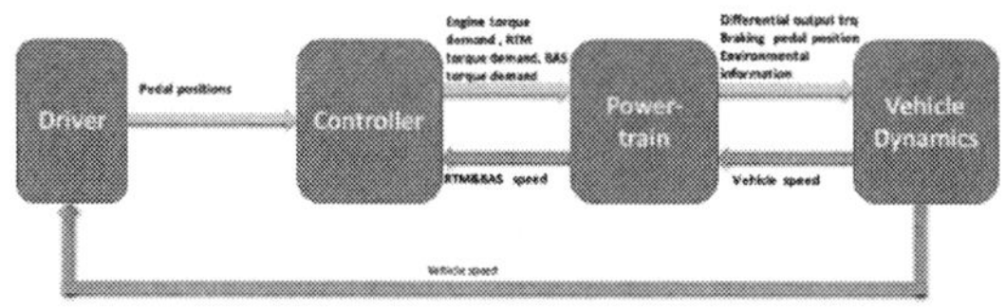

Figure 5: EcoCAR2 Model Structure

RTM and Super BAS are modelled using efficiency maps at different speed and torque, combined with inverter efficiency. These permanent-magnet synchronous motor (PMSM) M/G has a high-power density.

The PNGV model is used to capture the behavior of battery in this work as shown in Figure 6. In the equivalent circuit model, $OCV$and V_L are the open circuit and loaded voltage; R_0, R_1 and $R_2$are battery internal impedance and polarization resistances; C is the shunt capacitance around the R_1andR_2. The polarization currents can be induced by integration of the Eq. 1.

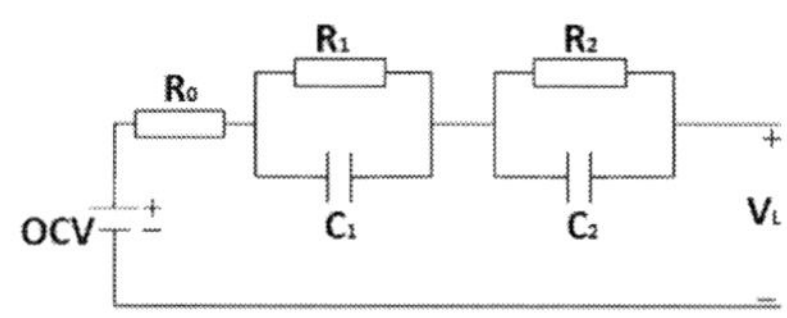

Figure 6: Battery Electric Circuit Model

$$\frac{dI_p}{dt} = \frac{I_L - I_p}{\tau} \qquad (1)$$

where τ is constant; I_L and I_p are the load and polarization current. The battery dynamics can be induced as:

$$\frac{OCV - V_L}{I_L} = R_0 + R_{p1} \times I_{p1}/I_L + R_{p2} \times \frac{I_{p2}}{I_L} \qquad (2)$$

Vehicle Dynamics Model

Typically, HEVs and PHEV models neglect the 3D multi-body dynamics of vehicles, instead, representing them as a particle mass with rolling resistance and aerodynamic drag, contributing to the power loss. The road load is then:

$$F_{RL} = F_{rr} + F_{ad} + F_{rg} \tag{3}$$

where F_{rr} is the rolling resistance of the wheel/tire, F_{ad} is the aerodynamic drag, and F_{rg} is the road grade resistance.

ENERGY MANAGEMENT STRATEGY

The commonly used optimal energy management strategy (EMS) for HEVs is aimed at properly propelling the vehicle with optimal operation mode and speed/torque allocation of ICE and M/G. Researchers at the US Argonne National Lab (ANL) compared two different powertrain configurations and different control strategy, and the impaction of control strategy on mean fuel consumption is presented [4]. Numerous EMS strategies have been developed for HEVs and extended to PHEV. With the ability to operate as an EV, the energy management strategy of PHEV, however, is quite different. Typically, its energy management strategy can be classified into two type: rule based and optimization based, as shown in Figure 7.

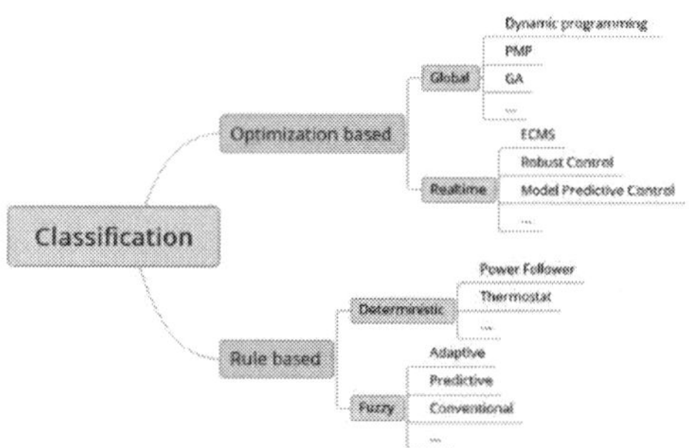

Figure 7: Types of HEV/PHEV EMS Strategies

Rule-Based Control Strategy

Thermostat Control Strategy

Heuristic rule-based control strategies that apply on a series HEV is straight forward, this method manage the power flow primitively, by turning on and off the engine periodically to maintain proper ESS State of Charge (SOC).

Power Follower Control strategy

The strategy follows the vehicle power load:

- Below certain vehicle speed, using M/G only (EV mode).
- When power demand is greater than the ICE can provide at a specify speed, the M/G kicks in to assist the ICE.
- When braking is applied, charge the ESS using regenerative power with assistance of mechanical brake.
- When ICE operates at an inefficient state, turn off the ICE or force it to operate at a more efficient operating point.
- If the SOC is below the threshold, increase ICE output.

The power follower strategy significantly outperforms simple thermostat control strategy. However, it is not optimal.

Fuzzy Rule Based Control

Fuzzy logic allows approximate reasoning, to make it easier to apply generic human experience in designing controller. The HEVs/PHEVs are typically a nonlinear, multi-domain and time varying complex system and it is hard to determine the exact state for the M/G to kick in. In [5], a fuzzy controller for parallel HEV was introduced. The result shows that the control is robust with considerably reduced emissions.

Optimization Based Control

Rule based energy management strategy is driven by experience and knowledge, and is not optimized. Due to the complexity of the PHEV powertrain, a global optimal solution is hard to achieve in real time. A close to optimal solution is normally used using the following schemes.

Dynamic programming (DP)

DP is a widely used method to solve the optimal control problem for a trip or for PHEV between two charges with the globally optimal solution. It cannot be applied to real time control due to the need of the entire trip information in advance and need of huge computation resources [6], but the method serves well for optimal control performance benchmark generation. For a vehicle condition represented as

$$x_{k+1} = f(x_k, u_k) \tag{4}$$

where k is an integer number, u_k is the control variable related to instantaneous fuel consumption and electric energy use at time k. Optimal control is then formulated as:

$$\min \; J(x_0, u) \tag{5}$$

where x_0 is the initial vehicle condition and u is a vector that contains all the control variables.

Equivalent Consumption Minimization Strategy (ECMS)

ECMS transfers the original global optimization problem into an easy to solve simple minimization problem for each instance of time using only past information. In the CS mode of PHEV, ESS is treated as an energy buffer to store or lease energy from the ICE dynamically while maintaining the ICE and the system to operate at peak energy efficiency. The virtual or equivalent fuel consumption of the M/Gs is [7]

$$\frac{dm_{ress}(t)}{dt} = \frac{s(t)}{Q_{lhu}} P_{batt}(t) \tag{6}$$

where $s(t)$ is the equivalence factor for converting used electrical energy into equivalent fuel consumption; $m_{ress}(t)$ is the equivalent fuel consumption. The sum of the virtual and actual instantaneous fuel consumption is then:

$$\frac{dm_{f,equ}(t)}{dt} = \frac{dm_f(t)}{dt} + \frac{dm_{ress}(t)}{dt} \times (1 - p(SOC) \tag{7}$$

where $\frac{dm_f(t)}{dt}$ is the fuel mass flow rate (unit: g/s), $p(SOC)$ is a correction term used to guarantee that the battery ESS does not overcharge or over discharge.

The optimal control problem is then formulated as:

$$\min \ \dot{m}_{f,equ}(t)$$

$$constrains: SOC_{\min} \leq SOC \leq SOC_{\max}$$

$$P_{\min}(t) \leq P(t) \leq P_{\max}(t)$$

The equivalence factor $s(t)$ plays an import role in ECMS. An optimal result could be obtained if the equivalence factor can be properly tuned, and future driving information is needed. Since its optimal value for different driving pattern is difficult to obtain, adaptive ECMS (A-ECMS) methods have been developed to eliminate the need of priori knowledge of the driving cycle or pattern. The ECMS method can achieve a suboptimal solution in real time by selecting the suitable factor as long as the driving pattern is given, but the accurate driving pattern is difficult to obtain due to varying driving conditions. This turn the focus of optimal control into driving data based pattern recognition issue. A-ECMS uses collected information to predict incoming driving pattern to tune the equivalence factor, and the technique consists of pattern recognition based schemes [8] and driving cycle prediction based schemes [9], [10]. Researchers at Ohio State U. compared three energy management strategies over various driving cycles, DP, ECMS with pre-optimized equivalence factor, and A-ECMS, showing A-ECMS' close performance to DP benchmark.

Pontryagin's Minimum Principle

PMP, that is quite similar to A-ECMS, has recently received the most attention [11]. The method does not require multiple equivalence factors by implicitly taking into account of different vehicle operation conditions. An adaptive PMP method was also introduced based on ESS SOC [7]

CONCLUSIONS

The architecture of PHEV powertrain system is illustrated using UVic EcoCAR2. The need as well as the pros and cons of different energy management strategy are briefly overviewed. The details of the tests of ECMS, RB, DP and PMP control strategies on the UVic EcoCAR2 model is beyond the scope of this paper. The overview serves as an introduction to PHEV powertrain and power control/energy management techniques.

ACKNOWLEDGMENT

Supports form Natural Resources of Canada, the EcoCAR program, and Transport Canada are gracefully acknowledged.

REFERENCES

1. Kaban, S., Nelford, J., Dong, Z., Dong, J. et al. 2012, "An Innovative 4WD PHEV Utilizing a Series-Parallel Multiple-Regime Architecture", *SAE Int. J. Alt. Power*, 1(2), pp. 419-437.
2. Cheng, R., Dong, J., and Dong, Z. 2013, "Modeling and Simulation of a Multiple-Regime Plug-In Hybrid Electric Vehicle", *ASME 2013 International Design Engineering Technical Conferences and Computers and Information in Engineering Conference*, Portland, USA, pp. V001T01A009.
3. Baumann B., M, Washington G., Glenn B., C, Rizzoni G. 2000, "Mechatronic design and control of hybrid electric vehicles", *IEEEASME Trans On*, 5(1), pp. 58-72.
4. Moawad A., Singh G., Hagspiel S., Fellah M., Rousseau A. 2009, "Impact of real world drive cycles on PHEV fuel efficiency and cost for different powertrain and battery characteristics", *The 24th International Electric Vehicle Symposium and Exposition EVS-24*, Stavanger, Norway, pp. 13-16.
5. Lee H-D, Sul S-K. 1998, "Fuzzy-logic-based torque control strategy for parallel-type hybrid electric vehicle", *IEEE Trans Ind Electron*, 45(4), pp. 625-632.
6. Kaban S., Dong Z., Crawford C. 2015, "Performance Modeling and Benchmark Analysis of an Advanced 4WD Series-Parallel PHEV Using Dynamic Programming", *2015 IEEE Vehicle Power and Propulsion Conference*, Montréal, Canada. pp. 1–6.
7. Onori S., Serrao L., Rizzoni G. 2016, *Hybrid Electric Vehicles*, Springer London, London
8. Jeon S., Jo S., Park Y., Lee J. 2000, "Multi-Mode Driving Control of a Parallel Hybrid Electric Vehicle Using Driving Pattern Recognition", *Journal of Dynamic Systems, Measurement, and Control*, 124(1), pp. 141-149.
9. Sciarretta A., Back M., Guzzella L. 2004, "Optimal control of parallel hybrid electric vehicles", *IEEE Trans Control Syst Technol*, 12(3), pp. 352-362.
10. Zhang X., Dong Z., Crawford C. 2011, "A Novel Method for Driving Condition Recognition Based on Compressed Sensing", *ASME 2013 International Design Engineering Technical Conferences and Computers and Information in Engineering Conference*, Portland, USA, pp. 1-14.
11. Dong J, Cheng R, Dong Z, Crawford C. 2013, "Fast Online-Implementable Sub-Optimal Energy Management of a Power-Split Hybrid Electric Vehicle", *ASME 2013 International Design Engineering Technical Conferences and Computers and Information in Engineering Conference*, Portland, USA, pp. 1-6.

Renewable Energy Systems 2

Proceedings of the 26th CANCAM
Victoria, British Columbia, Canada, May 29 – June 1, 2017

A REVIEW OF LITHIUM-ION BATTERY DEGRADATION MECHANISMS AND MODELING METHODS

Li Chen
Department of Mechanical Engineering
University of Victoria
Victoria, B.C., Canada
chenli@uvic.ca

Zuomin Dong
Department of Mechanical Engineering
University of Victoria
Victoria, B.C., Canada
zdong@uvic.ca

ABSTRACT

Electrified vehicles (EVs) with onboard battery Energy Storage System (ESS) provide promising solutions to improve energy efficiency and reduce environmental impacts and greenhouse gas emissions. However, batteries, including the widely used lithium-ion (Li-ion) batteries, are all suffering various degree of performance degradation and capacity deficient due to aging phenomena. Properly predicting and controlling battery's performance degradation and service life are essential to the life-cycle cost estimate, optimal design and energy management of EVs. This paper presents a systematic review of Li-ion battery's life degradation mechanism, and compares three different battery modeling methods for battery capacity and service life estimation to meet the stated needs.

KEYWORDS: Li-ion battery, degradation mechanism, battery modeling

INTRODUCTION

Environmental pollution and harmful effects of Greenhouse Gases (GHG) emission caused increased concerns globally. The transportation sector accounts for about 31% of total CO_2 emissions by burning gasoline and diesel in 2014. Electrified vehicles (EVs), including battery electric vehicles (BEVs), hybrid EVs (HEVs), plug-in hybrid EVs (PHEVs) and fuel cell vehicles (FCVs) have experienced quick developments and market expansion in recent years as promising solutions. Li-ion battery has been largely used in EVs to substitute or reduce internal combustion engines (ICEs). Metallic Li has many advantages as battery electrode material, low density (0.534 g/cm^3 at room temperature), high reduction voltage (-3.04V vs. standard hydrogen electrode) and high theoretical capacity (3860 mAh/g).

Even the technology has already satisfied the requirements for commercial vehicles, the price of Li-ion batteries and electric drive systems are still considerably more expensive than ICEs, which becomes a critical barrier for the production. The cost of the battery package occupies a heavy part of total cost in EV. As it was showed in literature[1], the optimistically estimates of the cost of Li-ions batteries might be declined from above US $1,000 per kWh to about US $410 per kWh from 2007 to 2014. Moreover, it could fall to about US $200 per kWh by 2020 according to an analysis report from Mckinsey [2]. But still, it would be hard to be competitive at market if the battery pack could not fall below US$150 per kWh. Giving the common estimation regarding the manufacturing volumes and assembly cost by OEMs, the rough cost of battery pack is among US $400 ~ $600 per kWh at present (Figure 1).

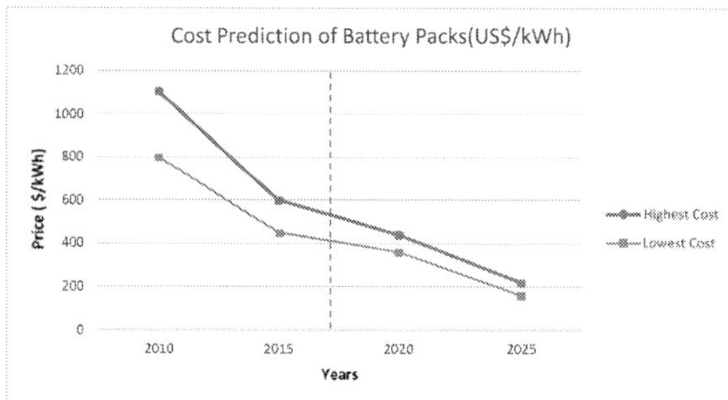

Figure 1: Cost prediction of Li-ion battery pack

Another critical problem for Li-ion batteries is the aging phenomena, caused by microscopic electrochemical reactions inside each single battery cell. Battery capacity and lifetime could be much shortened under inappropriate usage. The replacement of battery pack would aggravate the total vehicle ownership cost. Materials, manufacturing, operating temperature and other conditions could all affect battery performance deterioration. A better understanding on the battery aging process and the avoiding methods to prolong its lifetime are becoming more critical as the size of these batteries increases for improved functionality. This paper intends to make a thorough analysis on battery life degradation mechanisms and the available mathematical modeling methods for life estimation.

LI-ION BATTERY DEGRADATION MECHANISM REVIEW

The degradation mechanisms of Li-ion batteries highly depend up the materials used in anodes, cathodes, and electrolytes. Therefore a good understanding on available materials and their characteristics is vital to further reviews.

Lithiated carbon (LiC_6) is the most popular anode material. Layered crystal compound graphite has abundant natural reserve volume and stable performance for ion intercalation. LiC_6 has a relative low overpotential versus Li plate (about 0.5V vs. Li/Li+), which enables a more than 3.5 to 4.5V nominal voltage depending on the cathode material. The theoretical energy capacity of LiC_6 is about 372 mAh/g. Another anode material Lithium Titanate (LTO) has much higher cycle numbers that is ten times higher than graphite-based batteries. LTO has excellent cycling performance and negligible volume change [3]. However, the overpotential of LTO anode is about 1.55V against Li/Li+, which lowered overall voltage to about 2.4V. The theoretical capacity of LTO is only half of graphite anode (~170 mAh/g). New materials, such as transition-metal oxides (MO, where M could be Fe, Co, Ni, or Cu), graphene nanosheet (GNS) materials, have been introduced in laboratory but their technology is still immature for commercial product.

Positive electrode material includes a variety of lithium metal oxide compound. $LiFePO_4$ (or LFP) battery is the earliest massive productions used in EVs. LFP has an olivine structure, which can provide a flat potential of 3.45V vs. Li/Li+. The strong bonding of oxygen in phosphate group makes it stable to resist thermal runaway and have a safe performance at high temperature. $LiCoO_2$ (or LCO) can achieve high overpotential vs. Li/Li+ and provide a stable cycling voltage at 3 to 4.2 V [4]. $LiNiO_2$ (or LNO) has the highest specific energy capacity (about 275mAh/g in theory and 150mAh/g in practice). The synthesis process of LNO usually gains over-stoichiometric phase of extra Ni ions and cause low stability, poor cycling performance and rapid capacity fading. The spinal $LiMn_2O_4$ (or LMO) makes excellent reversibility and also has lower internal resistance, safer thermal stability, cheap cost and non-toxicity for environment. Doping and substituting small amount of Co, Mn or Al into Ni-based materials can largely improve the overall performances. The combinations of these metals are able to provide high energy and power density with stable thermal behavior.$LiNiMnCoO_2$ (or NMC) and $LiNiCoAlO_2$ (or NCA) are promising candidates for automotive industry with multiple advantages.

Performance decay due to side reactions in battery's lifetime is an inevitable process. Temperature, current, and state of charge (SOC) are all relevant to the decay rate. Regarding the working or storage states, battery degradation mechanisms perform different effects for both negative and positive electrodes.

Cyclic capacity decay starts right after the first charge/discharge process. Li-ions from electrolyte spontaneously react with active graphite anode and create solid-electrolyte layers (SEI). This process not only consumed available ions but also increased resistance. Ions have to penetrate SEI to insert into the crystal lattice. A sharp capacity decay can be observed from negative electrode [5]. Normally, more than adequate Li-ions and graphite are provided to fulfill SEI formation to improve total Coulombic efficiency. However, extra anode materials make cell's performance limited by positive electrode. Bourlot, et al. [6] compared fresh batteries with 1.5 years-aged ones produced by Johnson Controls - Saft, showed that the SEI layer on negative electrode stayed relatively stable while binder dissolutions and active metal Li were found on positive electrode, which implied positive electrode limited cycling performance. Furthermore, constantly periodic cycling may lead to continuous surface reactions. SEI layers become harder and thicker, causing impedance increment. Temperature is critical for SEI structure. When temperature raised up, SEI started to dissolve and becomes more permeable for solvent. Ramadass, et al. [7] analyzed Sony 18650 cells at elevated temperature and showed an increase of capacity losses. Itou and Ukyo [8] tested commercial 18650 cells with $LiNi_{0.8}Co_{0.15}Al_{0.05}O_2$ positive material and found out the resistance increased quickly than capacity decay. Another factor to affect cycle life is the fast charge rate. Large amount of Li-ions would accumulate on anode under high current density, causing metal Li plating along carbon surface. Dendritic Li might penetrate separator film. Inner short circuit will induce thermal runaway under some conditions.

When battery is on rest, chemical reactions happened only because of the thermodynamic instability between materials. Lithiated graphite has strong reactivity with the organic electrolyte. Side reactions, mainly on negative electrode, were believed to be the main reason of calendar life fading [6]. The reaction intensity is highly related to material properties, storage temperature, and battery open circuit voltage (or OCV) level. If battery stores at high temperature, secondary reactions will facilitate reactions and speed up corrosions. Mild or low temperature, on the other hand, can depress the ions diffusion from reacting. SOC level represents the ions proportion between cathode and anode. Broussely, et al. [9] tested a group of lithium-ion battery with LCO as positive material and revealed storage temperature is more critical than voltage (or SOC) level for calendar life degradation.

To conclude, negative electrode contributes some amount of capacity losses and impedance raises at beginning, while positive electrode continuously cause capacity loss during cycling. Storage deterioration mainly happens on negative electrode. Degradation mechanisms are different between anode and cathode. Main reasons for carbon-based anodes aging are: SEI formation and growth, loss of available lithium due to side reactions, lithium metal plating at low temperature and high rate current, etc. As for lithium metal oxide cathodes, wearing of active materials, compound structure changing, and electrolyte dissolving are relevant to performances decay.

LI-ION BATTERY SIMULATION MODELS

To quantify battery's state of health (SOH) during its deterioration process, researchers have proposed different modeling methods to simulate battery performance and capacity fading [10, 11]. Three most commonly used models are outlined in this section.

Empirical Model

Empirical model was the earliest developed mathematical model for battery life estimation. Based on Arrhenius equation, empirical model took advantage of measurement data to fit an equation (Eq. 1), which can relate available energy to temperature and activation energy.

$$Q = Ae^{\left(-\frac{E_a}{RT}\right)}t^z \quad (1)$$

where Q is power, t is time, z is the exponent of time and usually is 0.5. Parameters A and E_a need to be fitted by measurement data.

Empirical battery model is suitable for calendar life estimation rather than cycle life. Broussely, et al. [9] believed battery performance decay is due to the growth of SEI layer caused by irreversible side reactions during storage. By doing experiments on Ni-Co based batteries, they provided an equation by fitting the experimental data from stored temperature 10 to 60°C.

$$t = e^{\left(\frac{4661}{T}-14\right)}x^2 + e^{\left(\frac{4437}{T}-11.6\right)}x \quad (2)$$

where t is the calendar life starting from initial state, T is the storage temperature in K, x is the number of active Li-ion in moles. The calendar life degradation rate with respect to temperature was simulated in Figure 2:

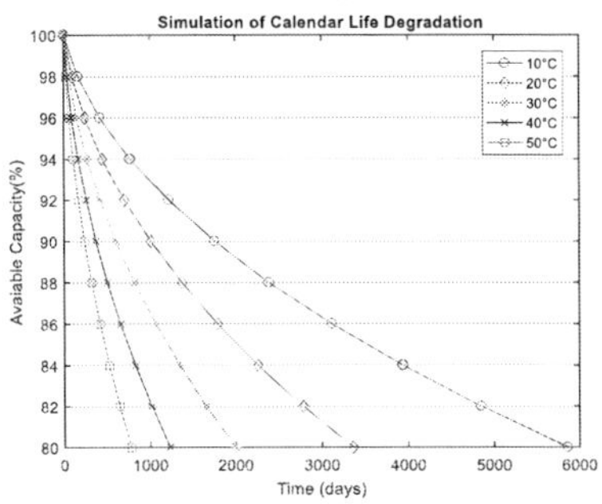

Figure 2: Simulation of calendar life degradation

Equivalent Circuit Model

Equivalent circuit model is the most widely used model in battery analysis. The physical battery is simplified as a circuit with a voltage source and several passive elements. This method adopted resistor-capacitor (RC) circuits to simulate the concentration polarization and electrochemical polarization process, so-called dual polarization model (or DP model). A typical DP model has an OCV source V_{oc}, an internal resistor R_i, and two RC circuits with R_1, C_1 and R_2, C_2 (as illustrated in Figure 3).

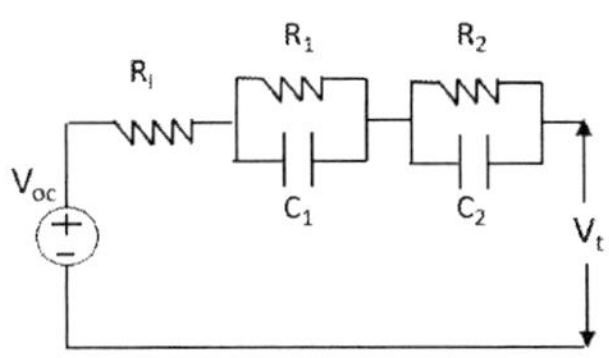

Figure 3: DP model

SOC is calculated by Coulomb counting method with a known nominal capacity Q:

$$S\dot{O}C = -\frac{I}{Q} \quad (3)$$

Open circuit voltage is a function of SOC, and can be fitted by test data:

$$V_{OC}(SOC) = a_0 + a_1(\mathrm{SOC}) + a_2(\mathrm{SOC})^2 + \cdots, a_i \in \mathbb{R} \quad (4)$$

The terminal voltage is derived by Kirchhoff laws:

$$V_t = V_{oc} - V_i - V_1 - V_2 \quad (5)$$

where the inner voltage drop for each section is:

$$\begin{aligned} V_i &= IR_i \\ \dot{V_1} &= -\frac{V_1}{R_1C_1} + \frac{I}{C_1} \\ \dot{V_2} &= -\frac{V_2}{R_2C_2} + \frac{I}{C_2} \end{aligned} \quad (6)$$

Parameters $(R_i, R_1, C_1, R_2, C_2)$ have to be fixed before using this model for calculation. Their values strictly depend on the measurement data. It requires optimization method to identify parameters, such as linear regression algorithm or genetic algorithm (GA). The best solution can be found by minimizing the differences between simulated terminal voltage V_t and the test data V_{data}.

$$\min \int (V_t - V_{data})^2 \quad (7)$$

The identified parameters for a 25Ah Li-NMC battery was listed below (Table 1):

Table 1: Identified parameters

R_i (Ω)	R_1 (Ω)	C_1 (F)	R_2 (Ω)	C_2 (F)
0.00204	0.00741	1837105	0.00077	292559

The accuracy of this model can be validated by using another test data. The sum of squared errors (SSE) of simulated SOC (SOC_{sim}) and measured SOC (SOC_m) is 0.31 in 4821s. The SSE of terminal voltage (between V_{sim} and V_m) is 1.1886. Comparisons can be seen in Figure 4.

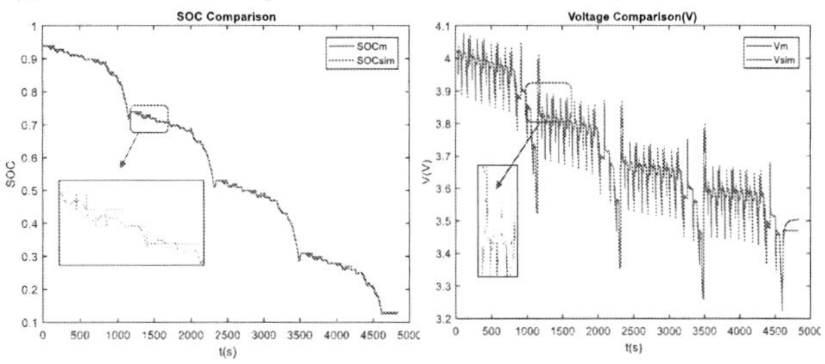

Figure 4: Simulated and measured results

The simulation errors from equivalent circuit model is quite large under high current density. One reason might be from model itself. Because there is no actual physical meanings for these resistors and capacitors inside battery. Another reason is caused by parameter identification method.

Electrochemical Model

Electrochemical model is based on the chemical reaction principles happened during battery charge/discharge process. The first full-order electrochemical model was proposed by Doyle, et al. [12] in 1993, so-called 'Doyle-Fuller-Newman' model (or DFN model). DFN model built a general Li-ion battery model by using concentration solution theories to capture ions' diffusion dynamics inside solid state and soluble electrolyte. Microscopic ions flow rate j^{Li} was coupled with macroscopic external current I. The Butler-Volmer kinetic equation is used to describe the relationship between overpotential and microscopic current density. Fick's law is used to describe the ions' diffusion process inside active solid particles. It is much more accurate however the modeling and computation are very time consuming [13].

Due to the complexity and expensive computational time, DFN model has not been widely used today. Recently, single particle model (SPM) was proposed by simplifying the DFN model into one solid particle on both electrodes to reduce the computational order [14]. With the assumptions of uniform electrolyte concentration inside battery, the accuracy under high current density cannot be guaranteed [15].

Pseudo 2-dimensional (P2D) model used a set of PDEs to calculate Li ions' diffusion along the thickness of battery and the radius of solid particle. These PDEs are highly nonlinear and coupled equations with boundary conditions [16]. Numerical solutions such as finite-difference method (FDM) has been adopted to solve it in software. Specific methods can be found in references [17]. The details of solving these PDEs are beyond the scope of this paper.

Due to its potential capability to better capture Li-ion battery's performance degradation mechanism. The authors' research team is working on using the electrochemical modeling in optimizing the ESS energy management and prolong its lifetime for PHEV/HEV.

CONCLUSIONS

Li-ion battery life degradation is a complex process which could be affected by battery materials, temperature, charge/discharge current, SOC, etc. The deterioration process under cycling and storage conditions has been discussed and the main mechanisms have been revealed. Moreover, different modelling methods have been presented for further studies.

REFERENCES

1. Nykvist, B., & Nilsson, M. 2015, "Rapidly falling costs of battery packs for electric vehicles", *Nature Climate Change, 5*(4), 329-332.
2. Hensley, R., Newman, J., Rogers, M., & Shahinian, M. 2012, "Battery technology charges ahead", *McKinsey Quarterly, 3*, 5-50.
3. Shenouda, A. Y., & Murali, K. 2008, "Electrochemical properties of doped lithium titanate compounds and their performance in lithium rechargeable batteries", *Journal of Power Sources, 176*(1), 332-339.
4. Wang, B., Bates, J., Hart, F., Sales, B., Zuhr, R., & Robertson, J. 1996, "Characterization of Thin - Film Rechargeable Lithium Batteries with Lithium Cobalt Oxide Cathodes", *Journal of the Electrochemical Society, 143*(10), 3203-3213.
5. Arora, P., White, R. E., & Doyle, M. 1998, "Capacity fade mechanisms and side reactions in lithium - ion batteries", *Journal of the Electrochemical Society, 145*(10), 3647-3667.
6. Bourlot, S., Blanchard, P., & Robert, S. 2011, "Investigation of aging mechanisms of high power Li-ion cells used for hybrid electric vehicles", *Journal of Power Sources, 196*(16), 6841-6846.
7. Ramadass, P., Haran, B., White, R., & Popov, B. N. 2002, "Capacity fade of Sony 18650 cells cycled at elevated temperatures: Part I. Cycling performance", *Journal of Power Sources, 112*(2), 606-613.
8. Itou, Y., & Ukyo, Y. 2005. "Performance of LiNiCoO 2 materials for advanced lithium-ion batteries", *Journal of Power Sources, 146*(1), 39-44.
9. Broussely, M., Herreyre, S., Biensan, P., Kasztejna, P., Nechev, K., & Staniewicz, R. 2001, "Aging mechanism in Li ion cells and calendar life predictions", *Journal of Power Sources, 97*, 13-21.
10. Arora, P., Doyle, M., & White, R. E. 1999, "Mathematical Modeling of the Lithium Deposition Overcharge Reaction in Lithium - Ion Batteries Using Carbon - Based Negative Electrodes", *Journal of the Electrochemical Society, 146*10), 3543-3553.
11. Brand, J., Zhang, Z., & Agarwal, R. K. 2014, "Extraction of battery parameters of the equivalent circuit model using a multi-objective genetic algorithm", *Journal of Power Sources, 247*, 729-737.
12. Doyle, M., Fuller, T. F., & Newman, J. 1993, "Modeling of galvanostatic charge and discharge of the lithium/polymer/insertion cell", *Journal of the Electrochemical Society, 140*(6), 1526-1533.
13. Doyle, M., & Newman, J. 1995, "The use of mathematical modeling in the design of lithium/polymer battery systems", *Electrochimica Acta, 40*(13), 2191-2196.
14. Smith, K., & Wang, C.-Y. 2006, "Solid-state diffusion limitations on pulse operation of a lithium ion cell for hybrid electric vehicles", *Journal of Power Sources, 161*(1), 628-639.
15. Santhanagopalan, S., Guo, Q., & White, R. E. 2007, "Parameter estimation and model discrimination for a lithium-ion cell", *Journal of the Electrochemical Society, 154*(3), A198-A206.
16. Di Domenico, D., Stefanopoulou, A., & Fiengo, G. 2010, "Lithium-Ion Battery State of Charge and Critical Surface Charge Estimation Using an Electrochemical Model-Based Extended Kalman Filter", *Journal of Dynamic Systems, Measurement, and Control, 132*(6), 061302. doi:10.1115/1.4002475
17. Zhang, L., Lyu, C., Wang, L., Luo, W., & Ma, K. 2013, "Thermal-electrochemical modeling and parameter sensitivity analysis of lithium-ion battery", *CHEMICAL ENGINEERING, 33*.

Proceedings of the 26th CANCAM
Victoria, British Columbia, Canada, May 29 – June 1, 2017

EXPERIMENTAL CONSIDERATIONS OF THE WAKE EFFECTS OF WIND TURBINES

Michael McKinnon
Mechanical and Mechatronics Engineering
University of Waterloo
Waterloo, Ontario, Canada
mmckinno@uwaterloo.ca

David A. Johnson
Mechanical and Mechatronics Engineering
University of Waterloo
Waterloo, Ontario, Canada
david.johnson@uwaterloo.ca

ABSTRACT

The wakes created by wind turbines have an effect on the downstream turbines and can affect the performance. The performance loss contributes to the production loss, blade loading, rotor performance, and fluctuating electrical output of the turbine. Therefore, experimental methods to determine characteristics, such as wake recovery, meandering, and behavior, are determined. Using the University of Waterloo's two turbines, experiments are conducted to determine these characteristics. The turbines include one outdoor turbine and one located in a large-scale controlled wind facility. Flow visualization using smoke will give qualitative and quantitative measurements of the wake behavior. Instrumentation such as a light ranging and detection (LiDAR) measurement device is used to measure the characteristics of the turbine wakes. The experiments have the advantage of studying the wake effects in a controlled environment and compare to the outdoor external environment.

KEYWORDS: wind turbine, wind energy, wake, LiDAR, wind farm density

INTRODUCTION

With more wind farms being developed and denser farms being created, the wake effects of turbines are becoming an increasingly important research topic. A rotating wind turbine produces a wake of turbulent air flow that affects the performance of downstream turbines. The wake can reduce the efficiencies of a downstream turbine and can even lead to wear on the components, increasing the maintenance required to keep the farm operational. Knowing the behavior of the wake will allow for better planning of wind farms by optimizing the spacing between turbines to ensure no significant losses occur from upstream wake effects. The spacing of the turbine would be determined by the recovery distance of the wake. Some of the negative effects the wake has on downstream turbines are loss of production, blade loading, rotor performance, and fluctuating electrical output [1].

Light detection and ranging (LiDAR) measurement systems can be used to measure the speed of particles in the air, such as dust and pollen. This system has been used in characterizing the wakes of turbines and the meandering of the wake. However challenges have arose due to varying wind conditions and changing orientation of the turbine as a result of unpredictable outdoor conditions. If a controlled environment, such as a large wind tunnel, is available it would be useful to study the wake of the turbine in such a facility using the LiDAR measurement device.

Therefore, to determine the characteristics of the wind turbine wakes, experimental methods are used to gain a practical understanding of the profile. This is explored with flow visualization using smoke and measurements with a LiDAR system. The University of Waterloo Wind Energy Group has two turbines that will be studied for wake characteristics. An outdoor turbine, a 35 kW Wenvor turbine, will allow an understanding of the wake with external contributions. The second turbine is in a large controlled wind facility that will give an idea of the wake characteristics without environmental conditions, and allow for a better understanding of the theoretical versus actual behavior of the wake.

EXPERIMENTAL FACILITIES

The Wenvor turbine is a 35 kW grid-connected turbine located at University facilities in the Region of Waterloo. It was manufactured by Wenvor Technologies and has two blades, 10 m in diameter, and a hub height of 30 m. The blades have a passive variable pitch system. The Wenvor turbine can be seen in Fig. 1. This turbine has been used for various research topics by the University of Waterloo Wind Group and has the ability to be lowered and raised to attach instrumentation and perform maintenance.

Figure 1: University of Waterloo Wenvor wind turbine

The University of Waterloo's Wind Generation facility is a large-scale indoor wind turbine testing facility that allows for controlled condition testing. The facility has six 75 kW variable frequency drive fans that produce variable airspeeds up to 11 m/s in the facility, shown in Fig. 2.

Figure 2: Wind Generation facility fans

A 4.5 m diameter turbine developed within the Wind Energy Group is used in the facility and can produce 3 kW of power in 11 m/s conditions. The turbine, called the UW WEG turbine, inside of the facility is shown in Fig. 3.

Figure 3: UW WEG wind turbine inside the Wind Generation facility

FLOW VIZUALIZATION

Since air flow is not visible by the naked eye, flow visualization is used to view the path of the air. Flow visualization techniques, such as smoke, can be used to get a visual of the behavior of the wake behind the wind turbine. If the smoke seeding is emitted from the turbine blade tips the wake can be visualized and quantitatively measured [3].

This method was performed on the UW WEG wind turbine inside the Wind Generation facility. A smoke emitter was attached to a single blade of the turbine. The wind speed was adjusted to 6 m/s and the rotational speed of the turbine was set to a constant 100 RPM using a controller. This resulted in a tip speed ratio of approximately 3.9. A progression of the blade tip smoke visualization as the blade rotates can be seen in Fig. 4. The smoke provides a visual of the flow coming off of the blade tip and how it propagates downstream. Further downstream of the turbine the smoke begins to dissipate into a less concentrated form.

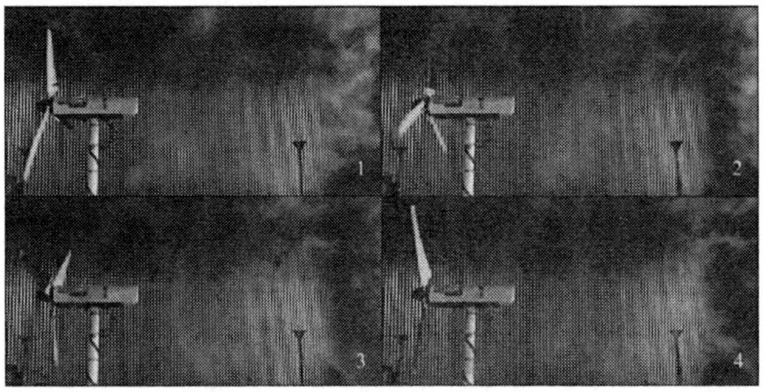

Figure 4: Tip smoke visualization progression of the UW WEG turbine for 6 m/s wind speed operating at 100 RPM

Looking at a single time frame of the tip smoke visualization in Fig. 5 it can be seen that the wake follows the typical helical vortex wake distribution. Vortex cylinders are shed from the blades of the turbine and then follow the helical path downstream of the turbine. Additionally, wake expansion can be seen from the experiment, however a frontal view of the flow will allow a better visualization of this characteristic.

Figure 5: Blade tip smoke visualization of the UW WEG turbine for 6 m/s wind speed operating at 100 RPM

From the angle of the smoke leaving the turbine blade, the helix angle, ϕ, also known as the flow angle at the blade tip, can be determined [4]. This angle can be used to determine the direction of the vortex cylinder parallel to the axis of rotor rotation. Once the vorticity strength and helix angle are known, the induced velocity at the rotor plane can be determined using Biot-Savart law. The helix angle from the experiment can be seen from a close-up image of the blade shown in Fig. 6.

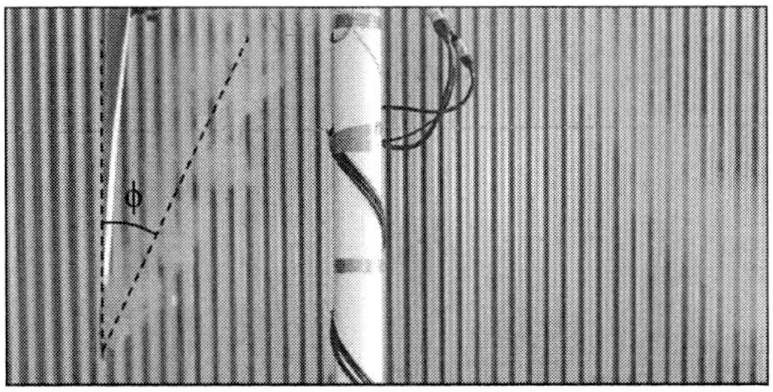

Figure 6: Helix angle of the UW WEG turbine for 6 m/s wind speed operating at 100 RPM

Another technique releases smoke upstream of the turbine to see the characteristics of the smoke through the turbine. This method will give a visual of the wind tip vortices being shed off of the turbine blade tips.

This method was also performed on the UW WEG wind turbine at 6 m/s and 100 RPM blade rotational speed. The result is shown in Figure 7. A tip vortex can be seen as the blade passes through the smoke. Another vortex is seen further downstream as is moves with the air flow. The distance between the vortices is related to the helical path of the vortex cylinders.

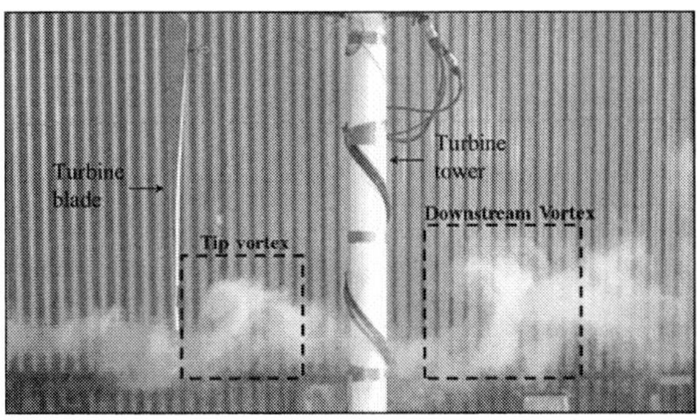

Figure 7: Upstream smoke visualization of the UW WEG turbine for 6 m/s wind speed operating at 100 RPM

TURBINE WAKE MEASUREMENTS

Using a ZephIR z150 LiDAR, as shown in Fig. 8, the velocity profile behind the turbines can be measured. The LiDAR is placed in the wake of the Wenvor turbine to gain information about the airspeed of particles in the wake. The measurements depend on the ambient wind speed and therefore require a large sample of data to analyze the effects from the varying speeds and direction. The ZephIR LiDAR has the capability to remain on site collecting data for up to a year.

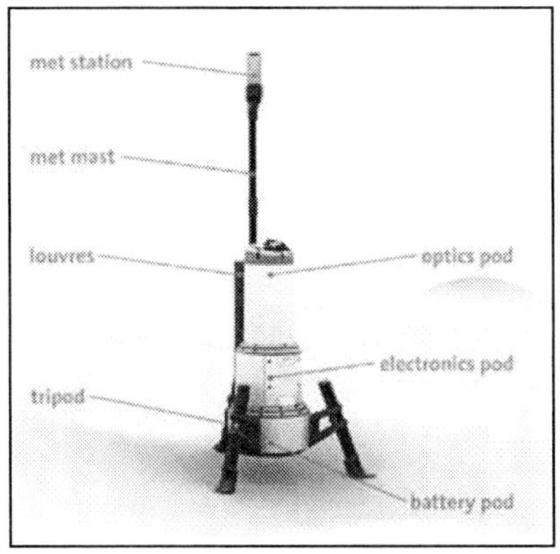

Figure 8: ZephIR z150 LiDAR [5]

Using a lens to minimize the measurement area, the LiDAR measures the wind speed on the wake of the UW WEG turbine in the Wind Generation facility. The Wind Generation facility gives the ability to obtain data at specific free stream airspeeds allowing faster collection of samples. This is one of the issues of most research using LiDAR measurement devices and is the advantage of a controlled wind facility [1]. In addition to the two turbines at the University of Waterloo, the LiDAR can potentially be used to measure existing large scale turbines situated in wind farms, providing access can be obtained.

CONCLUSIONS

The wake of a wind turbine is important for determining the downstream effect on other turbines and can be used when planning the layout of a wind farm. Characteristics, such as wake recovery, meandering, and behavior, are useful to determine the wake effects. To determine these characteristics several methods are used, including flow visualization with smoke and LiDAR measurements.

The University of Waterloo Wind Energy Group has two turbines that are used to study the effects of the turbine wake. One turbine, the Wenvor wind turbine, is located outdoors and is exposed to the variable effects of the wind and ambient conditions. The other turbine, the UW WEG turbine, is located within a large-scale controlled wind facility. The facility allows for the wind speed to be adjusted for speeds up to 11 m/s.

Smoke flow visualization is used to obtain a visual on the behavior of the wind turbine wakes. Two methods have been utilized for this visualization technique. The first attached a smoke emitter on a blade tip of the wind turbine to see how the flow coming off of the tips propagates downstream. The results of the experiment showed a helical wake distribution expected from theory. The helix angle can also be determined from the visualization to gain an understanding of the induced velocity at the rotor plane once the vorticity of the helical vortex wake is known.

The other method introduced smoke upstream of the turbine to visualize the wind tip vortices shedding from the blade tips of the turbine. The vortices were seen as the blade passed through the smoke introduced upstream and also seen propagating downstream of the turbine. The distance between the vortices related to the spacing of the helical vortex wake distribution seen in the first experiment.

Using a ZephIR z150 LiDAR, measurements for the airspeed in the wake of the turbines are determined. The LiDAR is placed in the wake of the Wenvor turbine to collect a large data sample for the varying airspeeds and conditions. Using a lens, the LiDAR has the ability to be placed inside the controlled wind facility to measure the wake of the UW WEG turbine.

The results in controlled conditions and the comparison to the external environment will give a better understanding of wind turbine wake effects to optimize wind farm density and turbine efficiencies.

ACKNOWLEDGMENTS

The authors acknowledge the help and support of the University of Waterloo Wind Energy Group for the research topic.

REFERENCES

1. Stevens, R., and Meneveau, C. 2017, "Flow Structure and Turbulence in Wind Farms", *Annual Review of Fluid Mechanics*, Vol(49), pp. 311-339.

2. Gallant, T. E., and Johnson, D. A. 2016, "In-blade angle of attack measurement and comparison with models", *Journal of Physics: Conf. Ser.*, 753, 072007.

3. Hand, M., Simms, D., Fingersh, D., Cotrell, J., Schreck, S., and Larwood, S. 2001, "Unsteady Aerodynamics Experiment Phase VI: Wind Tunnel Test Configurations and Available Data Campaigns", National Renewable Energy Laboratory, Golden, Colorado.

4. Burton, T., Sharpe, D., Jenkins, N., and Bossanyi, E. 2001, *Wind Energy Handbook*, John Wiley & Sons Ltd., Chichester, England.

5. Natural Power 2009, "ZephIR Operations & Maintenance Manual", Malvern, Worcs.

DESIGN AND ENERGY MODELLING OF A SOLAR HOME MECHANICAL SYSTEM USING TRNSYS

Kajen Ethirveerasingham, MASc Student
Mechanical and Industrial Engineering
Ryerson University
Toronto, Ontario, Canada
kethirve@ryerson.ca

Alan S. Fung, PhD, PEng, FCSME
Mechanical and Industrial Engineering
Ryerson University
Toronto, Ontario, Canada
alanfung@ryerson.ca

ABSTRACT

EcoStudio is a Toronto based team that is designing an affordable, efficient, comfortable and marketable Net-zero Energy Home (NZEH) for the Canadian climate. For this paper, the mechanical system was modelled using the most recent architectural design using TRNSYS 17. A base case was modelled using an efficient HRV, a two-stage variable capacity ASHP, and an EWH. This case showed high values of overheating, approximately 16.2% for the main floor and 12.8% for the top floor throughout an entire year. To reduce this, an upgraded case with a hybrid ASHPWH was used. Along with this upgrade, a BIPV/T system was integrated with the ASHP during the winter when ambient air is insufficient, these two additions are considered the upgraded case. Overheating did reduce, but not significantly. Energy savings for DHW load were also typical, similar to other findings. Further investigation must be done to identify better control strategies and technologies to better regulate zone temperature.

KEYWORDS: Solar home, NZEH, TRNSYS, Heat Pump, Energy modelling

INTRODUCTION

In Canada, natural gas is a convenient and economical fuel source used to heat buildings but, in return, generates greenhouse gas. Different solutions must be found in order to reduce the carbon footprint in Canadian homes, where 63% of energy usage in homes is strictly used for space heating and 17% for water heating [1]. There are however, several global initiatives to help develop these solutions, one such initiative is the Solar Decathlon competition that was started by the US Department of Energy. The goal of this competition is to have students and faculty from universities around the world to develop solar powered homes that would be affordable, comfortable and practical. Team Toronto (a consortium of Ryerson University, University of Toronto, and Seneca College) is one of such teams to have participated in this competition, although now going by the name EcoStudio, the goal to build an affordable net-zero energy home (NZEH) still stands.

Since the heating demand during the Canadian winters are high, EcoStudio had designed a home that is meant to utilize solar gains to reduce this demand. This is done by adding large windows on the south side, and having high insulation.

Two technologies that were investigated were air-source heat pump (ASHP) and an air-source heat pump water heater (ASHPWH). ASHP coefficient of performance (COP) however, typically decline during the winter [2]. To improve this, a building integrated photovoltaic thermal (BIPV/T) system, which has been thoroughly investigated by [2, 3], is to be considered to be integrated with the ASHP to improve its effectiveness.

Although there are a multitude of various combinations of equipment and systems, our initial investigation centers around an HVAC system design with a BIPV/T integrated ASHP along with an ASHPWH and a heat recovery ventilator (HRV). In this design, the ASHPWH is to provide DHW while also behaving similar to a thermal storage tank by attempting to extract the excess heat during periods of overheating in the winter. Simulations were performed in TRNSYS to gauge the effectiveness of such combination for the Canadian climate. The primary goal of the current system is to find a successful solution to prevent overheating, while also reducing the carbon footprint and attempting to converge to a net-zero status.

OBJECTIVE

The objective of this paper is to perform a case study with TRNSYS using the EcoStudio house design with a BIPV/T assisted ASHP and ASHPWH and compare these against a base case to identify any potential benefits. For space heating, the base case, will be a house with rack mounted PV panels with a standalone ASHP (the ASHP is not connected to any air preheating device); the upgraded case study is of an air based (BIPV/T) system with flow optimization coupled with an ASHP. The first case study represents the conventional installation of PV panels and standalone operation of ASHP's.

The second case will allow the air flow rate through the BIPV/T system to vary (according to a set criterion) and the air is diverted to the ASHP, to be used as an air source; this system promises better performance during the winter relative to the base case. For DHW heating, the base case will have a simple electric water heater (EWH), while the upgraded case will use a ASHPWH that will have electrical elements in situations where using the heat pump is not sufficient. Both cases will use an 85% efficient HRV as a standard for ventilation.

For the base case, the systems that will be used will consist of an ASHP with an 2.765 kW heating and 2.455 kW cooling capacity and EWH with an annual EF of 0.94 [4, 5]. The upgraded case will use a 5x5 panel BIPV/T array with the same ASHP and EWH swapped out for an ASHPWH with an energy factor (EF) of 2.78 in HP only mode [4].

MECHANICAL SYSTEM MODEL

The TRNSYS model used for the BIPV/T system was developed by Kamel and Fung and was based off of Type 567 BIPV systems [3]. The authors Kamel and Fung developed Type 289, an opaque backing BIPV/T system which is the model used in the upgraded case of the house [3]. The ASHP TRNSYS component Type 665 that the BIPV/T system was integrated with was also tested and validated using experimental data collected in an Archetype Sustainable House [5]. However, since the capacity of the ASHP is far too high for the EcoStudio house model, it will be sized down to 25% of its original capacity and performance to be used in the case studies. The sizing down to 25% of the original was decided as it provided an output that most closely matched the energy demand while maintaining relatively sufficient temperature control.

To model the ASHPWH, two TRNSYS components were required, Type 938 and 1237. Type 938 represents an air to water heat pump, where heat from the air is extracted and in return, cooling is provided to the zone that is used as the air source. For the EcoStudio house model, since the main floor will have the most overheating due to the large south facing windows, the ASHPWH will be using the main floor temperature as its source. In reality, the ASHPWH will result in localized cooling, however, it is possible to control how heat is extracted and cooling is distributed using ducting. The model and performance of the ASHP that will be connected to the water tank was verified using experimental data from a 50 gallon hybrid ASHPWH with a rated EF of 2.78 on HP only mode [4].

Type 1237 is a typical thermal storage tank that is modelled to contain water as the primary fluid that would be distributed. To model this tank as a EWH, two electric elements are added in the tank, one near the lower portion of the tank, and another higher up. Both these elements operate at 1500 W each and provide heating depending on the set point provided. For the base case, Type 1237 only used the electric elements to represent a EWH as the reference for DHW. The EWH was used in the base model had an annual EF of 0.94 [4].

The ventilator used is from the TRNSYS component library and uses air that is mixed from both floors to be directed into the HRV and recover heat from the exhaust air. For both the base case and upgraded case, the HRV parameters will be kept the same. The efficiency of the system is set at 85% and set to a flow rate of 300 CFM [6]. The ventilation was controlled through the building Type 56 TRNSYS component.

The building model Type 56 is a proprietary TRNSYS model that is compliant with ASHRAE 140-2001 [7]. For this study, due to the immense overheating, a feature known as coupling was used in the Type 56 component. Coupling defines an air flow through adjacent zones but does not account for any fan power. This feature was used to help mix the air coming through the adjacent wall in the atrium of the main floor to the top floor, then return through the staircase back to the main floor. This is to help moderate the temperature of the building while also circulating any air due to high insulation and low infiltration.

HOUSE MODEL

The design of the EcoStudio house has two floors with a combined floor space of 1200 ft^2. A primary goal of the design is to build a home that can maximize solar gains and efficiency while being aesthetically pleasing and marketable for urban communities. The most current design has several large windows on the south face of the home to help reduce the heating load, hypothetically. The angle of the south roof is 25° where BIPV/T system will be mounted. The current design has a very large top floor volume, which is approximately 50% larger than the main floor. Since the main floor also has a higher potential to overheat, air must be circulated to maintain a balance between the two floors.

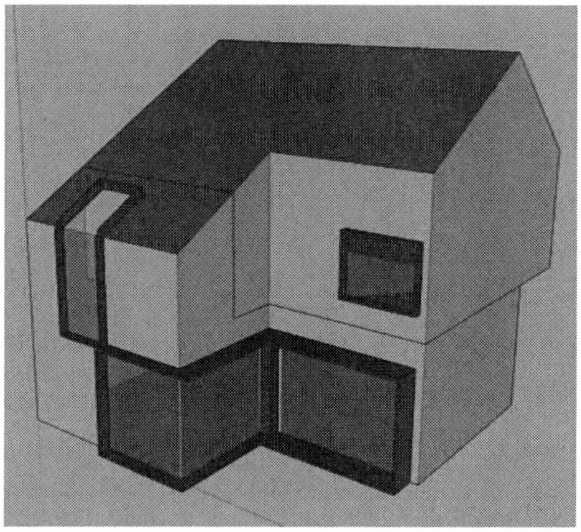

Figure 1: EcoStudio Solar home 3D model, southeast side

One unique feature of the building is that there is a small atrium in the first floor that is about 5 m high. This introduces the potential for thermal stratification if there is no circulation. The coupling that is added through the model helps to account for this issue. The house model is defined to have an infiltration rate of 0.5 ACH at 50Pa, which is equivalent to 0.025 ACH natural infiltration rate. Only internal shading is implemented during the cooling season and all other shading throughout the year is due to the facades built in place. The geometry and model of the house can be seen in Figure 1.

CONSTRUCTION

The external walls are made of R40 structurally insulated panels (SIP) and the external roof is R60 SIPs, where both have a layer of drywall for the interior surfaces. Internal walls are uninsulated studded walls using plywood and drywall that have an insulated of approximately R4.

The ground floor is R25 insulated 2x8 studs between plywood with 4 in concrete on top to be used as thermal mass. The high insulation is used to avoid losing heat during the winter, but must be optimized to account for the overheating due to solar gains.

INTERNAL GAINS

Since the goal of the EcoStudio is to build a NZEH, the internal gains that are input into the system cannot be standard values for residential homes, instead they must be values that would be typical for NZEHs. For this reason, the same modelling information used for internal gains from the reports for the past Solar Decathlon North House and Equilibrium Initiative EcoTerra House were used [8, 9]. Both use TRNSYS and as such, appliance internal gains values from these reports were used from the North House case study's highly efficient appliance list [8].

For the lighting, TRNSYS has an option to set very efficient lighting for a specified floor space. In these case studies, the total annual lighting load was approximately 701 kWh with a split of about 60% in the top floor and 40% in the main floor, which is close to the range of other similar sized energy efficient homes. This can be improved in the future with passive daylighting.

As for occupancy, internal TRNSYS setting were used to assume four adults where the occupancy scheduling was for a typical home where all persons work a day job. The lighting schedule was also designed to reflect this. The resulting total annual internal gains due to occupancy is to be 1223 kWh evenly split between both floors.

For the domestic hot water load, similar to the appliance heat gains, energy efficient water use had to be taken into account for the draw profiles. A daily hot water usage of 98.5 L was assumed, following similar standards to what the EcoTerra house had estimated as their usage [9]. The draw profile used was the same one that was used to validate the ASHPWH and normalized for down from 200 L/day to 98.5 L/day [4].

CONTROL STRATEGY

The system uses a very basic control system for the ASHP and ASHPWH. The ASHP uses a simple control matrix to deliver heating or cooling depending on demand. Since the ASHP is not strictly variable, values and dampers would have to be used. The control matrix seen in Table 1 shows that the energy is delivered depending on which zone(s) have active demand depending on season, i.e., cooling only during summer, heating only during winter. When both zones need heating or cooling, a fraction that was solved using peak demands of both zones with respect to the total demands was used. All other cases, the rooms would receive the full ASHP output. This is not ideal however, as a more complicated control should be designed in the future.

Table 1. Control Matrix for Heating and Cooling of Zones using ASHP

Zone	Both (Cooling/ Heating)	Main Floor Requires	Top Floor Requires	None Required
Main Floor	0.6 / 0.552	1	0	0
Top Floor	0.4 / 0.448	0	1	0

HOT WATER HEATING

For DHW heating, the control was set separately. For the base case, the electrical elements provided all the energy required as a temperature set point of 55°C with a 1°C dead band. For the ASHPWH in the upgraded case, the electrical elements had a set point of 45°C when there was insufficient energy in the home to be extracted by the HP, and the ASHPWH temperature set point was set similarly to the elements in the base case. These settings where only in place whenever the water tank required heating and when the thermostat of the main floor signaled the need for cooling. When the water required heating, but the temperature of the zone was below 20°C, the electric element temperature setting was changed to 55°C, so it would provide the full required heating. This is done because if the ASHPWH is left on when there is DWH load demand but no cooling demand, overcooling occurs, resulting in a higher heating demand for the ASHP.

SIMULATION RESULTS

The TRNSYS simulation used weather data for the Toronto Metro area, where the highest temperature is 33.9°C and the lowest at -22.65°C. The results show that the main floor only had a 10% higher heating demand than the top floor. However, for cooling the demand of the main floor was nearly double that of the top floor. This was consistent with both the base and upgraded case. Space heating load and cooling load did not improve significantly. This may be due to a lack of controls put in place to coordinate the ASHP with the ASHPWH, resulting in actions that may result in conflicting demand mentioned earlier, especially during the winter, where heating is required, but large amounts of overheating trigger the ASHPWH to function. The angle of the BIPV/T mounting is also poor to be able to fully utilize the solar energy at 25°. There were significant savings due to the ASHPWH however, as the electricity consumption for DHW load was reduced 54% due to a reduced usage of the electrical elements and the higher EF when in HP only mode. This is in line with other studies that have been done on ASHPWH [4, 10].

Table 2. Total Hours of Overheating or Overcooling depending on Reference Temperature Criteria for each Zone

Case	Zone	>24°C (hr)	>25°C (hr)	>26°C (hr)	>27°C (hr)	<22°C (hr)	<21°C (hr)	<20°C (hr)
Base	Main	1421.1	329.8	83.9	8.0	3475.5	1667.3	920.7
	Top	1123.0	286.9	67.9	7.7	2947.1	1709.0	943.8
Upgrade	Main	1185.1	248.2	65.3	1.0	3702.5	1678.8	928.4
	Top	855.1	206.6	47.5	0.0	2963.2	1720.6	952.9
% Reduced	Main	16.6%	24.7%	22.2%	87.5%	-6.5%	-0.7%	-0.8%
	Top	23.9%	28.0%	30.0%	100%	-0.5%	-0.7%	-1.0%

Unfortunately, the issue of overheating was not successfully tackled with the upgraded case as effectively as expected. Initially, it can be seen that in the base case, overheating occurs for 16.2% and 12.8% for the main and top floor respectively throughout an entire year. Although overheating was reduced, the values were not significant. As seen in Table 2, although the percentage decrease in overheating is fairly significant, in comparison, the increase in overcooling most likely due to the ASHPWH, is nearly equivalent in magnitude. Thus the benefits do not seem to outweigh with the consequences.

CONCLUSIONS

Conclusions that can be drawn from this simulation are many. A dedicated thermal storage tank should be used, since 98.5 L of hot water is not sufficient to fully extract the excess heat during periods of overheating. Better control must be devised, one where the ASHP and the ASHPWH are coordinated to work together to achieve optimal heating and cooling, while preventing any undesired consequences such as overcooling. Additional thermal mass should be considered, especially phase change materials (PCMs) to help buffer and offset the milder overheating and overcooling. Geometry should be refined, since a 25° angle roof is not optimal for PV/T, although aesthetics will be compromised; generally, a 45° angle is used in BIPV/T applications [3]. Windows and facades also must be designed to help optimize solar gains. Different ASHP technologies should also be investigated that is more appropriately sized and optimized for such a house. As the EcoStudio solar house model evolves, validation still must be completed to verify these results. Currently similar models are being developed by colleagues in EnergyPlus which will be used to verify future results.

ACKNOWLEDGMENTS

I would like to acknowledge all the contributions from past Team Toronto members and faculty and current members of EcoStudio in designing this home and for giving me the the opportunity to work on this project.

REFERENCES

[1] Natural Resources Canada, "Energy Efficiency Trends in Canada 1990 to 2004," Energy, no. March, pp. 1–48, 2006.

[2] R. S. Kamel, "Combined Building Integrated Photovoltaic-Thermal Collector with Air Source Heat Pump for Cold Climate," Ryerson Univ., Dissertation, no. July, 2015.

[3] R. S. Kamel and A. S. Fung, "Modeling, simulation and feasibility analysis of residential BIPV/T+ASHP system in cold climate—Canada," Energy Build., vol. 82, pp. 758–770, 2014.

[4] A. Amirirad, "Heat Pump Water Heater for Cold Climate - Canada," Ryerson Univ., MASc Thesis, 2016.

[5] A. A. Safa, "Performance Analysis of a Two-Stage Variable Capacity Air Source Heat Pump and a Horizontal Loop Coupled Ground Source Heat Pump System," Ryerson Univ., MASc Thesis, 2012.

[6] ASHRAE, ASHRAE Handbook 2013 Fundamentals, vol. 53, no. 9. 2013.

[7] S. A. Klein, "TRNSYS 17: A Transient System Simulation Program," Sol. Energy Lab. Univ. Wisconsin, Madison, USA, vol. 1, pp. 1–5, 2010.

[8] T. Saeid, "Technical Feasibility Study of Net-Zero Energy House for Canada - Case Study of Team North 2009 US DOE Solar Decathlon Competition," MASc Thesis, 2010.

[9] Sustainable Urbanism Initiative, "Sustainable Design in Toronto's Annex: Net Zero Energy Health Housing proposal on behalf of 'The Sustainable Urbanism Initiative'," no. January, 2007.

[10] Y. I. Sharaf-Eldeen, "Performance of Heat Pump Water Heater (HPWH) Systems and Their Impacts on Electric Utility Loads," J. Therm. Sci. Eng. Appl., vol. 2, no. 2, p. 21008, 2010.

Fluid Mechanics 2

Proceedings of the 26th CANCAM
Victoria, British Columbia, Canada, May 29 – June 1, 2017

IMPLEMENTATION OF LEES EDWARDS BOUNDARY CONDITION IN THREE DIMENSIONAL PARTICULATE FLOWS

Behnam Khalili
Mechanical Engineering Department
Shahid Bahonar University of Kerman
Kerman, Iran
Behnam.kh88@gmail.com

Ned Djilali
Mechanical Engineering Department
University of Victoria
Victoria, BC, Canada
ndjilali@uvic.ca

Mohammad Rahnama
Mechanical Engineering Department
Shahid Bahonar University of Kerman
Kerman, Iran
rahnama@uvic.ca

ABSTRACT

Direct numerical simulation of three dimensional particulate flows is a computationally demanding task especially for dense suspensions. Affordable computational domain can be achieved using Lees Edwards boundary conditions. The present paper deals with the implementation of Lees Edwards boundary condition in Lattice Boltzmann Simulations of a particle in shear flow. The particle motion simulations account for fluid-solid intercations using a smoothed profile method. Important issues of implementation of the Lees Edwards boundary and the problem of a particle crossing the shear boundary are discussed in detail. The method allows straightforward investigation of the effect of increasing particle volume fraction, and results in significantly improved physical representation compared to 2D simulations.

KEYWORDS: Lattice Boltzmann Method, Lees Edwards boundary condition, Smoothed profile method, particulate flow

INTRODUCTION

The rheological properties of a suspension are of both theoretical and practical importance and can be studied through shear flow of a suspension produced using solid walls. However, the existence of solid walls causes inhomogeneity in the suspension near the walls through various mechanisms such as shear-induced particle diffusion, which in turn affect the accurate prediction of rheological properties.

Lees Edwards Boundary Condition (LEBC) [1] is a method of generating shear flow without solid walls while particles may cross those boundaries. In fact, LEBC is an extension of a periodic boundary condition in which particles crossing a boundary experience both position and velocity changes due to boundary motion. As a main advantage of using the LEBC, the behaviour of suspensions can be studied with a finite number of particles in a computationally affordable domain. However, LEBC implementation is not straightforward in all computational procedures, especially those using the lattice Boltzmann method (LBM).

The lattice Boltzmann method is a computational technique used in fluid flow simulations, which has shown great success in predicting multi-phase flows [2]. Suspension simulation using the LBM has reached a state of such autonomy and robustness that dense suspensions can be accurately simulated. As a main challenge in suspension simulations, accurate prediction of fluid-solid interactions is of prime importance due to the subsequent dependence of both fluid and particle motions on such interactions. Among the various methods proposed for modelling solid-fluid interaction in the LBM, the smoothed profile method (SPM) [3] can be considered as efficient because it uses the same grid for both fluid and solid computations, resulting in a more computationally efficient simulation approach than other methods.

The LEBC was first implemented in an LBM of fluid flow by Wagner and Pagonabarraga [4]. They presented a method for obtaining unknown components of velocity distribution functions at the LEB based on the Galilean invariance of macroscopic velocity as fluid particles pass the LEB. Nearly all of the subsequent published LBM studies of particulate flows, have examined the Lees-Edwards boundary condition for two-dimensional flows [5,6]. Lorenz and Hoekstra [5] implemented Lees-Edwards boundary conditions for lattice Boltzmann simulations of particulate suspensions. Although such implementation was performed by Wagner and Pagonabarraga [8] for fluid-only systems, their discussion of the motion of particles passing through the Lees-Edwards boundary had not been previously reported. In a recent study, Zhou *et al.* [6] presented their implementation of the Lees-Edwards boundary condition in a node-based suspension method and demonstrated its essential Galilean invariance.

Implementation of the Lees-Edwards boundary condition in a three-dimensional particulate flow situation using the LBM requires special attention because particles are free to move in all three dimensions. The present paper deals with flow and particle simulation in a shear flow using Lattice Boltzmann Method and Smoothed Profile Method.

LATTICE BOLTZMANN METHOD COMBINED WITH SMOOTHED PROFILE METHOD

The lattice Boltzmann method is a solution procedure for the Boltzmann equation from which important macroscopic fluid properties such as velocity and pressure can be obtained. The main parameter of interest in the LBM is the velocity distribution function, f(x, v, t) which is a function of space, velocity, and time. With a single relaxation time, the LBM can be expressed as:

$$f_\alpha(\mathbf{x}+\mathbf{e}_\alpha\Delta t, t_n+\Delta t) = f_\alpha(\mathbf{x},t_n) - \frac{1}{\tau}\left[f_\alpha(\mathbf{x},t_n) - f_\alpha^{eq}(\mathbf{x},t_n)\right] + \frac{\omega_\alpha \Delta t}{c_s^2}(\mathbf{f}_H.\mathbf{e}_\alpha) \quad (1)$$

where α is the velocity direction, which can be any direction from 0 to 18 for D3Q19. The equilibrium distribution function, f^{eq}, is a distribution function that exists at equilibrium condition and can be expressed as:

$$f_\alpha^{eq} = \rho w_\alpha\left[1+3\frac{\mathbf{c}_\alpha.\mathbf{u}_f}{c_s^2}+\frac{9}{2}\frac{(c_\alpha.\mathbf{u}_f)^2}{c_s^4}-\frac{3}{2}\frac{\mathbf{u}_f.\mathbf{u}_f}{c_s^2}\right] \quad (2)$$

where τ is the relaxation time in Eq. (1), which is related to the fluid viscosity as $\tau = \upsilon/(c_s^2\Delta t) + 0.5$, and c_s is the speed of sound, equal to $1/\sqrt{3}$ in D3Q19. Macroscopic properties such as density and velocity ($\mathbf{u}_f$ in Eq. (2)) are obtained from the velocity distribution function as follows:

$$\rho(\mathbf{x},t) = \sum_{\alpha=0}^{8} f_\alpha(\mathbf{x},t) \quad (3)$$

$$\rho(\mathbf{x},t)\mathbf{u}_f(\mathbf{x},t) = \sum_{\alpha=1}^{8} \mathbf{c}_\alpha f_\alpha(\mathbf{x},t) \quad (4)$$

The smoothed profile method (SPM) is used to obtain fluid-solid interaction forces by defining a smoothed profile for the solid-fluid interface [4,5]. In a particulate flow, part of the domain is occupied by solid particles. In SPM, a function, φ, is defined such that it is equal to one inside of the particle and zero outside it, with a smoothly varying function from zero to one at the interface. Different forms have been proposed for the functions φ_i where suffix 'i' represents i^{th} particle, with the characteristics described above; among them, the following function was selected for the present study [3]:

$$\phi_i(x,t) = s(R-|x-R_i|)$$
$$s(x) = \begin{cases} 0 & x < -\zeta/2 \\ \frac{1}{2}\sin(\frac{\pi x}{\zeta}+1) & |x| < \zeta/2 \\ 1 & x > \zeta/2 \end{cases} \quad (5)$$

where ζ is the thickness of the interfacial region for the ith particle, in which φ_i varies from zero to one. The velocity of a point located inside a solid particle, $\mathbf{u}^p(\mathbf{x},t)$, is obtained from the kinematics of a rigid body as:

$$\phi(\mathbf{x},t)\mathbf{u}^p(\mathbf{x},t) = \sum_{i=1}^{N_P}\phi_i(\mathbf{x},t)\left[\mathbf{U}_i^p(t)+\boldsymbol{\omega}_i\times\{\mathbf{x}-\mathbf{R}_i^p(t)\}\right] \quad (6)$$

where $\mathbf{R}_i^p$, $\mathbf{U}_i^p$, and $\boldsymbol{\omega}_i$ are respectively the location, the velocity of the centre of mass, and the angular velocity of the ith particle. In SPM, a force is computed at each particle's internal node and implemented in LBM as an external hydrodynamic force (cf. Eq. 1) such that fluid velocity corresponds to that of solid particle at those nodes. This force appears as a result of the rate of change of momentum difference between fluid and solid at solid nodes. Mathematically speaking, for the ith particle, this statement can be written as:

$$\mathbf{f}_i^H(\mathbf{x},t_n) = \phi(\mathbf{x},t_n)\mathbf{f}_i^P(\mathbf{x},t_n) = \phi(\mathbf{x},t_n)\frac{(\mathbf{u}_i^p(\mathbf{x},t_n)-\mathbf{u}^f(\mathbf{x},t_n))}{\Delta t} \quad (2)$$

The total hydrodynamic force and torque exerted on the ith particle can be obtained from:

$$\mathbf{f}_i^p = \int \mathbf{f}_i^H(\mathbf{x},t)dv_i$$
$$\mathbf{N}_i^p = \int \mathbf{r}\times\mathbf{f}_i^H(\mathbf{x},t)dv_i$$

Knowing the resultant force and torque exerted on a solid particle, the particle acceleration, velocity, and location can be obtained from the equations of motion for a rigid body as:

$$M_i^p\frac{d\mathbf{U}_i^c}{dt} = \mathbf{F}_i^H + \mathbf{F}_i^{ext}$$
$$\mathbf{I}_i^p.\dot{\boldsymbol{\omega}}_i = \mathbf{T}_i^H + \mathbf{T}_i^{ext}$$

As mentioned earlier, any point in a particulate flow belongs either to a fluid domain, a particle domain, or an interfacial region ξ. Therefore, a general expression for the velocity at any point can be written as:

$$\mathbf{u}(\mathbf{x},t) = \phi(\mathbf{x},t)\mathbf{u}^p(\mathbf{x},t)+(1-\phi(\mathbf{x},t))\mathbf{u}^f(\mathbf{x},t) \quad (2)$$

where F represents the force applied, m is mass and a is the acceleration of the motion.

LEES EDWARDS BOUNDARY CONDITION

Lees and Edwards [4] proposed an extension of periodic boundary conditions for molecular dynamics to produce a shear flow without the effect of solid walls. This extension reduces the size of the domain in the shear direction by allowing solid particles to cross these boundaries as long as no wall effects influence their motion.

LEBC can be explained based on the concept of moving frames as shown in figure 1. As is observed in this figure, a shear flow can be divided into two rectangular domains of ABCD and A'B'C'D'. The first domain has top and bottom boundaries AD and BC respectively, moving with a constant velocity U_{LE}, but in opposite directions, whereas the corresponding boundaries of the second domain, A'D' and

C'D', move with velocities of U_{LE} and U_{LE} respectively, both in the positive x-direction. Two reference frames are attached to the domains and are shown as XY and X'Y'. Figure 3b shows the position of each domain after time t has passed. Assuming that A'B'C'D' moves with a velocity of 2 U_{LE}, it is identical to the ABCD frame. In other words, the top and bottom boundaries of A'B'C'D' move with velocity U_{LE} relative to frame X'Y' in this case, but in opposite directions, a situation similar to the velocity profile in ABCD. As a result, nodes located on the lower boundary B'C' of A'B'C'D' should represent their corresponding nodes on the lower boundary BC of ABCD. However, these corresponding nodes are located at different positions on BC. In other words, nodes located to the right of point B' move backward, whereas nodes located on the left side of BC move forward, as shown in Fig. 3b.

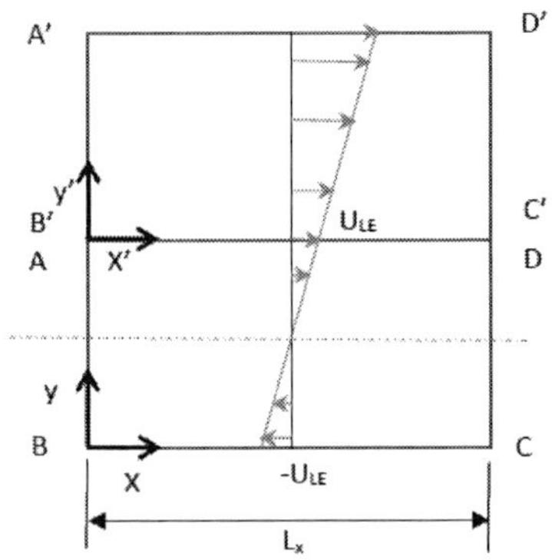

(a)

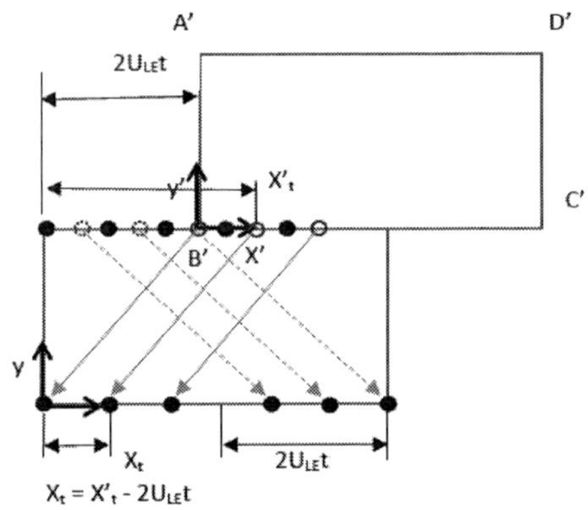

(b)

Figure 1. Lees-Edwards boundary condition described as a moving frame at (a) t=0 and (b) after time t. Filled black, unfilled, and dotted nodes belong to ABCD, A'B'C'D', and the domain to the left of A'B'C'D' (not shown).

Nodes located on top and bottom boundaries experience a change in both velocity and position in LEBC. These are used in streaming part of LB simulation for components of distribution function which cross LEB. A detailed discussion can be found in Ref. [5] for two dimensional simulations.

RESULTS

As a first test case, a shear flow with a particle placed at its center is investigated. This flow system has been studied by various authors in a two-dimensional flow [5,6,7]. Figure 2a shows the geometry of a particle placed at the center of a cubic domain under shear. Stating from zero velocity for all internal lattice nodes, a steady-state condition evolves for the case of shear produced by solid walls, at which the particle rotates with a constant angular velocity of $0.5\dot{\gamma}$ while no displacement occurs. This situation arises for a particle with small diameter compared to the domain dimensions. To validate the computational method proposed here, moving boundaries for such a flow configuration were implemented using LEBC with opposite velocities at top and bottom. A particle with a diameter of 10 lattice units was located at the center of the computational domain consisting of 66×66×66 lattice nodes, and its steady-state angular velocity was computed.

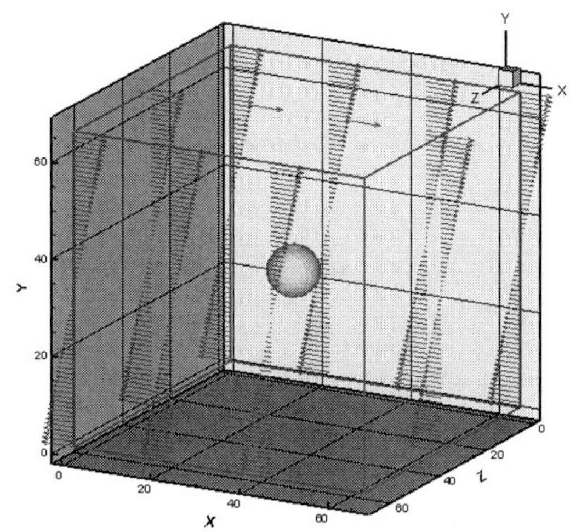

(a)

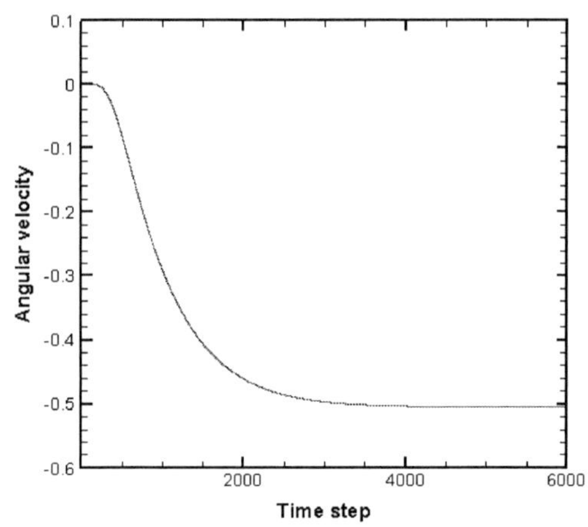

(b)

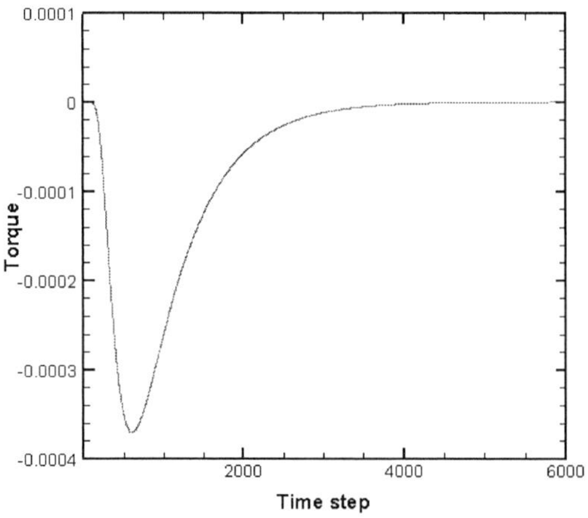

(c)

Figure 2. (a) Flow geometry of a particle under shear produced by LEBC, (b) variation of its dimensionless angular velocity and (c) torque with time.

Figures 2b and 2c show variation of the dimensionless z-component of angular velocity, $\omega/\dot{\gamma}$, and torque with time. It is observed that after 4000 time steps, a constant dimensionless angular velocity of 0.5 is appeared which corresponds to zero torque. Although such constant angular velocity is observed in this case, it does not happen for the same geometry but with larger particle. The bi-periodic LEBC implies other particles are present in the neighborhood of and interact with the main particle under study. Hwang et al. [8] studied the variation of particle angular velocity with diameter and showed that there were fluctuations in angular velocity over time. In this research, the motion of a spherical particle placed at the center of a cubic domain was studied under shear, as shown in Figure 2a, while its diameter was changed. Figure 3 shows variation of angular velocity with time for the cases of spheres with dimensionless diameters of $R = R_p/L = 0.25$ and 0.45 where 'L' is domain length. Higher values of angular velocity is observed for the present three dimensional computations compared to two dimensional ones. The reason for such behavior has its root in the different effect of LEBC in two and three dimensional computations. In three dimensional LEBC, there are particles placed periodically in z-direction in addition to x- and y-directions which affect main particle motion.

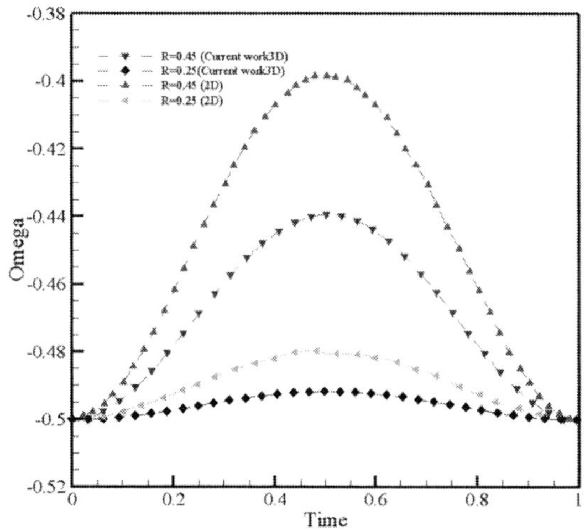

Figure 3.

To investigate the LEBC for particle motion in three dimensions, a cubic computational domain consisting of 100×100×100 lattice nodes was considered, with a spherical particle placed at its center. The particle diameter was selected as 10 lattice units. As shown in Figure 2a, the velocities at the top and bottom boundaries were selected as ULE = 0.0005 in lattice units, which corresponds to a shear rate of 1×10^{-5}. Computations were performed with a relaxation time of 1 and an SPM parameter $\xi=2$. The computations started with creation of a fluid-only shear flow for 5000 time steps. This number of time steps was required to reach a fully developed shear flow based on the authors' computations. Then a velocity field of (0.025, 0.005, 0.0) was superimposed on the fully developed shear flow, and a particle with the same velocity was placed at the center of the domain. Implementation of a transverse velocity of 0.005 made the particle move in a cross-flow direction and cross the LEB.

Figure 4 shows variation of particle velocity in the y-direction during the period in which the particle crosses LEB. Previous results of two-dimensional simulations for the same flow geometry [5,9,10] are also shown in this figure. The results obtained by Aidun, et. al. [9] and Ladd [10] show variation in y-component of velocity as the particle crosses LEB compared to those of Lorenz and Hoekstra [5] which show negligible increase in y-component of velocity. Such variations are not observed in the present three dimensional computations, and verify its accuracy compared to previous two dimensional simulations.

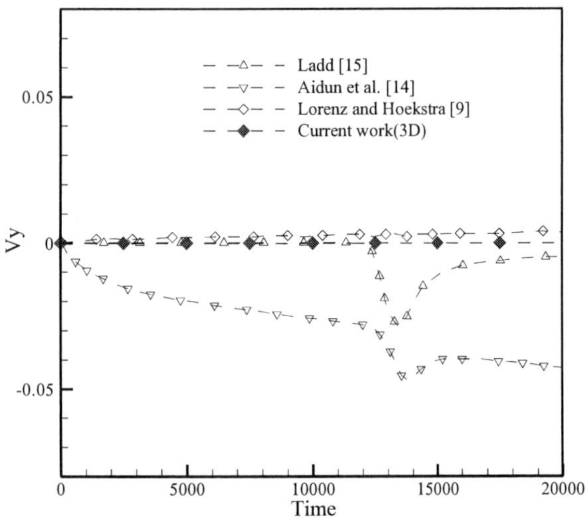

CONCLUSIONS

The LEBC was systematically implemented for the problem of a particle in shear flow placed at the center of the computational domain. As a particle crosses the LEB, it evolves in two parts requiring a consistent calculation of the forces and torques exerted on the two parts of a particle as it crosses one plane of the LEB. The present implementation is consistent and ensures in particular that the transverse components of the particle velocity does not vary spuriously due to the bi-periodic representation. The results show a constant angular velocity of $0.5\dot{\gamma}$ for small particles with a volume fraction less than 0.1, a result that accurately duplicates the theoretical value. However, increasing the particle diameter creates fluctuations in angular velocity over a certain period of motion as a result of the LEBC. This is in fact physically consistent and is a result of the proximity of other neighboring particles when the LEBC is implemented.

This work was primarily motivated by questions related to the fabrication of fuel cell catalyst layers [10] from colloidal inks suspensions where agglomeration plays an important role. The next phase of this work will focus on extending the method to account for inter-particle forces and its application to suspensions with a large number of particles.

ACKNOWLEDGMENTS

This work was supported in part by the CaRPE-FC Network (Catalysis Research for Polymer Electrolyte Fuel Cells).

REFERENCES

1. Lees, A.W. and Edwards, S.F., 1972, "The computer study of transport processes under extreme conditions", *J. Phys. C: Solid State Phys.*, Vol (5), pp 1921–1929.
2. Ladd, A.J.C., 2015, "Lattice Boltzmann methods for suspensions of solid particles", *Molecular Physics*, Vol (113), pp 2531–2537.
3. Jafari, S., Yamamoto, R, Rahnama, M., 2011, "Lattice-Boltzmann method combined with smoothed-profile method for particulate suspensions", *Physical Review E*, Vol (83), pp 026702-1.
4. Wagner, A.J. and Pagonabarraga, I., 2002, "Lees-Edwards boundary conditions for Lattice Boltzmann", *Journal of Statistical Physics*, Vol (107), pp 521–537.
5. Lorenz, E. and Hoekstra, A.G., 2009, "Lees-Edwards boundary conditions for lattice Boltzmann suspension simulations", *Physical Review E*, Vol (79), pp 036706-1.
6. Zhou, G., Wang, L., Wang, X., Gei, W., 2011, "Galilean-invariant algorithm coupling immersed moving boundary conditions and Lees-Edwards boundary conditions", *Physical Review E*, Vol (84), pp 066701-1.
7. Lee, Y.K., Ahn, K.H., Lee, S.J., 2014, "Local shear stress and its correlation with local volume fraction in concentrated non-Brownian suspensions: Lattice Boltzmann simulation", *Physical Review E*, Vol (90), pp 062317-1.
8. Hwang, W.R., Hulsen, M.A., and Meijer, H.E.H., 2004, "Direct simulation of particle suspensions in sliding bi-periodic frames", *J. Comput. Phys.* Vol (194), pp 742–772.
9. Aidun, C. K., Lu, Y., Ding, E., 1998, "Direct analysis of particulate suspensions with inertia using the discrete Boltzmann equation", *J. Fluid Mech.* Vol (373) pp 287.
10. Ladd, A. J. C., 1994, "Numerical simulations of particulate suspensions via a discretized Boltzmann equation. Part II. Numerical results", *J. Fluid Mech.* Vol (271), pp 311– 339.
11. Y.J. Choi Y.K., Djilali, N., 2016, "Direct numerical simulations of agglomeration of circular colloidal particles in two-dimensional shear flow", *Physics of Fluids*, Vol (28), 013304.

RHEOLOGY OF A PROPYLENE-GLYCOL ICE SLURRY

Charles Landa Onokoko
Génie mécanique
Université de Sherbrooke
Sherbrooke, QC, Canada
charleslanda@gmail.com

Nicolas Galanis
Génie mécanique
Université de Sherbrooke
Sherbrooke, QC, Canada
nicolas.galanis@usherbrooke.ca

Michel Poirier
CanmetEnergy
Natural Resources Canada
Varennes, QC, Canada
michel.poirier@canada.ca

Sébastien Poncet
Génie mécanique
Université de Sherbrooke
Sherbrooke, QC, Canada
sebastien.poncet@usherbrooke.ca

ABSTRACT

Friction factors for the flow of a particular propylene glycol ice slurry (freezing point -1.29 °C) obtained from measurements of pressure drops with different flowrates and ice fractions are compared with corresponding values from CFD simulations which assume Newtonian behaviour. The agreement between these sets of values is good for turbulent conditions but very poor for laminar flows. The experimental data is therefore used to obtain expressions for the yield stress and the viscosity consistency as a function of the ice fraction. When coupled with the Danish-Kumar laminar correlation the obtained Bingham rheological model produces friction factors in excellent agreement with the experimentally determined values.

KEYWORDS: ice slurry, Newtonian fluid, Bingham fluid, yield stress, viscosity consistency, ice mass fraction.

INTRODUCTION

Ice slurries are mixtures of small ice particles (typically 0.1 to 1 mm of diameter) and a carrier liquid (a mixture of water and an additive, such as glycol, alcohol or salt, which lowers the freezing temperature). They improve energy transport density and energy storage due to the combined effects of sensible and latent heat. Applications include cooling of buildings, food processing and the replacement of secondary refrigerants in ice rinks or supermarkets. Their thermophysical properties can be derived from linear weighting of the corresponding properties of the ice (essentially determined by the temperature) and the carrier liquid (which vary with temperature and additive concentration) [1].

The flow behaviour of ice slurries is complex. Thus, in horizontal pipes separation of the ice particles and carrier liquid occurs with any particle size at very low velocities and with large particles at high velocities. Various flow patterns (homogeneous, heterogeneous, sliding bed and stationary bed) are encountered and affect the hydrodynamics of the flow and the mechanism of heat transfer [1]. Kitanovski & Poredos [2] used a 1D approach to calculate the concentration distribution of ice and applied it to obtain the average ice slurry viscosity by integrating the Thomas equation. They concluded that for high average velocities and very low ice concentrations "the ice slurry viscosity is almost independent of velocity as for Newtonian fluids". On the other hand, ice slurries exhibit a non-Newtonian behavior for ice concentrations exceeding approximately 20 % but this threshold value is influenced by parameters such as the size of the ice particles and the nature of the additive. Several experimental studies have determined values of the effective viscosity of ice slurries and compared them to different rheological models (Bingham, Ostwald-de-Waele, Casson and Herschel-Bulkley) [3]. Simulations treating ice slurries as single-phase fluids have used the Bingham model for an ethanol ice slurry [4] and the Newtonian fluid approximation for an ethylene glycol ice slurry [5] with good results.

These, and other, diverse attempts to characterize the rheology of ice slurries indicate that their exact nature is not yet fully understood. In view of this situation the present study aims to add new data regarding the behavior of propylene glycol ice slurry for which non-Newtonian behavior has been reported even for low ice concentrations [6].

NEW RHEOLOGICAL MODEL

The ice slurry under consideration is an aqueous solution of propylene glycol whose measured freezing point is -1.29 °C. The pressure loss in a long horizontal pipe ($L_p/D \approx 177$) as well as the inlet and outlet density and temperature of the ice slurry were measured for flow rates between 1 and 38 kg/min and ice fractions of 5 to 20% by mass. Figure 1 compares the friction factor for an ice mass fraction of 15% determined from the experimental data with the corresponding numerical results from a 3D CFD model which assumes Newtonian behaviour [5] and with values determined from published correlations

[7, 8]. These comparisons (and similar results for other ice mass fractions) indicate a good agreement between the friction factors derived from measurements and the CFD predictions in the turbulent flow regime. In the laminar flow regime however there are significant differences between the friction factors derived from measurements and the CFD values which are close to the predictions of the Poiseuille expression. On the other hand, the results predicted by assuming a Bingham model are in very good agreement with the measured data for both flow regimes. The details of the calculations with the Bingham model are presented in the following paragraphs.

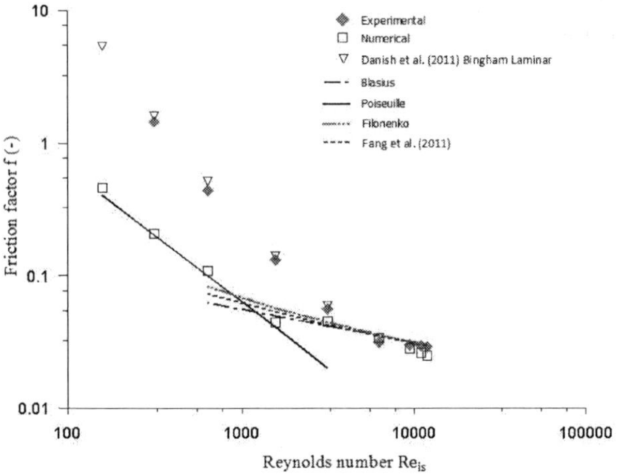

Figure 1: Friction factors in a smooth horizontal pipe (ice mass fraction = 15%)

Experimental pressure drops for laminar flow with three ice mass fractions (5, 15 and 20%) are plotted in Figure 2. In all cases, the curves which do not intercept the origin are linear. Therefore, these ice slurry flow situations can be fit by the Bingham model.

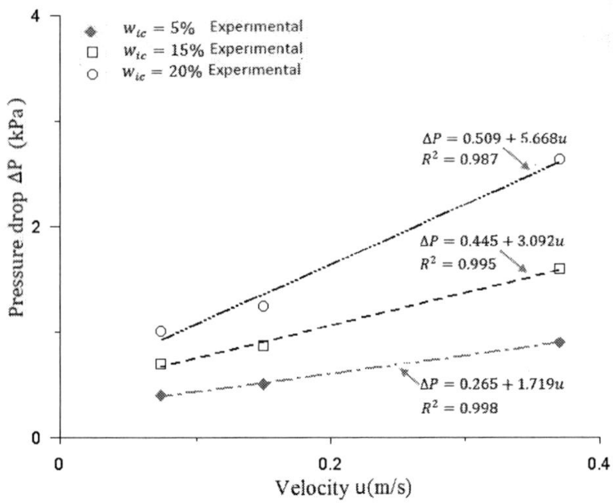

Figure 2: Measured pressure loss versus average inlet velocity for three ice mass fractions

Bingham fluids are characterized by two parameters: the yield stress τ_B and the viscosity consistency μ_B. Under steady state and constant temperature and pressure, the relationship between shear stress τ and shear rate γ for Bingham fluids is:

$$\tau = \tau_B + \mu_B \ \gamma \tag{1}$$

For the purposes of the present study the velocity from Figure 1 is converted into an apparent shear rate ($8u/D$) and the corresponding pressure drop into the wall shear stress ($\tau_w = D\Delta P/\ 4L_p$). The results of this operation are presented in Figure 3 together with the best linear correlation of the data. The agreement between the experimental values and the linear correlations is always very good.

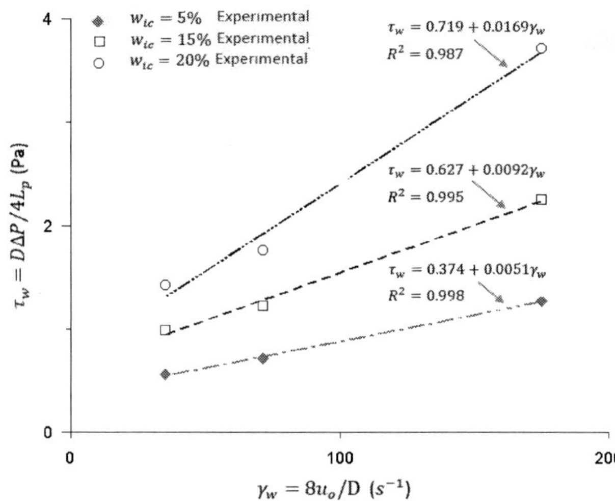

Figure 3: Apparent shear stress versus apparent shear rate for three ice mass fractions

From each line in Figure 3, we can extract the yield shear stress τ_B from the intercept point and the viscosity consistency μ_B which is equal to the slope. By applying the curve fitting techniques to these points, we obtain the following expressions for τ_B and μ_B as functions of the ice mass fraction w_{ic} (for ice mass fractions between 5% and 20%):

$$\tau_B = 1.534 \ {w_{ic}}^{0.471} \tag{2}$$

$$\mu_B = 0.0034 \ e^{7.98 w_{ic}} \tag{3}$$

Figure 4 illustrates the influence of the ice mass fraction on the yield stress and the viscosity consistency. The agreement between the experimental data and the proposed correlations is always very good.

The combination of Equations (1), (2) and (3) provides the general expression of the derived empirical Bingham model in terms of the ice mass fraction (applicable for laminar flow with w_{ic} between 5% and 20%):

$$\tau = 1.534 {w_{ic}}^{0.471} + (0.0034 e^{7.98 w_{ic}})\gamma \tag{4}$$

Figure 5 illustrates the effect of the ice mass fraction on the

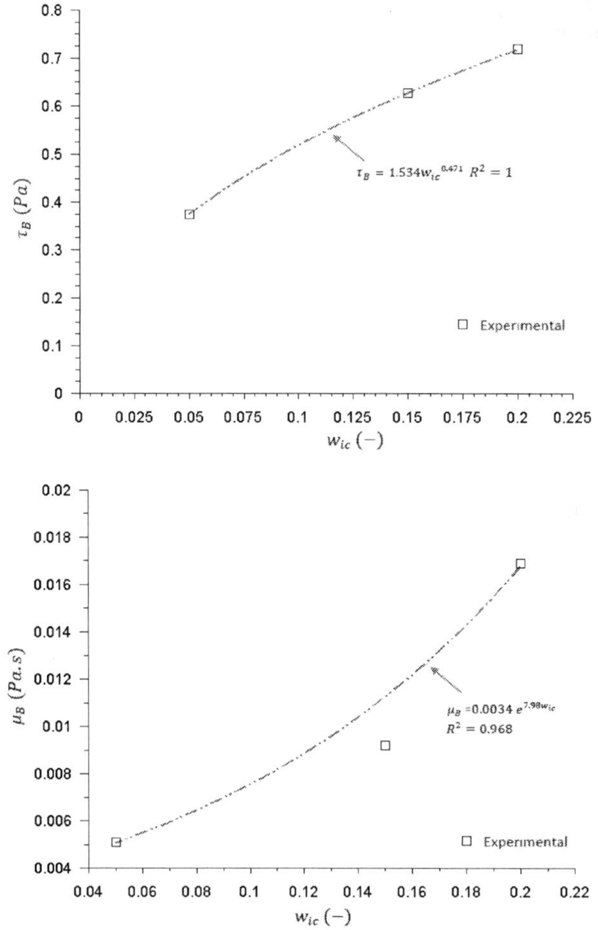

Figure 4: Effect of the ice mass fraction on the yield stress and the viscosity consistency

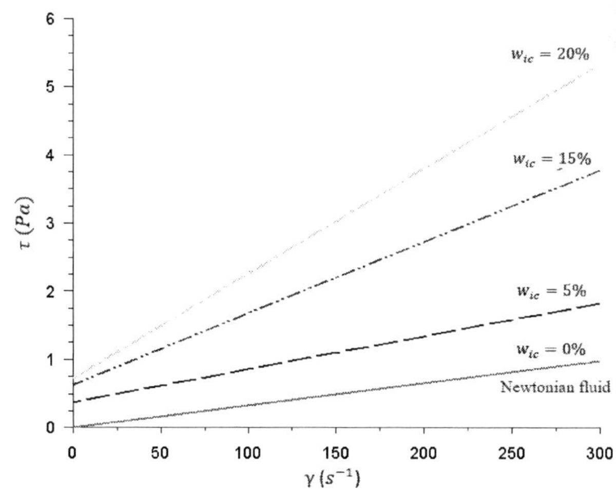

Figure 5: Effect of the ice mass fraction on the relation between shear stress and shear rate

relation between the shear stress and the shear rate for the Newtonian and Bingham models. These results reflect the fact that both the yield stress and the viscosity consistency increase with the ice mass fraction as shown in Figure 4.

FRICTION FACTOR COMPARISONS

Validation is obtained by comparing friction factors derived from the measurements with those predicted using Equation (4). For this purpose we apply the following explicit relation [9] for the laminar flow of Bingham fluids in smooth pipes:

$$f = 4K_1 \left\{ \begin{matrix} 1 - 10[K_2/(K_1^4 + 3K_1)]^4 \\ +6[K_2/(K_1^4 + 3K_1)]^3 + K_2/(K_1^4 + 3K_1) \end{matrix} \right\} \quad (5)$$

where $K_1 = (16/Re) + (16He/6Re^2)$ and $K_2 = (-16He^2/3Re^8)$. The Reynolds and Hedstrom numbers are calculated using the ice slurry density ρ and the values of τ_B, μ_B from Equations (2), (3) respectively ($Re = \rho uD/\mu_B$, $He = \tau_B \rho D^2/\mu_B$). To ensure that the flow is laminar this Reynolds number for each combination of measured flow rate (or, equivalently, average velocity) and ice mass fraction is compared with the corresponding critical value obtained from the following relation [10] which is valid for $He \leq 10^9$:

$$Re_c = 2100(1 + He/3600)^{0.35} \quad (6)$$

Table 1 gives the Bingham Reynolds number as a function of the flow mean velocity u_o for different ice mass fractions w_{ic} as well as the corresponding critical Reynolds Re_c and the Hedstrom number He for all the flow combinations used in the present study (see Figures 1-5). It confirms that all these flow combinations are indeed laminar since $Re < Re_c$.

Table 1: Values of the Bingham Reynolds number

u (m/s)	w_{ic} = 5% Re_c = 2752 He = 4195	w_{ic} = 15% Re_c = 2360 He = 1422	w_{ic} = 20% Re_c = 2241 He = 732
0.074	248	112	75
0.15	504	226	151
0.37	1242	556	374
0.74	2485	1115	747

It should be noted that these values of the Reynolds number were also used for the evaluation of the friction factors by all of the correlations mentioned in Figure 1.

The results of these calculations are shown in Figure 1 for w_{ic} = 15% and in Figure 6 for w_{ic} = 5% and 20%. The agreement between the experimentally derived values of the friction factor and those calculated by combining equations (2), (3) and (5) is excellent while the predictions of the Newtonian CFD model grossly underestimate these values. We conclude that Eq. (4) is valid for the present conditions.

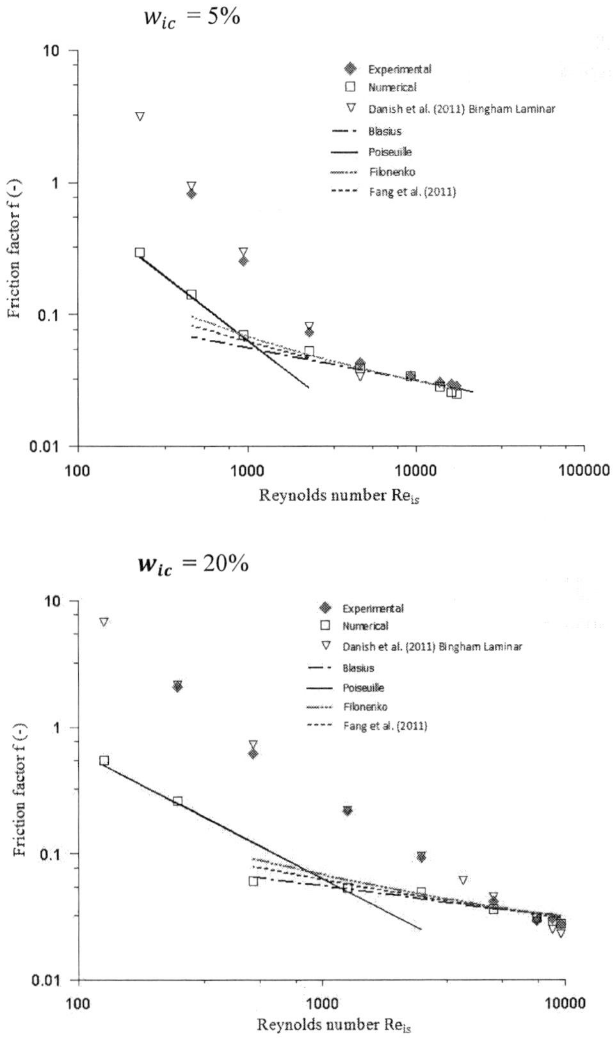

Figure 6: Friction factors in a smooth horizontal pipe (ice mass fraction = 5% and 20%)

CONCLUSIONS

Comparisons between friction factors derived from measurements with corresponding values obtained from CFD simulations indicate that the Newtonian behaviour is justified for turbulent flows of the propylene-glycol slurry under study. For laminar flow a Bingham formulation for the shear stress is developed and produces friction factors in excellent agreement with the experimentally derived values.

ACKNOWLEDGMENTS

The authors thank NSERC for its financial support and Natural Resources Canada for making available the instrumented experimental facility at CanmetEnergy Varennes.

REFERENCES

1. Kauffeld M., Kawaji M. & Egolf P. 2005, "*Handbook on Ice Slurries-Fundamentals and Engineering*", International Institute of Refrigeration (IIF/IIR), France.
2. Kitanovski A. & Poredos A. 2002, "Concentration distribution and viscosity of ice-slurry in heterogeneous flow", *International Journal of Refrigeration*, 25, pp. 827-835.
3. Kitanovski A., Vuarnoz D., Ata-Caesar D., Egolf P., Hansen T. & Doetsch C. 2005, "The fluid dynamics of ice slurry", *International Journal of Refrigeration*, 28, pp. 37-50
4. Niezgoda-Zelasko B. & Zalewski W. 2006, "Momentum transfer of ice slurry flows in tubes, modeling", *International Journal of Refrigeration*, 29, pp. 429-436.
5. Onokoko C. L. & Galanis N. 2013, "Stratification in isothermal ice-slurry pipe flow", *ASME Int. Mech. Eng. Congress & Expo*, paper IMECE2013-6380.
6. Mellari S., 2016, "Experimental investigations of ice slurry flows in horizontal pipe based on monopropylene glycol", *International Journal of Refrigeration*, 65, pp. 27-41.
7. Renaud-Boivin S., Poirier M. & Galanis N., 2012, "Experimental study of hydraulic and thermal behavior of an ice slurry in a shell and tube heat exchanger", *Experimental Thermal and Fluid Science*, 37, pp. 130-141.
8. Fang X., Xu Y. & Zhou, Z. 2011, "New correlations of single-phase friction factor for turbulent pipe flow and evaluation of existing single-phase friction factor correlations", *Nuclear Engineering and Design*, 241, pp. 897–902
9. Danish M., Kumar S. & Kumar Su. 2011, "Approximate explicit analytical expressions of friction factor for flow of Bingham fluids in smooth pipes using Adomian decomposition method", *Communications in Nonlinear Science and Numerical Simulation*, 16, pp. 239–251.
10. Swamee P. K. & Aggarwal N. 2011, "Explicit equations for laminar flow of Bingham plastic fluids", *Journal of Petroleum Science and Engineering*, 76, pp. 178–184

Proceedings of the 26th CANCAM
Victoria, BC, Canada, May 29 - June 1, 2017

A GPU-ACCELERATED SMOOTHED PARTICLE HYDRODYNAMICS ALGORITHM FOR FREE SURFACE FLOWS

Jiandong He
School of Aerospace
Beijing Institute of Technology
Beijing 100081, China
chrishe1900@gmail.com

Nima Tofighi
Department of Mechanical Engineering
University of Victoria
Victoria, BC V8W 2Y2, Canada
ntofighi@uvic.ca

Mehmet Yildiz
Faculty of Engineering
and Natural Sciences (FENS)
Sabanci University
Istanbul 34956, Turkey
meyildiz@sabanciuniv.edu

Juanmian Lei
School of Aerospace
Beijing Institute of Technology
Beijing 100081, China
leijm@bit.edu.cn

Afzal Suleman
Department of Mechanical Engineering
University of Victoria
Victoria, BC V8W 2Y2, Canada
suleman@uvic.ca

ABSTRACT

Smoothed Particle Hydrodynamics (SPH) is a meshless numerical method commonly used to simulate complex free surface flows. Due to large number of particles involved in each operation, SPH is computationally expensive. Recently, the computational capabilities of Graphics Processing Units (GPU) have been employed to accelerate the simulations. Here, we present and validate a weakly compressible SPH (WCSPH) algorithm suitable for GPU computing and compare its performance with the serial code.

Keywords: WCSPH, GPU, Free Surface Flows

INTRODUCTION

SPH is a Lagrangian meshless method introduced by Gingold and Monaghan [1] and Lucy [2] for astrophysics problems. Since then, SPH has been extended to simulate complex problems such as fluid dynamics [3–5] and solid mechanics [6]. SPH interpolates the parameters at Lagrangian particles using the values at neighboring particles through a kernel function. While this interpolation scheme makes SPH a very flexible method, it also introduces significant computational cost.

Use of parallel algorithms is a common method of accelerating SPH simulations. Here, we use GPU as the hardware of choice for parallelization. Specifically, we implement our code using Nvidia's Compute Unified Device Architecture (CUDA) interface on a GTX 960 graphics card. A detailed review and discussion of GPU's application in computational fluid dynamics may be found in [7]. Examples of using GPU accelerated SPH simulations are given in [8, 9].

In the current study, we employ the wall boundary condition described in [10] as it relies on a constant number of auxiliary particles. This facilitates the parallelization of the scheme. Additionally, we apply a density filter to recover a smoothed pressure field [11]. We simulate the breaking of water column and compare our results to experimental data of [12]. Our simulations show that use of GPU-parallel SPH code results in more than 20 times speed up compared to CPU-serial SPH code.

GOVERNING EQUATIONS

In this study, we assume the fluid to be non-viscous, governed by the following equations

$$\frac{d\rho}{dt} = \rho \nabla \mathbf{v}, \qquad (1)$$

$$\frac{d\mathbf{v}}{dt} = -\frac{\nabla p}{\rho} + \mathbf{g}, \qquad (2)$$

$$p(\rho) = \rho_0 g H \left[\left(\frac{\rho}{\rho_0} \right)^7 - 1 \right], \qquad (3)$$

where ρ, $\mathbf{v}$, t and $\mathbf{g}$ denote density, velocity, pressure, time and gravity. ρ_0 is the reference density while H is the initial water height.

Equations (1) and (2) are discretized as

$$\frac{d\rho_a}{dt} = \rho_a \sum_b \frac{m_b}{\rho_b} \mathbf{v}_{ab} \nabla_a W_{ab}, \qquad (4)$$

$$\frac{d\mathbf{v}_a}{dt} = -\sum_b m_b \left(\frac{p_a}{\rho_a^2} + \frac{p_b}{\rho_b^2} + \Pi_{ab} \right) \nabla_a W_{ab} + \mathbf{g}, \qquad (5)$$

where m denotes particle mass, $\mathbf{v}_{ab} = \mathbf{v}_a - \mathbf{v}_b$ and Π_{ab} is the artificial viscosity term [13]. W_{ab} is the kernel function between particles a and b where a Gaussian kernel with a cut-off radius of 3 particle spacing, h, is used. Wall boundary conditions are imposed through the method described in [10] while Shepard filter is applied every 25 iterations [11]. The continuity and momentum equations are advanced in time using a predictor-corrector scheme.

CUDA IMPLEMENTATION

A schematic of the solution procedure for both serial and parallel codes is shown in figure 1. For the parallel code, we carry out initialization procedures on host (CPU and RAM) and transfer the data to GPU memory to start the simulation. At this point, the GPU carries out all computations and the control and data are only transferred back to host for writing the output. The main operations carried out on particles by GPU are cell determination, sorting and time advancement.

In SPH method, each particle is assigned a number and needs to be aware of its neighboring particle numbers within the kernel cut-off radius, here $3h$. A common method to search for neighbors is to divide the computational domain into square cells. Assigning each particle to a cell is a process that benefits significantly from parallelization. As the flow evolves, the particles move and those with consecutive numbers may not fall within the same cell. To improve memory access, the particle numbers are reordered. This process is much more efficient when carried out in parallel. Finally, calculating the right hand side of equations (4) and (5) is significantly accelerated when each particle is assigned to a single thread on a GPU.

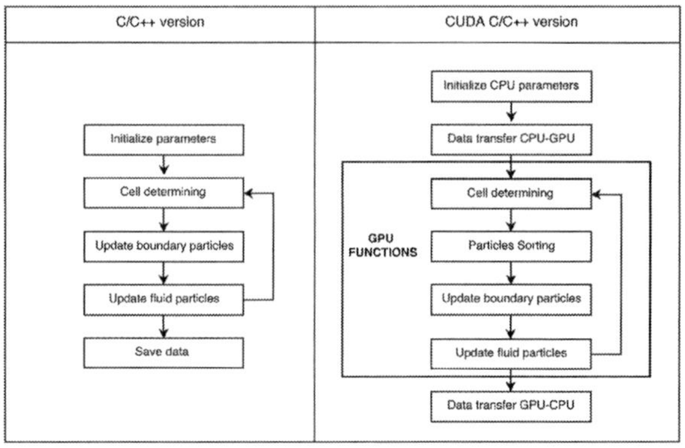

Figure 1: Flow diagrams of the C/C++ and CUDA C/C++ based SPH code.

DAM BREAK PROBLEM

Here we simulate a classical dam break problem. A water column of height $H = 1$ and width $W = 2$ is released from rest and allowed to break down under its own weight. The tank width is chosen as $L = 5.366$ to enable comparison with literature data [10, 12]. Figure 2 shows the snapshots of the interface profile colored by fluid pressure. The surface profile and pressure distribution observed here are in agreement with the SPH results reported in [10].

Pressure load on the right wall at $y/H = 0.2$ is plotted in figure 3. Compared to the experimental data of [12], the overall behavior is well captured and the results are quantitatively comparable.

The current simulation is run on an Intel Core i5-4460 CPU with an Nvidia GeForce GTX 960 GPU. For 24944 particles, each iteration of serial code takes $1.341s$ while the parallel code completes the same task in $0.054s$.

CONCLUSIONS

A GPU-parallel WCSPH scheme for free-surface flows is presented. The method relies on boundary conditions with static auxiliary particles and uses Shepard filter to obtain smooth pressure. The serial code is written in C++ and executed on an Intel Core i5-4460 CPU while the parallel code is written in CUDA C/C++ and is executed on an Nvidia GTX 960 graphics cards. The scheme is tested for dam break problem and the results are in quantitative agreement with experimental data. The parallel code is found to be 20 times faster than the serial counterpart.

ACKNOWLEDGMENTS

JH wishes to acknowledge the financial support from the program of China Scholarship Council under the grant number 201406030037. NT gratefully acknowledges financial support from Mitacs-Accelerate program under the grant number IT08354. MY acknowledges financial support provided by the Scientific and Technological Research Council of Turkey

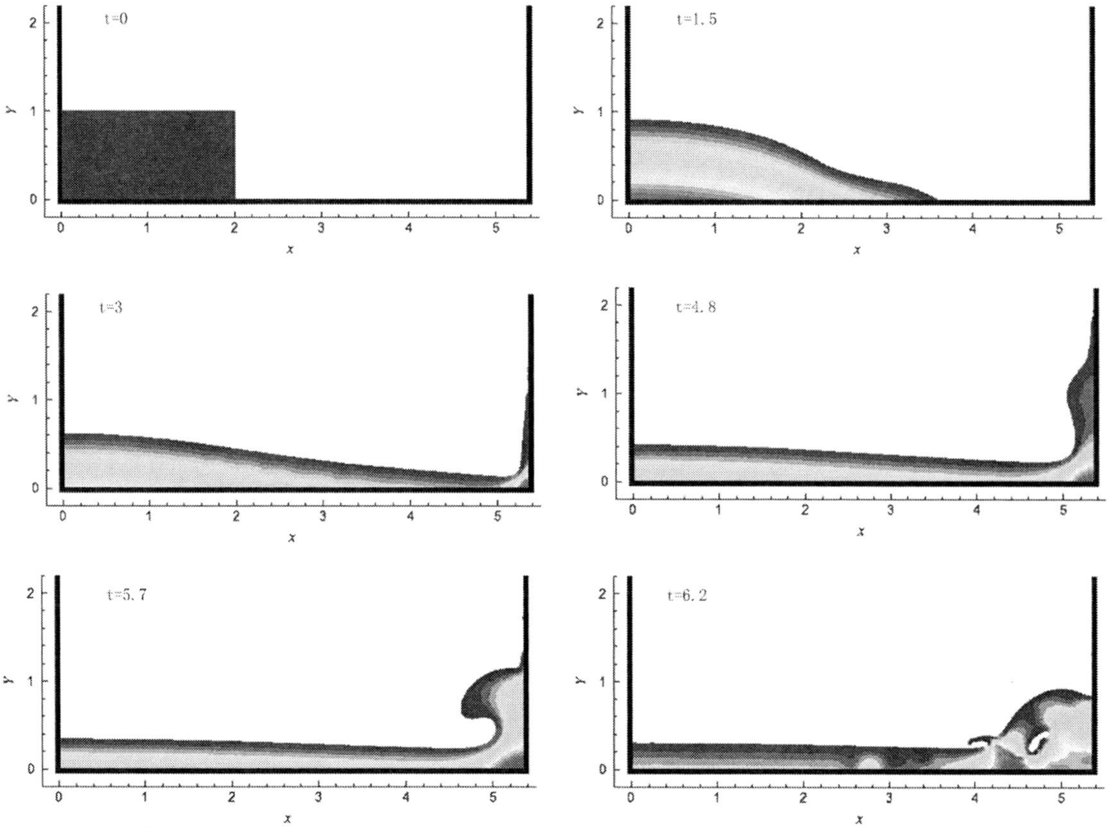

Figure 2: Snapshot of the evolution of the dam break simulation.

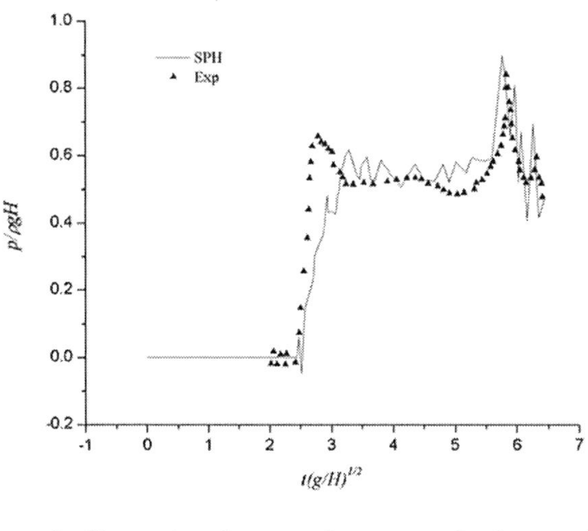

Figure 3: Comparison between the pressure loads over time measured experimentally and predicted by SPH.

(TUBITAK) for grant number 112M721. AS acknowledges the financial support provided by the Natural Sciences and Engineering Research Council (NSERC) for the Canada Research Chair in Computational and Experimental Mechanics.

REFERENCES

[1] Gingold, R.A., and Monaghan, J.J. 1977, "Smoothed particle hydrodynamics: theory and application to non-spherical stars," *Monthly Notices of the Royal Astronomical Society*, 181(3) pp. 375-389.

[2] Lucy, L.B. 1977, "A numerical approach to the testing of the fission hypothesis," *The Astronomical Journal*, 82, pp. 1013-1024.

[3] Monaghan, J.J. 1994, "Simulating free surface flows with SPH," *Journal of Computational Physics*, 110(2), pp. 399-406.

[4] Shao, S., and Lo, E.Y. 2003, "Incompressible SPH method for simulating Newtonian and non-Newtonian flows with a free surface," *Advances in Water Resources*, 26(7), pp.787-800.

[5] Farrokhpanah, A., Samareh, B. and Mostaghimi, J. 2015, "Applying Contact Angle to a Two-Dimensional Multiphase Smoothed Particle Hydrodynamics Model," *ASME Journal of Fluids Engineering*, 137(4), p. 041303.

[6] Gray, J.P., Monaghan, J.J. and Swift, R.P. 2001, "SPH elastic dynamics," *Computer Methods in Applied Mechanics and Engineering*, 190(49), pp. 6641-6662.

[7] Vanka, S.P. 2013, "2012 Freeman Scholar Lecture: Computational Fluid Dynamics on Graphics Processing Units," *ASME Journal of Fluids Engineering*, 135(6), p. 061401.

[8] Herault, A., Bilotta, G. and Dalrymple, R.A. 2010, "SPH on GPU with CUDA," *Journal of Hydraulic Research*, 48(S1), pp. 74-79.

[9] Dominguez, J.M., Crespo, A.J., Valdez-Balderas, D., Rogers, B.D. and Gómez-Gesteira, M., 2013, "New multi-GPU implementation for smoothed particle hydrodynamics on heterogeneous clusters," *Computer Physics Communications*, 184(8), pp. 1848-1860.

[10] Adami, S., Hu, X.Y. and Adams, N.A. 2012, "A generalized wall boundary condition for smoothed particle hydrodynamics," *Journal of Computational Physics*, 231(21), pp. 7057-7075.

[11] Colagrossi, A. and Landrini, M. 2003, "Numerical simulation of interfacial flows by smoothed particle hydrodynamics," *Journal of Computational Physics*, 191(2), pp. 448-475.

[12] Buchner, B. 2002, "Green water on ship-type offshore structures," PhD Dissertation, Delft University of Technology.

[13] Monaghan, J.J. and Gingold, R.A. 1983, "Shock simulation by the particle method SPH," *Journal of Computational Physics*, 52(2), pp. 374-389.

Fluid Mechanics and Heat Transfer

Characterization of frost formation on a cold surface

N. Niroomand[11]*, M. Fauchoux[12], E. Walia[2], C. J. Simonson[13]
[1]University of Saskatchewan, Department of Mechanical Engineering, Saskatoon, Saskatchewan, Canada
[11](Nan846@mail.usask.ca)
[12](melanie.fauchoux@usask.ca)
[13](carey.simonson@usask.ca)
[2] University of Saskatchewan, Department of Computer Science, Saskatoon, Saskatchewan, Canada
(Ewb178@mail.usask.ca)

ABSTRACT

This paper presents an experimental study on frost formation on a plate under natural convection conditions. Frost thickness, mass and density are measured at different surface temperatures as a function of time. It is found that lighter and thicker frost forms on a plate at lower surface temperature.

KEYWORDS: Frost, thickness, visualization, mass, density.

INTRODUCTION

Reducing greenhouse gas emissions is an important topic, due to global warming and climate change. Decreasing energy consumption for heating and cooling of buildings can play an important role in the reduction of greenhouse gas emissions. Air-to-air energy/heat exchangers are one way to reduce energy consumption for conditioning occupied buildings. In cold weather, however, frost can form inside the exchangers, and can adversely affect their performance [1][2]. Understanding the behavior of frost formation under different conditions is necessary to estimate the performance of a frosted exchanger.

Frost formation depends on several parameters, such as air temperature, air humidity ratio, surface temperature and surface properties. Many studies have been done on frost formation on cold impermeable plates. Cheng and Shiu [3] investigated the effects of velocity, temperature, and relative humidity of air on frost thickness on a cold surface, especially at the leading edge. Kim et al. [4] proposed a CFD-based model to predict frost distribution on a cold plate. They also calculated the frost thickness and frost surface temperature for a non-uniform frost density. Their theoretical results agreed well with the available experimental data. Alizadeh et al. [5] analyzed frost formation dynamics on surfaces with different wettability (i.e. hydrophilic and hydrophobic surfaces). They used a combination of high-speed photography and infrared thermography to detect the occurrence of freezing by monitoring the droplet temperature during the icing phase change. It was shown that surface energy can play an important role on the nucleation rate. They concluded that by controlling the surface properties (texture and chemistry), the ice nucleation can be delayed. Liu et al. [6] carried out an experimental and theoretical study on the influence of surface hydrophilicity on frost deposition on a vertical cold plate under natural convection conditions. They compared the size of initial water drops formed on a hydrophobic surface and that on a plain copper plate. They showed that water drops that formed on the copper plate have a larger size and shorter retention time compared to those formed on hydrophobic surfaces. Shin et al. [7] investigated the influence of surface energy (using surfaces with different hydrophilicity) on frost formation. They developed new test techniques to measure frost temperature, thickness, mass and thermal conductivity. They used four fine thermocouples located at different points, within 4 mm, above the cold surface to measure the frost temperature. They showed that surfaces with lower dynamic contact angles will result in frost with higher density and thermal conductivity after the first 120 min of test time.

In this study, frost formation on an aluminum plate under natural convection conditions is investigated. Mass, thickness and density of the frost layer are measured as a function of time. Frost thickness was measured by using a visualization method. It is found that different types of frost crystals form on a plate under different operating conditions which results in different behavior of frost growth. Understanding the process of frost deposition and the parameters that affect frost formation on a plate can help to develop methods to delay or avoid frost formation in exchangers.

EXPERIMENTS

In this study, properties of a frost layer deposited on an aluminum plate are measured experimentally. Experiments are conducted in an environmental chamber under natural convection conditions where the air temperature and humidity ratio can be maintained within ±0.2°C and ±2% RH (Figure 1). The aluminum plates, with dimensions of 4.5 cm by 4.5 cm, are placed on an isothermal block which is connected to a thermal bath, which provides cold refrigerant for reducing the surface to a desired temperature.

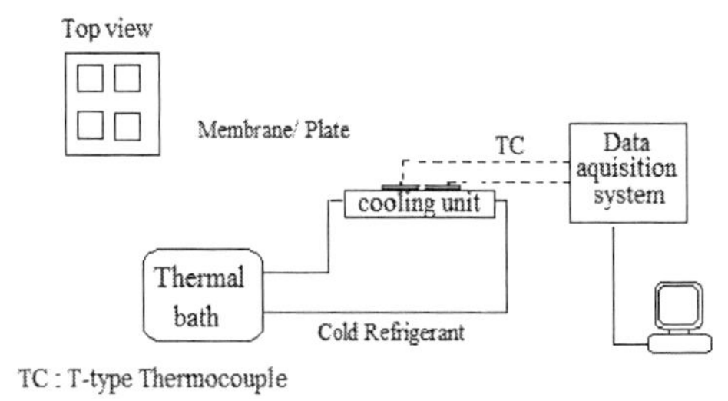

Figure 1: Schematic of the test setup in the environmental chamber.

The isothermal block, with the plates on it, is covered with a large acrylic box, which is open on one side to allow placement and removal of the aluminum plates during the tests. This box ensures that natural convection conditions exist inside the box. The temperature of the plate is measured using a T-type thermocouple mounted to the top of the plate. The surface temperature is continuously monitored, using a data acquisition system, every 30 seconds.

The thickness and mass of frost that grows on the surface of the aluminum plate are measured multiple times during each test to measure the accumulation of frost over time. Measurements are taken every 30 min. Each test is run for 120 min total.

The frost layer thickness is measured by analyzing images taken from the side view of the plate during frost buildup. The thickness of the plate is measured at the beginning of the test, when there is no frost, at four different locations along the length of the plate, from the side view picture of the plate. Then, when frost forms on the plate, the thickness of the plate, with frost on it, is measured periodically at the same locations during the test. The average of the difference between these two measurements at each time shows the frost layer growth. Sample pictures of the plate with and without frost can be seen in Figure 2.

The uncertainty of the measurement is calculated from systematic and precision uncertainties. To calculate systematic uncertainty in thickness measurement, the thickness of a standard plate is measured by the presented method and compared to its actual value, which is found to be equal to ±0.014 mm. The precision uncertainty is determined by calculating the thickness from several photos taken from the frost layer at a specific time and is equal to ±0.08 mm. Therefore, the total uncertainty for these measurements is equal to ±0.08 mm.

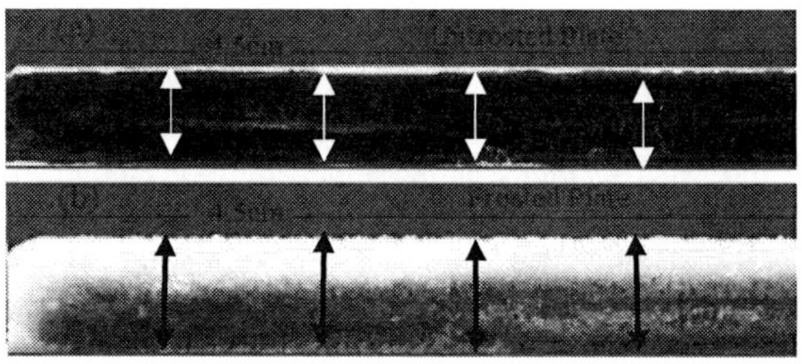

Figure 2: Sample photographs of the plate (a) with and (b) without frost.

The mass of frost is measured by removing the aluminum plate from the test surface using tweezers and placing it in a sealed plastic container. The mass of the container and frosted plate is measured using a mass balance. The tweezers are used to reduce the physical damage and melting of the frost due to contact or heat transfer. The sealed container protects the frost from evaporation. After the test, the frost is allowed to melt and evaporate completely, then plate in the container is weighed again. The difference between two readings is mass of the frost. The frost mass concentration is calculated by dividing the mass of the frost by the area of the plate. The systematic uncertainty in mass measurement was found by calibrating the mass balance and is equal to ±0.087 g. To calculate the precision uncertainty, four plates were placed on the cooling unit, then the mass of frost deposited on them were measured at the same time (30 min after the beginning of the test). The uncertainty in mass and mass concentration measurements were found to be equal to ±0.03 g and ±0.02 kg/m^2 respectively. The total uncertainty in mass measurement is found to be equal to $\pm 0.03\ kg/m^2$.

Frost density is calculated by dividing the frost mass concentration by the thickness of the frost at each specific time.

RESULTS AND DISCUSSION

The thickness and mass of the frost layer on the plate are compared to results from the literature in Figure 3, to verify the methodologies used in this paper. It can be seen that both frost thickness and mass data agree well with the results from the literature (Kandula) [8]. The operating conditions for the data from the literature (air Temperature = 22°C, surface temperature = -10°C, RH = 72%RH, forced convection) are different from this experiment (air temperature = 20°C, RH = 50%RH, surface temperature = -10°C, free convection), which may account for the small difference in results. A higher relative humidity will result in an increase in the frost mass and thickness layer. The similar trend between the experimental data measured in this paper and the data in the literature shows that the methodology can be used for measuring mass and thickness of the frost deposition in other cases.

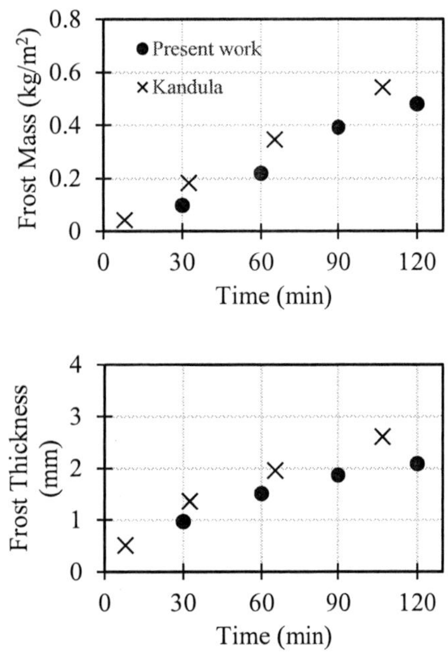

Figure 3: Comparison of frost mass and thickness from the present work with results in the literature.

Figure 4 shows the frost thickness, mass, density and at two different surface temperatures. It can be seen that the frost thickness, mass and density increase with time, however the rates at which the thickness and density increase reduce over time. The thickness of the frost layer is smaller when the surface temperature is higher, due to a lower potential for mass transfer between the air and the frost surface. The frost mass has a different behavior, heavier frost forms on the plate with a higher surface temperature. The reason for this is that the frost crystals deposited on the plate with a lower surface temperature are different from the frost crystals formed on the surface with a higher temperature. Frost crystals formed on the plate at lower temperature, have a feather shape, which results in a more porous layer of frost. Therefore, because there are more air voids in the frost layer at a lower surface temperature, the density of the frost layer is higher as can be seen in third figure. As density has an opposite relation with thermal conductivity, it means that frost with higher thermal conductivity will form on the plate at lower surface temperature.

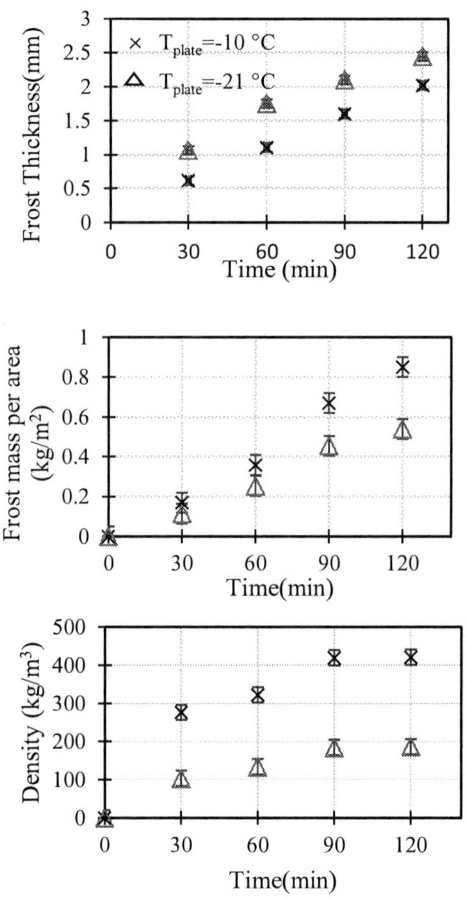

Figure 4: Effect of surface temperature on frost growth when T_{air}=20°C and RH=50%RH.

CONCLUSIONS

The present paper is the first step in a PhD research project to observe and characterize the frost formation on heat/energy exchanger surfaces. A test facility is developed to measure the thickness, mass and density of frost on a plate over time. Results from measuring frost thickness and mass on the plate are used to verify the experimental facility and methods. The following conclusions could be stated;

- Thicker and lighter frost forms on a plate with a lower surface temperature.
- Lower density, frost accumulates on a plate with a lower surface temperature.

REFERENCES

1. Rafati Nasr, M., Fathieh, F., Kadylak, D., Huizing, R., Besant, R. W., and Simonson, C. J. 2015, "Experimental methods of detecting frosting in cross-flow air-toair energy exchangers", *Experimental Thermal and Fluid Science*, 77, pp. 100-115.
2. Rafati Nasr, M., Fauchoux, M., Besant, R. W., and Simonson, C. J. 2014, "A review of frosting in air-to-air energy exchangers", *Renewable and Sustainable Energy Reviews*, *30*, pp. 538–554.
3. Cheng, C. H., and Shiu, C. C. 2002, "Frost formation and frost crystal growth on a cold plate in atmospheric air flow", *International Journal of Heat and Mass Transfer"*, 45(21) 4289–4303.
4. Kim, D., Kim, C., and Lee, K.-S. 2015, "Frosting model for predicting macroscopic and local frost behaviors on a cold plate", *International Journal of Heat and Mass Transfer*, 82, pp. 135–142.
5. Alizadeh, A., Yamada, M., Li, R., Shang, W., Otta, S., Zhong, S. and Blohm and M. L. 2012, . "Dynamics of Ice Nucleation on Water Repellent Surfaces", *Langmuir*, 28, pp.3180-3186.
6. Liu, Z., Zhang, X., Wang, H., Meng, S. and Cheng, S. 2007, "Influences of surface hydrophilicity on frost formation on a vertical cold plate under natural convection conditions", *Experimental Thermal and Fluid Science*, 31(7), pp. 789–794.
7. Shin, J., Tikhonov, A. V., and Kim, C. 2003, "Experimental Study on Frost Structure on Surfaces With Different Hydrophilicity: Density and Thermal Conductivity", *Journal of Heat Transfer*, 125(1), pp. 84-95.
8. Kandula, M. 2011, "Frost growth and densification on a flat surface in laminar flow with variable humidity", *Journal of heat and mass transfer,* 54, pp. 3719-3731.

Proceedings of the 26th CANCAM
Victoria, British Columbia, Canada, May 29 - June 1, 2017

CONDENSATION MODELING ON THE FIN TOP OF A MICRO-GROOVED HEAT PIPE

Mobin Alipour
Mechanical Engineering
Middle East Technical University
Ankara, Turkey
alipour.mobin@metu.edu.tr

Zafer Dursunkaya
Mechanical Engineering
Middle East Technical University
Ankara, Turkey
refaz@metu.edu.tr

ABSTRACT

Heat pipes are used as an alternative to forced cooling systems, especially for cooling current electronic devices which contain concentrated hot spots. Condensation occurs on the top of the fins and approximations to estimate the condensing mass is used in literature. In this study, four different approaches to condensation modeling in micro-grooved heat pipes are discussed. The effect of disjoining pressure, the reverse angles and temperature differences are studied.

Keywords: Condensation, micro-grooved heat pipe, disjoining pressure, micro region

INTRODUCTION

Electrical devices with high heat dissipation should be thermally managed with cooling systems in order to improve their life and performance. Because of the large amount of heat required for phase change, heat pipes can operate with lower temperature difference between heat addition and removal locations, in addition to lower maximum temperatures, an advantage in cooling of electronic components. Due to the ultra thin fluid layers encountered during condensation and evaporation, intermolecular forces and surface-liquid interactions have a significant impact on the behavior of the thin liquid film and consequently on the thermal performance of the heat pipe. In 2001, Zhang and Faghri [1] modeled the condensation film inside the groove and on the fin top of the rectangular shaped groove by using the VOF method. They stated that convection does not have a substantial effect on the condensation process, they also concluded that most of the condensation occurs on the fin top. In 2005, Jiao et al. [2] presented a hydrodynamic model for modeling the condensed liquid on the fin top with the assumption of a constant film thickness. Lefévre et al. [3, 4] presented a hydrodynamic model for condensation on the fin top without including the effect of disjoining pressure. The resulting fourth order nonlinear ordinary differential equation derived from the augmented Young-Laplace and two dimensional conservation of mass and momentum equations was solved numerically by using a Runge-Kutta scheme. Condensation mass flux was assumed to be constant along the fin top by using Clausius-Clapeyron equation. Do et al. [5] suggested a mathematical model for the condensation film on the fin top by using an integral method and fitting a 4th order polynomial for solving the nonlinear ordinary differential equation derived from the conservation of mass, momentum and Young-Laplace equations. The effect of disjoining pressure was omitted and capillary pressure assumed to be a function of the second derivative only. Effect of disjoining pressure on the behavior of the evaporation meniscus was discussed by some researchers, it was not considered as a part of condensation modeling. Derjaguin and Zorin [6] described the concept of the disjoining pressure and presented experimental data for polar liquids, which were formulated as a logarithmic function by Holm and Goplen [7] in 1979. Derjaguin and Zorin [6] also introduced the polynomial formula which was only a function of film thickness for calculating the disjoining pressure of the non-polar liquids. These formulae are commonly used in the literature to include the effects of disjoining pressure in the evaporation modeling for both polar and non-polar liquids [5, 8, 9, 10]. In 2004, Wu and Wang [11] expressed a new slope and curvature dependent disjoining pressure, which is modified for small variations of fluid film (i.e. small slopes). Biswal et al. [12] included the expression derived by Wu and Wang [11] in the modeling of the

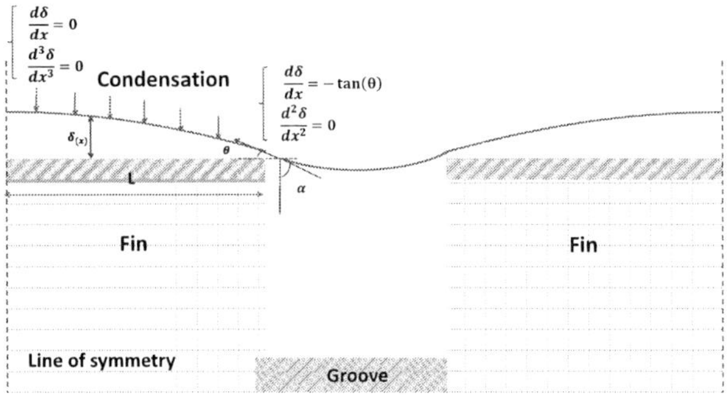

Figure 1: Schematic of liquid film on the fin top and its boundary conditions

evaporation meniscus. This study is the only one in which the effect of the slope and curvature dependent disjoining pressure on the evaporation is taken into consideration. There has not been any research related to analyzing the effect of disjoining pressure for the condensation film, with or without the slope dependency, up to present day. In this study condensation film is modeled by considering the conservation of mass and momentum along with the augmented Young-Laplace equation which is solved numerically while the effect of disjoining pressure is included in the calculation process. Beside this model, the 4th order polynomial fitting approach and the case in which the disjoining pressure is neglected, are taken into consideration and compared.

MODELING

In the condensation region, vapor is condensed both in the groove and on the fin top, but because of the thinner thickness of the liquid film on the fin top most of the heat is transferred in this part of the condensation region, hence majority of condensation occurs. Consequently, accurate modeling of the liquid film is expected to result in a better approximation of the thermal model. Schematic of the liquid film on the fin top is given in Figure 1. The film is assumed to be continuous at the point where the fin and the grove meet, rendering the continuity of first derivatives. The angle α shown in the figure is defined in this study as the "reverse angle" and the sum of the reverse angle and the contact angle inside the channel add up to 90°. Pressure difference between vapor and liquid is described by using the augmented Young-Laplace equation as Eq. (1), which is the function of both capillary and disjoining pressures.

$$P_v - P_l = P_c + P_d \tag{1}$$

Capillary pressure is related to the curvature and slope of the liquid film and can be written as Eq. (2).

$$P_c = \frac{\sigma \delta_{xx}}{\left(1+{\delta_x}^2\right)^{3/2}} \tag{2}$$

An expression for disjoining pressure was derived using the same approach used by Wu and Wang [11]. Although the approximation for small slopes was not used, therefore, in this study there are no limitations to the curvature of the liquid film in the expression for disjoining pressure. At equilibrium condition, total energy of the system should be at its minimum and its first derivative should be equal to zero. By expanding the function which gives the equilibrium condition, an expression for disjoining pressure is found [11] as,

$$P_d = -\frac{\partial E}{\partial \delta} + \frac{d}{dx}\left(\frac{\partial E}{\partial\left(d\delta/dx\right)}\right) \tag{3}$$

Excess energy is derived by integrating the total intermolecular potential ϕ with respect to x, from a cut-off distance D to the film thickness δ at any given point.

$$E = -\frac{\pi {n_f}^2 \beta_{ff}}{4\delta^2}\left\{\frac{1-\lambda}{3} + \frac{(1-\rho)\,{\delta_x}^2}{4}\left(1-\frac{1}{(1+{\delta_x}^2)^{1/2}}\right)\right\} \tag{4}$$

By substituting Eq. 4 in Eq. 3 the final form of disjoning pressure is derived as,

$$P_d = \pm\frac{B}{\delta^3}\left\{\alpha^4 - \frac{2}{3}{\delta_x}^2 - \frac{2{\delta_x}^4}{3\left(1+{\delta_x}^2\right)^{3/2}} + \frac{2{\delta_x}^2}{3\left(1+{\delta_x}^2\right)^{1/2}} + \frac{2}{3}\delta\delta_{xx} - \frac{\delta{\delta_x}^4\delta_{xx}}{(1+{\delta_x}^2)^{5/2}} + \frac{5\delta{\delta_x}^2\delta_{xx}}{3\left(1+{\delta_x}^2\right)^{3/2}} - \frac{2\delta\delta_{xx}}{3\left(1+{\delta_x}^2\right)^{1/2}}\right\} \tag{5}$$

where

$$B = \frac{3\pi {n_f}^2 \beta_{ff}(1-\rho)}{16}, \quad \alpha = \left[\frac{8(1-\lambda)}{9(1-\rho)}\right]^{1/4}, \quad \lambda = \frac{n_s \beta_{fs}}{n_f \beta_{ff}}, \rho = \frac{n_g \beta_{fg}}{n_f \beta_{ff}} \tag{6}$$

It should to be noted that β_{ff}, β_{fs} and β_{fg} represent the strengths of the van der Waals potentials between fluid-fluid, fluid- solid and fluid-gas molecules, respectively. Molecular densities of fluid, solid and gas are expressed by n_f, n_s and n_g. The liquid flow on the fin top can be assumed as a one dimensional fully developed flow by neglecting the inertia terms, which leads to the velocity profile,

$$u_{(y)} = \frac{1}{2\mu}\frac{dP_l}{dx}(y^2 - 2\delta y). \tag{7}$$

Mass flow rate is calculated by integrating the velocity distribution respect to y and multiplying it with density, which results in

$$m' = -\frac{1}{3\upsilon}\frac{dP_l}{dx}\delta^3. \tag{8}$$

Interfacial mass flux is applied by using the modified form of Wayner's equation which is adapted by Moosman and Homsy

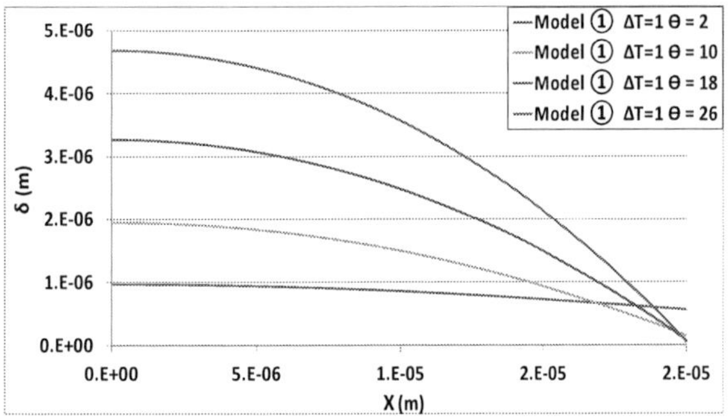

Figure 2: Variation of $\delta_{(x)}$ with reverse angle

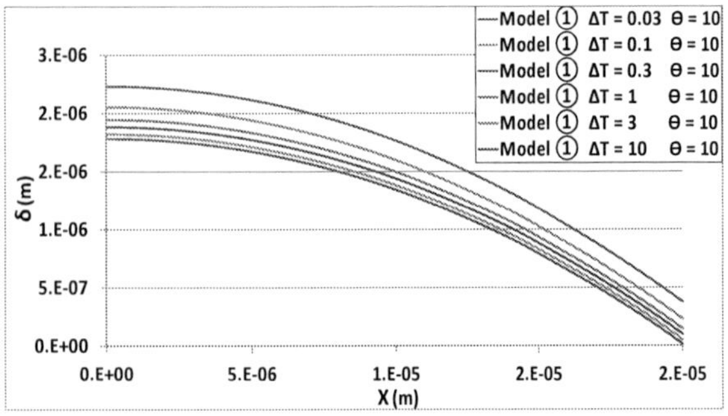

Figure 3: Variation of $\delta_{(x)}$ with temperature difference

[13] to be a function of a temperature jump between the temperatures of solid wall and vapor. It should be noted that the accommodation coefficient, c, is taken as unity in Eq. (10,11).

$$\dot{m}_e'' = \frac{a\left(T_w - T_v\right) - b\left(P_c + P_d\right)}{1 + a\delta h_{lv}/k_l} \tag{9}$$

$$a = \frac{2c}{2-c}\left(\frac{M}{2\pi R_u T_l v}\right)^{1/2}\frac{P_v M h_l v}{R_u T_v T_l v} \tag{10}$$

$$b = \frac{2c}{2-c}\left(\frac{M}{2\pi R_u T_l v}\right)^{1/2}\frac{P_v V_l}{R_u T_l v} \tag{11}$$

By balancing mass fluxes calculated by Equations (8) and (9), the governing differential equation is derived

$$-\frac{1}{3\nu}\delta^3\frac{d^2 P_l}{dx^2} - \frac{\delta^2}{\nu}\frac{d\delta}{dx}\frac{dP_l}{dx} = -\frac{a\left(T_w - T_v\right) - b\left(P_c + P_d\right)}{1 + a\delta h_{lv}/k_l} \tag{12}$$

SOLUTION METHODOLOGY

Eq. (12) is solved numerically by using a variable step size Runge-Kutta-Felhberg scheme. The problem has four boundary conditions shown in Figure 1. In order to solve the problem with Runge-Kutta-Felhberg approach, the problem should be converted into a form with simultaneous 1st order ODE's with corresponding initial conditions. Consequently, two boundary conditions that are located at the fin edge should be assumed as initial conditions at the symmetric line and solved by shooting method. A two dimensional Newton-Raphson approach is chosen for finding the roots of the shooting method. In addition to this study, one 4th order polynomial is also used in an approximation to the shape of the fluid film and is solved numerically by secant method. The results are obtained for four different models listed below:

- Model 1 : Including the slope and curvature dependent disjoining pressure and solving by Runge-Kutta-Felberg approach.
- Model 2 : Including capillary pressure, neglecting the disjoning pressure and solving by Runge-Kutta-Felberg approach.
- Model 3 : Fitting a 4th order polynomial to the fluid film and considering a simple form of the capillary pressure, function of second derivative only.
- Model 4 : Fitting a 4th order polynomial to the fluid film and considering the complete form of capillary pressure.

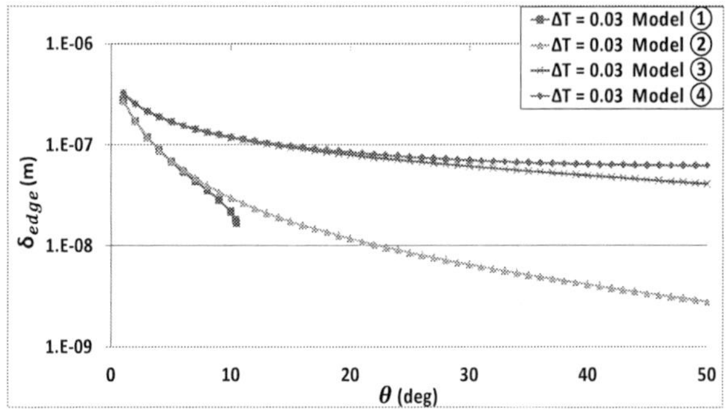

Figure 4: Variation of δ_{edge} with reverse angle for all models

RESULTS

Water is chosen as the operating fluid and vapor temperature is set to 373 K. Film thickness variation as a function of reverse angle and temperature difference is shown in Figures 2 and 3. It can be seen from Figure 2 that by decreasing the reverse angle, the film profile becomes flatter. Figure 3 shows that for smaller temperature differences the film is thinner. Comparison between the results of all four models for δ_{edge} as a function of reverse angles is presented in Figure 4. The results of Models 3 and 4 are very different from the results of Models 1 and 2. This results in a substantial difference in heat flux estimated by Models 3 or 4 vs. Models 1 or 2. The inclusion of the effect of disjoining pressure introduces a limitation to the results. Figure 4 shows that, in contrast to Model 1, Model 2 can predict results for all values of reverse angles. The predicitions of Model 1 on the other hand, are limited by the reverse angle. This limitation appears when the film is thin enough to render disjoining pressure dominant. The variation of this cut-off point with temperature difference

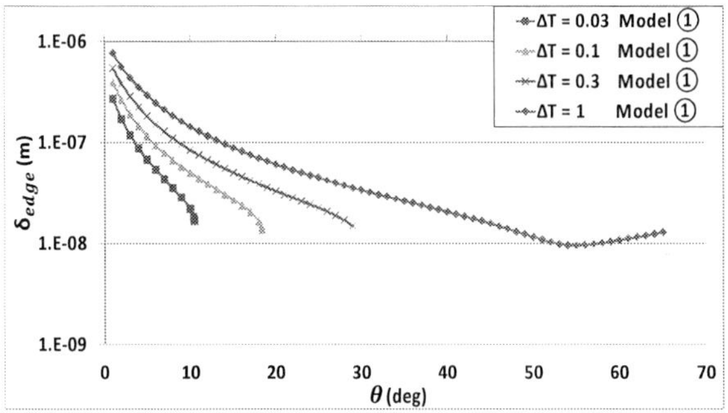

Figure 5: Variation of δ_{edge} with reverse angle

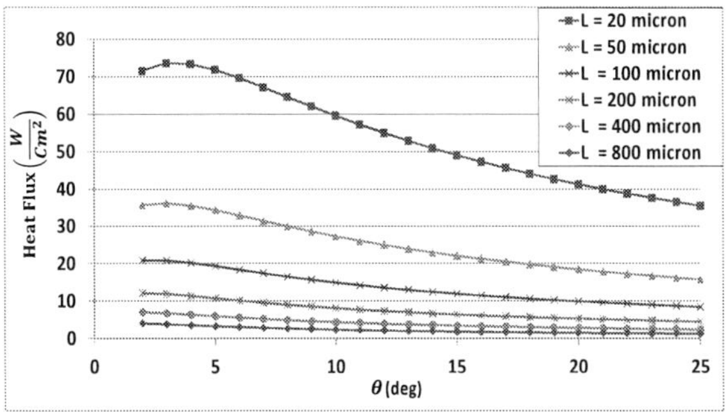

Figure 6: Variation of heat flux with reverse angle

is shown in Figure 5. Variation of heat flux with reverse angle for different fin thicknesses for Model 1 is presented in Figure 6. There is a maximum point for the heat flux in the reverse angles less than 10°. By decreasing the fin thickness heat flux increases rapidly, which suggests that increasing the number of fins – i.e. having more thinner fins than less wider fins – could result in enhanced condensation performance.

CONCLUSIONS

In this study four different models were used to assess the effect of capillary and disjoining pressures on condensation. The results of Models 3 and 4 which are based on solving the governing equation by fitting a 4th order polynomial to the fluid film profile predict results substantially different than the results of Models 1 and 2, where no approximation to the shape of the profile is made. This difference in profile shape, results in approximately 30% increase in the computed heat flux. Although the results of Models 1 and Model 2 are close up to a point and the maximum heat flux difference between them is approximately 5 percent, an important difference is the limitation that appears in Model 1 due to the effect of disjoining pressure.

ACKNOWLEDGMENTS

This study is financially supported by Turkish Scientific and Technical Research Council, under grant No. 213M351.

REFERENCES

[1] Faghri, A., Zhang, Y. 2001, "Numerical Simulation of Condensation on a Capillary Grooved Structure," *Numerical Heat Transfer, Part A: Applications*, Vol(39), pp. 227-243.

[2] Jiao, A. J., Riegler, R., Ma, H. B., Peterson, G. P. 2005, "Thin film evaporation effect on heat transport capability in a grooved heat pipe," *Microfluidics and Nanofluidics*, Vol(1), pp. 227-233.

[3] Lefvre, F., Rullire, R., Pandraud, G., Lallemand, M. 2008, "Prediction of the temperature field in flat plate heat pipes with micro-grooves - Experimental validation," *International Journal of Heat and Mass Transfer*, Vol(51), pp. 4083-4094.

[4] Lefvre, F., Lips, S., Bonjour, J. 2005, "Frontiers in Heat Pipes," *Microfluidics and Nanofluidics*, Vol(1).

[5] Do, K. H., Kim, S. J., Garimella, S. V. 2008, "A mathematical model for analyzing the thermal characteristics of a flat micro heat pipe with a grooved wick," *International Journal of Heat and Mass Transfer*, Vol(51), pp. 4637-4650.

[6] Derjaguin, B.V., Zorin, Z.M. 1957, "Title of the paper," *Proc. 2nd Int. Congr. Surface Activity*, London, England, pp. 145-152.

[7] Holm, F. W., Goplen, S. P. 1979, "Heat Transfer in the Meniscus Thin-Film Transition Region," *Journal of Heat Transfer*, Vol(101), pp. 543-547.

[8] Wayner, P. C., Kao, Y. K.,LaCroix, L. V. 1979, "Heat Transfer in the Meniscus Thin-Film Transition Region," *International Journal of Heat and Mass Transfer*, Vol(19), pp. 487-492.

[9] Kou, Z. H., Lv, H. T., Zeng, W., Bai, M L., Lv, J. Z. 2015, "Comparison of different analytical models for heat and mass transfer characteristics of an evaporating meniscus in a micro-channel," *International Communications in Heat and Mass Transfer*, Vol(63), pp. 49-53.

[10] Akkuş, Y., Dursunkaya, Z. 2016, "A new approach to thin film evaporation modeling," *International Journal of Heat and Mass Transfer*, Vol(101), pp. 742-748.

[11] Wu, Q., Wong, H. 2004, "A slope-dependent disjoining pressure for non-zero contact angles," *Journal of Fluid Mechanics*, Vol(506), pp. 157-185.

[12] Biswal, L., Som, S. K., Chakraborty, S 2013, "Thin film evaporation in microchannels with slope and curvature dependent disjoining pressure," *International Journal of Heat and Mass Transfer*, Vol(57), pp. 742-748.

[13] Moosman, S., Homsy, G.M. 1980, "Evaporating menisci of wetting fluids," *Journal of Colloid and Interface Science*, Vol(73), pp. 212-223.

EFFECT OF OUTDOOR AIR PRE-CONDITIONING ON ASHP PERFORMANCE USING EAHE

Deva S. Veylan
Architectural Science
Ryerson University
Toronto, Ontario, Canada
dveylan@ryerson.ca

Alan S. Fung
Mechanical and Industrial Engineering
Ryerson University
Toronto, Ontario, Canada
alanfung@ryerson.ca

ABSTRACT

A simplified model has been developed to estimate the increase in coefficient of performance (COP) that can be achieved by pre-conditioning outdoor air using an earth-air heat exchanger (EAHE) prior to the air being directed over the outdoor coil of a cold climate variable capacity air-source heat pump (ASHP). Data is compiled from a previously-developed TRNSYS model of the Toronto and Region Conservation Authority (TRCA) Archetype Sustainable House A (House A) at the Kortright Center in Woodbridge, Ontario, Canada and a TRNSYS model of an EAHE developed for this study. The EAHE model is used to estimate the effect of pre-conditioning the outdoor air at different flow rates for the ASHP outdoor coil. The effect of this pre-conditioning on the ASHP is studied using sensitivity and spreadsheet analysis to plot the effect on the COP of the ASHP. A moderate improvement to the COP of the ASHP is seen in the heating season, and a slight improvement in the cooling season. Results show an improvement in COP from 3.16 to 3.67 in the heating season, and an improvement in COP from 4.57 to 4.69 in the cooling season, for the typical Toronto climate.

KEYWORDS: earth air heat exchanger, variable capacity air source heat pump, coefficient of performance (COP), TRNSYS modeling, Archetype Sustainable House

INTRODUCTION

The Paris Agreement on Climate Change, which came into effect in late 2016, obligates Canada and the other signatories to reduce their greenhouse gas emissions to a level that will prevent an average global temperature increase of 2°C. A shift towards energy efficiency and renewable energy must be made in Canada's residential sector because fossil fuel reserves are finite and cannot be depleted without contributing to potentially catastrophic damage to our climate. According to the Intergovernmental Panel on Climate Change (IPCC), heat pumps are a key residential renewable energy technology that can contribute to a sustainable future [1].

Much of Canada sees temperatures that fall below -25°C in winter, and ASHPs operating at low winter temperatures struggle to provide the heat demanded, and do so at a much lower operating efficiency than at milder temperatures. Even a two-stage variable capacity ASHP designed for cold weather has a COP of 1.6 at -19°C, and below -24°C is less efficient than conventional electric heating and requires a supplemental heat source [2]. At these low temperatures, ground-source heat pumps (GSHPs) operate more efficiently due to the relatively constant and warmer ground temperatures rather than being subject to severe low air temperatures [3]. TRCA House A and the Archetype Sustainable House B have an ASHP and GSHP installed, respectively, and through studies performed by Fung, Safa, Seters, and Kumar in 2014 and 2015 the seasonal heating and cooling performance of the ASHP and GSHP are comparable [2]–[4].

At $34,500, GSHPs are more expensive to install compared to an ASHP at $14,500 [2]. The savings of approximately $20,000 realized by installing an ASHP could be used to install a 9 kWp solar photovoltaic (PV) system. This PV system could generate approximately 10,000 kWh of electricity per year and would allow a low-energy designed house to reach the net-zero standard currently promoted by the Canadian Home Builders' Association.

If a simple and cost-effective method can be developed to enhance the cold-weather performance of an ASHP, many residential and small commercial buildings could employ a relatively inexpensive advanced ASHP even in colder regions of Canada without needing a supplemental backup heating system to reach net-zero performance.

This paper discusses a novel approach of combining a two-stage variable capacity ASHP designed for cold weather performance with an EAHE using TRNSYS energy simulation software. The model developed simulates how the performance of an ASHP is affected by using an EAHE to pre-condition the outdoor air supplied to the ASHP.

TECHNOLOGIES USED IN THIS STUDY

This paper studies the potential improvement of cold-climate ASHP performance achieved by pre-conditioning air seen by the exterior coil by passing the air through a buried EAHE. A model developed for the House A shown in Fig. 1 is used to simulate this arrangement using TRNSYS 17 energy simulation software.

Figure 1: Archetype Sustainable House A

DESIGN CONCEPT

Energy efficiency upgrades can be costly, and to ensure economical viability, cost of installation and lifecycle costs must be optimally reduced. Any structure built in an environment with a winter frost line below grade must have foundation footings placed below the frost line to prevent frost heaving and differential settlement of the structure. In most Canadian locations, excavation must occur to allow footings to be poured. The most effective way of reducing the installation cost of an EAHE is to place it underground when the footing excavation occurred. In many new houses built in Canada, excavation for footings and slab is deeper than 6 feet below grade. Placing an EAHE on the outside perimeter of a foundation in the existing foundation excavation reduces or eliminates additional excavation costs when installing an EAHE. This paper discusses an EAHE system installed around the perimeter of an average-sized new home built in the greater Toronto area (GTA).

LENGTH OF EAHE

The estimated perimeter of an EAHE installed around a standard 2000 ft^2 new house in the GTA is 50 m. The EAHE design for this case study is limited to this size to conform with typical space constraints imposed by property lines.

DEPTH OF EAHE

Standard depth of basements in the GTA is 6 feet below grade, and accordingly the depth used in the design is 1.83 m below grade. This depth is shallower than the ideal identified by Brum et al. in 2013 [5]; however, burying the EAHE deeper would significantly increase the cost of installation.

SIZE OF EAHE

The effect of various pipe diameters is simulated in the EAHE TRNSYS model. The diameter of the EAHE piping alone does not have a significant effect on the heat transfer to the pre-conditioned air flowing through the system. According to the conclusions reached by Lee and Strand in 2008, in order to reduce pressure loss and flow rate through the pipes and increase the ΔT, a balance must be struck between pipe diameter and flow rate [4]. A pipe diameter of 8 inches was chosen to reduce the pressure drop through the EAHE that would be caused by a smaller diameter pipe, while still allowing a large ΔT between the inlet and outlet temperatures of the system. A commercially-available option in 8 inches diameter is also readily available from REHAU Inc. [6].

FLOW RATES THROUGH EAHE

Another factor affecting design is the flow rate through the EAHE system. A higher flow rate produces a greater pressure drop and lower ΔT. This pressure drop is overcome by the fan pulling air through the EAHE system. Minimizing the fan power draw is vitally important for the viability of the EAHE system, and therefore minimizing the pressure drop through the system is critical. De Paepe and Janssens show that an effective way to maximize ΔT and minimize pressure drop is to provide more pipes running in parallel [7]. For this reason, the design provides three 8-inch pipes running in parallel. The flow rate through each pipe will not exceed 1,200 CFM, which is one third of the total outdoor air flow rate required by the ASHP at peak capacity.

LIST OF VARIABLES

Fig. 2 shows the list of variables used in developing the EAHE model in TRNSYS.

Variables used in EAHE Design and TRNSYS Model	
Toronto	Location for Weather File
25	Nodes Along Pipe Axis
50 m	Pipe Length
0.1872 m	Pipe Inside Diameter
0.2032 m [6]	Pipe Outside Diameter
PVC [6]	Pipe Material
0.28 W/m*K [8]	Pipe Thermal Conductivity
0	Pipe/Soil Contact Resistance
1000 m	Depth of Groundwater Layer
10 m	Deep Earth Distance
1.005 kJ/kg*K [9]	Air Specific Heat
1.25 kg/m^3 [9]	Air Density
0.025 W/m*K [9]	Air Thermal Conductivity
0.000018 kg/m*s [9]	Air Viscosity
2.45 W/m*K [10]	Conductivity of Soil Layer
3200 kg/m^3	Density of Soil Layer
0.84 kJ/kg*K	Specific Heat of Soil Layer
10.1˚C [10]	Deep Earth Temp. (Toronto)
11.0˚C	Amplitude of Soil Temperature
5.0 CFM/W [11]	Efficiency of Fan

Figure 2: Variables Used in EAHX TRNSYS Model

LIST OF ASSUMPTIONS

Some assumptions were necessary in order to simplify the modeling of the EAHE. Further research is needed to eliminate all assumptions and model the effectiveness of the system to a high degree of accuracy. The following assumptions were made:

- Foundation wall thermal effects are ignored
- The three EAHE pipes do not thermally interact
- Properties of the air flowing through the EAHX, including density and viscosity are fixed at 25°C
- Some properties of the soil are unknown for the location, and are set to TRNSYS defaults
- Pressure drop through the pipe system is not modeled, and therefore fan efficiency is fixed
- Full TRNSYS modelling of the combined House A and EAHX is not performed. Some values are calculated approximations and can be modeled more precisely in further research
- Saturation of ground heat sink is ignored.

HOUSE A TRNSYS MODEL

A TRNSYS model of House A, built by Safa et al. in 2015, is used for modeling purposes [12]. The following information is collected from the House A model:

- Time
- Ambient temperature
- House heating and cooling loads
- Power demand for ASHP.

These outputs are gathered for both the heating and cooling seasons and are used in the spreadsheet calculations and analysis.

SPREADSHEET DESIGN

A spreadsheet is used to combine the data output from both the House A TRNSYS simulation and the EAHE TRNSYS simulation, and to calculate the interaction between the two for the heating and cooling seasons.

Time, ambient temperature, ASHP power draw, and house cooling and heating loads are the outputs collected from the House A TRNSYS simulation. A sensitivity analysis is performed on the EAHE model to determine the ΔT for various ambient temperature conditions and flow rates, and the results are shown in Figs. 3 and 4.

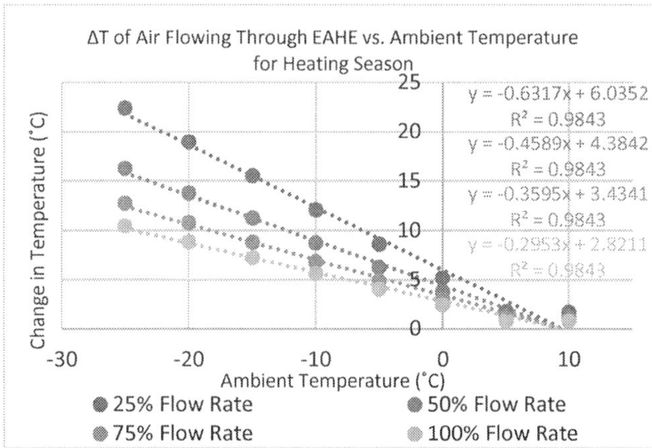

Figure 3: ΔT of Air Flowing Through EAHE vs. Ambient Temperature for Heating Season

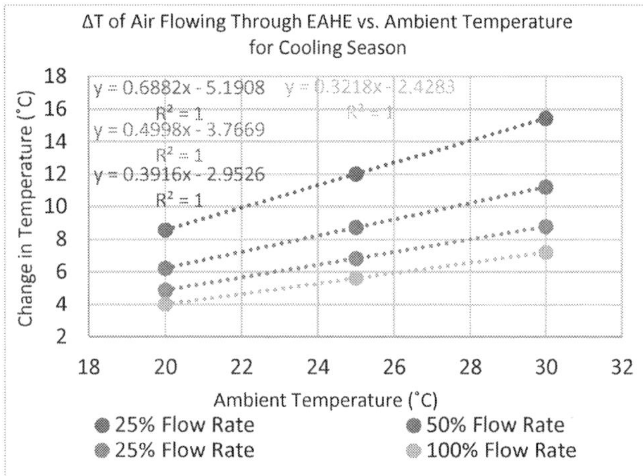

Figure 4: ΔT of Air Flowing Through EAHE vs. Ambient Temperature for Cooling Season

This sensitivity analysis is performed by simulating the ΔT between the inlet and outlet of the EAHE system at various outdoor temperatures (5°C intervals between -25°C and 30°C) and at 25%, 50%, 75% and 100% of the maximum flow rate experienced by the EAHE. The results of this sensitivity analysis are expressed as formulas used in the spreadsheet to interpolate the outlet temperature of the EAHE for the various ambient temperatures and flow rates experienced through the course of the year in the House A simulation.

CALCULATIONS

The following calculations are made to analyse the performance of the combined ASHP and EAHE system. Several calculations are used to combine the data gathered from both TRNSYS models and calculate the resulting change in performance of the combined system.

Firstly, the COP of the ASHP is calculated using Eqs. (1) and (2) from Safa et al. (2015) work and is based on the ambient temperature for the heating and cooling seasons [12]:

$$\text{Heating Season COP} = 3.7258 + 0.1158 \times T_{outdoor} \quad (1)$$

$$\text{Cooling Season COP} = -0.0568 \times T_{outdoor} + 6.6644 \quad (2)$$

Next, the heat through the ASHP coil is calculated for the heating and cooling seasons based on energy balance equations, Eqs. (3) and (4):

$$\text{Heating Season } Q_c \text{ (kW)} = \text{House Heating Load } Q_h \text{ (kW)} - \text{ASHP Power Draw (kW)} \quad (3)$$

$$\text{Cooling Season } Q_h \text{ (kW)} = \text{House Heating Load } Q_c \text{ (kW)} + \text{ASHP Power Draw (kW)} \quad (4)$$

From this, the fan flow rate through the outdoor coil of the ASHP is estimated, which determines the flow rate through the EAHE using Eq. (5):

$$Q = \dot{m} \times C_p \times \Delta T = V \times \rho \times C_p \times \Delta T \quad (5)$$

Where:
Q = *Energy Input from the Outdoor Coil (kW)*
$\dot{m}$ = *Mass Flow Rate (kg/s)*
C_p = *Heat Capacity (kJ/kg°C)*
T = *Temperature (°C)*
V = *Volumetric Flow Rate (m^3/s)*
ρ = *Density (kg/m^3)*

In the model developed for this study, the calculated air flow rate through the outdoor coil of the ASHP is equal to the flow rate through the EAHE. The ΔT through the EAHE is estimated by interpolation between the formulas in Figs. 3 and 4, and a new temperature flowing through the external coil of the ASHP is determined.

This change in temperature seen by the ASHP will reduce its electricity draw at a given ambient temperature, and improve the COP of the ASHP. The power consumption from the fan needed to draw air through the EAHX is added to the new lower power draw for the ASHP, and if a net reduction in power consumption is realized by the system, an improvement to performance has been achieved.

The efficiency of the additional fan needed to draw air through the EAHE is estimated based on work done by Iain Walker of Berkeley Lab [10]. Not all conditions will see a benefit from pre-conditioning the air seen by the external coil of the ASHP through the EAHE. A control mechanism is necessary to determine when it is appropriate to pre-condition the air through the EAHE and when the ASHP should receive ambient air. If the combined power draw of the ASHP and EAHE system is higher than the draw of the ASHP operating alone, the EAHE system should not be utilized. Similarly, if the ambient temperature in the heating season is above the ground temperature, then no performance improvement can be realized and the EAHE should not operate.

To determine the overall heating season COP, the total heating load from the house is divided by the ASHP and EAHE power consumption using an optimized control scheme. A similar calculation is performed for the cooling season. The COP of the combined system can then be compared to the original COP of the ASHP alone to gauge whether a performance improvement has been realized.

RESULTS

The efficiency of the fan used to pull air through the EAHE is a critical factor for the performance of the combined ASHP and EAHE system. If the additional fan draws too much power to run, it negates the benefit of the reduction in power draw from the ASHP by pre-conditioning the air.

Figs. 5 and 6 show how the efficiency of the additional fan affects the combined COP of the ASHP and EAHE system.

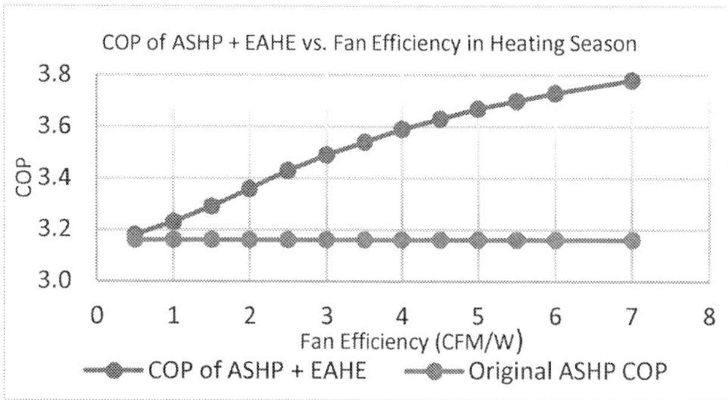

Figure 5: COP of ASHP + EAHE vs. Fan Efficiency in Heating Season

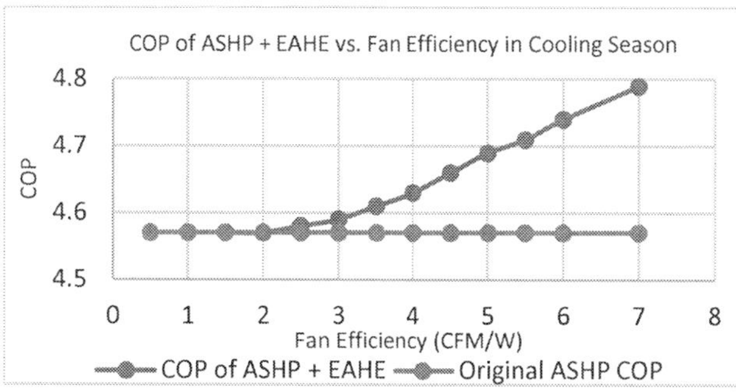

Figure 6: COP of ASHP + EAHE vs. Fan Efficiency in Cooling Season

Though it is not realistic, a very high efficiency fan is analogous to the ASHP itself drawing all of the air required through the EAHE. This is the maximum potential COP of the combined system. For the heating season, the maximum potential COP is 4.20, and for the cooling it is 5.78.

Using a more realistic efficiency value of 5 CFM/W for the additional fan [11], the COP for the combined ASHP and EAHE system is 3.67 compared to 3.16 for the ASHP alone in the heating season as shown in Fig. 7. This is an improvement of 16% and results in a reduction of 666.6 kWh/year of electricity in the typical Toronto climate.

For the cooling season, the combined system COP is 4.69, an improvement of 0.12 from the original COP of 4.57 for the ASHP alone as shown in Fig. 8. This translates into a reduction in electricity consumption of 11 kWh/year in the typical Toronto climate.

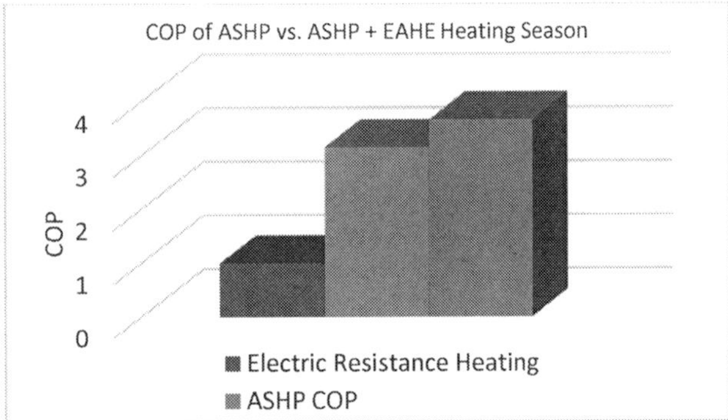

Figure 7: COP of ASHP vs. ASHP + EAHE Heating Season

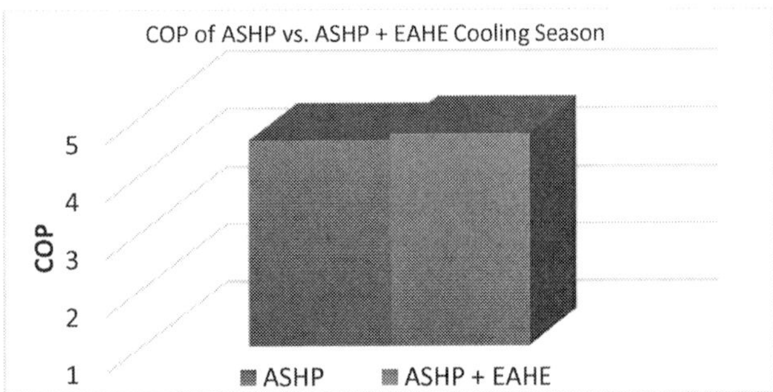

Figure 8: COP of ASHP vs. ASHP + EAHE Cooling Season

FURTHER RESEARCH

This case study is performed on a proof-of-concept basis and makes several assumptions and simplifications. Further research is recommended to eliminate these assumptions and improve the accuracy of the results, including:

- Research ground conditions for more accurate values
- Simulate the effect of the foundation wall
- Simulate the effect of insulation above the EAHE pipes to isolate them from the effects of ambient temperature
- Perform economic analysis, including payback period for the additional system costs
- Combine the House A ASHP and EAHE models into one TRNSYS model
- Run the model for different climate zones.

CONCLUSIONS

On a proof-of-concept basis, pre-conditioning air flowing through the outdoor coil of an AHSP using an EAHE is shown to improve the COP of an ASHP in both the heating and cooling seasons in Toronto. A moderate improvement in the COP from 3.16 to 3.67 is realized in the heating season, and a minor improvement in the COP of the ASHP from 4.57 to 4.69 is realized in the cooling season.

ACKNOWLEGMENTS

The authors would like to thank the Toronto and Region Conservation Authority (TRCA).

REFERENCES

1. United Nations, 2015 "Paris Agreement," *21st Conference of the Parties of the United Nations Framework Convention on Climate Change*, pp. 1–16.
2. Safa, A., Van Seters, T., and Fung, A. S. 2014, "Performance Assessment of a Variable Capacity Air Source Heat Pump and a Single-Capacity Horizontal Loop Coupled Ground Source Heat Pump System," *IEA Heat Pump Conference*, Montreal, Canada, pp. 1–14.
3. Safa, A., Fung, A. S., and Kumar, R. 2015, "Comparative thermal performances of a ground source heat pump and a variable capacity air source heat pump systems for sustainable houses," *Applied Thermal Engineering*, vol. 81, pp. 279–287.
4. Safa, A., Fung, A. S., and Kumar, R. 2015 "Heating and cooling performance characterisation of ground source heat pump system by testing and TRNSYS simulation," *Renewable Energy*, vol. 83, pp. 565–575.
5. Brum, R. D. S., Vaz, J., Rocha, L. A. O., Dos Santos, E. D., and Isoldi, L. A. 2013, "A new computational modeling to predict the behavior of Earth-Air Heat Exchangers," *Energy and Buildings*, vol. 64, pp. 395–402.
6. REHAU 2016, "REHAU ECOAIR ground-air heat exchange system"
7. De Paepe, M., and Janssens, A. 2003 "Thermo-hydraulic design of earth-air heat exchangers," *Energy and Buildings*, vol. 35, no. 4, pp. 389–397.
8. Hutcheon, N. B., and Handegord, G. O. P. 1995, *Building Science for a Cold Climate*. National Research Council of Canada.
9. "Air Properties," *The Engineering Toolbox*. [Online]. Available: http://www.engineeringtoolbox.com/air-properties-d_156.html [Accessed: 02-Nov-2016].
10. Alzahrani, W. S., 2014 "Experimental study of the performance of a vertical and a horizontal ground loops coupled to a ground source heat pump system."
11. Walker, I., "Blower Energy Use." Berkeley Lab
12. Safa, A., Fung, A. S., and Kumar, R. 2015, "Performance of two-stage variable capacity air source heat pump: Field performance results and TRNSYS simulation," *Energy and Buildings*, vol. 94, pp. 80–90.

Author Index

Z